Advances in Intelligent Systems and Computing

Volume 748

Series editor

Janusz Kacprzyk, Polish Academy of Sciences, Warsaw, Poland
e-mail: kacprzyk@ibspan.waw.pl

The series "Advances in Intelligent Systems and Computing" contains publications on theory, applications, and design methods of Intelligent Systems and Intelligent Computing. Virtually all disciplines such as engineering, natural sciences, computer and information science, ICT, economics, business, e-commerce, environment, healthcare, life science are covered. The list of topics spans all the areas of modern intelligent systems and computing such as: computational intelligence, soft computing including neural networks, fuzzy systems, evolutionary computing and the fusion of these paradigms, social intelligence, ambient intelligence, computational neuroscience, artificial life, virtual worlds and society, cognitive science and systems, Perception and Vision, DNA and immune based systems, self-organizing and adaptive systems, e-Learning and teaching, human-centered and human-centric computing, recommender systems, intelligent control, robotics and mechatronics including human-machine teaming, knowledge-based paradigms, learning paradigms, machine ethics, intelligent data analysis, knowledge management, intelligent agents, intelligent decision making and support, intelligent network security, trust management, interactive entertainment,Web intelligence and multimedia.

The publications within "Advances in Intelligent Systems and Computing" are primarily proceedings of important conferences, symposia and congresses. They cover significant recent developments in the field, both of a foundational and applicable character. An important characteristic feature of the series is the short publication time and world-wide distribution. This permits a rapid and broad dissemination of research results.

More information about this series at http://www.springer.com/series/11156

M. Tanveer · Ram Bilas Pachori
Editors

Machine Intelligence and Signal Analysis

 Springer

Editors
M. Tanveer
Discipline of Mathematics
Indian Institute of Technology Indore
Simrol, Madhya Pradesh
India

Ram Bilas Pachori
Discipline of Electrical Engineering
Indian Institute of Technology Indore
Simrol, Madhya Pradesh
India

ISSN 2194-5357   ISSN 2194-5365 (electronic)
Advances in Intelligent Systems and Computing
ISBN 978-981-13-0922-9   ISBN 978-981-13-0923-6 (eBook)
https://doi.org/10.1007/978-981-13-0923-6

Library of Congress Control Number: 2018943386

This Springer imprint is published by the registered company Springer Nature Singapore Pte Ltd.
The registered company address is: 152 Beach Road, #21-01/04 Gateway East, Singapore 189721, Singapore

Preface

Machine learning and signal processing are widely used approaches for solving real-world problems. Machine learning is a revolutionizing domain of public research which involves optimization and signal processing techniques. It is an interdisciplinary paradigm which covers a lot of areas of science and engineering. In order to provide better solutions, advanced signal processing techniques with optimal machine learning solutions are required. In the recent years, the techniques related to signal processing and machine learning have been frequently used for biomedical applications.

This book presents some important applications and improvements in the areas of machine learning and signal processing. The whole book is organized into 64 chapters. The focuses of these chapters are feature extraction, time frequency analysis, classification, and the diagnosis of various diseases. This book also includes some review on biomedical application related to various kinds of diseases such as neurological disorders and different types of cancers. The introductory material presented in these chapters gives future direction to interested researchers involved in these interdisciplinary domains. In brief, this book motivates the young researchers to involve and pursue research in these interdisciplinary areas of machine intelligence and signal processing.

We hope that the papers contained in this proceedings will prove helpful toward improving the understanding of machine learning and signal processing at the teaching and research levels and will inspire researchers to work in these interdisciplinary domains.

Indore, India
April 2018

M. Tanveer
Ram Bilas Pachori

Conference Organizing Committee

Organizing Chairs

M. Tanveer, IIT Indore, India
Ram Bilas Pachori, IIT Indore, India

Program Chair

Hem C. Jha, IIT Indore, India

Publicity Chairs

Hrishikesh V. Raman, IIIT, Sricity, India
Deepak Gupta, NIT Arunachal Pradesh, India
Santanu Manna, IIT Indore, India

International Advisory Committee

U. Rajendra Acharya, Ngee Ann Polytechnic, Singapore
Jim Schroeder, Florida Institute of Technology, USA
Johan Suykens, KU Leuven, Belgium
Palaniappan Ramaswamy, University of Kent, UK
Yuan-Hai Shao, Zhejiang University of Technology, China
Olivier Colliot, CNRS & INRIA, France
P. N. Suganthan, NTU, Singapore
Yuh-Jye Lee, NCTU, Taiwan
S. Suresh, NTU, Singapore
Shen-Shyang Ho, Rowan University, USA

David Hewson, University of Bedfordshire, UK
Hichem Snoussi, UTT, Troyes, France
Dante Mantini, KU Leuven, Belgium
Moonis Ali, Texas State University, San Marcos, USA
Girijesh Prasad, Ulster University, UK

National Advisory Committee

V. M. Gadre, IIT Bombay, India
Pradip Sircar, IIT Kanpur, India
B. K. Panigrahi, IIT Delhi, India
Pravat Mandal, NBRC, Gurgaon, India
G. Ramamurthy, IIIT, Hyderabad, India
S. Dandapat, IIT Guwahati, India
Suresh Chandra, IIT Delhi, India
Jayadeva, IIT Delhi, India
M. Abulaish, SAU, Delhi, India
S. Mitra, ISI, Kolkata, India
Omar Farooq, AMU, Aligarh, India
Asif Ekbal, IIT Patna, India
Reshma Rastogi, SAU, Delhi, India
Millie Pant, IIT Roorkee, India

Local Advisory Committee

Sk. Safique Ahmad, IIT Indore, India
Trapti Jain, IIT Indore, India
Anand Parey, IIT Indore, India
Niraj K. Shukla, IIT Indore, India
Md. Aquil Khan, IIT Indore, India
Aruna Tiwari, IIT Indore, India
Kapil Ahuja, IIT Indore, India
Amod C. Umarikar, IIT Indore, India
Devendra Deshmukh, IIT Indore, India
Swadesh Kumar Sahoo, IIT Indore, India

Technical Program Committee

Justin Dauwels, NTU, Singapore
Lalit Garg, University of Malta, Malta
Ayush Kumar, Stony Brook University, USA
A. T. Azar, Benha University, Egypt

Shamsollahi, Sharif University of Technology, Tehran, Iran
Nisar K. S., PSAU, Saudi Arabia
Pramod Gaur, Ulster University, UK
Anand Pandyan, Keele University, UK
Vidya Sudarshan, SUSS, Singapore
K. C. Veluvolu, Kyungpook National University, South Korea
Ratnadip Adhikari, Fractal Analytics, Bengaluru, India
Dinesh Bhati, AITR, Indore, India
Omar Farooq, AMU, Aligarh, India
Anand Parey, IIT Indore, India
P. Rajalakshmi, IIT Hyderabad, India
Rajib Kumar Jha, IIT Patna, India
Debi Prosad Dogra, IIT Bhubaneswar, India
Tony Jacob, IIT Guwahati, India
Varun Bajaj, IIITDM, Jabalpur, India
Anil Kumar, IIITDM, Jabalpur, India
Joyeeta Singha, LNMIIT, Jaipur, India
Pushpendra Singh, Bennett University, Noida, India
Deepak Mishra, IIST, Trivandrum, India
Shiv Ram Dubey, IIIT, Sricity, India
Hrishikesh Venkataraman, IIIT, Sricity, India
Abhishek Rawat, IITRAM, Ahmedabad, India
Kapil Gupta, NIT Uttrakhand, India
Manish Sharma, IITRAM, Ahmedabad, India
Shaik Rafi Ahamed, IIT Guwahati, India
Prashant Bansod, SGSITS, Indore, India
Dilip Singh Sisodia, NIT Raipur, India
Rajesh Bodade, MCTE, Mhow, India
Reshma Rastogi, SAU, New Delhi, India
K. V. Arya, ABV-IIITM, Gwalior, India
Vrijendra Singh, IIIT, Allahabad, India
Dheeraj Agarwal, MANIT Bhopal, India
Pooja Mishra, IIIT, Allahabad, India
Anil Kumar Vuppala, IIIT, Hyderabad, India
Syed Azeemuddin, IIIT, Hyderabad, India
Rajeev Sharma, IIT Indore, India
Preety Singh, LNMIIT, Jaipur, India
Umakant Dhar Dwivedi, RGIPT, Raebareli, India
Deepak Gupta, NIT Arunachal Pradesh, India
Prithwijit Guha, IIT Guwahati, India
M. Sabarimalai Manikandan, IIT Bhubaneswar, India
Sri Rama Murty, IIT Hyderabad, India
Navjot Singh, NIT Uttarakhand, India
Akanksha Juneja, NIT Delhi, India
Aruna Tiwari, IIT Indore, India

M. Tanveer, IIT Indore, India
Ram Bilas Pachori, IIT Indore, India
Surya Prakash, IIT Indore, India
Hem C. Jha, IIT Indore, India
Amod C. Umarikar, IIT Indore, India
Ravibabu Mulaveesala, IIT Ropar, India
Jyotindra Singh Sahambi, IIT Ropar, India
Manas Kamal Bhuyan, IIT Guwahati, India
Anil Kumar Sao, IIT Mandi, India
Nishchal K. Verma, IIT Kanpur, India
Nithin V. George, IIT Gandhinagar, India
Rishikesh D. Kulkarni, IIT Guwahati, India
Shubhajit Roy Chowdhury, IIT Mandi, India
Manisha Verma, IIT Gandhinagar, India
Rajesh Tripathy, S 'O' A University, Bhubaneswar, India
L. N. Sharma, IIT Guwahati, India
S. M. Shafiul Alam, NREL, Golden, CO, USA
Meenakshi Sood, JUIT, Solan, India
Vaibhav Gandhi, Middlesex University, UK
Mohammed Imamul Hassan Bhuiyan, BUET, Dhaka, Bangladesh
Siuly Siuly, Victoria University, Melbourne, Australia
Ankit Bhurane, IIIT, Nagpur, India
Vijay Bhaskar Semwal, IIIT, Dharwad, India
Hisham Cholakkal, Mercedes-Benz R&D, India
Keshav Patidar, Symbiosis University of Applied Sciences, Indore, India
Radu Ranta, CRAN—Université de Lorraine/CNRS, ENSEM, France
Le Cam Steven, ENSEM—CRAN UMR CNRS, Université de Lorraine, France
Amit Prasad, IIT Mandi, India
Babita Majhi, G. G. Vishwavidyalaya, Bilaspur, India
Shovan Barma, IIIT, Guwahati, India
Manoj Kumar Saxena, RRCAT, Indore, India
Neeraj Kumar Singh, University of Toulouse, France
Dipankar Deb, IITRAM, Ahmedabad, India
R. N. Yadav, MANIT Bhopal, India
Sanjeev Sharma, ABV-IIITM, Gwalior, India
Amalin Prince A., BITS Pilani—Goa Campus, Goa, India
Manoj Kumar, NIMHANS, Bangalore, India
Milind Padikar, Vehant Technologies, Noida, India

Acknowledgements

We would like to express our gratitude to all the researchers who provided their constant support and dedication to publish this book as a proceedings for international conference on Machine Intelligence and Signal Processing (MISP 2017).

We would also like to thank our colleagues who supported and encouraged us in spite of all the time it took us away from them. It was a long and difficult journey for them.

The editors of this book would like to thank all the authors for their contributions made in this book. The editors also want to thank the Springer team for their continuous assistance and support in the publication of this book.

The editors express their sincere gratitude to MISP 2017 speakers, reviewers, technical program committee members, international and national advisory committee, program and publicity chairs, local organizing committee, and institute administration, without whose support the quality and standard of the conference could not be maintained.

We thank all the participants who had presented their research papers and attended the conference. A special mention of thanks is due to our student volunteers including secretarial assistance Vipin Gupta, Bharat Richhariya, and Chandan Gautam for the spirit and enthusiasm they had shown throughout the duration of the event, without which it would have been difficult for us to organize such a successful event.

Contents

About the Editors

Dr. M. Tanveer is Assistant Professor and Ramanujan Fellow at the Discipline of Mathematics, Indian Institute of Technology Indore, India. Prior to that, he worked as Postdoctoral Research Fellow at Rolls-Royce@NTU Corporate Lab, Nanyang Technological University (NTU), Singapore. He served as Assistant Professor at the Department of Computer Science and Engineering, LNM Institute of Information Technology (LNMIIT), Jaipur, India. He received his Ph.D. degree in Computer Science from the Jawaharlal Nehru University, New Delhi, India, and his M.Phil. degree in Mathematics from Aligarh Muslim University, Aligarh, India. His research interests include support vector machines, optimization, applications to Alzheimer's disease and dementias, biomedical signal processing, and fixed-point theory and applications. He has been awarded competitive research funding by various prestigious agencies such as Department of Science and Technology (DST), Council of Scientific and Industrial Research (CSIR), and Science and Engineering Research Board (SERB). He is the recipient of 2017 SERB Early Career Research Award in Engineering Sciences and the only recipient of 2016 prestigious DST-SERB Ramanujan Fellowship in Mathematical Sciences. He is a member of the editorial review board of Applied Intelligence, Springer (International Journal of Artificial Intelligence, Neural Networks, and Complex Problem-Solving Technologies). He has published over 24 papers in reputed international journals.

Dr. Ram Bilas Pachori received his B.E. degree with honors in Electronics and Communication Engineering from Rajiv Gandhi Proudyogiki Vishwavidyalaya, Bhopal, India, in 2001; M.Tech. and Ph.D. degrees in Electrical Engineering from the Indian Institute of Technology (IIT) Kanpur, India, in 2003 and 2008, respectively. He worked as Postdoctoral Fellow at Charles Delaunay Institute, University of Technology of Troyes, France, during 2007–2008. He served as Assistant Professor at Communication Research Center, International Institute of Information Technology, Hyderabad, India, during 2008–2009. He served as Assistant Professor at Discipline of Electrical Engineering, IIT Indore, India, during 2009–2013. He worked as Associate Professor at Discipline of Electrical Engineering, IIT Indore, Indore, India, during 2013–2017 where presently he has been working as Professor since 2017. He worked as Visiting Scholar at Intelligent Systems Research Center, Ulster University, Northern Ireland, UK, during December 2014. His research interests are in the areas of biomedical signal processing, non-stationary signal processing, speech signal processing, signal processing for communications, computer-aided medical diagnosis, and signal processing for mechanical systems. He has more than 125 publications which include journal papers, conference papers, book, and chapters.

Detecting R-Peaks in Electrocardiogram Signal Using Hilbert Envelope

Y. Madhu Keerthana and M. Kiran Reddy

Abstract In this paper, the unipolar property of the Hilbert envelope is exploited for detecting R-peaks in electrocardiogram (ECG) signals. In the proposed method, first, the ECG signals are bandpass filtered to reduce various kinds of noises. Then, the Hilbert envelope of the bandpass filtered ECG signals is used to estimate the approximate locations of R-peaks. These locations are further processed to determine the correct R-peaks in the ECG signal. The performance of the proposed method is evaluated using 30 ECG records from the MIT-BIH arrhythmia database. Evaluation results show that the proposed method has a very less detection error rate of 0.31% with a high sensitivity and positive predictivity of 99.83 and 99.86%, respectively. Furthermore, the results indicated that the performance of the proposed method is much better compared to other well-known methods in the presence of noise/artifacts, low-amplitude, and negative QRS complexes.

1 Introduction

The electrophysiologic pattern of the heart muscles during each heartbeat is represented using electrocardiogram (ECG) signal. ECG signal processing is essential for the diagnosis of variability in heart rate, biometric systems, and cardiac diseases. In the ECG signal, the QRS complex represents the depolarization of ventricles of the heart. The maximum amplitude in the QRS complex is termed as the R-peak. The

Y. Madhu Keerthana (✉)
School of Engineering and Technology, Sri Padmavati Mahila Visvavidyalayam,
Tirupati, India
e-mail: madhu.keerthu@gmail.com

M. Kiran Reddy
Department of Computer Science and Engineering, Indian Institute
of Technology Kharagpur, Kharagpur, West Bengal, India
e-mail: kiran.reddy889@gmail.com

© Springer Nature Singapore Pte Ltd. 2019
M. Tanveer and R. B. Pachori (eds.), *Machine Intelligence and Signal Analysis*,
Advances in Intelligent Systems and Computing 748,
https://doi.org/10.1007/978-981-13-0923-6_1

distance between two R-peaks also referred as R-R interval is a measure of heart rate. Hence, accurate detection of R-peaks is essential for heartbeat analysis. However, this is often difficult due to the presence of noise sources such as 50/60 Hz power interference, baseline drift, motion artifacts, and muscle noise [1]. Also, ECG signals with low-amplitude R-peaks, tall sharp T waves, negative R-peaks, and sudden changes in R-peak amplitudes make the detection problem a more challenging task.

In literature, several approaches are proposed based on the derivatives [2], digital filters [3–6], wavelet transform (WT) [7, 8], neural networks [9], empirical mode decomposition (EMD) [10], K-nearest neighbor [11], area of curve [12], and Shannon energy [13, 14] for R-peak detection. The filtering-based methods are computationally efficient and hence commonly used for R-peak detection in real time. The most widely used filtering-based approach is the method proposed by Pan-Tompkins [4]. In [5], it is shown that the filtering-based approaches have very poor accuracy when the ECG signal contains negative R-peaks and noise. The WT-based [7] and EMD-based [10] methods are found to be more effective compared to other techniques, especially for ECG signals with unusual R-peaks and noise. However, selection of suitable mother wavelet and scales in case of WT-based method, and selection of appropriate intrinsic mode functions (IMFs) for the EMD-based method under noisy environment is very difficult. Also, these approaches have high computational costs due to complicated check up strategy and multiple threshold values, making them inapplicable in real time [13].

In this work, a new R-peak detection algorithm is presented. The proposed method uses the Hilbert envelope of the bandpass filtered ECG signals to identify the precise locations of R-peaks in the ECG signal. The proposed approach is simple as it does not require any complex operations like wavelet transform. The performance of the proposed method is evaluated using ECG records from the standard MIT-BIH arrhythmia database. The evaluation results show that the performance of the proposed method in the presence of noise and unusual R-peaks is significantly better than other popular approaches. The rest of the paper is organized as follows. The Hilbert transform and Hilbert envelope are discussed in brief in Sect. 2. Section 3 describes the proposed R-peak detection approach. Performance evaluation results are presented and discussed in Sect. 4. Finally, Sect. 5 concludes the paper.

2 Hilbert Envelope

The Hilbert transform (HT) is a linear operator which derives the analytic representation of a signal. The HT [15] $\tilde{y}(t)$ of a input signal x(t) is defined as

$$\tilde{y}(t) = H[x(t)] = \frac{1}{\pi} \int_{-\infty}^{\infty} \frac{x(\tau)}{t - \tau} d\tau \tag{1}$$

The Hilbert envelope h(t) of the signal x(t) is a magnitude function constructed from the input signal and its Hilbert transform as follows:

$$h(t) = \sqrt{x^2(t) + \tilde{y}^2(t)} \tag{2}$$

From (2), it is evident that h(t) is a positive function, and hence, it is always unipolar. In this work, the unipolar nature of Hilbert envelope is utilized in R-peak detection.

3 Proposed Method

The block diagram of the proposed R-peak detector is shown in Fig. 1. The first step in the proposed method is bandpass filtering. In general, ECG signals are often corrupted with distinct noises, like baseline drift, power-line interference, muscle noise, and motion artifacts [13]. The range of frequencies in which the energy of QRS complex is maximum is approximately 5–20 Hz [16]. The power-line interference (50 Hz or 60 Hz) is a noise source which is frequently encountered in the ECG signals [13]. Baseline drift (frequency <1 Hz) is caused due to respiration [17]. The high-grade noise frequency is usually above 30 Hz. Therefore, in this work, a third-order Butterworth bandpass filter with lower and upper cutoff frequencies 5 and 25 Hz, respectively, is used to attenuate noise and enhance the ECG signal. To combat phase distortion, the filter is applied in both the forward and reverse directions. The bandpass filtered output will be a bipolar signal, and thus a transformation needs to be used. The goal of transformation is to make all the signal values positive so that a single-sided thresholding mechanism can be used by emphasizing the QRS complexes. The squaring transformation is frequently used in the literature, but it significantly decreases the magnitude of the low-amplitude R-peaks [13]. Since Hilbert envelope is unipolar (discussed in Sect. 2), therefore here we study its effectiveness as a transformation technique. The performance of squaring and Hilbert envelope transformation techniques for an ECG signal containing R-peaks with low amplitude and negative polarity is illustrated in Fig. 2. From the figure, it can be observed that

Fig. 1 Block diagram of proposed method for R-peak detection

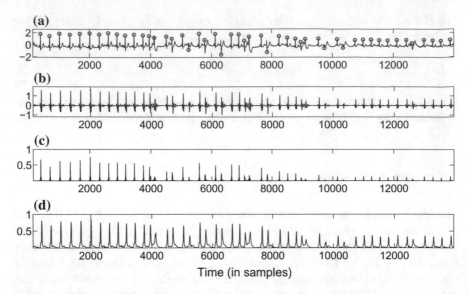

Fig. 2 Illustration of the performance of Hilbert envelope and squaring approach for the ECG signal with low amplitude and negative QRS complexes. **a** ECG signal, and **b** bandpass filtered ECG signal. Transformed bandpass filtered ECG signals with **c** squaring approach and **d** Hilbert envelope approach

with squaring approach the amplitude of the R-peak candidates, especially those with low amplitude is considerably diminished. On the other hand, Hilbert envelope magnifies the peaks which approximately represent R-peak instants. This indicates that the R-peak detection rate, especially in the presence of negative and low-amplitude R-peaks can be enhanced by using Hilbert envelope transformation. Therefore, the second step in the proposed method is to apply the Hilbert transform on the bandpass filtered ECG signal to extract Hilbert envelope. Then, in the third step, R-peak candidates are estimated by comparing the Hilbert envelope against a fixed amplitude and duration thresholds. Only those peaks with amplitude values greater than 0.2 and separated from their neighbors by a minimum distance of 140 samples are considered as R-peak candidates. These thresholds have been taken into consideration after experimenting with large data. It is observed that there is a slight difference between the estimated R-peaks locations and the correct instants of R-peaks. Therefore, the true R-peak instants are obtained by searching on the ECG signal for the largest amplitude within ±25 samples of each estimated R-peak location. Figure 3 demonstrates the proposed R-peak detector for a normal ECG signal.

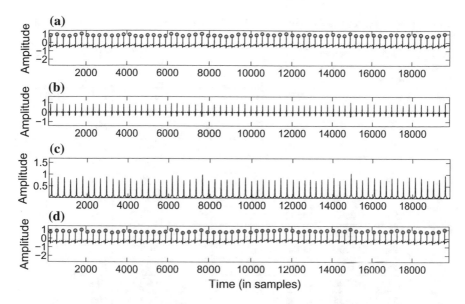

Fig. 3 Illustration of proposed method. **a** ECG signal taken from a record 100. On the ECG signal, manually annotated R-peaks (present in the database) are shown with red circles. **b** Bandpass filtered signal, **c** Hilbert envelope, and **d** output of R-peak detector. Peaks detected with proposed method are marked with blue circles

4 Results and Discussion

The performance of the proposed R-peak detection method is evaluated using 30 ECG records from the *defacto* standard MIT-BIH arrhythmia database [18], which contain noise, artifacts, low amplitude, and negative R-peaks. The ECG signals are sampled at 360 Hz. The manually annotated R-peaks present in the database serves as the ground truth. The length of each ECG record is 30 min, and hence, the total duration is 15 h. Table 1 summarizes the R-peaks detection results for the 30 ECG recordings using the proposed method. True positive (TP) implies that a R-peak is identified precisely with the proposed method, false positive (FP) implies that an erroneous peak is detected as R-peak, and false negative (FN) implies that a true R-peak is missed out. The performance is evaluated using three standard measures: detection error rate (DER), positive predictivity (+P), and sensitivity (Se) defined as

$$Se = \frac{TP}{TP + FN} * 100 \tag{3}$$

$$+P = \frac{TP}{TP + FP} * 100 \tag{4}$$

Table 1 Performance evaluation the proposed R-peak detection method using MIT-BIH arrhythmia database

ECG record no.	Total true R-peaks	Detected R-peaks	TP	FP	FN	SE (%)	+P (%)	DER (%)
100	2273	2273	2273	0	0	100	100	0.00
101	1865	1866	1861	5	4	99.79	99.73	0.48
103	2084	2084	2084	0	0	100	100	0.00
104	2229	2233	2218	15	11	99.51	99.33	1.17
105	2572	2583	2556	27	16	99.40	98.96	1.6
106	2027	2030	2027	3	0	100	99.85	0.15
108	1763	1748	1746	2	17	99.04	99.89	1.08
109	2532	2532	2532	0	0	100	100	0.00
112	2539	2539	2539	0	0	100	100	0.00
113	1795	1796	1795	1	0	100	99.95	0.05
115	1953	1953	1953	0	0	100	100	0.00
116	2412	2396	2392	4	20	99.17	99.83	1.00
117	1535	1535	1535	0	0	100	100	0.00
118	2278	2280	2278	2	0	100	99.92	0.08
121	1863	1861	1861	0	2	99.90	100	0.10
122	2476	2476	2476	0	0	100	100	0.00
123	1518	1518	1518	0	0	100	100	0.00
200	2601	2603	2598	5	3	99.89	99.81	0.3
202	2136	2126	2126	0	10	99.53	100	0.48
212	2748	2748	2748	0	0	100	100	0.00
213	3251	3248	3238	10	13	99.60	99.69	0.71
219	2154	2155	2154	1	0	100	99.95	0.04
220	2048	2045	2045	0	3	99.85	100	0.14
221	2427	2426	2426	0	1	99.96	100	0.04
223	2605	2606	2604	2	1	99.96	99.92	0.12
228	2053	2065	2050	15	3	99.85	99.27	0.87
230	2256	2257	2256	1	0	100	99.96	0.04
231	1571	1571	1571	0	0	100	100	0.00
232	1780	1780	1780	0	0	100	100	0.00
234	2753	2754	2753	1	0	100	99.97	0.03
Overall	66097	66087	65993	94	104	99.83	99.86	0.31

Table 2 Comparison of the R-peaks detectors for the specific ECG records from MIT-BIH arrhythmia database

Record no.	Number of false negative (FN) detections					Number of false positive (FP) detections				
	Bandpass filtering [4]	Low-pass filtering	EMD-based method [10]	WT-based method [7]	Proposed method	Bandpass filtering [4]	Low-pass filtering	EMD-based method [10]	WT-based method [7]	Proposed method
105	22	31	14	3	16	67	21	15	18	27
108	22	49	9	50	17	199	34	68	99	2
112	1	0	0	0	0	0	0	3	0	0
113	0	0	0	1	0	0	0	6	0	1
116	22	20	–	2	20	3	2	–	1	4
219	0	1	1	1	0	0	1	0	1	1
228	5	37	22	5	3	25	36	38	10	15
232	1	1	0	2	0	6	0	26	0	0
Total	73	138	46	64	56	300	94	176	129	50

$$DER = \frac{FP + FN}{TP} * 100 \tag{5}$$

There are totally 66097 R-peaks in the database, and the proposed method detected 65993 R-peaks correctly. Only 104 true R-peaks are missed out, and 94 false peaks are identified. Hence, the proposed method achieves a very high average sensitivity of 99.83%, and positive predictivity of 99.86% and very less detection error rate of 0.31%. In the MIT-BIH arrhythmia database, ECG records 105, 108, and 228 include artifact and high noise. Negative R-peaks and baseline drifts are also present in record 108. Abrupt changes and severe baseline drifts are present in records 112, 116, 219, and 228. Irregular rhythmic patterns can be observed in 219. Record 113 contains sharp and tall T waves. Long pauses are included in records 219 and 232. The total number of false detections (FN+FP) for these ECG recordings is more in all the existing algorithms. However, there is a significant improvement in R-peak detection in such signals using the proposed method. In Table 2, the proposed method is compared with the published results of the most popular R-peak detectors for the specific records. From the table, it can be seen that the bandpass filtering-based method [4] produced a maximum total detection failure of 373 (73 FN and 300 FP). The total false detections are 232 (138 FN and 94 FP) for the low-pass filtering-based method. The total number of false detections are 222 (46 FN and 176 FP) and 193 (64 FN and 129 FP) for the EMD-based method [10] and wavelet transform-based method [7], respectively. The proposed method produced a minimum total

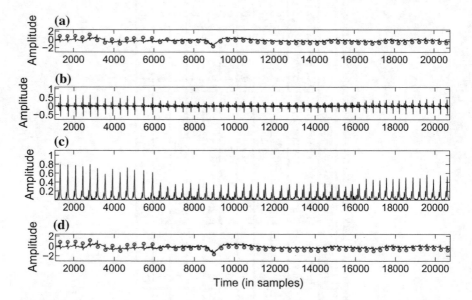

Fig. 4 Illustration of proposed method for a segment from ECG record 108. **a** ECG signal, **b** Bandpass filtered signal, and **c** Hilbert envelope. Manually annotated R-peaks are shown with red circles, and the R-peaks detected with the proposed method are shown with blue circles

detection failure of 106 (56 FN and 50 FP). The major improvement can be seen in case of the most difficult ECG record 108. For this particular record, the total number of false detections is only 19 with the proposed method while the next best being the EMD-based method with 77 false detections. For remaining records (apart from those shown in Table 2), the performance of all the methods is equally well. Figure 4 illustrates the proposed method for a segment from the ECG record 108. The ECG signal in Fig. 4(a) contains severe baseline drift, noise, low amplitude, and negative R-peaks. From Fig. 4(d), it can be seen that the proposed method detects R-peaks accurately even in such signals. Thus, the proposed method significantly outperforms the other well-known methods for the automatic detection of R-peaks in ECG signals.

5 Conclusion

This paper proposes a simple and efficient method for detecting R-peaks automatically in the ECG signals. First, different noises are removed by filtering the ECG signal with a zero-phase Butterworth bandpass filter. Then, the Hilbert transform is applied to the ECG signal to extract the Hilbert envelope. Finally, using suitable thresholds, the approximate R-peak instants are estimated, which are further processed to detect the true R-peaks. The proposed method is validated using 30 ECG records of the standard MIT-BIH arrhythmia database. The proposed method has a very less detection error rate of 0.31% and performs better compared to other existing methods in the case of ECG signals with noise, artifacts, baseline drifts, low amplitude, and negative R-peaks.

References

1. Gary Friesen, M., et al.: A comparison of the noise sensitivity of nine QRS detection algorithms. IEEE Trans. Biomed. Eng. 37(1), 85–98 (1990)
2. Yeh, Y.-C., Wang, W.-J.: QRS complexes detection for ECG signal: the difference operation method. Comput. Methods Programs Biomed. 91(3), 245–254 (2008)
3. Crema, C., et al.: Efficient R-peak detection algorithm for real-time analysis of ECG in portable devices. In: IEEE Sensors Applications Symposium (SAS), pp. 1–6 (2016)
4. Pan, Jiapu, et al.: A real-time QRS detection algorithm. IEEE Trans. Biomed. Eng. 32(3), 230–236 (1985)
5. Natalia Arzeno, M., et al.: Analysis of first-derivative based QRS detection algorithms. IEEE Trans. Biomed. Eng. 552, 478–484 (2008)
6. Patrick Hamilton, S., Willis Tompkins, J.: Quantitative investigation of QRS detection rules using the MIT/BIH arrhythmia database. IEEE Trans. Biomed. Eng. 33(12), 1157–1165 (1986)
7. Bouaziz F., et al.: Multiresolution wavelet-based QRS complex detection algorithm suited to several abnormal morphologies. IET Signal Process. 8(7), 774–782 (2014)
8. Elgendi, M., et al.: R wave detection using Coiflets wavelets. In: Proceedings of IEEE 35th Annual Northeast Conference on Bioengineering, Boston, pp. 1–2 (2009)

9. Abibullaev, B., Seo, H.D.: A new QRS detection method using wavelets and artificial neural networks. J. Med. Syst. **35**(4), 683–691 (2011)
10. Xing, H., Huang, M.: A new QRS detection algorithm based on empirical mode decomposition. In: Proceedings of IEEE International Conference on Bioinformatics and Biomedical Engineering, pp. 693–696 (2008)
11. Saini, I., et al.: QRS detection using K-Nearest Neighbor algorithm and evaluation on standard ECG databases. J. Adv. Res. **4**(4), 331–344 (2013)
12. Liao, Y., et al.: Accurate ECG R-peak detection for telemedicine. In: Proceedings of IEEE International Conference on Humanitarian Technology, pp. 1–5 (2014)
13. Sabarimalai Manikandan, M., Soman, K.P.: A novel method for detecting R-peaks in electrocardiogram (ECG) signal. Biomed. Signal Process. Control **7**(2), 118–128 (2012)
14. Zhu, H., Dong, J.: An R-peak detection method based on peaks of Shannon energy envelope. Biomed. Signal Process. Control **8**(5), 466–474 (2013)
15. Alexander Poularikas, D.: Transforms and Applications Handbook. CRC Press, Boca Raton (2010)
16. Nitish Thakor, V., et al.: Estimation of QRS complex power spectra for design of a QRS filter. IEEE Trans. Biomed. Eng. **30**(11), 702–706 (1984)
17. Poungponsri, S., Yu, X.-H.: An adaptive filtering approach for electrocardiogram (ECG) signal noise reduction using neural networks. Neurocomputing **117**, 206–213 (2013)
18. George Moody, B., Roger Mark, G.: The MIT-BIH arrhythmia database on CD-ROM and software for use with it. In: IEEE Proceedings Computers in Cardiology, pp. 185–188 (1990)

Lung Nodule Identification and Classification from Distorted CT Images for Diagnosis and Detection of Lung Cancer

G. Savitha and P. Jidesh

Abstract An automated computer-aided detection (CAD) system is being proposed for identification of lung nodules present in computed tomography (CT) images. This system is capable of identifying the region of interest (ROI) and extracting the features from the ROI. Feature vectors are generated from the gray-level covariance matrix using the statistical properties of the matrix. The relevant features are identified by adopting principle component analysis algorithm on the feature space (the space formed from the feature vectors). Support vector machine and fuzzy C-means algorithms are used for classifying nodules. Annotated images are used to validate the results. Efficiency and reliability of the system are evaluated visually and numerically using relevant measures. Developed CAD system is found to identify nodules with high accuracy.

1 Introduction

Medical imaging is a process of creating visual representation of the interior parts of the body, such as organs and tissues. It includes radiology which uses the imaging technologies like X-ray radiography, magnetic resonance imaging, ultrasound thermography, and computed tomography (CT) techniques such as positron emission tomography (PET), single photon emission computed tomography (SPECT), etc.

The imaging modalities provide vital information on abnormalities or disorders in the human body. Among different types of cancers, lung cancer is more severe and common; hence, cancer detection plays a vital role. Primary lung cancer originating

G. Savitha (✉) · P. Jidesh
National Institute of Technology Karnataka, Surathkal 575025, Karnataka, India
e-mail: gsavitha24@gmail.com

P. Jidesh
e-mail: jidesh@nitk.edu.in

© Springer Nature Singapore Pte Ltd. 2019
M. Tanveer and R. B. Pachori (eds.), *Machine Intelligence and Signal Analysis*,
Advances in Intelligent Systems and Computing 748,
https://doi.org/10.1007/978-981-13-0923-6_2

in lung cells manifests itself as tumors (nodules) in lungs. It is observed that 5-year survival rate for all the stages combined is only 16% [1]. Nodules have homogeneous soft-tissue attenuation on CT scans. Extensive research work has been carried out in computer-aided detection for nodules in chest CT scans [8, 13, 14, 16].

A dedicated CAD algorithm for detecting lung nodule is need of the hour. Generally, texture and intensity features are used for defining slice-based CAD system and performance evaluation is carried out by receiver operating characteristic curve (ROC) method [10]. Vessel suppression, intensity, and texture features are considered for detecting and segmenting the nodules by automatic scheme in [17]. A multilevel detection scheme adopting classification at both voxel and object level with focus on small volumes of interest produced by detector algorithm is introduced by [16].

The present study aims at developing an automated CAD system to identify the nodules present in lung CT images. It involves a process of denoising, segmentation, and feature identification followed by selection and classification.

2   Image Data Acquisition

Low dose CT scans collected from LIDC/IDRI database generated by Diagnostic Image Analysis Group, Department of Radiology and Nuclear Medicine, Radboud University Medical Center, Nijmegen, The Netherlands are used. 888 CT scans are available in LIDC-IDRI database.

3   Methodology

An automated CAD system for identifying the nodules present in lung CT images is tried to develop. The process involves suppression of noise in the raw input images followed by identification and segmentation of ROI from the processed images. Relevant features are identified and classified into two groups. They are analyzed to establish the reliability of the system. The overall design of the model is presented in Fig. 1.

3.1   Data Preprocessing

The raw images gather unwanted information during capture, storage, transmission, processing which are termed as noise. Hence, denoising is a vital preprocessing step in this model, since the segmentation fails to extract the ROI in a noisy environment. The medical images distorted by noise can be identified by one of the probability distribution functions such as Gamma, Poisson, Gaussian, and Rayleigh [7].

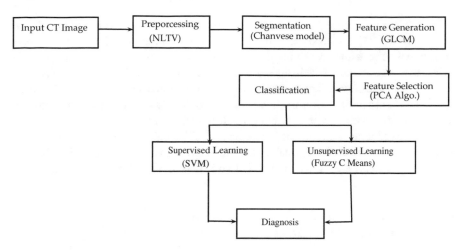

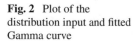

Fig. 1 Flowchart of the proposed CAD system

Fig. 2 Plot of the distribution input and fitted Gamma curve

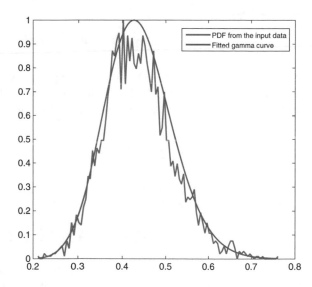

An analysis of noise distribution from the input image is done empirically. A low-oscillatory region (constant/homogeneous intensity region) from the input data is extracted as the variation in these regions is generally due to noise. It is observed that PDF (probability density function) of plotted distribution (evaluated from the histogram) is similar to Gamma noise. The noise parameters k (shape) and σ (scale) are estimated using the maximum likelihood estimators. The plot of the distribution using the estimated parameters fits well to the plot of the distribution from the data selected from low-oscillatory region. The two plots are presented in Fig. 2. Therefore, it could be concluded that the noise more likely follows Gamma distribution.

It has been observed from the literature that variety of filters or combination of filters are extensively used for denoising of nodular images. Wiener filter is used to remove noise while preserving the edges and fine details of nodular images [12]. Also, the combination of Weiner filter and median filter is used for denoising images.

Since it is observed that the noise follows a Gamma law, the noise is source-dependent and multiplicative in nature. Hence, generally used methods for denoising do not perform well in case of data-correlated distribution.

A nonlocal total variation minimization (NLTV) model is derived using the maximum a posteriori (MAP) estimator for Gamma distributed noise adopting the concepts from Aubert and Aujol method [2] and the same is introduced in the present study.

3.1.1 Nonlocal Total variation

A total variation (TV) regularization under nonlocal framework [6] for additive Gaussian noise can be stated as

$$\min_u \{\|\nabla_{NL}u\| + \lambda\|u - u_0\|_2^2\}, \tag{1}$$

where u and u_0 are the original and noisy images, respectively, λ denotes a regularization parameter, and $\|.\|$ is the TV norm. The above model takes the following form under a Gamma distributed noise setup (derived through a MAP estimator process):

$$\min_u \left\{ \int_\Omega \left(\|\nabla_{NL}u\| + \lambda \log u + \frac{u_0}{u} \right) dxdy \right\}. \tag{2}$$

The nonlocal gradient of a function $u : \Omega \longrightarrow \mathbb{R}$, for a pair of points or pixels $(x, y) \in \Omega \times \Omega$ is defined as

$$\nabla_{NL}u(x, y) = (u(y) - u(x))\sqrt{w(x, y)} : \Omega \times \Omega \longrightarrow \mathbb{R}, \tag{3}$$

where $w(x, y)$ defines the weight of edge between x and y. It is assumed that the $w : \Omega \times \Omega \longrightarrow \mathbb{R}$ is symmetric, i.e., $w(x, y) = w(y, x)$. Therefore, norm of the nonlocal (NL) gradient is defined as

$$|\nabla_{NL}u|(x) = \sqrt{\int_\Omega (u(y) - u(x))^2 w(x, y)dy} : \Omega \longrightarrow \mathbb{R}. \tag{4}$$

The projection scheme proposed in [3] is adopted for numerically solving the above-mentioned functional as most of the explicit schemes converge very slowly. The projection-based method converges faster than the explicit scheme.

3.2 Segmentation

Nodule location and lesion to be identified in the preprocessed images is achieved by segmentation process. This helps in treatment planning prior to radiation therapy.

Automatic segmentation of the nodules present in lungs is a primary requirement in CAD system. For example, a model has been developed to segment the thorax region from the background. Here, a gray-level distribution of the Wiener-filtered image is used to identify different regions [12].

The authors used pulmonary nodule segmentation algorithm to acquire accurate nodule segmentation for solid nodules [11]. A robust nodule segmentation is achieved by an efficient combination of morphological operations.

In this study, a contour-based model proposed in [4] is used to segment the region of interest as it is found suitable for segmenting images without proper edge details. A level set function ϕ defining a contour on the image can split and merge simultaneously as the topology of the image changes. Hence, more than one boundary can be defined with this approach. Multiple initial contours can be placed which helps in automatic segmentation. The energy functional E is given as

$$E(c_1, c_2, \phi) = \lambda_1 \int_{\Omega} (u(x) - c_1(x))^2 H(\phi(x)) dx + \lambda_2 \int_{\Omega} (u(x) - c_2(x))^2 (1 - H(\phi(x))) dx \quad (5)$$

in which $u(x)$ represents the pixel value, ϕ is level set function that represents the contour C in domain Ω. Here, $c_1(x)$ and $c_2(x)$ are the constants referring to average intensities inside and outside the curve. It is defined as

$$c_1(x) = \frac{\int_{\Omega} g_k(x - y)(u(y)H(\phi(y))) dy}{\int_{\Omega} g_k(x - y)H(\phi(y)) dy} \quad (6)$$

and

$$c_2(x) = \frac{\int_{\Omega} g_k(x - y)(u(y)(1 - H(\phi(y)))) dy}{\int_{\Omega} g_k(x - y)(1 - H(\phi(y))) dy}, \quad (7)$$

where g_k is the Gaussian kernel function and $g_k(x - y)$ is the weight assigned to each intensity $u(y)$ at y.

3.2.1 Morphological Operation

Morphological postprocessing is carried out to combine similar featured areas. It is observed that a disk structuring element of 4×4 pixel is found to give nodule size of 10–30 mm diameter as given by [9]. In the present study, "open" filter followed by "dilation" operation is adopted to make the segmented areas more clearly visible.

Table 1 Features considered

Autocorrelation	Contrast	Correlation	Cluster prominence
Cluster shade	Dissimilarity	Energy	Entropy
Homogeneity	Maximum probability	Sum of squares (variance)	Sum average
Sum variance	Sum entropy	Difference variance	Difference entropy
Information measure of correlation	Inverse difference normalized	Inverse difference moment normalized	

3.3 Feature Generation

Segmentation process is followed by identification of features required for classification of nodules. The classifier output determines the possibility of the given structure as nodule or non-nodule. Features such as intensity, texture, shape, and context of an image have been used to define lung nodules [9]. First-order and second-order statistical properties are calculated, which give relevant and distinguishable features without involving any computational transformations [5]. Gray-level covariance matrix (GLCM) performs well compared to other statistical properties in homogeneous images [15]. In the present study, second-order statistical properties are considered as it allows to use pixel neighborhood relationship in an image.

GLCM is created from a grayscale image. It calculates how often a pixel with gray-level value of i occurs whether horizontally, vertically or diagonally to adjacent pixel with value j. The 19 statistical properties generated from GLCM of an image given in Table 1 are considered as features for further process. However, the interdependence of these features may lead to bias results in identifying the nodules. Hence, it is important to choose relevant features from these and is achieved by principal component analysis (PCA) algorithm.

3.3.1 Feature Selection

Principal component analysis (PCA) is used for identifying fewer number of uncorrelated variables where specific features need to be identified in a large data set while retaining much of its variation. The components are ordered in such a way that first few components carry most of the variations of the original data. Mathematically this concept is achieved by solving the eigenvalue problem of square symmetric matrix. Eigenvalues are arranged in decreasing order and the corresponding eigenvectors form a new basis for the feature space. The first few eigenvalues highlight the most relevant feature pertaining to nodules present in an image and its corresponding eigenvectors determine the direction of uncorrelated variables. A Gaussian plot is drawn to identify relevant number of eigenvalues for identification of nodules. The eigenvalues which cover 90% area of the plot are taken into consideration. It is observed that only two to three prominent features among the 22 can be considered for further analysis.

3.4 Classification

Classification assigns the class labels for similar kind of data. Classes in the present analysis concentrate only on the nodules and non-nodules in CT lung images. For the purpose, support vector machine (SVM) and fuzzy C-means have been considered.
Support Vector Machine (SVM): SVM is one of the supervised classification technique which performs classification tasks by constructing hyperplanes in a multidimensional space that separates cases of different class labels. It works well on the linearly separable data sets. A clear distinction of classes in data sets is created by drawing hyperplanes which correspond to the line passing through the middle of two classes. Vectors(cases) that define hyperplane are called support vectors. Kernels can be linear, polynomial or radial basis function (RBF).
Fuzzy C-Means: It is an unsupervised learning classification technique where each piece of data can belong to more than one cluster. Each data is assigned to a class such a way that items in the same class are similar to each other. clusters are identified using the similarity measures such as distance, connectivity, and intensity based on the data or application. A number of clusters are chosen and each data is assigned to one of the clusters randomly. This method is repeated until the algorithm converges.

4 Results and Discussions

4.1 Data Preprocessing

NLTV method adopted in the present analysis for denoising is found to remove noise to the maximum extent while preserving fine details, edges, and texture.

Local Region-based Chan–Vese model adopted for the purpose of segmentation works on the principle of energy. Level set function ϕ is evolved to define the area enclosed in a contour. This technique identifies the prominent edges of the region bounded by the contour. The method adopted for the purpose of segmentation is found to define the region of interest (nodular structures) precisely.

Relevant ROI from the defined contours is extracted by applying morphological operators such as opening followed by dilation operator. In the process, small protuberances and other irrelevant connections are removed from the main region of interest. Figure 3 shows the process of preprocessing.

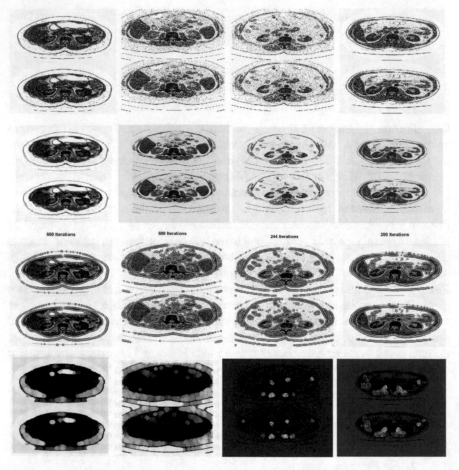

Fig. 3 Row 1: Original lung CT images. Row 2: Denoised images using NLTV. Row 3: Segmented lung images using Chan–Vese segmentation method. Row 3: Results of morphological operations

4.2   Feature Identification and Selection

The gray-level covariance matrix (GLCM) is used to compute second-order statistical features for all the images considered in this study. Dimensionality reduction model named principle component analysis is adopted for identifying prominent features among this pool of features.

The first few eigenvalues covering 90% area under the curve given in Fig. 4 depicts the prominent features to identify the nodules. This is achieved by plotting standard deviation (σ) along the abscissa on both sides of origin. It is observed that 3σ value covers 90% area under the curve, indicating that the first three eigenvalues (vectors) are the significant ones which can identify the nodules precisely.

Fig. 4 Gauss curve of eigenvalues

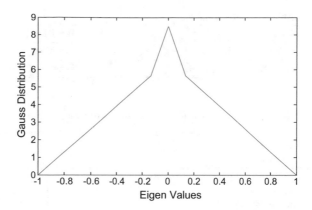

Table 2 Classification results of SVM

Number of images	True positive	True negative	False positive	False negative
40 (22 nodule, 18 non-nodule)	18	20	2	0
80 (44 nodule, 36 non-nodule)	33	43	1	3
120 (43 nodule, 77 non-nodule)	74	41	2	3
160 (83 nodule, 77 non-nodule)	73	79	4	4
200 (111 nodule, 89 non-nodule)	84	105	6	5

4.3 Classification

4.3.1 SVM Method

Out of randomly selected 500 CT lung images 300 images are considered as training data set. These images are assigned classes based on the computed feature values for annotated images. The testing data set are randomly grouped into five sets consisting of 40, 80, 120, 160 and 200 images. Each set is classified using SVM method. The result is given in Table 2.

It can be observed from Table 2 that the number of false prediction is minimum. The efficiency of the system is tested by computing the accuracy measures such as accuracy, precision, recall, specificity, error rate, F-measure for grouped images in the training set. The expressions for computing these parameters are given in Table 3. Table 4 presents the computed values for these parameters.

Table 3 Accuracy measures

Accuracy = $\frac{TP+TN}{(TP+TN+FP+FN)}$	Precision (P) = $\frac{TP}{(TP+FP)}$
Recall (sensitivity) (R) = $\frac{TP}{(TP+FN)}$	Specificity = $\frac{TN}{(TN+FP)}$
Error rate = $\frac{FP+FN}{(TP+TN+FP+FN)}$	F-measure = $\frac{2*(P*R)}{(P+R)}$

Table 4 Performance analysis of SVM classification method

Image number Images	Accuracy (percent)	Precision	Recall	Specificity	Error rate	F-measure
40 (22 nodule, 18 non-nodule)	95	0.90	1	0.90	0.05	0.94
80 (44 nodule, 36 non-nodule)	95	0.97	0.91	0.97	0.05	0.94
120 (43 nodule, 77 non-nodule)	95.8	0.97	0.96	0.95	0.04	0.96
160 (83 nodule, 77 non-nodule)	95	0.94	0.94	0.95	0.05	0.94
200 (111 nodule, 89 non-nodule)	94.5	0.93	0.94	0.94	0.05	0.93

It is observed from Table 4 accuracy in all the cases is found to be more than 94%. Specificity is on an average of 95%, sensitivity(recall) is in the order of 94%. Error rate presented in all is found to be in the range of 0.05%.

4.3.2 Fuzzy C-Means

Classification is done using fuzzy C-means algorithm for the same set of 200 test images used in SVM method. The classification results are presented in Table 5 and its performance analysis is presented in Table 6.

It can be seen from Table 6 that accuracy in all the cases is again more than 94%. Specificity and sensitivity are in the order of 94%. Error rate of this algorithm is found to be 0.05% similar to SVM.

Receiver operating characteristics (ROC) curves are plotted for both the classification methods considered in the present study. These curves are generally considered for studying the performance of system. They are presented in Fig. 5.

Table 5 Classification results of fuzzy C-means

Number of images	True positive	True negative	False positive	False negative
40 (22 nodule, 18 non-nodule)	18	20	2	0
80 (44 nodule, 36 non-nodule)	33	42	2	3
120 (43 nodule, 77 non-nodule)	71	42	1	6
160 (83 nodule, 77 non-nodule)	72	80	3	5
200 (111 nodule, 89 non-nodule)	83	107	4	6

Table 6 Performance analysis of fuzzy C-means algorithm

Image number Images	Accuracy (percent)	Precision	Recall	Specificity	Error rate	F-measure
40 (22 nodule, 18 non-nodule)	95	0.90	1	0.90	0.05	0.94
80 (44 nodule, 36 non-nodule)	93.7	0.94	0.91	0.95	0.062	0.92
120 (43 nodule, 77 non-nodule)	94.1	0.98	0.92	0.97	0.05	0.95
160 (83 nodule, 77 non-nodule)	95	0.96	0.93	0.96	0.05	0.94
200 (111 nodule, 89 non-nodule)	95	0.95	0.93	0.96	0.05	0. 94

It can be seen that the area under the curve is nearing one indicating that the classification is acceptable in both cases. However, they can be further compared using F-measure and accuracy curves.

The parameter values of the two classification techniques are plotted against number of samples and are given in Fig. 6.

From these graphs, it can be observed that both the classification methods give consistent results with marginal increase in the supervised classification method.

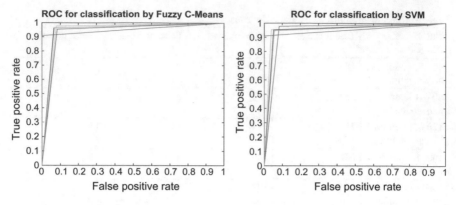

Fig. 5 ROC curve for SVM and fuzzy C-means

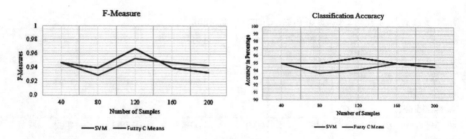

Fig. 6 Plot showing F-measure and classification accuracy

5 Conclusion

An automated CAD system for identification of lung nodules present in lung CT images is developed. This system includes the process such as denoising using the nonlocal TV, identifying the region of interest using Chan–Vese model, identification and selection of features using PCA, and classification using SVM and fuzzy C-means technique. Sensitivity of the CAD system developed using supervised (SVM) classification is in the order of 94% while for unsupervised (FCM) classification, it is 93%. Though supervised classification gives a slightly better result, unsupervised classification is found to give equally good result. Hence, the unsupervised classification method may be preferred which makes the system more consistent. CAD system developed in the study is found to be efficient in identifying the lung nodules in CT images without human intervention. Mathematical concepts adopted for identifying the lung nodules makes the developed CAD system more reliable. The processing time required for the developed CAD system is approximately 76 s on a normal machine.

References

1. American Cancer Society, Cancer Facts and Figures, America Cancer Society (2016). http://www.cancer.org/acs/groups/content/@research/documents/document/acspc-047079.pdf
2. Aubert, G., Aujol, J.F.: A variational approach to removing multiplicative noise. SIAM J. Appl. Math **68**(5), 925–946 (2008)
3. Chambolle, A.: An algorithm for total variation minimization and applications. J. Math. Imaging Vis. **20**(1), 89–97 (2004)
4. Chan, T.F., Vese, L.A.: Active contours without edges. IEEE Trans. Image Process. **10**(2), 266–277 (2001)
5. Chen, L., et al.: Speech emotion recognition: features and classification models. Digit. Signal Process. **22**(6), 1154–1160 (2012)
6. Gilboa, G., Osher, S.: Nonlocal operators with applications to image processing. Multiscale Model. Simul. **7**(3), 1005–1028 (2008)
7. Gravel, P., Beaudoin, G., De Guise, J.A.: A method for modeling noise in medical images. IEEE Trans. Med. Imaging **23**(10), 1221–1232 (2004)
8. Jacobs, C., et al.: Computer-aided detection of ground glass nodules in thoracic CT images using shape, intensity and context features. In: International Conference on Medical Image Computing and Computer-Assisted Intervention. Springer, Berlin (2011)
9. Jacobs, C., et al.: Automatic detection of subsolid pulmonary nodules in thoracic computed tomography images. Med. Image Anal. **18.2**, 374–384 (2014)
10. Joo, S., et al.: Computer-aided diagnosis of solid breast nodules: use of an artificial neural network based on multiple sonographic features. IEEE Trans. Med. Imaging **23**(10), 1292–1300 (2004)
11. Kuhnigk, J.-M., et al.: Morphological segmentation and partial volume analysis for volumetry of solid pulmonary lesions in thoracic CT scans. IEEE Trans. Med. Imaging **25**(4), 417–434 (2006)
12. Magdy, E., Zayed, N., Fakhr, M.: Automatic classification of normal and cancer lung CT images using multiscale AM-FM features. J. Biomed. Imaging **2015**, 11 (2015)
13. Messay, T., Hardie, R.C., Rogers, S.K.: A new computationally efficient CAD system for pulmonary nodule detection in CT imagery. Med. Image Anal. **14**(3), 390–406 (2010)
14. Murphy, K., et al.: A large-scale evaluation of automatic pulmonary nodule detection in chest CT using local image features and k-nearest-neighbour classification. Med. Image Anal. **13**(5), 757–770 (2009)
15. Soh, L.-K., Tsatsoulis, C.: Texture analysis of SAR sea ice imagery using gray level co-occurrence matrices. IEEE Trans. Geosci. Remote Sens. **37**(2), 780–795 (1999)
16. Tao, Y., et al.: Multi-level ground glass nodule detection and segmentation in CT lung images. In: International Conference on Medical Image Computing and Computer-Assisted Intervention. Springer, Berlin (2009)
17. Zhou, J., et al.: An automatic method for ground glass opacity nodule detection and segmentation from CT studies. In: 28th Annual International Conference of the IEEE Engineering in Medicine and Biology Society, 2006. EMBS'06. IEEE, 2006

Baseline Wander and Power-Line Interference Removal from ECG Signals Using Fourier Decomposition Method

Pushpendra Singh, Ishita Srivastava, Amit Singhal and Anubha Gupta

Abstract Analysis of electrocardiogram (ECG) signals helps us in detecting various abnormalities and diseases of heart. These signals commonly suffer from the problems of baseline wander and power-line interference. In this paper, we propose a new approach to eliminate such noises from ECG signals using the Fourier decomposition method. Simulation results are presented to show the efficacy of our method over previously used EMD-based methods. The proposed method has been shown to preserve shape characteristics of ECG signals of heart abnormalities.

Keywords Baseline wander and Power-line interference · ECG signal
Empirical mode decomposition · Fourier decomposition method · Linearly
independent non-orthogonal yet energy preserving (LINOEP) vectors

1 Introduction

Electrocardiogram (ECG) signals are used to examine the activity of human heart. It records the electrical activity of the heart over a period of time with the help

P. Singh (✉)
School of Engineering and Applied Sciences, Bennett University, Greater Noida, India
e-mail: pushpendrasingh@iitkalumni.org; spushp@gmail.com

I. Srivastava · A. Singhal
Jaypee Institute of Information Technology, Noida, India
e-mail: ishitasrivastava1996@gmail.com

A. Singhal
e-mail: singhalamit.iitd@gmail.com

A. Gupta
Indraprastha Institute of Information Technology-Delhi (IIIT-D), New Delhi, India
e-mail: anubha@iiitd.ac.in

© Springer Nature Singapore Pte Ltd. 2019
M. Tanveer and R. B. Pachori (eds.), *Machine Intelligence and Signal Analysis*,
Advances in Intelligent Systems and Computing 748,
https://doi.org/10.1007/978-981-13-0923-6_3

of electrodes placed on the skin. An ECG signal is used to detect any disease or abnormality in the functioning of heart, if present. Generally, for a normal ECG signal, the amplitude range is 10–5 mV and frequency range is 0.05–100 Hz [1]. In today's era, automated ECG analysis is very popular in telemedicine [2].

There are various problems that may arise while recording an ECG signal. It may get distorted due to the presence of various noises such as baseline wander (BLW), channel noise, power-line interference (PLI), physiological artifact, etc. Due to the presence of these noises, it becomes difficult to diagnose diseases and thus, appropriate treatment may be impacted [3]. In this paper, we focus on the removal of BLW noise and power-line interference to obtain clean ECG signals using the Fourier decomposition method (FDM) that is based on discrete Fourier transform (DFT) and finite impulse response (FIR) filtering.

PLI is a high-frequency noise of 50 or 60 Hz, while BLW is a low-frequency noise of usually [0–0.5] Hz. BLW is the effect where the base axis (X-axis) of any signal appears to "wander" or move up and down rather than be straight when viewed on a screen (like oscilloscope). This causes the entire signal to shift from its normal base. There are many methods that have been proposed to remove BLW and power-line interference from ECG signals. One of the most common methods for removing BLW is high-pass filtering with cutoff frequency of 0.7 Hz. However, the use of high-pass filter may lead to wrong diagnosis at times. Some other techniques include adaptive filters, notch filters, LMS algorithm, wavelet transform, projection operator-based approach, empirical mode decomposition (EMD) methods [4–7], robust sparse signal decomposition, cubic spline method, etc.

The problem of signal distortion in stress tests due to conventional adaptive filters that use complex detector has been highlighted in [8]. Removal these noises using notch filters is explored in [9]. In LMS algorithm, the problem of BLW is formalized as a mean-square-error regression, and low-complexity least mean squares (LMS) solutions have been proposed that comply with wearable device implementation requirements [10]. Similar blocks of samples are grouped in a matrix and then denoising is achieved by the shrinkage of its two-dimensional discrete wavelet transform coefficients in [11]. Projection operator approach includes noise subspace that is generated using sample functions of the first-order fractional Brownian motion fBm processes characterizing BLW noise. The orthogonal projection of noisy ECG signal onto the noise subspace provides an estimate of BLW noise [12, 13]. In EMD-based approach, the power-line interference is canceled by passing the first IMF of the Empirical mode decomposed noisy signal through FIR low-pass filter as discussed in [14–18].

In addition to EMD, there exist several methods for nonlinear and nonstationary signal decomposition and analysis such as synchrosqueezed wavelet transforms (SSWT) [19], variational mode decomposition (VMD) [20], eigenvalue decomposition (EVD) [21], empirical wavelet transform (EWT) [22], and statistical modeling-based approaches [13, 23]. These approaches are based on the perception that the Fourier methods are not suitable for nonlinear system and nonstationary data analysis. However, the recently proposed Fourier decomposition method (FDM) [24], which is based on the Fourier theory and zero-phase filtering approach, has established that the Fourier methods are indeed most suitable tools for nonlinear system and nonstationary data analysis. Some recent studies on FDM [25–29] demonstrate its usefulness for nonlinear and nonstationary signal analysis. In this work, we present baseline wander and power-line interference removal from ECG signals using FDM, and compare the results with widely used empirical mode decomposition (EMD) algorithm. We have used real-life ECG data to demonstrate the efficacy of the proposed method.

The rest of the study is organized as follows: Sect. 2 presents the overview of FDM and its application on BLW and PLI removal. Results and discussions are presented in Sect. 3. Finally, conclusions are presented in Sect. 4.

2 Fourier Decomposition Method, BLW, and Power-Line Interference Removal

The FDM is an adaptive signal decomposition method based on the well-established Fourier theory and zero-phase filtering (ZPF) approach [24]. It decomposes a signal, $x[n]$ of length N, into a set of, M, Fourier intrinsic band functions (FIBFs), $\{c_i[n]\}$, such that

$$x[n] = a_0 + \sum_{i=1}^{M} c_i[n], \tag{1}$$

where a_0 is average value of the signal, and $\{c_1[n], \ldots, c_M[n]\}$ are the amplitude–frequency modulated (AM-FM) FIBFs which are adaptive, local, complete, orthogonal, or LINOEP by virtue of decomposition [24].

A block diagram of the FDM, which is a DFT-based zero-phase filter bank to obtain orthogonal FIBFs, is shown in Fig. 1. The frequency response of the ith band of this filter bank is defined as [30]

$$\left.\begin{aligned} H_i[k] &= 1, \quad (K_{i-1}+1) \leq k \leq K_i \text{ and } (N - K_i) \leq k \leq (N - K_{i-1} - 1),\\ &= 0, \quad \text{otherwise,} \end{aligned}\right\} \tag{2}$$

where $1 \leq i \leq M$; $K_0 = 0$ and $K_M = N/2$ (or $K_M = (N-1)/2$ if N is odd), K_i is the upper cutoff frequency of ith band, and M is the number of desired frequency bands. A component $c_i[n]$ is obtained by the inverse DFT (IDFT) as

$$c_i[n] = \sum_{k=0}^{N-1} \left[H_i[k] X[k] \exp(j2\pi kn/N) \right], \tag{3}$$

where $X[k] = \frac{1}{N} \sum_{n=0}^{N-1} x[n] \exp(-j2\pi kn/N)$ is the DFT of signal $x[n]$.

Algorithm 1: The FDM algorithm [30] to obtain LINOEP vectors $\mathbf{c}_i$ by decomposition of a signal $\mathbf{x}$ such that $\mathbf{x} = \mathbf{c}_0 + \sum_{i=1}^{M} \mathbf{c}_i$ and $\mathbf{c}_i \perp \sum_{l=i+1}^{M} \mathbf{c}_l$. The PART A (PART B) of algorithm generates $\{\mathbf{c}_1, \ldots, \mathbf{c}_M\}$ in order of highest to lowest (lowest to highest) frequency components.

%PART A
$\mathbf{c}_0 = mean(\mathbf{x})$;
$\mathbf{x}_1 = \mathbf{x} - \mathbf{c}_0$;
for $i = 1$ *to* $M - 1$ **do**
  $\mathbf{y}_i = ZPHPF_i(\mathbf{x}_i, f_{ci})$;
  $\mathbf{r}_i = \mathbf{x}_i - \mathbf{y}_i$;
  $\alpha_i = \frac{\langle \mathbf{y}_i, \mathbf{r}_i \rangle}{\langle \mathbf{r}_i, \mathbf{r}_i \rangle}$;
  $\mathbf{c}_i = \mathbf{y}_i - \alpha_i \mathbf{r}_i$;
  $\tilde{\mathbf{c}}_{i+1} = (1 + \alpha_i)\mathbf{r}_i$;
  $\mathbf{x}_{i+1} = \tilde{\mathbf{c}}_{i+1}$;

$\mathbf{c}_M = \tilde{\mathbf{c}}_M$;

%PART B
$\mathbf{c}_0 = mean(\mathbf{x})$;
$\mathbf{x}_1 = \mathbf{x} - \mathbf{c}_0$;
for $i = 1$ *to* $M - 1$ **do**
  $\mathbf{y}_i = ZPLPF_i(\mathbf{x}_i, f_{ci})$;
  $\mathbf{r}_i = \mathbf{x}_i - \mathbf{y}_i$;
  $\alpha_i = \frac{\langle \mathbf{r}_i, \mathbf{y}_i \rangle}{\langle \mathbf{y}_i, \mathbf{y}_i \rangle}$;
  $\mathbf{c}_i = (1 + \alpha_i)\mathbf{y}_i$;
  $\tilde{\mathbf{c}}_{i+1} = \mathbf{r}_i - \alpha_i \mathbf{y}_i$;
  $\mathbf{x}_{i+1} = \tilde{\mathbf{c}}_{i+1}$;

$\mathbf{c}_M = \tilde{\mathbf{c}}_M$;

Fig. 1 A block diagram of the FDM [30], which is the DFT-based zero-phase filter bank, to decompose a signal $x[n]$ into a set $\{c_1[n], c_2[n], \ldots, c_M[n]\}$ of orthogonal desired frequency bands

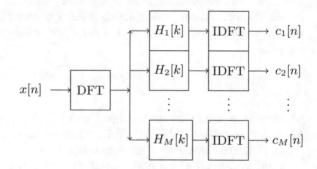

The FDM algorithm [30], which uses zero-phase noncausal finite impulse response (FIR) filters to decompose a signal into a set of LINOEP vectors, is summarized in Algorithm 1. For each iteration in this algorithm, $ZPHPF_i$ ($ZPLPF_i$) is zero-phase high (low) pass filter (e.g., FIR filtering using *filtfilt* function of MATLAB) with required cutoff frequency f_{ci}. The value of α_i is obtained such that $\mathbf{c}_i \perp \tilde{\mathbf{c}}_{i+1}$. As the filters are not ideal, i.e., non-brick wall frequency response, therefore $\mathbf{c}_i \not\perp \mathbf{c}_l$ for $i, l = 1, 2, \ldots, M - 1$, and only $\mathbf{c}_{M-1} \perp \mathbf{c}_M$. We use PART A (PART B) of algorithm to obtain $\{\mathbf{c}_1, \ldots, \mathbf{c}_M\}$ in the order of highest to lowest (lowest to highest) frequency components.

In order to eliminate BLW and PLI from ECG, we provide ECG signal as an input to FDM which decomposes the ECG data into a set of desired frequency bands. Bandlimited signals corresponding to BLW and PLI are removed, and by superposition of the rest of the signal components, a clean ECG signal is obtained. A block diagram of the proposed methodology is shown in Fig. 2.

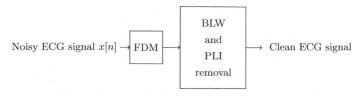

Fig. 2 A block diagram of the proposed baseline wander (BLW) and power-line interference (PLI) removal using FDM

3 Results and Discussions

The data for ECG signal of a subject suffering from MIT-Arrhythmia disease is available at physionet.org. We first consider the individual effects of baseline wander (BLW) and power-line interference (PLI) and then consider the combined effect of both these noises. Denoising is performed using FIR filtering and DFT-based FDM method.

We have added a low-frequency noise of [0–0.5] Hz to clean ECG signal to emulate the effect of baseline wander. The clean ECG signal and the BLW affected signal are shown in Fig. 3. We employ the DFT-based FDM method to obtain the clean signal. It is clearly visible from Fig. 4 that the proposed method successfully separates the ECG signal from BLW noise. The FIR-based FDM approach gives similar results on removing baseline wander noise, as discussed later in the paper.

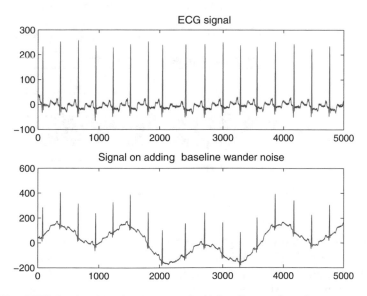

Fig. 3 Clean ECG signal and the signal corrupted with baseline wander noise

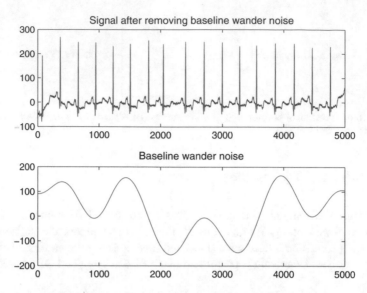

Fig. 4 DFT-based FDM approach to remove baseline wander noise

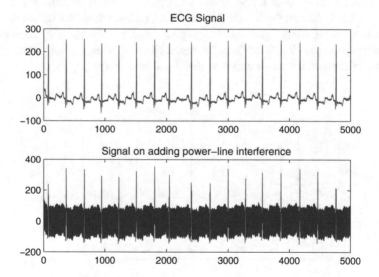

Fig. 5 ECG signal and the signal with power-line interference

Figure 5 depicts the clean signal and the signal obtained by adding a high-frequency noise of 50 Hz to produce an effect akin to PLI. The effect of PLI is removed using FIR-based FDM approach. The clean signal and PLI are separated out successfully, as shown in Fig. 6. Similar results are obtained by using the DFT-based FDM approach.

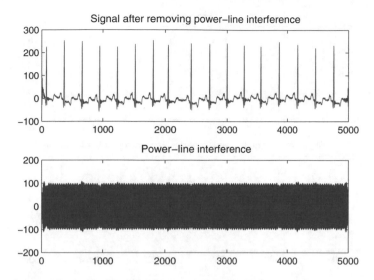

Fig. 6 FIR-based FDM approach to remove power-line interference

Lastly, we consider a noisy ECG signal such that it exhibits behavior of both high-frequency and low-frequency noise components. The low-frequency noise is BLW with a frequency of [0–0.5] Hz while the high-frequency noise is PLI with a frequency of 50 Hz. Figure 7 shows the clean and the noisy ECG signal. The denoising is performed using FIR-based and DFT-based FDM approaches and the results are compared with the EMD-based approach. It can be seen from Fig. 8 that the signal obtained after removing the noise using EMD-based method suffers from some distortions. Figure 9 shows the results for FIR-based FDM approach while the outcomes for the DFT-based FDM method are shown in Fig. 10. It is clearly visible that both the proposed methods are able to separate out the ECG signal from BLW and PLI affected signal. The obtained ECG signal does not have distortions and resembles the original ECG signal.

In order to present numerical comparison, we consider the input and output SNR defined as

$$\mathrm{SNR}_i = 10 \log \left(\frac{\sum_{n=0}^{N-1} x^2[n]}{\sum_{n=0}^{N-1} w^2[n]} \right), \quad \mathrm{SNR}_o = 10 \log \left(\frac{\sum_{n=0}^{N-1} x^2[n]}{\sum_{n=0}^{N-1} (x[n] - \tilde{x}[n])^2} \right), \quad (4)$$

respectively, where $x[n]$ is the original ECG signal, $w[n]$ is the sum of PLI and BLW noise, and $\tilde{x}[n]$ is the estimated ECG signal (i.e., after removing PLI and BLW noise) using the FDM and EMD approaches. A comparison of the FDM and EMD approaches using the input and output SNR criterion is shown in Fig. 11. This clearly demonstrates that the output SNR is best for FDM with DFT approach, followed by FDM with FIR filtering approach, and in both cases, the proposed method outperforms the widely used EMD algorithm.

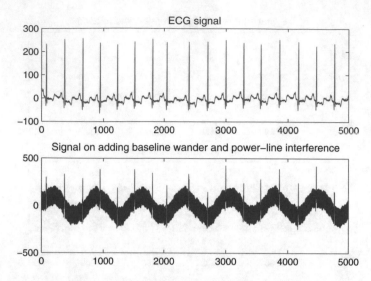

Fig. 7 ECG signal and the signal with both the noises: BLW and PLI

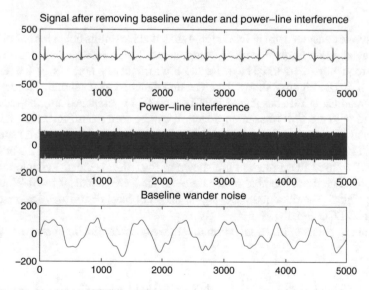

Fig. 8 Results obtained by EMD-based approach

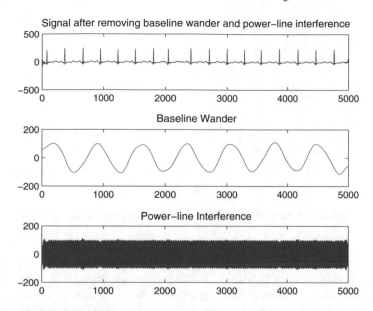

Fig. 9 Results obtained by FIR-based FDM approach

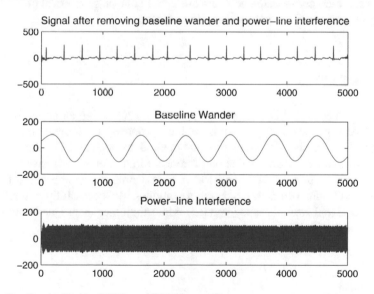

Fig. 10 Results obtained by DFT-based FDM approach

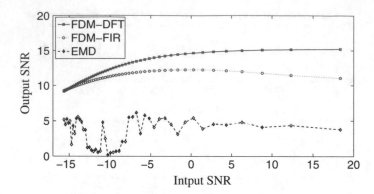

Fig. 11 A comparison of the FDM and EMD approaches using input and output SNR criterion

The performance of the FDM is better because it can decompose a signal into a set of desired frequency bands (i.e., FIBFs) which are orthogonal or LINOEP by construction. Thus, removal of low-frequency BLW, PLI, and high-frequency noise from a signal is easy, efficient, and effective. However, in the case of EMD, number of IMFs and their cutoff frequencies are not under control, so removal of BLW, PLI, and high-frequency noise from a signal is not effective.

4 Conclusion

An accurate analysis of ECG signal is imperative for making an appropriate decision on a person's health. In this work, we have proposed an efficient noise removal technique, based on Fourier decomposition method, to clean an ECG signal affected by baseline wander and power-line interference. Simulation results have been compared with the popularly used empirical mode decomposition approach. The proposed method outperforms the existing techniques by successfully extracting the clean ECG signal from its noisy version, while the other techniques may require manual intervention to obtain the clean signal.

References

1. Weiting, Y., Runjing, Z.: An improved self-adaptive filter based on LMS algorithm for filtering 50 Hz interference in ECG signals. In: Proceedings of the 8th International Conference on Electronic Measurement and Instruments, vol. 3, pp. 874–878 (2007)
2. Pettis, K.S., Savona, M.R., Leibrandt, P.N., Maynard, C., Lawson, W.T., Wagner, G.S.: Evaluation of the efficacy of hand held computer screens for cardiologists interpretations of 12-lead electrocardiograms. Am. Heart J. **138**, 765–770 (1990)

3. Islam, M.K., Haque, A.N.M.M., Tangim, G., Ahammad, T., Khondokar, M.R.H.: Study and analysis of ECG signal using MATLAB and LABVIEW as effective tools. Int. J. Comput. Electr. Eng. **4**, 404–408 (2012)
4. Huang, N.E., Shen, Z., Long, S., Wu, M., Shih, H., Zheng, Q., Yen, N., Tung, C., Liu, H.: The empirical mode decomposition and Hilbert spectrum for non-linear and non-stationary time series analysis. Proc. R. Soc. A **454**, 903–995 (1988)
5. Singh, P., Joshi, S.D., Patney, R.K., Saha, K.: The Hilbert spectrum and the Energy Preserving Empirical Mode Decomposition. arXiv:1504.04104 [cs.IT] (2015)
6. Singh, P., Joshi, S.D., Patney, R.K., Saha, K.: Some studies on nonpolynomial interpolation and error analysis. Appl. Math. Comput. **244**, 809–821 (2014)
7. Singh, P., Srivastava, P.K., Patney, R.K., Joshi, S.D., Saha, K.: Nonpolynomial spline based empirical mode decomposition. In: 2013 International Conference on Signal Processing and Communication (ICSC), pp. 435–440 (2013)
8. Velazquez, R.: An optimal adaptive filtering approach for stress-tests motion artifacts removal: application on an ECG for telediagnosis. In: Proceedings of the 6th International Conference on Signal Processing, vol. 2, pp. 1504–1507 (2002)
9. Shaik, B.S., Chakka, V.K., Goli, S., Reddy, A.S.: Removal of narroband interference (PLI in ECG signal) using Ramanujan periodic transform(RPT). In: Proceedings of the 2016 International Conference on Signal Processing and Communication (ICSC), pp. 233–237 (2016)
10. Argenti, F., Bamich, B., Giarre, L.: Regularized LMS methods for baseline wandering removal in wearable ECG devices. In: Proceedings of the 2016 IEEE Conference on Decision and Control (CDC), pp. 5029–5034 (2016)
11. Yadav, S.K., Sinha, R., Bora, P.K.: Electrocardiogram signal denoising using non-local wavelet transform domain filtering. IET Signal Process. **9**(1), 88–96 (2015)
12. Agrawal, S., Gupta, A.: Projection operator based removal of baseline wander noise from ECG signals. In: Proceedings of the 2013 Asilomar Conference on Signals, Systems and Computers, pp. 957–961 (2013)
13. Agrawal, S., Gupta, A.: Fractal and EMD based removal of baseline wander and powerline interference from ECG signals. Comput. Biol. Med. **43**(11), 1889–1899 (2013)
14. Zhao, Z., Chen, Y.: A new method for removal of baseline wander and power-line interference in ECG signals. In: Proceedings of the 2006 International Conference on Machine Learning and Cybernetics, pp. 4342–4347 (2006)
15. Bansod, P., Lambhate, R.: A new approach for removal of baseline wander in ECG signal using empirical mode decomposition and hurst exponent. In: Proceedings of the 2016 International Conference on Recent Advances and Innovations in Engineering (ICRAIE), pp. 1–6 (2016)
16. Wu, C., Zhang, Y., Hong, C., Chiueh, H.: Implementation of ECG signal processing algorithms for removing baseline wander and electromyography interference. In: Proceedings of the 2016 8th IEEE International Conference on Communication Software and Networks (ICCSN), pp. 118–121 (2016)
17. Anapagamini, S.A., Rajavel, R.: Removal of baseline wander from ECG signal using cascaded empirical mode decomposition and morphological functions. In: Proceedings of the 2016 3rd International Conference on Signal Processing and Integrated Networks (SPIN), pp. 769–774 (2016)
18. Anapagamini, S.A., Rajavel, R.: Removal of artifacts in ECG using empirical mode decomposition. In: Proceedings of the 2013 International Conference on Communication and Signal Processing, pp. 288–292 (2013)
19. Daubechies, I., Lu, J., Wu, H.T.: Synchrosqueezed wavelet transforms: an empirical mode decomposition-like tool. Appl. Comput. Harmon. Anal. **30**, 243–261 (2011)
20. Dragomiretskiy, K., Zosso, D.: Variational mode decomposition. IEEE Trans. Signal Process. **62**(3), 531–544 (2014)
21. Jain, P., Pachori, R.B.: An iterative approach for decomposition of multi-componentnon-stationary signals based on eigenvalue decomposition of the Hankel matrix. J. Franklin Inst. **352**(10), 4017–4044 (2015)
22. Gilles, J.: Empirical wavelet transform. IEEE Trans. Signal Process. **61**(16), 3999–4010 (2013)

23. Agrawal, S., Gupta, A.: Removal of baseline wander in ECG using the statistical properties of fractional Brownian motion. In: 2013 IEEE International Conference on Electronics, Computing and Communication Technologies (CONECCT) (2013)
24. Singh, P., Joshi, S.D., Patney, R.K., Saha, K.: The Fourier decomposition method for nonlinear and non-stationary time series analysis. Proc. R. Soc. A 20160871 (2017). https://doi.org/10.1098/rspa.2016.0871
25. Singh, P.: Some studies on a generalized Fourier expansion for nonlinear and nonstationary time series analysis. Ph.D. thesis, Department of Electrical Engineering, IIT Delhi, India (2016)
26. Singh, P.: Time-Frequency analysis via the Fourier Representation. arXiv:1604.04992 [cs.IT] (2016)
27. Singh, P., Joshi, S.D.: Some studies on multidimensional Fourier theory for Hilbert transform, analytic signal and space-time series analysis. arXiv:1507.08117 [cs.IT] (2015)
28. Singh, P.: LINOEP vectors, spiral of Theodorus, and nonlinear time-invariant system models of mode decomposition. arXiv:1509.08667 [cs.IT] (2015)
29. Singh, P., Joshi, S.D., Patney, R.K., Saha, K.: Fourier-based feature extraction for classification of EEG signals using EEG rhythms. Circuits Syst. Signal Process. 35(10), 3700–3715 (2016)
30. Singh, P.: Breaking the limits: Redefining the instantaneous frequency. Circuits Syst. Signal Process. (2017). https://doi.org/10.1007/s00034-017-0719-y

Noise Removal from Epileptic EEG signals using Adaptive Filters

Rekh Ram Janghel, Satya Prakash Sahu, Gautam Tatiparti
and Mangesh Kose

Abstract Electroencephalography (EEG) is a well-established clinical procedure which provides information pertinent to the diagnosis of various brain disorders. EEG waves are highly vulnerable to diverse forms of noise which pose notable challenges in the analysis of EEG data. In this paper, adaptive filtering techniques, namely, Recursive Least Squares (RLS), Least Mean Squares (LMS), and Shift Moving Average (SMA) filters, were applied to the collected EEG signals to filter noise from the EEG signal. Various fidelity parameters, namely, Mean Square Error (MSE), Maximum Error (ME), and Signal-to-Noise Ratio (SNR), were observed. Our method has shown better performance compared to previous filtering techniques. Overall, in comparison to the previous methods, this proposed strategy is more appropriate for EEG filtering with greater accuracy.

1 Introduction

Electroencephalography (EEG) is a brain signal processing technique that records the electrical activity of the brain from the scalp, the first recordings of EEG were made in the year 1929 by a German psychiatrist Hans Berger [1]. Human EEG reflects the electrical activity of the brain. Spontaneous EEG signals are classified into several

R. R. Janghel (✉) · S. P. Sahu · G. Tatiparti · M. Kose
National Institute of Technology, Raipur, Raipur, Chhattisgarh, India
e-mail: rrjanghel.it@nitrr.ac.in; rrj.iiitm@gmail.com

S. P. Sahu
e-mail: spsahu.it@nitrr.ac.in

G. Tatiparti
e-mail: gautam1993@gmail.com

M. Kose
e-mail: mangeshkose@gmail.com

© Springer Nature Singapore Pte Ltd. 2019
M. Tanveer and R. B. Pachori (eds.), *Machine Intelligence and Signal Analysis*,
Advances in Intelligent Systems and Computing 748,
https://doi.org/10.1007/978-981-13-0923-6_4

rhythms based on their frequencies, namely, δ rhythm (0.3–4 Hz), θ rhythm (4–8 Hz), α rhythm (8–13 Hz) and β rhythm (13–30 Hz) [2]. One of the principal advantages of electroencephalography compared to clinical methods is that it is continuous and does not need the patient's cooperation. An EEG is of use particularly at times when the brain is at risk by providing a sensitive indication of cerebral functioning. Such intervals are usually of long time spans, hence an extended EEG recording is needed. Usually, a 16 channel, 24 hr recording of EEG produces 500 MB of data which corresponds to 8500 pages of traditional EEG data, the processing of such large amounts of EEG data is a cumbersome task [3].

Early studies have shown evidence of this abnormal activity to be a convenient aid in detection of epilepsy and cerebral tumors [21, 28]. Nowadays EEG signals are used to get information relevant to the diagnosis, prognosis, and treatment of these abnormal conditions [4]. EEG data carries details regarding abnormalities in response to certain stimuli in the human brain. EEG signals are measured using electrodes placed on the scalp and have small amplitudes of the order of 20 μV [5].

EEG signals are multichannel signals, which reflect the response of stimulation or a task known as Evoked Potentials [18]. Pure EEG signals without artifacts improve the quality of visual inspection. Artifacts make the diagnosis of diseases unfeasible due to resemblance between signals and pathological signals of interest [6].

EEG artifacts are often caused due to ocular, heart or muscular activities [7]. The blinking and movement of eyes produce electric potential which propagates over the scalp producing EOG artifacts in the electroencephalogram (EEG), these artifacts make the analysis of the EEG data unfeasible [8, 19, 20]. EMG artifacts can be eliminated by letting the patient find a comfortable position before the EEG signals are recorded, and by refraining from any tasks which require oral gesture or any form of physical activity. EEG signals are often contaminated due to interference by the 50 Hz line frequency because of the wires, the ignition of light of fluorescents is captured by the electrodes, which causes artificial spikes which can cause hindrance in the analysis of the EEG signals [9].

The human brain is one of the most complex systems in the universe. The largest portion of the human brain known as the cortex is divided into four parts, namely, the frontal lobe, the temporal lobe, the parietal lobe, and the occipital lobes; the frontal lobe is responsible for conscious thoughts, the functioning of olfactory organs, and the analysis of complicated stimuli such as recognition of faces is controlled by the temporal lobe [23–27]; the parietal lobe integrates information obtained from different senses; finally, the occipital lobe is responsible for the sense of sight [10, 23].

Epilepsy is a chronic brain disorder in which the nervous system of the patient gets affected, As per a 2015 report by the World Health Organisation, around 50 million worldwide suffer from epilepsy, around 90% of them live in developing countries and three fourths of these people do not get the necessary treatment [11], seizures occur due to abnormal electrical discharge from neurons, seizures can differ from less than once per year to several times per day [12, 22, 29]. Epilepsy is of two major kinds: focal and non-focal epileptic seizures [11–16].

2 Proposed Methodology

2.1 Data

The EEG data used for this research has been taken from CHB-MIT scalp EEG database the database can be obtained from the PhysioNet website : http://physionet.org/physiobank/database/chbmit/ [17].

This database consists of 916 h of continuous scalp EEG data that was sampled with a sampling frequency of 256 Hz, the data was documented from 23 pediatric patients at the Children's Hospital Boston and 1 adult at Beth Israel Deaconess Medical Centre. The patients encountered 173 events found to be clinical seizures while the recordings were being made [13]. In this paper, only 10 records of the EEG signal, each of 10 s (2560 samples), have been used.

2.2 Methodology

Shift Moving Average (SMA): SMA is a filtering algorithm in which each output is the average of some input data points. Number of input data points is equal to window size M. The SMA algorithm is implemented using Eq. (1),

$$b(n) - (a(n) + a(n-1) + \cdots + a(n-M+1))/M \qquad (1)$$

here, $b(n)$ denotes the output from SMA algorithm at point n. a is the input signal. The window size is M and should be taken odd (sometimes the value of $b(n)$ is calculated at the center of the window) [5].

Adaptive Filters: Adaptive filters employ a computational algorithm which seeks to build a correlation among signals. Adaptive filters are commonly defined by four characteristics: first one being the signals that are being filtered, second one being the composition which specifies computation and selection of input and output signals, third one being the parameter (weight vector) of construction that can be remodeled to build and adjust the filter's input–output correspondence, and fourth one being the algorithm that specifies the way parameters are tuned at each iteration.

The most popular algorithms to design adaptive filter are LMS and RLS algorithms. The steps and calculations of both the algorithms are given below: LMS algorithm starts with Eq. (2),

$$b(n) = W^T * A \qquad (2)$$

where $y(n)$ is the output signal which is the convolution of weight vector W and X which is the input signal, W is a weight vector $W = [w(0), w(1), \ldots, w(M-1)]$ having length equal to filter length M. $A = [a(n), a(n-1), \ldots, a(n-M-1)]$ vector of input signal (Noisy Signal) with length equal to filter length. After calculation of the output value b, error is calculated as $e(n) = b(n) - d(n)$.

$d(n)$ represents the desired signal also called as the reference signal (in noise cancelation model estimation of noise is selected). $e(n)$ is the value of error at point n. The main equation for the updation of weight vector of the filter is given in Eq. (3),

$$W(n + 1) = W(n) + 2\mu e(n)a(n) \tag{3}$$

where μ is the step size which is used to manage the convergence speed of the algorithm. In RLS algorithm, weights of adaptive filters are updated through Eq. (4),

$$w(n) = Q_D(n)Y_D(n) \tag{4}$$

In Eq. (4), Q_D and Y_D represent the deterministic correlation matrix of the input signal and the deterministic cross-correlation vector between the input and desired signals, respectively. Q_D and Y_D are determined using Eqs. (5) and (6),

$$Q_D = \frac{1}{\lambda}\left[Q_D(n - 1) - \frac{Q_D(n - 1)a(n)a^T(n)Q_D(n - 1)}{\lambda + a^T(n)Q_D(n - 1)a(n)} \right] \tag{5}$$

and

$$Y_D(n) = \lambda Y_D(n - 1) + d(n)a(n) \tag{6}$$

Q_D (−1) is initialized as δI, here δ represents inverse of the input signal power, λ and I represent the forgetting factor and identity matrix, respectively. Simulations: Input signal to the filters is prepared by adding noise to EEG signal. EMG Noise is used to contaminate the pure EEG signal (Fig. 1).

3 Simulations and Results

Various simulations were performed for the evaluation of the proposed method. The performance of methods was observed with the help of several conventionally used metrics. The simulations were carried out in MATLAB 8.5 environment and the performance was compared qualitatively and quantitatively.

3.1 Qualitative Evaluation

In this portion, the performance of EEG denoising methods is assessed qualitatively and the methods under consideration are compared by visual inspection. The signals resulting from the proposed methods can be observed in the figures below (Figs. 2, 3, 4, 5, 6 and 7).

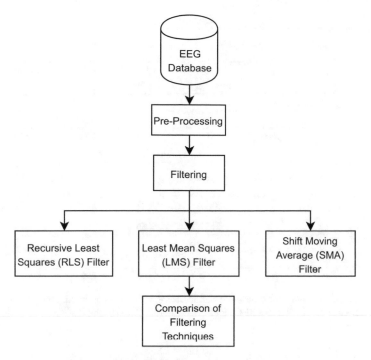

Fig. 1 Flowchart for the proposed methodology

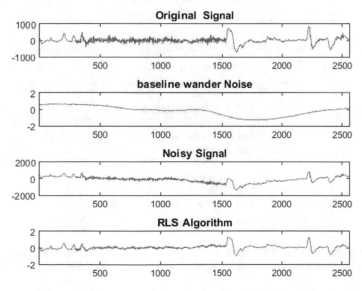

Fig. 2 Record chb01 and −10bB baseline wander noise, first plot is of the initial signal, next plot is of the noise, third plot is of noisy EEG signal, and the last plot is of EEG after filtering through RLS

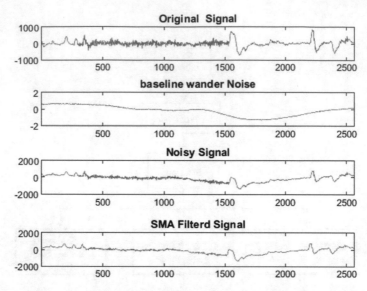

Fig. 3 Record chb01 and −10bB baseline wander noise, first plot is of the initial signal, next plot is of the noise, third plot is of noisy EEG signal, and the last plot is of the EEG after filtering through SMA

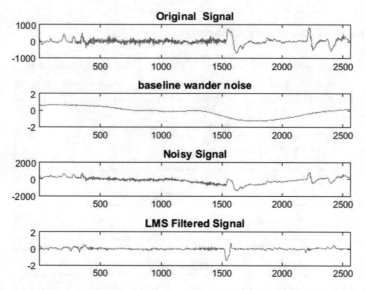

Fig. 4 Record chb01 and −10bB baseline wander noise, first plot is of the initial signal, next plot is of the noise, third plot is of noisy EEG signal, and the last plot is of EEG after filtering through LMS

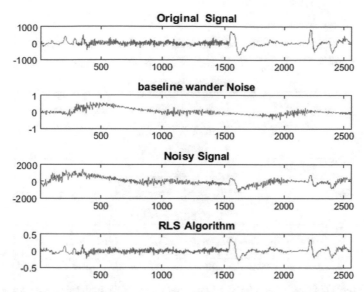

Fig. 5 Record chb01 and −10bB EMG artifact, first plot is of the initial signal, next plot is of the noise, third plot is of noisy EEG signal, and the last plot is of EEG after filtering through RLS

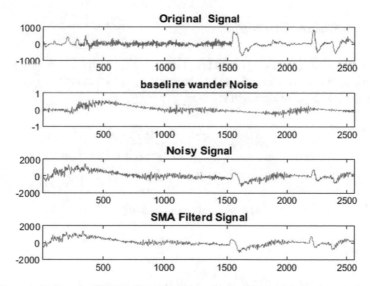

Fig. 6 Record chb01 and −10bB EMG artifact, first plot is of the initial signal, next plot is of the noise, third plot is of noisy EEG signal, and the last plot is of EEG after filtering through SMA

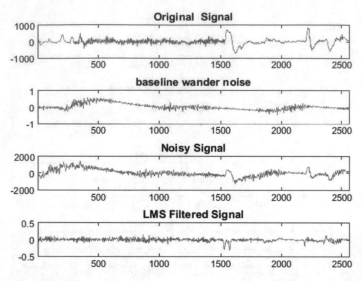

Fig. 7 Record chb01 and −10bB EMG artifact, first plot is of the initial signal, next plot is of the noise, third plot is of noisy EEG signal, and the last plot is of EEG after filtering through LMS

3.2   Quantitative Evaluation

Now, the performance of our method is compared quantitatively with other methods using three metrics, which are Signal-to-Noise Ratio (SNR) in dB, Maximum Error (ME), and Mean Square Error (MSE)

$$SNR_{dB} = \sum_{i=1}^{N} \frac{a(n)^2}{[b(n) - a(n)]^2} \tag{7}$$

$$ME = max(a(n) - b(n))^2 \tag{8}$$

$$MSE = \frac{1}{N} \sum_{i=1}^{N} (b(n) - a(n))^2 \tag{9}$$

Here, $a(n)$ represents the actual signal and $b(n)$ is the signal obtained after filtering; N represents the number of EEG samples under evaluation. At a particular SNR, for a denoising method to be considered better, it is desirable to have higher SNR_{dB}, and lower MSE and ME (Tables 1, 2, 3, 4, 5 and 6).

Table 1 Average SNR of 10 records for EMG artifact

Filtering technique	Pre-filtering	Post-filtering
LMS	−15.22594	−16.50382
RLS	−15.22594	0.0045
SMA	−15.22594	−1.25815

Table 2 Average MSE of 10 records for EMG artifact

Filtering technique	Pre-filtering	Post-filtering
LMS	0.19725	2.77808
RLS	0.19725	0.04069
SMA	0.19725	0.09266

Table 3 Average ME of 10 records for EMG artifact

Filtering technique	Pre-filtering	Post-filtering
LMS	0.02492	0.32772
RLS	0.02492	0.32756
SMA	0.02492	0.07404

Table 4 Average SNR of 10 records for baseline wander noise

Filtering technique	Pre-filtering	Post-filtering
LMS	−8.30957	−15.76885
RLS	−8.30957	0.01361
SMA	−8.30957	−1.75961

Table 5 Average MSE of 10 records for baseline wander noise

Filtering technique	Pre-filtering	Post-filtering
LMS	0.12136	2.77808
RLS	0.12136	0.04065
SMA	0.12136	0.03787

Table 6 Average ME of 10 records for baseline wander noise

Filtering technique	Pre-filtering	Post-filtering
LMS	0.67316	0.32778
RLS	0.67316	0.32713
SMA	0.67316	1.10862

4 Conclusion

EEG signals can be used to discriminate between normal and epileptic states of the brain. In this work, we have applied adaptive filtering techniques, namely, Least Mean Squares (LMS), Recursive Least Squares (RLS), and Shift Moving Average (SMA), to filter out baseline wander noise from the noisy EEG signal. We have shown that the adaptive filter based on RLS algorithm performs the best among all reported filters in the study as it gives the highest improvement in SNR up to 0.01361 for baseline wander noise.

References

1. Majumdar, K.: Human scalp EEG processing: various soft computing approaches. Appl. Soft Comput. J. **11**(8), 4433–4447 (2011)
2. Zhou, W., Gotman, J.: Removal of EMG and ECG artifacts from EEG based on wavelet transform and ICA. In: 26th Annual International Conference of the IEEE Engineering in Medicine and Biology Society, vol. 1, pp. 392–395 (2004)
3. Agarwal, R., Gotman, J., Flanagan, D., Rosenblatt, B.: Automatic EEG analysis during long-term monitoring in the ICU. Electroencephalogr. Clin. Neurophysiol. **107**(1), 44–58 (1998)
4. Gevins, A.S., Yeager, C.L., Diamond, S.L., Spire, J., Zeitlin, G.M., Gevins, A.H.: Automated analysis of the electrical activity of the human brain (EEG): A progress report. Proc. IEEE **63**(10), 1382–1399 (1975)
5. Selvan, S., Srinivasan, R.: Removal of ocular artifacts from EEG using an efficient neural network based adaptive filtering technique. IEEE Signal Process. Lett. **6**(12), 330–332 (1999)
6. Priyadharsini, S.S., Rajan, S.E.: An efficient soft-computing technique for extraction of EEG signal from tainted EEG signal. Appl. Soft Comput. J. **12**(3), 1131–1137 (2012)
7. Repov, G.: Dealing with Noise in EEG Recording and Data Analysis Spoprijemanje s umom pri zajemanju in analizi EEG signala, pp. 18–25 (2010)
8. Cuong, N.T.K., et al.: Removing Noise and Artifacts from EEG Using Adaptive Noise Cancelator and Blind Source Separation, pp. 282–286 (2010)
9. Guruvareddy, A.: Artifact removal from EEG signals. Int. J. Comput. Appl. **77**(13), 9758887 (2013)
10. Fonseca, M.J., Member, S., Alarc, S.M.: Emotions Recognition Using EEG Signals: A Survey, vol. 3045, pp. 120 (2017)
11. Acharya, U.R., Molinari, F., Sree, S.V., Chattopadhyay, S., Ng, K.H., Suri, J.S.: Automated diagnosis of epileptic EEG using entropies. Biomed. Signal Process. Control **7**(4), 401408 (2012)
12. WHO Report on Epilepsy http://www.who.int/mediacentre/factsheets/fs999/en/ as seen on 07.08.2017
13. Shoeb, A., Guttag, J.: Application of Machine Learning To Epileptic
14. Egiazarian, K.: Automatic Removal of Ocular Artifacts in the EEG without an EOG Reference Channel Automatic Removal of Ocular Artifacts in the EEG without an EOG Reference Channel, no. July 2017 (2006)
15. Khammari, H., Anwar, A.: A spectral based forecasting tool of epileptic seizures. Int. J. Comput. Sci. Issues **9**, no. 3 3–3, pp. 337–346 (2012)
16. Kim, S.G., Yoo, C.D., Nguyen, T.Q.: Alias-free subband adaptive filtering with critical sampling. IEEE Trans. Signal Process. **56**(5), 18941904 (2008)
17. CHB-MIT Scalp EEG Database PhysioNet, https://physionet.org/

18. Ahirwal, M.K., Kumar, A., Singh, G.K.: EEG/ERP adaptive noise canceller design with controlled search space (CSS) approach in cuckoo and other optimization algorithms. In: IEEE/ACM Trans. Comput. Biol. Bioinf. **10**(6), 1491–1504 (2013)
19. Jung, T.-P., et al.: Removing electroencephalographic artifacts by blind source separation. Psychophysiology **37**(2), S0048577200980259 (2000)
20. Puthusserypady, S., Ratnarajah, T.: H adaptive filters for eye blink artifact minimization from electroencephalogram. IEEE Sig. Proc. Lett. **12**(12), 816819 (2005)
21. Acharya, U.R., Vinitha Sree, S., Swapna, G., Martis, R.J., Suri, J.S.: Automated EEG analysis of epilepsy: a review. Knowl.-Based Syst. **45**, 147165 (2013)
22. Pijn, J.P., Velis, D.N., van der Heyden, M.J., DeGoede, J., van Veelen, C.W., Lopes da Silva, F.H.: Nonlinear dynamics of epileptic seizures on basis of intracranial EEG recordings. Brain Topogr. **9**(4), 24970 (1997)
23. Mantini, D., Perrucci, M.G., Del Gratta, C., Romani, G.L., Corbetta, M.: Electrophysiological signatures of resting state networks in the human brain. Proc. Natl. Acad. Sci. U. S. A. **104**(32), 131705 (2007)
24. Theiler, J.: On the evidence for low dimensional chaos in an epileptic electroencephalogram. Phys. Lett. A. **196**(94), 335341 (1995)
25. He, P., Wilson, G., Russell, C.: Removal of ocular artifacts from electro-encephalogram by adaptive filtering. Med. Biol. Eng. Comput. **42**(3), 407412 (2004)
26. Salido-Ruiz, R.A., Ranta, R., Louis-Dorr, V.: EEG montage analysis in blind source separation. IFAC Proc. **7**(PART 1), pp. 389–394 (2009)
27. Winterhalder, M., et al.: Spatio-temporal patient-individual assessment of synchronization changes for epileptic seizure prediction. Clin. Neurophysiol. **117**(11), 2399–2413 (2006)
28. Bhati, D., Sharma, M., Pachori, R.B., Gadre, V.M.: Time frequency localized three-band biorthogonal wavelet filter bank using semidefinite relaxation and nonlinear least squares with epileptic seizure EEG signal classification. Digit. Signal Process. A Rev. J. **62**, 259–273 (2017)
29. Singh, P., Joshi, S.D., Patney, R.K., Saha, K.: Fourier-based feature extraction for classification of EEG signals using EEG rhythms, circuits. Syst. Signal Process. **35**(10), 3700 3715 (2016)
30. Normal Brain Waves EEG stock vector. Image of anatomy - 29444815. [Online]. https://www.dreamstime.com/royalty-free-stock-photo-normal-brain-waves-eeg-image29444815. Accessed 06 Dec 2017

Time–Frequency–Phase Analysis for Automatic Detection of Ocular Artifact in EEG Signal using S-Transform

Kedarnath Senapati and Priya R. Kamath

Abstract Artifacts are unwanted components in the EEG signals which may affect the EEG signal reading, thereby not allowing the signal to be interpreted properly. One of the most common artifacts is the ocular artifact. This artifact arises due to the movement of the eye including eye blink. In most cases, detection of ocular artifacts in EEG signals is done by skilled professionals who are small in number. This paper proposes a new approach of automatic detection of ocular artifacts using the phase information present in the S-transform (ST) of EEG signal. S-transform of a signal provides absolutely referenced phase information of the signal in addition to time–frequency information. A time delay exists between the signals recorded by electrodes placed at different distances from the point of origin of the artifact. This time delay translates to phase delay in the frequency domain. The phase information of the EEG signal recorded from different electrodes placed in the frontal region is used to detect the artifacts which are generated near the region where the eye is located.

1 Introduction

EEG signals are complex signals that record the brain activity. The EEG signals are measured by either placing electrodes on the scalp or surgically placing electrode on the exposed surface of the subject's brain. Most commonly, these EEG recordings are recorded for a few hours to analyze the behavior of the brain. EEG can provide var-

K. Senapati (✉) · P. R. Kamath
National Institute of Technology Karnataka, Surathkal, Mangalore, India
e-mail: kedar@nitk.edu.in

P. R. Kamath
e-mail: priyarkamath@gmail.com

© Springer Nature Singapore Pte Ltd. 2019
M. Tanveer and R. B. Pachori (eds.), *Machine Intelligence and Signal Analysis*,
Advances in Intelligent Systems and Computing 748,
https://doi.org/10.1007/978-981-13-0923-6_5

ious information about nervous system-related disorders like epilepsy, Alzheimer's disease, dementia, etc.

Since the EEG is recorded over a long period of time, the patient may move about and also look around causing muscle and ocular artifact, respectively, in the EEG signals. This may adversely affect the EEG signal reading.

For the medical examiners to view pure EEG signals, such artifacts need to be removed. A lot of work has been done to remove the EEG artifacts from the signal. Some methods suggest the use of independent component analysis (ICA) and its variations [1]. However, there are reports suggesting that the usage of ICA affects the phases of the signals. Also, artifact detection techniques need human intervention to determine which of the independent component contains the artifact so as to separate it from the rest of the EEG signals [4, 5]. Efforts are being made by researchers to avoid the human intervention in ICA driven techniques [9]. Some other removal techniques using discrete wavelet transform (DWT), discrete cosine transform (DCT) have also been suggested in [7, 10, 18]. Ocular artifact removal is also attempted by many researchers using S-transform and its modified versions [10, 16]. However, these techniques do not make use of the phase information in the EEG signals. Phase is a critical part of the EEG signal. The phase of EEG signals determines the signal interdependency. The relation between EEG signal and its phase has been investigated by many researchers [3].

This study aims at automatic detection of ocular artifacts using phase information extracted from S-transform of EEG signals.

1.1 S-Transforms

S-transform(ST) is a modified version of wavelet transforms, in the sense that it uses a window function to obtain time–frequency resolution. Fourier transform of a signal can be obtained from the ST of the signal. Thus, unlike wavelet transforms, the ST has a direct relation to the Fourier transform of the signal. The phase information provided by the ST is referenced at the time point t = 0. This absolutely referenced phase enables analysis of the signal using the phase information content of the signal. In addition, ST can be used to derive the instantaneous frequency of the signal. The S-transform of a continuous signal x(t) is defined as [15]

$$X_S(t, f) = \int_{-\infty}^{\infty} x(\tau) \frac{|f|}{\sqrt{2\pi}} e^{\frac{-f^2(t-\tau)^2}{2}} e^{-i2\pi ft} d\tau \tag{1}$$

The Fourier transform $X(f)$ of the signal $x(t)$ can be obtained by summing the S-transform $X_S(t, f)$ over τ as follows:

$$X(f) = \int_{-\infty}^{\infty} X_S(\tau, f) d\tau \tag{2}$$

Equation (2) can also be rewritten using Eq. (1) as follows:

$$X(f) = \int_{-\infty}^{\infty} \int_{-\infty}^{\infty} x(t) \frac{|f|}{\sqrt{2\pi}} e^{\frac{-f^2(t-\tau)^2}{2}} e^{-i2\pi ft} dt d\tau \tag{3}$$

The original signal can be recovered from $X(f)$ using inverse Fourier transform

$$x(t) = \int_{-\infty}^{\infty} X(f) e^{i2\pi ft} df \tag{4}$$

The above equation can also be written as

$$x(t) = \int_{-\infty}^{\infty} \left\{ \int_{-\infty}^{\infty} X_S(\tau, f) d\tau \right\} e^{i2\pi ft} df \tag{5}$$

Equation (5) can be viewed as an inverse S-transform through the inverse Fourier transform.

For discrete time signals $x(kt)$, its S-transform is defined as

$$X_S \left(jT, \frac{n}{NT} \right) = \sum_{m=0}^{N-1} X \left(\frac{m+n}{NT} \right) e^{\frac{-2\pi^2 m^2}{n^2}} e^{\frac{i2\pi mj}{N}} \tag{6}$$

and for n = 0,

$$X_S(jT, 0) = \frac{1}{N} \sum_{m=0}^{N-1} x \left(\frac{m}{NT} \right) \tag{7}$$

where j, m and n = 0, 1, 2, ..., N−1. The discrete inverse of S-transform is given by

$$x(kt) = \sum_{n=0}^{N-1} \left\{ \frac{1}{N} \sum_{j=0}^{N-1} X_S \left(jT, \frac{n}{NT} \right) \right\} e^{\frac{i2\pi nk}{N}} \tag{8}$$

2 Methodology

The proposed method uses the phase information inherent to the ST of a signal for detecting ocular artifact.

Like wavelets, the S-transforms provides multi-resolution time–frequency analysis using a Gaussian window which is very similar to the window function used in Morlet wavelets. Apart from this, the ST also provides absolutely referenced phase information which the wavelet transform fails to provide. A detailed comparison between S-transform and wavelet transform can be found in [17]. There have been numerous recent developments in regard to localization of time–frequency distribution of signals which can be found in [11–14] and the references therein.

2.1 Time–Frequency–Phase Analysis of EEG Signal

The phase information present in a neural signal consists of two components. While the first corresponds to the inherent phase of the signal and can be computed using the gain function, the second component corresponds to the phase change due to propagation delay of the signal between two electrodes [3].

In this study, time, frequency, and phase analysis of EEG signal has been carried out using ST. The signal due to eye movements generated near one electrode affects the signals at neighboring electrodes. However, upon measuring the signals at two neighboring electrodes, a small time delay exists which corresponds to the propagation delay. The propagation delay is indicated by phase change in the corresponding ST of the signals. This phase change information can be used to detect the origin of the signal in a patient. For example, the phase information present in the signal can be used to find the origin of the ictal spikes in epilepsy patients and can also be used to determine if the seizure was due to a focal discharge or generalized [8]. However, it was also pointed out that if the ictal events are very close to each other such that the Gaussian window function used in the ST cannot discriminate between the ictal spikes, or if the noise in the recorded EEG signal is high, the ST may not be useful. The phase between the two signals is measured as phase synchrony. Lower values of phase synchrony indicate low correlation between the signals [6].

Since the ST localizes spectral components in time, the cross-correlation of specific events in two spatially separated S-transforms gives the phase difference, and hence, the phase synchrony can be estimated.

Cross-ST of two signals $x(t)$ and $y(t)$ is defined as follows [2]:

$$CrossST(t, f) = X_S(t, f).Y_S(t, f)^* \tag{9}$$

where $()^*$ denotes the complex conjugate.

The phase of the cross-ST is given by [2]

$$arg(CrossST(t, f)) = \phi_x(t, f) - \phi_y(t, f) \tag{10}$$

where ϕ_x and ϕ_y are phases of the signals $x(t)$ and $y(t)$, respectively. The above equation shows that the phase difference (phase synchrony) between the two signals is equal to cross-power spectrum.

It is not very hard to show that, if a signal $x(t)$ in time domain is shifted by time t_1, then the phase of the signal in its S-transform domain is shifted by $2\pi f t_1$ which is also observed in Fourier transform. This can be explained as follows.

If a signal originating near an electrode consists of ocular artifact, its neighboring electrodes also record this signal activity with some propagation delay. This propagation delay translates to phase delay in the transform domain. Also, it is known that higher values of phase synchrony indicate higher correlation between the signals. One can use electrodes close to the eye (example FP1, F3, F7 electrodes for the left eye and FP2, F4, F8 for the right eye in 10-20 system positioning) to obtain the phase plot. This phase plot is obtained using cross-ST.

In order to check if the ocular artifact originated from the left eye, signals recorded at electrodes FP1, F3, and F7 are considered. High values of phase synchrony between the signal pairs at electrodes pairs (Fp1 and F3) and (FP1 and F7) at the same point of time indicates that an artifact generated near electrode FP1 has propagated to F3 and F7. This segment of the signal is considered to contain an ocular artifact if it lies in the frequency band of 0–16 Hz and the signal at electrode FP1 has relatively high amplitude.

An algorithm is suggested below to detect ocular artifacts generated by left eye using the electrodes FP1, F3, and F7.

Algorithm 1 Algorithm to detect ocular artifacts using electrodes Fp1, F3 and F7

Input: EEG segments at electrodes Fp1, F3, F7
Output: Ocular artifact zones along the time axis *(artifactDetected)*
1: **procedure** ARTIFACT- DETECT
2:　　$S1(t, f) = ST(EEG$ *segments at electrodes Fp1)*
3:　　$S2(t, f) = ST(EEG$ *segments at electrodes F3)*
4:　　$S3(t, f) = ST(EEG$ *segments at electrodes F7)*
5:　　$crossST_{FP1F3} = S1(t, f).S2(t, f)^*$
6:　　$crossST_{FP1F7} = S1(t, f).S3(t, f)^*$
7:　　for all rows i and columns j of the matrices, select $0 < frequency < 16Hz$
8:　　**if** $abs(S1(i, j)) >= threshold_1$ and $abs(angle(crossST_{FP1F3}(i, j))) >= threshold_2$ and $abs(angle(crossST_{FP1F7}(i, j))) >= threshold_3$ **then**
9:　　　　*artifactDetected(j)* ← 1
10:　　**if** *artifactDetected(j) = 1* **then**
11:　　　　**for k=j:1 do**
12:　　　　**if** $crossST_{FP1F3}(i, k) > threshold_2$ and $crossST_{FP1F7}(i, k) > threshold_3$ **then**
13:　　　　　　*artifactDetected(k)* ← 1
14:　　　　**end for**

3　Results and Discussion

In this study, 19 electrodes are used to collect the EEG data, from a healthy human subject, with a sampling rate of 256 Hz. The signal is preprocessed using a low pass filter with a cutoff frequency of 30 Hz followed by normalization. Normalization removes any unwanted bias during signal acquisition. The signal consists of eye-blink artifacts and the signals at electrodes labeled FP1, F3, and F7 using average reference montage is considered which are shown in Fig. 1a, b, c, respectively. The S-

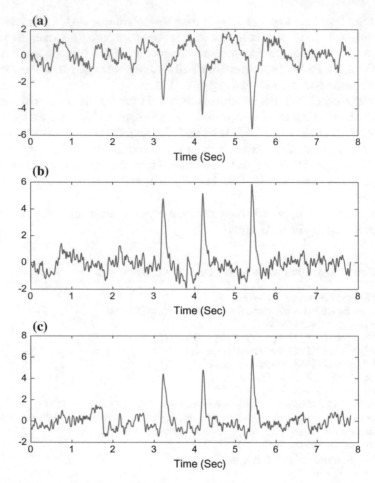

Fig. 1 EEG signal at electrodes **a** FP1, **b** F3, and **c** F7 using average montage

transforms of these signals are displayed in Fig. 2a, b, c. The amplitude of S-transform is color-coded from blue to red as shown.

The phase synchrony between signals at electrodes (FP1 and F3) and (FP1 and F7) are displayed in Fig. 3a, b, respectively. These phase synchrony values are color-coded from blue to red. It is found that the phase synchrony values are high at eye blinks. Also, the signal zones where phase synchrony values are high for the signal at electrode FP1 with respect to signal at neighboring electrodes (F3 and F7) are considered as ocular artifacts. The ocular artifact zone for the signal captured by the electrode FP1 is shown in Fig. 4.

Three threshold values are used in the algorithm. Here, $theshold_1$ indicates the minimum amplitude of the signal that can be considered as an eye blink. This value of $theshold_1$ is the sum of mean and standard deviation of the signal.

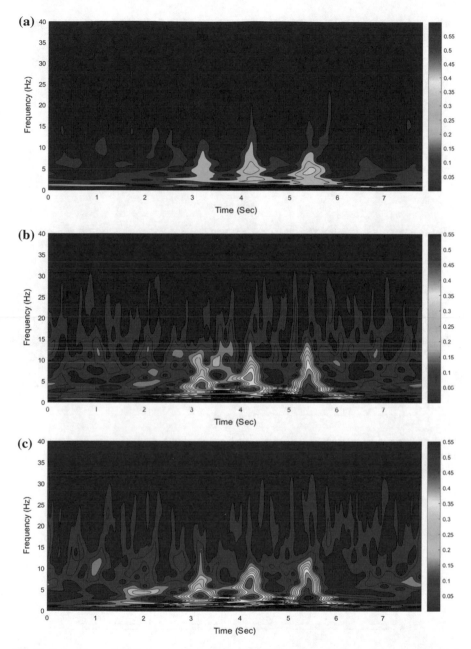

Fig. 2 S-transform of EEG signal at electrodes **a** FP1, **b** F3, and **c** F7. The graphs are color-coded to indicate the amplitude values of the signal components, which varies continuously from blue via green, yellow to red. The blue and red indicate the low and high values of the amplitude, respectively

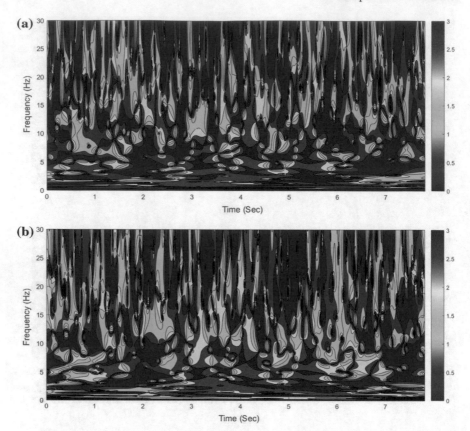

Fig. 3 Phase synchrony between the signals at electrodes **a** FP1 and F3, and **b** FP1 and F7. The color bar with blue via yellow to red indicates phase difference (in radian) between the signals. Higher phase difference is indicated in red

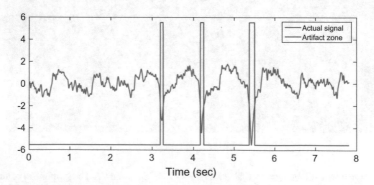

Fig. 4 The signal at electrode FP1: red vertical boxes denote the detected ocular artifact zone due to eye blink

The other two thresholds, namely, $threshold_2$ and $threshold_3$, determine the minimum phase difference that should exist between the two signals for interpreting the signal as an artifact. Both these values (in radian) are suitably chosen to be unity. Note that the threshold values can be made to change adaptively.

4 Conclusion

Artifacts are unwanted components in an EEG signal. In this paper, an attempt has been made to automatically detect the eye blink ocular artifact from the EEG signal using the phase information present in the S-transform of EEG signal. For a signal originating from a given region, a phase difference due to propagation delay exists between neighboring electrodes. This phase information has been used while in determining artifacts in the signal.

As a future work, aforementioned detection technique could be extended to detect other artifacts in EEG signals and recognize the area of origin for such an artifact so that they can be successfully removed from the EEG signal to obtain artifact-free signals.

This research involves no more than minimal risk to the human subjects who have participated in the experiment to acquire the EEG data. Also, consents were taken from the subjects and were provided with additional pertinent information before and after they had participated in this experiment. And also, no research ethics committee is involved in this process. So the statement on ethical approval is not required.

References

1. Akhtar, M.T., Mitsuhashi, W., James, C.J.: Employing spatially constrained ICA and wavelet denoising, for automatic removal of artifacts from multichannel EEG data. Signal Process. **92**(2), 401–416 (2012)
2. Assous, S., Boashash, B.: Evaluation of the modified-S transform for time-frequency synchrony analysis and source localisation. EURASIP J. Adv. Signal Proc. **2012**, 49 (2012)
3. da Silva, F.H.L.: Event-related neural activities: what about phase? Prog. Brain Res. **159**, 3–17 (2006)
4. Devuyst, S., Dutoit, T., Ravet, T., Stenuit, P., Kerkhofs, M., Stanus, E.: Automatic processing of EEG-EOG-EMG artifacts in sleep stage classification. In: 13th International Conference on Biomedical Engineering, pp. 146–150. Springer, Berlin (2009)
5. Jung, T., Makeig, S., Humpries, C., Lee, T., McKeown, M.J., Iragui, V., Sejnowski, T.J.: Removing electroencephalographic artifacts by blind source separation. Psychophysiology **37**, 163–178 (2000)
6. Kreuz, T.: Measures of neuronal signal synchrony. Scholarpedia **6**(12), 11922 (2011). http://www.scholarpedia.org/article/Measures_of_neuronal_signal_synchrony
7. Krishnaveni, V., Jayaraman, S., Aravind, S., Hariharasudhan, V., Ramadoss, K.: Automatic identification and removal of ocular artifacts from EEG using wavelet transform. Meas. Sci. Rev. **6**(4), 45–57 (2006)

8. Pinnegar, C.R., Khosravani, H., Federico, P.: Time-frequency phase analysis of ictal EEG recordings with the S-transform. IEEE Trans. Biomed. Eng. **56**(11), 2583–2593 (2009)
9. Raduntz, T., Scoutena, J., Hochmuthb, O., Meffert, B.: EEG artifact elimination by extraction of ICA-component features using image processing algorithms. J. Neurosci. Methods **243**(147), 84–93 (2015)
10. Senapati, K., Routray, A.: Comparison of ICA and WT with S-transform based method for removal of ocular artifact from EEG signals. J. Biomed. Sci. Eng. **4**(05), 341 (2011)
11. Sharma, M., Bhati, D., Pillai, S., Pachori, R.B., Gadre, V.M.: Design of time? frequency localized filter banks: transforming non-convex problem into convex via semidefinite relaxation technique. Circuits Syst. Signal Process. **35**(10), 3716–3733 (2016)
12. Sharma, M., Achuth, P., Pachori, R., Gadre, V.: A parametrization technique to design joint time-frequency optimized discrete-time biorthogonal wavelet bases? Signal Process. **135**, 107–120 (2017)
13. Sharma, M., Dhere, A., Pachori, R.B., Gadre, V.M.: Optimal duration-bandwidth localized antisymmetric biorthogonal wavelet filters. Signal Process. **134**, 87–99 (2017)
14. Singh, p, Joshi, S.D., Patney, R.K., Saha, K.: The fourier decomposition method for nonlinear and non-stationary time series analysis. Proc. R. Soc. A. (The Royal Society) **473**(2199), 20160871 (2017)
15. Stockwell, R.G., Mansinha, L., Lowe, R.P.: Localization of the complex spectrum: the S transform. IEEE Trans. Signal Process. **44**(4), 998–1001 (1996)
16. Upadhyay, R., Padhy, P.K., Kankar, P.K.: Ocular artifact removal from EEG signals using discrete orthonormal Stockwell transform. In: 2015 Annual IEEE India Conference (INDICON), pp. 1–5. IEEE (2015)
17. Ventosa, S., Simon, C., Schimmel, M., Daobeitia, J.J., Mnuel, A.: The S-transform from a wavelet point of view. IEEE Trans. Signal Process. **56**(7), 2771–2780 (2008)
18. Zikov, T., Bibian, S., Dumont, G.A., Huzmezan, M., Ries, C.R.: A wavelet based de-noising technique for ocular artifact correction of the electroencephalogram. In: Engineering in Medicine and Biology, 2002. 24th Annual Conference and the Annual Fall Meeting of the Biomedical Engineering Society EMBS/BMES Conference, 2002. Proceedings of the Second Joint, Vol. 1, pp. 98–105. IEEE (2002)

An Empirical Analysis of Instance-Based Transfer Learning Approach on Protease Substrate Cleavage Site Prediction

Deepak Singh, Dilip Singh Sisodia and Pradeep Singh

Abstract Classical machine learning algorithms presume the supervised data emerged from the same domain. Transfer learning on the contrary to classical machine learning methods; utilize the knowledge acquired from the auxiliary domains to aid predictive capability of diverse data distribution in the current domain. In the last few decades, there is a significant amount of work done on the domain adaptation and knowledge transfer across the domains in the field of bioinformatics. The computational method for the classification of protease cleavage sites is significantly important in the inhibitors and drug design techniques. Matrix metalloproteases (MMP) are one such protease that has a crucial role in the disease process. However, the challenge in the computational prediction of MMPs substrate cleavage persists due to the availability of very few experimentally verified sites. The objective of this paper is to explore the cross-domain learning in the classification of protease substrate cleavage sites, such that the lack of availability of one-domain cleavage sites can be furnished by the other available domain knowledge. To achieve this objective, we employed the TrAdaBoost algorithm and its two variants: dynamic TrAdaBoost and multisource TrAdaBoost on the MMPs dataset available at PROSPER. The robustness and acceptability of the TrAdaBoost algorithms in the substrate site identification have been validated by rigorous experiments. The aim of these experiments is to compare the performances among learner. The experimental results demonstrate the potential of dynamic TrAdaBoost algorithms on the protease dataset by outperforming the fundamental and other variants of TrAdaBoost algorithms.

1 Introduction

The extracellular matrix (ECM) [1] is the noncellular segment exhibited inside all tissues and organs, and provides not only fundamental physical framework for the cell constituents but additionally starts off and evolved critical biochemical and biome-

D. Singh (✉) · D. S. Sisodia · P. Singh
National Institute of Technology, Raipur, Raipur 492001, Chhattisgarh, India
e-mail: dsingh.phd2016.cs@nitrr.ac.in

© Springer Nature Singapore Pte Ltd. 2019
M. Tanveer and R. B. Pachori (eds.), *Machine Intelligence and Signal Analysis*,
Advances in Intelligent Systems and Computing 748,
https://doi.org/10.1007/978-981-13-0923-6_6

chanical signals which will be required for tissue morphogenesis, separation, and homeostasis. The ECM degradation is the most crucial factor for the tissue repair and remodeling. Arthritis, fibrosis, cancer, nephritis, encephalomyelitis, and chronic ulcers are the cause of ECM abnormal functioning [2].

Matrix metalloproteinase (MMP), also called matrixins, is one of the essential proteinases which are implicated in ECM degradation [3]. The removal and degradation of ECM molecules are the primary functions of matrixins from the tissue. The normal physiological activities of MMPs are specifically regulated at the level of transcription, activation of the precursor zymogens, interaction with specific ECM components, and inhibition by endogenous inhibitors.

The objective of understanding the physiological role of protease is to recognize the behavior and natural functioning with respect to substrates. Proteases are enzymes that carry out the split process from the particular cleavage sites within the substrate [4]. These proteases primarily depend on the active sites which are a sequence of amino acids with defined positions. In addition, the cleavage sites are dependent on the physical co-location, three-dimensional structures of the protease, and the matching substrates.

The use of the computational method and machine learning algorithms in bioinformatics are crucial tools to alleviate the time and cost in contrast to a manual process. The challenge of predicting the cleavage site from substrate associated with MMP can be fulfilled by machine learning techniques. Supervised learning algorithms such as Naïve Bayes [5], support vector machines (SVM) [6], decision tree [7], etc. have already shown its potential in solving the cleavage site prediction problem.

In a typical supervised scenario, it is considered to have an adequate amount of annotated training data so that the training of the learner model can be accomplished successfully. In many biological applications, the availability of supervised training data is confined. Sometimes, acquiring this data will be compromised by paying a huge cost. This problem of unavailability in supervised data results in few instances for the training process. These limited training instances may lead to overfitting model which becomes a significant bottleneck in practice. As a consequence, the learner model will perform poorly when given new instances. To solve the overfitting problem, various machine learning techniques have been developed. One solution could be the use of transfer learning [8], which refers to a learning framework that utilizes the knowledge from other similar domains.

Transfer learning beliefs in the reusability of knowledge from the auxiliary domains where the feasible amount of labeled data is present. However, it cannot be used directly. The reusability of knowledge is achieved by comparing the auxiliary and target problem domains of data such that a common useful knowledge between them can be extracted. This common knowledge will boost the learning performance in the target domain. Transfer learning has been successfully applied in many different applications over the last two decades. This includes the problems in text mining [9], signal processing, pattern recognition [10], bioinformatics [11], and ubiquitous computing.

Inadequacy in the training data, overfitted model, and inability to learn from the existing domain are the issues identified in the conventional machine learning strategies. In response to the problems, instance-based transfer learning algorithms are employed in this work. Instance-based algorithms work on reweighting and attenuating the differences in source- and target-domain marginal probability. We have used TrAdaBoost [12] and their variants [13, 14] for the MMP substrate cleavage site prediction.

In the rest of the paper, Sect. 2 briefs about the past work done, Sect. 3 enlightens the instance-based transfer learning approach, Sect. 4 describes the workflow of the desired objective, Sect. 5 lists the experiment setup and simulation results along with the discussion, and Sect. 6 concludes the work.

2 Related Work

The prediction of protease substrate cleavage sites based on computational methods had a variety of successful attempts made by the researcher over the last decade. The conventional machine learning algorithms usually train their models from a supervised set of peptides data. These supervised data have cleavage site information. The learner model exploits the training data and extracts the useful features for the construction of the predictive model. There are various machine learning approaches practiced in the peptide cleavage site classification in the past [15, 16]. These methods take labeled substrate peptide sequences as the input to the classifiers, and the trained classifiers can predict cleavage sites with maximum accuracies based on different data distributions of the datasets.

A computational model based on the amino acid sequence for the protease prediction built on the experimental data is developed and named as PoPS [17]. Additionally, it ranks likely cleavages within a single substrate and also among the proteomes. SitePrediction [18] is a tool for identifying potential cleavage site with a feature of a wide range of proteinases acceptability and usage. Cascleave (termed) is used to predict both classical and nontypical caspase cleavage sites [19]. Caspase belongs to the family of cysteine proteases that play essential roles in vital biological processes. Further, the second version is proposed with an amendment of classifying the granzyme cleavage sites too [20]. Pripper is another tool that uses three different classifiers for predicting caspase cleavage sites from protein sequences in the FASTA format. The prediction results were enhanced due to the combination of three different learners [21]. CaSPredictor is a tool which incorporated a PEST-like index and the position-dependent amino acid matrices for the classification of caspase cleavage sites in individual proteins as well as in protein datasets [22]. GraBCas is yet another tool that provides a position-based specific scoring scheme for the prediction of cleavage sites for granzyme B and caspases [23]. There are a lot of alternative methods that are available as well but they focus on specific families of substrate cleavage sites [6, 24, 25].

Among the MMP site prediction techniques, the most powerful framework is PROSPER which has been used to predict 24 different proteases [26]. Use of support vector regression (SVR) as a classifier based on a variety of sequence-based features in the system helps to predict the sites highly accurate. PROSPER is the only tool that can predict MMP-2, -3, -7, and -9 substrate cleavage sites. The prediction method of CleavPredict [27] utilizes the MMP-specific position weight matrices (PWMs) derived from statistical analysis of high-throughput phase that displays experimental results. This tool has the ability to classify the 11 different MMPs substrate cleavage sites. The work of classifying the MMP substrate cleavage sites is insufficient consequently; the accurate prediction of different MMP substrate cleavage sites remains an outstanding need and challenging problem. These results highlight the need for new and efficient means of improved prediction of MMP substrate cleavage sites that are highly desired [11].

3   Transfer Boost-Based Transfer Learning

To understand, let us define the terms domain and task. A domain $\mathfrak{D}$ is specified by two entities, a feature space $\mathcal{X}$, and a marginal probability distribution $\mathcal{P}(x)$, where $\mathcal{X} = \{x_2, x_2, \ldots, x_n\} \in \mathcal{X}$. For example, if the machine learning application is protease substrate cleavage site classification and each peptide character is treated as a feature, this feature is then encoded by some encoding technique, the encoded vector is feature, x_i is the ith feature vector (instance) corresponding to the ith protease site, n is the number of feature vectors in $\mathcal{X}$, and $\mathcal{X}$ is a particular learning sample.

For a given domain $\mathfrak{D}$, a task $\mathcal{T}$ is specified by two entities, a label space $\mathcal{Y}$ and a predictive function $f(\cdot)$, which predicts the label y_i from the feature vector x_i where $x_i \in \mathcal{X}$ and $y_i \in \mathcal{Y}$. Referring to the protease substrate cleavage site classification application, $\mathcal{Y}$ is the set of labels and in this case it contains $\{-1, 1\}$, y_i takes on a value of -1 if the site is uncleavage or 1 if the site is cleavage, and $f(x)$ is the learner that predicts the label value for the protease x_i.

From the definitions above, a domain $\mathfrak{D} = \{\mathcal{X}, \mathcal{P}(x)\}$ and a task $\mathcal{T} = \{\mathcal{Y}, f(x)\}$. Now, source-domain data is defined as D_s where $D_s = \{(x_{s1}, y_{s1}), \ldots, (x_{sn}, y_{sn})\}$, where $x_{si} \in \mathcal{X}_s$ is the ith data instance of D_s and $y_{si} \in \mathcal{Y}_s$ is the respective class label for x_{si}. In a similar way, the target-domain data is defined as D_T where $D_T = \{(x_{T1}, y_{T1}), \ldots, (x_{Tn}, y_{Tn})\}$ where $x_{Ti} \in \mathcal{X}_T$ is the ith data instance of D_T and $y_{Ti} \in \mathcal{Y}_T$ is the respective class label for x_{Ti}. Further, the source task is denoted as T_s, the target task as T_T, the source predictive function as $f_s(\cdot)$, and the target predictive function as $f_T(\cdot)$. In the cleavage site classification, D_s can be a set of protease term together with the relevant class label -1 or 1 from the one data source (MMP$_9$) and D_T is from the other data source (MMP$_3$).

The formal definition of transfer learning can be given as source-domain D_s with a related source task T_s, and a target domain D_T with a related task T_T, transfer learning is the process of predicting the target domain or minimizing the empirical error of a function $f_T(\cdot)$ by using the related information from D_s and T_T, where $D_s \neq D_T$ or $T_s \neq T_T$. From the definition of transfer learning, if $D_s \neq D_T$ it means either the $x_s \neq x_T$ or $\mathcal{P}(x_s) \neq \mathcal{P}(x_T)$ else if $T_s \neq T_T$ then it is possible that y_s

$\neq y_T$ or conditional probability distribution between the source and target domains is varying.

Transfer learning problem can be solved by the various different strategies. One could be minimizing the marginal distribution differences between the source and target domains, and other could be minimizing the conditional distribution difference between them. Trying for both marginal and conditional distributions, minimization is also being adopted by many solutions. These strategies are categorized into four methodologies: Instance-based, feature-based, parameter–based, and relational-based. In this study, we preferred instance-based approaches over the other techniques because of the better performance during marginal distribution differences among source and target domains.

Instance-based transfer learning approach utilizes the source-domain instances for transferring the knowledge. In other words, the source-domain data cannot be reused directly; there are specific data that can still be reused together with a few labeled data in the target domain. So the instances of the source domain are reweighted for the sake of minimizing the marginal distribution differences. Further, these reweighted instances are used for the prediction of target-domain test data.

The idea of AdaBoost is further extended to form TrAdaBoost algorithm, with an objective of solving the transfer learning problems. TrAdaBoost assumes data to be homogeneous. To be precise, there have to be an equal number of features and labels in both the source and target domains, but the difference in the distributions of the data in the two domains could exist. In addition, TrAdaBoost works on the basic principle of having some common data between source and target domains that have similar marginal distribution. However, the uncommon data between both the domains may weaken the predictive model. To deal with such situation, TrAdaBoost iteratively reweights the source-domain data by reducing the weight of uncommon (bad) source data while increasing the weight of common (good) source data to participate more in target-domain prediction. For each round of iteration, TrAdaBoost boosts the base classifier on the computed; the higher the weight of good data, the lesser the weight of bad data. The empirical error is computed only on the target data. Furthermore, the weight update strategy during the incorrectly classified instances of target-domain data in TrAdaBoost is similar to the AdaBoost algorithm, while employing a different strategy from AdaBoost to update the incorrectly classified source instances in the source domain.

Dynamic TrAdaBoost is another algorithm with an integrated dynamic cost to resolve the issues in the boosting-based transfer algorithm, TrAdaBoost. This issue causes source instances to converge before they can be used for transfer learning. Multisource TrAdaBoost is another enhancement on the TrAdaBoost with the capability of adding more than one source-domain data. The selection of weak classifier from the source that appears to be the most closely related to the target is primarily adopted by the multi-source transfer boosting.

4   Transfer Learning in Substrate–Protease Cleavage Site

Matrix metalloprotease substrate cleavage site prediction is a binary classification problem, i.e., being classified as either a cleavage or non-cleavage site $\{1, -1\}$. In this paper, we applied the three transfer AdaBoost learning techniques, to solve the complex task of predicting substrate cleavage sites of different proteases. TrAdaBoost, dynamic TrAdaBoost, and multisource TrAdaBoost are efficient classification algorithms suitable for solving classification in transfer-domain problems.

TrAdaBoost is an extension of the AdaBoost algorithm used to deal with the homogenous transfer learning problems. Although there is the presence of a difference in the distributions of source and target domains, there exist some source data that may be useful in learning the target domain which is the primary assumption of the TrAdaBoost algorithm. The algorithm works on the inductively based transfer learning settings where some labeled target data are required for the training. Therefore, we consider five percent of the total target data to be used in training phase.

A formal representation of the framework is shown in Fig. 1. As it can be seen from the figure, the source data are carried from the three different sources: MMP-2, MMP-3, and MMP-9, whereas the target data carried only from one MMP-7 source. When the dataset of one MMP is in the target domain, the remaining three were used as the source-domain data. In addition, we also took five percent of the target data in the training phase. Three out of four MMP sites and the few labeled target-domain data are collectively treated as training data.

The training data are formed due to the combination of multiple individual sources which may cause the presence of redundant entries. These redundant entries may lead to overfitting the learner model; so to overcome this problem, the removal of such entries is mandatory. The preprocessing method of removing such redundant entries is then carried out using CD-HIT program with 70% sequence identity threshold.

The protease sequences are of varying lengths and to extract the cleavage and non-cleavage sites, we truncated the sequence into 16 amino acid character lengths using a sliding window. This 16 amino acid sequence is then passed through the encoding schemes. Sequence-encoding schemes play an important role in determining the predictive performance of machine-learning-based models. Here, we use BLOSUM sequence-encoding schemes for training TrAdaBoost models. The BLOSUM62 matrix was used to extract primary sequence information. A vector of $L \times 21$ elements was used to represent each segment in our datasets, where L is the length of the segment and 21 represents the 20 standard amino acids and an additional one representing non-conserved amino acid residues. Therefore, the BLOSUM-derived features for a segment of length $L = 16$ comprised a $16 \times 21 = 336$-dimensional vector.

The encoded vectors are then used as training data in the TrAdaBoost models, and the target-domain encoded vectors are used as testing data.

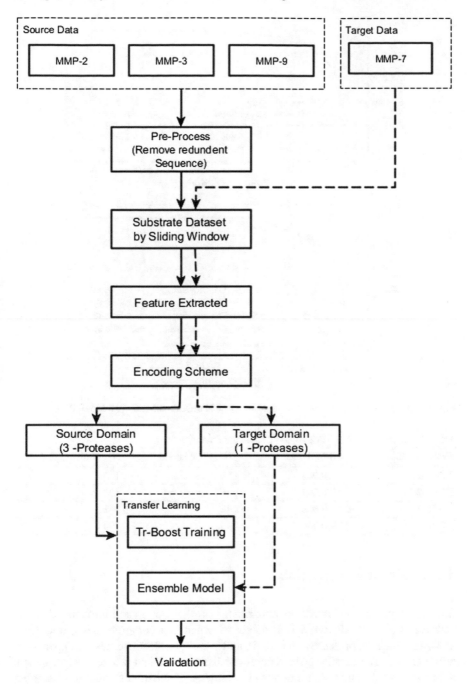

Fig. 1 Workflow of the proposed model

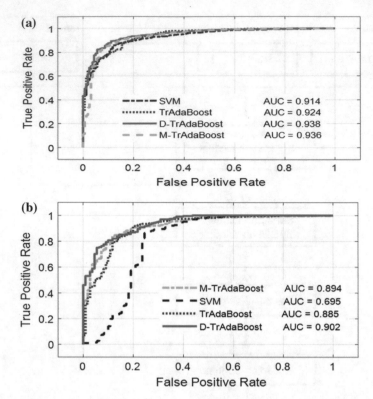

Fig. 2 ROC curves of TrAdaBoost variants and base learner on **a** MMP-2, **b** MMP-3

Table 1 Characteristics of MMP-specific substrate dataset

Protease name	MEROPS ID	No. of substrates	No. of cleavage sites
MMP-2	M10.003	35	115
MMP-3	M10.005	44	132
MMP-7	M10.008	42	142
MMP-9	M10.004	43	290

5 Results and Discussion

In this study, we conducted our experiment on the four metalloprotease datasets extracted from the MEROPS [28] database which is a comprehensive, integrated knowledgebase for proteases, substrates, and inhibitors; all substrates are experimentally verified. The details of the dataset are listed in Table 1 where rows represent the type of MMP and column depicts the number of substrates and a total number of sites present in the substrates. MMP-9 is a higher number of sites carrying data, whereas MMP-2 carries the lowest number of sites.

Table 2 Predictive performance of base learner and TrAdaBoost variants

MMP TYPE	Measures	SVM	TrAdaBoost	D-TrAdaBoost	Multisource TrAdaBoost
MMP-2	Accuracy	85.39	85.41	86.45	**86.78**
	Sensitivity	72.72	70.45	**72.41**	71.48
	F-Measure	71.33	70.14	71.89	**72.17**
	Specificity	89.61	88.65	**90.12**	87.23
MMP-3	Accuracy	88.10	87.89	**88.32**	84.23
	Sensitivity	81.92	81.46	**82.85**	80.24
	F-Measure	77.71	74.23	78.03	**78.69**
	Specificity	89.38	87.26	90.35	**90.73**
MMP-7	Accuracy	84.86	86.45	**86.75**	86.49
	Sensitivity	71.05	72.65	**73.12**	71.46
	F-Measure	70.13	68.94	71.45	**73.14**
	Specificity	89.74	88.75	89.65	**89.97**
MMP-9	Accuracy	83.33	84.25	84.46	**87.28**
	Sensitivity	63.88	65.12	**68.94**	64.13
	F-Measure	65.71	**68.41**	64.12	62.67
	Specificity	89.81	88.47	89.44	**90.03**

To ensure proper machine-learning-based model training and performance, we followed the fivefold cross-validation model. We evaluated accuracy, sensitivity, specificity, F-score, and area under the curve for the performance comparison among the classifiers. The performance measures are the average values obtained by the mean of 10 runs. When the substrate dataset of an MMP was used as the dataset in the target domain, the remaining three MMP datasets were used as the dataset in the source domain.

Table 2 provides a statistical summary of the average accuracy on MMP-specific substrate datasets used in this study. The first column indicates the name of the dataset that is considered for the test (target-domain) data. The source data consist of the remaining three MMP substrate datasets. Following observations were drawn from the table:

- TrAdaBoost and their variants are performing well compared to the baseline classifier. Among the TrAdaBoost, the average accuracy of the D-TrAdaBoost in the two MMP-3 and MMP-7 sites has superior performance, while in MMP-2 and MMP-9, multisource TrAdaBoost has good accuracy rate.
- Average sensitivity on the referred datasets is noted as the second entry in the table. The observation shows the significant improvement in the mean sensitivity when D-TrAdaBoost is employed.
- The F-scores of TrAdaBoost are relatively low as compared to the base learner. However, the multisource TrAdaBoost values vary from 0.72 to 0.78.

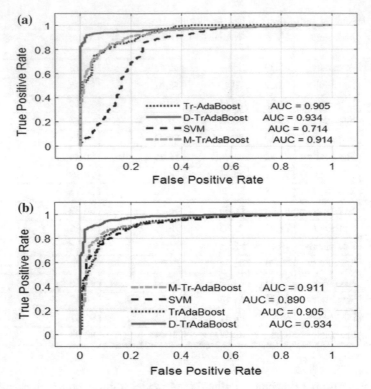

Fig. 3 ROC curves of TrAdaBoost variants and base learner on **a** MMP-7, **b** MMP-9

- The compelling improvement made by the multisource TrAdaBoost in the specificity value of the three MMP datasets indicates the stability when the number of sites is higher.

The ROC and AUC are the two important measures that are widely evaluated for the biomedical data. In Figs. 2 and 3, ROC of each classifier with the achieved AUC values is listed for all the four MMP sites. The curves represent various classifiers used, where black represents the base learner, red denotes the TrAdaBoost, green depicts the multisource TrAdaBoost, and blue represents the dynamic TrAdaBoost. The observation from all the ROC curves and AUC values in all the dataset is that the dynamic TrAdaBoost has outperformed during the classification of MMP-7 and MMP9 sites, while in the case of MMP-2 and MMP-3, there is a minor improvement in comparison to other two TrAdaBoost variants. Overall, the performance of TrAdaBoost and its variants with respect to the base learner is superior in the MMP substrate sites.

6 Conclusion

The optimal use of auxiliary or archival domain data for the prediction of the relevant current domain data is the prime objective of this paper. In this work, we evaluate the TrAdaBoost transfer learning algorithms for the prediction of MMP substrate cleavage sites and confirm its efficiency by applying these techniques to acquire the knowledge from the three-source domain such that the prediction of cleavage sites of one MMP in the target-domain improves. The four MMP datasets from MEROPS had been preprocessed and truncated into 16-character protease sequence. This protease is further encoded by the BLOSUM matrix encoding technique. The various combinations of source and target MMPs were formed and considered during the experiment. The experiment on voluminous and complex data with three TrAdaBoost learners indicates that transfer learning is robust and particularly useful when limited training target data are available for predicting cleavage sites of MMPs. Overall, the experimental results suggest that dynamic TrAdaBoost provides a useful alternative for the characterization of cleavage sites of MMP substrate. For future, the work can be extended in the transductive domain where the target-domain data are unavailable for training.

References

1. Lu, P., Takai, K., Weaver, V.M., Werb, Z.: Extracellular matrix degradation and remodeling in development and disease. Cold Spring Harb. Perspect. Biol. 3(12), 1–24 (2011)
2. Coussens, L.M., Fingleton, B., Matrisian, L.M.: Matrix metalloproteinase inhibitors and cancer: trials and tribulations. Science 295(5564), 2387–2392 (2002)
3. Cieplak, P., Strongin, A.Y.: Matrix metalloproteinases—from the cleavage data to the prediction tools and beyond. In: Biochimica et Biophysica Acta (BBA)—Molecular Cell Research, pp. 1–12, Jan 2017
4. Rögnvaldsson, T., Etchells, T.A., You, L., Garwicz, D., Jarman, I., Lisboa, P.J.G.: How to find simple and accurate rules for viral protease cleavage specificities. BMC Bioinform. 10, 149 (2009)
5. Yousef, M., Nebozhyn, M., Shatkay, H., Kanterakis, S., Showe, L.C., Showe, M.K.: Combining multi-species genomic data for microRNA identification using a Naïve Bayes classifier. Bioinformatics 22(11), 1325–1334 (2006)
6. Wee, L.J.K., Tan, T.W., Ranganathan, S.: CASVM: web server for SVM-based prediction of caspase substrates cleavage sites. Bioinformatics 23(23), 3241–3243 (2007)
7. Tan, A.C., Gilbert, D.: An empirical comparison of supervised machine learning techniques in bioinformatics, vol. 19, no. Apbc (2009)
8. Pan, S.J., Yang, Q.: A survey on transfer learning. IEEE Trans. Knowl. Data Eng. 22(10), 1345–1359 (2010)
9. Glorot, X., Bordes, A., Bengio, Y.: Domain adaptation for large-scale sentiment classification: a deep learning approach. In: Proceedings of the 28th International Conference on Machine Learning, no. 1, pp. 513–520 (2011)
10. Iqbal, M., Xue, B., Al-Sahaf, H., Zhang, M.: Cross-domain reuse of extracted knowledge in genetic programming for image classification. IEEE Trans. Evol. Comput. PP(99), 1 (2017)
11. Wang, Y., et al.: Knowledge-transfer learning for prediction of matrix metalloprotease substrate-cleavage sites. Sci. Rep. 1–15 (2017)

12. Dai, W., Yang, Q., Xue, G.-R., Yu, Y.: Boosting for transfer learning. In: Proceedings of the 24th international conference on Machine learning—ICML '07, pp. 193–200 (2007)
13. Al-Stouhi, S., Reddy, C.K.: Adaptive boosting for transfer learning using dynamic updates. Lecture Notes in Computer Science (Including Subseries Lecture Notes in Artificial Intelligence and Lecture Notes in Bioinformatics), LNAI, vol. 6911, no. PART 1, pp. 60–75 (2011)
14. Yao, Y., Doretto, G.: Boosting for transfer learning with multiple sources. In: 2010 IEEE Computer Society Conference on Computer Vision and Pattern Recognition, pp. 1855–1862 (2010)
15. Chen, C.T., Yang, E.W., Hsu, H.J., Sun, Y.K., Hsu, W.L., Yang, A.S.: Protease substrate site predictors derived from machine learning on multilevel substrate phage display data. Bioinformatics **24**(23), 2691–2697 (2008)
16. Barkan, D.T., et al.: Prediction of protease substrates using sequence and structure features. Bioinformatics **26**(14), 1714–1722 (2010)
17. Boyd, S.E., Garcia de la Banda, M., Pike, R.N., Whisstock, J.C., Rudy, G.B.: PoPS: a computational tool for modeling and predicting protease specificity. In: Proceedings/IEEE Computational Systems Bioinformatics Conference, CSB. IEEE Computational Systems Bioinformatics Conference, no. Csb, pp. 372–381 (2004)
18. Verspurten, J., Gevaert, K., Declercq, W., Vandenabeele, P.: SitePredicting the cleavage of proteinase substrates. Trends Biochem. Sci. **34**(7), 319–323 (2009)
19. Song, J., et al.: Cascleave: towards more accurate prediction of caspase substrate cleavage sites. Bioinformatics **26**(6), 752–760 (2010)
20. Wang, M., Zhao, X.M., Tan, H., Akutsu, T., Whisstock, J.C., Song, J.: Cascleave 2.0, a new approach for predicting caspase and granzyme cleavage targets. Bioinformatics **30**(1), 71–80 (2014)
21. Piippo, M., Lietzén, N., Nevalainen, O.S., Salmi, J., Nyman, T.A.: Pripper: prediction of caspase cleavage sites from whole proteomes. BMC Bioinform. **11**(1), 320 (2010)
22. Garay-Malpartida, H.M., Occhiucci, J.M., Alves, J., Belizário, J.E.: CaSPredictor: a new computer-based tool for caspase substrate prediction. Bioinformatics **21**(SUPPL. 1), 169–176 (2005)
23. Backes, C., Kuentzer, J., Lenhof, H.P., Comtesse, N., Meese, E.: GraBCas: a bioinformatics tool for score-based prediction of Caspase- and granzyme B-cleavage sites in protein sequences. Nucleic Acids Res. **33**(SUPPL. 2), 208–213 (2005)
24. Dönnes, P., Elofsson, A.: Prediction of MHC class I binding peptides, using SVMHC. BMC Bioinform. **3**, 25 (2002)
25. Widmer, C., Toussaint, N.C., Altun, Y., Kohlbacher, O., Rätsch, G.: Novel machine learning methods for MHC class I binding prediction. Lecture Notes in Computer Science (Including Subseries Lecture Notes in Artificial Intelligence and Lecture Notes in Bioinformatics), LNBI, vol. 6282, pp. 98–109 (2010)
26. Song, J., et al.: PROSPER: an integrated feature-based tool for predicting protease substrate cleavage sites. PLoS ONE **7**(11) (2012)
27. Kumar, S., Ratnikov, B.I., Kazanov, M.D., Smith, J.W., Cieplak, P.C.: CleavPredict: a platform for reasoning about matrix metalloproteinases proteolytic events. PLoS ONE **10**(5), 1–19 (2015)
28. Rawlings, N.D., Barrett, A.J., Finn, R.: Twenty years of the MEROPS database of proteolytic enzymes, their substrates and inhibitors. Nucleic Acids Res. **44**(D1), D343–D350 (2016)

Rényi's Entropy and Bat Algorithm Based Color Image Multilevel Thresholding

S. Pare, A. K. Bhandari, A. Kumar and G. K. Singh

Abstract Colored satellite images are difficult to segment due to their low illumination, dense features, uncertainties, etc. Rényi's entropy is a famous entropy criterion that provides excellent outputs in bi-level thresholding based segmentation. But such method suffers lack of accuracy, inefficiency, and instability when extended to perform color image multilevel thresholding. Therefore, a new color image multilevel segmentation strategy based on Bat algorithm and Rényi's entropy is proposed in this paper to determine the optimal threshold values more efficiently. The experiments are conducted on four real satellite images and two well-known test images at different threshold levels. The study shows that the proposed algorithm obtains good quality and adequate segmented results more effectively as compared to other multilevel thresholding algorithms such as Rényi's-PSO and Otsu-PSO.

Keywords Color images · Multilevel thresholding · Rényi's entropy
Bat algorithm

S. Pare (✉) · A. Kumar
Indian Institute of Information Technology Design and Manufacturing,
Jabalpur, Jabalpur 482005, Madhya Pradesh, India
e-mail: shreya.pare9@gmail.com

A. Kumar
e-mail: anilkdee@gmail.com

A. K. Bhandari
National Institute of Technology, Patna, Patna 800005, Bihar, India
e-mail: bhandari.iiitj@gmail.com

G. K. Singh
Indian Institute of Technology Roorkee, Roorkee 247667, Uttarakhand, India
e-mail: gksngfee@gmail.com

© Springer Nature Singapore Pte Ltd. 2019   71
M. Tanveer and R. B. Pachori (eds.), *Machine Intelligence and Signal Analysis*,
Advances in Intelligent Systems and Computing 748,
https://doi.org/10.1007/978-981-13-0923-6_7

1 Introduction

Color image segmentation is a fundamental step in image processing which distinguishes an image into distinct regions in such a way that the pixels belonging to same class have more homogeneity than other class. Color images have more information than the grayscale images due to the presence of different frames (red, green, and blue). Therefore, segmented color images are more suitable than the original images in performing object recognition, image representation, and other image processing tasks. Segmentation using entropy-based thresholding has become very popular in past some decades. Basically, image thresholding can be classified as bi-level and multilevel thresholding. In bi-level thresholding, an image is classified into two classes object and background using a single-threshold value. Multilevel thresholding partitions an image into several classes using more than two threshold levels.

Due to the developments in information theory, entropy-based thresholding criterion using between-class variance [1], moment-preserving principle [2], maximum entropy criterion [3], fuzzy approach [4], cross entropy [5], and Rényi's entropy [6] have gained wide popularity in bi-level thresholding. However, the above approaches suffer from lack of accuracy, inefficiency, and instability when extended to perform multilevel thresholding of color images. Moreover, owing to the increasing application of satellite images in astronomical sciences, geoscience, and geographical studies, satellite image segmentation has become a demanding field. However, due to the dense features and complexities, it is necessary to perform multilevel thresholding on such images to extract remote sensing information for image classification and retrieval. Consequently, a more practical approach to satellite segmentation is multilevel thresholding. In this paper, Rényi's entropy criterion is used as an objective function that has been studied previously for the multilevel segmentation of grayscale images [6–8] and in multispectral image segmentation [9]. Due to the successful results of Rényi's entropy-based previous implementations, it is exploited in this paper for the first time to perform multilevel thresholding of colored satellite images and other natural images.

To improve the performance of conventional entropy-based thresholding functions, several nature-inspired optimization algorithms based on multilevel segmentation techniques have been developed [10–13]. Also, multilevel thresholding approaches for satellite images using differential evolution [9], genetic algorithm [14], artificial bee colony [15], Cuckoo search [16–18], and other nature-inspired algorithms [19–22] have also been presented.

A new stochastic global search method, bat algorithm, has been used in diverse applications [23–25]. Bat algorithm has reported to give excellent performance in multilevel thresholding of grayscale images using Kapur's and Otsu method [26, 27], and fuzzy entropy [28]. Yet, it can be depicted from the literature that the Bat algorithm has not been used for the segmentation of any color or satellite image. Thus, it can be used due to its inevitable capability in finding the best solutions.

Due to the increase in the application of satellite images in geoscience and geographical studies, segmentation of such images has become important to perform further image processing tasks. But, because of dense features and low illumination of such images, it is challenging to acquire an accurate segmentation. Consequently, in this paper, a new multilevel thresholding technique has been developed using Rényi's entropy criterion implemented by Bat algorithm to search optimal thresholds and to perform a computationally faster and adequate segmentation of color satellite and natural images. The performance of the proposed algorithm is compared against PSO algorithm based on Renyi's entropy and Otsu method as an objective function on the set of four real satellite images and two natural images.

The rest of the paper is organized as follows: Section 2 gives a brief explanation of Rényi's entropy method. Section 3 discusses Bat algorithm. Section 4 presents the proposed methodology. In Sect. 5, simulation results and discussion are presented. Finally, conclusion is drawn in Sect. 6.

2 Rényi's Entropy-Based Multilevel Thresholding

For an image I with L gray levels, consider a set of discrete probability distributions p defined as

$$p = (p_1, p_2, ..., p_n) \in \Delta_n \tag{1}$$

where

$$\Delta_n = \{(p_1, p_2, ..., p_n)| p_i \geq 0, \; i = 1, 2, \ldots, n, n \geq 2, \sum_{i=1}^{n} p_i = 1\}$$

For additively independent events, Rényi's entropy can be defined as [6, 9]

$$H_\alpha[P] = \frac{1}{1-\alpha} \log_2 \left(\sum_{i=1}^{n} p_i^\alpha \right) \tag{2}$$

where order of entropy is a positive integer α. When α reaches unity, it becomes a limiting case of Rényi's entropy. Divide the total distribution for an image into N classes. Then, the priori Rényi's entropy for every distribution can be defined as [6, 9]

$$H_\alpha[C_1] = \frac{1}{1-\alpha}\left[\ln\sum_{i=0}^{t_1}\left(\frac{P(i)}{P(C_1)}\right)^\alpha\right],$$

$$H_\alpha[C_2] = \frac{1}{1-\alpha}\left[\ln\sum_{i=t_1+1}^{t_2}\left(\frac{P(i)}{P(C_2)}\right)^\alpha\right],\ldots,$$

$$H_\alpha[C_N] = \frac{1}{1-\alpha}\left[\ln\sum_{i=t_N+1}^{L-1}\left(\frac{P(i)}{P(C_N)}\right)^\alpha\right], \tag{3}$$

where

$$P(c_1) = \sum_{i=0}^{t_1} P(i),$$

$$P(c_2) = \sum_{i=t_1+1}^{t_2} P(i),\ldots$$

$$P(c_N) = \sum_{i=t_N+1}^{L-1} P(i) \tag{4}$$

$P(i)$ shows normalized histogram. Rényi's entropy-based thresholding approach obtains the optimal threshold ($t^* = \{t_1, t_2,\ldots,t_N\}$) by maximizing H_R:

$$H_R = H[c_1] + H[c_2] + \cdots + H[c_N] \tag{5}$$

$$t^* = \arg\max(H[c_1] + H[c_2] + \cdots + H[c_N]) \tag{6}$$

The application of the multilevel Rényi's entropy is limited since its computation complexity becomes $O(L^N - 1)$ due to the use of exhaustive search process to maximize H_R. As can be seen, when this method is extended to solve multilevel thresholding problem, the computation complexity increases as the thresholding level is increased. Thus, to reduce the computation complexities, heuristic search approach based on Bat algorithm is implemented to search the optimal multilevel threshold values using multilevel Rényi's entropy as an objective function.

3 Bat Optimization Algorithm

Bat algorithm is based on the echolocation of the bats through which they detect prey and avoid obstacles [23]. In the bat algorithm, bats navigation depends on time delay from emission to reflection. The three idealized rules which are the basis of Bat algorithm are as follows [28]:

1. Bats use echolocation to measure distance and they know the difference between the obstacles and the prey in some own magical way.
2. Bats randomly fly with velocity v_i at position x_i using a fixed frequency f_{min}, varying wavelength λ, and loudness A_0 to prey. Bats are able to adjust λ of their emitted pulse. Bats are capable to adjust pulse emission rate r from 0 to 1 based on the proximity of target, where the absence of any emission is denoted by 0 and the maximum rate of pulse emission is shown by 1.
3. The loudness can vary from a large positive value L_0 to a minimum value L_{min}.

Virtual bats are used in the algorithm to deal with the practical problems. New solutions and velocities at step t are updated in a multidimensional search space defined by [27, 28]

$$x_i^{t+1} = x_i^t + v_i^t \tag{7}$$

$$v_i^{t+1} = v_i^t + (x_i^t - x_{best})f_i \tag{8}$$

$$f_i = f_{min} + (f_{max} - f_{min})\beta \tag{9}$$

where, a vector $\beta \in [0, 1]$ generated randomly using uniform distribution. v^{t+1} and x^{t+1} are the velocity and position vectors of ith bat at tth and $(t+1)$th iteration, respectively. After performing a comparison of all solutions generated by N bats, a global optimum solution, x_{best} is located. The frequency f_i adjusts the velocity change. The bat frequency is randomly considered between f_{min} and f_{max}. In local search, current best solution is detected. Afterward, a new solution is generated locally through random walk using

$$x_{new} = \begin{cases} x_{best} + \varepsilon A^t, & \text{if } rand > r_i^t \\ x_i^t, & \text{otherwise} \end{cases} \tag{10}$$

where x_{new} and x_{old} are the solutions generated using (7). $\varepsilon \in [0, 1]$ is a randomly generated number through Gaussian distribution, which is the average loudness of all bats at tth iteration. $rand \in [0,1]$ is a uniform random number. When the new solutions are improved, loudness L_i and the rate of emission r_i are correspondingly updated along with t which shows that bats are nearer to the optimal solution. Here, $L_{min} = 0$ indicates that bat has just discovered the prey and for some time it will stop emitting. Loudness function $L^{(t+1)}$ and pulse emission function $r^{(t+1)}$ are defined as [27, 28]:

$$L_i^{t+1} = \alpha L_i^t \tag{11}$$

$$r_i^{t+1} = r_i^0[1 - e^{-\gamma t}] \tag{12}$$

where α is similar to the cooling factor of the simulating annealing algorithm. For any value of the constants $\alpha \in [0, 1]$ and $\gamma > 0$, $L_i^t \to 0$, $r_i^t \to r_i^0$, as $t \to \infty$.

Table 1 Parameters adopted for Bat and PSO algorithms

Algorithm	Parameters	Value
Bat	Number of bats and iterations	20 and 300
	Rate of pulse emission	[0, 1]
	Loudness value	[1, 2]
PSO	Number of particles and iterations	20 and 300
	Cognitive, social, and neighborhood acceleration (C_1, C_2, C_3)	3, 2, 1

4 Proposed Algorithm

In this paper, an effective method based on multilevel Rényi's entropy and BAT algorithm is used to perform multilevel segmentation of the colored satellite images at various thresholding levels. The Rényi's entropy is maximized to determine the optimal threshold values. The flowchart of the proposed algorithm is shown in Fig. 1.

5 Results and Discussion

The performance of the proposed algorithm and the other state-of-the-art techniques such as Rényi's-PSO and Otsu-PSO has been evaluated on the set of four colored satellite images and two well-known test images each of size 256×256 shown in Fig. 2. The optimal thresholds have been determined at 2, 5, and 8 levels. The control parameters set for the bat and PSO algorithms have been shown in Table 1. The results are evaluated in terms of peak signal-to-noise ratio (PSNR), mean square error (MSE), structural similarity index (SSIM), feature similarity index (FSIM), and execution time in seconds (CPU time) [18]. The best results among 10 different runs are documented in Tables 2, 3 and 4. The visual results are shown in Figs. 3, 4, 5, 6 and 7.

Table 2 shows the PSNR and MSE values for the predefined threshold levels ($m = 2$, 5, and 8) to measure the effectiveness of the algorithms. It can be depicted from the table that the proposed algorithm obtains better PSNR and MSE values than other compared methods for most of the cases. Moreover, Rényi's-PSO obtains better values than the Otsu-PSO. It is noticeable that an increase in the threshold levels causes improvement in the PSNR and MSE values of the algorithms.

Table 3 shows the best segmentation metrics, SSIM, and FSIM values which have been used to measure the accuracy of algorithms. The obtained values prove that the proposed algorithm is significantly better than Otsu-PSO and Rényi's-PSO methods for all the test images including satellite images. Again, the SSIM and FSIM values become significantly higher at highest thresholding level which shows better segmentation.

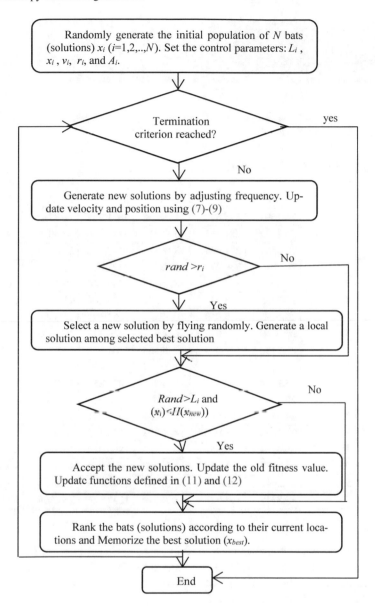

Fig. 1 Flowchart of the proposed algorithm

Fitness values can be acquired by maximizing the objective criterions. For the assessment of the solution quality and optimization ability, best fitness values obtained by different algorithms have been compared in Table 4. The values indicate that the proposed algorithm obtains best fitness values than the other approaches for most of the experiments. To measure the computation efficiencies of the algorithms,

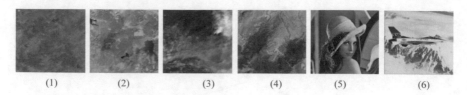

| | (1) | (2) | (3) | (4) | (5) | (6) |

Fig. 2 Original images (1) Satellite image 1, (2) Satellite image 2, (3) Satellite image 3, (4) Satellite image 4, (5) Lena, and (6) aircraft

Table 2 Comparison of PSNR and MSE

I	m	Otsu-PSO		Renyi-PSO		Proposed algorithm	
		PSNR	MSE	PSNR	MSE	PSNR	MSE
1	2	24.4391	235.572	11.04658	5801.223	14.9213	942.1113
	5	24.1147	221.587	17.57312	2143.096	15.4079	949.9231
	8	25.4391	211.844	22.09159	1417.471	21.4643	680.825
2	2	24.5071	230.777	11.63228	4671.006	11.5333	2086.552
	5	25.4242	187.804	18.25517	1982.724	15.2248	1995.483
	8	25.7585	173.33	22.69722	1352.888	15.9397	1766.928
3	2	24.5473	228.619	10.16721	2661.598	11.3496	1936.054
	5	25.0749	203.018	18.69922	1880.823	16.5035	1466.439
	8	25.6118	179.874	22.08654	1405.016	18.9082	1091.107
4	2	24.5026	230.685	11.48154	3934.487	18.9854	1934.487
	5	25.1754	197.555	16.27865	1807.851	15.7276	1807.851
	8	25.2711	193.873	15.5798	1320.515	18.6186	1320.515
5	2	24.5547	228.53	13.46137	2677.926	14.8959	1977.926
	5	24.9842	208.086	17.20793	1830.138	14.5878	1830.138
	8	25.2754	1.95289	20.63137	1366.315	16.3649	1366.315
6	2	24.1338	251.029	14.03987	3064.212	14.4842	1964.212
	5	24.2842	242.473	16.61647	1542.233	15.3459	1542.233
	8	25.8319	174.922	21.73319	1233.82	15.6594	1233.82

CPU time (in seconds) has been compared in Table 4. CPU time acquired by Otsu-PSO seems to be lowest, while Otsu-PSO and Rényi's-PSO use almost equivalent time to compute the optimal thresholds which can be attributed to the formulation of Rényi's entropy described in Sect. 2.

To visually compare the segmentation results of the algorithms, Figs. 3, 4, 5, 6 and 7 show the segmented outputs for different thresholding levels. It can be interpreted that at lowest levels, the results are not satisfactory due to inaccurate distribution of the homogenous pixels into different classes, whereas, on increasing the thresholding level, the segmented results of the test images have become visibly better showing improved segmentation quality. Still, the proposed algorithm represented in Fig. 5 has gained most accurate and high-quality segmented colored satellite and natural images than Otsu-PSO and Rényi's-PSO as shown in Fig. 3 and Fig. 4, respectively.

Table 3 Comparison of SSIM and FSIM

I	m	Otsu-PSO		Renyi-PSO		Proposed algorithm	
		SSIM	FSIM	SSIM	FSIM	SSIM	FSIM
1	2	0.86531	0.59201	0.8426	0.54709	0.914687	0.668472
	5	0.96927	0.74933	0.97136	0.767379	0.914555	0.710499
	8	0.96111	0.79101	0.98953	0.902672	0.987177	0.849137
2	2	0.91072	0.5469	0.8659	0.654419	0.859336	0.627234
	5	0.97171	0.82103	0.97342	0.888174	0.946729	0.717173
	8	0.96176	0.83319	0.99041	0.955532	0.954483	0.743934
3	2	0.93303	0.60803	0.817005	0.532278	0.873627	0.587298
	5	0.95945	0.7018	0.974851	0.773762	0.959427	0.719249
	8	0.95804	0.75893	0.988399	0.881559	0.967805	0.769684
4	2	0.86315	0.56198	0.899214	0.600715	0.847285	0.55874
	5	0.97363	0.78217	0.9798	0.826486	0.957091	0.722486
	8	0.94897	0.71772	0.991463	0.934342	0.949925	0.717273
5	2	0.88483	0.6467	0.932039	0.722866	0.920854	0.707125
	5	0.95378	0.77188	0.975259	0.865156	0.968091	0.804456
	8	0.95689	0.76156	0.989151	0.914179	0.98064	0.873449
6	2	0.74089	0.59609	0.939524	0.770326	0.935509	0.747843
	5	0.93374	0.72498	0.9915	0.905451	0.972896	0.767956
	8	0.97992	0.83112	0.996083	0.946228	0.9887	0.889663

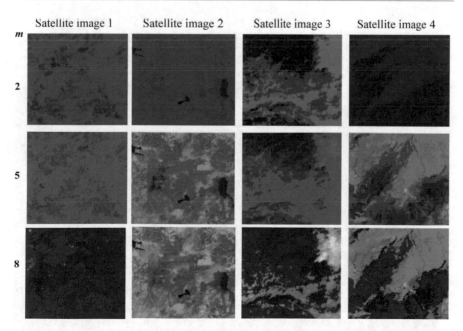

Fig. 3 Otsu-PSO based multilevel thresholding results for satellite images

Table 4 Comparison of fitness and CPU time (in seconds)

I	m	Otsu-PSO		Renyi-PSO		Proposed algorithm	
		Fitness	CPU	Fitness	CPU	Fitness	CPU
1	2	15.6606	4.06942	17.9575	16.38876	11.9442	20.584245
	5	19.95	4.03719	24.8563	20.61638	22.0525	20.640623
	8	27.2038	4.07242	32.5875	23.1849	30.3053	22.361312
2	2	28.4667	4.72252	48.4538	17.33141	73.4377	21.442789
	5	38.3604	4.72778	58.1408	21.42008	81.4994	21.548843
	8	46.5904	4.69461	66.2572	24.0755	83.3176	23.562375
3	2	15.4109	4.49236	28.177	17.91192	18.1163	21.099683
	5	24.8966	4.52346	37.9643	21.84455	37.682	23.035318
	8	32.7969	4.65028	45.6819	23.82585	51.4421	22.761302
4	2	24.6817	4.5952	30.5878	17.61524	90.1798	21.453026
	5	33.8869	4.63191	40.0358	21.5958	98.4291	21.863828
	8	41.2767	4.70657	47.6286	24.28583	105.929	23.693896
5	2	12.7164	4.42334	15.809	17.91026	13.6226	20.347612
	5	21.2688	4.56089	24.6184	21.96785	23.4862	21.408829
	8	25.7873	4.46443	28.3139	23.96744	27.5972	23.146076
6	2	11.5594	4.58355	14.9118	17.49183	31.1621	20.108488
	5	20.1623	4.53321	24.3735	20.91297	73.583	21.699044
	8	26.2805	4.7461	23.7509	31.38971	78.8795	23.449232

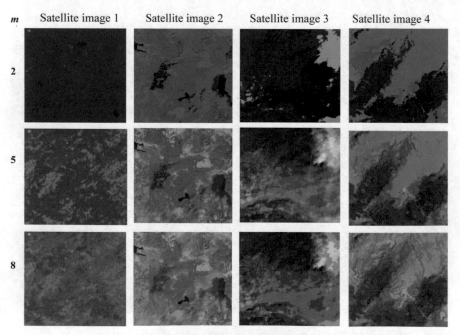

Fig. 4 Rényi-PSO based multilevel thresholding results for satellite images

| m | Satellite image 1 | Satellite image 2 | Satellite image 3 | Satellite image 4 |

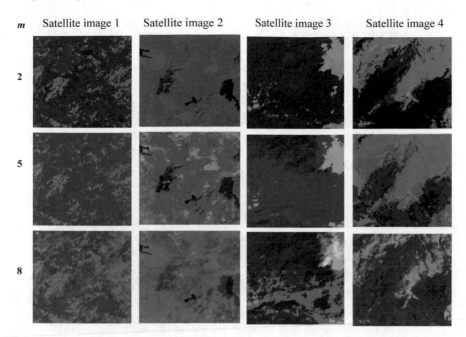

Fig. 5 Proposed algorithm (Renyi-Bat) based multilevel thresholding results for satellite images

Based on the segmentation results, it can be concluded that despite having slow convergence, proposed Bat-Rényi's algorithm has better segmentation results in terms of accuracy, quality, and robustness. Moreover, the proposed algorithm appears to be significantly feasible in performing multilevel thresholding of the colored satellite images due to better explorational and exploitational properties of Bat algorithm.

6 Conclusion

In this paper, an optimal multilevel segmentation algorithm has been proposed that employs Bat algorithm based Renyi's entropy function for determining the best threshold values for the segmentation of color images. Different colored satellite and natural images have been used to evaluate the merits of the proposed algorithm. Experimental results have been compared with some other methods such as PSO-based Otsu entropy and PSO-based Renyi's entropy algorithms. The objective and subjective results on different test images depict that the proposed method is more efficient and feasible for multilevel thresholding based segmentation of different color images including satellite image.

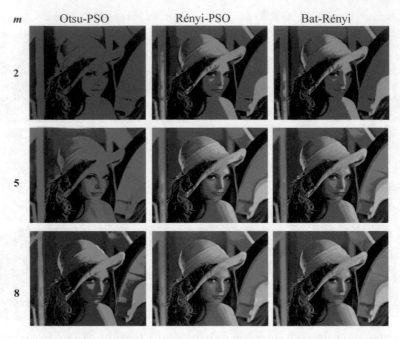

Fig. 6 Multilevel thresholding results for colored natural images using Otsu-PSO, Rényi-PSO, and Rényi-Bat algorithms at 2-level, 5-level, and 8-level

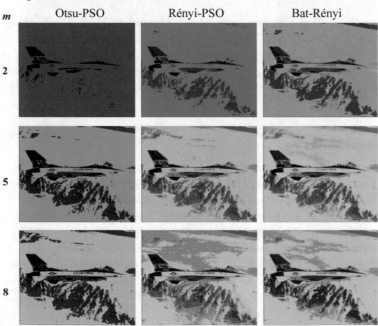

Fig. 7 Multilevel thresholding results for colored natural images using Otsu-PSO, Rényi-PSO, and Rényi-Bat algorithms at 2-level, 5-level, and 8-level

References

1. Otsu, N.: Threshold selection method from gray-level histograms. IEEE Trans. Syst. Man Cybern. **9**(1), 62–66 (1979)
2. Tsai, W.H.: Moment-preserving thresolding: a new approach. Comput. Vis. Gr. Image Process. **29**(3), 377–393 (1985)
3. Kapur, J.N., Sahoo, P.K., Wong, A.K.: A new method for gray-level picture thresholding using the entropy of the histogram. Comput. Vis. Gr. Image Process. **29**(3), 273–285 (1985)
4. Lim, Y.W., Lee, S.U.: On the color image segmentation algorithm based on the thresholding and the fuzzy c-means techniques. Pattern Recogn. **23**(9), 935–952 (1990)
5. Li, C.H., Lee, C.K.: Minimum cross entropy thresholding. Pattern Recogn. **26**(4), 617–625 (1993)
6. Sahoo, P., Wilkins, C., Yeager, J.: Threshold selection using Renyi's entropy. Pattern Recogn. **30**(1), 71–84 (1997)
7. Sahoo, P.K., Arora, G.: A thresholding method based on two-dimensional Renyi's entropy. Pattern Recogn. **37**(6), 1149–1161 (2004)
8. Wang, S., Chung, F.L.: Note on the equivalence relationship between Renyi-entropy based and Tsallis-entropy based image thresholding. Pattern Recogn. Lett. **26**(14), 2309–2312 (2005)
9. Sarkar, S., Das, S., Chaudhuri, S.S.: Hyper-spectral image segmentation using Rényi's entropy based multi-level thresholding aided with differential evolution. Expert Syst. Appl. **50**, 120–129 (2016)
10. Sarkar, S., Das, S., Chaudhuri, S.S.: A multilevel color image thresholding scheme based on minimum cross entropy and differential evolution. Pattern Recogn. Lett. **54**, 27–35 (2015)
11. Sağ, T., Çunkaş, M.: Color image segmentation based on multiobjective artificial bee colony optimization. Appl. Soft Comput. **34**, 389–401 (2015)
12. Beevi, S., Nair, M.S., Bindu, G.R.: Automatic segmentation of cell nuclei using Krill Herd optimization based multi-thresholding and localized active contour model. Biocybern. Biomed. Eng. **36**(4), 584–596 (2016)
13. Rajinikanth, V., Couceiro, M.S.: RGB histogram based color image segmentation using firefly algorithm. Proc. Comput. Sci. **46**, 1449–1457 (2015)
14. Pare, S., Bhandari, A.K., Kumar, A., Singh, G.K., Khare, S.: Satellite image segmentation based on different objective functions using genetic algorithm: a comparative study. In: IEEE International Conference on Digital Signal Processing (DSP), pp. 1–13. IEEE (2015)
15. Bhandari, A.K., Kumar, A., Singh, G.K.: Modified artificial bee colony based computationally efficient multilevel thresholding for satellite image segmentation using Kapur's, Otsu and Tsallis functions. Expert Syst. Appl. **42**(3), 1573–1601 (2015)
16. Bhandari, A.K., Singh, V.K., Kumar, A., Singh, G.K.: Cuckoo search algorithm and wind driven optimization based study of satellite image segmentation for multilevel thresholding using Kapur's entropy. Expert Syst. Appl. **41**(7), 3538–3560 (2014)
17. Pare, S., Kumar, A., Bajaj, V., Singh, G.K.: A multilevel color image segmentation technique based on cuckoo search algorithm and energy curve. Appl. Soft Comput. **47**, 76–102 (2016)
18. Pare, S., Kumar, A., Bajaj, V., Singh, G.K.: An efficient method for multilevel color image thresholding using cuckoo search algorithm based on minimum cross entropy. Appl. Soft Comput. **61**, 570–592 (2017)
19. Bhandari, A.K., Kumar, A., Singh, G.K.: Tsallis entropy based multilevel thresholding for colored satellite image segmentation using evolutionary algorithms. Expert Syst. Appl. **42**(22), 8707–8730 (2015)
20. Bhandari, A.K., Kumar, A., Chaudhary, S., Singh, G.K.: A novel color image multilevel thresholding based segmentation using nature inspired optimization algorithms. Expert Syst. Appl. **63**, 112–133 (2016)
21. Pare, S., Bhandari, A.K., Kumar, A., Singh, G.K.: An optimal color image multilevel thresholding technique using grey-level co-occurrence matrix. Expert Syst. Appl. **87**, 335–362 (2017)
22. Pare, S., Bhandari, A.K., Kumar, A., Bajaj, V.: Backtracking search algorithm for color image multilevel thresholding. Signal Image Video Process 1–8 (2017)

23. Yang, X.S.: A new metaheuristic bat-inspired algorithm. In: Nature Inspired Cooperative Strategies for Optimization (NICSO 2010), pp. 65–74 (2010)
24. Hasançebi, O., Teke, T., Pekcan, O.: A bat-inspired algorithm for structural optimization. Comput. Struct. **128**, 77–90 (2013)
25. Hasançebi, O., Carbas, S.: Bat inspired algorithm for discrete size optimization of steel frames. Adv. Eng. Softw. **67**, 173–185 (2014)
26. Alihodzic, A., Tuba, M.: Bat algorithm (BA) for image thresholding. In: Recent Researches in Telecommunications, Informatics, Electronics and Signal Processing, pp. 17–19 (2013)
27. Alihodzic, A., Tuba, M.: Improved bat algorithm applied to multilevel image thresholding. The Sci. World J. (2014)
28. Ye, Z.W., Wang, M.W., Liu, W., Chen, S.B.: Fuzzy entropy based optimal thresholding using bat algorithm. Appl. Soft Comput. **31**, 381–395 (2015)

Excitation Modeling Method Based on Inverse Filtering for HMM-Based Speech Synthesis

M. Kiran Reddy and K. Sreenivasa Rao

Abstract In this paper, we propose a novel excitation modeling approach for HMM-based speech synthesis system (HTS). Here, the excitation signal obtained via inverse filtering is parameterized into excitation features, which are modeled using HMMs. During synthesis, the excitation signal is reconstructed by modifying the natural residual segments in accordance with the target source features generated from HMMs. The proposed approach is incorporated into the HTS. Subjective evaluation results indicate that the proposed method enhances the quality of synthesis and is better than the traditional pulse and STRAIGHT-based excitation models.

1 Introduction

Among several speech synthesis systems, the most widely used one is the HMM-based speech synthesis system (HTS) [12]. HTS is flexible enough to generate different voice qualities with reduced complexity. HTS relies on source-filter representation for modeling the speech. The excitation or source is produced due to vocal fold vibration. The cascade of resonators formed in the vocal tract refers to filter. Efficient modeling of excitation is necessary to synthesize good quality speech. The basic HTS framework provides only pulse sequences and white noise as voiced and unvoiced excitation, respectively. These excitation signals are significantly different from the natural excitation. Hence, buzziness can be perceived in the speech

M. Kiran Reddy (✉) · K. Sreenivasa Rao
Department of Computer Science and Engineering, Indian Institute
of Technology Kharagpur, Kharagpur, West Bengal, India
e-mail: kiran.reddy889@gmail.com

K. Sreenivasa Rao
e-mail: ksrao@iitkgp.ac.in

© Springer Nature Singapore Pte Ltd. 2019
M. Tanveer and R. B. Pachori (eds.), *Machine Intelligence and Signal Analysis*,
Advances in Intelligent Systems and Computing 748,
https://doi.org/10.1007/978-981-13-0923-6_8

synthesized with HTS. For overcoming this problem, excitation models which can generate excitation signals close to the natural ones are needed to be developed.

Few attempts have been made in the literature to improve the quality of synthesis by using a subtle excitation model. Yoshimura et al. [14] proposed the first excitation model in which band-pass filtered pulse train and white noise are added together to generate voiced excitation. An excitation modeling approach for HTS based on speech transformation and representation using adaptive interpolation of the weighted spectrum (STRAIGHT) [5] is proposed in [15]. Mia et al. proposed closed-loop training-based mixed excitation model for HTS [8]. In this model, excitation is generated by employing the pulse trains and state-dependent filters. Natural glottal pulses modified as per the parameters generated by HMMs were utilized to synthesize excitation signal in [11]. The drawback in this approach is the estimation of glottal flow, which is a very complex problem. A source model based on pitch-scaled spectrum of excitation is presented in [13]. In [2, 3], Drugman et al. proposed a deterministic plus stochastic model (DSM) of excitation. Here, the first eigenvector obtained from PCA analysis of pitch-synchronous residual frames and white noise modified both in frequency and time are utilized to obtain excitation signal. In [1], the source is generated by concatenating the segments of residual signal around epochs and the segments modeled by amplitude envelope. Recently, deterministic plus noise (DPN) model-based approach is proposed in [9, 10].

The aim of this paper is to enhance the quality of HTS via efficient excitation modeling. In the proposed approach, source parameters obtained from each frame of the excitation signal are modeled by HMMs. At the time of synthesis, the spectral properties of the voiced excitation signals generated using natural residual segments are modified as per target source features to incorporate the time-varying characteristics in the real excitation. Speech is synthesized by exciting the mel-generalized log spectrum approximation (MGLSA) filter with resulting excitation signal. Here, the words excitation, residual, and source are used interchangeably. Section 2 illustrates the proposed excitation model. Section 3 gives details about speech synthesis using the proposed approach. In Sect. 4, the listening tests results are presented. Section 5 concludes the paper.

2 Proposed Source Model

In the proposed model, the excitation signal is parameterized in such a way that the essential characteristics of the real excitation signal are represented accurately. Figure 1 presents the various steps in the proposed method. First, the speech signal is inverse filtered to acquire the excitation signal. Next, using rectangular window the excitation signal is divided into frames of 25-ms with 5-ms frameshift. Then, the two excitation parameters: energy and source spectrum are obtained from every frame. Tenth-order LP analysis is used to estimate the source spectrum, which captures the spectral behavior of the source signal in every frame. The spectral distortion is reduced by transforming the LP coefficients into LSF coefficients. Besides excitation

Fig. 1 Proposed excitation
model

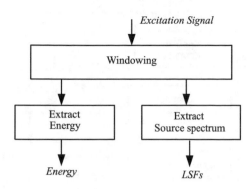

parameters, F_0 and 34th order mel-generalized cepstral (MGC) coefficients [15] are
obtained from the speech signals. The extracted F_0, excitation parameters, and MGC
coefficients are modeled using HMMs.

3 Synthesizing Speech Using Proposed Method

The proposed method is integrated into the framework of HTS to synthesize speech.
The openly available HTS toolkit [4] is used for implementing the HTS. Figure 2
shows the HTS framework incorporating the proposed excitation model. In HTS, the
two important steps are training and synthesis. During training, F_0 and vocal tract
components are obtained from the speech signals available in the database. Thirty-
fourth order mel-generalized cepstral (MGC) coefficients represent the vocal tract
component. F_0 is estimated using the pitch estimation method proposed in [6]. The
speech signal is inverse filtered using MGLSA filter to obtain the excitation signal.
A set of excitation parameters, namely, energy and source spectrum are estimated
from every frame (frame size and frameshift are 25 and 5 ms, respectively) of the
excitation signal using the proposed method. The extracted parameters are modeled
using HMMs. During synthesis, from the text input, a sentence HMM is constructed
according to the generated label sequence. MGC coefficients, F_0, and excitation
parameters (energy and LSFs) are obtained from the sentence HMM. The LSF coefficients are converted back to LPC coefficients. The excitation signal is reconstructed
using the generated parameters in two steps. First, voiced excitation signal is created
using the reference segment. Second, the excitation signal is inputted into an infinite
impulse response (IIR) filter for modifying its spectral properties to match the target
source spectrum. Here, we consider a single two-pitch period long and GCI centered
residual segment extracted from the real excitation of a sustained vowel as the reference segment. The selected segment acts as a representative sample of the excitation
signal for the given speaker and is normalized both in pitch and energy [9]. The
example of a reference segment is shown in Fig. 4. The reference residual segment is
expected to be more natural than a train of impulses. The steps in the synthesis stage
are shown in Fig. 3. The excitation signals are generated separately for voiced and

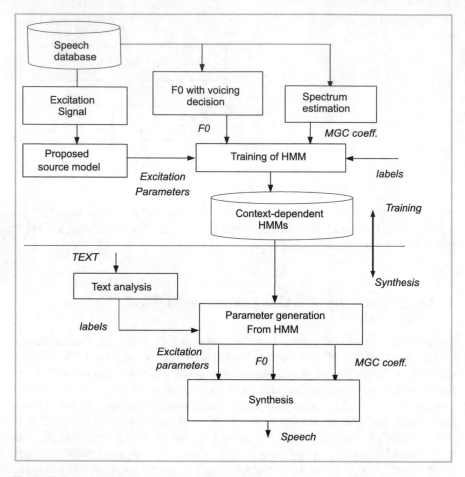

Fig. 2 HTS incorporating proposed excitation model

unvoiced speech. The reference segment is the basis for voiced excitation. First, the reference segment is resampled as per the target pitch period. Next, the segments are overlap added pitch-synchronously to reconstruct the entire excitation of a frame. The spectrum of the reconstructed excitation signal is different from that of the target spectrum. Therefore, its spectrum is modified according to the target spectrum obtained from HMMs to compensate for the differences between the spectra. Finally, the target energy obtained from HMMs is imposed on the excitation signal. In case of unvoiced speech, the energy modified white Gaussian noise is considered as the excitation. Synthetic speech is generated by filtering the obtained excitation signal with MGLSA filter.

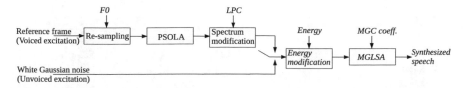

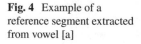

Fig. 3 Block diagram showing different steps in the synthesis stage

Fig. 4 Example of a
reference segment extracted
from vowel [a]

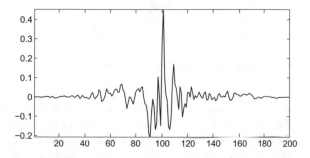

4 Subjective Evaluation

The performance of the proposed method is evaluated using two speakers (male (AWB) and female (SLT)) from CMU Arctic database [7]. For training about 1000 phonetically balanced English utterances are used for each speaker. The test set consists of 20 sentences that are not present in the training data. Fifteen research scholars with sufficient speech processing background participated in the subjective evaluation. The proposed method is compared with two widely used excitation modeling approaches, namely, STRAIGHT-based HTS [5] and pulse-based HTS. In pulse-based HTS, pulse sequence controlled by the pitch is used to excite voiced speech, and white noise is used to excite unvoiced speech. As a result, the speech synthesized with the pulse-HTS contains buzziness. Pulse-HTS is the standard method for testing effectiveness of the proposed excitation models. In the case of STRAIGHT-HTS [15], the excitation is modeled by the five a-periodic measures in five spectral sub-bands: [0-1], [1-2], [2-4], [4-6] and [6-8] kHz besides fundamental frequency. Mixed excitation made up of a pulse train and a noise element weighted by band-pass filtered parameters are used to create the excitation in STRAIGHT-HTS. Evaluation is done using preference tests and comparative mean opinion scores (CMOS) [2]. In CMOS test, the research scholars listened to two synthesized speech samples (one from proposed method and one from STRAIGHT-HTS or pulse-HTS) and indicated their overall preference on a 7-point scale which ranges from −3 to 3. To avoid bias, the speech utterances were played in arbitrary order to the scholars. The CMOS score is positive when the participants prefer proposed method over other methods. The participants' preference to other methods over proposed method is indicated by a negative CMOS score. The results of CMOS test taking into account

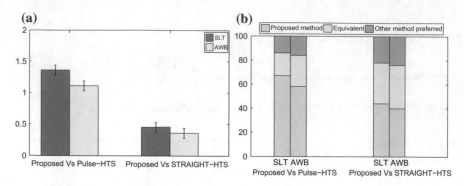

Fig. 5 **a** CMOS scores and **b** preference scores

the 95% confidence intervals are shown in Fig. 5a. Comparing with pulse-HTS and STRAIGHT-HTS, the average CMOS scores are around 1.2 and 0.4. The scores are relatively higher for the female speaker than that of the male speaker. This indicates that the speech synthesized by proposed approach has a superior quality compared to other approaches. In the case of preference tests, the scholars have to prefer the one that is better among a pair of synthesized speech samples. They have the option to prefer both as equal if they feel that the speech utterances sound similar. Figure 5b shows the preference scores. For both male and female speakers, the scholars preferred the proposed model significantly over other approaches. The use of natural residual segments in the proposed approach has upgraded synthesis quality.

5 Conclusion

In this paper, a novel source or excitation modeling method for HMM-based speech synthesis system is proposed. In the proposed method, the excitation parameters obtained from each frame of the residual signal are modeled by HMMs. At the time of synthesis, a single natural residual segment is used to reconstruct realistic excitation signals from the HMM generated parameters. Speech is synthesized by filtering the reconstructed excitation signal with MGLSA filter. The proposed method is simple and computationally less intensive. Subjective evaluation results show that the speech synthesized with proposed method has a better quality than two popular source modeling approaches. In future, the proposed method may be extended for synthesizing different voice qualities such as creaky voice. Also, by incorporating the proposed method in DNN-based speech synthesis system, the improvement in the quality of synthesis may be analyzed.

References

1. Cabral, J.P.: Uniform concatenative excitation model for synthesizing speech without voiced/unvoiced classification. In: Proceedings of Interspeech, pp. 1082–1086 (2013)
2. Drugman, T., Dutoit, T.: The deterministic plus stochastic model of the residual signal and its applications. IEEE Trans. Audio Speech Lang. process. **20**(3), 968–981 (2012)
3. Drugman, T., Raitio, T.: Excitation modeling for HMM-based speech synthesis: breaking down the impact of periodic and aperiodic components. In: Proceedings of International Conference on Audio, Speech and Signal Processing (ICASSP), pp. 260–264 (2014)
4. HMM-based speech synthesis system (HTS). http://hts.sp.nitech.ac.jp/
5. Kawahara, H., Masuda-Katsuse, I., de Cheveigne, A.: Restructuring speech representations using a pitch-adaptive time-frequency smoothing and an instantaneous frequency-based F0 extraction: possible role of a repetitive structure in sounds. Speech Commun. **27**, 187–207 (1999)
6. Kiran Reddy M., Sreenivasa Rao, K.: Robust pitch extraction Method for the HMM-Based speech synthesis system: IEEE Signal Process. Lett. **24**(8), 1133–1137 (2017)
7. Kominek, J., Black, A.: The CMU arctic speech databases. In: Proceedings of ISCA Speech Synthesis Workshop, pp. 223–224 (2004)
8. Maia, R., Toda, T., Zen, H., Nankaku, Y., Tokuda, K.: An excitation model for HMM-based speech synthesis based on residual modeling. In: Proceedings of ISCA Speech Synthesis Workshop, pp. 131–136 (2007)
9. Narendra, N.P., Sreenivasa Rao, K.: A deterministic plus noise model of excitation signal using principal component analysis for parametric speech synthesis. In: Proceedings of International Conference on Audio, Speech and Signal Processing (ICASSP), pp. 5635–5639 (2016)
10. Narendra, N.P., Kiran Reddy M., Sreenivasa Rao K.: Excitation modeling for HMM-based speech synthesis based on principal component analysis. In: proceedings of IEEE National Conference on Communication (NCC), pp. 1–6 (2016)
11. Raitio, T., Suni, A., Yamagishi, J., Pulakka, H., Nurminen, J., Vainio, M., Alku, P.: HMM-based speech synthesis utilizing glottal inverse filtering. IEEE Trans. Audio, Speech, Lang. Process. **19**(1), 153–165 (2011)
12. Tokuda, K., et al.: Speech synthesis based on hidden Markov models. Proc. IEEE **101**(5), 1234–1252 (2013)
13. Wen Z., Tao J., Hain H.-U.: Pitch-scaled spectrum based excitation model for HMM-based speech synthesis. In: Proceedings of IEEE international conference on Signal Processing (ICSP), pp. 609–612 (2012)
14. Yoshimura, T., Tokuda, K., Masuko, T., Kobayashi, T., Kitamura, T.: Mixed-excitation for HMM-based speech synthesis. In: Proceedings of the Eurospeech, pp. 2259–2262 (2001)
15. Zen, H., Toda, T., Nakamura, M., Tokuda, K.: Details of Nitech HMM-based speech synthesis system for the Blizzard Challenge 2005. IEICE Trans. Inform. Syst. **E90-D**, 325–333 (2007)

Efficient Methodology for Estimation of Vibration Thresholds for Electrical Machines

D. Ganga and V. Ramachandran

Abstract This paper discusses about deducing the impacts of dynamic operating conditions of the DC machine on the shaft vibration pattern by proposing a statistical classification based signal processing technique, which estimates the signal oscillations at multiple amplitude levels. This analysis parameterizes the vibration signal with the oscillation information so that the non-stationary amplitude pattern of the vibration has been extracted efficiently which enables effective fixation of vibration thresholds for safe and reliable operation of machines in industrial environments. The variations in the pattern of the non-stationary vibration signal have been identified by using the signal transition matrix. The proposed technique determines the vibration thresholds at different machine operating conditions from the decomposed signal oscillations by clustering and enumerating the set of denser oscillation levels as characterized levels of vibration which distinguish the incipient changes in the machine conditions efficiently. The technique on implementation has traced the dynamic changes in the real-time rotor shaft vibration of DC shunt motor during starting from stall condition and sudden load changes with and without external disturbances. The efficiency of the technique has been outlined by comparing its performance with the outcomes of widely adopted non-stationary signal feature extraction methods of Joint Time-Frequency Analysis (JTFA) such as Short Time Fourier Transform and Gabor Transform, executed with different settings of the parameters namely type of window, window length and time steps. The effectiveness of the proposed technique in extracting the variations of the shaft vibration in a simple, faster and detailed manner has been elucidated through comparative analysis.

Keywords Vibration threshold · Electrical machines · Time-frequency analysis
Signal processing

D. Ganga (✉)
NIT Nagaland, Chumukedima, Dimapur 797103, India
e-mail: gangaadhan@gmail.com

V. Ramachandran
College of Engineering Guindy, Anna University, Chennai 600025, India
e-mail: rama5864@gmail.com

© Springer Nature Singapore Pte Ltd. 2019   93
M. Tanveer and R. B. Pachori (eds.), *Machine Intelligence and Signal Analysis*,
Advances in Intelligent Systems and Computing 748,
https://doi.org/10.1007/978-981-13-0923-6_9

1 Introduction

Vibration thresholds determined in terms of amplitude and frequency have been the decision factors in condition monitoring of rotating machinery. Accordingly, the techniques employed for fixing thresholds fall into three major domains of time, frequency, and time-frequency to extract informative signal features. The time domain techniques of RMS and Crest Factor providing power and impulse properties of the signal have been proved to be effective only when the fault has developed substantially. Later the stochastic processing techniques namely probability density function, mean, variance and kurtosis that abridge the amplitude variations of the signals have been used widely in fault diagnosis to determine the abnormal conditions [1]. Monitoring of faults in certain system components such as gears, bearings used methods based on spectrum and cepstrum which identify faults with reference to vibration frequency determined using Fourier Transform. The vibration signals when processed using Fourier Transformation gives the better frequency information of the signal for fault detection under stationary conditions, whereas under dynamic conditions the spectrum response is highly smeared and does not provide any valid reference level for condition monitoring. As a measure of overcoming this drawback, time-frequency analysis has been adopted as a solution for non-stationary feature extraction which uses time-frequency localization in signal by using window. In this, the signal variations are captured in the windowed signal, which is mainly dependent on the window function that has been selected. While determining the features of unknown signal, selection of appropriate window poses high challenge and is prone to loss of signal features in case incorrect choice. The wavelet transform attempts to solve this problem of time-frequency resolution by introducing scalable windows formed by mother wavelet as a base function, whose selection is crucial and tedious again [2]. By Heisenberg's uncertainty principle, obtaining high resolution in both time and frequency simultaneously is impossible. These constraints in the extraction of frequency and amplitude information from a non-stationary vibration signal calls for alternate analysis methods for fixation of effective condition monitoring thresholds. Though vibration signals of rotating machines are highly non-stationary, several processing methods have been adopted to extract the signal features since vibrations are always potential indicators of machine conditions. The machines often undergoing different operating modes possess dynamic vibration behavior which calls for optimum threshold fixation, in order to evade the implications of false alarms and ensure safe operations. This paper addresses this challenge in determining absolute vibration thresholds adaptive to the machine operating conditions by extracting the oscillation count of the signal at different classified levels of amplitude. The pattern of the non-stationary vibration signal has been tracked in a detailed manner to determine the signal features of amplitude in conjunction with oscillations for setting the precise threshold limits corresponding to operating conditions for reliable condition monitoring. The paper has been organized in sections to present the proposed technique. Section 1 gives an introduction on condition monitoring thresholds and techniques used. Section 2 reviews about the vibration based condition monitoring methods. Section 3 presents the proposed technique with explanation and algorithm. Section 4 briefly explains the experimental set up. Section 5 furnishes the results of real-time implementation. Section 6 compares the performance of the proposed technique with time-frequency techniques. Section 7 highlights significance of the condition monitoring thresholds and the advantages of the proposed technique.

2 Relevant Works

The earlier works report that alarm levels generated with improper analysis create chaos in machine maintenance due to false status identifications. Diego Galar et al., have cited that single valued thresholds provided by the manufactures are not suitable for fault identification under non-stationary operations, environmental changes and aging. Hence the signal features extracted using data-driven or physics methods are compared with the normal features for making optimum decisions on fault occurrence. The discrepancies in the features compared have been used to decide upon false alarms and actual risks. Various signal and process analysis methods are applied to obtain reliable thresholds for efficient monitoring. The thresholds maintained for a limited period of time or positions of a manufacturing cycle are said to provide optimum monitoring [3]. Straczkiewicz M. et al. [4] have specified that the number of parameters considered in any condition monitoring system are high and various efforts are said to be undertaken to distinguish more critical alarms to be addressed in priority. Hence introduction of Violation Priority Coefficient (VPC) calculated from the violations of individual features measured over a period a time has been considered as a potential indicator for decision making. Such decision-making process relies on the thresholds that determine the warning and alarm levels. An incorrectly fixed thresholds result in overwhelming condition monitoring. In order to alleviate this, diagnosis of reliable thresholds adaptive to the operating conditions by using effective techniques need to be focused for efficient decision-making. The vibration analysis has been widely considered to assess the state of machine due to its high precision. With respect to time-varying conditions such as load or speed changes the vibration signals are non-stationary and the characterization of the signal is a challenge for condition assessment and classification of machine state. The authors have used time-frequency extraction method, whose response is expressed in terms of Linear Frequency Cepstral Coefficients (LFCC) and Spectral Sub-Band Centroids (SSC) as a measure of reducing the loss in the estimation of dynamic features. The requirement for specifying variety of dynamic features has been emphasized for better condition monitoring under non-stationary operations [5]. Various types of condition indicators based on statistical measures and analysis algorithms used for the representation of specific failure modes of gears, bearings and shafts have been discussed by Junda Zhu et al. The application of Time Synchronous Averaging and Time Synchronous Resampling algorithms, as an effort of determining stabilized condition monitoring indicators for wind turbines has been demonstrated [6]. The application of time-frequency analysis to capture the bearing damages caused due to bearing current has been discussed by Aurelien Prudhom et al. The degradation in the bearings of induction motor is detected from the vibration signal frequency evolving over time in the time-frequency map of the STFT. This analysis has discriminated the bearings subjected to similar failures and failures of other components [7]. In STFT localization of time with high and low frequency components requires a shorter and longer time window respectively and since the window function is constant over the time the technique has set back in real-time non-stationary signals analysis in capturing the characteristics such as impulses and frequency variations. The adaptive window of the wavelet transform decomposes signal into low frequency approximations and high frequency details but with less time and frequency resolutions respectively, whereas wavelet packet transforms give high time and frequency resolutions. Unlike wavelet transform, wavelet packet transform decomposes the detail signals also iteratively. Though it provides better results, selection of wavelets matching the signal characteristics is a challenge. The other techniques based on bilinear transformation such as Wigner-Ville distribution, Wigner higher order, L-class and S-class distributions suffer from cross term interferences [8]. The threshold calculations in condition monitoring have more importance but are not given due consideration. Thousands of false alarms are generated due to adoption of default threshold levels. Gaussian distribution of raw vibration data is used to calculate mean and standard deviation of the data. The nature of vibration is assessed from the range of values falling in between the multiples of standard deviation. Generally, the real-time data do not take Gaussian distribution always and hence the calculations lead to false alarms. In such approach, different distributions such as Weibull probability distribution, generalized extreme value probability distribution, extreme value probability distribution, inverse Gaussian probability distribution are used instead of Gaussian distribution to characterize the vibration data for threshold fixation [9]. Various such works portray different methodologies developed for condition monitoring, where focus is more effective for diagnosis of abnormalities from the available data nature than threshold estimation adaptive to operating condition. Unless thresholds are estimated precisely, the criticality of the abnormal conditions could not be realized to the fullest extent.

3 Proposed Technique for Threshold Estimation

The fixed thresholds adopted need to be alternated with adaptive alarms specific to operating nature of the individual machine, by using suitable analysis technique for accurate and quick condition monitoring of electrical machines. Hence a new statistical classification based signal decomposition and clustering technique has been proposed for signal feature extraction and precise vibration threshold estimation. The proposed technique tracks the vibration signal transitions between the classes at multiple levels of amplitudes and computes the number of oscillations between the levels. While classifying the signal, the amplitude is segmented into a desired set of classes of equal and desired width. The transition matrix (T) containing signal transitions that has resulted out of this statistical classification with 'n' number of classes, used to compute the oscillation nature of the vibration is as follows:

$$\mathbf{T} = \begin{bmatrix} ST_{11} & \cdots & ST_{1n} \\ \vdots & \ddots & \vdots \\ ST_{n1} & \cdots & ST_{nn} \end{bmatrix} \tag{1}$$

- ST_{ij} is the number of signal transitions from *class i* to *class j*
- n is the total number of classes

In some of the previous works, the transition matrix has been reported to predict the state transition probabilities, whereas in the proposed vibration analysis technique, number of signal transitions has been considered to calculate the oscillations between every class and other classes so as to extract the threshold levels of vibration. The transition of the signal from a lower class to higher class and vice versa are accounted as positive and negative slope respectively in the transition matrix. Thus all the diagonal elements of matrix become zeros in this case. The analysis has been carried out by considering the transition matrix and progressing through the upper diagonal matrix row-wise and lower diagonal matrix column-wise elements or its vice versa. The n column vectors of upper diagonal matrix give the signal transitions from a class corresponding to the row to class of the diagonal element and n row vectors of the lower diagonal matrix represent the signal transitions from class of the diagonal element to lower class corresponding to column.

3.1 Algorithm

The algorithm below measures the oscillations in the real-time non-stationary vibration signal at multiple class levels using statistical classification of the signal amplitude and transition matrix obtained.

Step 1: Find the maximum amplitude (Acc_{max}) and minimum amplitude (Acc_{min}) of the shaft acceleration signal after the removal of outliers, if required, and fix them as initial and end points for classification.

Step 2: Choose the amplitude resolution for classification as per monitoring requirements and assign it as the width of each class (Class Width)

Step 3: Calculate the range of classification as $Range = (\text{Acc}_{\text{max}}) - (\text{Acc}_{\text{min}})$. Classify the signal into classes of equal width as per Eq. 2.

$$\text{No. of classes} = \frac{Range}{\text{Class Width}} \tag{2}$$

Step 4: From the signal transition matrix ($\mathbf{T}$), obtain Upper Diagonal matrix $\mathbf{T}^{\text{UD}}$ and Lower Diagonal Matrix $\mathbf{T}^{\text{LD}}$. Compare each column vector of $\mathbf{T}^{\text{UD}}$ with corresponding row vector of $\mathbf{T}^{\text{LD}}$. For every pair of elements, segregate the lowest or equal values as in Eq. 3 and store OSC_{ji} as row element of new Oscillation Matrix (**OM**). Also calculate the difference between every pair of elements, if unequal as in Eq. 4.

$$OSC_{ji} = \min(ST_{ij}, ST_{ji}) \tag{3}$$

$$DS = \text{Sub}(ST_{ij}, ST_{ji}) \cdots DS_{ij} if > 0,$$
$$DS_{ji} if < 0.. \tag{4}$$

Step 5: The resulting **OM** will be a lower diagonal matrix with matrix elements of every row giving the distribution of oscillation between class corresponding to the row and other lower classes.

Step 6: If the values determined by Eq. 4 is positive, record the resulting positive difference, DS_{ij} (Eq. 4) in a new matrix **PS**. Elements of **PS** represent the Excess Positive Slopes (EPS) inferring the existence of oscillations which have started from a lower class i, but have not ended at the same lower class.

Step 7: In case of negative values, store the results of Eq. 4, DS_{ji} in a new matrix **NS**. The elements of **NS** give the Excess Negative Slopes (ENS) which represent the existence of oscillations ending at the corresponding class j but have not started from the same lower class. Thus the movement of the non-stationary signal between dissimilar amplitudes could be traced from Eq. (5).

$$(DS > 0 \rightarrow EPS = DS_{ij}) \wedge (ENS = DS_{ji}) \tag{5}$$

Using the results of Oscillation Matrix (**OM**), the strategy for condition monitoring threshold estimation has been proposed based on the oscillation density of the similar amplitude oscillations happening between every class and its lower classes. The density of oscillations (OD_k) (Eq. 8) between any class k and its lower classes is obtained from the measure of oscillations pertaining to a class OC_k (Eq. 6) and Total Oscillations (Eq. 7) as shown below.

$$OC_k = \sum_{j=1}^{j=k-1} OM_{kj} \forall 1 \leq k \leq n \tag{6}$$

$$Total \text{ Oscillations} = \sum_{k=1}^{k=n} OC_k, \tag{7}$$

where,

OM_{kj} is oscillations between class k and j
n   is the number of classes
k   is the class of observation (Row index)
j   is the column index of the Oscillation Matrix (OM)

$$\%OD_k = \frac{OC_k}{\text{Total Oscillations}} \times 100 \tag{8}$$

The above-derived features of OM are applied to classify the safe vibration thresholds specific to the operating conditions of electrical machines by making following observations:

- Determination of dominant classes possessing comparatively higher number of oscillations with lower classes. The set of classes which cumulatively have an oscillation density of 90–95% have been clustered and labeled as upper threshold class cluster as shown in Eq. 9.
  Upper Threshold Class Cluster =

$$\left\{ Class\ UT_1,\ Class\ UT_2 \ldots Class\ UT_p \Big| 90\% \leq \left(\sum_{c=1}^{c=p} \%OD_{ClassUT_c} \right) \leq 95\% \right\}, \quad (9)$$

where p is the number of upper threshold classes identified.
- The oscillation distribution of every upper threshold class $Class\ UT_p$, with that of its lower classes, is extracted from the corresponding row of Oscillation Matrix **OM**, and distribution of the oscillation density along the lower classes is determined as in Eq. 10.

$$\%OD_{Lck} = \frac{OM_{kj}}{OC_k} \times 100 \,\forall\, 1 \leq j \leq (k-1), \quad (10)$$

where Lck represent each Lower Class of Upper Threshold Class k.
- Clustering of the lower classes having cumulative Oscillation Density of 65 and above percent as Lower Threshold Class Cluster (Eq. 11).
  Lower Threshold Class Cluster =

$$\left\{ Class\ LT_1,\ Class\ LT_2 \ldots Class\ LT_q \Big| \left(\sum_{c=1}^{c=p} \%OD_{LC\ ClassLT_q} \right) \geq 65\% \right\}, \quad (11)$$

where q is the number of lower threshold classes identified.

This condition monitoring strategy framed with Upper and Lower Threshold clusters identifies the machine's vibration thresholds adaptive to the operating conditions very precisely. The shift in the Total Oscillations, Upper and Lower Threshold Classes identify the change of operating conditions very quickly and characterizes the effect of the changes on machine vibration so effectively even though the vibration signal is highly non-stationary. The intensity of abnormalities can be measured from the range of deviations. Instead of predefined thresholds, this identification of cluster of classes as upper and lower thresholds adaptive to the operating nature of the machines enables accurate fixation of the vibration reference levels in the field conditions. Thus, the proposed methodology of adaptive thresholds brings out effective condition monitoring for electrical machines which are normally operated at different environmental and operating conditions and eliminates false identification of failures.

4   Experimental Set up

The algorithm is implemented on the real-time shaft vibration signals acquired from DC motor coupled to AC generator at different operating conditions of DC motor: i. starting to no load speed at standalone condition and in the presence of the mechanical disturbance caused by running a three-phase induction motor under constant speed in the neighborhood of DC machine ii. loading of DC motor at standalone condition. The tri-axial accelerometer (ADXL345) [10] mounted on the rigid structure supporting the DC motor's rotor shaft senses the shaft vibration which is acquired by data acquisition system, my-RIO [11] in fast data transfer mode of I2C (400 kHz) with Output Data Rate (ODR) and Bandwidth as 800 Hz and 400 Hz respectively. The ADXL345 is used in 13-bit resolution at measurement range of ±16 g with sensitivity of 256 LSB/g. Hall effect transducers are used to convert the higher levels of generated AC voltage and input DC armature current into lower levels suitable for data acquisition. The motor input current, voltage generated and the shaft accelera-tion signals of 3,96,000 samples are acquired using myRIO at the sampling rate of 2000 samples per second and the physical values are deduced by converting the raw values into actual values.

5   Real-Time Analysis Results of the Algorithm

The parameters of shaft acceleration signal pertaining to different operating modes have been extracted and the calculations made as per the classification requirements of the proposed algorithm are furnished in Table 1.

The algorithm implemented on the shaft acceleration signals of different operating conditions using LabVIEW (DIAdem) has resulted in the Oscillation Matrices (**OM**)

Table 1   Vibration signal classification parameters at different operating conditions

Operating mode	Initial and end point of classification	Range	Class width	No. of classes
Starting to no load speed (Standalone condition)	Acc_{max}: 368 Acc_{min}: −286	654	54	12
Starting to no load speed (Disturbance condition)	Acc_{max}: 453.6 Acc_{min}: −296.3	750	53.5	14
Load changes (Standalone condition)	Acc_{max}: 389 Acc_{min}: −292	681	52.5	13

Table 2 Oscillations during starting to no load speed (standalone condition)

Class k	Class mean	No. of oscillations between every class and its lower classes												OC_k	% OD_k
		1	2	3	4	5	6	7	8	9	10	11	12		
1	−259.05	0	–	–	–	–	–	–	–	–	–	–	–	0	0.0
2	−204.47	0	0	–	–	–	–	–	–	–	–	–	–	0	0.0
3	−149.88	0	4	0	–	–	–	–	–	–	–	–	–	4	0.1
4	−95.29	0	14	12	0	–	–	–	–	–	–	–	–	26	0.5
5	−40.70	3	24	81	126	0	–	–	–	–	–	–	–	234	4.9
6	13.88	4	44	143	229	290	0	–	–	–	–	–	–	710	14.9
7	68.46	20	108	224	384	308	135	0	–	–	–	–	–	1179	24.7
8	123.05	4	130	268	480	261	74	27	0	–	–	–	–	1244	26.0
9	177.64	3	48	203	309	187	92	10	0	0	–	–	–	852	17.8
10	232.23	0	11	59	134	94	58	15	0	0	0	–	–	382	8.0
11	286.82	0	3	30	30	36	29	6	1	0	0	0	–	135	2.8
12	341.41	0	0	6	2	3	2	1	0	0	0	0	0	14	0.3
	Total Oscillations													4780	100

Table 3 Total oscillations and threshold class clusters of DC Motor shaft vibration

Operating condition	Total oscillations	Upper threshold class cluster	Lower threshold class cluster
Starting to no load Speed (Standalone mode)	4780	{13.88, 68.46, 123.05, 177.64, 232.23}	{−149.88, −95.29, −40.70}
Starting to no load Speed (With external disturbance)	6221	{−1.68, 51.88, 105.46, 159.03, 212.61}	{−162.41, −108.84, −55.26}
Loading (Standalone mode)	7810	{−3.79, 48.67, 101.13, 153.60, 206.07}	{−161.18, −108.72, −56.25}

providing the details of oscillations between every class and its lower classes. The computed values of **OM,** percentage density of oscillations (%OD_k) corresponding to the operating mode of starting to no load speed at standalone condition alone has been furnished in Table 2 for illustration.

The results of Total Oscillations, Upper and Lower Threshold Class Clusters of DC motor's shaft vibration signal calculated for every operating condition based on the respective Oscillation Matrix (**OM**) are furnished in Table 3. From the results tabulated in Table 3, it is observed that the proposed analysis identifies the presence of external disturbance, change of operating conditions so elegantly from the variations in the number of Total Oscillations that are evaluated. The shifts in the amplitude levels of denser oscillations caused by different operating conditions on the vibration signals are revealed precisely which lead to the formation of different threshold class clusters adaptive to the operation conditions. The observed thresholds during loading at standalone condition and starting to no load speed at disturbance condition elucidate that the external disturbance during starting to no load speed condition has created a vibration effect on the machine shaft equivalent to that of loading at standalone condition, which is unrevealed in the electrical measurements. This information extracted helps to prescribe the field operational constraints for machines in real-time applications so that the machine performance and lifetime can be improved.

6 Comparative Study with Joint Time-Frequency Analysis Techniques

The Joint Time-Frequency analysis [12] is implemented on the same set of vibration signals of different operating conditions using LabVIEW. The results of Short Time Fourier Transform (STFT) applied with different windows and window lengths are shown in Fig. 1 with details of window and its length and dominant frequencies determined with time index. The rectangular window chosen for length of 6000 and time step of 1500 has captured only frequency of 25 Hz in both normal and disturbance conditions, and has not distinguished the change at all. However, the increase in the window length to 24,000 with overlap of 6000, has shown an additional frequency of 125 Hz and 75 Hz in normal and disturbance condition respectively. Well, having set these levels as thresholds for condition determination, further analysis carried out with window length of 85000 with overlap of 56000, shows the existence of 25–160 Hz in normal condition and 25–150 Hz in disturbance condition. This overrides previous threshold of 125 Hz in normal condition (160 Hz) and 75 Hz of disturbance condition (150 Hz) which leads to the condition of false alarms. This exploration has been done on the vibration signals of normal and disturbance conditions based on same window and different window lengths.

An alternate perception of using different windows for the same window length has been outlined below:

Out of the four windows chosen for STFT analysis, only Flat top and Gaussian identify closely similar pattern of frequencies for all window lengths in normal condition, whereas the same windows show different patterns during disturbance condition. On the other hand, considering the 75 Hz determined by both windows (for length of 6000) as threshold will cause a false alarm for 125 Hz present in the normal condition, which is identified by the window length of 24,000. If 125 Hz detected at normal condition for the window length of 24,000 has been fixed as threshold, then Flat top misses to detect 150 Hz in the disturbance signal which is determined by the Gaussian window. From the examination, it is observed that STFT using Gaussian window finds the presence of peak frequencies of 125 and 140 (in Hz) under normal condition and 120, 125, 130 and 150 (in Hz) under disturbance in which demarcating deviation is very marginal. Moreover, the performance Gaussian window is not so convincing when compared to Flat top when window length of 85000 is chosen for disturbance condition.

The results of alternate analysis made using Gabor Transform on the same signal are listed below:

- Gabor Transform finds the frequency content ranging between 0 and 150 Hz at different time instants in normal condition which is against the 150 Hz uncaptured by Gaussian window in normal condition.
- Under disturbance, the frequencies from 0 to 270 Hz have been observed which also remain unidentified.

Such inconsistency in the features extracted by STFT and Gabor Transform for the same vibration signal creates ambiguity in the aspect of threshold fixation. In order to fix thresholds and make comparisons using thresholds, an intensive feature extraction is required. The time-frequency methods which are characterized by time-frequency localization stand good in detecting a specific range of frequencies which depend on window or wavelet selection. However, for threshold determination comprehensive information of the real-time vibration signal is required to detect the abnormalities at the earliest. Though the proposed technique in the current version

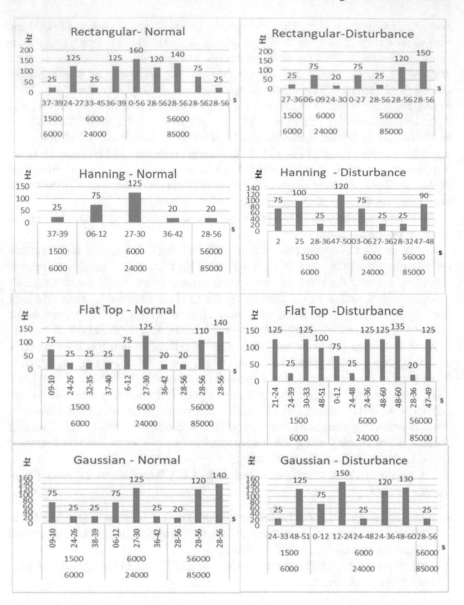

Fig. 1 Time-frequency details of vibration signal during starting to no load speed (STFT)

lacks time-frequency localization, it is observed that the features extracted are more detailed to identify and validate the changes in the nonstationary vibration signal due to disturbance and operating condition precisely than time-frequency techniques which depend on selection of type of window and length as major factors. Hence the information obtained in terms of oscillations at classified amplitude levels and density based threshold class clusters using proposed technique are observed to provide detailed reference for analysis of nonstationary vibration signal at any of the operating conditions.

7 Conclusion

Industrial condition monitoring systems generally have enormous monitoring parameters, for which threshold calculations are more important. But due consideration has not been given so far for threshold determination and focus is more on fault diagnosis. More researches made in this domain with conventional and improved time-frequency localization techniques are intended for detecting pronounced or developing faults than threshold determination. On the other hand, techniques based on signal amplitude encompass statistical measures such that the determined signal characteristics could reflect only major changes. Hence, the time-frequency techniques are opted comparatively more than the amplitude based methods. This work contributes an analysis method which traces the vibration oscillations at quantized amplitudes as per the selected class width and also gives the distribution of the vibration oscillations at all the quantized levels. Applying these extracted features, implementation of density-based clustering has been carried out for determination of vibration thresholds under dynamic operating conditions for effective decision-making in condition monitoring. This outpaces the amplitude-based methods by providing piecewise amplitude information of the entire signal and threshold class clusters based on the actual oscillation density of vibration pattern. In that way, the proposed technique identifies i. effects of even tolerable environmental disturbances on machine's vibration ii. distinguishes the vibration levels changing with respect to operating conditions very precisely and iii. extracts the features both in terms of amplitude and oscillations for any length of the signal. This has been validated on the real-time vibration signals collected from DC machine shaft under normal, disturbed and loaded conditions. The detailed changes observed in the vibration pattern, effectively discriminates the machine conditions and reveals the characteristics which are not estimated with electrical measurements. The electrical machines having same technical specifications show different behavior within the maximum specified limits of factory conditions when subject to different environmental and operational conditions. In such dynamic conditions, the contemporary condition monitoring could identify the incipient vibration changes with limited capabilities due to the non-stationary nature of the vibration signal. In this work, significant condition monitoring indicators have been derived through qualitative analysis carried out on the Oscillation Matrix. The variations in the intensity of vibratory oscillations along with the nature of signal transitions are distinctly observed under dynamic operating conditions of DC motor. The features are so detailed and versatile in nature that they can be used efficiently for threshold determination, incipient abnormality identification and damage detection. Nevertheless, the signal features of Oscillations, Excess Positive Slopes and Excess Negative Slopes extracted from the transition matrix have large scope for analysis and not limited only to the illustrated threshold strategy.

References

1. Vishwakarma, M., Purohit, R., Harshlata, V., Rajput, P.: Vibration analysis & condition monitoring for rotating machines: a review. In: Materials Today: Proceedings of 5th International Conference of Materials Processing and Characterization (ICMPC 2016), pp. 2659–2664. Elsevier Ltd. (2017)
2. Al-Badour, F., Sunar, M., Cheded, L.: Vibration analysis of rotating machinery using time–frequency analysis and wavelet techniques. Mech. Syst. Signal Process. **25**, 2083–2101 (2011)
3. Galar, D., Sandborn, P., Kumar, U., Johansson, C.-A.: SMART: integrating human safety risk assessment with asset integrity. In: Dalpiaz, G., D'Elia, G., Rubini, R., Cocconcelli, M., Chaari, F., Haddar, M., Zimroz, R., Bartelmus, W. (eds.) Proceedings of the Third International Conference on Condition Monitoring of Machinery in Non-Stationary Operations CMMNO 2013, LNME, vol. 7, pp. 37–59. Springer, Heidelberg (2014)
4. Straczkiewicz, M., Barszcz, T., Jablonski, A.: Detection and classification of alarm threshold violations in condition monitoring systems working in highly varying operational conditions. J. Phys: Conf. Ser. **628**, 1–8 (2015)
5. Cardona-Morales, O., Alvarez-Marin, D., Castellanos-Dominguez, G.: Condition monitoring under non-stationary operating conditions using time–frequency representation-based dynamic features. In: Dalpiaz, G., D'Elia, G., Rubini, R., Cocconcelli, M., Chaari, F., Haddar, M., Zimroz, R., Bartelmus, W. (eds.) Proceedings of the Third International Conference on Condition Monitoring of Machinery in Non-Stationary Operations CMMNO 2013, LNME, vol. 7, pp. 441–451. Springer, Heidelberg (2014)
6. Zhu, J., Nostrand, T., Spiegel, C., Morton, B.: Survey of condition indicators for condition monitoring systems. In: Proceedings of Annual Conference of the Prognostics and Health Management Society, pp. 1–13 (2014)
7. Prudhom, A., Antonino-Daviu, J., Razik, H., Climente-Alarcon, V.: Time-frequency vibration analysis for the detection of motor damages caused by bearing currents. Mech. Syst. Signal Process. **84 Part A**, 747–762 (2017)
8. Feng, Z., Liang, M., Chu, F.: Recent advances in time–frequency analysis methods for machinery fault diagnosis: a review with application examples. Mech. Syst. Signal Process. **38**, 165–205 (2013)
9. Jablonski, A., Barszcz, T., Bielecka, M., Breuhaus, P.: Modeling of probability distribution functions for automatic threshold calculation in condition monitoring systems. Measurement **46**, 727–738 (2013)
10. www.analog.com/media/en/technical-documentation/data-sheets/ADXL345.pdf
11. http://www.ni.com/pdf/manuals/376047c.pdf
12. National Instruments: LabVIEW-Joint Time-Frequency Analysis Toolkit Reference Manual, Part Number 320544D-01 (1998)

Comparison Analysis: Single and Multichannel EMD-Based Filtering with Application to BCI

P. Gaur, G. Kaushik, Ram Bilas Pachori, H. Wang and G. Prasad

Abstract A brain–computer interface (BCI) aims to facilitate a new communication path that translates the motion intentions of a human into control commands using brain signals such as magnetoencephalography (MEG) and electroencephalogram (EEG). In this work, a comparison of features obtained using single channel and multichannel empirical mode decomposition (EMD) based filtering is done to classify the multi-direction wrist movements-based MEG signals for enhancing a brain–computer interface (BCI). These MEG signals are presented as a dataset 3 as part of the BCI competition IV. These single channel and multichannel EMD methods decompose MEG signals into a group of intrinsic mode functions (IMFs). The mean frequency measure of these IMFs has been used to combine these IMFs to obtain enhanced MEG signals which have major contributions from the low-frequency band (<15 Hz). The shrinkage covariance matrix has been computed as a feature set. These features have been used for the classification of MEG signals into multi-direction wrist movements using the Riemannian geometry classification method. Significant improvement of $>8\%$ in the test stage using the multichannel EMD-based filtering and $>4\%$ when compared with single channel EMD method and BCI competition winner, respectively. This analysis offers evidence that the multichannel EMD-based filtering has the potential to be used in online BCI systems which facilitate a broad use of noninvasive BCIs.

P. Gaur (✉) · G. Kaushik · G. Prasad
Intelligent Systems Research Centre, Ulster University, Derry, UK
e-mail: gaur-p@email.ulster.ac.uk

R. B. Pachori
Discipline of Electrical Engineering, Indian Institute of Technology Indore, Indore, India

H. Wang
School of Computing and Mathematics, Ulster University, Jordanstown, UK

© Springer Nature Singapore Pte Ltd. 2019
M. Tanveer and R. B. Pachori (eds.), *Machine Intelligence and Signal Analysis*,
Advances in Intelligent Systems and Computing 748,
https://doi.org/10.1007/978-981-13-0923-6_10

1 Introduction

Most people suffering from severe motor disabilities, particularly those who are totally paralyzed may need a communication pathway which do not need muscle control. Many studies use brain signals such as magnetoencephalography (MEG) or electroencephalogram (EEG) which serves as a basis for this new communication pathway called a brain–computer interface (BCI) system. It aims to facilitate a new communication path that translates the motion intentions of a human into control commands for an output device using brain signals such as EEG or MEG [3, 21]. However, these electrophysiological signals have low signal-to-noise ratio (SNR) due to external interferences such as electrical power line, etc. and other artefacts resulting from muscle movement, electromyogram (EMG) or eye movements, electrooculogram (EOG) interferences resulting in degraded classification accuracy. It is important to remove these interferences in the preprocessing step to achieve high accuracy in classification problems. To handle this problem, a number of studies have been carried out wherein research groups studied common spatial pattern (CSP) [12] and some extension methods based on the CSP algorithm [1]. Other research groups have utilized the space of symmetric positive definite matrices (SPDM) and computed covariance matrices as a feature set and then further classified using the Riemannian geometry framework [2].

In recent times, empirical mode decomposition (EMD) has shown potential as a very promising decomposition technique to analyse EEG signals [10]. It has been used for the classification of epileptic EEG signals [17, 18] and motor imagery BCI classification problems [5]. However, the single channel EMD method has some potential issues such as mode-mixing and frequency localization problems. To address these issues different variants have been proposed as an extension to single channel EMD method namely, ensemble EMD (EEMD) which adds white Gaussian noise (WGN) [22] but is computationally expensive, and multivariate EMD (MEMD) [13, 20] which is a multichannel extension of EMD. These methods adaptively decompose the MEG/EEG signal into a set of intrinsic mode functions (IMFs). Some of these IMFs contain noise and residuals as well. The important challenge is to identify the IMFs which are of interest and the remaining IMFs may be discarded. To address this issue, a single channel EMD-based filtering (EMDBF) [5] and a multichannel EMD-based filtering (MEMDBF) [8] method have been proposed. These filtering techniques identified the IMFs which are of interest and discarded the rest of them based on a statistical measure namely, mean frequency. Mean frequency computation has been carried out using Fourier spectrum [15]. Then, the identified IMFs are summed to enhance the EEG/MEG signals. These techniques have shown potential to classify the motor imagery EEG signal into two and four classes [5, 7, 8]. In addition, these methods have been studied to classify the MEG signals to address the multi-class classification problem [6, 9, 11]. In this paper, a comparative study of features extracted after applying the EMDBF and MEMDBF algorithms is done to classify MEG signals recorded during hand movements in four directions. A block diagram of the proposed methodology is shown in Fig. 1.

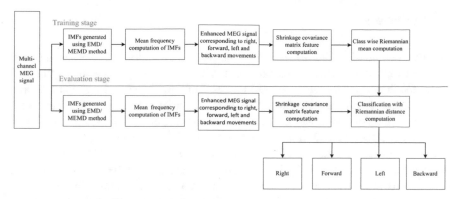

Fig. 1 Block diagram of the proposed methodology

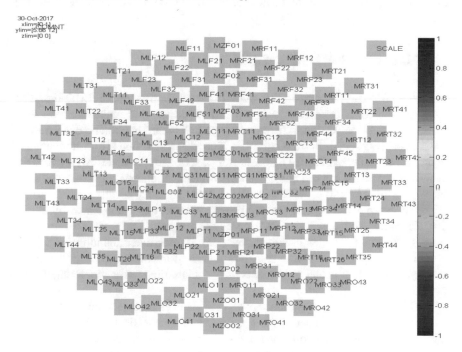

Fig. 2 MLC21, MLC22, MLC23, MLC32, MLC31, MLC41, MLC42, MZC01, MZC02 and MRC41 channels are used for the present work

The remaining paper is organized as follows: Sect. 2 provides the details of dataset used for our study, and Sect. 3 discusses the background of EMD and MEMD related decomposition techniques, respectively. Section 4 presents the analysis of results obtained by comparing the MEMDBF with EMDBF and BCI competition winners. Finally, Sect. 5 describes the conclusions of this analysis.

2 Materials

The BCI competition IV dataset 3 contains MEG signals for four classes, namely: right, forward, left and backward movements. These MEG signals from 10 channels above the motor areas have been used for the study as shown in Fig. 2. The dataset contains data on two subjects S01 and S02. Each subject contains MEG signals for one training session and one test session. A training set contains 40 trials of each class giving a total of 160 trials for S01 and 160 trials in subject S02, respectively. In evaluation session, there are 74 trials in S01 and 73 trials in S02 to be classified into four different classes [16]. These signals are sampled at 400 Hz. Each trial of the right, forward, left and backward movements MEG signals contain 400 sample points giving a trial length of 1 s. This dataset has been extensively studied by different research groups in a number of studies [6, 9, 11, 16]. For more details refer to [19].

3 EMD and MEMD Algorithm

This article describes a comparative study between single channel EMD-based filtering [5] and MEMD-based filtering [7, 8] to classify the multi-direction wrist movements-based MEG signals for enhancing a BCI is carried out. The EMD method [10] breaks MEG signals into a set of IMFs. The MEMD method [13] simultaneously decomposes multichannel MEG signals into a set of multivariate IMFs (MIMFs). These IMFs/ MIMFs can be considered narrowband, amplitude and frequency modulated (AM-FM) signals.

The recorded MEG signals are highly non-stationary and non-linear by nature. They also suffer from low SNR. Also, there is a strong possibility that they may contain disturbances from electromyography (EMG), electrooculography (EOG) and power line, etc. [14]. Therefore, the MEG signals of interest pertaining to a particular cognition task or actual movements may contain high noise which may lead to erroneous results.

The EMD was proposed by Huang et al. [10]. This method decomposes the signal in the time domain into a set of multiple IMFs. These IMFs are AM-FM type of signals. The mathematical equation for EMD method is expressed as

$$IM(t) = \sum_{p=1}^{k} U_p(t) + S_k(t) \tag{1}$$

where $IM(t)$ presents the original signal in time domain, $U_p(t)$ gives the pth IMF and $S_k(t)$ denotes the residue. Although this method decomposes the signal into a set of IMFs, it suffers from the mode-mixing problem [13]. Since the method does not account for any a priori about the information for the signal. This gives us different number of IMFs which leads to losing some of the cross-channel information present across the channels, and even the frequency components are not localized in the frequency domain. To address this issue, an MEMD method has been proposed which simultaneously does the decomposition of all channels. As a result, the number of IMFs obtained remains the same, and cross-channel information is also utilized. The mathematical expression for MEMD method is given as below:

$$IT(t) = \frac{1}{p} \sum_{p=1}^{k} e^{\theta_p}(t) \tag{2}$$

where $e^{\theta_p}(t)$ gives the envelope curves for multivariate data in all directions vectors and p denotes the length of the vectors. More details may be obtained from [13].

As discussed earlier, these decomposition techniques break the signal into a set of IMFs. Some of the IMFs contain the noise and the residual as well. Hence, this is needed to identify the IMFs which provide the actual information pertaining to any tasks. Henceforward, a filtering method is required to filter the noise out without deterioration of the original signal. Recently, Gaur et al. has proposed a single channel [5] and multichannel filtering [7, 8] technique to handle this issue. These filtering techniques have been built as an enhancement to EMD and MEMD techniques. These filtering techniques first identify the IMF to obtain the reconstructed signal based on signal of interest. They are identified based on the mean frequency measure.

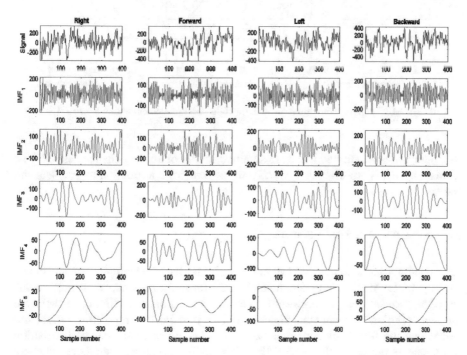

Fig. 3 The MEG signals from channel LC21 for wrist movement to right, forward, left and backward directions and first three IMFs generated

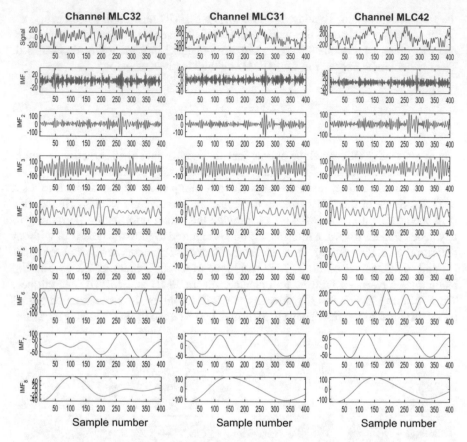

Fig. 4 The MEG signals from three channels for right wrist movement and obtained IMFs

Secondly, these identified IMFs are summed up to obtain the enhanced signal. The remaining IMFs are discarded which provide a major contribution to artefacts and noise.

4 Results and Discussions

The decomposed components (IMFs) for the EMD-based decomposition are shown in Fig. 3. The MEG signals for channel LC21 for multi-direction wrist movements in right, forward, left and backward directions and its obtained first three IMFs are shown. It is clearly evident that the IMFs obtained in this decomposition suffer from localizing the frequency components. Also, it has different frequency distribution components present in the same IMFs. If the IMF$_5$ is selected then there may be a strong possibility that we may end up losing some MEG data which actually can help to achieve better feature separability at a later stage.

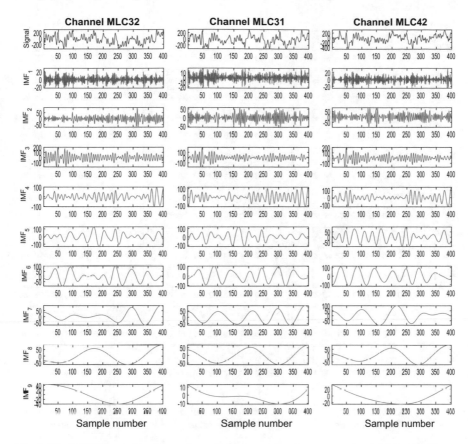

Fig. 5 The MEG signals from three channels for forward wrist movement and obtained IMFs

Figures 4 and 5 show the obtained MIMFs from MEMD-based decomposition method for right hand and forward wrist movements. Three channels MLC32, MLC31 and MLC42 are randomly selected to plot the decomposition mechanism.

Although in the actual study all of the ten channels have been considered and decomposed simultaneously. It is clearly evident that the same frequency distribution component is obtained in the same MIMFs as shown in Figs. 4 and 5. This decomposition helps to utilize the cross-channel information. If a particular MIMFs is/are selected, then it is possible to gain high localized frequency without losing information. Similarly, Figs. 6 and 7 display the MIMFs obtained by applying MEMD-based decomposition method for left hand and backward wrist movements.

In this work, the shrinkage covariance matrix (SHCM) is computed from the enhanced MEG signals obtained from single channel and multi channel filtering methods. Let F_1, F_2, F_3, ..., F_f denote the f feature vectors. The unbiased estimator of the mean is given as

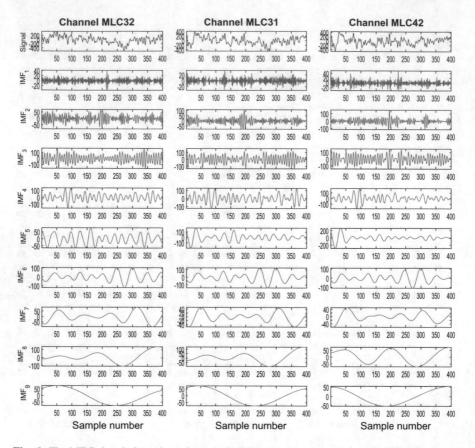

Fig. 6 The MEG signals from three channels for left wrist movement and obtained IMFs

$$\hat{M} = \frac{1}{f} \sum_{i=1}^{f} F_i \tag{3}$$

Also, the unbiased estimator of the covariance matrix is denoted as

$$\hat{C} = \frac{1}{f-1} \sum_{i=1}^{f} (F_i - \hat{M})(F_i - \hat{M})^T \tag{4}$$

To account for the estimation error, $\hat{C}$ is substituted by

$$C(i) = (1 - \gamma)\hat{\sigma} + \gamma \upsilon I \tag{5}$$

with a tuning parameter $(\gamma) \in [0, 1]$. υ denotes the average eigenvalue. More details may be obtained from [4].

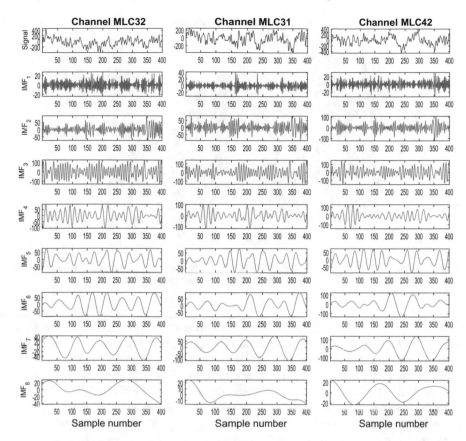

Fig. 7 The MEG signals from three channels for backward wrist movement and obtained IMFs

Table 1 Classification accuracies with the proposed method when evaluated on BCI competition IV dataset 3

Subject	MEMDBF	EMDBF	Winner1	Winner2	Winner3	Winner 4
S01	52.7	40.54	**59.5**	31.1	16.2	23.0
S02	49.31	43.83	34.3	19.2	31.5	17.8
Average	51.00	42.18	46.90	25.15	23.85	20.4
Std	2.4	2.33	17.82	8.41	10.82	3.68

Table 1 shows the classification accuracy when evaluated on BCI competition IV dataset 3 using EMD-based filtering and MEMD-based filtering and comparison with BCI competition winners [19]. Here, Std denotes the standard deviation in Table 1 along with classification accuracy comparison with other BCI competition winners. The classification accuracy has been computed for subjects S01 and S02. The main highlights observed based on the classification accuracy are as follows: (1) The average classification accuracy computed with the EMD-based filtering and MEMD-

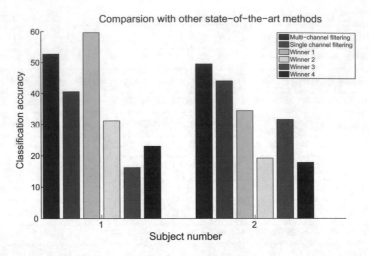

Fig. 8 The bar graphs depict the comparison of classification accuracy with BCI IV competition winners for two subjects. The data is taken from BCI Competition IV dataset 3

based filtering for both subjects gives a minimum standard deviation of 2.33 and 2.4 as compared to BCI competition winners [19]. (2) Subject S02 gives the maximum classification accuracy of 49.31% which is higher than (>5%) with EMDBF method and >15% with the BCI competition winner [19]. (3) As the higher classification is achieved in the multi-class classification problem using the MEMDBF technique, thus the features are more separable as compared to EMDBF method. These filtering techniques have been served as a preprocessing step. It should be noted that no complexity has been introduced at the feature extraction and classification steps.

Figure 8 shows the graph plot for the comparison of classification accuracies between MEMDBF, EMDBF [6] and BCI competition winners [19]. The improvement obtained in the average classification accuracy across the two subjects is illustrated with the bar graphs. The performance improvement for the BCI dataset IV dataset 3 is shown for the evaluation session. There is a significant improvement in average classification accuracy for subject S02 using the MEMDBF method in the evaluation session.

5 Conclusion

A comparison analysis of the multi-direction wrist movement MEG signals feature discrimination using the EMD-based filtering and MEMD-based filtering methods has been undertaken. The shrinkage covariance matrix has been computed as a feature set to classify MEG signals into multiple classes of right, forward, left and backward movements. These features have been calculated from the enhanced MEG signal obtained from EMD-based filtering and MEMD-based filtering methods. Consequently, their capability to classify the MEG signals into multiple classes has been explored based on the computed classification accuracy and standard deviation. The minimum standard deviation has been obtained across subjects using these filtering technique using MEMDBF. Features obtained from MEMDBF provides superior discrimination capability which leads to higher classification accuracy. Thus, using MEMDBF shows good potential in classifying the MEG signals into multiple classes. Future work may involve studying different features such as frequency domain and time–frequency domain features and to employ suitable classifiers, such as support vector machine (SVM) or artificial neural network (ANN) or k-nearest neighbour (KNN) to classify multi-direction wrist movements MEG signals where MEMDBF may be utilized as a preprocessing step. Thus, this study may aid researchers to develop an improved method that would be helpful in rehabilitation for people suffering from strokes.

References

1. Arvaneh, M., Guan, C., Ang, K.K., Quek, C.: Optimizing spatial filters by minimizing within-class dissimilarities in electroencephalogram-based brain-computer interface. IEEE Trans. Neural Netw. Learn. Syst. **24**(4), 610–619 (2013)
2. Barachant, A., Bonnet, S., Congedo, M., Jutten, C.: Multiclass brain-computer interface classification by Riemannian geometry. IEEE Trans. Biomed. Eng. **59**(4), 920–928 (2012)
3. Blankertz, B., Dornhege, G., Krauledat, M., Müller, K.R., Curio, G.: The non-invasive Berlin brain-computer interface: fast acquisition of effective performance in untrained subjects. NeuroImage **37**(2), 539–550 (2007)
4. Blankertz, B., Lemm, S., Treder, M., Haufe, S., Müller, K.R.: Single-trial analysis and classification of ERP components - a tutorial. NeuroImage **56**(2), 814–825 (2011)
5. Gaur, P., Pachori, R.B., Wang, H., Prasad, G.: An empirical mode decomposition based filtering method for classification of motor-imagery EEG signals for enhancing brain-computer interface. In: International Joint Conference on Neural Networks, pp. 1–7 (2015)
6. Gaur, P., Prasad, G., Wang, H., Pachori, R.: An MEG based BCI for classification of multi-direction wrist movements using empirical mode decomposition. MEG UK 2016, York, UK (2016)
7. Gaur, P., Pachori, R.B., Wang, H., Prasad, G.: A multivariate empirical mode decomposition based filtering for subject independent BCI. In: 27th Irish Signals and Systems Conference (ISSC), pp. 1–7. IEEE (2016)
8. Gaur, P., Pachori, R.B., Wang, H., Prasad, G.: Enhanced Motor Imagery Classification in EEG-BCI using Multivariate EMD based Filtering and CSP Features. In: Proceedings of the Sixth International Brain-Computer Interface Meeting: BCI Past, Present, and Future (2016)
9. Gaur, P., Bornot, J., Prasad, G., Wang, H., Pachori, R.: Decoding of Multi-direction wrist movements using multivariate empirical mode decomposition. In: MEG UK 2017, University of Oxford, UK (2017)
10. Huang, N.E., Shen, Z., Long, S.R., Wu, M.C., Shih, H.H., Zheng, Q., Yen, N.C., Tung, C.C., Liu, H.H.: The empirical mode decomposition and the Hilbert spectrum for nonlinear and non-stationary time series analysis. Proc. R. Soc. Lond A: Math. Phys. Eng. Sci. **454**, 903–995 (1998)
11. Kaushik, G., Gaur, P., Prasad, G., Wang, H., Pachori, R.: An MEG based multi direction wrist movements analysis using empirical mode decomposition and multivariate empirical mode decomposition. In: MEG UK 2017. University of Oxford, UK (2017)

12. Lotte, F., Guan, C.: Regularizing common spatial patterns to improve BCI designs: unified theory and new algorithms. IEEE Trans. Biomed. Eng. **58**(2), 355–362 (2011)
13. Park, C., Looney, D., Ahrabian, A., Mandic, D.P., et al.: Classification of motor imagery BCI using multivariate empirical mode decomposition. IEEE Trans. Neural Syst. Rehabil. Eng. **21**(1), 10–22 (2013)
14. Pfurtscheller, G., Neuper, C., Flotzinger, D., Pregenzer, M.: EEG-based discrimination between imagination of right and left hand movement. Electroencephalogr. Clin. Neurophysiol. **103**(6), 642–651 (1997)
15. Phinyomark, A., Thongpanja, S., Hu, H., Phukpattaranont, P., Limsakul, C.: The usefulness of mean and median frequencies in electromyography analysis. In: Computational Intelligence in Electromyography Analysis-A Perspective on Current Applications and Future Challenges, InTech, (2012)
16. Sardouie, S.H., Shamsollahi, M.B.: Selection of efficient features for discrimination of hand movements from MEG using a BCI competition IV data set. Front. Neurosci. **6** (2012)
17. Sharma, R., Pachori, R.B.: Classification of epileptic seizures in EEG signals based on phase space representation of intrinsic mode functions. Expert Syst. Appl. **42**(3), 1106–1117 (2015)
18. Sharma, R., Pachori, R.: Automated classification of focal and non-focal EEG signals based on bivariate empirical mode decomposition. Biomedical Signal and Image Processing in Patient Care, IGI Global (2017)
19. Tangermann, M., Müller, K.R., Aertsen, A., Birbaumer, N., Braun, C., Brunner, C., Leeb, R., Mehring, C., Miller, K.J., Müller-Putz, G.R., et al.: Review of the BCI competition IV. Front. Neurosci. **6** (2012)
20. ur Rehman, N., Park, C., Huang, N.E., Mandic, D.P.: EMD via MEMD: multivariate noise-aided computation of standard EMD. Adv. Adapt. Data Anal. **5**(02), 1350007 (2013)
21. Wolpaw, J.R., Birbaumer, N., McFarland, D.J., Pfurtscheller, G., Vaughan, T.M.: Brain-computer interfaces for communication and control. Clin. Neurophysiol. **113**(6), 767–791 (2002)
22. Wu, Z., Huang, N.E.: Ensemble empirical mode decomposition: a noise-assisted data analysis method. Adv. Adapt. Data Anal. **1**(01), 1–41 (2009)

A Two-Norm Squared Fuzzy-Based Least Squares Twin Parametric-Margin Support Vector Machine

Parashjyoti Borah and Deepak Gupta

Abstract A two-norm squared fuzzy-based least squares version of twin parametric-margin support vector machine is proposed to reduce the effect of outliers and noise by assigning fuzzy membership values to each training data samples. Further, by considering two-norm squared the slack variable multiplied to the fuzzy membership values makes the objective function strongly convex. Here, we substitute the inequality constraints of the primal problems with equality constraints to solve the two primal problems instead of solving two quadratic programming problems which eliminates the need of external optimization toolbox and provides with a lower computational cost. A performance comparison of the proposed method with twin support vector machine, least squares twin support vector machine, twin parametric-margin support vector machine and least squares twin parametric-margin support vector machine is presented in this paper.

Keywords Support vector machine · Fuzzy · Quadratic programming problem
Least squares · Parametric-margin

1 Introduction

In the recent years, support vector machine (SVM) [21] has gained immense popularity because of its high generalization ability and its structural risk minimization property. The idea of SVM lies on maximizing the margin between two parallel separating hyperplanes by solving a quadratic programming problem (QPP) and defining the classifier which is equidistant from the two separating hyperplanes. Since its introduction, SVM has been widely and effectively applied to many real-life appli-

P. Borah (✉) · D. Gupta
Computer Science & Engineering, NIT Arunachal Pradesh, Yupia, India
e-mail: parashjyoti@hotmail.com

D. Gupta
e-mail: deepakjnu85@gmail.com

© Springer Nature Singapore Pte Ltd. 2019   119
M. Tanveer and R. B. Pachori (eds.), *Machine Intelligence and Signal Analysis*,
Advances in Intelligent Systems and Computing 748,
https://doi.org/10.1007/978-981-13-0923-6_11

cation areas such as object detection and recognition [15, 16], handwritten character recognition [1, 4], face recognition [6, 7], speech recognition [3, 22], etc.

In the last few decades many other algorithms based on the classical SVM have been proposed that utilizes the properties of SVM incorporated with other computational techniques or ideas. Lagrangian support vector machine (LSVM) [11], v-support vector machine (v-SVM) [18], least squares support vector machine (LSSVM) [20] and fuzzy support vector machine (FSVM) [10] are some variants of the classical SVM. Based on the v-SVM, Hao [5] has proposed parametric-margin v-support vector machine (par-v-SVM) for heteroscedastic noise structure.

Although SVM has high generalization ability, its huge training cost $O(n^3)$ is one of its challenging issues. Different researchers have come up with ideas to resolve this issue of the classical SVM. Mangasarian and Wild [12] proposed generalized eigenvalue proximal support vector machine (GEPSVM) that unlike SVM tries to find two non-parallel hyperplanes for each class of data. Recently, Khemchandani and Chandra [8] proposed twin support vector machine (TWSVM) which also finds two non-parallel hyperplanes closer to one class and as far possible from the other class. The computational complexity of TWSVM is approximately four times lesser than the classical SVM as it solves two smaller sized QPPs instead of solving one larger sized QPP. Kumar and Gopal [9] proposed least squares version of TWSVM, termed as least squares twin support vector machine (LSTSVM) that solves the primal formulations and is faster as compared to TWSVM. Further, Peng [14] proposed twin parametric-margin support vector machine (TPMSVM) which finds two flexible parametric-margin hyperplanes able of taking care of heteroscedastic noise structure. Shao et al. [19] further combined the concepts of LSTSVM and TPMSVM and proposed a least squares model for TPMSVM and named it LSTPMSVM.

In this paper, a two-norm squared fuzzy-based least squares twin parametric-margin support vector machine is proposed which is a TWSVM based model based on the idea of FSVM and TPMSVM. We name this proposed model as fuzzy least squares twin parametric-margin support vector machine (FLSTPMSVM). In this method fuzzy membership values are computed and assigned to each training data samples. Data samples get membership values according to their degree of belongingness to their respective classes and that makes outliers and noise samples get lower membership values based on the fuzzy membership function used. Thus, the membership values multiplied with the slack variables makes the decision surface less sensitive towards outliers and noise present in real-world datasets. Further, by taking the two-norm squared the slack variable multiplied with the membership value makes the objective function strongly convex. Moreover, as in TPMSVM, the proposed model takes care of heteroscedastic noise structure by defining two flexible parametric-margin hyperplanes. Furthermore, as in case of LSSVM, the inequality constraints of the primal formulation are replaced by equality constraints that make the proposed FLSTPMSVM computationally faster. FLSTPMSVM finds the solutions from its primal formulations and therefore does not need any optimization toolbox as in case of TWSVM and TPMSVM. To validate the proposed method, in this paper, we have further presented a performance comparative analysis of FLSTPMSVM with TWSVM, LSTSVM, TPMSVM, and LSTPMSVM.

2 Background

All the vectors in this paper are considered as column vectors. $||.||$ calculates the two-norm of a vector. y^t represents the transpose of a vector y. Let $X_1 \in R^{l_1 \times n}$ denote the training points belonging to the positive (+1) class, $X_2 \in R^{l_2 \times n}$ is the matrix of training samples of the negative (−1) class and $X = [X_1; X_2] \in R^{l \times n}$ where $l = l_1 + l_2$.

2.1 Twin Support Vector Machine

Twin support vector machine (TWSVM) [8] is an SVM-based binary classification technique which tries to find two non-parallel hyperplanes closer to the samples of one class and as far possible from the samples of the other class. TWSVM solves two smaller sized QPPs instead of solving a single large sized QPP. The hyperplanes for the nonlinear case are defined as $K(x^t, X^t)w_1 + b_1 = 0$ and $K(x^t, X^t)w_2 + b_2 = 0$ which can be obtained by solving the following pair of QPPs,

$$\min \frac{1}{2}||K(X_1, X^t)w_1 + e_1 b_1||^2 + C_1 e_2^t \xi$$
$$\text{subject to: } -(K(X_2, X^t)w_1 + e_2 b_1) + \xi \geq e_2, \ \xi \geq 0 \tag{1}$$

and

$$\min \frac{1}{2}||K(X_2, X^t)w_2 + e_2 b_2||^2 + C_2 e_1^t \eta$$
$$\text{subject to: } (K(X_1, X^t)w_2 + e_1 b_2) + \eta \geq e_1, \ \eta \geq 0 \tag{2}$$

where, C_1 and C_2 are penalty parameters, $e_1 \in R^{l_1}$ and $e_2 \in R^{l_2}$ are vectors of $1's$, $\xi \in R^{l_1}$ and $\eta \in R^{l_2}$ are slack variables. Here, $K(x^t, D^t) = (k(x, x_1), \ldots, k(x, x_l))$ is a row vector in R^l, such that $k(x, x_i) = \varphi(x) \times \varphi(x_i)$ for $i = 1 \ldots l$, where $\varphi(.)$ is the mapping function from input space to a higher dimensional feature space.

The dual QPPs corresponding to Eqs. (1) and (2) are derived by finding the Lagrangian function and applying the Karush–Kuhn–Tucker (KKT) necessary and sufficient conditions as,

$$\max e_2^t \alpha_1 - \frac{1}{2} \alpha_1^t Q(P^t P)^{-1} Q^t \alpha_1$$
$$\text{subject to: } 0 \leq \alpha_1 \leq C_1 \tag{3}$$

and

$$\max e_1^t \alpha_2 - \frac{1}{2} \alpha_2^t P(Q^t Q)^{-1} P^t \alpha_2$$
$$\text{subject to: } 0 \leq \alpha_2 \leq C_2 \tag{4}$$

where $\alpha_1 \in R^{l_1}$ and $\alpha_2 \in R^{l_2}$ are vectors of Lagrangian multipliers, $P = [K(X_1, X^t) \ e_1]$ and $Q = [K(X_2, X^t) \ e_2]$. The hyperplane parameters w_1, w_2, b_1 and b_2 are computed by solving the above dual QPPs for α_1 and α_2, then from the below augmented matrices as $\begin{bmatrix} w_1 \\ b_1 \end{bmatrix} = -(P^t P + \varepsilon I)Q^t \alpha_1$ and $\begin{bmatrix} w_2 \\ b_2 \end{bmatrix} = (Q^t Q + \varepsilon I)P^t \alpha_2$, where I is an identity matrix of dimension $(l + 1) \times (l + 1)$ and $\varepsilon > 0$ is a real number as defined in [8].

A new data point $x \in R^n$ is assigned to class ± 1 based on the result of the following equation:

$$class \ i = \min |K(x^t, X^t)w_i + b_i|, \text{ for } i = 1, 2 \tag{5}$$

Here, $|.|$ is the perpendicular distance of the test sample from each hyperplanes.

2.2 Twin Parametric-Margin Support Vector Machine

In twin parametric-margin support vector machine (TPMSVM) [14] two non-parallel flexible parametric-margin hyperplanes $\varphi(x)^t w_1 + b_1 = 0$ and $\varphi(x)^t w_2 + b_2 = 0$ are considered in the feature space. The QPPs for TPMSVM are given below,

$$\min \tfrac{1}{2}||w_1||^2 + \tfrac{v_1}{l_2} e_2^t (\varphi(X_2)w_1 + e_2 b_1) + \tfrac{C_1}{l_1} e_1^t \xi$$
$$\text{subject to: } \varphi(X_1)w_1 + e_1 b_1 + \xi \geq 0, \ \xi \geq 0 \tag{6}$$

and

$$\min \tfrac{1}{2}||w_2||^2 + \tfrac{v_2}{l_1} e_1^t (\varphi(X_1)w_2 + e_1 b_2) + \tfrac{C_2}{l_2} e_2^t \eta,$$
$$\text{subject to: } \varphi(X_2)w_2 + e_2 b_2 - \eta \leq 0, \ \eta \geq 0 \tag{7}$$

where $v_1, v_2 > 0$ are scalars for determining the penalty weights. The corresponding Wolfe duals of Eqs. (6) and (7) are,

$$\max \ -\tfrac{1}{2}\alpha_1^t K(X_1, X_1^t)\alpha_1 + \tfrac{v_1}{l_2} e_2^t K(X_2, X_1^t)\alpha_1$$
$$\text{subject to: } 0 \leq \alpha_1 \leq \tfrac{C_1}{l_1} e_1, \ e_1^t \alpha_1 = v_1 \tag{8}$$

and

$$\max \ -\tfrac{1}{2}\alpha_2^t K(X_2, X_2^t)\alpha_2 + \tfrac{v_2}{l_1} e_1^t K(X_1, X_2^t)\alpha_2$$
$$\text{subject to: } 0 \leq \alpha_2 \leq \tfrac{C_2}{l_2} e_2, \ e_2^t \alpha_2 = v_2 \tag{9}$$

respectively. The hyperplane parameters w_1, w_2, b_1 and b_2 are computed as, $w_1 = \varphi(X_1)^t\alpha_1 - \frac{v_1}{l_2}\varphi(X_2)^t e_2$, and $w_2 = -\varphi(X_2)^t\alpha_2 + \frac{v_2}{l_1}\varphi(X_1)^t e_1$, $b_1 = -\frac{1}{|N_1|}\sum_{i \in N_1} \varphi(X_i)^t w_1$ and $b_2 = -\frac{1}{|N_2|}\sum_{i \in N_2} \varphi(X_i)^t w_2$ where, N_i is the index set of samples satisfying $\alpha_i \in \left(0, \frac{c_i}{l_i}\right)$ for $i = 1, 2$. The classifier for TPMSVM is given as,

$$class\ i = sign\left(\sum_{i=1,2}\left(\frac{w_i}{||w_i||}\right)^t \varphi(x) + \sum_{i=1,2}\left(\frac{b_i}{||w_i||}\right)\right), \ i = 1, 2 \qquad (10)$$

2.3 Least Square Twin Support Vector Machine

Least squares twin support vector machine (LSTSVM) [9] considers two-norms squared of the slack variables instead of one-norm as in case of TWSVM and TPMSVM, and replaces the inequality constraints of the primal QPPs by equality contraints. The two non-parallel hyperplanes for nonlinear LSTSVM $K(x^t, X^t)w_1 + b_1 = 0$ and $K(x^t, X^t)w_2 + b_2 = 0$ are computed using the following primal QPPs:

$$\min \frac{1}{2}||K(X_1, X^t)w_1 + e_1b_1||^2 + \frac{C_1}{2}\xi^t\xi$$
$$\text{subject to:} -(K(X_2, X^t)w_1 + e_2b_1) + \xi = e_2 \qquad (11)$$

and

$$\min \frac{1}{2}||K(X_2, X^t)w_2 + e_2b_2||^2 + \frac{C_2}{2}\eta^t\eta$$
$$\text{subject to:} (K(X_1, X^t)w_2 + e_1b_2) + \eta = e_1 \qquad (12)$$

By substituting the constraints of Eqs. (11) and (12) into their respective objective functions, the QPPs (11) and (12) are reformulated as,

$$\min \frac{1}{2}||K(X_1, X^t)w_1 + e_1b_1||^2 + \frac{C_1}{2}||K(X_2, X^t)w_1 + e_2b_1 + e_2||^2 \qquad (13)$$

and

$$\min \frac{1}{2}||K(X_2, X^t)w_2 + e_2b_2||^2 + \frac{C_2}{2}||-K(X_1, X^t)w_2 - e_1b_2 + e_1||^2 \qquad (14)$$

By rearranging the Eqs. (13) and (14) in matrix form, the solutions to the above QPPs can be obtained as $\begin{bmatrix} w_1 \\ b_1 \end{bmatrix} = -\left(Q^t Q + \frac{1}{C_1} P^t P\right)^{-1} Q^t e_2$ and $\begin{bmatrix} w_2 \\ b_2 \end{bmatrix} = \left(P^t P + \frac{1}{C_2} Q^t Q\right)^{-1} P^t e_1$. The classifier for nonlinear LSTSVM is similar to TWSVM as given in Eq. (5).

2.4 Least Squares Twin Parametric-Margin Support Vector Machine

Least Squares twin parametric-margin support vector machine (LSTPMSVM) [19] merges the concepts of LSSVM and TPMSVM to find two non-parallel flexible parametric-margin hyperplanes $K(x^t, X^t)w_1 + b_1 = 0$ and $K(x^t, X^t)w_2 + b_2 = 0$ by solving the following pair of primal QPPs:

$$\min \frac{1}{2}(||w_1||^2 + b_1^2) + \nu_1 e_2^t(K(X_2, X^t)w_1 + e_2 b_1) + \frac{C_1}{2}\xi^t\xi$$
$$\text{subject to:} (K(X_1, X^t)w_1 + e_1 b_1) + \xi = e_1 \tag{15}$$

and

$$\min \frac{1}{2}(||w_2||^2 + b_2^2) - \nu_2 e_1^t(K(X_1, X^t)w_2 + e_1 b_2) + \frac{C_2}{2}\eta^t\eta$$
$$\text{subject to:} (K(X_2, X^t)w_2 + e_2 b_2) + \eta = -e_2 \tag{16}$$

By substituting the equality constraints into their objective functions and rearranging in matrix form, the solution to the equations are obtained as $\begin{bmatrix} w_1 \\ b_1 \end{bmatrix} =$ $(C_1 P^t P + I)^{-1}(C_1 P^t e_1 - \nu_1 Q^t e_2)$ and $\begin{bmatrix} w_2 \\ b_2 \end{bmatrix} = (C_2 Q^t Q + I)^{-1}(C_2 Q^t e_2 + \nu_1 P^t e_1)$. The classifier for nonlinear LSTPMSVM is similar to the classifier of TPMSVM given in Eq. (10).

3 Proposed Two-Norm Squared Fuzzy-Based Least Squares Twin Parametric-Margin Support Vector Machine

In this section, a fuzzy-based least squares twin support vector machine (FLSTPMSVM) is proposed that considers the square of two-norm of slack vectors multiplied to the membership matrix as discussed below. FLSTPMSVM is a TWSVM based model which is inspired by the works of Lin and Wang [10], Suykens and Vandewalle [20] and Hao [5].

3.1 Linear FLSTPMSVM

In linear case, FLSTPMSVM determines the two non-parallel hyperplanes in the input space given by the equations $x^t w_1 + b_1 = 0$ and $x^t w_2 + b_2 = 0$. The formulations for FLSTPMSVM can be obtained by solving the following pair of optimization problems.

$$\min \frac{1}{2}(||w_1||^2 + b_1^2) + v_1 e_2^t (X_2 w_1 + e_2 b_1) + \frac{C_1}{2}||S_1 \xi||^2$$
$$\text{subject to: } (X_1 w_1 + e_1 b_1) + \xi = e_1 \tag{17}$$

and

$$\min \frac{1}{2}(||w_2||^2 + b_2^2) - v_2 e_1^t (X_1 w_2 + e_1 b_2) + \frac{C_2}{2}||S_2 \eta||^2,$$
$$\text{subject to: } -(X_2 w_2 + e_2 b_2) + \eta = e_2 \tag{18}$$

where S_1 and S_2 are diagonal matrices of membership values of dimension $l_1 \times l_1$ and $l_2 \times l_2$ respectively for positive and negative class. Considering the Eq. (17), the equality constraint substitutes the slack variables in the objective function to obtain the following minimization problem,

$$\min \frac{1}{2}(||w_1||^2 + b_1^2) + v_1 e_2^t (X_2 w_1 + e_2 b_1) + \frac{C_1}{2}||S_1(-X_1 w_1 - e_1 b_1 + e_1)||^2 \tag{19}$$

Now, setting the gradients of Eq. (19) to 0 with respect to w_1 and b_1, we obtain

$$w_1 + v_1 X_2 e_2^t - C_1 X_1^t S_1^t S_1 (-X_1 w_1 - e_1 b_1 + e_1) = 0 \tag{20}$$
$$b_1 + v_1 e_2^t e_2 - C_1 e_1^t S_1^t S_1 (-X_1 w_1 - e_1 b_1 + e_1) = 0 \tag{21}$$

The Eqs. (20) and (21) can be arranged in matrix form as follows:

$$C_1 \begin{bmatrix} X_1^t S_1^t S_1 X_1 + \dfrac{I}{C_1} & X_1^t S_1^t S_1 e_1 \\ e_1^t S_1^t S_1 X_1 & e_1^t S_1^t S_1 e_1 + \dfrac{1}{C_1} \end{bmatrix} \begin{bmatrix} w_1 \\ b_1 \end{bmatrix} + v_1 \begin{bmatrix} X_2^t \\ e_2^t \end{bmatrix} e_2 - C_1 \begin{bmatrix} X_1^t \\ e_1^t \end{bmatrix} S_1^t S_1 e_1 = 0$$

which is equivalent to

$$\begin{bmatrix} w_1 \\ b_1 \end{bmatrix} = \left(C_1 \begin{bmatrix} X_1^t \\ e_1^t \end{bmatrix} S_1^t S_1 [X_1 \ \ e_1] + I \right)^{-1} \left(C_1 \begin{bmatrix} X_1^t \\ e_1^t \end{bmatrix} S_1^t S_1 e_1 - v_1 \begin{bmatrix} X_2^t \\ e_2^t \end{bmatrix} e_2 \right) \quad (22)$$

The Eq. (22) can be written as,

$$\begin{bmatrix} w_1 \\ b_1 \end{bmatrix} = (C_1 G^t S_1^t S_1 G + I)^{-1} (C_1 G^t S_1^t S_1 e_1 - v_1 H^t e_2) \quad (23)$$

where, $G = [X_1 \ \ e_1]$ and $H = [X_2 \ \ e_2]$. Following an exact similar way, solutions to w_2 with b_2 are obtained from (18) as,

$$\begin{bmatrix} w_2 \\ b_2 \end{bmatrix} = (C_2 H^t S_2^t S_2 H + I)^{-1} (-C_2 H^t S_2^t S_2 e_2 + v_2 G^t e_1) \quad (24)$$

For a new data sample $x \in R^n$, the classifier for FLSTPMSVM is defined as,

$$class\ i = sign \left(\sum_{i=1,2} \left(\frac{w_i}{||w_i||} \right)^t x + \sum_{i=1,2} \left(\frac{b_i}{||w_i||} \right) \right), \text{ for } i = 1, 2 \quad (25)$$

3.2 Nonlinear FLSTPMSVM

In nonlinear case, the two separating hyperplanes in the feature space, given by $K(x^t, X^t) w_1 + b_1 = 0$ and $K(x^t, X^t) w_2 + b_2 = 0$, are obtained by solving the following primal QPPs:

$$\min \frac{1}{2}(||w_1||^2 + b_1^2) + v_1 e_2^t (K(X_2, X^t) w_1 + e_2 b_1) + \frac{C_1}{2}||S_1 \xi||^2$$
$$\text{subject to: } (K(X_1, X^t) w_1 + e_1 b_1) + \xi = e_1 \quad (26)$$

and

$$\min \frac{1}{2}(||w_2||^2 + b_2^2) - v_2 e_1^t (K(X_1, X^t) w_2 + e_1 b_2) + \frac{C_2}{2}||S_2 \eta||^2$$
$$\text{subject to: } - (K(X_2, X^t) w_2 + e_2 b_2) + \eta = e_2 \quad (27)$$

Following the exact similar steps as in the linear case, the solutions to (26) and (27) are obtained as follows:

$$\begin{bmatrix} w_1 \\ b_1 \end{bmatrix} = (C_1 P^t S_1^t S_1 P + I)^{-1}(C_1 P^t S_1^t S_1 e_1 - v_1 Q^t e_2) \tag{28}$$

and

$$\begin{bmatrix} w_2 \\ b_2 \end{bmatrix} = (C_2 Q^t S_2^t S_2 Q + I)^{-1}(-C_2 Q^t S_2^t S_2 e_2 + v_2 P^t e_1) \tag{29}$$

In both (29) and (30), one can see that inverse of one matrix of dimension $(n + 1) \times (n + 1)$ needs to be computed. The computational costs of (29) and (30) can further be reduced by using Sherman–Morrison–Woodbury (SMW) formula [2] as,

$$\begin{bmatrix} w_1 \\ b_1 \end{bmatrix} = \left(I - Y^t \left(\frac{I}{C_1} + YY^t \right)^{-1} Y \right)(C_1 Y^t S_1 e_1 - v_1 Q^t e_2) \tag{30}$$

and

$$\begin{bmatrix} w_2 \\ b_2 \end{bmatrix} = \left(I - Z^t \left(\frac{I}{C_2} + ZZ^t \right)^{-1} Z \right)(-C_2 Z^t S_2 e_2 + v_2 P^t e_1), \tag{31}$$

where $Y = S_1 P$ and $Z = S_2 Q$. Further, the classifier of FLSTPMSVM for nonlinear case is obtained as,

$$class\ i = sign\left(\sum_{i=1,2} \left(\frac{w_i}{||w_i||} \right)^t K(x^t, X^t) + \sum_{i-1,2} \left(\frac{b_i}{||w_i||} \right) \right), \text{ for } i = 1, 2 \tag{32}$$

4 Numerical Experiment

To validate the performance of the proposed method, a performance analysis of the proposed FLSTPMSVM with TWSVM, LSTSVM, TPMSVM, and LSTPMSVM is conducted and presented in this paper. The experiments are performed on one synthetic dataset and 8 publicly available real-world datasets. All the experiments are performed on a PC with 64 bit, 3.40 GHz Intel© Core™ i7-3770 CPU and 4 GB RAM, running Windows 7 operating system. The software package used is MATLAB R2008a along with MOSEK optimization toolbox for TWSVM and TPMSVM, available at https://www.mosek.com.

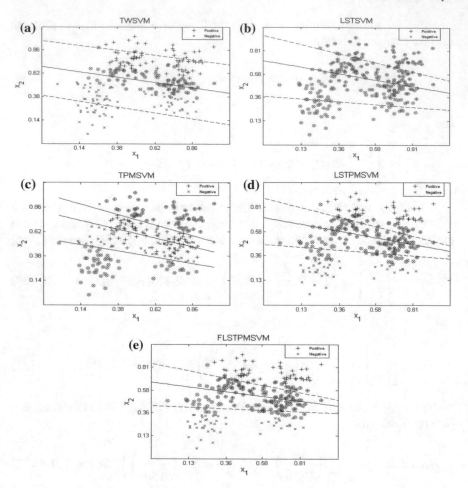

Fig. 1 Results for linear case on Ripley's dataset

The datasets are normalized to the range [0, 1] before experiments are performed on them. The results for linear as well as nonlinear case of all the considered algorithms are analyzed in this experimental. For nonlinear case the popular Gaussian kernel is used which is given by $K(x_i, x_j) = \exp(-\frac{\|x_i - x_j\|^2}{2\sigma^2})$. The optimum kernel parameter σ is chosen by using tenfold cross validation technique on training data and from the set $\{2^{-5}, \ldots, 2^5\}$. In a similar way, the optimum values of the penalty parameters C_i and the terms ν_i/C_i, for $(i = 1, 2)$ are selected from the sets $\{10^{-5}, \ldots, 10^5\}$ and $\{0.1, \ldots, 0.9\}$ respectively. In our experiment, to compute the membership weights of the training samples a centroid based fuzzy membership function is used which is given as, $S_j(i, i) = 1 - \frac{|x_j^{(cen)} - x_i|}{r_j + \delta}$ for $i = 1 \ldots l_j, \ j = 1, 2$

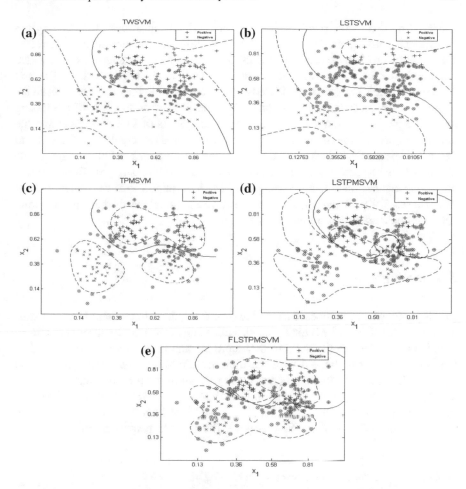

Fig. 2 Results for nonlinear case on Ripley's dataset

where S_j is a diagonal matrix of membership values of samples of class j, $x_j^{(cen)}$ and r_j are respectively the centroid and the radius of class j.

The performance is evaluated based on classification accuracy. In this experiment, for plotting the classifiers of all the methods the Ripley's synthetic dataset [17] is used which consists of 250 training samples and 1000 samples for testing in R^2. The classifiers of linear and nonlinear cases are shown as continuous lines in Fig. 1a–e and Fig. 2a–e respectively. In these figures, the hyperplanes associated to each class are depicted using broken lines and the support vectors are circled around them. The classification results along with optimum parameter values and training time are presented in Table 1 and Table 2 for linear and nonlinear cases respectively.

Further, the experiment is further extended to eight real-world benchmark datasets namely, Breast Cancer, Breast Tissue, BUPA Liver, German, Haberman, Monk3,

Tic-Tac-Toe and Wine, publicly available at the UCI repository [13]. Among the considered datasets, six datasets are of binary class and two are of more than two classes. In cases of non-binary datasets, the majority class is considered as the positive class and the other classes altogether are considered as the negative class. The first 40% of the whole dataset is considered as training data and the next 60% is used for testing. Further, a statistical result analysis is performed on tenfolds of the testing data. The performance results and optimum parameters of FLSTPMSVM and the considered algorithms for linear and nonlinear cases are tabulated in Table 3 and Table 4 respectively. Tables 5 and 6 presents the average ranks of all the algorithms based on their classification accuracies. To summarize our experiment, from Tables 3 and 4 one can see that the proposed FLSTPMSVM outperforms the others in most of the cases. Tables 5 and 6 show that the proposed model has the best average rank for both linear as well as nonlinear cases.

5 Conclusion

In this paper, we have proposed a novel two-norm squared fuzzy-based least squares TPMSVM (FLSTPMSVM) for binary classification problems. By introducing fuzzy membership values to the training data samples and taking two-norm squared of the slack variable multiplied with the membership values, our proposed method becomes less sensitive to outliers and noise present in the training data. Moreover, needless to solve a pair of QPP as in TWSVM and TPMSVM, to solve two primal problems builds a faster computational model of FLSTPMSVM as in case of LSTSVM and LSTPMSVM. Furthermore, the flexible parametric-margin hyper-

Table 1 Classification results on Ripley's dataset using linear kernel

Dataset train size test size	TWSVM accuracy (C) time	LSTSVM accuracy (C) time	TPMSVM accuracy $(C, \nu/C)$ time	LSTPMSVM accuracy $(C, \nu/C)$ time	FLSTPMSVM accuracy $(C, \nu/C)$ time
Ripley 250×2 1000×2	87.5 $(10^{\wedge}0)$ 0.0164	**89.8** $(10^{\wedge}-5)$ 0.0002	86 $(10^{\wedge}-1, 0.2)$ 0.0226	88.9 $(10^{\wedge}2, 0.8)$ 0.0002	89.5 $(10^{\wedge}1, 0.3)$ 0.0007

Table 2 Classification results on Ripley's dataset using Gaussian kernel

Dataset train size test size	TWSVM accuracy (C, μ) time	LSTSVM accuracy (C, μ) time	TPMSVM accuracy $(C, \mu, \nu/C)$ time	LSTPMSVM accuracy $(C, \mu, \nu/C)$ time	FLSTPMSVM accuracy $(C, \mu, \nu/C)$ time
Ripley 250×2 1000×2	88.9 $(10^{\wedge}0, 2^{\wedge}-1)$ 0.1228	90.3 $(10^{\wedge}0, 2^{\wedge}-1)$ 0.1113	90 $(10^{\wedge}-3, 2^{\wedge}-3, 0.4)$ 0.1212	89.4 $(10^{\wedge}0, 2^{\wedge}-3, 0.9)$ 0.1178	**90.9** $(10^{\wedge}0, 2^{\wedge}-3, 0.2)$ 0.1187

Table 3 Classification results on UCI benchmark datasets using linear kernel

Dataset train size test size	TWSVM accuracy (C) time	LSTSVM accuracy (C) time	TPMSVM accuracy (C, ν/C) time	LSTPMSVM accuracy (C, ν/C) time	FLSTPMSVM accuracy (C, ν/C) time
Breast cancer 279×9 420×9	97.1031 ± 2.891 (10^0) 0.0461	$\mathbf{98.3333} \pm 1.9602$ (10^1) 0.0003	97.7723 ± 2.7134 $(10^{-2}, 0.2)$ 0.0721	97.8571 ± 2.0848 $(10^0, 0.9)$ 0.0016	96.1905 ± 5.7451 $(10^{-1}, 0.1)$ 0.0013
Breast tissue 42×9 64×9	81.6667 ± 25.0924 (10^{-5}) 0.0109	96.6667 ± 7.0273 (10^0) 0.0002	79.5833 ± 23.7699 $(10^{-5}, 0.1)$ 0.0119	$\mathbf{98.3333} \pm 5.2705$ $(10^1, 0.6)$ 0.0001	$\mathbf{98.3333} \pm 5.2705$ $(10^2, 0.1)$ 0.0003
BUPA liver 137×6 208×6	52.679 ± 17.596 (10^0) 0.0146	56.9048 ± 27.2732 (10^1) 0.0002	56.5799 ± 16.8386 $(10^{-1}, 0.1)$ 0.017	43.5952 ± 23.3938 $(10^1, 0.9)$ 0.0003	$\mathbf{59.7619} \pm 17.9527$ $(10^4, 0.5)$ 0.0005
German 399×24 601×24	71.1359 ± 5.6243 (10^{-5}) 0.0621	74.8743 ± 6.2609 (10^0) 0.0007	70.3004 ± 7.9538 $(10^{-2}, 0.6)$ 0.0957	74.8743 ± 7.1794 $(10^1, 0.9)$ 0.0018	$\mathbf{75.7131} \pm 6.3618$ $(10^1, 0.1)$ 0.0031
Haberman 122×3 184×3	63.728 ± 7.8055 (10^{-5}) 0.0108	73.9474 ± 7.4769 (10^0) 0.0001	59.2026 ± 15.7231 $(10^{-3}, 0.4)$ 0.0147	72.3684 ± 7.9113 $(10^2, 0.8)$ 0.0002	$\mathbf{74.4737} \pm 6.7864$ $(10^5, 0.2)$ 0.0003
Monk3 221×7 333×7	65.6408 ± 16.263 (10^{-5}) 0.0222	79.287 ± 8.7684 (10^{-2}) 0.0003	65.7754 ± 17.9751 $(10^2, 0.9)$ 0.0239	68.7344 ± 12.084 $(10^1, 0.5)$ 0.0002	$\mathbf{81.7023} \pm 7.2182$ $(10^2, 0.9)$ 0.0007
Tic-Tac-Toe 383×9 575×9	56.9295 ± 7.009 (10^{-5}) 0.0527	65.5445 ± 5.7387 (10^1) 0.0004	57.8958 ± 5.8605 $(10^{-1}, 0.6)$ 0.0841	65.5445 ± 5.7387 $(10^0, 0.8)$ 0.0003	$\mathbf{67.2958} \pm 8.6392$ $(10^{-1}, 0.8)$ 0.0022
Wine 71×13 107×13	$\mathbf{100} \pm 0$ (10^0) 0.0078	$\mathbf{100} \pm 0$ (10^{-1}) 0.0002	95.8929 ± 6.4137 $(10^{-3}, 0.4)$ 0.0111	99.0909 ± 2.8748 $(10^1, 0.7)$ 0.0002	96.1818 ± 6.8273 $(10^1, 0.2)$ 0.0004

Table 4 Classification results on UCI benchmark datasets using Gaussian kernel

Dataset train size test size	TWSVM accuracy (C, μ) time	LSTSVM accuracy (C, μ) time	TPMSVM accuracy $(C, \mu, \nu/C)$ time	LSTPMSVM accuracy $(C, \mu, \nu/C)$ time	FLSTPMSVM accuracy $(C, \mu, \nu/C)$ time
Breast cancer 279×9 420×9	97.8571 ± 3.0635 $(10^{\wedge}-1, 2^{\wedge}2)$ 0.2955	97.381 ± 2.6203 $(10^{\wedge}-1, 2^{\wedge}4)$ 0.2479	97.619 ± 2.5097 $(10^{\wedge}-2, 2^{\wedge}0, 0.1)$ 0.2487	97.381 ± 2.3677 $(10^{\wedge}-5, 2^{\wedge}-2, 0.1)$ 0.2353	$\mathbf{98.0952} \pm 2.459$ $(10^{\wedge}1, 2^{\wedge}1, 0.1)$ 0.2486
Breast tissue 42×9 64×9	$\mathbf{98.3333} \pm 5.2705$ $(10^{\wedge}-2, 2^{\wedge}1)$ 0.0197	96.6667 ± 7.0273 $(10^{\wedge}-1, 2^{\wedge}1)$ 0.0061	92.1429 ± 10.7099 $(10^{\wedge}-3, 2^{\wedge}-1, 0.2)$ 0.0168	$\mathbf{98.3333} \pm 5.2705$ $(10^{\wedge}-5, 2^{\wedge}3, 0.2)$ 0.0093	$\mathbf{98.3333} \pm 5.2705$ $(10^{\wedge}3, 2^{\wedge}1, 0.1)$ 0.0072
BUPA liver 137×6 208×6	60.0476 ± 12.6516 $(10^{\wedge}0, 2^{\wedge}-1)$ 0.0718	$\mathbf{61.9762} \pm 13.1584$ $(10^{\wedge}0, 2^{\wedge}-1)$ 0.0632	61.9048 ± 15.9225 $(10^{\wedge}-1, 2^{\wedge}0, 0.1)$ 0.0614	47.8571 ± 28.1371 $(10^{\wedge}-5, 2^{\wedge}1, 0.9)$ 0.0628	60.5 ± 18.575 $(10^{\wedge}4, 2^{\wedge}0, 0.4)$ 0.0609
German 399×24 601×24	73.5574 ± 7.2328 $(10^{\wedge}-5, 2^{\wedge}2)$ 0.5827	75.2077 ± 6.1087 $(10^{\wedge}0, 2^{\wedge}3)$ 0.5345	71.5628 ± 4.252 $(10^{\wedge}-1, 2^{\wedge}0, 0.3)$ 0.5125	74.3743 ± 6.6354 $(10^{\wedge}-3, 2^{\wedge}3, 0.9)$ 0.4948	$\mathbf{75.7158} \pm 6.0469$ $(10^{\wedge}5, 2^{\wedge}4, 0.1)$ 0.5392
Haberman 122×3 184×3	70.6725 ± 7.2402 $(10^{\wedge}0, 2^{\wedge}-2)$ 0.0678	75.0585 ± 8.4172 $(10^{\wedge}0, 2^{\wedge}1)$ 0.0508	70.1462 ± 7.188 $(10^{\wedge}-3, 2^{\wedge}-1, 0.5)$ 0.0492	74.4737 ± 6.2607 $(10^{\wedge}-4, 2^{\wedge}2, 0.9)$ 0.0487	$\mathbf{75.5556} \pm 8.2117$ $(10^{\wedge}1, 2^{\wedge}0, 0.3)$ 0.0527
Monk3 221×7 333×7	94.5811 ± 5.2579 $(10^{\wedge}-1, 2^{\wedge}1)$ 0.1631	$\mathbf{96.3993} \pm 4.1591$ $(10^{\wedge}0, 2^{\wedge}0)$ 0.1562	94.5989 ± 4.4114 $(10^{\wedge}-1, 2^{\wedge}0, 0.2)$ 0.1547	94.9109 ± 3.6902 $(10^{\wedge}-4, 2^{\wedge}1, 0.3)$ 0.1599	$\mathbf{96.3993} \pm 4.1591$ $(10^{\wedge}4, 2^{\wedge}1, 0.1)$ 0.1724
Tic-Tac-Toe 383×9 575×9	98.6056 ± 1.1065 $(10^{\wedge}-1, 2^{\wedge}0)$ 0.5314	$\mathbf{99.3073} \pm 0.8943$ $(10^{\wedge}-1, 2^{\wedge}0)$ 0.4718	90.7653 ± 4.6272 $(10^{\wedge}0, 2^{\wedge}0, 0.1)$ 0.4435	98.784 ± 1.4314 $(10^{\wedge}-4, 2^{\wedge}0, 0.1)$ 0.44	98.784 ± 1.4314 $(10^{\wedge}5, 2^{\wedge}0, 0.1)$ 0.4712
Wine 71×13 107×13	95.3636 ± 6.5056 $(10^{\wedge}-5, 2^{\wedge}-1)$ 0.0237	$\mathbf{97.2727} \pm 6.1359$ $(10^{\wedge}-2, 2^{\wedge}-1)$ 0.0183	96.2727 ± 4.819 $(10^{\wedge}-3, 2^{\wedge}1, 0.5)$ 0.0214	95.3636 ± 4.8947 $(10^{\wedge}-1, 2^{\wedge}-1, 0.1)$ 0.0164	$\mathbf{97.2727} \pm 4.3913$ $(10^{\wedge}1, 2^{\wedge}-1, 0.6)$ 0.0189

Table 5 Average ranks of all the algorithms on UCI datasets using linear kernel

Dataset	TWSVM	LSTSVM	TPMSVM	LSTPMSVM	FLSTPMSVM
Breast cancer	4	1	3	2	5
Breast tissue	4	3	5	1.5	1.5
BUPA liver	4	2	3	5	1
German	4	2.5	5	2.5	1
Haberman	4	2	5	3	1
Monk3	5	2	4	3	1
Tic-Tac-Toe	5	2.5	4	2.5	1
Wine	1.5	1.5	5	3	4
Average rank	3.9375	2.0625	4.25	2.8125	**1.9375**

Table 6 Average ranks of all the algorithms on UCI datasets using Gaussian kernel

Dataset	TWSVM	LSTSVM	TPMSVM	LSTPMSVM	FLSTPMSVM
Breast cancer	2	4.5	3	4.5	1
Breast tissue	2	4	5	2	2
BUPA liver	4	1	2	5	3
German	4	2	5	3	1
Haberman	4	2	5	3	1
Monk3	5	1.5	4	3	1.5
Tic-Tac-Toe	4	1	5	2.5	2.5
Wine	4.5	1.5	3	4.5	1.5
Average rank	3.6875	2.1875	4	3.4375	**1.6875**

planes of the proposed FLSTPMSVM makes it suitable for heteroscedastic error structure. From the comparative performance analysis presented in this paper one can see that FLSTPMSVM delivers comparable or better classification accuracy as compared to the other reported algorithms.

References

1. Choisy, C., Belaid, A.: Handwriting recognition using local methods for normalization and global methods for recognition. In: Proceedings of Sixth International Conference on Document Analysis and Recognition, 2001, pp. 23–27. IEEE (2001)
2. Golub, G.H., Van Loan, C.F.: Matrix Computations. JHU Press (2012)
3. Gordan, M., Kotropoulos, C., Pitas, I.: Application of support vector machines classifiers to visual speech recognition. In: Proceedings of 2002 International Conference on Image Processing, vol. 3, pp. III-III. IEEE (2002)
4. Gorgevik, D., Cakmakov, D., Radevski, V.: Handwritten digit recognition by combining support vector machines using rule-based reasoning. In: Proceedings of the 23rd International Conference on Information Technology Interfaces 2001 (ITI 2001), 19 June 2001, pp. 139–144. IEEE (2001)

5. Hao, P.Y.: New support vector algorithms with parametric insensitive/margin model. Neural Netw. **23**(1), 60–73 (2010)
6. Heisele, B., Ho, P., Poggio, T.: Face recognition with support vector machines: Global versus component-based approach. In: Proceedings of Eighth IEEE International Conference on Computer Vision, 2001 (ICCV 2001), vol. 2, pp. 688–694. IEEE (2001)
7. Huang, J., Blanz, V., Heisele, B.: Face recognition using component-based SVM classification and morphable models. In: Pattern Recognition with Support Vector Machines, pp. 531–540 (2002)
8. Khemchandani, R., Chandra, S.: Twin support vector machines for pattern classification. IEEE Trans. Pattern Anal. Mach. Intell. **29**(5), 905–910 (2007)
9. Kumar, M.A., Gopal, M.: Least squares twin support vector machines for pattern classification. Expert Syst. Appl. **36**(4), 7535–7543 (2009)
10. Lin, C.F., Wang, S.D.: Fuzzy support vector machines. IEEE Trans. Neural Netw. **13**(2), 464–471 (2002)
11. Mangasarian, O.L., Musicant, D.R.: Lagrangian support vector machines. J. Mach. Learn. Res. **1**, 161–177 (2001)
12. Mangasarian, O.L., Wild, E.W.: Multisurface proximal support vector machine classification via generalized eigenvalues. IEEE Trans. Pattern Anal. Mach. Intell. **28**(1), 69–74 (2006)
13. Murphy, P.M., Aha, D.W.: UCI repository of machine learning databases. Department of Information and Computer Science, University of California, Irvine, CA
14. Peng, X.: TPMSVM: a novel twin parametric-margin support vector machine for pattern recognition. Pattern Recognit. **44**(10), 2678–2692 (2011)
15. Pittore, M., Basso, C., Verri, A.: Representing and recognizing visual dynamic events with support vector machines. In: Proceedings International Conference on Image Analysis and Processing, 1999, pp. 18–23. IEEE (1999)
16. Pontil, M., Verri, A.: Support vector machines for 3D object recognition. IEEE Trans. Pattern Anal. Mach. Intell. **20**(6), 637–646 (1998)
17. Ripley, B.D.: Pattern Recognition and Neural Networks. Cambridge University Press (2007)
18. Schölkopf, B., Smola, A.J., Williamson, R.C., Bartlett, P.L.: New support vector algorithms. Neural Comput. **12**(5), 1207–1245
19. Shao, Y.H., Wang, Z., Chen, W.J., Deng, N.Y.: Least squares twin parametric-margin support vector machine for classification. Appl. Intell. **39**(3), 451–464 (2013)
20. Suykens, J.A., Vandewalle, J.: Least squares support vector machine classifiers. Neural Process. Lett. **9**(3), 293–300 (1999)
21. Vapnik, V.N., Vapnik, V.: Statistical Learning Theory. Wiley, New York (1998)
22. Wan, V., Campbell, W.M.: Support vector machines for speaker verification and identification. In: Neural Networks for Signal Processing X, 2000. Proceedings of the 2000 IEEE Signal Processing Society Workshop, vol. 2, pp. 775–784. IEEE (2000)

Human Gait State Prediction Using Cellular Automata and Classification Using ELM

Vijay Bhaskar Semwal, Neha Gaud and G. C. Nandi

Abstract In this research article, we have reported periodic cellular automata rules for different gait state prediction and classification of the gait data using Extreme Machine Leaning (ELM). This research is the first attempt to use cellular automaton to understand the complexity of bipedal walk. Due to nonlinearity, varying configurations throughout the gait cycle and the passive joint located at the unilateral foot-ground contact in bipedal walk resulting variation of dynamic descriptions and control laws from phase to phase for human gait is making difficult to predict the bipedal walk states. We have designed the cellular automata rules which will predict the next gait state of bipedal steps based on the previous two neighbor states. We have designed cellular automata rules for normal walk. The state prediction will help to correctly design the bipedal walk. The normal walk depends on next two states and has total eight states. We have considered the current and previous states to predict next state. So we have formulated 16 rules using cellular automata, eight rules for each leg. The priority order maintained using the fact that if right leg in swing phase then left leg will be in stance phase. To validate the model we have classified the gait Data using ELM (Huang et al. Proceedings of 2004 IEEE international joint conference on neural networks, vol 2. IEEE, 2004, [1]) and achieved accuracy 60%. We have explored the trajectories and compares with another gait trajectories. Finally we have presented the error analysis for different joints.

Keywords Cellular automata (CA) · Human gait · Bipedal control · Humanoid robot · Extreme learning machine (ELM) · Pseudo-inverse

V. B. Semwal (✉)
Department of CSE, Indian Institute of Information Technology,
Dharwad, Dharwad, Karnataka, India
e-mail: vsemwal@iiitdwd.ac.in; vsemwal@gmail.com

N. Gaud
Institute of Computer Science, Vikram University, Ujjain, MP, India

G. C. Nandi
Robotics & AI Lab, Indian Institute of Information Technology,
Allahabad, Allahabad, UP, India
e-mail: gcnandi@iiita.ac.in

© Springer Nature Singapore Pte Ltd. 2019
M. Tanveer and R. B. Pachori (eds.), *Machine Intelligence and Signal Analysis*,
Advances in Intelligent Systems and Computing 748,
https://doi.org/10.1007/978-981-13-0923-6_12

1 Introduction

The bipedal walk is the combination of eight different sub-phases during normal walk. In running, we observed that it is a combination of four different sub-phases [2]. The Human walk is inherently unstable and non linear due to high degrees of nonlinearity, high dimensionality, under actuation (in swing phase), over actuation (in stance phase). The modern robotics industries have given boost the development of such robot which can walk in unconstrained environment similar to human. The modern robots are not capable enough to walk effectively in unconstraint environment. The next state predication for robotic walk is very difficult. The bipedal robot generally used to suffer from the singularity configuration. Using pseudo-inverse, we used to deal such configuration [3].

The equation of linear system can be represented in following form:

$$Ax = y \tag{1}$$

$$A \ is \ singluar \ if \ \|A\| = 0 \tag{2}$$

$$Pseudo \ Inverse x = \left(A^t A\right)^{-1} A^t y \tag{3}$$

This research is attempts to predict the state of robot walking. The walking pattern is very unique to each human being [4]. The human acquired this behavior through learning. The gait study is widely using in Biometric identification [3, 5], artificial limbs generation [6], Robotics walk [7, 8] and development of modern data driven computational walking model which can walk similar to human [9].

We have presented the cellular automata model as generalized predictive model. The model will be able to predict the next state based on current state and previous state. It will help the robot to plan the next state. Total 8! Permutation is possible. The model is able to predict the state on any terrain. In this case we have considered only states in terms of joints angle value we would not referred the terrain. Cellular automata are mathematical tool [10] for modeling a system that evolves with certain set of rule. It merges the specifications for discrete switching logic and continuous dynamics behaviors of any dynamical system. So, it is an appropriate model for predicting a human bipedal trajectory in terms of CA because a bipedal locomotion trajectory is also a combination of continuous and discrete phases [11] and a stable walk can be obtained using CA mathematical model. This bipedal trajectory very precisely can be designed using cellular automata.

The paper is organized into four parts as following. The next section is methodology section which is description of cellular automata rule and ELM algorithm description. The third section is results section. The fourth section is verification of results and final section is conclusion, discussion and future scope.

2 Methodology

To modeling the system we have used a bottom-up approach. We have first constructed the atomic component and modeled their behavior using Cellular automaton. We have merged all atomic component together and converted into composite components. The interactions can be formulated between composite and atomic components using the well-defined semantics of algebra of connectors and mainly the causal chain type of interaction. Later, we assigned the priorities which impose the certain ordering on the type of interactions. The priority also can help to avoid the deadlock. Before, we develop the cellular automata and their design, it is important to understand the different sub-phases of one complete gait. Mostly the gait data is used in the medical and health care sector. The gait cycle usually takes 1–1.2 s to complete. The gait is a time series data and broadly it has two phases Stance and Swing. The gait further can be dived into eight sub-phases [12, 13]. As gait is time series data, so each sub-phases takes certain percentage of entire gait cycle. The swing phases generally takes 40% of whole gait cycle and stance phase takes 60% of complete gait cycle. The percentage wise division of human gait is divided into following sub-phases [14–16]:

Stance phase:

1. Initial Contact—IC[0–2%]
2. Loading Response—LR[2–10%]
3. Mid-Stance—MS[10–30%]
4. Terminal Stance   TS[30 50%]

   Swing phase:

5. Pre-Swing—PS[50–60%]
6. Initial Swing—IS[60–73%]
7. Mid Swing—MS[73–87%]
8. Terminal Swing—TS[87–100%]

The initial contact sub-phase accounts for a very small percentage of the complete gait cycle, hence it is merged with loading response phase without loss of generality and we called the new merged state as initial contact. Similarly, we have merged the pre-swing and initial swing in one combined state named it as initial swing [17].

2.1 Design of Atomic Components for CA

The left and right legs can be decomposed into three atomic components named hip, knee and ankle. So, we have total six atomic components, i.e., each leg should have three atomic component. For two legs we will have two composite components. To express the nature of each atomic component is given by a six states (three for each swing and stance states) in cellular automaton. In our cellular automata model the states starts from stance phase which lead for automatically swing phase for right leg [18]. All shifting from one state to other happens in synchronization. So, the phase order during leg moment will be the left leg goes into swing phase and right will go in stance phase [19].

2.2 Developing Cellular Automata (CA)

Here we have written 16 CA rules to determine the state of atomic components of one leg with the help of second leg. It will be among one of the eight states so there will be a total of 16 rules. 1000 can be seen as two parts $1 + 000$ (leg + Sub-phase) which means right leg is in initial contact.

We have assumed binary state of movement of atomic components of a leg (Ankle, Knee, Hip) is either in motion or in rest. So we consider binary stage 0 and 1 for each component. 0 represents atomic components are in rest and 1 represents atomic components are in motion [19]. Since, there are three atomic components and each have two state either 0 or 1. So, there will be a total of eight states. During human locomotion each leg passes through eight sub-phases [20].

CA is discrete dynamic systems. CAs are said to be discrete because they operate in finite space and time and with properties that can have only a finite number of states. CAs are said to be dynamic because they exhibit dynamic behaviors. Equation 4 is the representation of state prediction. Where S(t) represents the current state. S is the set of all possible discrete states of our gait model for us it is 8. And Eq. 5 is the prediction of next state from the previous two states.

S: Finte set of state, i.e., discreter variable

$$S = \{IC, LR, MS, TST, PSW, ISW, MSW, TSW\} \tag{4}$$
$$S(t + 1) = \{S(t), S(t - 1)\} \tag{5}$$

Consider the state $S = \{1, 2, 3, 4, 5, 6, 7, 8\}$ so we have assumed eight discrete states here. In this work we have considered eight discrete state as eight neighbors. Equation 5 is the state prediction in case of Normal Walk and brisk walk. Each leg which is considered as atomic component passes through the three discrete states named initial contact, mid stance, terminal stance in synchronization. The phase transfer between the left and right leg happens alternatively. The swing phase gets complete once the left leg returns back to the stance phase [21]. The atomic components of the left leg are shown here ankle, knee, and hip. We have assumed binary state of movement of atomic components of a leg (Ankle, Knee, Hip) is either in motion or in rest. So we consider binary stage 0 and 1 for each component. 0 represents atomic components are in rest and 1 represents atomic components are in motion. Since, there are three atomic components and each have two state either 0 or 1. So, there will be a total of eight states (Refer Table 1).

Table 1 Binary state representation of bipedal Gait 8 states

Number	7	6	5	4	3	2	1	0
Neighborhood	111	110	101	100	011	010	001	000
Rule result	TS	MS	IS	PS	TS	MS	LR	IC

2.3 CA Rules

Here we have written 16 CA (Eqs. 6–14) rules to determine the state of atomic components of one leg with the help of second leg. All the states are represented using 4-bit stream. First bit represent the leg that if the fourth bit is zero it represents left leg whereas if the fourth bit is 1 it represents the right leg. Other three bits represents the sub-phases of that leg. It will be among one of the eight states so there will be a total of 16 rules. 1000 can be seen as two parts $1 + 000$ (leg + Sub-phase) which means right leg is in initial contact. The neighbor row represents the state or phase of another leg whereas the rule row depicts the state of atomic components of that leg. Set of rules to determine the state of locomotion. Cellular automata rule Rule-8, universal, generalizes Rule for left and right leg during normal walk. Following are the states relation between left and right leg.

$$Left\_Leg\_Stance \rightarrow Right\_Leg\_Swing \qquad (6)$$
$$Left\_Leg\_Swing \rightarrow Right\_Leg\_Stance \qquad (7)$$
$$Left\_Leg\_IC \rightarrow Right\_Leg\_PSw \qquad (8)$$
$$Left\_Leg\_MS \rightarrow Right\_Leg\_Msw \qquad (9)$$
$$Left\_Leg\_TS \rightarrow Right\_Leg\_TSw \qquad (10)$$
$$Left\_Leg\_PSw \rightarrow Right\_Leg\_LR \qquad (11)$$
$$Left\_Leg\_ISw \rightarrow Right\_Leg\_MS \qquad (12)$$
$$Left\_Leg\_MSw \rightarrow Right\_Leg\_TS \qquad (13)$$
$$Left\_Leg\_TSw \rightarrow Right\_Leg\_IC \qquad (14)$$

Tables 1, 2 and 3 are the binary state representation of bipedal gait and it is the novel contribution. It will help to predict the next state of robot. Figure 1 is the transaction of leg state using cellular automata. The unique approach to model the human gait presented here. It is able to model the normal human gait within a negotiable degree of error. Here we have written 16 CA rules to determine the state of atomic components of one leg with the help of second leg. All the states are represented using 4-bit stream. First bit represent the leg that if the fourth bit is zero it represents left leg whereas if the fourth bit is 1 it represents the right leg. Other

Table 2 Cellular automata state prediction for left leg

Number	7	6	5	4	3	2	1	0
Neighbor	0111	0110	0101	0100	0011	0010	0001	0000
Rule result	1011	1010	1001	1000	1111	1110	1101	1100

Table 3 Cellular automata state prediction for left leg

Number	15	14	13	12	11	10	9	8
Neighbor	1011	1010	1001	1000	1111	1110	1101	1100
Rule result	0111	0110	0101	0100	0011	0010	0001	0000

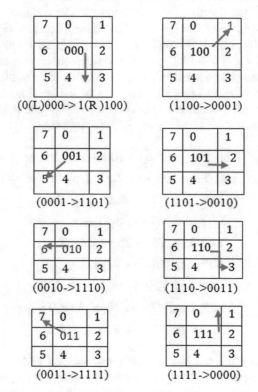

Fig. 1 Transaction of leg state using CA

three bits represents the sub-phases of that leg. It will be among one of the eight states so there will be a total of 16 rules. 1000 can be seen as two parts 1 + 000(leg + Sub-phase) which means right leg is in initial contact. The neighbor row represents the state or phase of another leg whereas the rule row depicts the state of atomic components of that leg.

CAs are discrete dynamic systems.

- CAs are said to be discrete because they operate in finite space and time and with properties that can have only a finite number of states.
- CAs are said to be dynamic because they exhibit dynamic behaviors.

2.4 Basic ELM Classifier

Hung et al. [1, 3] has proposed the ELM, which is fast and we called learning without iteration tuning. For given non-constant piece wise continuous function g, if continuous target function f(x) can be approximated by SLFNS with some adjustable weights.

Given a training set $\{(a_i, b_i)|a_i \in R^d, b_i \in R^m, i = 1, …, N\}$, hidden node output function G(a, b, x), and the number of hidden nodes L,

Step 1—Randomly hidden node parameter (a_i, b_i), $i = 1,…, N$.
Step 2—Output of hidden layer $H = [h(x_1), h(x_2)… h(x_N)]^T$
Step 3—The out weight β.

3 Experiments Results and Verification

In this paper, we will be looking on the results and outcome of the work done using cellular automata for the state prediction of bipedal walk. For this we have divided the human gait cycle data into different sub-phases for different joint angle right and left ankle, knee, and hip joints. These are the equations for all the six joints that are left hip, left knee, left ankle, and right hip, right knee and right ankle corresponding to each sub-phase of the gait cycle [22, 23].

Figure 2 is classification accuracy of cellular automata based gait using ELM, SVM, and linear Regression. ELM based classification has our performed with 65% to other classifiers. Similarly it has proved the cellular automata based Gait state predication is giving a better walking pattern. Figure 3 is the stick diagram to validate the cellular automata rules. In this model we have applied the joint trajectories. We have taken the input thigh and stride length. It is showing the state predicted through cellular automata is correct. Figure 4 is the different gait states prediction of Gait2354 model of Opensim. Figure 5 shows the hip joints limit cycle to validate the stability of state generated trajectory of cellular automata. The limit cycle is used to measure the stability of model. The trajectories generated using the model for left and right hip is depicted in Fig. 4. It is following the cycle which justify that the hip joints trajectories are stable. Figure 5 is the comparison of trajectories of our model with other well established trajectories and our previous work generated trajectories [24, 25]. The generated trajectories are very close to other established joints trajectories.

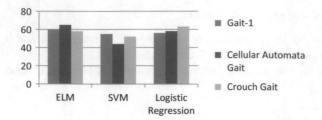

Fig. 2 Classification result of different gait using ELM, SVM, and linear regression

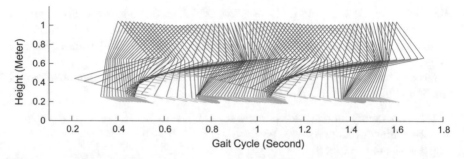

Fig. 3 Stick diagram for both leg gait pattern of cellular automata model

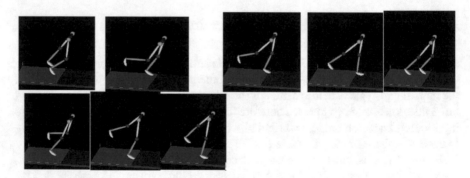

Fig. 4 The transaction and state prediction of leg state using cellular automata

4 Error Analysis

In this section we have validated the trajectories generated through cellular automata. To validate the correctness of model, it is important to perform the error analysis of predicated state using cellular automata model. The analytical human data base model and (stable for 4°) and the Cellular Automata Model (stable for 4°). We are now going to calculate the error in Analytical model with respect to Human data. We

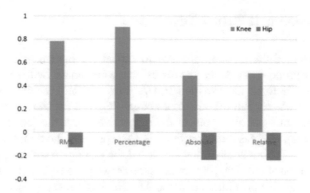

Fig. 5 Correlation between various error with Spearmen's error

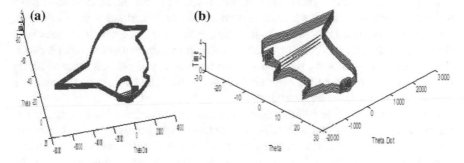

Fig. 6 **a** Cellular automata model knee LC **b** Cellular automata model hip LC

have then plotted this output in Fig. 6. From the figure, it is clearly visible that the mean percentage error has the maximum correlation for both knee and hip. Hence, this error is suitable for training the neural network for finding the error pattern recognition [26].

In this section we have compared the limit cycle behavior between Hip and Knee joints for the cellular automata generated trajectories. We have identified that the base model is unstable below some degree. For comparison, we have used walking along some degrees of inclination. As mentioned in the earlier section, the limit cycle behavior clearly shows that the system is unstable at some degrees of inclination. Now as seen from the Fig. 6 the limit cycle behavior shows that the same system is stable at some degrees angle of inclination [27]. This comparison clearly shows while the base model fails to justify stable walking by a real subject under certain conditions, the cellular model fully supports the same.

5 Conclusion and Future Discussion

We have presented the unique approach to model the human gait using cellular automata. The proposed model is able to design the human gait with some negotiable degree of error. Here we have written 16 CA rules to determine the state of atomic components of one leg with the help of second leg. All the states are represented using 4-bit stream. First bit represent the respective leg. If the fourth bit is zero it represents left leg whereas if the fourth bit is 1 it represents the right leg. Other three bits represents the sub-phases of that leg. It will be among one of the eight states so there will be a total of 16 rules. 1000 can be seen as two parts 1 + 000 (leg + Sub-phase) which means right leg is in initial contact. The Neighbor row represents the state or phase of another leg whereas the rule row depicts the state of atomic components of that leg. The cellular automata based model help to predict the next state of robot walking. It will generate the trajectories which we have verified using limit cycle curve and comparison with other trajectories. The covariance matrix is calculated about the parameter to prove how the cellular automata generated trajectories is close to other normal walk trajectories. We have achieved classification accuracy 60% through ELM for cellular automata based walk. The ELM-based classifier it best among all classifier and very fast.

Future Discussion: The cellular automata can be utilized for many other learning behaviors like push recover [27]. The behavior can be explored far better and can be implemented on bipedal robot model. The state prediction is unconstrained environment is very tough. It requires the human intelligence and neuron muscular co-ordination.

References

1. Huang, G.-B., Zhu, Q.-Y., Siew, C.-K.: Extreme learning machine: a new learning scheme of feedforward neural networks. In: Proceedings of 2004 IEEE International Joint Conference on Neural Networks, vol. 2. IEEE (2004)
2. Semwal, V.B., et al.: Design of vector field for different subphases of gait and regeneration of gait pattern. IEEE Trans. Autom. Sci. Eng. **PP**(99), 1–7 (2016)
3. Huang, G.-B., Chen, L., Siew, C.K.: Universal approximation using incremental constructive feedforward networks with random hidden nodes. IEEE Trans. Neural Netw. **17**(4), 879–892 (2006)
4. Raj, M., Semwal, V.B., Nandi, G.C.: Bidirectional association of joint angle trajectories for humanoid locomotion: the restricted Boltzmann machine approach. Neural Comput. Appl. 1–9 (2016)
5. Semwal, V.B., Raj, M., Nandi, G.C.: Biometric gait identification based on a multilayer perceptron. Robot. Auton. Syst. **65**, 65–75 (2015)
6. Mukhopadhyay, S.C.: Wearable sensors for human activity monitoring: a review. IEEE Sens. J. **15**(3), 1321–1330 (2015)
7. Zhang, Z., Hu, M., Wang, Y.: A survey of advances in biometric gait recognition. In: Biometric Recognition, pp. 150–158. Springer, Berlin, Heidelberg (2011)
8. Gupta, J.P., et al.: Human activity recognition using gait pattern. Int. J. Comput. Vis. Image Process. (IJCVIP) **3**(3), 31–53 (2013)

9. Semwal, V.B.: Data Driven Computational Model for Bipedal Walking and Push Recovery. https://doi.org/10.13140/rg.2.2.18651.26403

10. Raj, M., Semwal, V.B., Nandi, G.C.: Hybrid model for passive locomotion control of a biped humanoid: the artificial neural network approach. Int. J. Interact. Multimed. Artif. Intell. (2017)

11. Semwal, V.B., et al.: Biologically-inspired push recovery capable bipedal locomotion modeling through hybrid automata. Robot. Auton. Syst. **70**, 181–190 (2015)

12. Huang, G.-B., et al.: Extreme learning machine for regression and multiclass classification. IEEE Trans. Syst. Man Cybern. Part B (Cybernetics) **42**(2), 513–529 (2012)

13. Wang, C., Zhang, J., Wang, L., Pu, J., Yuan, X.: Human identification using temporal information preserving gait template. IEEE Trans. Pattern Anal. Mach. Intell. **34**(11), 2164–2176 (2012)

14. Wang, L., Tan, T., Ning, H., Hu, W.: Silhouette analysis-based gait recognition for human identification. IEEE Trans. Pattern Anal. Mach. Intell. **25**(12), 1505–1518 (2003)

15. Semwal, V.B., et al.: An optimized feature selection technique based on incremental feature analysis for bio-metric gait data classification. Multimed. Tools Appl. 1–19 (2016)

16. Nag, A., Mukhopadhyay, S.C., Kosel, J.: Wearable flexible sensors: a review. IEEE Sens. J. **17**(13), 3949–3960 (2017)

17. Sinnet, R.W., Powell, M.J., Shah, R.P., Ames, A.D.: A human-inspired hybrid control approach to bipedal robotic walking. In: 18th IFAC World Congress, pp. 6904–6911 (2011)

18. Semwal, V.B., Nandi, G.C.: Toward developing a computational model for bipedal push recovery: a brief. Sens. J IEEE **15**(4), 2021–2022 (2015)

19. Parashar, A., Parashar, A., Goyal, S.: Push recovery for humanoid robot in dynamic environment and classifying the data using K-Mean. Int. J. Interact. Multimed. Artif. Intell. **4**(2), 29–34 (2016)

20. Semwal, V.B., Nandi, G.C.: Generation of joint trajectories using hybrid automate-based model: a rocking block based approach. IEEE Sens. J. **16**(14), 5805–5816 (2016)

21. Semwal, V.B., Katiyar, S.A., Chakraborty, P., Nandi, G.C.: Biped model based on human Gait pattern parameters for sagittal plane movement. In: 2013 International Conference on Control, Automation, Robotics and Embedded Systems (CARE), pp. 1–5, 16–18 Dec 2013

22. Raj, M., Semwal, V.B., Nandi, G.C.: Multiobjective optimized bipedal locomotion. Int. J. Mach. Learn. Cybern. 1–17 (2017)

23. Nandi, G.C., et al.: Modeling bipedal locomotion trajectories using hybrid automata. In: Region 10 Conference (TENCON), 2016 IEEE. IEEE (2016)

24. Semwal, V.B., Chakraborty, P., Nandi, G.C.: Biped model based on human gait pattern parameters for sagittal plane movement. In: IEEE International Conference on Control, Automation, Robotics and Embedded Systems (CARE), pp. 1–5 (2013)

25. Semwal, V.B., Nandi, G.C.: Robust and more accurate feature and classification using deep neural network. Neural Comput. Appl. **28**(3), 565–574

26. Semwal, V.B., Nandi, G.C.: Study of humanoid push recovery based on experiments. In: IEEE International Conference on Control, Automation, Robotics and Embedded Systems (CARE) (2013)

27. Semwal, V.B., Chakraborty, P., Nandi, G.C.: Less computationally intensive fuzzy logic (type-1)-based controller for humanoid push recovery. Robot. Auton. Syst. **63**, Part 1, 122–135 (2015)

Redesign of a Railway Coach for Safe and Independent Travel of Elderly

Dipankar Deb, Tirthankar Deb and Manish Sharma

Abstract The elderly population have specific needs while venturing beyond their homes and traveling without an accompanying care provider. In this paper, we look at specific technological and ergonomic innovations required so as to make travel safe and comfortable for the elderly with some user-centered design modifications. Without loss of generality, we consider the redesign of a railway coach and propose alterations in the entry and suggest introduction of medical care desks and automated basic healthcare consoles. The coaches are also proposed to be equipped with advanced sensors to study and monitor any ailments (new or existing) through the evaluation of gait of the elderly passengers.

Keywords Technology for health care · Usability · Assisted travel · Adaptive console interface

1 Introduction

The population worldwide is aging, and since disabilities increase with age, the demand for accessible transport as and when one desires, is expected to increase. Mobility is important for daily activities and quality of life, but owing to diseases

D. Deb (✉) · M. Sharma
Institute of Infrastructure Technology Research and Management,
Ahmedabad 380026, Gujarat, India
e-mail: dipankardeb@iitram.ac.in; ddeb30@gmail.com

M. Sharma
e-mail: manishsharma.iitb@gmail.com

T. Deb
Department of Pharmacology, Kalpana Chawla Government Medical College,
Karnal 132001, Haryana, India
e-mail: tirthdeb@gmail.com

© Springer Nature Singapore Pte Ltd. 2019   147
M. Tanveer and R. B. Pachori (eds.), *Machine Intelligence and Signal Analysis*,
Advances in Intelligent Systems and Computing 748,
https://doi.org/10.1007/978-981-13-0923-6_13

such as dementia and strokes, mobility impairment usually impedes travel due to lack of accessibility at different point of time during journey. According to National survey (Jan-June 2004), 8% of older Indians were confined to their home or bed [1]. Nonavailability of suitable travel options may be a reason for such restrictions. Providing accessible transportation for elderly and people with limited mobility will significantly alter the future quality of life.

Governments around the world should ensure that infrastructure and services are accessible to those with limited mobility, or visual, hearing, and other impairments, the prevalence of which generally increases with age. Proliferation of mobile applications and interconnected sensors offers a variety of new channels for safeguarding the travel of older persons by delivering services related to health, security or environmental hazards during travel. Little information is available regarding the risk behaviors and the health of elderly travelers, during travel, compared to their younger counterparts. Due to their more complex medical background and decreasing immunity, elderly travelers are more prone to various health risks and would seek medical care more intensively during and after travel [2]. Generally, a growth in older population causes a decrease in regular transit use relative to total population growth. Increased elderly disability rates lead to predicted increases in paratransit ridership. However, when they use transit services they may face significant challenges which cause unsafe movement at different point of time during journey [3]. Certain disabilities and chronic diseases are most frequent among older people. Psychological problems, impaired memory, rigidity of outlook, and dislike of change are some of the mental changes found in the aged [4].

India will experience a 270% growth in the senior population (age 60 and above) by 2050. The significant number of aging population underscores the urgency in eliminating age-related discrimination, promoting and protecting the dignity of the elderly, and facilitating their full participation in the travel industry and related infrastructure. Under such changing demographics, a transportation planner has to ensure equal opportunity for travel in the conventional public transport systems like rail service. Acting on a Public Interest Litigation (PIL), a division bench of Delhi High Court on Jan 22, 2014, directed the Ministry of Railways to employ one trained doctor along with the support staff, besides medicines and equipment necessary for life support, in each long-distance train. It is not cost-effective to do employ a resident doctor in each train and even if it is done in many cases specialists may be required. This paper proposes a cost-effective alternative.

The Persons with Disabilities (Equal Opportunities, Protection of Right and Full Participation) Act, 1995 (PDA) was promulgated by the Indian Government to ensure equal opportunities to person with disabilities and their full participation in nation-building. According to census 2001, there are 0.219 billion persons with disabilities in India who constitute 2.13% of the total population. Indian Railways is the world's fourth largest railway network, transporting 8.4 billion passengers annually. However, basic amenities for the elderly with mobility impairment is lacking. Chatterjee (1999) carried out a survey of Indian travelers wherein out of 158 (35%) reporting illness during their visit, 20 (4.4%) had fever for more than 5 days necessitating a visit to a doctor [5].

According to National Sample Survey Organization (2004), 5.2% of persons aged 60 years and above live alone. For such people, special care has to be taken when they move out, and user-friendly provisions are needed in various public transport systems. Presently there are no coaches specially designed and commissioned for the elderly who often have limited mobility and are less aware of their surroundings. Interconnected (with network connectivity) devices at different locations of the train can aid and/or alert others of their urgent needs. Recent initiatives have taken place to develop anthropometric datasets for specific populations with currently available digital human modeling technology [6]. However, there is a need for a comprehensive national effort toward improved tools for development of a robust anthropometric dataset that informs the design community and shapes design practice in the coming decades. Each coach that would board the elderly or disabled would host various forms of monitoring devices, noninvasive measuring devices/systems, injury preventions, blood parameters, etc., that would provide on-demand data to the care provider who may be on board in a different coach or at major stations in the route of the train.

2 Anomaly Diagnosis in Traveling Elderly

Given the impending growth of older population in the coming decades, health systems in the developing infrastructure should prepare to address their travel-related health concerns. Morbidities in elderly restrict their movement both due to current health status as well as fear of sudden exacerbation during journey. Provision of immediate medical care designed to address specific health ailments may provide necessary confidence to such elderly people to travel by railways, even without an accompanying caregiver.

Provision of live and automated assessment of sickness condition during the course of the journey to address specific health ailments may provide necessary confidence to such elderly people to travel by railways, even without an accompanying caregiver. Yeung et al. (2005) estimated the willingness to pay (WTP) to prevent travel health problems in Hong Kong, based on the questionnaire addressing 26 typical health problems ranging from diarrhea to genital ulcers, and found that 77% of the sample are willing to pay from their own pockets indicating possible cost sharing. Furthermore, those who took precautionary measures to protect themselves against health risks had higher WTP values, suggesting that an economical way to protect the population against travel health risk is to improve attitudes regarding health problems [7]. Presently, Indian Railways provides fare concession of 40% to males (60 years and above) and 50% to females (58 years and above). Reduced amount of concession can make these additional medical facilities economically viable. During the past two decades, there have been a number of studies to examine the relationship between various illnesses and traveling [8, 9]. Risk behavior associated with travel and its impact on developing communicable diseases were also reported in other studies. Moderately and severely immune-compromised travelers

are at increased risk of developing a serious health problem during or after a trip
in a tropical country and should be well informed about such specific risks. Travel
medicine health professionals favor effective preventive measures for such travelers
and envisage standby antibiotic treatment [10].

Elderly patients who wish to avail the medical care facilities on board may be
asked to opt for this option while booking ticket, so that they can be allotted berths in
the specially designed medical care coach. The same software used for booking ticket
can have a format for entry of health information of the person including baseline
records of vital parameters like blood pressure (BP) and heart rate, last investigation
reports like fasting blood sugar, postprandial blood sugar, hemoglobin (Hb), blood
cell counts, blood group, ECG (if any positive finding), serum creatinine, urine test
reports, or any other significant result. Moreover, details of present prescription
including medications, dosage, and advice may be recorded while booking. As shown
in Fig. 1, the different parts of the medical console such as ticket scanner, ECG device
holder, glucometer with strips, electronic sphygmomanometer, nebuliser with sterile

Fig. 1 Medical console for
onboard diagnostics

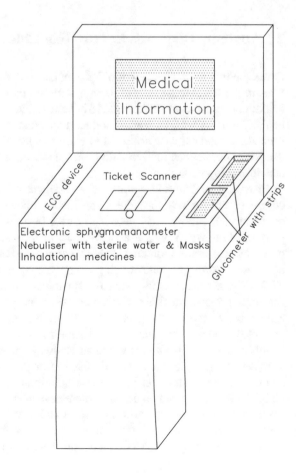

water, etc., are located appropriately for better usability considering that the attendant may need to quickly attend to the elderly with an appropriate device or item.

These consoles would provide live wellness health data when an elderly passenger scans his/her ticket which identifies the stored baseline health parameters, while boarding the train. In case of any change in prescription or new investigation reports done between booking of ticket and boarding the train, it may be informed to the attendant at the console after boarding, which can be entered there. However, a basic fitness status for traveling may be assured by demanding a fit to travel certificate by a registered medical practitioner while boarding, so that travel of severely ill patients alone may be avoided as they cannot be taken care of with only the immediate medical care facilities provided in coach. Railways should have tie-up with a large cross section of healthcare providers so that the baseline health status data of the senior travelers is recorded in the console beforehand or is identified when the ticket is scanned. At any time of travel, the passenger can check the vital parameters, and the console will indicate if anything is untoward and call in emergency care if needed at the next station. In order to identify add-on devices that can be provided in railway coaches for senior citizens, one must evaluate the most likely incidents related to their existing disease conditions. For the present work, we consider only those ailments and disabilities prevalent among the elderly that need immediate care while traveling and which can be addressed without a medical doctor in immediate attendance.

Serious adverse events can be prevented by recognizing and responding to early vital signs [11]. Adaptive console system for elderly people to monitor various form of health parameters is desired. We propose a centrally located console system in the coach so that people can check the vital signs of medical problems. Vital signs are (body temperature, pulse rate, rate of breathing, and blood pressure) useful in detecting or diagnosing the medical conditions. The console also checks the basic blood parameters which convert into various forms of medical terms related to health like sugar level, depression level, etc. The console would have access to health records of all individual elderly passengers provided by themselves directly or through their health insurance agencies during ticketing. This inbuilt medical record helps determining adaptively for each elderly passenger their requisite dosage for any medicine to be taken should their real-time health parameters go beyond certain thresholds. Such information is observed by the trained attendant who would administer the dosages accordingly.

Researchers have found that blood test results above a specified threshold could precisely and reliably predict the probability of individuals responding to the treatments. To anticipate various medical conditions, robust algorithms that can be validated need to be developed. Also, to be formulated is a mobile application which will store all the data through Bluetooth/ wireless network and send to the medical officer if a passenger suddenly suffers from some serious health issue during journey. The medical officer can then guide the attendant for further treatment on call or the subject can be attended to by a doctor at the next station. The above-stated tasks are accomplished on a modified version of an about to be condemned coach of Indian Railways, and using real data of human health parameters received and analyzed at real time and also real data of ergonomics-based studies in public transport.

3   Design of Railway Coach for Assisted Travel

Usability studies on all technologies for health care are important. Assistive devices for elderly with physical disabilities such as wheelchairs, walkers, canes, crutches, prosthetic devices, and orthodontic devices enable accessibility and enhances their quality of travel and safety. Technological interventions for the elderly for improved quality of life especially when they are on move, is the need of the hour. Advanced sensing and wireless interconnectivity across various add-on devices can provide individual autonomy and independence to persons with disabilities. We propose a holistic development by creating an enabling environment through application of advanced sensing and control in the railway coach.

The proposed add-on equipment to the latest railway coaches would be cost-effective and affordable and are aimed toward improving functionality, assisted safety measures, methods for learning, etc. With ubiquitous wireless network connectivity now available with Indian Railways, it is possible to interconnect the devices to monitor the health conditions and gait of the elderly. Figure 2 describes a specially designed coach for assisted travel. There are 42 berths in this coach, two less than an Air-Conditioned 2-Tier Coach of Indian Railways.

Two berths have been removed to accommodate the specially trained attendant who operates a fully equipped console near his desk. Another console is proposed by removing one of the four usual lavatories, and this will be primarily used for urine, stool, and related tests that need close access to a lavatory. All the lavatories in this coach will be wheelchair enabled and equipped with non-slip floors and sufficient number of handles. The elderly in this coach would only be issued the lower berths to facilitate their ease of use.

Medical care desk is shown in Fig. 2 and the occupancy status is prominently displayed in three different colored lights for watchful observance. Shaded circle represents blue colored bulb to tell the attendant that passenger is present in the berth, the white circle represents green colored bulb indicating that the passenger is not present in the berth, and the half dark and half white circle indicates a red colored bulb triggering a concern to the attendant that the passenger is not in the berth for an extended period of time. In Fig. 2 it is clear that one of the lavatories is occupied by a passenger who is missing from berth for an extended period, and another passenger is missing from the berth for extended period of time but not in the lavatories, indicating that he/she may have fallen anywhere in the coach. Two other lavatories are vacant.

Falls are one of leading injury causes among older travelers. One with limited mobility or impaired balance could request the attendant for a walking aid. The two hinged doors of the coach facing each other are provided with sensors to detect falls and alert the attendant. Display on top of these doors already provides occupancy status of the lavatories and can be modified to also include the information about upcoming stations and estimated time to reach there. Falls in elderly may be further precipitated during railway travel due to the jerking movement of coaches, specially in case of those with hearing and visual impairment. Since urinary incontinence,

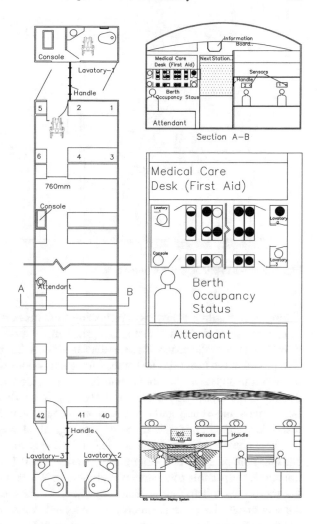

Fig. 2 Specially designed coach for elderly

urgency, increased frequency of micturition are common in elderly, frequent movement to the lavatory from the individual berth is necessitated, and therefore the risk of fall may be one of the factors discouraging elderly persons from choosing railways for travel. Structural modification of coaches with facility of horizontal and perpendicular bars, modified hangers in the passage to hold while moving from berth to lavatory, and similar special stuff in lavatories meant for elderly to hold while inside the chamber, will definitely infuse confidence.

As already shown in Fig. 2, and an enlarged view in Fig. 3 below, the sensors are installed on all the upper berths to monitor the presence of elderly passenger in the lower berth facing the sensor. These sensors obtain the body postures. Various image processing techniques such as template matching, skeleton extraction, posture recognition, background deduction, etc., are used to detect falls [12]. These sensors

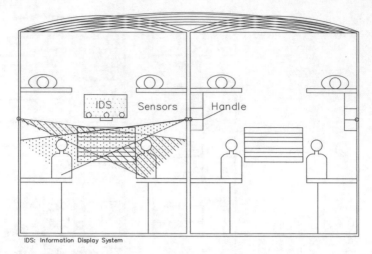

IDS: Information Display System

Fig. 3 Sensors to remotely monitor the elderly

determine if the elderly passenger is missing from his berth beyond a certain duration of time which is decided and stored a priori in the system. If an elderly passenger falls anywhere in the coach, the attendant is alerted. Such sensors (not cameras) are also placed in the lavatories and at the top of the two hinged doors facing each other, for the same objective. Additional handles are provided in the region between hinged doors and the lavatories for ease of movement. An elderly person who is at risk of falling or having a heart attack can avail of an equipment that is part of the emergency response system in this state of the art redesigned railway coach.

These equipment have a call service provision at the click of a button, a receiver console that plugs into a mobile jack, and a small battery-operated push-button transmitter attachable to a pendant or bracelet. When the elderly passenger pushes the transmitter's help button from anywhere in the coach, it signals the receiver console to call the system's emergency response center which need not necessarily be physically in the train. This provision is useful for the issues unaddressable by the designated attendant.

4 Design Evaluation

Feasibility testing with prospective users was performed among a group of 15 Elderly persons (9 Males and 6 Females) in the age group of 53–75 who presently live alone in an Indian metropolis. The questionnaire used for this study is summarized below:

1. Are you suffering from any disease? If yes, mention the name of the disease
2. Did you suffer from any acute exacerbation of your disease in last 1 year?
3. Do you think acute exacerbation of your disease can be prevented?
4. How frequently do you visit a doctor?
5. How frequently do you get blood investigations done?
6. Are you on any regular medication? If yes, mention the name of the medication
7. Do you depend on any other person for administering your medications?
8. Do you use any device to administer your medications? If yes, please specify. Do you need some other persons' help to use the device?
9. When did you last travel by long-distance train?
10. Mention your three important concerns while planning your travel by train
11. Did you need an attendant in your last journey by train? If yes, mention two reasons why you needed an attendant to accompany.
12. Are you ready to travel alone by train if medical care to prevent exacerbations of your morbidities is provided in coach during travel?
13. Are you ready to travel alone if facility to perform emergency blood tests in case of exacerbation of your disease is provided in coach during travel?
14. Are you ready to travel alone by train if emergency medications needed in case of exacerbation of your disease are provided in coach during travel?
15. Are you ready to travel alone by train with certain structural modifications in coach taking into consideration common health concerns of the elderly?
16. If you have any specific suggestion toward modification or facilities to be provided in railway coach for safe elderly travel, please mention.

They have all traveled a minimum of 8 hours by train alone. Out of these, 13 individuals have pre-existing medical conditions such as gallbladder stone, chest pain, abdomen pain, joint pain, diabetes, Parkinson's disease, vision difficulty, and hypertension. The respondents visit a doctor at least once in 6 months. The eldest respondent (aged 75 years) has not traveled by train for 3 years; others have at least once in part year. All but two participants stated that they needed an attendant to accompany them in their last train journey and all 15 participants expressed their willingness to travel alone by train if medical care facilities to prevent exacerbations of their morbidities are provided in the railway coach during travel. All of them also stated that they would be encouraged to travel alone by train if certain structural modifications were done in railway coach taking into considerations common health concerns of the elderly.

5   Concluding Remarks

In conclusion, there will be a need to expand the capacity to respond to the growing need for ubiquitous medical care over the coming years which will arise from the aging population. At the same time, greater policy efforts may be needed to prevent or postpone as much as possible health and disability problems among elderly people. Provision of real-time anomaly diagnosis may encourage the elderly to travel independently by trains. The requisite technological upgrades have been suggested. As a future task, we would incorporate automatic detection of falls of the elderly in the railway coach through more advanced wireless sensors. This work would need multiple motion detectors which would sense any movement and if there is any change of orientation of the person in the horizontal plane so as to avoid false alarms due to sitting or climbing down.

References

1. Rao, K.V.: Morbidity, health care and condition of the aged. Chronicle of higher education, national sample survey organization ministry of statistics and program, New Delhi, Report No. 507 (60/25.0/1), (2006)
2. Alon, D., Shitrit, P., Chowers, M.: Risk behaviors and spectrum of diseases among elderly travelers: a comparison of younger and older adults. J. Travel Med. 17(4), 250–255 (2010)
3. Rodman, W., Berez, D., Moser, S.: The National mobility management initiative: State DOTS connecting specialized transportation users and rides. Boston, MA, Project No. 20-65 Task 60 (2016)
4. Dhar, H.L.: Specific problems of health in elderly. Indian J. Med. Sci. 47(12), 285–92 (1993)
5. Chatterjee, S.: Compliance of Malaria Chemoprophylaxis among travelers to india. J. Travel Med. 6(1), 7–11 (1999)
6. Chakrabarti, D.: Indian anthropometric dimensions for ergonomic design practice. National Institute of Design (1997)
7. Yeung, R., Abdullah, A.S.M., McGhee, S.M., Hedley, A.J.: Willingness to pay for preventive travel health measures among Hong Kong Chinese residents. J. Travel Med. 12, 66–71 (2005)
8. Abdullah, A.S.M., Hedley, A.J., Fielding, R.: Assessment of travel associated health risk in travelers from Hong Kong. In: Fifth International Conference on Travel Medicine, Geneva, (1997)
9. Abdullah, A.S.M., McGhee, S.M., Hedley, A.J.: Health risk during travel: a population based study amongst Hong Kong Chinese. Ann. Trop. Med. Parasitol. 95(1), 105–110 (2001)
10. Dekkiche, S., de Vallière, S., D'Acremont, V., Genton, B.: Travel-related health risks in moderately and severely immunocompromised patients: a case-control study. J. Travel Med. 23(3) (2016)
11. Kyriacos, U., Jelsma, J., Jordan, S.: Monitoring vital signs using early warning scoring systems: a review of the literature. J. Nurs. Manag. 19(3), 311–30 (2011)
12. Mirmahboub, B., Samavi, S., Karimi, N., Shirani, S.: Automatic monocular system for human fall detection based on variations in Silhouette area. IEEE Trans. Biomed. Eng. 6(2), 427–436 (2013)

A Neuro-Fuzzy Classification System Using Dynamic Clustering

Heisnam Rohen Singh, Saroj Kr Biswas and Biswajit Purkayastha

Abstract Classification task provides a deep insight into the data and helps in better understanding and effective decision-making. It is mostly associated with feature selection for better performance. Various techniques are used for classification; however, they provide poor explanation and understandability. Neuro-fuzzy techniques are most suitable for better understandability. In the neuro-fuzzy system, the features are interpreted with some linguistic form. In these existing neuro-fuzzy systems, numbers of linguistic variables are produced for each input. This leads to more computational, limited explanation, and understandability to the generated classification rules. In this, a neuro-fuzzy system is suggested for rule-based classification and the novelty lies in the way significant linguistic variables are generated, and it results in better transparency and accuracy of classification rules. The performance of the proposed system is tested with eight benchmark datasets from UCI repository.

Keywords Classification · Neural network · Fuzzy logic · Neuro-fuzzy system Clustering · Linguistic variable selection

1 Introduction

Technology is rapidly evolving and with it, the amount of data is continuously increasing. The enormous amount of data generates massive attention in the information industry, for tuning the data into useful information and knowledge. The information and knowledge extracted are appropriate for the organization to monitor and improve their business strategies, market analysis, the scientific, and medical research. How-

H. R. Singh (✉) · S. K. Biswas · B. Purkayastha
NIT, Silchar 788010, Assam, India
e-mail: rohenheisnam87@gmail.com

S. K. Biswas
e-mail: bissarojkum@yahoo.com

B. Purkayastha
e-mail: biswajit_purkayastha@hotmail.com

© Springer Nature Singapore Pte Ltd. 2019   157
M. Tanveer and R. B. Pachori (eds.), *Machine Intelligence and Signal Analysis*,
Advances in Intelligent Systems and Computing 748,
https://doi.org/10.1007/978-981-13-0923-6_14

ever, its challenge is to extract knowledge from data because of the enormous size, noise, and unstructured manner. Various data mining tasks are established to extract the hidden information and knowledge within the enormous data. One of the data mining tasks that provide a deep insight into the data and help in better understanding and effective decision-making is classification [1]. The chance of irrelevant, redundant, and noisy features or attributes increases with the increase in the dimension of data [2]. Because of the dimension, the learning method used to analyze the data might not work well. A common method for reducing the dimension is to remove the insignificant and unimportant features and retailing only the important features. This process of reducing the dimension with the capability to produce significant or better results is called as feature selection. This process not only reduces the dimension of the data but also for the computational cost improves the accuracy and understandability of the systems [3].

Various methods [4–7] are used for feature selection. This feature selection process enhanced the performance of system [8] and it might also happen to lower the performance based on the system [9].

The ANNs are suitable for extracting knowledge from data. The knowledge in ANN is represented by the values of synaptic weights. They exhibit black-box nature as they do not give any explanation and description of the underlying systems. Fuzzy systems are capable of modeling nonlinear, uncertain, and complex system, and explain these systems in an easily understandable manner using fuzzy IF-THEN rules. However, they cannot acquire knowledge directly from data. The fuzzy logic and ANN techniques are complementary, and it is advantageous to use these techniques in combination among themselves rather than exclusively as a neuro-fuzzy system [9, 10]. The neuro-fuzzy system (NFS) hybridization of them can overcome each other limitations. This hybrid system provides human-like logical reasoning using the deep learning and connection of neural network. The learning process of the underlying system in NFS is done by sophisticated layer-by-layer learning of the ANN, and the information and knowledge are represented by the IF-THEN rules of the fuzzy model. NFSs are utmost suitable for extracting and representing knowledge in a data-driven environment.

Various feature selection and classification systems using neuro-fuzzy techniques have been continually growing for better performance. In [11], the authors proposed a neuro-fuzzy system in which the feature is evaluated and indexed based on the aggregation compactness and separation of the individual classes. The membership functions were used for the evaluation of it. Li et al. [12] proposed a fuzzy neural system capable of classification of patterns and the selection of features. In this, the network selected the important features while maintaining the maximum recognition rate. In [13], a neuro-fuzzy classification system and feature reduction were proposed based on triaxial accelerometer. A modified clustering algorithm was used to construct the network structure. The feature reduction methods are done by linear discriminate analysis. In [14], a neuro-fuzzy classifier was proposed based on quantum membership functions were used for multilevel activation functions. This model also proposed a learning algorithm based on quantum fuzzy entropy. In [15, 16], neuro-fuzzy systems for classification were proposed in which the membership matrix was developed using the information of the degree of belongingness of a pattern in a class. The proposed models were divided into three steps: first is fuzzification, in which a membership matrix is generated consisting of the degree of membership for each pattern to different classes. The second step is building the MLPBPN. Finally, the defuzzification process utilized MAX operation for classification. Cetisli [17, 18] used the concept of linguistic hedges of membership functions to design a classification system. He also used the same concept for feature selection. In this model, the importance of the fuzzy sets for fuzzy rules is determined by the linguistic hedges. The linguistic hedges and other parameters of the proposed network are trained by gradient training algorithm. In [19], the authors proposed a neuro-fuzzy classifier which worked on medical big data. The proposed system is an extension of [17] in which the linguistic hedge is used to select features for dimension reduction. In this system, the feature selection is done using hedge value. The classification is done in two ways: Linguistic hedge neuro-fuzzy classifier with and without feature selection.

In [20], a classification neuro-fuzzy system proposed in this online feature selection was also done. The network learns and automatically generates the classification rules from the data. The indifferent/ bad features are removed using modulation function. The rules are easily read from the network. In [21], a neuro-fuzzy system was proposed in which significant fuzzy sets are selection for classification. It detects the void in the representation of domain and removed insignificant fuzzy sets representing the missing domain. The architecture of this is similar to the [20]. This model is capable of learning the important fuzzy sets and pruned redundant and insignificant fuzzy sets, so that the size and number of rules are reduced.

The neuro-fuzzy schemes [22, 23] were proposed which is the extension of [20, 21] and these schemes are more simplified, interpretable, and used only three linguistic variables for each input feature. However, these schemes were not a generalized way of interpreting the input. Biswas et al. [24] proposed an interpretable neuro-fuzzy classifier without losing knowledge. The proposed model extends the rule generation mechanism of [22, 23]. The classification rules are generated based on the important features which are determined by the frequencies of the top linguistic features. The work of [22, 23] is extended in [25], in which the authors proposed a generalized way of representing the features with linguistic variables. For it, a dynamic clustering algorithm based on class-based grouping was proposed. The information generated by the clustering algorithm was also utilized in the neuro-fuzzy network. However, unfortunately, this system produces numbers of linguistic variables which degrade the understandability underlying system. Napook et al. [26] further extended [25] by proposing a modified dynamic clustering algorithm. In this, the authors determined the proper significant linguistic representation of the input features. In this model, fuzzy union operation is used to improve the quality of linguistic variables generated from clustering process. The proposed method also used golden section search method to determine the number of linguistic variable in the rules.

The linguistic variables representing the features are produced with a different approach. If a feature is represented by a large number of the linguistic variables, then it ultimately hammered the understandability and increased computational complexity. The main idea to reduce and produce significant linguistic variables is by increasing the elements in each cluster produced in preprocessing phase. For this purpose, the proposed system modified the threshold of the proposed clustering algorithm. This is the novelty of the proposed work.

2 Proposed System

The crucial part of the proposed neuro-fuzzy system is representing input features with linguistic forms. It is divided into three main phases which include preprocessing, training, and rules generation. In preprocessing, the significant linguistic variables and the parameters of the neuro-fuzzy network are established. These system parameters are tuned and updated by training the network. Finally, classification rules are generated from the trained network. The different phases are shown in Fig. 1.

2.1 Preprocessing Phase

The main focus of this research is representing the input feature with significant linguistic variables for proper interpretability. The linguistic variables used to represent feature are determined in this phase. Here, a dynamic clustering algorithm (DCA) is proposed for clustering which produces much less significant linguistic variables. The detailed algorithm is given below. Suppose x_i represents the ith input feature values.

1. It begins by arranging the feature x_1 in ascending order.
2. A cluster is created and the first element in the order is assigned to it.
3. For remaining data points, the following step is repeated:

Fig. 1 Proposed system

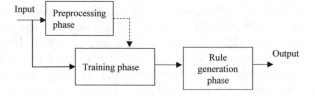

   a. If the new point belongs to the same class as the previous point, then the point is put in the same cluster of the previous point.

   b. If the class of new point is different, then the point is assigned to a new cluster.

4. A threshold (T_{x^i}) is generated for x_i which determines the existence of cluster.

$$T_{x^i} = \frac{\sum_{j=1}^{N_{x^i}} M_j}{N_{x^i}} + C_n \qquad (1)$$

where M_j, N_{x^i}, and C_n denote the number of members in cluster j, the number of clusters of x_i, and number of class, respectively.

5. The clusters whose number of elements is less than T_{x^i} are eliminated.
6. The elements of the eliminated cluster are merged to the nearest cluster. The distance is calculated as

$$d[c_j, c_k] = abs(c_j - c_k) \qquad (2)$$

where c_j: is the centroid eliminated cluster and c_k is centroids of the other clusters.

7. The parameters of each cluster, i.e., mean (c) and variance (σ), are calculated.
8. The steps 1–8 are repeated for others features.

2.2 Training Phase

The proposed neuro-fuzzy model consists of four layers—input, membership, binary transformation, and output layer as shown in Fig. 2. This layer performs a clear and compelling task.

Layer I—input layer
The node acts as buffer for the features and its number is same as number of features. The output y_i of ith node of this layer is given as

$$y_i = x_i \qquad (3)$$

where x_i is the ith input feature.

Layer II—membership layer
The memberships of the features to each linguistic variable are determined in this layer using Gaussian function, and its parameters are produced during the preprocessing phase. The membership of jth linguistic variable of input feature x_i is given by (4).

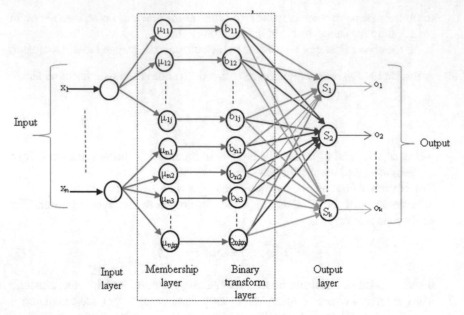

Fig. 2 The network structure

$$\mu_{ij} = \begin{cases} 0, & \text{if } \sigma_{ij} = 0 \text{ and } x_i \neq \bar{x}_{ij} \\ e^{-\left(\frac{1}{2}\frac{(x_i - \bar{x}_{ij})^2}{\sigma_{ij}^2}\right)}, & \text{if } \sigma_{ij} \neq 0 \\ 1, & \text{if } \sigma_{ij} = 0 \text{ and } x_i = \bar{x}_{ij} \end{cases} \tag{4}$$

where x_i is an input feature and $\bar{x}_{ij}$ and σ_{ij} denote the mean and standard deviation of jth cluster of ith feature.

Layer III—binary transform layer

In this layer, the linguistic variable with the maximum membership value, among all the linguistic variables from an input feature, is assigned 1 and the others are assigned 0 as shown in (5). The number of nodes is same as that of the previous layer. This layer also represents the antecedent part of the classification rules.

$$y_z = \begin{cases} 1; & \text{if } \mu_{ij} \text{ is maximum of } i\text{th feature} \\ 0; & \text{otherwise} \end{cases} \tag{5}$$

where μ_{ij} is the membership of jth linguistic variable of ith input feature.

Layer IV—output layer

This layer node represents the consequent part of the rule. Sigmoid activation function is used in this, and the output of the kth node is given by (5).

$$O_k = \frac{1}{(1 + e^{-S_k})}, \text{ where } S_k = \sum_{i=1}^{n} \sum_{j=1}^{N_i} w_{kj} b_{ij} \tag{6}$$

where n is the number of features and N_i is the total number of linguistic variables.

2.2.1 Learning

The important linguistic variables of the classification rules are represented by the link between the binary transform and output layer. During learning, these weights are adjusted by backpropagation algorithm to distinguish the linguistic variables.

The error $e_k(z)$ of some node k in output layer for iteration z is given by (7).

$$e_k(z) = D_k(z) - O_k(z) \tag{7}$$

where $D_k(z)$ is the k node desired output and $O_k(z)$ is k node actual output. The error delta value of the k node is

$$\delta_k(z) = e_k(z)(1 - O_k)O_k \tag{8}$$

The weights are updated as

$$w_{ki}(z + 1) = w_{ki}(z) + \eta \delta_k(z) b_{ij} \tag{9}$$

where η is the learning rate.

2.3 Rule Generation Phase

The neuro-fuzzy classification rules are generated based on top N linguistic variables that are selected, i.e., linguistic variables with higher values of weights are in classification rules. After selection of linguistic variables, "Simple OR" approach [22, 23] is used for rule creation. Inconsistent rules are removed to produce the final rule-based system (Tables 1, 2 and 3).

Table 1 Collection of data point of ith feature

Feature i	Class
x_{i1} (6)	c1
x_{i2} (8)	c2
x_{i3} (10)	c1
x_{i4} (11)	c2
x_{i5} (14)	c1

Table 2 Clusters generated

Feature i	Class	Cluster
x_{i1} (6)	c1	1
x_{i2} (8)	c2	2
x_{i3} (10)	c1	3
x_{i4} (11)	c2	4
x_{i5} (14)	c1	5

Table 3 Another sample of clusters

Feature i	Class	Cluster
x_{i1}(5)	c2	1
x_{i2}(6)	c2	
x_{i3}(7)	c1	2
x_{i4}(8)	c2	
x_{i5}(9)	c1	3
x_{i6}(9)	c2	

3 Brief Explanation of the Proposed System

The proposed neuro-fuzzy system is similar to [25, 26]. The main focus of this research is interpreting each input feature with significant linguistic features so that the classification rules are understandable. One major problem in the existing neuro-fuzzy systems is the linguistic variables used to represent the features. The preprocessing phase used a clustering technique to determine the proper linguistic variables. The clustering algorithm in [25] produced a number of linguistic variables to interpret the features. With the increase of it, the underlying system understandability is reduced, i.e., it becomes harder and harder for a human to understand [27]. In [26], the problem of linguistic variables is tackled using the concept of fuzzy union operation which merges linguistic variables with same means. In this proposed system, the problem of numbers of linguistic variables is tackled by increasing the elements in each cluster so that less number of clusters is produced. In [25, 26], the number of elements of each cluster is governed by the threshold. So, in this, the equation of the threshold is modified by introducing the term C_n that represents the number of classes. To explain the model considered, the collection of data points is shown in Table 5 and they are already sorted.

When assigned, the clusters shown in Table 2 are obtained.

The threshold of clusters [25, 26] gives its a value of 1, so the cluster remains as it is. In some other scenario, the following can be obtained.

Here, some of the clusters contain only elements of different classes. Such cluster is not good. So the threshold is modified such that at least two elements from a class come into each cluster.

Table 4 Linguistic variables for each feature of liver disorder dataset

Feature	No. of linguistic variables
1	17
2	29
3	31
4	56
5	51
6	23

Table 5 Selected linguistic variables

Class	Top 5				
1	50	49	173	132	44
2	134	40	145	42	34

4 An Illustrative Example

The proposed system is explained using liver disorder dataset from UCI repository. 90% and remaining 10% of total patterns are used for training and testing, respectively. The linguistic variables generated by the proposed clustering algorithm used to interpret each input feature are shown in Table 4.

Using the information generated by the clustering, neuro-fuzzy network consisting of 6, 207, 207, and 2 nodes in input, fuzzification, binary transformation, and output layer are established. After training, five most important linguistic variables are taken as shown in Table 5.

Then, the rule for classification is generated as follows:

If f3 is lv50 or f3 is lv49 or f5 is lv173 or f4 is lv132 or f2 is lv44 then Class 1
Else if f4 is lv134 or f2 is lv40 or f5 is lv145 or f2 is lv42 or f2 is lv186 then Class 2
Else no class

Here, f3 represents third feature of the dataset and lv50 represents 50th linguistic variable of all linguistic variables (or fourth linguistic variable of the third feature). The accuracy for tenfold validation is found as 64.7% for liver disorder database.

5 Experiment and Results

The proposed model is trained and tested with eight benchmark UCI datasets whose details are given in Table 6.

The comparison of the neuro-fuzzy models is conducted in terms of accuracy and other performance measures given in Table 7. In these comparisons, all parameters such as a number of the epoch, learning rate, etc. are kept same. For comparing the accuracy, tenfold cross-validation is used. In these equations, true-positive (TP) is the number of positive instances correctly predicted, false-positive (FP) is the number of positive instances incorrectly predicted, false-negative (FN) is the number

Table 6 Details of experimental datasets

Dataset	Feature type	Size	Feature	No. of class
Breast cancer	Real	680	9	2
Heart	Categorical, real	270	13	2
Ionosphere	Integer, real	350	33	2
Liver disorder	Categorical, integer, real	345	6	2
Swine flu	Integer	250	11	2
Sonar	Real	208	60	2
Iris	Real	150	4	3
Seed	Real	150	7	3

Table 7 Details of various performance measures

Precision	$\frac{TP}{TP+FP}$
FP-rate	$\frac{FP}{FP+TN}$
recall	$\frac{TP}{TP+TN}$
F-measure	$\frac{2*precision*recall}{precision+recall}$
MCC	$\frac{TP \times TN - FP \times FN}{\sqrt{(TP+FP)(TP+FN)(TN+FP)(TN+FN)}}$

of negative instances incorrectly predicted, and true-negative (TN) is the number of negative instances correctly predicted.

It is expected from a good classifier to have higher value for these performance measures except for FP-rate.

5.1 Result and Analysis

The overall comparison of accuracies of the proposed system and other systems, ENF [25] and ADCNF [26] are shown in Table 6. In this, the accuracies of the systems are shown for different linguistic variables.

Golden selection search (GSS) is also used in [26] to determine the linguistic memberships in the final classification rule. The accuracy of ADCNF with GSS [26] is shown in Table 7 for the abovementioned datasets.

The proposed system gives better accuracy than others for dataset breast cancer, heart, sonar, swine flu, and seed datasets for all linguistic variables. For ionosphere, iris, and liver disorder datasets, the proposed system produced better results than other models [25, 26] for 10 and 5 linguistic variables. The proposed system accuracy for ionosphere with 15 linguistic variables is same with ADCNF. The proposed system accuracies for liver disorder and iris, with 15 linguistic variables, are same with others models.

Fig. 3 Comparison of linguistic variables

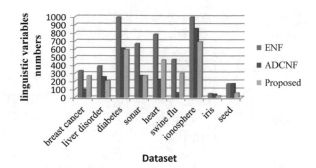

The linguistic variables generated for each feature by various models are compared and shown in Fig. 3. The linguistic variables are represented in y-axis. The linguistic variables for all datasets are significantly reduced for our model as compared to [25]; however, it is slightly higher than [26] for the datasets breast cancer, heart, and swine flu (Tables 8 and 9).

The neuro-fuzzy classifier is evaluated with different performance measures using onefold and is shown in Table 10.

The proposed system performance measures for the liver disorder, heart, ionosphere, swine flu, sonar, and seed datasets are better than [25, 26]. In the breast cancer and iris data except for precision, the proposed system is better than [25, 26]. The precision of the proposed system for breast cancer is better than [25, 26]. The precision of the proposed system for iris dataset is same with other existing systems.

6   Conclusion

In neuro-fuzzy systems, the proper interpretation of the input features in the linguistic form is essential for better understanding of the underlying system. Here, a neuro-fuzzy classification system is proposed. And the innovation lies in the way of generating the linguistic variables for each feature. The superiority of the proposed system than the existing systems is established from experimental results and analysis with 8 (Eight) datasets is taken from UCI repository. This superiority is because of the significant linguistic variables generated by the proposed system to interpret the features. The proposed system can be used for feature selection and classification task with great accuracy and understandability.

Table 8 Accuracy for classification using tenfold cross-validation

Dataset	System	Linguistic variables		
		15	10	5
Breast cancer	ADCNF [26]	88.9	90.9	90.2
	ENF [25]	92.3	90.7	88.1
	Proposed	92.9	92.8	90.3
Heart	ADCNF [26]	59.2	53.3	48.5
	ENF [25]	59.2	56.7	53.7
	Proposed	65.6	61.9	57.8
Ionosphere	ADCNF	47.4	49.4	50
	ENF	45.1	47.7	51.4
	Proposed	47.4	50.5	53.4
Liver disorder	ADCNF [26]	53.8	53.3	50.5
	ENF [25]	53.8	50.3	47.6
	Proposed	53.8	55.3	53.8
Swine Flu	ADCNF [26]	72.4	78.4	77.6
	ENF [25]	88.4	85.6	76.4
	Proposed	89.2	89.6	80
Sonar	ADCNF [26]	62.5	59.3	59.6
	ENF [25]	63.2	56.8	52.1
	Proposed	63.7	60.3	65.5
Iris	ADCNF [26]	92	90	90
	ENF [25]	92	90	87
	Proposed	92	92	94
Seed	ADCNF [26]	68	64	61
	ENF [25]	68	68	52
	Proposed	78	70	79

Table 9 Accuracy for classification using ADCNF with GSS

Dataset	ADCNF with -GSS [26]
Breast cancer	85.6
Heart	55.9
Ionosphere	38.3
Liver disorder	53.3
Swine Flu	78
Sonar	61.3
Iris	92
Seed	60

Table 10 Performances measure

Dataset	System (%)	Precision (%)	FP-rate (%)	Recall (%)	F-measure (%)	MCC
Breast cancer	ADCNF [26]	90.91	15.79	83.33	86.96	0.782
	ENF [25]	90.63	15.79	82.86	86.57	0.779
	Proposed	92.86	23.26	72.22	81.25	0.771
Liver disorder	ADCNF [26]	94.44	88.24	53.13	68.00	0.147
	ENF [25]	93.75	89.47	46.88	62.50	0.107
	Proposed	95.24	84.62	64.52	76.92	0.221
Heart	ADCNF [26]	76.47	10.00	92.86	83.87	0.582
	ENF [25]	60.00	50.00	21.43	31.58	0.130
	Proposed	80.00	45.45	28.57	42.11	0.449
Ionosphere	ADCNF [26]	88.89	88.24	51.61	65.31	0.013
	ENF [25]	89.47	87.50	54.84	68.00	0.039
	Proposed	96.30	62.50	83.87	89.66	0.478
Swine flu	ADCNF [26]	68.75	22.22	84.62	75.86	0.403
	ENF [25]	71.43	27.27	76.92	74.07	0.423
	Proposed	72.22	0.00	100.00	83.87	0.551
Sonar	ADCNF [26]	76.92	66.67	50	60.60	0.140
	ENF [25]	90.91	58.82	50	64.52	0.468
	Proposed	91.67	56.25	55	68.75	0.501
Iris	ADCNF [26]	80	20	66.67	72.73	0.632
	ENF [25]	80	40	50	61.54	0.478
	Proposed	80	10	80	80	0.7
Seed	ADCNF [26]	83.33	13.33	71.47	76.92	0.725
	ENF [25]	50	14.28	85.71	63.16	0.273
	Proposed	85.71	7.14	85.71	85.71	0.786

References

1. Guyon, I., Elisseeff, A.: An introduction to variable and feature selection. J. Mach. Learn. Res. **3**(1), 1157–1182 (2003)
2. Chang, C., Verhaegen, P.A., Duflou, J.R.: A comparison of classifiers for intelligent machine usage prediction. Intell. Environ. (IE) 198–201 (2014)
3. Kohavi, R.: A study of cross-validation and bootstrap for accuracy estimation and model selection. Int. Joint Conf. Artif. Intell. (IJCAI) 1137–1145 (1995)
4. Puch, W., Goodman, E., Pei, M., Chia-Shun, L., Hovland, P., Enbody, R.: Further research on feature selection and classification using genetic algorithm. In: International Conference on Genetic algorithm, 557–564 (1993)
5. Inza, I., Larranaga, P., Sierra, B.: Feature selection by bayesian networks: a comparison with genetic and sequential algorithm. Approx. Reason. **27**(2), 143–164 (2001)

6. Ledesma, S., Cerda, G., Avina, G., Hernandez, D., Torres, M.: Feature selection using artificial neural networks. In MICAI 2008. Adv. Artif. Intell. **5317**, 351–359 (2008)
7. Cover, T.M., Hart, P.E.: Nearest neighbor pattern classification. IEEE Trans. Inf. Theory **13**(1) (1967)
8. Ngai, E.W.T., Xiu, L.: Chau, DCK.: Application of data mining techniques in customer relationship management: a literature review and classification. Expert Syst. Appl. **36**(2), 2592–2602 (2009)
9. Jang, J.R.: ANFIS: adaptive-network-based fuzzy inference system. IEEE Trans. Syst. Man Cybern. **23**(3), 665–685 (1993)
10. Nauck, D.: Neuro-fuzzy classification studies in classification. Data Anal. Knowl. Organ. 287–294 (1998)
11. De, R.K., Basak, J., Pal, S.K.: Neuro-fuzzy feature evaluation with theoretical analysis. Neural Netw. **12**(10), 1429–1455 (1999)
12. Li, R.P., Mukaidono, M., Turksen, I.B.: A fuzzy neural network for pattern classification and feature selection. Fuzzy Sets Syst. **130**(1), 101–108 (2002)
13. Yang, J.Y., Chen, Y.P, Lee, G.Y., Liou, S.N., Wang, J.S.: Activity recognition using one triaxial accelerometer: a neuro-fuzzy classifier with feature reduction. Entertain. Comput. ICEC. **4740**, 395–400 (2007)
14. Chen, C.H., Lin, C.J., Lin, C.T.: An efficient quantum neuro-fuzzy classifier based on fuzzy entropy and compensatory operation. Soft. Comput. **12**(6), 567–583 (2008)
15. Ghosh, A., Shankar, B.U., Meher, S.K.: A novel approach to neuro-fuzzy classification. Neural Networks. **22**, 100–109 (2009)
16. Ghosh, S., Biswas, S., Sarkar, D., Sarkar, P.P.: A novel Neuro-fuzzy classification technique for data mining. Egypt. Inf. J. **15**(3), 129–147 (2014)
17. Cetisli, B.: Development of an adaptive neuro-fuzzy classifier using linguistic hedges: Part 1. Expert Syst. Appl. **37**, 6093–6101 (2010)
18. Cetisli, B.: The effect of linguistic hedges on feature selection: Part 2. Expert Syst. Appl. **37**, 6102–6108 (2010)
19. Azar, A.T., Hassanien, A.E.: Dimensionality reduction of medical big data using neural-fuzzy classifier. Soft. Comput. **19**(4), 1115–1127 (2015)
20. Chakraborty, D., Pal, N.R.: designing rule-based classifiers with on-line feature selection: a neuro-fuzzy approach. In: Pal, N.R., Sugeno, M. (eds.), AFSS. LNAI, vol. 2275, 251–259 (2002)
21. Sen, S., Pal, T.: A neuro-fuzzy scheme for integrated input fuzzy set selection and optimal fuzzy rule generation for classification. Premi LNCS **4815**, 287–294 (2007)
22. Eiamkanitchat, N., Umpon, T., Sansanee, A.: A novel neuro-fuzzy method for linguistic feature selection and rule-based classification. In: Proceedings of the in 2nd International Conference on Computer and Automation Engineering (ICCAE), pp. 247–252 (2010)
23. Eiamkanitchat, N., Umpon, T.: Colon tumor microarray classification using neural network with feature selection and rule based classification. Adv. Neural Netw. Res. Appl. **67**, 363–372 (2010)
24. Biswas, S.K., Bordoloi, M., Singh, H.R., Purkayasthaya, B.: A neuro-fuzzy rule-based classifier using important features and top linguistic features. Int. J. Intell. Inf. Technol. (IJIIT) **12**(3), 38–50 (2016)
25. Wongchomphu, P., Eiamkanitchat, N.: Enhance neuro-fuzzy system for classification using dynamic clustering. In: Proceedings of the in 4th Joint International Conference on Information and Communication Technology, Electronic and Electrical Engineering (JICTEE), pp.1–6 (2014)
26. Napook, P., Eiamkanitchat, E.: The adaptive dynamic clustering neuro-fuzzy system for classification. Inf. Sci. Appl. **339**, 721–728 (2015)
27. Jin, Y.: Fuzzy modeling of high-dimensional systems: complexity reduction and interpretability improvement. IEEE Trans. Fuzzy Syst. **8**(2), 212–221 (2000)

Evaluating the Performance of Signal Processing Techniques to Diagnose Fault in a Reciprocating Compressor Under Varying Speed Conditions

Vikas Sharma and Anand Parey

Abstract An inefficient detection of a fault in a reciprocating compressor (RC) by a signal processing technique could lead to high energy losses. To achieve a high-pressure ratio, RCs are used in such pressure-based applications. This paper evaluates the performance of nonstationary signal processing techniques employed for monitoring the health of an RC, based on its vibration signal. Acquired vibration signals have been decomposed using empirical mode decomposition (EMD) and variational mode decomposition (VMD) and compared respectively. Afterward, few condition indicators (CIs) have been evaluated from decomposed modes of vibration signals. Perspectives of this work are therefore detailed at the end of this paper.

Keywords Condition indicator · Mode decomposition · Signal processing
Reciprocating compressor

1 Introduction

Reciprocating compressors (RCs) are used in almost every industry for compressing the fluid. The detection of the failure of either suction or discharge valve is highly important. An incorrect indication of failure by signal processing techniques could result in both energy and material loss.

Vibration signals of the RC are observed to be nonstationary in nature [1, 2]. Techniques like wavelet transform (WT), empirical mode decomposition (EMD), and variational mode decomposition (VMD) are suitable for nonstationary signal processing of mechanical systems. However, In WT, the signal is decomposed using filter bands, thereby restricting to have mono-frequency components. The choice of basis function also limits WT. On the other hand, both EMD and VMD are

V. Sharma (✉) · A. Parey
Discipline of Mechanical Engineering, Indian Institute of Technology Indore, Indore, India
e-mail: s.vikasiiti@gmail.com

A. Parey
e-mail: anandp@iiti.ac.in

© Springer Nature Singapore Pte Ltd. 2019
M. Tanveer and R. B. Pachori (eds.), *Machine Intelligence and Signal Analysis*,
Advances in Intelligent Systems and Computing 748,
https://doi.org/10.1007/978-981-13-0923-6_15

specialized in decomposition and do not require any basis function. They both decompose the signal into n band components. Besides this, EMD offers some limitations like mode mixing, sensitivity to noise and sampling, boundary distortion. Furthermore, it lacks mathematical proof. Authors like Parey and Pachori [3], Sharma and Parey [4], Deng et al. [5], and Zhong and Shixiong [6] have attempted to address the problems of EMD. Methods such as ensemble empirical mode decomposition [7], and empirical wavelets [8] were partially being successful in solving such problems.

For diagnosing the faults in RC, a technique using the rational Hermite interpolation for enveloping to enhance the performance of fault diagnosis has been used [9]. Wang et al. [10] proposed an EMD coherence-based fault diagnosis approach to remove the interference components from cylinder signal. In his earlier work, an approach using autocorrelation function of intrinsic mode functions (IMFs) was used for fault diagnosis of RC [11]. A very few research articles have been found for EMD-based health monitoring of RC.

Monitoring the health of the RC based on vibration signals by various signal processing techniques other than EMD has also been observed in the literature [11–17]. Furthermore, based on vibration signals, Qin et al. [18] presented a new approach by denoising via basis pursuit, feature extraction via wave matching to diagnose different compressor valve faults. Tran et al. [19] proposed an approach by associating the Teager-Kaiser energy operator for diagnosing different compressor valve faults using vibration, pressure, and current signals. For small RC, Yang et al. [20] presented a comparison of the wavelet transform of raw noise and a vibration signal. Cui et al. [21] to diagnose valve faults using different information entropy features extracted from vibration signals as inputs. However, each of these signal processing techniques has some limitations in highlighting the health of RC.

In addition to this, condition indicators (CI) have been used to characterize the vibration signal generated because of appearing fault phenomenon. By means of these CIs, level of the fault has been exhibited [22]. CIs like root mean square (RMS), skewness, kurtosis, and entropy based on wavelet have been used to study the behavior of vibration signals of RC [15, 20, 21].

Furthermore, the application of VMD has not been explored to state the health of an RC. Nevertheless, VMD has proved its capability in fault diagnosis of gearbox [23–25], bearing [26–28], and other mechanical applications [29–31]. It has also been found that the VMD could effectively restrain the mode mixing problem of EMD. Thus to address the aforementioned gap, vibration-based health monitoring of RC has been carried out in the present article. VMD has been used to study the behavior of RC. Subsequently, a comparison of EMD and VMD has been presented.

The paper is structured as follows: the main steps of the procedure of EMD and VMD are discussed in Sect. 2. The CIs used in this study have been also mentioned in subsequent part of Sect. 2, respectively. Section 3 illustrates the description of the test facility. The experimental findings, followed by the discussions, are presented in Sect. 4. Conclusions of the present study have been addressed in Sect. 5.

2 Theoretical Background

To process the nonstationary vibration signals acquired from any machine, the most important feature is the adaptive capability of the signal processing technique. Both EMD and VMD are adaptive decompositions [32]. In this section, both the techniques are explained sequentially. Afterward, the CIs used in this study have been mentioned.

2.1 EMD

This technique does not require information about the signal periodicity, linearity, or about stationary behavior. EMD decomposes a multicomponent signal $x(t)$ into mono-component signals which are called as IMFs which are sets of the band-limited functions. The main steps of the method are briefly presented here: for a complete summary, [33] can be referred. Each IMF should satisfy two conditions:

(i) In the complete signal, the number of extrema (extreme values) and zero-crossings must either equal or at most differ by one.
(ii) At any point, the mean value of the envelope defined by local maxima and the local minima is zero.

Therefore, in accordance with the definition of EMD, a nonstationary signal $x(t)$ can be expressed as a linear sum of IMFs and the residual components. Mathematically,

$$x(t) = \sum_{n=1}^{N} IMF_n(t) + r_n(t) \tag{1}$$

where $IMF_n(t)$ symbolizes the nth intrinsic mode function and $r_n(t)$ the residual component. The EMD algorithm can be further summarized as follows:

(1) Extraction of all extrema of $x(t)$.
(2) Interpolation between minima and maxima for obtaining two envelopes $L_{min}(t)$ and $L_{max}(t)$.
(3) Compute the mean: $a(t) = (L_{max}(t) + L_{max}(t))/2$.
(4) Extract the detail $h(t) = x(t) - a(t)$.
(5) Test if $h(t)$ is an IMF using below conditions:

    (a) If it yes, procedure from the step (1) about the residual signal $f(t) = x(t) - h(t)$ is repeated.
    (b) If not, then replace $x(t)$ with $h(t)$ and repeat the procedure from step (1).

The $IMF_1(t)$ contains highest frequency component and lower frequency components will appear in subsequent IMFs in such a way that the lowest frequency component lies in $IMF_N(t)$. These frequency components change with the variation of signal $x(t)$. The residue $r_n(t)$ shows the central tendency of the signal.

2.2   Variational Mode Decomposition (VMD)

The VMD method [34] decomposes a real-valued signal $r(t)$ into K narrowband components $s_k(t)$. It also computes the center frequencies ω_k where $k = 1, 2, \ldots,$ K. In order to obtain these narrowband components and their corresponding center frequencies, this method formulates constrained optimization problem [34].

Before formulation of the optimization problem, the Hilbert transform (HT) is applied to the components $s_k(t)$ in order to compute the unilateral frequency spectrum. After that, modulation property is used to shift the frequency spectrum of these components based on the estimated center frequencies ω_k. The bandwidth was estimated through the H^1 Gaussian smoothness of the demodulated signal [34]. Now, the constrained optimization problem is formulated as [34]:

$$\min_{\{s_k\}, \{\omega_k\}} \left\{ \sum_k \left\| \partial_t \left[\left(\delta(t) + \frac{j}{\pi t} \right) * s_k(t) \right] e^{-j\omega_k t} \right\|_2^2 \right\} \tag{2}$$

such that $\sum_k s_k(t) = r(t)$.

The Lagrangian multiplier (λ) has been applied to convert the constrained optimization problem (2) into an unconstrained optimization problem. The unconstrained optimization problem can be expressed as [34]

$$L(\{s_k\}, \{\omega_k\}, \lambda) := \alpha \sum_k \left\| \partial_t \left[\left(\delta(t) + \frac{j}{\pi t} \right) * s_k(t) \right] e^{-j\omega_k t} \right\|_2^2$$

$$+ \left\| r(t) - \sum_k s_k(t) \right\|_2^2 + \left\langle \lambda(t), r(t) - \sum_k s_k(t) \right\rangle \tag{3}$$

In (3), parameter α is the penalty factor. The estimated narrowband component and corresponding center frequency during $n + 1$ iteration can be computed as follows [34]:

$$\hat{s}_k^{n+1}(\omega) = \frac{\hat{r}(\omega) - \sum_{i \neq k} \hat{s}_i(\omega) + \frac{\hat{\lambda}(\omega)}{2}}{1 + 2\alpha(\omega - \omega_k)^2} \tag{4}$$

$$\omega_k^{n+1} = \frac{\int_0^\infty \omega |\hat{s}_k(\omega)|^2 d\omega}{\int_0^\infty |\hat{s}_k(\omega)|^2 d\omega} \tag{5}$$

In (4) and (5), $\hat{r}(\omega)$, $\hat{s}_k(\omega)$, $\hat{s}_k^{n+1}(\omega)$ and $\hat{\lambda}(\omega)$ represent the Fourier transform of $\hat{r}(t)$, $\hat{s}_k(t)$, $\hat{s}_k^{n+1}(t)$ and $\hat{\lambda}(t)$, respectively. The update in λ can be expressed by the following expression [34]:

$$\hat{\lambda}^{n+1}(\omega) = \hat{\lambda}^{n}(\omega) + \tau \left[r(\omega) - \sum_k \hat{s}_k^{n+1}(\omega) \right] \quad (6)$$

Here, τ represents the dual ascent.

In VMD method, the tolerance of convergence (*tol*) parameter is used for controlling the relative errors corresponding to the estimated narrowband components. Interestingly, the expression of $\hat{s}_k^{n+1}(\omega)$ in (4) contains Wiener filter structure for denoising [34]. The $\hat{s}_k^{n+1}[n]$ can be computed from the real part of the inverse Fourier transform applied on $\hat{s}_k^{n+1}(\omega)$. The advantages of VMD technique are as follows:

(1) The relative error is largely independent of the harmonic's frequency and is controlled by tolerance level.
(2) The VMD achieves good tones separation [34].

2.3 Condition Indicators (CIs)

The brief descriptions of the CIs used in this study are mentioned as follows:

(1) Root mean square (R.M.S)—It is defined as the square root of the mean of the sum of the squares of the signal samples. It reflects the vibration amplitude and energy of the signal in time domain [35].

$$RMS = \sqrt{\frac{1}{N} \sum_{i=1}^{N} (x_i)^2} \quad (8)$$

(2) Kurtosis—It is used to reflect the presence of impulses in the signal. It is the fourth-order normalized moment of a given signal and is given by [35]

$$k = \frac{N \sum_{i=1}^{N} (x_i - \bar{x})^4}{\left(\sum_{i=1}^{N} (x_i - \bar{x})^2 \right)^2} \quad (9)$$

where x is the original sampled and I is the number of samples. A signal consisting exclusively of Gaussian distributed data will have a kurtosis of approximately 3.

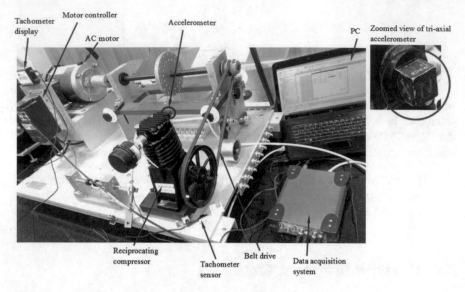

Fig. 1 Experimental test rig for RC, zoomed view of accelerometer

3 Experimental Investigations

3.1 Test Setup

To evaluate the performances of both signal processing techniques, viz., EMD and VMD, the experimental analysis has been carried out. The vibration signals were recorded from machinery fault simulator (MFS) as shown in Fig. 1. The test rig comprises motor, power transmitting belt drive, and a single stage RC. The discharge of RC connected to a tank by a hosepipe. The RC used in this study is a single-piston half-horsepower, with bore diameter and stroke length of 50.8 mm and 38.1 mm, respectively. The compressor is driven by the belt drive from the main shaft. Further, the test rig is equipped with two optical tachometers: one for the main shaft speed, and the other one for the compressor piston speed. As shown in zoomed view of Fig. 1, an accelerometer was mounted on the top of the compressor to acquire vibration signals. Figure 2, shows the suction and discharge valves of the RC.

The severity of the leakage of the RC valve has been illustrated in Fig. 3; the following severity of faults has been considered to examine the proposed approach in this study. The details of RC valve are listed in Table 1. The severity level of leakage in the discharge valve is made with the help of screw.

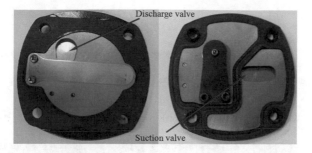

Fig. 2 Suction and discharge valves

Fig. 3 Leakages in valve fault, **a** no leak, **b** partial leak, **c** full leak

Table 1 Discharge valve specification

Type of valve	Finger type
Area of discharge valve (mm^2)	145.14
Lift of valve (mm)	6.15

3.2 Measurement Conditions

As the prime object of the proposed work deals with the performance of signal processing technique in diagnosing the severity of RC valve, signal acquisition for all the RC was performed by increasing the gap as discussed in Sect. 3.1 and Fig. 3. The vibration signals were acquired at a rate of 51.2 kHz and 10 datasets were acquired for each arrangement. The speed of the motor was fluctuated from 2200 to 2300 rpm in a random way. This makes the operating speed of RC to fluctuate from 8.40 Hz to 8.80 Hz.

4 Result and Discussions

The vibration signals have been acquired for the different levels of the leak in the exhaust of an RC compressor as shown in Fig. 4. The impulses are generated due to movement of the external valve, as the leaf of the external valve strikes the screw and generates an impulse. From Fig. 4, the amplitude of the occurred impulses found

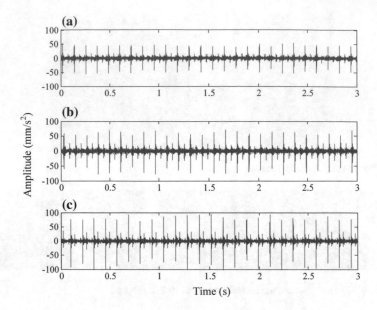

Fig. 4 Time-domain vibration signal of RC, **a** no leak, **b** partial leakage, **c** full leakage

Table 2 Performance of CIs for raw vibration signals

Valve conditions	Condition indicators	
	RMS	Kurtosis
No leak	5.7770	33.7907
Partial leak	5.2696	37.1752
Full leak	7.5126	62.4593

increased for different levels of discharge valve leaks. Further, RMS and kurtosis have been evaluated for the time-domain vibration signals as mentioned in Table 2.

Table 2 highlights that RMS fails to respond toward fault when used for raw vibration signals. On the other hand, the value of kurtosis is found to be very high. These results mislead about the health of the RC. Therefore, the time-domain vibration signals were further decomposed by both EMD and VMD. The responses of EMD and VMD have been compared to illustrate the effectiveness of the respective techniques.

The decomposition of the acquired vibration signals into IMFs has been shown in Fig. 5. Subsequently, the values of CIs have been evaluated and presented in Table 3. The behavior of RMS has been found decreasing for increasing leakage in the discharge valve of RC.

By comparing the IMFs decomposed with the EMD, it could be discovered that the main energy of the original signal was contained within the first six IMFs. Therefore, the first six IMFs have been explored to extract the fault features by means of CIs.

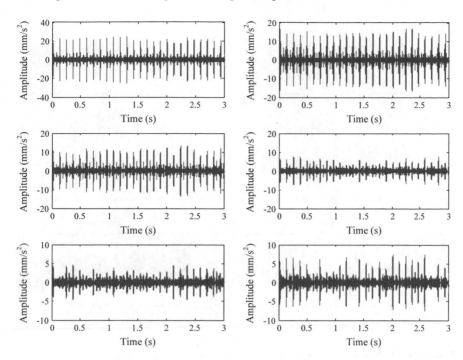

Fig. 5 First six IMFs decomposed by EMD

Table 3 Performance of CIs for IMFs decomposed by EMD

IMF	No leak		Partial leak		Full leak	
	RMS	Kurtosis	RMS	Kurtosis	RMS	Kurtosis
IMF1	3.3848	18.1290	3.1386	23.6219	3.8418	15.1290
IMF2	2.8406	10.2375	2.3584	12.7136	3.1406	7.2375
IMF3	2.0764	11.5770	2.1571	14.3324	2.7779	14.2470
IMF4	1.4305	9.5997	1.5235	10.1109	1.9905	19.1997
IMF5	1.0621	7.0330	1.1145	7.9261	2.0621	5.1330
IMF6	1.2888	8.4178	1.1003	5.3345	2.2888	10.4178

Theoretically, when there is a leakage, the leaf will have more deformation due to cantilever behavior. Because of the pulsation force of the compressed fluid on the leaf, it will make an impact during the discharge of the fluid. An increase in the amplitude of impulses must be reflected by an increase in RMS. But, from the observed readings of both RMS and kurtosis, it shows misleading results for EMD decomposed vibration signals.

To employ VMD for decomposing signals, VMD has three parameters to be set, which are the decomposing mode number K, balancing parameter of the data fidelity enforcement τ and moderate bandwidth constraint α. In this paper, K was set to be 6, α was set to the default value 2000, while τ was set as 0 to avoid distortion. The

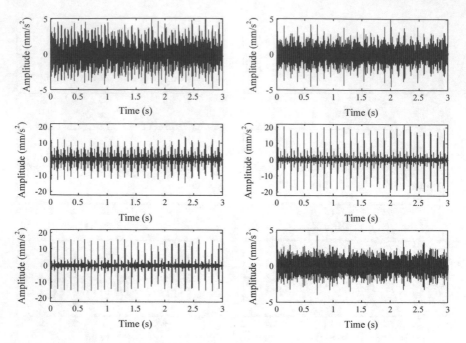

Fig. 6 First six VMFs decomposed by VMD

Table 4 Performance of CIs for IMFs decomposed by VMD

VMFs	No leak		Partial leak		Full leak	
	RMS	Kurtosis	RMS	Kurtosis	RMS	Kurtosis
VMF1	1.3570	3.2589	1.3030	12.9263	1.5120	42.8469
VMF2	0.9872	3.5032	0.9763	23.1967	1.1890	43.5499
VMF3	2.5222	4.0469	2.1215	19.2918	2.3637	43.7532
VMF4	2.5599	3.5616	2.5440	22.8995	3.1895	28.3453
VMF5	1.8411	2.9618	1.5202	11.4123	3.3227	32.6521
VMF6	0.8053	3.4178	0.8004	13.4882	1.2202	39.1796

VMFs decomposed by VMD are shown in Fig. 6 and the corresponding CIs are presented in Table 4, respectively.

It has been noticed that the kurtosis of no leak RC vibration signal is around 3. This value does not indicate fault/ presence of peaks. On the contrary, for partial leak and full leak, the value of kurtosis is found very large comparatively. Moreover, the RMS remains insensitive for both VMD and EMD decomposed components.

Figure 5 shows the first six IMFs by EMD result for the vibrations signal acquired from an RC. The mode function decomposed with VMD for the same signal is shown in Fig. 6. Comparing the decomposing results of EMD and VMD, it is obvious that EMD fails to separate the basic mode components with different timescales, while

VMD achieves better decomposition result in terms of improved performance of CIs under fluctuating speed of RC.

5 Conclusion

In this paper, VMD has been used to analyze the nonstationary signals that give information about the leakage in the discharge valve of RC under fluctuating speed conditions. The performance of VMD has been also compared to EMD.

Under fluctuating speed and presence of leakage in discharge, VMD showed efficient separation of the different modes that correspond to the variation in speed and the effect of impulses due to leak.

CIs like RMS and kurtosis when used on raw vibration signals. They fail to characterize the level of leak present in signal. A similar behavior has been noted for IMFs decomposed by EMD. However, when kurtosis computed for VMFs, i.e., the decomposition with VMD, it turns positive and detects the level of leaks.

References

1. Zhao, H., Wang, J., Han, H., Gao, Y.: A feature extraction method based on HLMD and MFE for bearing clearance fault of reciprocating compressor. Measurement **89**, 34–43 (2016)
2. Chen, G. J., Zou, L. Q., Zhao, H. Y., & Li, Y. Q. (2016). An improved local mean decomposition method and its application for fault diagnosis of reciprocating compressor. J. Vibroeng. **18**(3)
3. Parey, A., Pachori, R.B.: Variable cosine windowing of intrinsic mode functions: Application to gear fault diagnosis. Measurement **45**(3), 415–426 (2012)
4. Sharma, V., Parey, A.: Frequency domain averaging based experimental evaluation of gear fault without tachometer for fluctuating speed conditions. Mech. Syst. Signal Process. **85**, 278–295 (2017)
5. Deng, Y., Wang, W., Qian, C., Wang, Z., Dai, D.: Boundary-processing-technique in EMD method and Hilbert transform. Chin. Sci. Bull. **46**(11), 954–960 (2001)
6. Zhong, C., Shixiong, Z.: Analysis on end effects of EMD method. J. Data Acquis. Process. **1**, 025 (2003)
7. Wang, T., Zhang, M., Yu, Q., Zhang, H.: Comparing the applications of EMD and EEMD on time–frequency analysis of seismic signal. J. Appl. Geophys. **83**, 29–34 (2012)
8. Hu, J., Wang, J., Xiao, L.: A hybrid approach based on the Gaussian process with t-observation model for short-term wind speed forecasts. Renew. Energy, 670–685 (2017)
9. Li, Y., Xu, M., Wei, Y., Huang, W.: Diagnostics of reciprocating compressor fault based on a new envelope algorithm of empirical mode decomposition. J. Vibroeng. **16**(5), 2269–2286 (2014)
10. Wang, L., Zhao, J.-L., Wang, F.-T., Ma, X.-J.: Fault diagnosis of reciprocating compressor cylinder based on EMD coherence method. J. Harbin Instit. Technol. (New Series) **19**(1), 101–106 (2012)
11. Wang, L., Zhao, J.-L., Wang, F.-T., Ma, X.-J.: Fault diagnosis of reciprocating compressors valve based on cyclostationary method. J. Donghua Univ. (Engl. Edit.) **28**(4), 349–352 (2011)
12. Guerra, C.J., Kolodziej, J.R.: A Data-Driven Approach for Condition Monitoring of Reciprocating Compressor Valves. J. Eng. Gas Turbines Power **136**(4), 041601 (2014)

13. Duan, L., Wang, Y., Wang, J., Zhang, L., Chen, J.: Undecimated lifting wavelet packet transform with boundary treatment for machinery incipient fault diagnosis. Shock Vib. **2016** (2015)
14. Pichler, K., Lughofer, E., Pichler, M., Buchegger, T., Klement, E.P., Huschenbett, M.: Fault detection in reciprocating compressor valves under varying load conditions. Mech. Syst. Signal Process. **70**, 104–119 (2016)
15. Wang, J., Zhang, Y., Duan, L., Wang, X.: Multi-domain sequential signature analysis for machinery intelligent diagnosis. In: 10th International Conference on Sensing Technology (ICST), pp. 1–6. IEEE (2016)
16. Duan, L., Zhang, Y., Zhao, J., Wang, J., Wang, X., Zhao, F.: A hybrid approach of SAX and bitmap for machinery fault diagnosis. In: International Symposium on Flexible Automation (ISFA), pp. 390–396. IEEE, Aug 2016
17. Wang, Y., Gao, A., Zheng, S., Peng, X.: Experimental investigation of the fault diagnosis of typical faults in reciprocating compressor valves. Proc. Inst. Mech. Eng. Part C J. Mech. Eng. Sci. **230**(13), 2285–2299 (2016)
18. Qin, Q., Jiang, Z.N., Feng, K., He, W.: A novel scheme for fault detection of reciprocating compressor valves based on basis pursuit, wave matching and support vector machine. Measurement **45**(5), 897–908 (2012)
19. AlThobiani, F., Ball, A.: An approach to fault diagnosis of reciprocating compressor valves using Teager-Kaiser energy operator and deep belief networks. Expert Syst. Appl. **41**(9), 4113–4122 (2014)
20. Yang, B.S., Hwang, W.W., Kim, D.J., Tan, A.C.: Condition classification of small reciprocating compressor for refrigerators using artificial neural networks and support vector machines. Mech. Syst. Signal Process. **19**(2), 371–390 (2005)
21. Cui, H., Zhang, L., Kang, R., Lan, X.: Research on fault diagnosis for reciprocating compressor valve using information entropy and SVM method. J. Loss Prev. Process Ind. **22**(6), 864–867 (2009)
22. Zouari, R., Antoni, J., Ille, J.L., Sidahmed, M., Willaert, M., Watremetz, M.: Cyclostationary modelling of reciprocating compressors and application to valve fault detection. Int. J. Acoust. Vib. **12**(4), 116–124 (2007)
23. Zhao, C., Feng, Z.: Application of multi-domain sparse features for fault identification of planetary gearbox. Measurement **104**, 169–179 (2017)
24. Li, Z., Jiang, Y., Wang, X., Peng, Z.: Multi-mode separation and nonlinear feature extraction of hybrid gear failures in coal cutters using adaptive nonstationary vibration analysis. Nonlinear Dyn. **84**(1), 295–310 (2016)
25. Mahgoun, H., Chaari, F., Felkaoui, A.: Detection of gear faults in variable rotating speed using variational mode decomposition (VMD). Mech. Ind. **17**(2), 207 (2016)
26. Zhang, M., Jiang, Z., Feng, K.: Research on variational mode decomposition in rolling bearings fault diagnosis of the multistage centrifugal pump. Mech. Syst. Signal Process. **93**, 460–493 (2017)
27. An, X., Tang, Y.: Application of variational mode decomposition energy distribution to bearing fault diagnosis in a wind turbine. Trans. Inst. Meas. Control 0142331215626247 (2016)
28. Zhao, H., Li, L.: Fault diagnosis of wind turbine bearing based on variational mode decomposition and Teager energy operator. IET Renew. Power Gener. **11**(4), 453–460 (2016)
29. Liu, J., Wang, G., Zhao, T., Zhang, L.: Fault diagnosis of on-load tap-changer based on variational mode decomposition and relevance vector machine. Energies **10**(7), 946 (2017)
30. Huang, N., Chen, H., Cai, G., Fang, L., Wang, Y.: Mechanical fault diagnosis of high voltage circuit breakers based on variational mode decomposition and multi-layer classifier. Sensors **16**(11), 1887 (2016)
31. An, X., Zeng, H.: Pressure fluctuation signal analysis of a hydraulic turbine based on variational mode decomposition. Proc. Inst. Mech. Eng. Part A J. Pow. Energy **229**(8), 978–991 (2015)
32. Fu, W., Zhou, J., Zhang, Y.: Fault Diagnosis for Rolling Element Bearings with VMD Time-Frequency Analysis and SVM. In: Fifth International Conference on Instrumentation and Measurement, Computer, Communication and Control (IMCCC). IEEE, pp. 69–72 (2015)

33. Huang, N.E., Shen, Z., Long, S.R., Wu, M.C., Shih, H.H., Zheng, Q., Liu, H.H.: The empirical mode decomposition and the Hilbert spectrum for nonlinear and non-stationary time series analysis. Proc. R. Soc. Lond. Math. Phys. Eng. Sci. **454**(1971), 903–995 (1998)
34. Dragomiretskiy, K., Zosso, D.: Variational mode decomposition. IEEE Trans. Signal Process. **62**(3), 531–544 (2014)
35. Sharma, V., Parey, A.: A review of gear fault diagnosis using various condition indicators. Proc. Eng. **144**, 253–263 (2016)

CA-DE: Hybrid Algorithm Based on Cultural Algorithm and DE

Abhishek Dixit, Sushil Kumar, Millie Pant and Rohit Bansal

Abstract Optimization problems can be articulated by numerous practical problems. These glitches stance a test for the academics in the proposal of proficient procedures skilled in digging out the preeminent elucidation with the slightest computing cost. In this study, we worked on differential evolution and cultural algorithm, conglomerates the features of both the algorithms, and proposes a new evolutionary algorithm. This jointure monitors the complex collaboration amalgam of two evolutionary algorithms, where both are carried out in analogous. The novel procedure termed as CA-DE accomplishes an inclusive inhabitant that is pooled among both metaheuristics algorithms concurrently. The aspect of the recycled approval action in credence space is to update the information of the finest individuals with the present information. This collective collaboration arises among both the algorithms and is presented to mend the eminence of resolutions, ahead of the individual performance of both the algorithms. We have applied the newly proposed algorithm on a set of six standard benchmark optimization problems to evaluate the performance. The comparative results presented demonstrate that CA-DE has an encouraging accomplishment and expandable conducts while equated with new contemporary advanced algorithms.

Keywords Nature-inspired computation (NIC) · Cultural algorithm (CA) Evolutionary computation (EC) · Differential evolution (DE)

A. Dixit (✉) · S. Kumar
Amity University, Noida, India
e-mail: abhishekdixitg@gmail.com

S. Kumar
e-mail: kumarsushiliitr@gmail.com; skumar21@amity.edu

M. Pant
Indian Institute of Technology Roorkee, Roorkee, India
e-mail: millifpt@iitr.ac.in

R. Bansal
Rajiv Gandhi Institute of Petroleum Technology, Noida, India
e-mail: rohitbansaliitr@gmail.com

© Springer Nature Singapore Pte Ltd. 2019
M. Tanveer and R. B. Pachori (eds.), *Machine Intelligence and Signal Analysis*,
Advances in Intelligent Systems and Computing 748,
https://doi.org/10.1007/978-981-13-0923-6_16

1 Introduction

Various nonfictional complications are feasibly articulated by means of optimization problems. Henceforth, a substantial volume of investigation is being dedicated to find out the resolution of the comprehensive optimization problems. Optimization problems usually can be articulated as follows:

$$Minimize : f\left(\vec{X}\right), \vec{X} = \{x_1, x_2, \ldots, x_D\}, x_i \in [x_1, x_u], \tag{1}$$

where objective function f(x) to be optimized over $\vec{X}$, dimensionality of the given problem can be termed as D and x_u, x_1 of parameter x_i is designated as upper and lower bounds correspondingly.

With regard to resolve these types of problems, numerous nature-inspired approaches have been recommended for instance Evolutionary Computation (EC) [1], Genetic Programming (GA) [2], Practical Swarm Optimization (PSO) [3], Differential Evolution (DE) [4], Simulated Annealing (SA) [5], and Tabu search (TS) [6]. CA delivers an overriding instrument in lieu of the resolution of refined complications and has been efficaciously smeared to numerous optimization problems [7–11]. CA is also characterized as a model which is cultural evolutionary having two constituents, specifically a population space and belief space, accompanied by a bundle of communication conventions which match up with collaboration among the two spaces. Though a problem faced by all nature-inspired algorithms is the probable for untimely or early merging [12–15], to resolve and overcome this problem, investigators typically augment the performance of their algorithms by the mixing of novel and well-organized effective procedures which mend the evolutionary algorithm or amalgamate other evolutionary algorithms to develop a hybrid algorithm [16–19]. For instance, in [20, 21], a repeated native exploration technique is fused using CA which integrates mutual data as of predefined multiple-independent population. This information is then distributed among the subdivision of populations by definite time breaks to stake their performance and utilize that knowledge to distress the development of upcoming peer group.

Throughout the optimization progression, almost all the hybridization methods experience the dearth of steadiness among the exploration and exploitation segments. In [22], there is a proposal to an optimal narrowing hypothesis for investigation and manipulation in optimization. In this paper, the encouragement of the task performed is to deliver a novel algorithm that can utilize the strong points of individual algorithm so as to enhance the performance for resolving various optimization problems. A new hybrid algorithm is proposed in this paper that is CA-DE which incorporate the competencies of both the evolutionary algorithms: CA and the eminent DE. Accept () function in cultural algorithm chooses the finest entities which are cast off to bring up to date the certainty space information bases by means of the Update () function. Most of which, to control the evolvement of the subsequent generation of the inhabitants, the Influence () function chooses the information bases. The most important basis for exploration in cultural algorithm is topographical knowledge which is generally grown in a top-down hierarchical way. This functionality of CA can be supported in newly proposed algorithm by DE. DE established consequent to the dissemination of the whole population by arranging an identical grounds of exploratory information, and therefore brands a seamless fusion to augment the search stages throughout the optimization procedure. The performance of CA-DE algorithm is authenticated and equated with six benchmark sets of functions.

The framework of the remaining part of this paper is in this way. Segments 2 and 3 discuss both CA and DE algorithms. The proposed algorithm is explained for the hybridization CA and DE in Sect. 4. Section 5 refers to the standard problems that are used to scrutinize the proposed algorithm's effectiveness, the constraint sets. Section 6 discusses the experiment outcomes together with evaluations of the methodology by means of other contemporary algorithms. To finish, Sect. 7 précises the inferences of the proposed.

2 Cultural Algorithm

Cultural algorithm introduced by Reynolds is an extension of genetic algorithm which is an evolutionary standard model that comprises the features like certainty, populace cosmoses, and a communication network. An additional module of the CA algorithm is known as cultural information foundations which can be inherent in the confidence space and are recycled to impact the exploration of the population entities. Culture comprises the habits, knowledge, beliefs, customs, and morals of a fellow of civilization. Culture does not be existent autonomous of the environment and can interrelate with the environment via positive or negative response sequences. The accentuation of the calculation is the knowledgebase information structure that registers assorted learning assortments grounded on the idea of the issue. For example, the structure might be utilized to record the best applicant arrangement found and additionally summed up data about ranges of the pursuit space that are relied upon to result. This social learning is found by the populace-based developmental inquiry, and is thus used to impact consequent ages. The acknowledgment work compels the correspondence of information from the populace to the learning base.

Figure 1 describes the basic pseudocode of the CA. The figure shows how the process is executed in each generation. First, the $Obj()$ function evaluates individuals in the population space, and the $Accept()$ function selects the best individuals that are used to update the belief space knowledge sources using the function $Update()$. The $Influence()$ function then uses these modified knowledge sources to influence the evolution of the next generation of the population.

Advantages of CA:

- It diminishes the requirement for juvenile people to squander vitality by bypassing experimentation emphases normally required to secure data about the earth.

Fig. 1 Pseudocode of
cultural algorithm

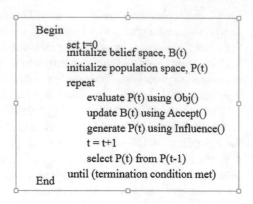

```
Begin
    set t=0
    initialize belief space, B(t)
    initialize population space, P(t)
    repeat
        evaluate P(t) using Obj()
        update B(t) using Accept()
        generate P(t) using Influence()
        t = t+1
        select P(t) from P(t-1)
    until (termination condition met)
End
```

- CA allows the broadcast of additional info as compared to any individual genome might practically encompass.
- It gives inhabitants with stability and flexibility as it is proficient of persevering further than the lifecycle of individual and it can be transferred faster.

3 Differential Evolution

DE proposed by storm and price [6] is a nature-inspired algorithm which is cast off to resolve various optimization problems established on numerous selections of mutation and crossover approaches. The major steps of DE are mutation, crossover, and selection that are repetitively enforced on to the population of individual candidate solutions. As the first step in mutation, a novel candidate is calculated, known as mutant vector through the support of the unsystematic vectors and target vectors. As a result, recombination integrates effective elucidations from the earlier generation. Donar vector components then go into the trial vector with probability CR. The last step is selection where two candidate solutions are compared and lowermost value function is confessed to subsequent generation. This information is well-matched as compared to the CAs topographic knowledge basis and can provide its effort for an improved search throughout the exploration. DE is well-thought-out to be an operative optimization algorithm and a strong procedure for providing solutions to numerous real-world problems [23]. By and itself DE can be a probable basis for topographic information which can be utilized by the CA to investigative search. Six mutation strategies were carried out in the novel DE [24] as termed underneath

$$DE/rand/1 : V_i^G = X_{r1}^G + F \times \left(X_{r2}^G - X_{r3}^G \right) \tag{2}$$

$$DE/rand/2 : V_i^G = X_{r1}^G + F \times \left(X_{r2}^G - X_{r3}^G \right) + F \times (X_{r4}^G - X_{r5}^G) \tag{3}$$

$$DE/best/1 : V_i^G = X_{best}^G + F \times \left(X_{r1}^G - X_{r2}^G \right) \tag{4}$$

$$DE/best/2 : V_i^G = X_{best}^G + F \times \left(X_{r1}^G - X_{r2}^G \right) + F \times (X_{r3}^G - X_{r4}^G) \tag{5}$$

$$DE/current - to - rand/1 : V_i^G = X_i^G + F \times \left(X_{r1}^G - X_i^G\right) + F \times (X_{r2}^G - X_{r3}^G) \quad (6)$$

$$DE/current - to - best/1 : V_i^G = X_i^G + F \times \left(X_{best}^G - X_i^G\right) + F \times (X_{r1}^G - X_{r2}^G) \quad (7)$$

where $V_i = \{v_i^1, v_i^2, \ldots, v_i^D\}$ known as the mutant vector and it is developed for every individual X_i in the populace interstellar r_1, r_2, r_3, r_4, r_5. Epitomize arbitrary and conjointly special integers created contained by the collection $[1, NP]$. Population size is represented as NP. Scaling factor is F which characterizes a real numeral for monitoring the scrabbling fraction of the target vector. X_{best} is the resolution with the finest fitness set up thus far.

A uniform crossover can be described as follows:

$$u_i = \left\{ \begin{array}{l} v_i, \; if \,(with\, probability\, of\; CR) \, or \, (j = j_{rand}) \\ x_i, \; otherwise \end{array} \right\} \quad (8)$$

where $j = 1, 2, \ldots, D$, and u_i is the developed descendants. CR $[0, 1]$ is the user-defined value known as crossover rate. CR supports the DE in governing the derivative quota of the modified vector obsessed by the sample vector.

Advantages of DE:

- Aptitude to knob nonlinear, non-differentiable, and multimodal cost functions.
- Manage parallelly the cost functions which are computation concentrated.
- Easier in implementation and understanding.
- Having very good property in convergence.

Disadvantages:

- Parameter tuning is important.
- Same parameters may not guarantee the global optimum solution.

4 Propose CA-DE

The objective of the proposed algorithm is to deliver an elementary structure for the hybridization which utilizes a vibrant value function to standardize the collaboration among both the algorithms. This benefits to decrease the likelihood of untimely conjunction and better guide the search in the direction of favorable regions. The straightforwardness of the outline ropes its prospective leeway to various different kinds of algorithms and variation in factors. The projected algorithm, CA-DE, succeeds in an all-inclusive population space that is mutual in between CA and DE at the equivalent interval. The population space P of the CA-DE algorithm at generation t is denoted as

$$X_i^g = \langle x_{i,1}^g, x_{i,2}^g, \ldots, x_{i,D}^g \rangle, \, i = 1, 2, \ldots, NP, \quad (9)$$

where NP is the population size with dimensionality D of the problem. In the beginning, the population have to shield the search space by lower and upper bounds, X_i and x_u, correspondingly as below:

$$x_{i,j} = x_{l,j} + rand(0, 1).(x_{u,j} - x_{l,j}), j = 1, 2, \ldots, D \qquad (10)$$

At the first-generation g_0 and after the initial population P_0 is evaluated, the initial participation ratio $\emptyset_T^{g_0}$ of each technique T, where $T \in [DE, CA]$, is calculated as below:

$$\emptyset_T^{g_0} = \frac{NP}{|T|} \qquad (11)$$

where $|T|$ is the number of practices cast off in the fusion procedure. In our close, we considered it as 2. Following are the main steps of proposed algorithm.

The quality function $QF_{T_k}^g$ can be well-defined for any cast-off module algorithm as

$$QF_{T_k}^g = \frac{\Delta f_{T_{best}}^{+,g} - \Delta f_{T_k}^{+,g}}{\Delta f_{T_{best}}^{+,g}} \qquad (12)$$

where $\Delta f_{T_{best}}^{+,g}$ is the variance in the volume of fitness of the best performing technique in the last two generations as computed in Eq. (13).

$$\Delta f_{T_k}^+ = \sum_{i=0}^{n_{S,T_k}^{g-1}-1} a_{f^+,T_k}^{g-1}(X_i), \qquad (13)$$

where $a_{f^+,T_k}^{g-1}(X_i)$ is the amount of fitness increment of T_k at generation $g-1$. For the current generation, every technique's participation ratio is rationalized as per the above quality function. From the results obtained by comparing the quality function of both the techniques, the best function is utilized to monitor a higher amount of individuals in the subsequent generation.

For any component algorithm, the participation ratio is defined as

$$\emptyset_{T_k}^g = QF_{T_k}^g \cdot NP \qquad (14)$$

1. Generate initial overall shared population P_0 of NP individual
2. Generate initial parameters of each technique T where $T \in [DE, CA]$,
3. Compute initial participation ratio $\emptyset_{T_k}^{g_0}$ of each subcomponent algorithm $\rightarrow \emptyset_T^{g_0} = \frac{NP}{|T|}$
4. **While** termination criteria not met **Do**

   a. Compute quality function QF_g^T of each T

Table 1 Constraint parameters

Parameter	Setting value
Dimension (D)	20
Population size (NP)	100
Number of Iterations	1000
Mutation probability	0.05
Elitism parameter	10
Habitat modification probability	1
Crossover rate	0.1
Scaling factor (F)	0.1

   b. Update participation ratio $\emptyset_g^T$ of each T according to QF_g^T
   c. **For** every technique t_k in T **Do**
       i. Generate novel entities as of in-progress population P_g using technique t_k
     ii. Calculate novel entities
    iii. Add novel entities to an secondary population
   d. **End For**
   e. Combine auxiliary populations of techniques to generate P_g
5. **End while**

5 Experiments

In command to approve the authorization of CA-DE, six benchmark functions publicized in the appendix are used. Keeping in mind the end goal to show the overwhelming nature of our expected CA-DE methodology, we compare our approach with the CA and DE algorithm. This algorithm is implemented in MATLAB and conducted in 2.10 GHz, Intel® Core™ i3-2310 M and 4 GB of RAM. Maximum iteration is considered to be 1000 as the termination criteria (Table 1).

6 Simulation Results and Analysis

The suggested CA-DE procedure is compared with cultural algorithm and classical DE optimization algorithm to assess the effectiveness of algorithm. The comparison outcomes obtained are shown in Figs. 1, 2, 3, 4, 5 and 6 on six different benchmark functions. Figures 1, 2, 3, 4, 5 and 6 present the best cost comparison on 1000 iterations. Figures clearly illustrate that accomplishment of anticipated procedure is enhanced than the CA and DE algorithm. In each of the six figures, our proposed algo-

Fig. 2 Best cost comparison on sphere function

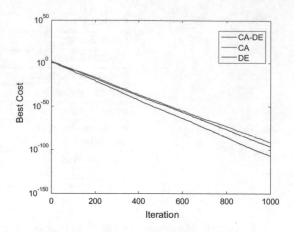

Fig. 3 Best cost comparison on quartic function

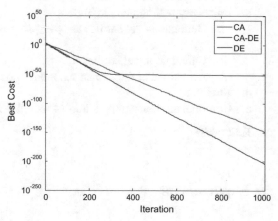

rithm outperforms with the existing CA and DE algorithm. By cautiously observing Figs. 1, 2, 3, 4, 5 and 6, it can be seen that in the start of the CA congregates quicker as compared to CA-DE, while CA-DE is competent to mend its result gradually for a lengthy course. Main purpose can be that CA is having respectable manipulation, however, deficiencies in the examination. Nevertheless, for CA-DE a new migration operator, its steadiness can affect the capability of the examination and the manipulation. Table 2 explains the statistical results obtained on six benchmark functions. We can see that CA-DE performs well in solving the problem and the algorithm has a good capability of escaping from local optima (Fig. 7).

7 Conclusion

An innovative hybrid algorithm CA-DE is introduced in this paper which amalgamate the competencies of two individual algorithms: a modified acknowledged variety of both individual algorithms. CA and the eminent DE. Accept () function in cultural

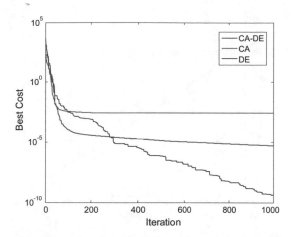

Fig. 4 Best cost comparison on Colville function

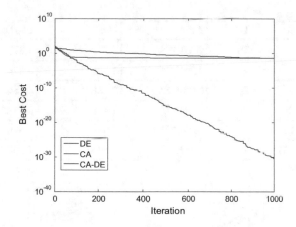

Fig. 5 Best cost comparison on Schwefel_1_2 function

algorithm chooses the finest entities which cast off to bring up to date the certainty space information bases by means of the Update () function. Most of which, to control the evolvement of the subsequent generation of the inhabitants, the Influence () function chooses the information bases. The most important basis for exploration in cultural algorithm is topographical knowledge which is generally grown in a top-down hierarchical way. This functionality of CA can be supported in newly proposed algorithm by DE. This collaboration is assisted in the succeeding way.

The efficiency of CA-DE algorithm was formerly matched with six individual CA and DE algorithms using six sets of benchmark functions. Its performance contrary to the individual CA and DE algorithms was categorized as #1 via experimental results.

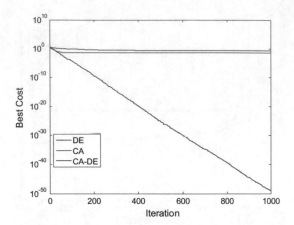

Fig. 6 Best cost comparison on Schwefel_2_21 function

Table 2 Statistical results

F	D	DE	CA	CA-DE
f1 (Sphere)	20	$7.67\mathrm{E}^{-97}$	$7.27\mathrm{E}^{-83}$	$1.37\mathrm{E}^{-107}$
f2 (Quartic)	20	$1.38\mathrm{E}^{-149}$	$1.49\mathrm{E}^{-52}$	$2.87\mathrm{E}^{-205}$
f3 (Colville)	20	$5.03\mathrm{E}^{-06}$	$2.60\mathrm{E}^{-03}$	$4.15\mathrm{E}^{-10}$
f4 (Schwefel_1_2)	20	$3.41\mathrm{E}^{-02}$	$1.43\mathrm{E}^{-02}$	$3.56\mathrm{E}^{-31}$
f5 (Schwefel_2_21)	20	$2.25\mathrm{E}^{-01}$	$3.57\mathrm{E}^{-02}$	$9.61\mathrm{E}^{-50}$
f6 (Schwefel_2_22)	20	$2.39\mathrm{E}^{-26}$	$9.54\mathrm{E}^{-07}$	$1.20\mathrm{E}^{-35}$

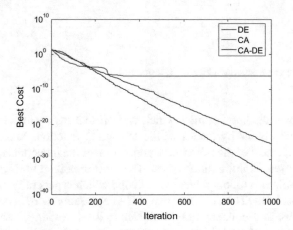

Fig. 7 Best cost comparison on Schwefel_2_22 function

References

1. Rechenberg, I.: Cybernetic solution path of an experimental problem, Royal Aircraft Establishment Library Translation, No. 1122, Aug 1965
2. Holland, J.H.: Adaption in Natural and Artificial Systems. University of Michigan Press, Ann Arbor, MI (1975)
3. Kennedy, J., Eberhart, R.: Particle swarm optimization. In: Proceedings of the IEEE International Conference on Neural Networks, vol. 4, pp. 1942–1948 (1995)
4. Storn, R., Price, K.V.: Differential evolution-a simple and efficient heuristic for global optimization over continuous spaces. J. Global Optim. **11**, 341–359 (1997)
5. Demsar, J.: Statistical comparisons of classifiers over multiple data sets. J. Mach. Learn. Res. **7**, 1–30 (2006)
6. Glover, F.: Heuristic for integer programming using surrogate constraints. Decis. Sci. **8**(1), 156–166 (1977)
7. Peng, B., Reynolds, R.G.: Cultural algorithms: knowledge learning in dynamic environments. In: Proceedings of the IEEE Congress on Evolutionary Computation, pp. 1751–1758 (2004)
8. Kim, Y., Cho, S.-B.: A hybrid cultural algorithm with local search for traveling salesman problem. IEEE International Conference on Robotics and Automation (CIRA), pp. 188–192 (2009)
9. Awad, N.H., Ali, M.Z., Suganthan, P.N., Reynolds, R.G.: CADE: a hybridization of cultural algorithm and differential evolution for numerical optimization. Inf. Sci. (2016)
10. Alia, M.Z., Awadc,N.H., Suganthanc, P.N., Reynolds, R.G., Lin, C.-J., Chen, C.-H., Lin, C.-T.: A modified cultural algorithm with a balanced performance for the differential evolution frameworks. Sci. Direct Knowl.-Based Syst. 73–86 (2016)
11. Sun, Y., Zhang, L., Gu, X.: A hybrid co-evolutionary cultural algorithm based on particle swarm optimization for solving global optimization problems. Neurocomputing **98**, 76–89 (2012)
12. Reynolds, R.G.: An introduction to cultural algorithms. In: Proceedings of the Annual Conference on Evolutionary Programming, pp. 131–139 (1994)
13. Xue, X., Yao, M., Cheng, R.: A Novel Selection Operator of Cultural Algorithm. Knowl. Eng. Manag. **123**, 71–77 (2012)
14. He, J., Xu, F.: Chaotic-search-based cultural algorithm for solving unconstrained optimization problem. Model. Simul. Eng. **2011**, 1–6 (2011)
15. Guo, Y.-N., Cheng, J., Cao, Y.-Y., Lin, Y.: A novel multi-population cultural algorithm adopting knowledge migration. Soft. Comput. **15**, 897–905 (2011)
16. Oleiwi, B.K., Roth, H., Kazem, B.I.: A hybrid approach based on ACO and GA for multi objective mobile robot path planning. Appl. Mech. Mater. **527**, 203–212 (2014)
17. Cai, Y., Wang, J.: Differential evolution with hybrid linkage crossover. Inf. Sci. **320**, 244–287 (2015)
18. Mahi, M., Baykan, Ö.K., Kodaz, H.: A new hybrid method based on particle swarm optimization, antcolony optimization and 3-opt algorithms for traveling salesman problem. Appl. Soft Comput. **30**, 484–490 (2015)
19. Das, P.K., Behera, H.S., Panigrahi, B.K.: A hybridization of an improved particle swarm optimization and gravitational search algorithm for multi-robot path planning. Swarm Evolut. Comput. (2016)
20. Nguyen, T.T., Yao, X.: An experimental study of hybridizing cultural algorithms and local search. Int. J. Neural Syst. **18**, 1–18 (2008)
21. Zheng, Y.-J.: A hybrid neuro-fuzzy network based on differential biogeography-based optimization for online population classification in earthquakes. Ling, H.-F., Chen, S.-Y., Xue, J.-Y.: IEEE Trans. Fuzzy Syst. **23**(4) (2015)
22. Chen, J., Xin, B., Peng, Z., Dou, L., Zhang, J.: Optimal contraction theorem for exploration-exploitation tradeoff in search and optimization. IEEE Trans. Syst. Man Cybern. Part A Syst. Hum. **39**, 680–691 (2009)

23. Zhang, J., Avasarala, V., Sanderson, A.C., Mullen, T.: Differential evolution for discrete optimization: AN EXPERIMENTAL Study on combinatorial auction problems. In: Proceedings of the IEEE World Congress on Computational Intelligence, Hong Kong, China, pp. 2794–2800 (2008)
24. Qin, K., Suganthan, P.N.: Self-adaptive differential evolution algorithm for numerical optimization. In: Proceedings of the IEEE Congress on Evolutionary Computation, pp. 1785–1791 (2005)

Optimal Design of Three-Band Orthogonal Wavelet Filter Bank with Stopband Energy for Identification of Epileptic Seizure EEG Signals

Dinesh Bhati, Ram Bilas Pachori and Vikram M. Gadre

Abstract We design three-band orthogonal wavelet filter bank using unconstrained minimization of stopband energies of low-pass, band-pass, and high-pass filters. The analysis polyphase matrix of the orthogonal filter bank is represented by the parameterized structures such that the regularity condition is satisfied by the designed perfect reconstruction filter bank (PRFB). Dyadic and householder factorization of the analysis polyphase matrix is employed to impose perfect reconstruction, orthogonality, and regularity order of one. Three-band orthonormal scaling and wavelet functions are generated by the cascade iterations of the regular low-pass, band-pass, and high-pass filters. The designed three-band orthogonal filter bank of length 15 is used for feature extraction and classification of seizure and seizure-free electroencephalogram (EEG) signals. The classification accuracy of 99.33% is obtained from the designed filter bank which is better than the most of the recently reported results.

Keywords Three-band filter bank · Regularity · Householder factorization
Dyadic factorization · EEG signal classification

D. Bhati (✉)
Department of Electronics and Communication Engineering, Acropolis Institute
of Technology and Research, Indore, India
e-mail: bhatidinesh@gmail.com; dineshbhati@acropolis.in

R. B. Pachori
Discipline of Electrical Engineering, Indian Institute of Technology Indore,
Indore 453552, India
e-mail: pachori@iiti.ac.in

V. M. Gadre
Department of Electrical Engineering, Indian Institute of Technology Bombay,
Mumbai, India
e-mail: vmgadre@ee.iitb.ac.in

© Springer Nature Singapore Pte Ltd. 2019
M. Tanveer and R. B. Pachori (eds.), *Machine Intelligence and Signal Analysis*,
Advances in Intelligent Systems and Computing 748,
https://doi.org/10.1007/978-981-13-0923-6_17

1 Introduction

Wavelet filter banks have been found extremely useful for time–frequency analysis of naturally occurring nonstationary signals such as speech, radar, electroencephalogram (EEG), electrocardiogram (ECG), image signals, etc. [1, 2]. Wavelet filter bank extracts the features of a given signal by decomposing it into its uncorrelated wavelet subband signals. It is well known that wavelet multiresolution analysis can be implemented using perfect reconstruction filter bank (PRFB) that satisfy certain regularity conditions. Chen et al. [3] has given sufficient conditions for the cascade iterations of a low-pass filter to converge to a regular smooth scaling function for M-band wavelet filter banks. Regularity of a filter bank is the measure of number of vanishing moments of the M-band wavelets [4], which is essential for approximation of polynomials [5]. The accuracy of polynomial approximation [6, 7] and the decay of wavelet coefficients [8] depend on the degree of regularity or the smoothness of the scaling function. As the regularity of the filter bank increases, it becomes more and more smoother, and the number of derivatives increases. The multiresolution decomposition of the signal into approximation and details is used in the applications such as feature extraction, signal interpolation, approximation, compression, and denoising [6, 7, 9].

Recently, Bhati et al. [10, 11] employed three-band wavelet filter banks for feature extraction and classification of epileptic seizure and seizure-free EEG signals. They proposed novel methods for design of time–frequency localized three-band wavelet filter banks and evaluated their performance in classification of seizure and seizure-free EEG signals. They propose novel metrics for mean and variances in time and frequency domains for a signal from the samples of its Fourier transform and use them for design of time–frequency localized three-band wavelet filter banks [12]. They have shown that time–frequency localized three-band wavelet filter banks outperform many other existing techniques for classification of seizure and seizure-free EEG signals [13]. Design of optimal PRFBs are studied by Patil et al. [14] and Vaidyanathan [15] wherein stopband and passband errors of the filters are minimized. Up to the best of our knowledge, the performance of stopband energy optimized three-band orthogonal wavelet filter bank has not been studied for classification of EEG signals. In this paper, we design optimal three-band orthogonal wavelet filter bank of length 15 by minimizing the sum of the stopband energies of the low-pass, band-pass, and high-pass filters with respect to the free parameters of the filter bank. We further use the designed filter bank for classification of seizure and seizure-free EEG signals.

The rest of the paper is organized as follows. The design of M-band orthogonal wavelet filter bank using householder and dyadic factorization is discussed in Sect. 2. In Sect. 3, An optimal three-band orthogonal wavelet filter bank is designed and used for classification of seizure and seizure-free EEG signals. Section 4 presents the conclusion of the paper.

2   Parameterized M-Band Wavelet Filter Banks

An analysis bank of M-band wavelet filter bank is used for decomposition of signal in its wavelet subbands and reconstructing the signal from its wavelet subbands using synthesis bank [10]. Let the filters of the analysis and synthesis filter bank of the canonical form are represented by $\mathbf{h}(z)$ and $\mathbf{f}(z)$ respectively, where $\mathbf{h}(z)$ and $\mathbf{f}(z)$ are given by

$$\mathbf{h}(z) = [H_0(z) \ H_1(z) \ \ldots \ H_{M-1}(z)]^T$$

$$\mathbf{f}(z) = [F_0(z) \ F_1(z) \ \ldots \ F_{M-1}(z)]^T$$

The analysis polyphase matrix and the synthesis polyphase matrix are represented by $\mathbf{E}(z)$ and $\mathbf{R}(z)$ respectively. Analysis bank filter coefficients $\mathbf{h}(z)$ can be determined using the expression given below [15]:

$$\mathbf{h}(z) = \mathbf{E}(z^M)\mathbf{e}(z) \tag{1}$$

where
$$\mathbf{e}(z) = [1 \ z^{-1} \ \ldots \ z^{-(M-1)}]^T \tag{2}$$

Perfect reconstruction (PR) condition ensures the reconstruction of the input signal within a delay. It imposes the condition that all the filters of the PRFB must be the spectral factors of the Mth band filter and paraunitary analysis and synthesis polyphase matrices [15]. Following are the equivalent conditions for an orthogonal or paraunitary filter bank.

In an M-band, orthogonal, real coefficient filter bank, it can be shown that the following conditions are equivalent [4]:

$$< h_i[k], h_j[Mn + k] >= \delta[i - j]\delta[n], \quad i, j = 0, 1 \ldots, M - 1 \tag{3}$$

$$\mathbf{R}(z) = \mathbf{E}^T(z^{-1}) \tag{4}$$

For $M = 3$ and $W^k = e^{-\frac{2\pi jk}{3}}$, it can be shown that the orthogonality condition (3) implies [15]

$$\mathbf{H}^T{}_m(z^{-1})\mathbf{H}_m(z) = 3I \tag{5}$$

where

$$\mathbf{H}_m(z) = \begin{bmatrix} H_0(z) & H_1(z) & H_2(z) \\ H_0(zW) & H_1(zW) & H_2(zW) \\ H_0(zW^2) & H_1(zW^2) & H_2(zW^2) \end{bmatrix}$$

The above expression represents the wavelet orthonormality condition [15]. The diagonal terms in Eq. (5) implies that the three filters $H_0(z)$, $H_1(z)$, and $H_2(z)$ are power complementary filters and the sum of the stopband energies of the three filters can be minimized to increase the attenuation of all the filters in their respective stopbands [15]. We use parametric forms for analysis and synthesis polyphase matrices $E(z)$ and $R(z)$ and the free parameters of the filter bank are optimized to minimize the sum of the stopband energies of the three analysis filters. Parameterized householder and dyadic factorization of polyphase matrices are employed to implement three-band orthogonal wavelet filter bank. In the subsequent subsections, we describe householder and dyadic factorizations for implementation of three-band orthogonal wavelet filter bank with regularity order of one.

2.1 Householder Factorization

A unitary matrix can be represented as product of householder matrices [15]. Let $p_0 \in \mathbb{R}^M$ such that $||p_0|| = 1$. If U represents a unitary matrix, its householder transformation $H[p_0]$ is given by [3]

$$H[p_0] = I - 2p_0 p_0^H \qquad (6)$$

The householder transformation aligns the first column of U with e_0, the desired coordinate axis [15]. The transformation can be represented by

$$H[p_0]U = \begin{bmatrix} e^{j\theta} & 0^T \\ 0 & T \end{bmatrix} \qquad (7)$$

where T represents a unitary matrix [3, 15].

2.2 Degree One Dyadic Factorization of Polyphase Matrix

Dyadic factorization of a degree-N paraunitary polyphase matrix $E(z)$ is given by [15]

$$E(z) = V_N(z)V_{N-1}(z) \ldots V_1(z)E_0 \qquad (8)$$

where E_0 and $V_N(z)$ are unitary matrix and dyadic block respectively. Let $v_m \in \mathbb{R}^M$, the degree one paraunitary and dyadic building block $V_m(z)$ is given by [15]

$$V_m(z) = I - v_m v_m^H + z^{-1} v_m v_m^H \qquad (9)$$

$$||v_m|| = 1 \qquad (10)$$

In order to impose the constraint (10) on $\mathbf{v_m}$, we choose the following parametrization [16]:

$$\mathbf{v_m}(\theta_1, \theta_2) = [\cos(\theta_1)\cos(\theta_2) \quad \cos(\theta_1)\sin(\theta_2) \quad \sin(\theta_1)]^T$$

2.3 Regularity

An M-band paraunitary filter bank (PUFB) is said to be K-regular if its scaling low-pass filter has a zero of multiplicity K at the Mth roots of unity $e^{\pm \frac{j2\pi m}{M}}$ for $m = 1, 2 \ldots, M-1$ [8]. Equivalently, an analysis polyphase matrix $\mathbf{E}(z)$ satisfies the condition of regularity order of one, if and only if the first row of the unitary matrix $\mathbf{E_0}$ has identical elements or $\mathbf{E_0}$ represents a householder matrix [3]. Thus problem reduces to find $\mathbf{p_0}$ such that $\mathbf{E_0} = \mathbf{H}[\mathbf{p_0}]$ is having regularity order of one. It can be shown that $\mathbf{p_0} = [a, b, b]^T$ imposes regularity order of one on to the matrix $\mathbf{E_0} = \mathbf{H}[\mathbf{p_0}]$. The parameters a and b are given by [3]

$$a = \sqrt{(\sqrt{M} - s)/(2\sqrt{M})} \tag{11}$$

$$b = -s/\sqrt{2(M - s\sqrt{M})} \tag{12}$$

where s can be 1 or -1. Further, it can be shown that

$$\mathbf{F_0 1_M} = [c_0 \ 0 \ 0]^T \tag{13}$$

where $c_0 = s\sqrt{M}$ [3]. For $s = 1$ and $M = 3$ the matrix $\mathbf{E_0}$ is given by [3]

$$\mathbf{E_0} = \begin{bmatrix} -0.5773 & -0.5773 & -0.5773 \\ -0.5773 & 0.7886 & -0.2113 \\ -0.5773 & -0.2113 & 0.7886 \end{bmatrix} \tag{14}$$

Note that regularity order of degree one of the polyphase matrix $\mathbf{E}(z)$ is completely determined by the unitary matrix $\mathbf{E_0}$ irrespective of the filter length and the McMillan degree N [6].

3 Results and Discussion

The paraunitary analysis polyphase matrix $\mathbf{E}(z)$ ensures that analysis filters are power complementary. The parameters of the filter bank are optimized to minimize the sum of the stopband energies C_{stop} of the analysis low-pass, high-pass, and band-pass filters. The sum of stopband energies C_{stop} is given by [15]

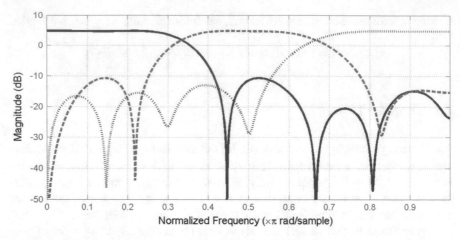

Fig. 1 Frequency response of the designed three-band filter bank

$$C_{\text{stop}} = \int_{\pi/3}^{\pi} |H_0(e^{j\omega})|^2 d\omega + \int_{0}^{\pi/3} |H_1(e^{j\omega})|^2$$

$$+ \int_{2\pi/3}^{\pi} |H_1(e^{j\omega})|^2 d\omega + \int_{0}^{2\pi/3} |H_2(e^{j\omega})|^2 d\omega \quad (15)$$

Starting from the initial guess for the free parameters, in each optimization iteration, the eight parameters [12] of the filter bank are optimized to minimize C_{stop} [17]. Several iterations of the optimization are run to obtain the sufficiently small value of the cost function C_{stop}. Table 1 shows the filter coefficients of the filters of the designed frequency response optimal filter bank. Figure 1 shows the frequency response of the filters. The scaling and wavelet functions are shown in Fig. 2. Eight cascade iterations are used to generate the scaling and wavelet functions.

The performance of the designed filter bank is evaluated in classification of seizure and seizure-free EEG signals. We have used seizure and seizure-free EEG signals from the Bonn university dataset [18]. The database contains 100 seizure signals represented by E class, 200 seizure-free EEG signals represented by C and D classes and 200 signals collected from healthy patients represented by A and B classes. In this paper, we have taken 100 seizure signals and 200 seizure-free EEG signals to evaluate the classification performance of the designed filter bank. The seven subband signals for each EEG signal is computed using the designed three-band filter bank at the third level of wavelet decomposition [10]. The subband norm of each subband signal is computed and a set of subband norms is used to form a feature vector to represent the given EEG signal [10]. The feature vectors for all the EEG signals from both the classes are computed. Multilayer perceptron neural network (MLPNN) [10] is used for classification of seizure and seizure-free EEG signals. Tenfold cross-validation [10, 19–21] is used to compute the classification accuracy of the designed filter bank. The classification accuracy obtained is 99.33%.

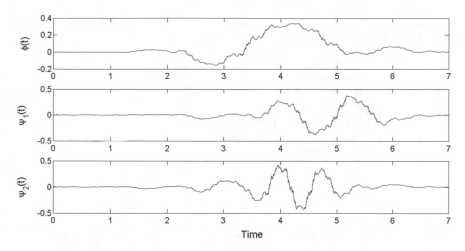

Fig. 2 Scaling and wavelet functions generated from designed three-band filter bank

Table 1 Filter coefficients of the designed three-band filter bank

$h_0[n]$	$h_1[n]$	$h_2[n]$
−0.016725530731057	−0.001253281108132	0.018828688837509
−0.043272317657348	−0.003242490722542	0.048713611397403
0.013708395221402	0.001027200453609	−0.015432162496709
0.093414073525234	0.045196454379951	0.031131021896882
0.252214888131053	0.117721658600813	0.068683781396342
0.075296605985903	−0.025664241910253	−0.196470529479625
−0.223843382240440	−0.029599310668852	0.129172243926326
−0.514319758584217	−0.095779736949852	0.175399763398251
−0.564570551784084	−0.378599620042335	−0.501234094590463
−0.469604882809573	0.276275122637543	0.607946293753909
−0.221205916697883	0.514958120390501	−0.476000092704774
0.010385146676252	−0.366690822590486	0.222608661914611
0.039409453066210	−0.501943850645696	0.001596886180187
−0.050767164381231	0.255017583275893	−0.028121928892408
−0.112169865289099	0.192577214899839	−0.086822144537441

Table 2 shows the classification accuracy obtained from various other methods for classification of seizure and seizure-free EEG signals. It shows that the designed filter bank performs better than most of the existing methods for EEG signal classification.

Bhati et al. [10] have shown that time–frequency localized three-band wavelet filter banks classify seizure and seizure-free EEG signals with classification accuracy

Table 2 Classification accuracy (%) obtained from the designed three-band filter bank and other methods for classification of seizure and seizure-free EEG signals

S.No.	Signal transformation techniques and feature extraction methods	Classifier used	Classification accuracy (%)	Year, Reference
1	Power spectral density and Higuchi fractal dimension, Hjorth parameters, and Petrosian fractal dimension	Probabilistic neural network	97	2008, [22]
2	Linear prediction	Energy of error signal	94	2010, [23]
3	Modeling error energy and fractional linear prediction	Least squares support vector machine	95.33	2014, [24]
4	95% Second-order difference plot (SODP) of intrinsic mode functions (IMFs) and 95% confidence ellipse area obtained from the SODP	Multilayer perceptron neural network	97.75	2014, [25]
5	Phase space representation (PSR) of IMFs and combination of 95% confidence ellipse area obtained from 2D PSR	Least squares support vector machine	98.67	2014, [26]
6	Histogram matching scores obtained from local binary patterns	Classifier based on nearest neighbor criterion	98.33	2015, [27]

(continued)

Table 2 (continued)

S.No.	Signal transformation techniques and feature extraction methods	Classifier used	Classification accuracy (%)	Year, Reference
7	Histogram of key point-based local binary pattern	Least squares support vector machine	99.45	2016, [28]
8	Subband norms obtained from time–frequency localized three-band synthesis filter bank	Multilayer perceptron neural network	99.33	2016, [10]
9	Tunable-Q wavelet transform and multi-scale entropy measure	Least squares support vector machine	99.5	2017, [29]
10	Fractal dimension of subband signals of analytic time–frequency flexible wavelet transform	Least squares support vector machine	98.67	2017, [30]
11	Frequency response optimized three-band orthogonal wavelet filter bank and subband norm	Multilayer perceptron neural network	99.33	This work

of 99.33%. In this paper, we have shown that even stopband energy optimal three-band wavelet filter bank of length 15 performs equivalent to time–frequency localized three-band wavelet filter banks in classifying seizure and seizure-free EEG signals with classification accuracy of 99.33%.

4 Conclusion

We designed optimal three-band orthogonal wavelet filter bank by minimizing stopband energies of low-pass, band-pass, and high-pass filters. Orthogonality and regularity order of one are imposed using householder and dyadic factorization of analysis polyphase matrix. The designed filter bank of length 15 is used for classification of seizure and seizure-free EEG signals. The designed frequency response optimal filter bank classify seizure and seizure-free EEG signals with classification accuracy of 99.33% which is better than the most of the recently reported results. The classification method proposed in the paper should be studied on huge database with EEG recordings of long duration before it is used for clinical purposes.

References

1. Cvetkovic, D., Übeyli, E.D., Cosic, I.: Wavelet transform feature extraction from human PPG, ECG, and EEG signal responses to ELF PEMF exposures: a pilot study. Digit. Signal Process. **18**(5), 861–874 (2008)
2. Lee, D.T., Yamamoto, A.: Wavelet analysis: theory and applications. Hewlett Packard J. **45**, 44–44 (1994)
3. Chen, Y.-J., Oraintara, S., Amaratunga, K.S.: Dyadic-based factorizations for regular paraunitary filterbanks and M-band orthogonal wavelets with structural vanishing moments. IEEE Trans. Signal Process. **53**(1), 193–207 (2005)
4. Vetterli, M., Herley, C.: Wavelets and filter banks: theory and design. IEEE Trans. Signal Process. **40**(9), 2207–2232 (1992)
5. Strang, G., Nguyen, T.: Wavelets and Filter Banks. Wellesley Cambridge Press, Wellesley (1996)
6. Oraintara, S.: Regular linear phase perfect reconstruction filterbanks for image compression. Ph.D. dissertation, Boston Univ., Boston, MA (2000)
7. Mallat, S.: A Wavelet Tour of Signal Processing, vol. 7. Academic, New York (1999)
8. Steffen, P., Heller, P.N., Gopinath, R.A., Burrus, C.S.: Theory of regular M-band wavelet bases. IEEE Trans. Signal Process. **41**(12), 3497–3511 (1993)
9. Akansu, A., Haddad, R.: Multiresolution Signal Decomposition: Transforms, Subbands and Wavelets. Academic, New York (2001)
10. Bhati, D., Sharma, M., Pachori, R.B., Gadre, V.M.: Time-frequency localized three-band biorthogonal wavelet filter bank using semidefinite relaxation and nonlinear least squares with epileptic seizure EEG signal classification. Digit. Signal Process. **62**, 259–273 (2017)
11. Bhati, D., Pachori, R.B., Gadre, V.M.: A novel approach for time-frequency localization of scaling functions and design of three-band biorthogonal linear phase wavelet filter banks. Digit. Signal Process. **69**, 309–322 (2017)
12. Bhati, D., Sharma, M., Pachori, R.B., Nair, S.S., Gadre, V.M.: Design of time-frequency optimal three-band wavelet filter banks with unit Sobolev regularity using frequency domain sampling. Circuits Syst. Signal Process. **35**(12), 4501–4531 (2016)
13. Pachori, R.B., Sircar, P.: EEG signal analysis using FB expansion and second-order linear TVAR process. Signal Process. **88**(2), 415–420 (2008)
14. Patil, B.D., Patwardhan, P.G., Gadre, V.M.: Eigenfilter approach to the design of one-dimensional and multidimensional two-channel linear-phase fir perfect reconstruction filter banks. IEEE Trans. Circuits Syst. I Regul. Pap. **55**(11), 3542–3551 (2008)
15. Vaidyanathan, P.P.: Multirate Systems and Filter Banks. Pearson Education, India (1993)
16. Lizhong, P., Wang, Y.: Parameterization and algebraic structure of 3-band orthogonal wavelet systems. Sci. China Ser. A Math. (Springer) **44**, 1531–1543 (2001)
17. Nocedal, J., Wright, S.: Numerical Optimization. Springer Science & Business Media, New York (2006)
18. Andrzejak, R.G., Lehnertz, K., Mormann, F., Rieke, C., David, P., Elger, C.E.: Indications of nonlinear deterministic and finite-dimensional structures in time series of brain electrical activity: dependence on recording region and brain state. Phys. Rev. E **64**(6), 061907 (2001)

19. Pachori, R.B.: Discrimination between ictal and seizure-free EEG signals using empirical mode decomposition. Res. Lett. Signal Process. **2008**, Article ID 293056, 1–5 (2008)
20. Bhattacharyya, A., Gupta, V., Pachori, R.B.: Automated identification of epileptic seizure EEG signals using empirical wavelet transform based Hilbert marginal spectrum. In: 22nd International Conference on Digital Signal Processing, London, UK (2017)
21. Sharma, R.R., Pachori, R.B.: Time-frequency representation using IEVDHM-HT with application to classification of epileptic EEG signals. IET Sci. Meas. Technol. **12**, 72–82 (2017)
22. Bao, F.S., Lie, D.Y.-C., Zhang, Y.: A new approach to automated epileptic diagnosis using EEG and probabilistic neural network. In: 20th IEEE International Conference on Tools with Artificial Intelligence, vol. 2, pp. 482–486 (2008)
23. Altunay, S., Telatar, Z., Erogul, O.: Epileptic EEG detection using the linear prediction error energy. Expert Syst. Appl. **37**(8), 5661–5665 (2010)
24. Joshi, V., Pachori, R.B., Vijesh, A.: Classification of ictal and seizure-free EEG signals using fractional linear prediction. Biomed. Signal Process. Control. **9**, 1–5 (2014)
25. Pachori, R.B., Patidar, S.: Epileptic seizure classification in EEG signals using second-order difference plot of intrinsic mode functions. Comput. Methods Programs Biomed. **113**(2), 494–502 (2014)
26. Sharma, R., Pachori, R.B.: Classification of epileptic seizures in EEG signals based on phase space representation of intrinsic mode functions. Expert Syst. Appl. **42**(3), 1106–1117 (2015)
27. Kumar, T.S., Kanhangad, V., Pachori, R.B.: Classification of seizure and seizure-free EEG signals using local binary patterns. Biomed. Signal Process. Control. **15**, 33–40 (2015)
28. Tiwari, A.K., Pachori, R.B., Kanhangad, V., Panigrahi, B.K.: Automated diagnosis of epilepsy using key-point-based local binary pattern of EEG signals. IEEE J. Biomed. Health Informatics **21**(4), 888–896 (2017)
29. Bhattacharyya, A., Pachori, R.B., Upadhyay, A., Acharya, U.R.: Tunable-Q wavelet transform based multiscale entropy measure for automated classification of epileptic EEG signals. Appl. Sci. **7**(4), 385 (2017)
30. Sharma, M., Pachori, R.B., Acharya, U.R.: A new approach to characterize epileptic seizures using analytic time-frequency flexible wavelet transform and fractal dimension. Pattern Recognit. Lett. **94**, 172–179 (2017)

Identification of Epileptic Seizures from Scalp EEG Signals Based on TQWT

Abhijit Bhattacharyya, Lokesh Singh and Ram Bilas Pachori

Abstract In this work, we propose a method for epileptic seizure detection from scalp electroencephalogram (EEG) signals. The proposed method is based on the application of tunable-Q wavelet transform (TQWT). The long duration scalp EEG signals have been segmented into one-second duration segments using a moving window-based scheme. After that, TQWT has been applied in order to decompose scalp EEG signals segments into multiple sub-band signals of different oscillatory levels. We have generated two-dimensional (2D) reconstructed phase space (RPS) plot of each of the sub-band signals. Further, the central tendency measure (CTM) has been applied in order to measure the area of the 2D-RPS plots. These computed area measures have been used as features for distinguishing seizure and seizure-free EEG signal segments. Finally, we have used a feature-processing technique which clearly discriminates epileptic seizures in the scalp EEG signals. The proposed method may also find application in the online detection of epileptic seizures from intracranial EEG signals.

Keywords TQWT · RPS · CTM · Scalp EEG signal · Epileptic seizure detection

A. Bhattacharyya (✉) · L. Singh · R. B. Pachori
Discipline of Electrical Engineering, Indian Institute of Technology Indore, Indore, India
e-mail: phd1401202001@iiti.ac.in

L. Singh
e-mail: mt1602102005@iiti.ac.in

R. B. Pachori
e-mail: pachori@iiti.ac.in

© Springer Nature Singapore Pte Ltd. 2019
M. Tanveer and R. B. Pachori (eds.), *Machine Intelligence and Signal Analysis*,
Advances in Intelligent Systems and Computing 748,
https://doi.org/10.1007/978-981-13-0923-6_18

1 Introduction

Epilepsy is caused by irregular synchronous firing of a cluster of neurons inside the human brain. It affects nearly 60 million people in the world [25]. The electroencephalogram (EEG) signals have been found useful for detection and diagnosis of epileptic seizure events. However, the manual monitoring of hours long EEG signals is a tedious and time-consuming task which may be prone to human error. In addition, the background noise and artifacts significantly degrade the recorded EEG signal quality. Thus, automatic and real-time seizure detection system may facilitate epileptologists in their diagnosis [2, 21].

In the literature, there exist several methods for the efficient detection of epileptic seizure events from long duration EEG signals. In [14], a seizure detection method was presented using wavelet-based band-pass finite-impulse response filter. The EEG signals were used in order to generate foreground sequence by determining the median value of squared EEG signal epochs. Then, the foreground to its background ratio was computed and comparison was made with predefined threshold value for the detection of EEG seizures. The authors obtained 100% sensitivity and specificity when evaluating their method in a short intracranial EEG signal database. The temporal pattern (TP) filter was utilized for the detection of EEG seizure onset patterns in [13]. The authors developed TP filter by computing the weighted sum of the relevant TPs which were chosen from both preictal and ictal durations of patients EEG signals. The abstruse seizure patterns from scalp EEG signals were found by employing the built TP filter. In [8], the time–frequency representation of the selected reference seizure segments were used in order to train the self-organizing map (SOM) based neural network. After that, test feature vectors were compared with reference SOM neural network vectors and the epileptic seizures were detected with rule-based decision. The method presented in [8] was studied on a vast EEG database in [7] and authors achieved 92.8 % sensitivity with 1.35 per hour false detection rate. In [22], an EEG seizure detector was designed and authors evaluated the performance of their method on Children's Hospital Boston-Massachusetts Institute of Technology (CHB-MIT) scalp EEG database. The two seconds duration EEG signals epochs were passed through a filter bank in the frequency range of 0.5–25 Hz with eight filters and energy features were computed from the output signal of each filter. The method was able to detect 96% of the tested seizures with a median false detection rate of two in 24-hour duration EEG recording. The time, frequency, and time–frequency domain features and nonlinear features were used in order to detect epileptic seizure in [12]. They designed a binary classifiers network, followed by morphological filtering, and the average sensitivity of 89.01% was achieved with 25% training data. In [26], authors developed an automatic seizure detection method based on projection of seizure-free and seizure EEG signals epochs in high dimensional space. They reported 88.27% sensitivity with 25% training data, using double-layered classifier architecture. The scaled versions of EEG signals and their sub-bands were mapped in two-dimensional (2D) space in order to generate texture image in [16]. The gray level co-occurrence matrix was applied for image texture analysis and extracted multivariate textural features in order to classify seizure-free and seizure EEG signals epochs. In this work, 70.19% average sensitivity was obtained with support vector machine classifier. In [2], authors proposed a multivariate technique based on empirical wavelet transform (EWT) for patient-specific epileptic seizure detection. They extracted three distinct features from the joint amplitude envelope of multichannel scalp EEG signals and processed the features with a novel feature-processing technique. They reported average sensitivity of 97.91% with random forest (RF) classifier in classifying seizure-free and seizure EEG signal segments. In a recent work [3], authors proposed an EEG seizure detection method based on EWT-based Hilbert marginal spectrum (HMS). They computed energy and entropy features from the HMSs of different oscillatory levels in order to classify seizure and seizure-free EEG signals using RF classifier. The method was studied on a short intracranial EEG database and obtained 100% sensitivity rate with 50% training rate.

In this work, we have proposed a new method for epileptic seizure detection from EEG signals based on tunable-Q wavelet transform (TQWT). The TQWT was applied in [4, 5, 10, 11] for the analysis of epileptic EEG signals. A single channel out of 23 available channels has been selected in this work for EEG seizure detection. We have applied a rectangular moving window of one-second duration in order to obtain nonoverlapping segments of the long duration EEG signals.

After that, TQWT has been applied in order to decompose EEG signals segments into sub-bands. Then, the obtained sub-bands have been projected into 2D reconstructed phase space (RPS) and further, the area span of the 2D RPS plots has been computed using central tendency measure (CTM). The computed area features have been processed using a feature-processing method in order to clearly discriminate epileptic seizure event in long-duration EEG signals. Figure 1 presents the block

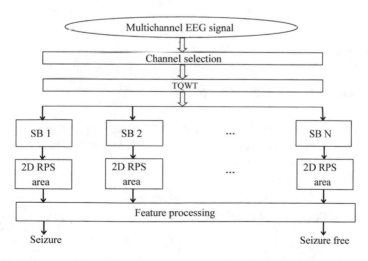

Fig. 1 Block diagram of TQWT-based epileptic seizure detection method

diagram of the proposed EEG seizure detection method based on TQWT. In the figure, the sub-band signals are denoted using SB 1, SB 2, and so on.

The rest of the paper is organized as follows: Sect. 2 discusses the database used in this work and channel selection procedure, Sect. 3 discusses the TQWT method and 2D RPS area computation. The obtained results are presented and discussed in Sect. 4. Finally, Sect. 5 concludes the work.

2 Database Used and Channel Selection

The scalp EEG signals used in this study are available in the PhysioNet CHB-MIT database [9, 23]. The EEG records were acquired from 23 pediatric patients. The female patients were of age between 1.5 and 19 years and male patients belonged from the age group of 3–22 years. The EEG signals have sampling rate of 256 Hz with 16-bit resolution. The 23 common channels were used (FP1-F7, F7-T7, T7-P7, P7-O1, FP1-F3, F3-C3, C3-P3, P3-O1, FP2-F4, F4-C4, C4-P4, P4-O2, FP2-F8, F8-T8, T8-P8, P8-O2, FZ-CZ, CZ-PZ, P7-T7, T7-FT9, FT9-FT10, FT10-T8, and T8-P8) with international 10–20 system electrode positioning protocol for recording the EEG signals. In the current study, we have considered 6 out of 23 patients. For each of the considered patients, two long-duration EEG records where seizure events occurred at least once have been considered. The considered patients' information and their EEG signals details have been presented in Table 1. We have selected the channel with minimum standard deviation from the available list of 23 channels. The channel having minimum standard deviation was found useful in [2] for efficient detection of epileptic seizures in EEG signals.

3 Method

3.1 Tunable-Q Wavelet Transform

The TQWT uses the concept of two-channel filter bank operation [18]. The frequency response corresponding to low-pass filter $\Gamma_0(\omega)$ and high-filter $\Gamma_1(\omega)$ are mathematically expressed as follows [18]:

$$\Gamma_0(\omega) = \begin{cases} 1, & \text{if } |\omega| \leq (1-\delta)\pi, \\ \Theta\left(\frac{\omega+(\delta-1)\pi}{\gamma+\delta-1}\right), & \text{if } (1-\delta)\pi < |\omega| < \gamma\pi, \\ 0, & \text{if } \gamma\pi \leq |\omega| \leq \pi. \end{cases} \tag{1}$$

$$\Gamma_1(\omega) = \begin{cases} 0, & \text{if } |\omega| \leq (1-\delta)\pi, \\ \Theta\left(\frac{\gamma\pi - \omega}{\gamma + \delta - 1}\right), & \text{if } (1-\delta)\pi < |\omega| < \gamma\pi, \\ 1, & \text{if } \gamma\pi \leq |\omega| \leq \pi, \end{cases} \tag{2}$$

where δ and γ denote the high-pass and low-pass scale factors, respectively, of the two-channel filter bank. The $\Theta(\omega)$ denotes the Daubechies filter frequency response. It should be noted that high-pass scale factor $(0 < \delta \leq 1)$ and low-pass scale factor $(0 < \gamma < 1)$ satisfy the condition $(\gamma + \delta) > 1$. The Quality factor (Q), maximum number of sub-bands (J_{max}) and redundancy parameter (R) are mathematically expressed as [18],

$$Q = \frac{2-\delta}{\delta}; \quad J_{max} = \left\lceil \frac{\log(\delta N/8)}{\log(1/\gamma)} \right\rceil, \quad R = \frac{\delta}{1-\gamma}; \tag{3}$$

where N is the length of the analyzed signal. The higher value of Q factor is desirable for the analysis of oscillatory signals, where lower Q value suitable for analyzing signals with transient nature. The redundancy parameter R localizes the wavelet in the time domain by maintaining its shape. In this work, R has been fixed to three and total nine sub-bands $(J = 8)$ have been extracted from each EEG signal segment. We have segmented long duration scalp EEG signals into one-second duration segments and each of the EEG signals segments have been decomposed into sub-bands using TQWT.

3.2 Area Computation from Reconstructed Phase Space

In this work, we have plotted the RPS of the sub-band signals obtained after decomposition with TQWT for the detection of EEG seizures. Let $x[p]$ is the analyzed signal where $p = 1, 2, \ldots P$, a reconstruction has been made by using time-delay vector $X[n]$ determined from the scalar quantity $x[p]$. In the multidimensional phase space, the vector $X[n]$ has been represented by delayed versions of $x[p]$ which is expressed as follows [6, 20]:

$$X[n] = \{x[n], x[n+\tau], \ldots, x[n+(d-1)\tau]\}, \tag{4}$$

where $n = 1, 2, \ldots, P - (d-1)\tau$.

$X[n]$ represents the point on trajectory of the phase space at time n; d denotes the embedding dimension which is the count of phase space coordinates, τ is the reconstruction time delay. In this work, we have fixed the value τ to 1 as considered in [24], and the value of d has been fixed to two.

In this work, we have measured the area span of the sub-band signals using CTM. The concept of CTM has been used in previous studies for epileptic seizure detection in [1], for the detection of bend generated error in [17]. The CTM denotes the ratio

Table 1 Information of considered patients and their EEG signals

Patient index	Age-gender	Number of seizure events	Seizure and seizure-free durations in seconds
1	11-Female	2	82 and 7118
2	11-Male	2	93 and 4467
7	14.5-Female	2	183 and 17943
17	12-Female	2	207 and 6993
19	19-Female	2	156 and 7056
24	NR-NR	3	125 and 7075

of points lying inside the circle with radius r to the total number of points of the 2D RPS projection. Let $x[k]$ has total P number of points and r is considered as the radius of the circle, then CTM is mathematically expressed as [6, 15, 19]

$$\text{CTM} = \frac{\sum_{i=1}^{P_1} s[c_i]}{P_1} \tag{5}$$

$$s[c_i] = \begin{cases} 1, & \text{if } \left((x[i])^2 + (x[i + \tau])^2\right)^{\frac{1}{2}} < r \\ 0, & \text{otherwise,} \end{cases} \tag{6}$$

where $P_1 = P - (d - 1)\tau$, $1 \leq i \leq P_1$, and $P_1 \leq P$.

We have computed the area πr^2 as feature from the 2D RPS plots for the discrimination of seizure-free and seizure EEG signals. After this, the computed area features have been passed through a feature-processing step (combination of moving average filter and Hadamard product) which was proposed in [2]. In this work, we have considered 100 s duration for moving average filter.

4   Results and Discussion

In this section, we have presented the effectiveness of our proposed method in finding epileptic events from long duration scalp EEG signals. Figure 2a shows the seizure EEG signal segment and its corresponding sub-band signals, whereas Fig. 2b presents the seizure-free EEG signal segment and its sub-band signals. From the figures, it is clear that seizure EEG signal segment has higher amplitude compared to seizure-free EEG signal. This may be a crucial information in distinguishing seizure and seizure-free EEG signal segments. The obtained sub-band signals have been projected into 2D RPS plane. In Fig. 3a, b we have presented the plots of 2D RPS of sub-bands of

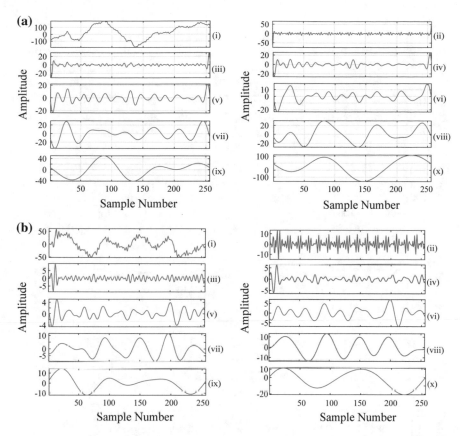

Fig. 2 a Plots of (i) seizure EEG signal segment, (ii)–(x) sub-bands of seizure EEG signal segment obtained using TQWT ($Q = 1$, $R = 3$, $J = 8$). **b** Plots of (i) seizure-free EEG signal segment, (ii)–(x) sub-bands of seizure-free EEG signal segment obtained using TQWT ($Q = 1$, $R = 3$, $J = 8$)

seizure and seizure-free EEG signal segments, respectively. It can be observed that 2D RPS plot of sub-bands corresponding to seizure EEG signals cover more area as compared to the sub-bands of seizure-free EEG signal. This can be considered as a visual information in order to distinguish seizure-free and seizure EEG signal segments. We have used this 2D RPS area corresponding to 50% CTM for detecting epileptic seizures in this work. For each of the considered six patients, we have randomly chosen two long-duration records for EEG seizure detection. Figure 4a–c present the plots of epileptic seizure detection for patients 1, 2, and 7, respectively, whereas Fig. 5a–c present the plots of epileptic seizure detection for patients 17, 19, and 24, respectively, when Q is considered as 1.

It is clear from the figures that our method detects all the seizure events for all the considered records. The EEG seizure onset and offset timings as per the annotations are marked with arrows in the figures. It should be noted that, though

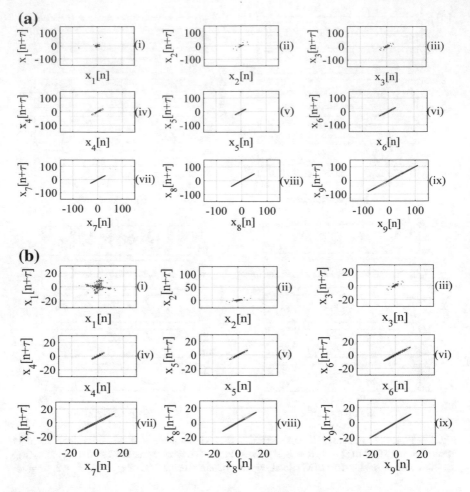

Fig. 3 **a** 2D RPS plots ((i)–(ix)) of sub-bands obtained from seizure EEG signal segment. **b** 2D RPS plots ((i)–(ix)) of sub-bands obtained from seizure-free EEG signal segment

our method has detected all the seizure events, we have not been able to detect the exact starting and ending times of the epileptic events. For patient 17, our proposed method has provided one false detection as shown in Fig. 5a (right column). This is due to the fact that seizure-free EEG signals segments also looks like seizure EEG signal segments over the interval of false detection. Figure 6 presents the plot of EEG seizure detection when Q has been considered as 5 and 10 (by keeping other parameters of TQWT fixed), respectively, in order to observe the changes in the EEG seizure detection performance of our method. It is clear from the figure that higher values of Q factor give poor performance as seizure segments are not enhanced and the method encounters multiple false detections. It should be noted that our algorithm

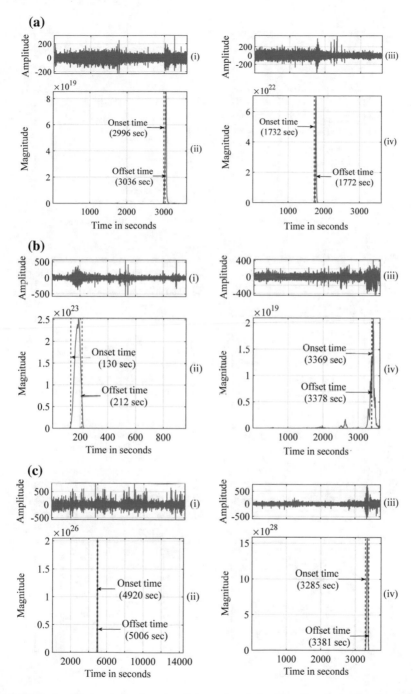

Fig. 4 Plots of epileptic seizure detection for **a** patient 1, **b** patient 2, and **c** patient 7 using the proposed method with $Q = 1$

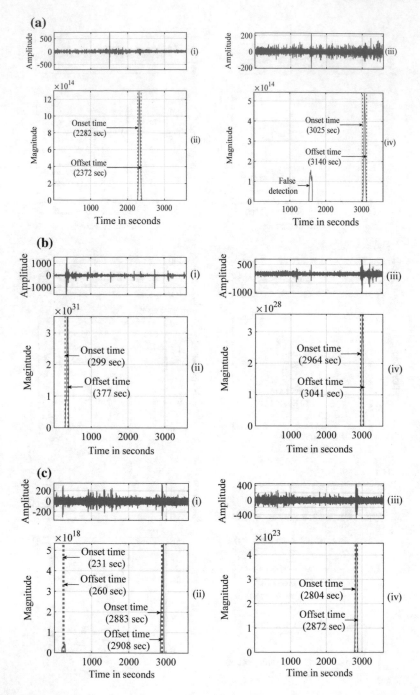

Fig. 5 Plots of epileptic seizure detection for **a** patient 17, **b** patient 19, and **c** patient 24 using the proposed method with $Q = 1$

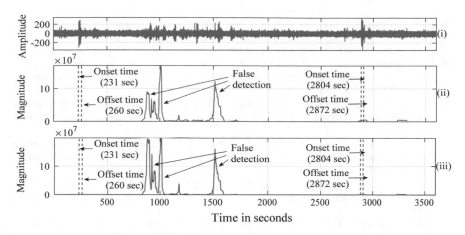

Fig. 6 Plot of epileptic seizure detection of patient 24, for (ii) $Q = 5$ and (iii) $Q = 10$

is extremely fast and only takes 39.2 s to find seizure events in one hour long EEG record on an Intel (R) Core (TM) i5-3470 CPU (3.20 GHz) with 12 GB RAM. The implementation of the proposed method has been performed in MATLAB.

5 Conclusion

We have proposed a new method of epileptic EEG seizure detection. The TQWT has been used for decomposing EEG signal segments into sub-band signals. We have projected the obtained sub-band signals into two-dimensional reconstructed phase space plane followed by computing the area span of the phase space plane. The computed area features were processed with a feature-processing technique. The proposed method has been studied for six patients of epilepsy. The proposed method clearly detects the epileptic seizure events in scalp EEG records. The performance of the proposed method degraded with higher values of quality factor (Q). The proposed method has very fast execution time can be applied for real-time epileptic seizure detection.

References

1. Bajaj, V., Pachori, R.B.: Epileptic seizure detection based on the instantaneous area of analytic intrinsic mode functions of EEG signals. Biomed. Eng. Lett. **3**(1), 17–21 (2013)
2. Bhattacharyya, A., Pachori, R.B.: A multivariate approach for patient-specific EEG seizure detection using empirical wavelet transform. IEEE Trans. Biomed. Eng. **64**(9), 2003–2015 (2017)

3. Bhattacharyya, A., Gupta, V., Pachori, R.B.: Automated identification of epileptic seizure EEG signals using empirical wavelet transform based Hilbert marginal spectrum. In 22nd International Conference on Digital Signal Processing, London, United Kingdom (UK) (2017)
4. Bhattacharyya, A., Pachori, R.B., Rajendra Acharya, U.: Tunable-Q wavelet transform based multivariate sub-band fuzzy entropy with application to focal EEG Signal Analysis. Entropy **19**(99) (2017)
5. Bhattacharyya, A., Pachori, R.B., Upadhyay, A., Acharya, U.R.: Tunable-Q wavelet transform based multiscale entropy measure for automated classification of epileptic EEG signals. Appl. Sci. **7**(385) (2017)
6. Bhattacharyya, A., Sharma, M., Pachori, R.B., Sircar, P., Acharya, U.R.: A novel approach for automated detection of focal EEG signals using empirical wavelet transform. Neural Comput. Appl. pp. 1–11 (2017)
7. Gabor, A.J.: Seizure detection using a self-organizing neural network: validation and comparison with other detection strategies. Electroencephalogr. Clin. Neurophysiol. **107**(1), 27–32 (1998)
8. Gabor, A.J., Leach, R.R., Dowla, F.U.: Automated seizure detection using a self-organizing neural network. Electroencephalogr. Clin. Neurophysiol. **99**(3), 257–266 (1996)
9. Goldberger, A.L., Amaral, L.A.N., Glass, L., Hausdorff, J.M., Ivanov, P.C., Mark, R.G., Mietus, J.E., Moody, G.B., Peng, C., Stanley, H.E.: Physiobank, physiotoolkit, and physionet. Circulation **101**(23), e215–e220 (2000)
10. Gupta, V., Bhattacharyya, A., Pachori, R.B.: Classification of seizure and non-seizure EEG signals based on EMD-TQWT method. In: 22nd International Conference on Digital Signal Processing, London, United Kingdom (UK) (2017)
11. Hassan, A.R., Siuly, S., Zhang, Y.: Epileptic seizure detection in EEG signals using tunable-Q factor wavelet transform and bootstrap aggregating. Comput. Methods Programs Biomed. **137**, 247–259 (2016)
12. Kiranyaz, S., Ince, T., Zabihi, M., Ince, D.: Automated patient-specific classification of long-term electroencephalography. J. Biomed. Inform. **49**, 16–31 (2014)
13. O'Neill, N.S., Koles, Z.J., Javidan, M.: Identification of the temporal components of seizure onset in the scalp EEG. Can. J. Neurol. Sci. **28**(3), 245–253 (2001)
14. Osorio, I., Frei, M.G., Wilkinson, S.B.: Real-time automated detection and quantitative analysis of seizures and short-term prediction of clinical onset. Epilepsia **39**(6), 615–627 (1998)
15. Pachori, R.B., Bajaj, V.: Analysis of normal and epileptic seizure EEG signals using empirical mode decomposition. Comput. Methods Programs Biomed. **104**(3), 373–381 (2011)
16. Samiee, K., Kiranyaz, S., Gabbouj, M., Saramäki, T.: Long-term epileptic EEG classification via 2D mapping and textural features. Expert Syst. Appl. **42**(20), 7175–7185 (2015)
17. Saxena, M.K., Raju, S.D.V.S.J., Arya, R., Pachori, R.B., Kher, S.: Instantaneous area based online detection of bend generated error in a Raman optical fiber distributed temperature sensor. IEEE Sens. Lett. **1**(4), 1–4 (2017)
18. Selesnick, I.W.: Wavelet transform with tunable Q-factor. IEEE Trans. Signal Process. **59**(8), 3560–3575 (2011)
19. Shah, M., Saurav, S., Sharma, R., Pachori, R.B.: Analysis of epileptic seizure EEG signals using reconstructed phase space of intrinsic mode functions. In: 9th International Conference on Industrial and Information Systems (ICIIS) 2014, pp. 1–6 (2014)
20. Sharma, R., Pachori, R.B.: Classification of epileptic seizures in EEG signals based on phase space representation of intrinsic mode functions. Expert Syst. Appl. **42**(03), 1106–1117 (2015)
21. Sharma, R.R., Pachori, R.B.: Time-frequency representation using IEVDHM-HT with application to classification of epileptic EEG signals. IET Sci. Meas. Technol. (2017)
22. Sheb, A., Guttag, J.: Application of machine learning to epileptic seizure detection. In: 27th International Conference on Machine Learning, Haifa, Israel (2010)
23. Shoeb, A.H.: Application of machine learning to epileptic seizure onset detection and treatment. Ph.D. thesis, Ph.D dissertation, Massachusetts Institute of Technology, Cambridge, MA, USA (2009)

24. Takens, Floris, et al.: Detecting strange attractors in turbulence. Lect. Notes Math. **898**(1), 366–381 (1981)
25. Witte, H., Iasemidis, L.D., Litt, B.: Special issue on epileptic seizure prediction. IEEE Trans. Biomed. Eng. **50**(5), 537–539 (2003)
26. Zabihi, M., Kiranyaz, S., Rad, A.B., Katsaggelos, A.K., Gabbouj, M., Ince, T.: Analysis of high-dimensional phase space via poincaré section for patient-specific seizure detection. IEEE Trans. Neural Syst. Rehabil. Eng. **24**(3), 386–398 (2016)

A Teaching–Learning-Based Particle Swarm Optimization for Data Clustering

Neetu Kushwaha and Millie Pant

Abstract The present study proposes TLBO-PSO an integrated Teacher–Learning-Based Optimization (TLBO) and Particle Swarm Optimization (PSO) for optimum data clustering. TLBO-PSO algorithm searches through arbitrary datasets for appropriate cluster centroid and tries to find the global optima efficiently. The proposed TLBO-PSO is analyzed on a set of six benchmark datasets available at UCI machine learning repository. Experimental result shows that the proposed algorithm performs better than the other state-of-the-art clustering algorithms.

Keywords Teaching–learning-based optimization · K-means · Clustering
Particle swarm optimization

1 Introduction

Clustering [1, 2] refers to the process of grouping similar data objects into a number of clusters or group [3]. This process can be applied in different areas such as document classification [4], marketing, bioinformatics, image segmentation [5], recommendations systems [6, 7], etc. Traditionally, the two methods for solving clustering problems are partitional clustering and hierarchal clustering [8, 9]. In this paper, focus is mainly on partitional clustering to find all clusters simultaneously. Partitional-based clustering algorithms include k-means, K-medoids [10], and PAM. k-means is one of the most widely used unsupervised learning algorithms introduced by Hartigan [11]. It is a partitional clustering algorithm in which whole dataset is divided into a certain number of groups (i.e., Disjoint clusters) selected a priori. The standard k-means algorithm can be described as follows:

N. Kushwaha (✉) · M. Pant
Department of Applied Science and Engineering, Indian Institute of Technology Roorkee,
Roorkee 247667, India
e-mail: neetumits@gmail.com

© Springer Nature Singapore Pte Ltd. 2019
M. Tanveer and R. B. Pachori (eds.), *Machine Intelligence and Signal Analysis*,
Advances in Intelligent Systems and Computing 748,
https://doi.org/10.1007/978-981-13-0923-6_19

1. Initialize k cluster centroid (randomly, if necessary).
2. Assignment step: Assign each data point to the cluster that has the closest centroid measured with specific distance metric. Most researchers use Euclidean distance as distance metric.

For each data point i in the dataset,

$$C_i := arg \left\| \min_j X_l - P_j \right\|$$

where X_l is the l-th data vector and P_j is the centroid vector of cluster j.

3. Centroid estimation: Update each cluster centroid value to the mean value of that cluster.

$$P_j = \frac{1}{|C_i|} \sum_{\forall x_l \in P_j} X_l$$

where X_l is the subset of data points that belongs to cluster j and $|C_i|$ is the number of data vectors in cluster j.

4. Repeat steps 2 and 3 until termination criteria reached.

Advantage of using k-means clustering algorithm is its simplicity and efficiency that can run even on a large amount of dataset [12, 13]. However, the drawback of k-means algorithm is that it easily gets affected by initial centroids and get stuck in local optimum [14, 15]. As we change the value of initial centroids, algorithm produces different solutions. Also, there are chances that the algorithm will get stuck in a local optimum solution. To overcome the shortcomings of k-means algorithm, different heuristic or evolutionary algorithms have been proposed. This type of clustering finds the optimal solution while avoiding the local optimum. These algorithms include but not are limited to Genetic Algorithms (GA) [16], Particle Swarm Optimization (PSO) [17, 18], Simulated Annealing (SA) [19], Artificial Bee Colony (ABC) [20, 21], and Ant Colony Optimization (ACO) [22].

In past, several years, many hybridized clustering algorithms based on metaheuristic methods have been introduced. Maulik and Bandyopadhyay [16] developed a data clustering technique based on genetic algorithm. The algorithm is tested on real-life and synthetic datasets to evaluate its performance. Another approach based on genetic k-means was proposed by Krishna and Murty [23]. They defined the basic mutation operator specific for clustering problem.

Van and Engelbrecht [18] utilized PSO to handle data clustering. It applies k-means to seed the initial swarm. Then, PSO is used to refine the clustering result. The results of these methods are compared with k-means and it was concluded that the proposed algorithm has low quantization error as compared to k-means and gave a better clustering.

Hybridization of Nelder–Mead simplex search and PSO was presented by Fan et al. [24] called as NM-PSO for global optimization on benchmark functions. Kao et al. [25] used this hybrid technique for data clustering problem. The authors proposed a hybridized algorithm called K-NM-PSO that combines k-means, Nelder–Mead simplex search (NM), and PSO technique for data clustering. To further improve the NM–PSO, such an algorithm applies k-means clustering first to seed the initial swarm of NM–PSO. It obtains the better data clustering result and is more robust.

But still the rate of convergence is not sufficient for searching global optima. To overcome this problem, Chuang et al. (2011) introduced a new data clustering algorithm called ACPSO [26] which combined accelerated convergence rate strategy [27] with Chaotic map Particle Swarm Optimization (CPSO). ACPSO algorithm finds better clustering of arbitrary data as compared to other previous clustering algorithms. The authors conducted experiment on six experimental datasets and demonstrate that performance of the ACPSO is better than other algorithms.

As we know that premature convergence is the main problem in all evolutionary algorithms [26], each of these algorithms performs well in certain cases and none of them are dominating one another. The key reason for hybridizing one algorithm with other is to take advantage of strengths of each individual algorithm while simultaneously overcoming its main limitations. In this paper, the hybrid algorithm combining TLBO and PSO is proposed to solve complex data clustering problem. Such approach can enhance the performance of the algorithm to find the optimal solution.

The rest of this paper is organized as follows. Section 2 presents a brief introduction of PSO and TLBO. Section 3 describes the proposed work with flowchart. The dataset description and experimental results are listed in Sect. 4. Brief conclusion of the paper is presented in Sect. 5.

2 Preliminaries

2.1 Particle Swarm Optimization

The Particle Swarm Optimization (PSO) algorithm proposed by Eberhart and Kennedy [28, 29] is inspired by the behavior of species which stay in group (swarm) and interact with each other. The members of the swarm are called particles. Each particle has a fitness value which represents a potential solution to the optimization problem. The core objective of PSO is used to find the optimal or sub-optimal of an objective function through the social interaction and information sharing between the particles.

Working of PSO can be explained as follows: Each particle is associated with a velocity vector v_i which measures the direction in which the particle moves and the position vector x_i which consists of Dim dimensions. The velocity of each particle is updated using two best positions, personal best position ($pbest_i$) and global best

position (*gbest*). The pbest$_i$ represents the personal best position, obtained so far
by particle i and gbest represents the global best position, obtained so far, by any
particle of the swarm. In every iteration (or generation), the particles move from one
position to another according to their velocity and update their position in the swarm.
While moving from one position to another, a particle evaluates different prospective
solutions of the problem. The process of updating velocities (see Eq. 1) and positions
(see Eq. 2) is continued iteratively till a stopping criterion is met:

$$v_i^j(t+1) = w * v_i^j(t) + r_1 * c_1 * \left(pbest_i^j(t) - x_i^j(t) \right) + r_2 * c_2 * \left(gbest^j(t) - x_i^j(t) \right) \quad (1)$$

$$x_i^j(t+1) = x_i^j(t) + v_i^j(t+1) \tag{2}$$

where i $= 1, 2, \ldots, N$, j $= 1, 2, \ldots,$ Dim, c_1, c_2 are weighted factors, also called
the cognitive and social parameter, respectively; r_1 and r_2 are random variables uniformly distributed within $(0, 1)$; and t denotes the iteration counter. In the original
PSO, c_1 and c_2 were termed as acceleration constants, w represents inertia weight,
a high inertia weight means enhances the search globally while low inertia weight
emphasizes on the local search.

PSO is an iterative process and terminates when a predefined condition is satisfied.
PSO can be divided into two versions: global and local version. In global version,
each particle updates its velocity using best position in the entire swarm means by
using global best position while in local version, position is updated by using best
positions in the group.

PSO is easy to implement and performs well on optimization problems but like
other metaheuristic algorithms it may easily get trapped into local optimum resulting
in a premature convergence while solving complex and high-dimensional multimodal
problem, which has many local optimum [30].

2.2 Teaching–Learning-Based Optimization

Teaching–Learning-based Optimization (TLBO) algorithm proposed by Rao et al.
[31] is used to simulate the classical teaching–learning process. Similar to other
nature-inspired algorithms, TLBO is a population-based algorithm based on the
influence of teacher on learners in class and learner's interaction between each other.
TLBO algorithm is divided into two phases: Teacher phase and learner phase.

Teacher phase:

In teacher phase, teacher provides knowledge to the learners to increase the mean
grades of the class. Let $F(x)$ be the objective function with Dim-dimensional variables, the ith student can be represented as $X_i = [x_{i1}, x_{i2}, \ldots x_{iDim}]$. To obtain a
new set of improved learners difference between the existing and the new mean rep-
resented by $Mean = \left(\sum_{i=1}^{N} x_{i,1}, \sum_{i=1}^{N} x_{i,2}, \ldots, \sum_{i=1}^{N} x_{i,N} \right) / N$ are added to the existing
population (N). A student having highest fitness value is taken as a teacher for the

current iteration itr and is represented as $x_{teacher}$. The position of each learner is updated by Eq. (3)

$$x_{i,new} = x_{i,old} + r * (x_{teacher} - T_F * Mean) \qquad (3)$$

where $x_{i,old}$ and $x_{i,new}$ are the ith learner's sold and new positions. r is a uniformly distributed random number in the range [0, 1]. T_F is the teaching factor which is decided randomly with equal probability according to Eq. (4)

$$T_F = round[1 + r(0, 1)\{2 - 1\}] \qquad (4)$$

Value of T_F can be either 1 or 2.

Learner phase: In this phase, learner enhances his or her knowledge from the different learners. Learner x_i randomly selects another learner $x_j (j \neq i)$ for enhancing his or her knowledge and the learning process of learners can be expressed by Eqs. (5) and (6):

$$x_{i,new} = x_{i,old} + r * \left(x_i - x_j\right) if f(x_i) \leq f(x_j) \qquad (5)$$
$$x_{i,new} = x_{i,old} + r * \left(x_j - x_i\right) if f(x_i) \geq f(x_j) \qquad (6)$$

where $x_{i,new}$ is accepted if it gives a better function value.

TLBO algorithm termination: The algorithm is terminated after Max_itr iterations are completed.

3 Proposed Algorithm

The proposed algorithm is a simple integration of TLBO and PSO. It starts with TLBO and the set of refined solutions obtained through it is processed through PSO. The proposed TLBO-PSO is customized for generating the optimum clusters. In TLBO-PSO, each particle represents a possible candidate solutions for the clustering data points. Therefore, a population contains multiple candidate solution for the data clustering. Every particle in the population is represented by a matrix $X_i = \{X_1, X_2, \ldots, X_k\}$ where k represents the number of clusters.

Initial population: All particles in the swarm randomly choose K different data points as the initial centroid.

Fitness evaluation: The objective function for the proposed clustering algorithm is the sum of intracluster distance.

Sum of the intracluster distance: The distances between the data point in a cluster and the centroid of that cluster is calculated and summed up for each cluster as shown in Eq. (7). Clearly, the smaller the sum of the intracluster distances is, the higher the quality of the clustering algorithm.

$$intra_{sum} = \sum_{i=1}^{k} X_{p,i} - P_i \tag{7}$$

where $X_{p,i}$ represents the data points X_p belongs to cluster i and P_i is the centroid of cluster i.

The detailed steps of TLBO-PSO can be described as follows:

1. Do the parameter setting: k denotes the number of clusters, N denotes the number of particles in the swarm, Max_itr denotes maximum iterations.
2. Run TLBO and obtain the initial k clusters.
3. Initialize the swarm particles: The outcome of TLBO will serve as the initial solution of PSO. Remaining particles in the swarm randomly choose k different data points from the dataset as the initial centroid for the particular particle.
4. Run PSO algorithm:

   (a) Calculate the distance of each data point to the centroid particles and assign these data points that are closest to those centroids.

   $$C_j = \{X_i : X_i - P_j < X_i - P_i \nabla j, 1\,1 < j < k\}$$

   (b) Calculate the fitness value of each particle using the sum of intracluster distance metric based on Eq. (7).
   (c) The values of $pbest$ and $gbest$ are evaluated and replaced if better particle's previous best position and global best position is obtained.
   (d) Update the velocity and position of each particle in the swarm using Eqs. (1) and (2).

5. Repeat step 3; current iteration number reaches the predetermined maximum iteration number which is reached.
6. Output the last $gbest$ particle value as the final solution of the TLBO-PSO algorithm.

The flowchart of the proposed algorithm is shown in Fig. 1.

4 Experimental Results

The proposed clustering algorithm is evaluated on six benchmark datasets using sum of intracluster distance metric. TLBO-PSO is implemented in MATLAB and run on a PC with i7 core and 8 GB RAM memory. The maximum iteration is set to 200 for all the algorithms.

Algorithms used for comparison are k-means, PSO, GA, KGA, NM–PSO, K–PSO and K–NM–PSO, CPSO, ACPSO, and TLBO. Each of the clustering algorithms was run 30 times independently and their mean standard deviation average best so far value for the result was calculated.

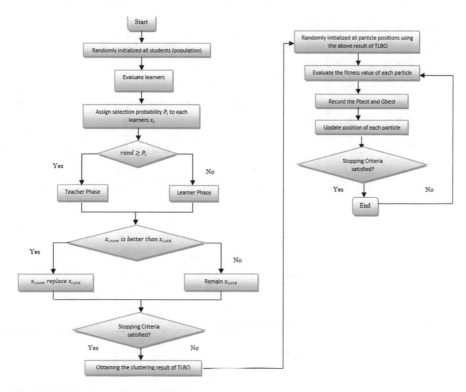

Fig. 1 The flowchart of TLBO-PSO

Parameter setting: The parameter N is based on the dataset and it is defined as $k * Dim$ where k is the number of cluster and Dim is the dimension of the dataset. The size of population is set to $3N$. The different parameters used for the TLBO-PSO are as follows: $xmin$ and $xmax$ are the maximum and minimum values from each dimension in the dataset. Inertia weight w is set as 0.72 and $c1$, $c2$ are set to 1.49445. The number of iteration is equal to 1000.

Datasets: To validate the proposed algorithm, six benchmark datasets, named, Iris, CMC (Contraceptive Method Choice), Crude Oil, Cancer, Vowel, and Wine were derived from the Department of Information and Computer Science (http://arc hive.ics.uci.edu/ml/index.php) at UCI machine learning repository. The description of each dataset is provided in Table 1. The iris dataset is shown in Fig. 2. It shows that the dataset contains three clusters of uneven-sized clusters indicated in red, green, and blue color.

Table 1 Characteristics of the datasets

S. no	Dataset	Instances (n)	Number of classes (k)	Number of features (d)
1.	Vowel	871	6	3
2.	Iris	150	3	4
3.	Crude oil	56	3	5
4.	CMC	1473	3	9
5.	Cancer	683	2	9
6.	Wine	178	3	13

Fig. 2 Original dataset (Iris dataset)

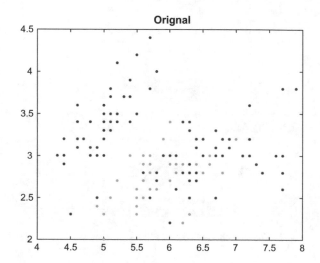

4.1 Results

Table 2 provides the objective function values obtained from the ten clustering algorithms for the datasets described in Table 1. The values reported are average of the sums of intracluster distances over 30 runs. Standard deviation is given in parentheses to indicate the spanning range of values that the algorithms produced and also the best solution of fitness. The smaller (larger) the intracluster distance value, the better (worse) is the efficiency of clustering algorithm. Average sum of intracluster distance obtained by the proposed algorithm is better as compared to other clustering algorithms in all six datasets. It can be seen from results that TLBO-PSO offers better optimized solutions than TLBO or PSO. In terms of the best distance, ACPSO, TLBO-PSO, TLBO and CPSO both have a best intracluster distance in cancer and CMC dataset as compared to proposed algorithm, which means they may achieve a global optimum.

Using the sum of intracluster distance metric, CPSO, NM–PSO, and k-means are shown to need more iteration to achieve the global optimum, while the other

Table 2 Intracluster distances for real-life datasets

Dataset	Criteria	GA	k-means	NM-PSO	KGA	CPSO	ACPSO	K-PSO	K-NM-PSO	PSO	TLBO	TLBO-PSO
Vowel	Average	390088.2	159242.8	151983.9	149358.4	151337	149051.8	149375.7	149141.4	168477	149580	**149470**
	(Std)	N/A	−916	−4386.43	N/A	−491.43	−67.27	−155.56	−120.38	−3715.73	776.6784	−578.3025
	Best	383484.1	149422.2	149240.0	149356.01	148996.5	148970.8	149206.1	149005	163882	148970	148970
Iris	Average	135.4	106.05	100.72	97.1	96.9	96.66	96.76	96.67	103.51	96.7119	**96.6555**
	(Std)	N/A	−14.11	−5.82	N/A	−0.303	−0.001	−0.07	−0.008	−9.69	0.1645	−1.35E−14
	Best	124.13	97.33	96.66	97.1	96.56	96.66	96.66	96.66	96.66	96.6555	96.6555
Crude oil	Average	308.16	287.36	277.59	278.97	277.24	277.24	277.77	277.29	285.51	278.24	**277.2165**
	(Std)	N/A	−25.41	−0.37	N/A	−0.038	−0.04	−0.33	−0.095	−10.31	−0.034	−0.022
	Best	297.05	279.2	277.19	278.97	277.21	277.21	277.45	277.15	279.07	277.22	277.2107
CMC	Average	N/A	5693.6	5563.4	N/A	5552.23	**5532.2**	5532.9	5532.7	5734.2	**5532.2**	**5532.2**
	(Std)	N/A	−473.14	−30.27	N/A	−004	−0.01	−0.09	−0.23	−289	0.0417	−2.63E−11
	Best	N/A	5542.2	5537.3	N/A	5552.19	5532.19	5532.88	5532.4	5538.5	5532.2	5532.2
Cancer	Average	N/A	2988.3	2977.7	N/A	2964.49	2964.42	2965.8	2964.7	3334.6	**2964.4**	**2964.4**
	(Std)	N/A	−0.46	−13.73	N/A	−0.12	−0.03	−1.63	−0.15	−357.66	3.77E−12	−3.34E−12
	Best	N/A	2987	2965.59	N/A	2964.4	2964.39	2964.5	2964.5	2976.3	2964.4	2964.4
Wine	Average	N/A	18061	16303	N/A	16292.9	16292.31	16294	16293	16311	16294	**16292**
	(Std)	N/A	−793.21	−4.28	N/A	−0.78	−0.03	−1.7	−0.46	−22.98	1.0394	−0.217
	Best	N/A	16555.6	16292	N/A	16292.19	16292.18	16292	16292	16294	16292	16292

clustering algorithms PSO and GA may get stuck at a local optimum, depending on the choice of the initial cluster centers.

5 Conclusions

The present study proposes TLBO-PSO, an integrated TLBO and PSO approach for dealing with data clustering problems. The efficiency of proposed TLBO-PSO is analyzed on six benchmark datasets, and the results obtained are compared with ten state-of-the-art clustering algorithms. It is observed that the proposed TLBO-PSO outperforms other algorithms in most of the cases. However, a drawback of the proposed algorithm is that it is not applicable when numbers of clusters are not known a priori. The authors are working on this issue as a future work.

References

1. Kant, S., Ansari, I.A.: An improved K means clustering with Atkinson index to classify liver patient dataset. Int. J. Syst. Assur. Eng. Manag. **7**(1), 222–228 (2016)
2. Cornuéjols, A., Wemmert, C., Gançarski, P., Bennani, Y.: Collaborative clustering: why, when, what and how. Inf. Fusion **39**, 81–95 (2018)
3. Han, X., Quan, L., Xiong, X., Almeter, M., Xiang, J., Lan, Y.: A novel data clustering algorithm based on modified gravitational search algorithm. Eng. Appl. Artif. Intell. **61**, 1–7 (2017)
4. Prakash, J., Singh, P.K.: Particle swarm optimization with K-means for simultaneous feature selection and data clustering. In: 2015 Second International Conference Soft Computing Machine Intelligence, pp. 74–78 (2015)
5. Zhang, C., Ouyang, D., Ning, J.: An artificial bee colony approach for clustering. Expert Syst. Appl. **37**(7) 4761–4767 (2010)
6. Kant, S., Mahara, T.: Merging user and item based collaborative filtering to alleviate data sparsity. Int. J. Syst. Assur. Eng. Manag. 1–7 (2016)
7. Kant, S., Mahara, T.: Nearest biclusters collaborative filtering framework with fusion. J. Comput. Sci. (2017)
8. Everitt, B.S., Landau, S., Leese, M., Stahl, D.: Cluster Anal. **14** (2011)
9. Xu, R., Ii, D.W.: Surv. Clust. Algorithms **16**(3), 645–678 (2005)
10. Khatami, A., Mirghasemi, S., Khosravi, A., Lim, C.P., Nahavandi, S.: A new PSO-based approach to fire flame detection using K-medoids clustering. Expert Syst. Appl. **68**, 69–80 (2017)
11. Hartigan, J.A.: Clust. algorithms. Wiley Publ. Appl. Stat. 1–351. 175 AD
12. Jain, A.K.: Data clustering: 50 years beyond K-means. Pattern Recognit. Lett. **31**(8), 651–666 (2010)
13. Macqueen, J.: Some methods for classification and analysis of multivariate observations. Proc. Fifth Berkeley Symp. Math. Stat. Probab. **1**(233), 281–297 (1967)
14. Peña, J., Lozano, J., Larrañaga, P.: An empirical comparison of four initialization methods for the K-means algorithm. Pattern Recognit. Lett. **20**(10), 1027–1040 (1999)
15. Celebi, M.E., Kingravi, H.A., Vela, P.A.: A comparative study of efficient initialization methods for the k-means clustering algorithm. Expert Syst. Appl. **40**(1), 200–210 (2013)
16. Maulik, U., Bandyopadhyay, S.: Genetic algorithm-based clustering technique. Pattern Recognit. **33**, 1455–1465 (2000)

17. Chen, C.-Y., Ye, F.: Particle swarm optimization algorithm and its application to clustering analysis. In: 2004 IEEE Conference on Networking, Sensing Control, no. 1, pp. 789–794 (2004)
18. van der Merwe, D.W., Engelbrecht, A.P.: Data clustering using particle swarm optimization. In: 2003 Congress on Evolutionary Computation, CEC'03, pp. 215–220 (2003)
19. Selim, S.Z., Alsultan, K.: A simulated annealing algorithm for the clustering problem. Pattern Recognit. **24**(10), 1003–1008 (1991)
20. Ozturk, C., Hancer, E., Karaboga, D.: Dynamic clustering with improved binary artificial bee colony algorithm. Appl. Soft Comput. J. **28**, 69–80 (2015)
21. Banharnsakun, A.: A MapReduce-based artificial bee colony for large-scale data clustering. Pattern Recognit. Lett. **93**, 78–84 (2017)
22. Shelokar, P.S., Jayaraman, V.K., Kulkarni, B.D.: An ant colony approach for clustering. Anal. Chim. Acta **509**(2), 187–195 (2004)
23. Krishna, K., Narasimha Murty, M.: Genetic K-means algorithm. IEEE Trans. Syst. Man Cybern. Part B **29**(3), 433–439 (1999)
24. Fan, S.S., Liang, Y., Zahara, E.: Hybrid simplex search and particle swarm optimization for the global optimization of multimodal functions. Eng. Optim. **36**(4), 401–418 (2004)
25. Kao, Y.-T., Zahara, E., Kao, I.-W.: A hybridized approach to data clustering. Expert Syst. Appl. **34**(3), 1754–1762 (2008)
26. Chuang, L.-Y., Hsiao, C.-J., Yang, C.-H.: Chaotic particle swarm optimization for data clustering. Expert Syst. Appl. **38**(12), 14555–14563 (2011)
27. Chuanwen, J., Bompard, E.: A self-adaptive chaotic particle swarm algorithm for short term hydroelectric system scheduling in deregulated environment. Energy Convers. Manag. **46**(17), 2689–2696 (2005)
28. Eberhart, R., Kennedy, J.: A new optimizer using particle swarm theory. In: Proceedings of the Sixth International Symposium on Micro Machine and Human Science, MHS'95, pp. 39–43 (1995)
29. Shi, Y., Eberhart, R.: A modified particle swarm optimizer. In: Proceedings of 1998 IEEE World Congress on Evolutionary Computation. In: 1998 IEEE International Conference on Computational Intelligence, pp. 69–73 (1998)
30. Chen, W.N., et al.: Particle swarm optimization with an aging leader and challengers. IEEE Trans. Evol. Comput. **17**(2), 241–258 (2013)
31. Rao, R.V., Savsani, V.J., Vakharia, D.P.: Teaching–learning-based optimization: a novel method for constrained mechanical design optimization problems. Comput. Des. **43**(3), 303–315 (2011)

A New Method for Classification of Focal and Non-focal EEG Signals

Vipin Gupta and Ram Bilas Pachori

Abstract In this paper, we have proposed a new methodology based on the empirical mode decomposition (EMD) for classification of focal electroencephalogram (FE) and non-focal electroencephalogram (NFE) signals. The proposed methodology uses EMD along with Sharma–Mittal entropy feature computed on Euclidean distance values from K-nearest neighbors (KNN) of FE and NFE signals. The EMD method is used to decompose these electroencephalogram (EEG) signals into amplitude modulation and frequency modulation (AM–FM) components, which are also known as intrinsic mode functions (IMFs) then the KNN approach-based Sharma–Mittal entropy feature has been computed on these IMFs. These extracted features play significant role for the classification of FE and NFE signals with the help of least squares support vector machine (LS-SVM) classifier. The classification step includes radial basis function (RBF) kernel along with tenfold cross-validation process. The proposed methodology has achieved classification accuracy of 83.18% on entire Bern-Barcelona database of FE and NFE signals. The proposed method can be beneficial for the neurosurgeons to identify focal epileptic areas of the patient brain.

Keywords EEG · EMD · KNN · IMF · LS-SVM

V. Gupta (✉) · R. B. Pachori
Discipline of Electrical Engineering, Indian Institute of Technology Indore,
Indore 453552, India
e-mail: vipingupta@iiti.ac.in

R. B. Pachori
e-mail: pachori@iiti.ac.in

© Springer Nature Singapore Pte Ltd. 2019   235
M. Tanveer and R. B. Pachori (eds.), *Machine Intelligence and Signal Analysis*,
Advances in Intelligent Systems and Computing 748,
https://doi.org/10.1007/978-981-13-0923-6_20

1 Introduction

In human being every function is controlled by the brain. These functions mainly depend on neurons present in the different parts of the brain. An epileptic seizure is the result of synchronous neuronal activity in the brain and the occurrence of at least one epileptic seizure is known as epilepsy [10]. Epilepsy is mainly of two types, the first one is generalized epilepsy which affects the whole part of the brain and another one is focal or partial epilepsy, in which a small neurons group is thought to give rise to localized epileptic discharge [11]. There are approximately 20% and 60% patients suffered from generalized and focal epilepsy among the world population, respectively [18]. The main reason behind the study of focal epilepsy is its development of drug resistance [18]. Therefore, the only way to cure this kind of epilepsy is the surgical removal of the affected brain parts.

The electroencephalogram (EEG) signal is a well-known way to represent the electrical activity of the brain. These signals consist information related to the neurological disorder of the brain. The recordings of these EEG signals are possible in two manners, the first one is scalp [5] and another one is intracranial EEG recordings. The intracranial EEG recording is utilized for the localization of focal brain area because scalp EEG may fail to show ictal changes in seizures arising from a small and deeply situated focal epileptic zone [18]. In [2], the analysis of focal electroencephalogram (FE) signals shows that these signals are more rejected for both randomness and nonlinear-independence test but more stationary in comparison to non-focal electroencephalogram (NFE) signals. These EEG signals require a signal processing-based automated method for the classification purpose because the identification of these long duration EEG signals is very cumbersome and time-consuming process by the neurologists. However, there are various automated methods for the classification of FE and NFE signals in the literature [6, 9, 13, 20–25].

The average sample entropies and average variance of instantaneous frequencies features computed from intrinsic mode functions (IMFs) extracted by empirical mode decomposition (EMD) have been fed to least squares support vector machine (LS-SVM) classifier with radial basis function (RBF) kernel. The maximum classification accuracy achieved in this approach is 85% with 50 FE and 50 NFE signals of Bern-Barcelona database [20]. In another work [21], EMD and LS-SVM are also utilized with average entropies features like as Renyi wavelet, Shannon wavelet, Tsallis wavelet, fuzzy, permutation, and phase with a classification accuracy of 87% on the same 50 EEG signals database. A new approach for the classification based on the integrated index and discrete wavelet transform (DWT) with different classifiers, such as probabilistic neural network, K-nearest neighbors (KNN), LS-SVM, and fuzzy have been discussed in [22]. The obtained classification accuracy in this method is 84% with separated integrated index on 50 FE and 50 NFE signals. In [9], the methods EMD and DWT have been also used with log-energy entropy feature and KNN classifier. The classification accuracy achieved for this method is 89.4% on 3750 FE and 3750 NFE signals. In another work [6], the empirical wavelet transform (EWT) has been utilized to decompose FE and NFE signals into rhythms and then the areas obtained from the reconstructed phase space plot of rhythms corresponding to different central tendency measures are used as features with LS-SVM classifier. The achieved classification accuracies are 90% and 82.53% for 50 and 750 of FE and NFE signals, respectively. In [23], the classification is based on the time–frequency orthogonal wavelet filter banks and the achieved classification accuracy is 94.25% for entire Bern-Barcelona FE and NFE signals database. A multivariate fuzzy entropy computed on sub-bands obtained from tunable-Q wavelet transform (TQWT) has been used in [8] and this method achieved a classification accuracy of 84.67% with LS-SVM classifier on entire EEG database of Bern-Barcelona. The classification of FE signals have been also discussed in [13], with flexible analytic wavelet transform and the obtained classification accuracy is 94.41% with full EEG database of Bern-Barcelona. The TQWT method-based decision support system is proposed in [24] which provided a classification accuracy of 95% for the classification of FE and NFE signals for the same full database. In [25], the authors have utilized Fourier-based mean–frequency and root-mean-square bandwidth features for the classification of 50 and 750 of FE and NFE signals. The achieved classification accuracies for 50 and 750 of EEG signals database are 89.7% and 89.52%, respectively.

In this work, our main objective is to explore Sharma–Mittal entropy with KNN approach feature in the EMD domain which can help to decompose difference signals obtained from FE pair and NFE pair signals into IMFs. The performance of the extracted features are evaluated by Kruskal–Wallis statistical test [17] with the help of three different parameters namely gamma (γ), delta (δ), and nearest neighbor (K) required for entropy computation [27]. The block diagram of the proposed method can be seen in Fig. 1. The proposed method requires difference operation on bivariate FE and NFE signals, EMD method-based decomposition, Sharma–Mittal entropy computation, Kruskal–Wallis statistical test, and LS-SVM-based classifier for classification of FE and NFE signals.

The rest of the paper is discussed as follows: Sect. 2 comprises the database description and the description of the proposed methodology, which includes EMD method, feature extraction process, and the classifier studied. Results and discussions are provided in Sect. 3 and then the conclusion of this paper is presented in Sect. 4.

2 Methodology

The database description, EMD method, feature extraction, and LS-SVM classifier are explained as follows.

2.1 Database Description

The EEG signals are obtained from a publicly available Bern-Barcelona EEG database [28]. The intracranial EEG signals of focal epilepsy are recorded from five

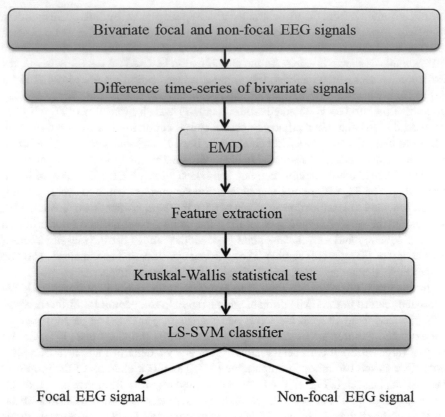

Fig. 1 Block diagram of the proposed method for classification of FE and NFE signals

patients at the Department of Neurology of the University of Bern, Bern, Switzerland. There are 7500 total pairs of EEG signals in which 3750 NFE pairs and 3750 FE pairs which are sampled at a frequency of 512 Hz for 20 s time duration corresponding to 10,240 samples. The database includes a pair of EEG signals which are depicted by "x" and "y" time-series. The plots of "x" and "y" time-series of a FE and NFE signals are represented in Fig. 2a–d, respectively.

In this study, we have considered the difference of "x" and "y" time-series ("x-y") of NFE and FE signals as suggested in [9] for the classification purpose and the plots of these time-series are shown in Fig. 2e, f.

2.2 Empirical Mode Decomposition

The EMD is an adaptive method for analyzing signals which can be represented by the sum of zero-mean amplitude modulation and frequency modulation (AM–FM)

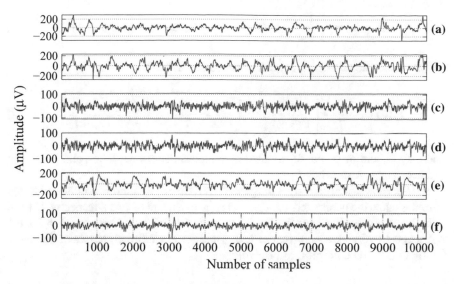

Fig. 2 Plots of **a** "x" time-series of FE **b** "y" time-series of FE **c** "x" time-series of NFE **d** "y" time-series of NFE **e** "x-y" time-series of FE **f** "x-y" time-series of NFE signals

components known as IMFs and each IMF has limited bandwidth [14]. An IMF needs to satisfy following two conditions [14]· (1) the number of extreme and zero-crossing must equal or differ at most by one. (2) the mean value of the upper and lower envelopes which are formed by local maxima and local minima, at any point should be zero.

IMF is derived from a signal using an iterative process of finding the upper and lower envelopes. The detailed description of this iterative process is explained in [14].

In [9, 20–22], the EMD method is shown significant for the classification of FE and NFE signals. This EMD method has been also utilized in [1, 4, 12, 16] for the detection and classification of epileptic seizure EEG signals.

2.3 Feature Extraction

Feature extraction is an important step for the identification of patterns present in the signals. In this work, we have used values of Euclidean distance based on the KNN approach from the IMFs, which are extracted after decomposition of difference EEG signals. These values of the distance are further considered for the Sharma–Mittal entropy computation. The Sharma–Mittal entropy can be mathematically expressed as [19, 27]:

$$\hat{E}_{\gamma,\delta}(v) = \frac{1}{1-\delta} \left[\hat{I}_{\gamma}(v)^{\frac{1-\delta}{1-\gamma}} - 1 \right] \quad \forall (\gamma > 0, \gamma \neq 1, \delta \neq 1) \tag{1}$$

For one-dimensional data, $\hat{I}_{\gamma}(v)$ can be expressed as [27]:

$$\hat{I}_{\gamma}(v) = \frac{T-1}{T} \left[\frac{\pi^{\frac{1}{2}}}{\Gamma(3/2)} \right]^{1-\gamma} \left[\frac{\Gamma(K)}{\Gamma(K+1-\gamma)} \right] \sum_{t=1}^{T} \frac{[\rho_K(t)]^{(1-\gamma)}}{(T-1)^{\gamma}} \tag{2}$$

where T, K, and $\rho_K(t)$ are the number of samples, nearest neighbor, and Euclidean distance values, respectively.

The Sharma–Mittal entropy is the generalized case of Shannon, Tsallis, and Renyi entropies when the parameters $(\gamma, \delta \longrightarrow 1)$, $(\gamma = \delta)$, and $(\delta = 1)$, respectively [27].

In this study, we have considered the values of K, γ, and δ between 3 to 10, 0.1 to 2, and 0.1 to 2 with a step sizes of 1, 0.1, and 0.1 excluding the cases of Shannon, Tsallis, and Renyi entropies, respectively.

2.4 Least Squares Support Vector Machine

The LS-SVM is the least square formulation of support vector machine (SVM) for the classification of two class problem. The LS-SVM requires solving a set of linear equations as compared to SVM which requires solution of quadratic programming problem [26]. The LS-SVM classifier can be expressed as [13, 20]:

$$F(x) = \text{sign} \left[\sum_{m=1}^{L} \alpha_m \psi_m K(\phi, \phi_m) + b \right] \tag{3}$$

where α_m, ϕ_m, ψ_m, b, and $K(\phi, \phi_m)$ are the Lagrangian multiplier, input data, class label for two different classes, bias, and kernel function, respectively.

In this paper, we have used RBF kernel function, which is mathematically expressed in [15] and the parameter of this kernel function is selected between the interval of 0.5 and 2.5 with a step size of 0.1. This kernel function with LS-SVM classifier is also used in [13, 23, 24] for the classification of FE and NFE signals.

3 Results and Discussion

The EEG signals used in this work contain two different time-series namely "x" and "y". First of all, the difference of time-series is evaluated from FE and NFE signals. This technique is previously used in [9] for increasing the discrimination between features. Therefore, we have obtained 3750 difference time-series corresponding to

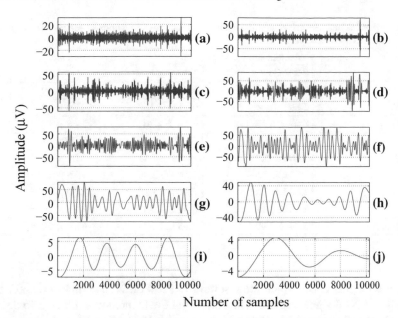

Fig. 3 Plots of **a** IMF1 **b** IMF2 **c** IMF3 **d** IMF4 **e** IMF5 **f** IMF6 **g** IMF7 **h** IMF8 **i** IMF9 and **j** IMF10 from "x-y" time-series of FE signal

Table 1 p-value and mean ± standard deviation of Sharma Mittal entropy features corresponding to 50 FE and 50 NFE signals

IMF$_s$	Mean ± standard deviation of features		p-value
	Focal class	Non-focal class	
IMF1	3.94 ± 1.91	5.45 ± 2.63	8.69×10^{-04}
IMF2	5.62 ± 2.35	7.39 ± 3.10	5.82×10^{-04}
IMF3	7.75 ± 3.39	10.17 ± 3.70	1.87×10^{-04}
IMF4	10.77 ± 4.10	11.58 ± 4.18	0.21
IMF5	13.32 ± 4.51	11.89 ± 5.36	0.08
IMF6	14.61 ± 5.75	12.13 ± 5.49	0.03
IMF7	12.79 ± 5.64	10.47 ± 4.51	0.06
IMF8	8.43 ± 6.12	6.81 ± 3.50	0.36
IMF9	4.41 ± 4.19	3.37 ± 2.39	0.21
IMF10	2.35 ± 1.40	2.95 ± 1.64	0.01

Table 2 p-value and mean ± standard deviation of Sharma–Mittal entropy features corresponding to 3750 FE and 3750 NFE signals

IMFs	Mean ± standard deviation		p-value
	Focal class	Non-focal class	
IMF1	4.15 ± 1.99	4.65 ± 2.54	1.86×10^{-35}
IMF2	5.89 ± 2.34	6.71 ± 3.18	4.29×10^{-59}
IMF3	8.07 ± 3.13	9.66 ± 4.26	1.68×10^{-91}
IMF4	11.18 ± 4.13	11.36 ± 4.96	0.0188
IMF5	13.66 ± 5.04	11.88 ± 5.41	2.01×10^{-48}
IMF6	14.33 ± 5.55	12.40 ± 5.94	1.75×10^{-54}
IMF7	12.02 ± 5.24	10.74 ± 5.32	2.30×10^{-25}
IMF8	7.56 ± 4.78	6.37 ± 3.73	3.86×10^{-22}
IMF9	3.99 ± 3.99	2.95 ± 2.30	6.19×10^{-30}

each EEG signal class, which are resulting in 7500 difference time-series for both the FE and NFE signal classes. Thereafter, the EMD method is utilized to decompose these difference time-series which provides IMFs. The EMD is an adaptive decomposition method so we have decomposed these difference EEG signals into 10 common IMFs for the analysis and classification purpose of 50 FE and 50 NFE signals. Plots of 10 common IMFs are depicted in Fig. 3a–j and 4a–j of difference FE and NFE signals, respectively. The analysis and classification are mainly based on the feature extraction process in which three different entropy evaluation parameters (γ, δ, and K) are considered. We have considered the variation of these three different parameters which are γ, δ, and K for feature evaluation and the limits of these parameters have been already discussed in feature extraction process. The values of Euclidean distance are computed from these obtained IMFs with KNN approach and then these values are further considered for Sharma–Mittal entropy evaluation with different selected values of γ, δ, and K. The classification of these extracted features is carried out with the help of LS-SVM classifier with RBF kernel. The tenfold cross-validation [15] procedure is also used in this proposed method by which we have obtained more appropriate results. The performance of the classification step is computed using three different parameters, such as accuracy (ACC), sensitivity (SEN), and specificity (SPE) [3] and the maximum achieved value of these parameters are 88%, 98%, and 78% for the classification of 50 FE and 50 NFE signals (small dataset), respectively. This experiment is further extended for the 7500 EEG signals to verify the efficacy of proposed method in term of classification accuracy. However, the 3750 FE and 3750 NFE signals have only nine IMFs in common. Therefore, we have used nine IMFs for the classification of 3750 FE and 3750 NFE signals. The selected values of γ, δ, and K are shown in Table 3 corresponding to maximum accuracy obtained from the small dataset part. We have also applied the Kruskal–Wallis statistical test [7, 17] on computed features from 50 and 3750 of FE and NFE signals database. This hypothesis test shows the discrimination between FE and NFE signal

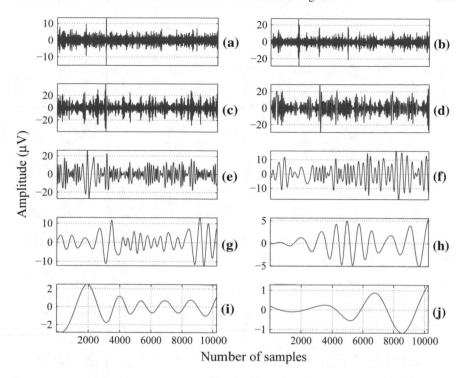

Fig. 4 Plots of **a** IMF1 **b** IMF2 **c** IMF3 **d** IMF4 **e** IMF5 **f** IMF6 **g** IMF7 **h** IMF8 **i** IMF9 and **j** IMF10 from "x-y" time-series of NFE signal

Table 3 Classification performance parameters (ACC, SEN, and SPE) with feature parameters (γ, β, and K) and kernel parameters on 50 and 3750 of FE and NFE signals

Focal and non-focal EEG signals	ACC	SEN	SPE	RBF kernel parameter	Feature parameters		
					γ	δ	K
50	88	98	78	1	1.5	0.5	10
3750	83.12	85.78	80.45	0.9			

classes on the basis of p-value. The Tables 1 and 2 represent the p-values with mean and standard deviations of the features corresponding to 50 and 3750 of FE and NFE signals, respectively. The lower value of p shows greater the discrimination ability between both classes of EEG signals. The box plots corresponding to p-values of 50 and 3750 FE and NFE signals are shown in Figs. 5 and 6, respectively. The achieved classification accuracy in the proposed method is 83.18% for 3750 FE and 3750 NFE signals. The obtained classification accuracies from both the performed studies are also presented in Table 3.

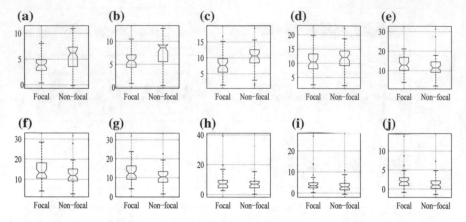

Fig. 5 Box plots corresponding to Sharma–Mittal entropy feature computed from IMF1-IMF10 depicted in **a–j** for 50 FE and 50 NFE signals

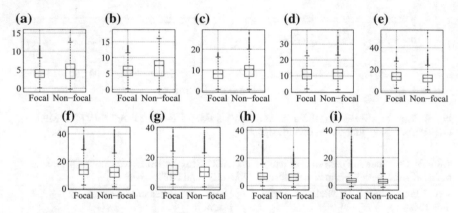

Fig. 6 Box plots corresponding to Sharma–Mittal entropy feature computed from IMF1-IMF9 depicted in **a–i** for 3750 FE and 3750 NFE signals

4 Conclusion

In this work, we have proposed the Sharma–Mittal entropy feature based on the KNN approach for classification of FE and NFE signals. The features show significant discrimination on IMFs from FE and NFE signals in terms of p-value. The LS-SVM classifier with RBF kernel has provided a classification accuracy of 83.18% with 85.78% sensitivity for the classification of 3750 FE and 3750 NFE signals. In comparison to the other methodologies present in the literature for the classification of FE and NFE signals, our methodology is new and involved only one entropy with lesser dimensionality of features set on full database. The proposed methodology can be utilized for the identification of focal epileptic zones from the human brain.

The real-time implementation and study on very large database recorded for long duration are required in this proposed method before applying it for clinical purpose.

Acknowledgements This work was supported by the Council of Scientific and Industrial Research (CSIR) funded Research Project, Government of India, Grant No. 22/687/15/EMR-II.

References

1. Altunay, S., Telatar, Z., Erogul, O.: Epileptic EEG detection using the linear prediction error energy. Expert Syst. Appl. **37**(8), 5661–5665 (2010)
2. Andrzejak, R.G., Schindler, K., Rummel, C.: Nonrandomness, nonlinear dependence, and nonstationarity of electroencephalographic recordings from epilepsy patients. Phys. Rev. E **86**, 046206 (2012)
3. Azar, A.T., El-Said, S.A.: Performance analysis of support vector machines classifiers in breast cancer mammography recognition. Neural Comput. Appl. **24**(5), 1163–1177 (2014)
4. Bajaj, V., Pachori, R.B.: Epileptic seizure detection based on the instantaneous area of analytic intrinsic mode functions of EEG signals. Biomed. Eng. Lett. **3**(1)(1), 17–21 (2013)
5. Bhattacharyya, A., Pachori, R.B.: A multivariate approach for patient specific EEG seizure detection using empirical wavelet transform. IEEE Trans. Biomed. Eng. **64**(9), 2003–2015 (2017)
6. Bhattacharyya, A., Sharma, M., Pachori, R.B., Sircar, P., Acharya, U.R.: A novel approach for automated detection of focal EEG signals using empirical wavelet transform. Neural Comput. Appl. **29**(8), 47–57 (2018)
7. Bhattacharyya, A., Gupta, V., Pachori, R.B.: Automated identification of epileptic seizure EEG signals using empirical wavelet transform based Hilbert marginal spectrum. In: 22nd International Conference on Digital Signal Processing August 23-25, London, United Kingdom. IEEE, 1 5 (2017)
8. Bhattacharyya, A., Pachori, R.B., Acharya, U.R.: Tunable Q wavelet transform based multivariate sub-band fuzzy entropy with application to focal EEG signal analysis. Entropy **19**(3) (2017)
9. Das, A.B., Bhuiyan, M.I.H.: Discrimination and classification of focal and non-focal EEG signals using entropy-based features in the EMD-DWT domain. Biomed. Signal Process. Control **29**, 11–21 (2016)
10. Fisher, R.S., Boas, W.E., Blume, W., Elger, C., Genton, P., Lee, P., Engel, J.: Epileptic seizures and epilepsy: definitions proposed by the international league against epilepsy (ILAE) and the international bureau for epilepsy (IBE). Epilepsia **46**(4), 470–472 (2005)
11. Gloor, P., Fariello, R.G.: Generalized epilepsy: some of its cellular mechanisms differ from those of focal epilepsy. Trends Neurosci. **11**(2), 63–68 (1988)
12. Gupta, V., Bhattacharyya, A., Pachori, R.B.: Classification of seizure and non-seizure EEG signals based on EMD-TQWT method. In: 22nd International Conference on Digital Signal Processing August 23-25, London, United Kingdom. IEEE, 1–5 (2017)
13. Gupta, V., Priya, T., Yadav, A.K., Pachori, R.B., Acharya, U.R.: Automated detection of focal EEG signals using features extracted from flexible analytic wavelet transform. Pattern Recogn. Lett. **94**, 180–188 (2017)
14. Huang, N.E., Shen, Z., Long, S.R., Wu, M.C., Shih, H.H., Zheng, Q., Yen, N.-C., Tung, C.C., Liu, H.H.: The empirical mode decomposition and the Hilbert spectrum for nonlinear and non-stationary time series analysis. In: Proceedings of the Royal Society of London A: Mathematical, Physical and Engineering Sciences, vol. 454, pp. 903–995. The Royal Society (1998)
15. Khandoker, A.H., Lai, D.T.H., Begg, R.K., Palaniswami, M.: Wavelet-Based feature extraction for support vector machines for screening balance impairments in the elderly. IEEE Trans. Neural Syst. Rehabil. Eng. **15**(4), 587–597 (2007)

16. Martis, R.J., Acharya, U.R., Tan, J.H., Petznick, A., Yanti, R., Chua, C.K., Ng, E.Y.K., Tong, L.: Application of empirical mode decomposition (EMD) for automated detection of epilepsy using EEG signals. Int. J. Neural Syst. **22**(6), 1250027, 1–16 (2012)

17. McKight, P.E., Najab, J.: Kruskal-Wallis test. Corsini Encyclopedia of Psychology. Wiley, Hoboken (2010)

18. Pati, S., Alexopoulos, A.V.: Pharmacoresistant epilepsy: from pathogenesis to current and emerging therapies. Clevel. Clin. J. Med. **77**, 457–467 (2010)

19. Sharma, B.D., Mittal, D.P.: New non-additive measures of entropy for discrete probability distributions. J. Math. Sci **10**, 28–40 (1975)

20. Sharma, R., Pachori, R.B., Gautam, S.: Empirical mode decomposition based classification of focal and non-focal EEG signals. In: International Conference on Medical Biometrics, pp. 135–140, Shenzhen (2014)

21. Sharma, R., Pachori, R.B., Acharya, U.R.: Application of entropy measures on intrinsic mode functions for the automated identification of focal electroencephalogram signals. Entropy **17**(2), 669–691 (2015)

22. Sharma, R., Pachori, R.B., Acharya, U.R.: An integrated index for the identification of focal electroencephalogram signals using discrete wavelet transform and entropy measures. Entropy **17**, 5218–5240 (2015)

23. Sharma, M., Dhere, A., Pachori, R.B., Acharya, U.R.: An automatic detection of focal EEG signals using new class of time-frequency localized orthogonal wavelet filter banks. Knowl.-Based Syst. **118**, 217–227 (2017)

24. Sharma, R., Kumar, M., Pachori, R.B., Acharya, U.R.: Decision support system for focal EEG signals using tunable-Q wavelet transform. J. Comput. Sci. **20**, 52–60 (2017)

25. Singh, P., Pachori, R.B.: Classification of focal and nonfocal EEG signals using features derived from Fourier-based rhythms. J. Mech. Med. Biol. **17**(04), 1740002, 1–16 (2017)

26. Suykens, J.A.K., Vandewalle, J.: Least squares support vector machine classifiers. Neural Process. Lett. **9**(3), 293–300 (1999)

27. Szabó, Z.: Information theoretical estimators toolbox. J. Mach. Learn. Res. **15**(1), 283–287 (2014)

28. The Bern-Barcelona EEG database (2013). http://ntsa.upf.edu/downloads

Analysis of Facial EMG Signal for Emotion Recognition Using Wavelet Packet Transform and SVM

Vikram Kehri, Rahul Ingle, Sangram Patil and R. N. Awale

Abstract Emotion recognition has been improved recently and effectively used in medical and diagnostic areas. Automatic recognition of facial expressions is an important application in human–computer interface (HCI). This paper proposed techniques for recognizing three different facial expressions such as happiness, anger, and disgust. Facial signals were recorded using two-channel wireless data acquisition system. Recorded facial EMG signals from zygomatic and corrugator face muscles were set up in four steps: Feature extraction, features selection, classification, and emotion recognition. The features have been extracted using wavelet packet transform method and feed to support vector machine for the classification of three different facial emotions. Finally, the proposed methodology gives classification accuracy 91.66% on 12 subjects.

Keywords Facial electromyogram (FEMG) · Human–computer interface (IICI) Wavelet packet transform (WPT) · Support vector machine (SVM)

1 Introduction

Researches related to FEMG signal processing to study emotion have started since the first half of last century. Researches on FEMG signal processing mainly focusing on the computation of signal processing and data analysis have been blooming in recent years. Several techniques and computational methods have been used in facial expression recognition [1–4]. Except for FEMG method, other techniques are video analysis and facial image by image processing technique [5]. The FEMG signal

V. Kehri (✉) · R. Ingle · S. Patil · R. N. Awale
Department of Electrical Engineering, Veermata Jijabai Technological Institute, Mumbai, India
e-mail: vakehri@el.vjti.ac.in; vakehri@vjti.org.in

R. Ingle
e-mail: rringle@el.vjti.ac.in

R. N. Awale
e-mail: rnawale@el.vjti.ac.in

© Springer Nature Singapore Pte Ltd. 2019
M. Tanveer and R. B. Pachori (eds.), *Machine Intelligence and Signal Analysis*,
Advances in Intelligent Systems and Computing 748,
https://doi.org/10.1007/978-981-13-0923-6_21

is nonlinear, small amplitude, and low-frequency signal. FEMG is information of electrical potentials generated by facial muscles cells [6]. Invasive (needle electrode) and noninvasive (surface electrodes) are the two methods of electrode placement for recording EMG signal [7].

Facial expression is considered as a source of information to understand person's affective state [8]. A facial expression focuses on two major muscles, corrugator and zygomaticus in the face. Zygomaticus muscles give the information about the expression of cheeks and mouth, whereas corrugator muscle is responsible for eyebrows upward producing upright wrinkle on face above the eyebrows [9]. The amplitude and frequencies range for FEMG are 0.1–15 mV and between 5 and 499 Hz.

In early days, FFT was used for the analysis of EMG datasets, but FFT suffers from noise sensitivity, since EMG signals are nonstationary and parametric methods are not used fully for frequency decomposition of these signals [10]. Another method is short-time Fourier transform (STFT) which provides resolution in short window of time for all frequencies. The performance of STFT is mainly dependent on selecting a proper length of the desired segment of the signal [8]. To overcome the problems of STFT and FFT, a powerful tool that is wavelet packet transform (WPT) can be applied to extract the features of FEMG signal [10].

There are several methods used for FEMG classification based on pattern recognition such as SVM, neural networks, neuro-fuzzy, probabilistic, FCM, and SFCM. Many literatures emphasize the success of neural networks in FEMG classification where MLP is applied to recognize time-domain features where LDA performed better with frequency-domain features [11]. SVM classifier has been widely used in medical diagnostic decision support system because of the belief that these have more accurate classification accuracy power.

2 Materials and Methods

2.1 Subjects

For this work, FEMG signals were recorded from twelve physical fit participants (8 male and 4 female) within the range of 20–35 age groups. Each participant was requested to perform three unique types of facial emotion such as happiness, anger, and disgust. Each emotion was recorded for three times from each participant. So the total numbers of emotion classes were 108. If subjects become aware of the recording than their facial expression may neglect its authenticity. So to avoid this situation, subjects were asked about the true emotions that they had felt during recording.

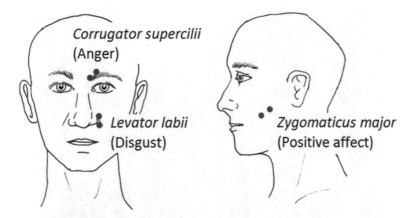

Fig. 1 Facial muscles involved in facial emotion activity

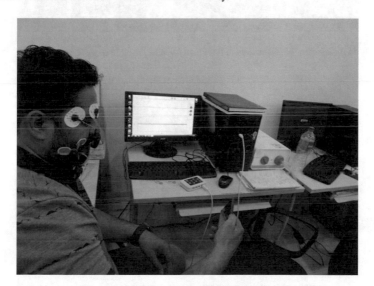

Fig. 2 Experimental setup for recording facial EMG signal

2.2 Experiment Setup

EMG signal depends on many factors such as the orientation of fibers with respect to electrodes, skin resistance, and space between muscle fibers and electrodes [12]. In this experiment, two electrodes are fixed on the zygomaticus and two are placed in corrugator face muscle at a distance of 1–2 cm. Figure 1 shows the location of facial muscles responsible for facial emotion activity. Figure 2 shows the real-time experiment setup, Fig. 3depict wireless data acquisition system and Fig. 4 represent FEMG signal acquired form zygomaticus muscle.

Fig. 3 Myon aktos-mini wireless data acquisition system

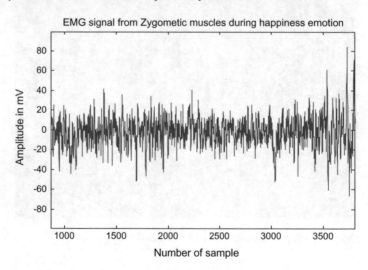

Fig. 4 FEMG from zygomaticus muscle during happiness emotion

Precise skin arrangement is essential to acquire accurate signal and circumvent noise. The area which is selected for signal recording must be cleaned and sweat free [13]. Conductive cream is applied to the electrodes before using them for experiment.

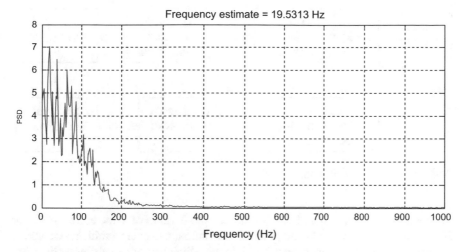

Fig. 5 Power spectral density of FEMG during happiness emotion

2.3 EMG Signal Recording and Data Acquisition

For recording facial signal, Myon aktos-mini data acquisition device is used. This is a high-speed wireless data acquisition device with two channels. It contains wireless transmitter, receiver, data acquisition kit, and biofeedback software. The subject can sit, lie down, or freely move around during any task. There is no restriction of movement, even when walking or running.

A signal is recorded for sampling frequency of 1000 Hz. Before recording subjects were asked to sit comfortable and relax, it is also very important to record rest signal and then actual active signal. Each recording was done for a period of 10 s. The bandpass filter is used to eliminate high and low frequencies with a frequency range of 10–300 Hz. Figure 5 shows the power spectral density of EMG signal acquired from zygomatic face muscle.

Ethical approval has been taken from the ethical committee constituted by the institute. Consent of all the subject has been taken prior to the experiments.

2.4 Block Diagram of the Proposed Method

An experiment was performed to acquire FEMG signal. From the acquired signal, it can be noticed that EMG signals have the ability to recognize emotional activity. The recorded signal was then decomposed into multilevel decomposition using WPT method to extract useful details and information that was unseen in recorded signal. After that output is given to the extraction algorithm for feature extraction, before processing, feature selection is proposed to reduce feature dimension filtered.

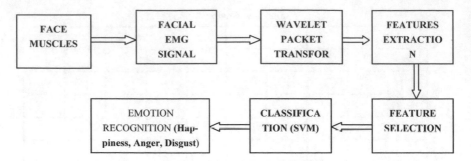

Fig. 6 Block diagram of the proposed method

Selection of features is based on specific threshold value of selected features. For this work, we have selected mean, standard deviation, and energy statistical features. The extracted features were categorized into three facial emotion subclasses by using support vector machine (SVM) classifiers. Figure 6 depicts the block diagram of the proposed work.

2.5 Feature Extraction

For feature extraction, we apply WPT method which generates wavelet packet that shows different frequency sub-bands. Active part of FEMG data contains 20,000 samples with a sampling frequency of 1000 Hz. This active part of FEMG data is decomposed by three levels of decomposition based on wavelet family "db4". Three levels of WPT decomposition produce 1256 wavelet coefficients. Each coefficient has frequency sub-band related to it. The wavelet packet coefficients span the frequency scale of 0–20 Hz. Out of every frequency sub-band, we extracted three different features such as mean, energy, and standard deviation. Figure 7 shows three levels of WPT decomposition producing 1256 samples. Figure 8 shows Linear SVM Classification.

2.6 Support Vector Machine (SVM)

The concept of machine learning enables the machine to learn and perform tasks by developing algorithms and methodologies. Machine learning combines with statistics in many aspects. Many techniques and methodologies were introduced for machine learning quest among which SVM proves to give a better and promising result for classification [14].

SVMs are a set of related supervised learning techniques which interpret data, identified patterns, and used statistical classification analysis [14]. This method was

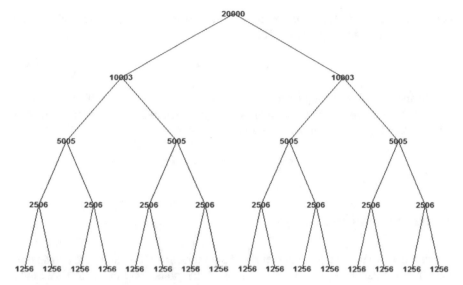

Fig. 7 Wavelet decomposition

Fig. 8 Simple kind of
support vector machine

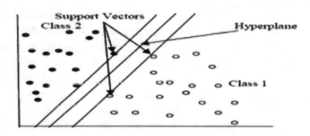

first proposed by Vapnik and Chervonenkis in 1965 and then upgraded for classification in 1992. In recent times, SVM classifier is used in various research works for
pattern recognition and FEMG signal classification [15].

SVM utilizes the idea of a hyperplane where a finest separating hyperplane in
high-dimensional feature space of training data that are mapped by a nonlinear kernel
function. The power of learning and generalization is improved by the use of nonlinear
kernel function [16]. This classification method gives good results compared to other
statistical models [14].

To find a hyperplane, this classified the input space with an extreme boundary.
The optimum hyperplane is established as [17]

$$\mathbf{Za_i + c} \geq +1 \ \mathbf{b_i} = +1 \tag{2.4.1}$$

$$\mathbf{Za_i + c} \leq +1 \ \mathbf{b_i} = -1 \tag{2.4.2}$$

where a_i is the ith input vector, b_i is the class label of the ith input, and Z represents
weight vector. The weight vector Z is always normal to the hyperplane. In the above

equation, c represents bias. The optimal hyperplane parallel to the two margins which is found by the below equation [17]:

$$Za_i + c \leq +1 \tag{2.4.3}$$

The margins found by input vectors are known as support vectors. If the problem is not linearly separable, then by applying kernel function to the input vectors, problem will become linearly separable in a transformed space.

$$Q(y_i, y_j) = \#(y_i)\#(y_j) \tag{2.4.4}$$

3 Results and Discussion

In this work, we have acquired facial EMG signal from zygomaticus and corrugators face muscles using two-channel electrodes. After processing of facial EMG data, features were extracted using WPT method. Mean, energy, and standard deviation features are obtained from the decomposed signal for 12 different subjects.

Twelve different subjects generate 12 groups of facial EMG datasets and each group contains three sample sets for each emotion class. Thus, total numbers of emotion sample sets were 108. Out of these sample sets, 72 sample sets were used to train classifier. The remaining 36 sample sets were used as test samples for the classification of facial emotion (see results in Table 1). Nonlinear SVM were used as classifier for the classification of remaining sample sets.

The WPT method to eliminate feature vectors integrates with support vector machine classifier technique that gives 91.66% accurate recognition rate. On the other hand, we used four sensors of data acquisition system which gives good recognition rate as compared to 1 or 2 channel data acquisition system. Table 2 analyzes the different research works and results in this particular area.

Table 1 Classification results for three types of emotion

Types of emotion	Number of samples	Test samples number identified correctly	Correct identification rate
Happiness	12	12	91.66%
Anger	12	11	
Disgust	12	10	

Table 2 A summary of some of the facial expression databases classification

References	No. of channels	Types of classifier (s)	Selected feature (s)	Classification result (s) (%)
Mohammad Rezazadeh et al. [18]	–	FCM	RMS	92.6
Bernardo et al. [19]	3	Minimum Distance	RMS, MEAN PDS	94.44
Hamedi et al. [20]	2	FCM	RMS	90.8
Sinha and Parsons [21]	2	LDA	–	86
Proposed work	4	SVM	Mean, Standard deviation, Energy	91.66

4 Conclusion and Future Scope

Emotion recognition is a tactful method of understanding human character. This research work mainly emphasizes a technique for recognizing three different types of facial emotion such as happiness, anger, and disgust. The facial signals were recorded using four channels Myon aktos-mini system. For feature extraction, WPT method was applied, which gives better results and decreases the data processing time compared to others features extraction methods. In this work, SVM classifier is used for facial emotion recognition, which has high conversion rate and simple. In training dataset, total 24 groups of EMG signals from each class were selected and the resultant 91.66% accuracy has been obtained. In future work, by increasing numbers of samples in the training dataset, we can get more improve classification efficiency. The application of facial EMG recognition has great perspective in medical engineering fields. Automatic facial emotion recognition can be used in muscular dystrophy and designing human–machine interaction systems. Most of the classification methods which have already used in this particular area suffer from low recognition rate and improper electrode placement. In future work, more diverse and advanced approach to data preprocessing, feature extraction, and combination of classifiers should be used in order to achieve excellent results.

Ethical Approval and Consent to Participate
Ethical Approval is taken from the Ethical committee constituted by this institute. The committee consists of following persons:

Written content has already taken from all participants. This consent is kept at the R&D Cell of the authors' institution. This can be made available at any time during the review process.

Sr. no.	Name of member	Address	Designation in ethical committee	Wheather affiliated to institute
01	Dr. Arya Desh Deepak	Department of Biochemistry, MGM Medical College, Kamothe, New Mumbai	Chairman	No
02	Dr. K. S. Yadav	Associate Professor, Department of Biochemistry, D. Y. Patil Medical College and Hospital, New Mumbai	Member	No
03	Mr. Nitin Yashwante	MSW, Social Worker at D. Y. Patil Medical College and hospital	Medical social worker	No
04	Ms. Pranjal Nayer	Assistant Professor, College of Law, New Mumbai	Legal person	No
05	Dr. Sharukh Tare	Management Teaching Institute	Common man's representative	No
06	Dr. Neeraj Rawani	Professor, Department of Psychiatry, Terna Medical College, New Mumbai	Member	No
07	Dr. Raval Awale	Professor, Electrical Engineering Department, VJTI Mumbai	Member secretary	Yes

References

1. Ang, L.B.P., Belen, E.F., Bernardo, R.A., Boongaling, E.R., Briones, G.H., Coronel, J.B.: Facial expression recognition through pattern analysis of facial muscle movements utilizing electromyogram sensors. In: TENCON 2004 IEEE Region 10 Conference, vol. 3, pp. 600–603 (2004)
2. Haag, A., Goronzy, S., Schaich, P., Williams, J.: Emotion recognition using bio-sensors: first steps towards an automatic system. Affect. Dialogue Syst. (2004)
3. Firoozabadi, M.S.P, Oskoei, M.R.A., Hu, H.: A human-computer interface based on forehead multi-channel bio-signals to control a virtual wheelchair, Tarbiat Modares University, Tehran, Iran (2008)
4. Huang, C.N., Chen, and H.Y. Chung: The review of applications and measurements in facial electromyography. J. Med. Biol. Eng. **25**, 15–20 (2004)
5. Buenaposada, J., Mu nez, E., Baumela, L.: Recognising facial expressions in video sequences. Pattern Anal. Appl. (2008)
6. Kale, S.N., Dudul, S.V.: Intelligent noise removal from EMG signals using focused time lagged recurrent neural network. Appl. Comput. Intell. Soft Comput. (2009)
7. Hargrove, L., Englehart, K., Hudgins, B.: A comparison of surface and intramuscular myo-electric signal classification. IEEE Trans. Biomed. Eng. **54**(5), 847–853 (2007)
8. Hamedi, M., Salleh, S.H., Sweev, T.T., Kamarulafizam: Surface electromyography-based facial expression recognition in bi-polar configuration. J. Comput. Sci. **7**(9), 1407–1415. ISSN 1549-3636 © 2011 Science Publications (2011)
9. Huang, C.-N., Chen, C.-H., Chung, H.-Y.: The review of applications and measurements in facial electromyography. J. Med. Biol. Eng. **25**(1), 15–20 (2004)
10. Englehart, K., Hudgin, B., Parker, P.: A wavelet based continuous classification scheme for multi-function myoelectric control. IEEE Trans. Biomed. Eng. **48**, 302–311 (2001)

11. Chu, J.U., Moon, I., Lee, Y.J., Kim, S.K., Mun, M.S.: A supervised feature-projection-based real time EMG pattern recognition for multifunction myoelectric hand control. Trans. Mech. IEEE/ASME **12**, 282–290 (2007)
12. Thulkar, D., Hamde, S.T.: Facial electromyography for characterization of emotions using lab VIEW. In: IEEE International Conference on Industrial Instrumentation and Control (IClC) (2015)
13. Mohammad Rezazadeh, I., Wan, X., Wang, R., Firoozabadi, M.: Toward affective handsfree human machine interface approach in virtual environments-based equipment operation training. In: 9th International Conference on Construction Applications of Virtual Reality, 5–6 Nov 2009
14. Oskoei, M.A., Hu, H.: Application of support vector machines in upper limb motion classification using myoelectric signals. In: Proceedings of the IEEE International Conference on Robotics and Biomimetics. IEEE Xplore Press, Sanya, pp. 388–393, 15–18 Dec 2007
15. Oskoei, M.A., Hu, H.: Support vector machine-based classification scheme for myoelectric control applied to upper limb. IEEE Trans. Biomed. Eng. (2008)
16. Foster, I., Kesselman, C.: The Grid: Blueprint for a New Computing Infrastructure. Morgan Kaufmann, San Francisco (1999)
17. Bayram, K.S., Kizrak, M.A., Bolat, B.: Classification of EEG signals by using support vector machines. In: IEEE International Symposium on Innovations in Intelligent Systems and Applications (INISTA), pp. 1–3. IEEE (2013)
18. Mohammad Rezazadeh, I., Wan, X., Wang, R., Firoozabadi, M.: Toward affective handsfree human machine interface approach in virtual environments-based equipment operation training. In: 9th International Conference on Construction Applications of Virtual Reality, 5–6 Nov 2009
19. Ang, L.B.P., Belen, E.F., Bernardo, R.A., Boongaling, E.R., Briones, G.H., Coronel, J.B.: Facial expression recognition through pattern analysis of facial muscle movements utilizing electromyogram sensors. In: TENCON 2004 IEEE Region 10 Conference, volume C (2004)
20. Hamedi, M., Salleh, S.-H., Swee, T.T., Kamarulafizam.: Surface electromyography-based facial expression recognition in bi-polar configuration. J. Comput. Sci. **7**(9), 1407–1415 (2011)
21. Sinha, R., Parsons, O.A.: Multivariate response patterning of fear. Conf. Cognit. Emot **10**(2), 173–198 (1996)

Machine Learning for Beach Litter Detection

Sridhar Thiagarajan and G. Satheesh Kumar

Abstract People from economically weaker sections find consolation in doing unskilled tasks which are easily available, though a few of them are certainly not for humans. They have to be replaced with robots since these jobs fall into the categories of dull, dirty, difficult, and dangerous jobs. Exclusion of human involvement in the demeaning tasks and provision of hygienic environments around the beach areas would contribute to a better opportunity for a country's tourism. The task of implementation of beach cleaning with robots throws many technical challenges, a few of which are addressed in this research work. Machine learning has influenced the progress and outcomes for various domains of engineering and science including statistics. Even in social domains the impact of advent of Machine learning has been felt by not just the end users, but also the researchers. In this work, different methods for classifications of beach litter are proposed and evaluated. A dataset is collected and the classifiers are evaluated based on various metrics. An appropriate classifier is then selected based on these metrics, and the system is used on a beach cleaning robot.

Keywords Machine learning · Classification · HOG · ConvNets
Data augmentation · SVM · Robotics · Beach

1 Introduction

Technology has helped people at all levels irrespective of their social or economic background. World over, robots [1–4] have been used for unknown terrain navigation and path-planning applications and have met success of varying degrees. Though the reach of technology is only limited by the delay of implementation, certain gray areas

S. Thiagarajan (✉) · G. Satheesh Kumar
SSN College of Engineering, Chennai, India
e-mail: thiagarajan14111@mech.ssn.edu.in

G. Satheesh Kumar
e-mail: satheeshkumarg@ssn.edu.in

© Springer Nature Singapore Pte Ltd. 2019
M. Tanveer and R. B. Pachori (eds.), *Machine Intelligence and Signal Analysis*,
Advances in Intelligent Systems and Computing 748,
https://doi.org/10.1007/978-981-13-0923-6_22

are yet to receive the attention of researchers, like the task of beach cleaning. Among a few beach cleaning robots that have been proposed and implemented some of those robots [5, 6] are for heavy duty application which might consume large amount of power and hence not allowing the usage of renewable energy. And as in [7–9], since they are wirelessly driven because of the human intervention required, there is a need for robots with autonomous capability.

Though machine learning algorithms have been actively researched [10–16] for replacement of humans from dirty and difficult tasks, beach cleaning deserves a thorough research from its implementation perspective. The parameters that govern the working of the algorithm are causatively dependant on the dynamic inputs fed into the robots. The task of cleaning the beaches of plastic bottles and cups is one which engages enormous manpower and consumes a lot of time. Hence, we propose to use a machine learning system for the identification of this litter. Coupled with a robot, this ensures that enormous manpower is not spent on menial tasks. In order to form a holistic solution to the beach litter problem to be foolproof and long lasting, we have defined the objectives of the research work as follows:

- To design, develop, and analyze cooperative robots and self-learning algorithms for cleaning applications over the beaches;
- To implement coordinated navigation incorporating obstacle avoidance and path-planning and perform experimental investigations with the developed algorithm; and
- To implement robot-to-robot communication for effective operation and perform onsite real-time experiments and modifications.

According to the scope of this paper, results and discussions in this paper are limited to the second part of the first objective. The following sections describe the methodology adopted to implement the objective, followed by results and discussions.

2 Methodology

Identification of debris on the beach stretch could be achieved in many ways. Given the scale of operation image processing seems to be a viable option on the outset. Given the three-dimensional nature of the debris and the randomness in shape, size, and conditions of the debris and their infinite possibilities, application of machine learning seems to be the only way out. This would bestow the robot with a capability to self-learn based on the experience they gain while they work. Debris found on the beach could be broadly classified into two types:

- visible (to naked eye) debris and
- hidden or buried debris.

The proposed methodology is meant for identifying and clearing the visible debris. Examples of visible debris include paper and plastic cups, bottles, corn cobs, other

(a) **(b)** **(c)**

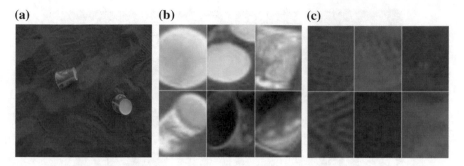

Fig. 1 **a** A sample image from the collected data; **b** Samples of positive instances extracted from the images; **c** Sample negative instances extracted from the images

paper, and plastic wastes. Biodegradable objects like twigs, leaves, etc. are excluded as they are not a major concern. Shells and living things like crabs and insects are also to be spared or avoided. One way of locating and clearing the buried debris is to extensively sweep every inch of the beach once with a robot having autonomous capability checking for debris at a depth of at least 5 cm. If implemented, this operation could be repeated only once every year and hence the cost involved would be controlled.

2.1 Data Preparation

135 images were collected at varying lighting conditions, some of them containing litter (Fig. 1). Positive and negative instances were extracted from them, using an image labeling tool. Each instance is a 32 × 32 RGB image. Once these instances are extracted, we apply data augmentation techniques on them to increase the size of the dataset (Fig. 2). Hearst et al. [10] showed that these techniques help bridge the gap between robot vision and computer vision. These include random rotations and pixel intensity changes and addition of random noise. Finally, the dataset consists of 400 positive samples and 800 negative samples. We show using a validation accuracy versus data size plot that these augmentations to increase the dataset size improve the machine learning system.

2.2 Classification

We formulate this problem as a classification problem. We choose not to pose this problem as purely segmenting out the sand, as this would not allow us to avoid picking up stones/beach reptiles. We evaluate four different classifiers on our dataset: Support vector machine (SVM) [11], histogram of oriented gradients + SVM [12],

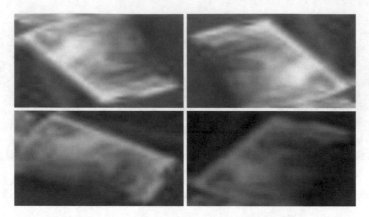

Fig. 2 Examples of augmentations, rotation, and pixel intensity transformations

Table 1 Parameters of HOG computation

Parameter	Value
Cell size	(3, 3)
Block size	(6, 6)
Block stride	(2, 2)
nBins	9
Window stride	(6, 6)

a fully connected network [13], and a convolutional neural network [14]. In the final robot, model was chosen keeping in mind runtime constraints and the metrics evaluated of the classifier. A Gaussian blur operation was performed on the image as preprocessing. Non-Maximum suppression [15, 16] was applied to bounding boxes to detect unique locations of litter.

(a) ***Linear SVM***: 32 × 32 × 3 images were fed directly to the classifier to evaluate its performance. As expected, with just raw pixels, it performs poorly. This was taken as baseline performance of the machine learning.

(b) ***HOG+SVM***: Histogram of oriented gradients was computed for the color image. This gives the SVM contextual information. Parameters of HOG computation are listed in Table 1.

(c) ***Fully connected neural network***: Two layers fully connected network with ReLU activation were used, with 64 and 32 neurons in each layer, with cross-entropy loss.

(d) **Convolutional neural network**: Two convolutional layers with 32 filters each, followed by fully connected layer with 16 neurons. Cross-entropy loss was used.

3 Model Selection

The machine learning system for the cleaning robot needs to be chosen primarily based on two factors.

(a) **Recall**: In litter detection, recall is the most important factor compared to other metrics like precision/accuracy. This is because it is acceptable to classify sand wrongly as litter (low precision), but not acceptable to wrongly classify litter as sand (low recall), as the latter classification results in the robot failing to pick up litter. Recall of various classifiers can be seen in Table 2.

(b) **Runtime**: The machine learning system being part of an autonomous robot has to have relatively low runtime and memory requirements. Runtime of various classifiers is evaluated as shown in Fig. 3.

Table 2 Results on test data for various classifiers

Classifier	Test accuracy (%)	Test precision	Test recall
Linear SVM	93.41	0.9096	0.8833
HOG + Linear SVM	97.11	0.9562	0.9259
Two layers fully connected network with ReLU activations	98.30	0.9462	0.9932
Two layers convolutional neural network, 16 filters, 16 neurons in final fully connected layer	99.87	0.9944	1

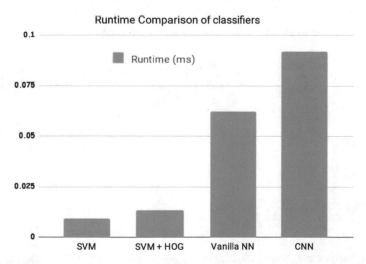

Fig. 3 Runtime comparison of various classifiers on an image (ms). Image is reshaped to 160 × 160, and then sliding window of stride 12 is applied. Note: in SVM + HOG, HOG is computed for whole preprocessed image once, and then pipeline is applied. Tests are conducted on an Intel I7-4510U CPU @ 2.00 GHz machine

The convolutional neural network has the highest recall among the classifiers under consideration, but has a runtime that is 3x that of the fully connected neural network. The fully connected network has close to perfect recall and has a relatively lower runtime. The Support Vector Machine (SVM), and SVM + Histogram of Oriented Gradients (HOG) classifiers do not have satisfactory recall rates. Hence, the fully connected network is chosen as our final model.

4 Performance Versus Data Size

The decision whether to obtain more training examples or to improve/change the machine learning model is often a crucial one, especially in cases where it is time-consuming to obtain more data. An important test that is done to decide this is by looking at the performance of our classifier as a function of data size. If it is noted that more data does not improve the accuracy, it is pointless to obtain more training samples. Hence, we evaluated the performance of one of our classifiers (SVM + HOG) as a function of the training set size, (Fig. 4). It is clear that the performance has saturated, and further training data will not help classification accuracy. Hence, we made the decision to not collect further training examples.

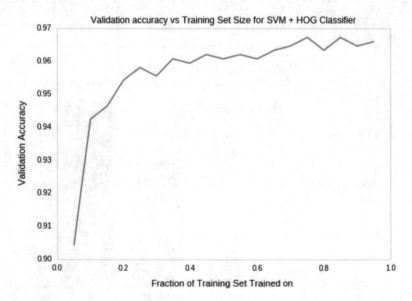

Fig. 4 Validation accuracy versus fraction of training set trained on for SVM + HOG classifier

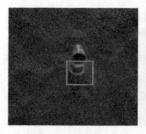 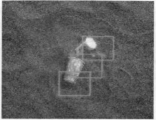 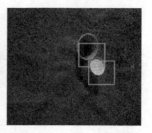

Fig. 5 Sample bounding box detections after non-maximum suppression

5 Sample Results

Figure 5 shows the output of the machine learning classifier on a test image. As stated in 2.2, sliding window classifier is used on the Gaussian blurred image, and finally non-maximum suppression is performed.

6 Conclusion and Future Work

A dataset containing litter (cups) was collected, and positive and negative instances were extracted. We formulated this as a classification problem rather than a pure segmentation of sand problem, in order to avoid picking up rocks/beach reptiles, etc. Classification algorithms were proposed and trained on this dataset. Metrics like accuracy and recall were evaluated on the train and test data and this was used to compare the algorithms. Runtime evaluation was done, and this combined with recall was determined to be the most important. The fully connected neural network was chosen to be the final model to be deployed on the robot. We plan to include further classes of litter in our dataset and expect similar performance from the algorithms. This machine learning system combined with localization algorithms like visual odometry will be combined and deployed on a robot, in order to successfully perform beach cleaning operation.

References

1. Ichimura, T., Nakajima, S.-I.: Development of an autonomous beach cleaning robot Hirottaro. In: IEEE International Conference on Mechatronics and Automation (ICMA), 7–10 Aug 2016. https://doi.org/10.1109/icma.2016.7558676
2. Wattanasophon, S., Ouitrakul, S.: Garbage collection robot on the beach using wireless communications. Int. Proc. Chem. Biol. Environ. Eng. **66**, 92–96 (2014)
3. Pinheiro, P., Cardozo, E., Wainer, J., Rohmer, E.: Cleaning task planning for an autonomous robot in indoor places with multiples rooms. Int. J. Mach. Learn. Comput. **5**(2), 86–90 (2015)

4. Nagasaka, Y., Saito, H., Tamaki, K., Seki, M., Kobayashi, K., Taniwaki, K.: An autonomous rice transplanter guided by global positioning system and inertial measurement unit. J. Field Robot. **26**, 537–548 (2009)
5. Fedio, C., Panico, G.: http://www.cbc.ca/news/canada/ottawa/robot-grrl-bowie-beach-cleaning-1.3604109. Accessed 12 Apr 2017
6. Dronyx: http://www.dronyx.com/solarino-beach-cleaner-robot/ (2016). Accessed 12 Apr 2017
7. Cartonerdd: http://www.instructables.com/id/Robot-Crab-Beach-Cleaning/ (2016). Accessed 12 Apr 2017
8. Le, A.T., Rye, D.C., Durrant-Whyte, H.F.: Estimation of track-soil interactions for autonomous tracked vehicles. In: IEEE International Conference on Robotics and Automation (1997). https://doi.org/10.1109/robot.1997.614331
9. Arkin, R.C.: Motor schema-based mobile robot navigation. Int. J. Robot. Res. **8**(4), 92–112 (1989)
10. Hearst, M.A., Dumais, S.T., Osuna, E., Platt, J., Scholkopf, B.: Support vector machines. IEEE Intell. Syst. Appl. **13**(4), 18–28 (1998). https://doi.org/10.1109/5254.708428
11. Dalal, N., Triggs, B.: Histograms of oriented gradients for human detection. IEEE Computer Society Conference on Computer Vision and Pattern Recognition (CVPR'05), vol. 1, no. 1, pp. 886–893, San Diego, CA, USA (2005). https://doi.org/10.1109/cvpr.2005.177
12. D'Innocente, A., Carlucci, F.M., Colosi, M., Caputo, B.: Bridging between computer and robot vision through data augmentation: a case study on object recognition. In: International Conference on Computer Vision Systems, pp. 1–10 (2017). arXiv:1705.02139
13. Krizhevsky, A., Sutskever, I., Hinton, G.E.: ImageNet classification with deep convolutional neural networks. In: proceedings of NIPS 2012 Advances in Neural Information Processing Systems, vol. 25, pp. 1–10 (2012)
14. Dollár, P., Appel, R., Belongie, S., Perona, P.: Fast feature pyramids for object detection. IEEE Trans. Pattern Anal. Mach. Intell. **36**(8), 1532–1545 (2014). https://doi.org/10.1109/tpami.2014.2300479
15. Ichimura, T., Nakajima, S.: Development of an autonomous beach cleaning robot Hirottaro. In: Proceedings of the IEEE International Conference on Mechatronics and Automation, pp. 868–872
16. Browne, M., Ghidary, S.S.: Convolutional neural networks for image processing: an application in robot vision. In: Gedeon, T.D., Fung, L.C.C. (eds.) AI 2003: Advances in Artificial Intelligence. AI 2003. Lecture Notes in Computer Science, vol. 2903, pp. 641–652, Springer, Berlin, Heidelberg (2003). https://doi.org/10.1007/978-3-540-24581-0_55

Estimation of Sampling Time Offsets in an N-Channel Time-Interleaved ADC Network Using Differential Evolution Algorithm and Correction Using Fractional Delay Filters

M. V. N. Chakravarthi and Bhuma Chandramohan

Abstract Higher sampling rates are essential in any communication system at present due to the demand for higher data rates. Such high sampling rates can be achieved with time-interleaved analog-to-digital converters (TI-ADCs). Even though TI-ADCs are faster, the sampling time offset is a setback. The sampling time offsets present in ADCs generate nonuniform samples. In the reconstruction process, these nonuniform samples might produce an erroneous signal. In this work, estimation of these sampling time offsets is performed using differential evolution algorithm. The proposed algorithm efficiently detects the sampling time offsets with minimum number of iterations. The estimated sampling time offsets are used to reconstruct the signal using fractional delay filters. Performance of the proposed algorithm is tested by considering various signals, i.e., speech signal, sinusoidal, and amplitude modulated signal (AM). Signal-to-noise ratio (SNR) and signal-to-noise distortion ratio (SNDR) are calculated for the signals. The results are compared with the existing works and noteworthy improvement with the proposed algorithm is demonstrated.

Keywords Time-interleaved ADC · Sampling time offset
Differential evolution · Fractional delay filter

M. V. N. Chakravarthi
Department of Electronics & Communication Engineering,
Acharya Nagarjuna University, Guntur, India
e-mail: chakrimvn@yahoo.com

B. Chandramohan (✉)
Department of Electronics & Communication Engineering,
Bapatla Engineering College, Bapatla, India
e-mail: chandrabhuma@yahoo.co.in

© Springer Nature Singapore Pte Ltd. 2019
M. Tanveer and R. B. Pachori (eds.), *Machine Intelligence and Signal Analysis*,
Advances in Intelligent Systems and Computing 748,
https://doi.org/10.1007/978-981-13-0923-6_23

1 Introduction

The sampling of high-frequency signals requires high switching rates of ADCs and is difficult with a single ADC. Such high sampling frequencies can be achieved with time-interleaved ADCs (TI-ADCs). The sampling frequency of each ADC is F_s/N, where F_s is the overall sampling frequency required and N is the number of ADCs. However, one of the limitations of TI-ADCs is the sampling time offsets present in the ADCs. These sampling time offsets generate nonuniform samples and result in an erroneous signal. Many algorithms are proposed for the estimation and correction of these sampling time offsets. Comb filters are used for offset cancellation and channel gain equalization in [1]. The calibration technique [2] makes use of the autocorrelation characteristics of the input signal to estimate mismatch errors. The correction scheme uses an improved fractional delay filter based on Farrow structure.

A novel background fast convergence algorithm for timing skew is proposed in [3]. A technique to obtain the derivative information of the analog input directly and use it to calibrate the timing skew and input bandwidth mismatch errors in TI-ADCs was reported [4]. A simple sampling clock skew correction technique can adjust the inherent delay in the input signal sampling network rather than the sampling clock [5]. A band-limited pseudorandom noise sequence is employed as the desired output of the TI-ADC and simultaneously is also converted into an analog signal, which is injected into the TI-ADC as the training signal during the calibration process. Adaptive calibration filters with parallel structures are used to optimize the TI-ADC output to the desired output [6]. Gain, offset, and timing mismatches are estimated and corrected, and input signal applied is either complex exponential or sinusoidal signals [7]. The error estimation based on the correlationship between the output samples of every two adjacent sub-ADCs and calibration based on the Taylor series expansions is proposed [8]. A single standard version of the time-interleaved pipeline ADC, which meets Bluetooth specifications while minimizing power consumption and mismatch issues inherent to the architecture, is reported [9]. A correction subsystem, which consists of $M-1$ adaptive FIR filters whose function is to compensate the error phase on the undesired alias signals, is presented by [10]. The sampling time error detection is achieved with Hilbert transform filter and the error is given as feedback to an adaptive filter used to correct the error [11].

A digital background calibration technique based on interpolation is proposed to reduce the channel timing error effects, specifically the timing clock skew effects [12]. The sampling and reconstruction of a signal in TIADCs involves multirate signal processing which is discussed in detail in [13]. In this work, differential evolution algorithm is used to estimate the sampling time offsets. These offsets are corrected using fractional delay filters. This paper is divided into six sections. In Sect. 2, the structure of time-interleaved ADCs TI-ADCs is covered. Section 3 discusses sampling time offsets present in TI-ADCs and their effect on reconstruction of the signal. The estimation of sampling time offsets by using differential evolution algorithm is covered in Sect. 4. Corrections of the sampling time offsets using fractional delay filters and results are covered in Sects. 5 and 6 respectively.

Fig. 1 N-Channel TI-ADC
network

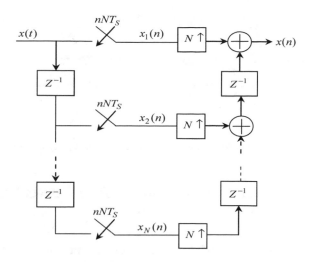

2 Time-Interleaved ADC Network

The sampling process of a signal can be performed by using more than one ADC
to achieve higher sampling frequencies. The successive ADCs are time-interleaved
by one sampling period (T_S). As shown in Fig. 1, an N-channel time-interleaved
ADC network consists of N-ADCs. The sampling period of each ADC is NT_S. The
samples of these ADCs are up-sampled, delayed, and then added to reconstruct the
signal with a sampling period of T_S. When a signal $x(t)$ is sampled by TI-ADC
network, the sampled signals from these ADCs can be represented as $x_i(n)$. After
reconstruction, the signal obtained is given by $x(n)$.

3 Sampling Time Offsets in TI-ADC Network

In a TI-ADC network, during the sampling process, sampling time offsets are present
in the ADCs due to inaccurate switching speeds. These offsets result in the generation of nonuniform samples after reconstruction. The time interval between two
successive samples is not uniform as shown in Fig. 2. The sampling time offsets are
denoted by Δt_i. These offsets are represented as fractions of sampling time T_S.

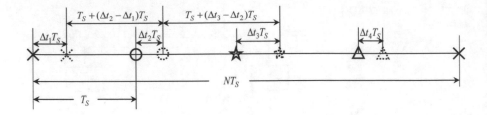

Fig. 2 Nonuniform samples generated due to sampling time offsets

4   Estimation of Sampling Time Offsets Using Differential Evolution Algorithm

To recover the uniform samples from the nonuniform samples, first the sampling time offsets should be estimated and then correction should be performed. The present work deals with estimation of these sampling time offsets and applying correction mechanism. The estimation of the sampling time offsets is performed using differential evolution (DE) algorithm. The proposed algorithm makes use of a sinusoidal test signal $(x_{test}(t))$ given by Eq. (1), for estimation of the offsets. As shown in Fig. 3, the erroneous samples from the i-th ADC are represented by $x_{test\_ei}(n)$, given by Eq. (2). These samples are fed to DE algorithm using which it performs estimation of the sampling time offsets $\left(\Delta t_i'\right)$.

$$x_{test}(t) = \sin(\Omega t) \tag{1}$$

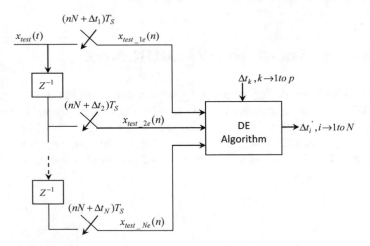

Fig. 3 Estimation of sampling time offset using DE

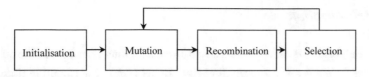

Fig. 4 Block diagram of DE algorithm

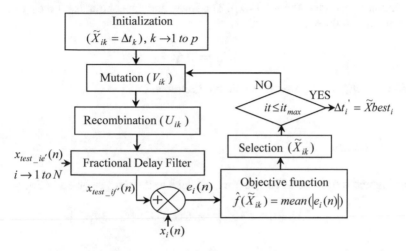

Fig. 5 Differential evolution algorithm flow graph

where Ω is the analog frequency in rad/s given by $\Omega = 2\pi F$ and F is the frequency in $cycles/s$.

$$x_{test\_ei}(n) = \sin \omega(nN + i - 1 + \Delta t_i) \tag{2}$$

where ω is the digital frequency in $rad/sample$ given by $\omega = 2\pi f$ and f is the frequency in $cycles/sample$.

The DE algorithm involves the steps shown in Fig. 4.

1. Initialisation.
2. Mutation.
3. Recombination.
4. Selection.

The flow graph for the algorithm is shown in Fig. 5.

4.1   Initialization

Since the algorithm is used for the estimation of sampling time offsets, the optimized parameter $\left(\tilde{X}_{ik}\right)$ is taken as sampling time offset (Δt_k). The sampling time offsets are represented as fractions of the sampling time period (T_S) and are initialized with a population represented as $k \rightarrow 1\, to\, p$. The lower and upper bounds for the parameter are defined as $0.00 \leq \tilde{X}_{ik} \leq 0.99$.

$$\tilde{X}_{ik} = \Delta t_k \tag{3}$$

The subscript (i) represents the channel number, i.e., $i \rightarrow 1\, to\, N$.

4.2   Mutation

The weighted difference of two vectors is added to the third as given in Eq. (4).

$$V_{ik} = \tilde{X}best_i + \tilde{F}\left(\tilde{X}_i(rand_1(p)) - \tilde{X}(rand_2(p))\right) \tag{4}$$

where $\tilde{F}$ is called as the mutation factor which is a constant selected in between 0 and 2 and $rand_1(p), rand_2(p)$ are the random number from $[1, 2, \ldots, p]$. The vector V_{ik} is called as the donor vector.

4.3   Recombination

Recombination incorporates successful solutions from the previous generation. The trial vector U_{ik} is developed from the elements of the target vector $\tilde{X}_{ik}$, and the elements of the donor vector, V_{ik} as given Eq. (5). Elements of the donor vector enter the trial vector with probability CR (Cross-over Ratio).

$$U_{ik} = \begin{cases} V_{ik}, rand(p) \leq CR\, or\, k = randi(p) \\ \tilde{X}_{ik}, rand(p)\, CR > and\, k \neq randi(p) \end{cases} \tag{5}$$

4.4   Selection

The parameter is optimized by the minimization of an objective function ($\hat{f}$), given by Eq. (6).

$$\hat{f}(\tilde{X}_{ik}) = mean\,(\,|\,e_i(n)\,|\,) \tag{6}$$

$$e_i(n) = x_{test\_if'}(n) - x_i(n) \tag{7}$$

Here $x_{test\_if'}(n)$ is the signal obtained after filtering $x_{test\_ie'}(n)$ with a fractional delay filter. The fractional delay is set as $\tilde{X}_{ik}$. After determining the objective function ($\hat{f}$) with the initial parameter values $\tilde{X}_{ik}$, the best parameter for the ith channel is determined. $\tilde{X}best_i$ is the parameter value for which objective function ($\hat{f}$) is minimum. The target vector is $\tilde{X}_{ik}$ compared with the trial vector V_{ik} and the one with the lowest objective function value ($\hat{f}$) is selected as given in Eq. (8). Mutation, recombination and selection continue until some stopping criterion is reached. The stopping criterion here is the number of iterations, $it \le it_{max}$.

$$\tilde{X}_{ik} = \begin{cases} U_{ik}, & if\ \hat{f}(U_{ik}) \le \hat{f}(\tilde{X}_{ik}) \\ \tilde{X}_{ik}, & otherwise \end{cases} \tag{8}$$

After the termination criterion is met the estimated sampling time offset for the ith channel is taken as $\Delta t_i' = \tilde{X}best_i$.

5 Correction of Sampling Time Offsets

Once the sampling time offsets are estimated, they are used for correction and reconstruction of uniform samples from the nonuniform samples is done. The reconstruction method is as shown in Fig. 6. If an arbitrary signal $x(t)$ is sampled with a sampling time offset of Δt, then the samples so obtained are erroneous. The correct sample values can be obtained by delaying the signal by Δt, which is illustrated in Fig. 6. For reconstruction, the input signal is delayed by a fractional delay which is equal to the sampling time offset present in the ADC.

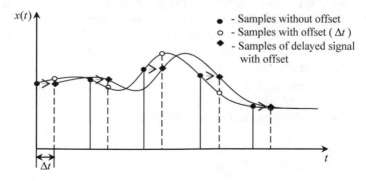

Fig. 6 Correction of sampling time offset by delaying the input signal

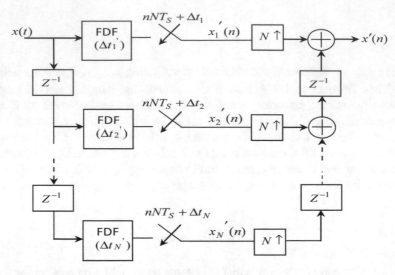

Fig. 7 Correction of sampling time offsets using fractional delay filters

As shown in Fig. 7, the fractional delay filter (FDF) is placed in each channel path to introduce a fractional delay $(\Delta t_i')$ estimated by the DE algorithm. The reconstructed signal is $x'(n)$ which is almost equal to signal $x(n)$ obtained in TI-ADC network without sampling time offsets.

6   Results

6.1   Estimation of Sampling Time Offsets

The sampling time offsets are estimated using DE algorithm with mutation factor $\tilde{F} = 0.8$ and cross-over ratio $CR = 0.4$. The number of channels N is taken as 4. The sampling time offsets (Δt_i) introduced into the four-channel TI-ADC network are 0.00, 0.02, 0.01, and 0.015. The sampling time offsets estimated are exactly equal to the actual offsets introduced, i.e., $\Delta t_i' = \Delta t_i$. The termination criterion is the number of iterations (it_{max}) and is taken as 10. The initialisation of the parameter $\tilde{X}_{ik}$ is done with a population (p) of 20. The convergence of the estimated sampling time offsets $\Delta t_i'$ toward the actual offsets Δt_i is given in Fig. 8.

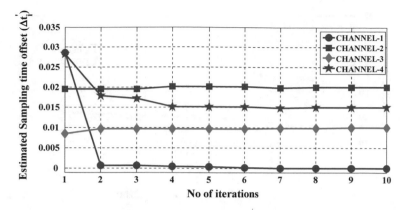

Fig. 8 Convergence of sampling time offsets of four channels

7 Correction of Sampling Time Offsets

The correction of the sampling time offsets is performed by using a six-tap fractional delay filter. The correction is performed with sinusoidal, speech, and amplitude modulated (AM) signals as input signals. For a sinusoidal signal, signal-to-noise ratio (SNR) and signal-to-noise distortion ratio (SNDR) versus normalized frequency (f) plot is shown in Fig. 9. For AM signal, the message and the carrier signal frequencies taken are $F_m = 100\,\mathrm{Hz}$, $F_c = 3\,\mathrm{kHz}$, respectively. The SNR versus sampling frequency, $F_S(\mathrm{kHz})$ plot, is shown in Fig. 10. Plots depicting the spectra of original speech signal, speech signal from TI-ADC network before correction, and speech signal from TI-ADC network after correction are shown in Figs. 11, 12, and 13, respectively.

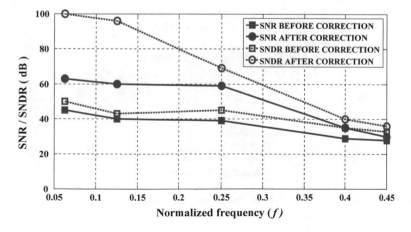

Fig. 9 SNR/SNDR versus normalized frequency

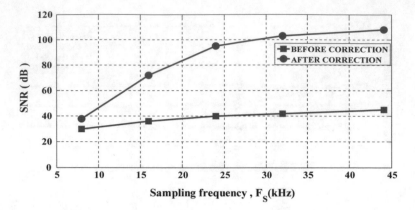

Fig. 10 SNR versus sampling frequency

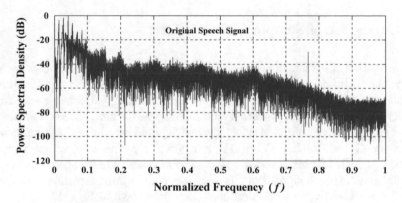

Fig. 11 PSD of the original speech signal

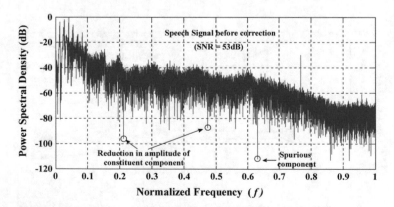

Fig. 12 PSD of the speech signal from TI-ADC network before correction

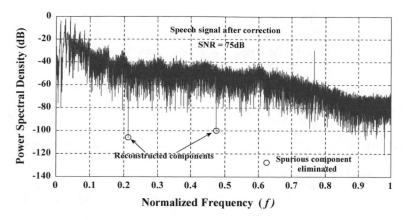

Fig. 13 PSD of the speech signal from TI-ADC network after correction

Table 1 Performance comparison

	Ref. [2]	Ref. [3]	Ref. [4]	Proposed work
Channels	4	8	4	4
Mismatch type	Timing	Timing	Timing	Timing
Frequency range	0–0.45	0–0.44	0–0.45	0–0.45
Filters	1 (5 taps)	7 (8 taps)	–	1 (6 taps)
Max SNR and *SNDR* (dB)	48.9 and 50	47.3	49.8	63 and 100 (sine)
				75 (speech)
				108 (AM)

The speech signal is a 1 kHz signal with a sampling frequency of 44.1 kHz. After correction, the amplitude of constituent components is improved and also spurious component is also eliminated. The SNR before and after correction are 53 dB and 75 dB, respectively, with an improvement of over 20 dB.

As shown in Table 1, the performance comparison is done with respect to the signal-to-noise ratio (SNR) and signal-to-noise and distortion ratio (SNDR). With the proposed algorithm, an improvement in the SNR and SNDR for a sinusoidal, speech, and AM input signals is shown.

8 Conclusion

A differential evolution algorithm based algorithm is proposed for accurately estimating sampling time offsets. They are corrected using fractional delay filters at the reconstruction level. Performance is compared with the existing algorithms in terms of SNR and SNDR. The correction provided by the fractional delay filters offered better SNR and SNDR as given in Table 1. The algorithm is tested using a sinusoidal signal, speech signal, and amplitude modulated signal.

References

1. Beydoun, Ali, Nguye, Van-Tam, Naviner, L., Loumeau, P.: Optimal digital reconstruction and calibration for multichannel time interleaved ADC based on comb-filters. Int. J. Electron. Commun. **67**, 329–339 (2013)
2. Chen, Hongmei, Pan, Yunsheng, Yin, Yongsheng, Lin, Fujiang: All-digital background calibration technique for timing mismatch of time interleaved ADCs. Integr. VLSI J. **57**, 45–51 (2017)
3. Li, D., Zhu, Z., Zhang, L., Yang, Y.: A background fast convergence algorithm for timing skew in time-interleaved ADCs. Microelectron. J. **47**, 45–52 (2015)
4. Benwei, Xu, Chiu, Yun: Comprehensive background calibration of time-interleaved analog-to-digital converters. IEEE Trans. Circuits Syst. I Regul. Pap. **62**, 1306–1314 (2015)
5. Prashanth, D., Seung, H: A sampling clock skew correction technique for time-interleaved SAR ADCs. In: GLVLSI'16 Proceedings of the 26th edition on Great Lakes Symposium on VLSI, pp. 129–132 (2016)
6. Liu, S.J., Qi, P.P., Wang, J.S., Zhang, M.H., Jiang, W.S.: Adaptive calibration of channel mismatches in time-interleaved ADCs based on equivalent signal recombination. IEEE Trans. Instrum. Meas. **63**, 277–286 (2014)
7. Qin, J., Liu, G.M., Guo, M.G.: Adaptive calibration method for timing mismatch error in time-interleaved ADC system. Chin. J. Sci. Instrum. **34**, 2371–2375 (2013)
8. Yao, Y., Yan, B., Li, G., Lin, S.: Adaptive blind compensation of timing mismatches in four-channel time-interleaved ADCs. In: 2013 International Conference on Communications, Circuits and Systems (ICCCAS), pp. 232–234 (2013)
9. Carvajal, W., Van Noije, W.: Time-interleaved pipeline ADC design: a reconfigurable approach supported by optimization. In: SBCCI Proceedings of the 24th Symposium on Integrated Circuits And Systems Design, pp. 17–22 (2011)
10. Camarero, D., Naviner, J.-F.: Digital background and blind calibration for clock skew error in time-interleaved analog-to-digital converters. In: SBCCI'04 Proceedings of the 17th Symposium on Integrated Circuits and System Design, pp. 228–232 (2004)
11. Jamal, S.M., Fu, D., Singh, M.P., Hurst, P.J., Lewis, S.H.: Calibration of sample-time error in a two-channel time- interleaved analog-to-digital converter. IEEE Trans. Circuits Syst. I **51**, 130–139 (2004)
12. Jin, H., Lee, E.K.F.: A digital-background calibration technique for minimizing timing-error effects in TI-ADCs. IEEE Trans. Circuits Syst. II Analog Digit. Signal Process. **47**, 603–613 (2000)
13. Vaidyanathan, P.P.: Multirate Systems and Filter Banks. Pearson, India, (1993)

Double Density Dual-Tree Complex Wavelet Transform-Based Features for Automated Screening of Knee-Joint Vibroarthrographic Signals

Manish Sharma, Pragya Sharma, Ram Bilas Pachori
and Vikram M. Gadre

Abstract Pathological conditions of knee-joints change the attributes of vibroarthrographic (VAG) signals. Abnormalities associated with knee-joints have been found to affect VAG signals. The VAG signals are the acoustic/mechanical signals captured during flexion or extension positions. The VAG feature-based methods enable a non-invasive diagnosis of abnormalities associated with knee-joint. The VAG feature-based techniques are advantageous over presently utilized arthroscopy which cannot be applied to subjects with highly deteriorated knees due to osteoarthritis, instability in ligaments, meniscectomy, or patellectomy. VAG signals are multicomponent nonstationary transient signals. They can be analyzed efficiently using time–frequency methods including wavelet transforms. In this study, we propose a computer-aided diagnosis system for classification of normal and abnormal VAG signals. We have employed double density dual-tree complex wavelet transform (DDDTCWT) for sub-band decomposition of VAG signals. The L_2 norms and log energy entropy (LEE) of decomposed sub-bands have been computed which are used as the discriminating features for classifying normal and abnormal VAG signals. We have used fuzzy Sugeno classifier (FSC), least square support vector machine (LS-SVM), and sequential minimal optimization support vector machine (SMO-SVM)

M. Sharma (✉)
Department of Electrical Engineering, Institute of Infrastructure, Technology,
Research and Management (IITRAM), Ahmedabad 380026, India
e-mail: manishsharma.iitb@gmail.com

P. Sharma
Department of Electronics and Communication Engineering,
Acropolis Institute of Technology and Research, Indore, India
e-mail: pragyasharma1512@gmail.com

R. B. Pachori
Discipline of Electrical Engineering, Indian Institute of Technology Indore,
Indore 453552, India
e-mail: pachori@iiti.ac.in

V. M. Gadre
Discipline of Electrical Engineering, Indian Institute of Technology Bombay,
Mumbai, India
e-mail: vmgadre@ee.iitb.ac.in

© Springer Nature Singapore Pte Ltd. 2019
M. Tanveer and R. B. Pachori (eds.), *Machine Intelligence and Signal Analysis*,
Advances in Intelligent Systems and Computing 748,
https://doi.org/10.1007/978-981-13-0923-6_24

classifiers for the classification with tenfold cross-validation scheme. This experiment resulted in classification accuracy of 85.39%, sensitivity of 88.23%, and a specificity of 81.57%. The automated system can be used in a practical setup in the monitoring of deterioration/progress and functioning of the knee-joints. It will also help in reducing requirement of surgery for diagnosis purposes.

Keywords Vibroarthrographic (VAG) signals · Analytic complex wavelet transform · Computer-aided diagnosis system · Support vector machine (SVM)

1 Introduction

One of the most commonly injured and vulnerable joints in the human body are knee-joints. Aging, activities that involve physical stress, such as exercises, sports, etc., accidental injuries, hereditary causes and obesity may lead to the damage of ligaments and articular cartilage of knee-joints, causing osteoarthritis, and other such knee-related problems [1]. The osteoarthritis (OA) is a chronic disease that causes pain, swelling, restricted range of motion, and stiffness in the joints. The OA has been reported to be one of the leading cause of disability in older adults [2]. The report on "Global Burden of Disease Study" [3] states that approximately 10–15% of people over age 60 suffering from some level of OA, with knee-joint as most commonly affected area. Women are more likely to be affected by OA than men [3]. Furthermore, musculoskeletal disorders attribute to 6.8% disability-adjusted life years (DALYs) all over the world [4]. United nation's report mentions that by year 2050, 20% of the world population will be over 60. Among them, an estimated 20% (130 million) will have some symptoms of OA and one-third (40 million) of whom will be disabled severely [5]. Thus, early detection and diagnosis is an important aspect in treatment and recovery of knee-joint OA. Currently available procedures involve both invasive and noninvasive approaches, such as physical examination, magnetic resonance imaging (MRI), X-ray, and computer tomography (CT) but they do not facilitate early diagnosis. Arthroscopy provides more information and is so far best available method for identification of knee-joint disorders. However, it is semi-invasive and is not practical in the case of people with highly deteriorated knee-joints.

The vibroarthrographic (VAG) signals are auditory vibrations emitted by mid-patella during flexion and extension movements and are supposed to represent pathological states of the knee-joint [1]. These signals are multicomponent and nonstationary exhibiting that joint surfaces may not join same way at a given position and time. Various signal processing and pattern detection algorithms have been employed to analyze VAG signals and to compute the features related to roughness, softening, breakdown, or lubrication state of the articular cartilage surfaces. Various pattern identification and signal processing techniques have been introduced previously utilizing VAG signals [6–8]. These methods include a range of time-domain, frequency domain, and time–frequency domain approaches that employ linear and nonlinear features for identification of abnormalities.

In the present work, we introduce a new approach for screening of VAG signals using double density dual-tree complex wavelet transform (DDDTCWT). The L_2 norm and log energy entropy (LEE) of sub-bands (SBs) of VAG signals are exploited as discriminating features. The raw VAG signals were first decomposed using DDDTCWT into sub-bands, then L_2 norm and LEE of the SBs are calculated as features. These features were then ranked using student's t-test and fed to classifiers to identify abnormal and normal knee-joints. The classification was achieved using fuzzy Sugeno classifier (FSC), least squared support vector machine (LS-SVM), and sequential minimal optimization support vector machine (SMO-SVM).

This paper is organized in following manner: Sect. 2 explains process with which the database was acquired and methodology to process the obtained data. This section discusses noise removal, signal transformation, feature calculation, selection, and ranking and classification methods. Results are discussed in Sect. 3 and concluded in Sect. 4.

2 Method and Material

Figure 1 depicts the stages to develop the automated identification system for the screening of VAG signals. Initially, raw VAG data is acquired and filtered to eliminate any noise and contaminations. Then, these processed VAG signals are decomposed into sub-bands of varying frequencies using DDDTCWT. The L_2 norm and LEE were calculated for these obtained SBs and used as feature vectors to identify normal and abnormal knee-joints. So obtained features were ranked using Student's t-test [9] and fed to the classifiers. In this work, FSC, LS-SVM, and SMO-SVM were employed for classification of abnormal and normal VAG signals

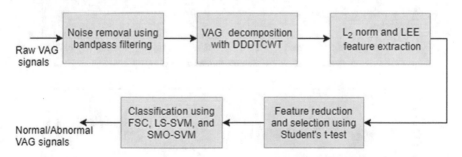

Fig. 1 Proposed algorithm steps for automated classification of normal and abnormal VAG signals

2.1   Dataset/Raw VAG Signals

The database used in this study consists of 89 signals, 51 among them obtained from normal subjects and 38 from subjects suffering from any knee-joint disorder. The approval for experimental protocol was given by the conjoint health research ethics board of the University of Calgory. Each of the volunteers positioned themselves on the rigid table with accelerometer to record the VAG signal. The accelerometer was of model 3115a, Dytran (Chatsworth, CA). It was placed over mid-patella of the knee to record VAG signals as the person swung their leg in ranges full flexion (Approximately 135 degree) and full extension (0 degree) and back again to full flexion state in 4 seconds [10, 11]. The subjects were instructed verbally for flexion, extension, and again flexion movements making approximately first-half of the recording to be the extension and second-half to be the flexion. The transducer utilized in this practice had a 3-dB bandwidth of 0.66–12 kHz, and sensitivity of 10 mV/G at a frequency of 100 Hz. These recorded VAG signals were prefiltered (10–1000 Hz), amplified, and then converted to digital at 2 kHz of the sampling rate (Fig. 2).

2.2   Denoising and Filtering

During leg movement and recording of VAG signals, random noise elements expected to affect the signal, even with the high-efficiency accelerometer sensor. Signal-to-noise ratio of VAG signals is not known beforehand. We have used Butterworth filter of sixth order [12].

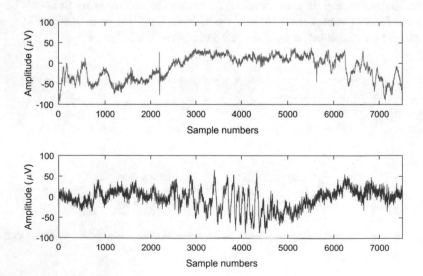

Fig. 2 The abnormal and normal raw VAG signals

2.3   Signal Transformation

The VAG signals exhibit random, multicomponent, and nonstationary characteristics. Recording these signals need leg movement in extension and flexion, which involves surface interaction in knee-joint area. Such surface interaction and time of connection in the joint may be different for each movement, thus producing nonstationary signals. The signal generated from a source may propagate through different tissue channels before being recorded by at mid-patella. Furthermore, contact of femoral condyle and patella surface avail different contact sources producing multifrequency signals at different frequencies. Such nonstationary and nonlinear signals can not be analyzed using the Fourier transform owing to the fact that the Fourier transform does not retain time-characteristics of the signals. Wavelet transform [13–17] and its variant transforms give a powerful means to deal with such nonstationary biomedical signals like VAG and electroencephalogram (EEG) [9, 12, 18–20] .

In this work, we have utilized DDDTCDWT introduced by Ivan W. Selesnick [21]. This transform can be considered as a combination of double-density discrete wavelet transform (DWT) and dual-tree DWT. Both of these transforms have their own advantages and characteristics, thus, a blend of these two provides a powerful tool for signal processing applications. Moreover, complex DWT offers better direction selectivity and it is shift invariant [22–24]. The complex DWT utilizes complex-valued analytic filter to decompose real and imaginary parts of transformed signals. The transform coefficients, both real and imaginary, represent amplitude and phase information. This separative quality is very useful in de-noising and signal processing applications. The dual-tree DWT can be employed as complex-transform, thereby, we get advantage of having better selectivity and shift invariance. Also, the double density DWT can add more design freedom.

The DDDTCWT is 4-times more expansive and uses two scaling functions and four different wavelets. All these wavelets have first two wavelet pair are offset by each other by half, and the remaining pair is used as approximate Hilbert transform duo [21, 25]. The DDDTCWT has two wavelets in dominating orientation, which can be interpreted as real and imaginary parts for a particular direction. Figure 3 represents the design flow of the DDDTCWT. We have used 4-level of decomposition in this study.

2.4   Feature Extraction

In this study, we computed L_2 norm and LEE of all 5 SBs obtained after 4-level VAG signal decomposition. We used these features as characteristics of normal and abnormal classes of VAG signals. There were total 10 SBs computed corresponding to each of the aforementioned features. The mathematical expression of the features are as follows:

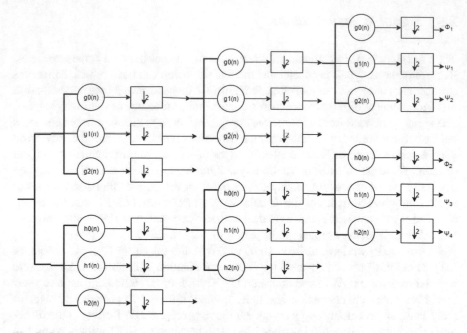

Fig. 3 Block diagram-based representation of DDDT CDWT method

$$L_2 norm = \sqrt[2]{\sum_n |g(n)|^2} \tag{1}$$

$$\log E = \sum_n \log_e |g(n)|^2 \tag{2}$$

where, $g(n)$ is the sub-bands obtained after decomposition, L_2 norm is level-2 norm of signal, and $\log E$ represents the LEE [26–29] of SB $g(n)$.

After computing the features for each VAG sub-band, feature ranking, and feature selection are followed, which are explained in the subsequent section.

2.5 Feature Ranking and Feature Selection

In this work, the decomposition of 4-levels was employed which provided with 5 SBs for each of the features. Thus, there were total 10 available SBs representing L_2 norm and LEE at different levels of decomposition.

However, not all the features prove effectively discriminatory to identify abnormal and healthy knee-joints. Thus, features are ranked in the order of their discriminative abilities which is determined by student's t-test [30]. The t-test ranking computes statistical significance of the features by calculating t-values of features. The t-values imply better significance for that particular feature. Features were then arranged in the order of their t-value ranking.

Having achieved feature ranking, the next step is feature selection. In this step, classification is performed by taking first highest significant feature set, then collective performance of first and second feature sets is calculated and so on. The feature set for which maximum accuracy is achieved is fed to the final classification step and the features for which classification accuracy degrades, are discarded. In our work, optimum performance was observed for only seven features. Thus, feature ranking and feature selection provide with optimal feature subset for optimum classification results. Feature selection reduces computational complexity by running for only selective features, and so increasing overall speed of classification algorithm.

2.6   Classification and Cross-Validation Process

Final step is to feed the computed features in appropriate classifier. In this work, we have observed maximum performances with FSC, LS-SVM, and SMO-SVM.

The FSC classifiers is based on the fuzzy interference system (FIS). It employs cluster information for optimally modeling the data behavior by utilizing minimum number of rules. The FIS uses fuzzy logic, instead of boolean logic, to map input to output, and thus provides base for pattern identification [31, 32].

Other choices for classifiers were LS-SVM [33] and SMO-SVM [26], both are variants of support vector machines (SVMs). The SVMs are supervised algorithms for classification and pattern identification and are widely used in data mining applications. The SVMs approach involves identifying optimum hyperplane by maximizing the distance between the classes. For this purpose, kernel-based SVMs were introduced by Boser et al. [34] and Cristianini et al. [35]. Gaussian radial basis function (RBF) was employed as a kernel in this work. A parameter σ was used to vary the width of Gaussian RBF between 1 to 20 with a step-size of 0.1. Maximum performance was observed at 2.4 (for LS-SVM) and 2.2 (for SMO-SVM) values of σ. Furthermore, cross-validation (tenfold) [36] ensured resilient classification model and avoid any overfitting. Final accuracy was computed as an average accuracy of all accuracies available after each fold of cross-validation.

3   Results and Discussion

In this section, the classification results, performance of the proposed model, and comparison with various other studies are presented.

The Kruskal–Wallis test [9] uses null hypothesis to test statistical significance by calculating p-values corresponding to each feature. The p-values signify probability of whether the null hypothesis is true. The hypothesis is accepted or rejected based on the p-values being greater or smaller than significance level. Smaller p-values indicate better discriminatory ability of a feature. Table 1 gives p-values calculated using the Kruskal–Wallis test.

While computing features, we calculated absolute value of wavelet coefficients from tree 1 and tree 2, representing real and imaginary parts of decomposition flow. The first SB depicts approximate wavelet coefficients and remaining four SBs are detailed wavelet coefficients.The p-values were calculated for absolute values of wavelet coefficients as shown in Table 1.

It can be observed that p-values corresponding to L_2 norm and LEE features of SB 1 have considerably smaller levels and represent the most significant features.

The box-plots analyze and visualize the identification ability of features for two classes. The Fig. 4 shows box-plots corresponding to sub-bands for L_2 norm and LEE. It is evident that not all features have good discrimination ability as can be observed from their box-plots (Table 2).

Table 3 represents the performance of the algorithm in terms of classification accuracy (CA), classification sensitivity (CSen), and classification specificity (CSpe)

Table 1 Computed p-values for the features

Sub-band	L_2 norm	LEE
1	2.9093e-05	7.8143e-04
2	0.0049	0.0304
3	0.2259	0.0745
4	0.4554	0.2323
5	0.0078	0.0122

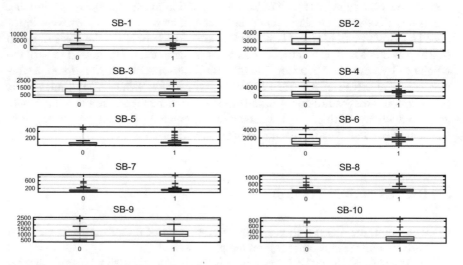

Fig. 4 Boxplots for all 10 features: The notations 1 and 0 represent SB of abnormal and normal VAG signals, respectively

Table 2 Feature Statistics: (mean ± standard deviation) of L_2 norm and LEE corresponding to 10 SBs

SB	L_2norm		LEE	
	Abnormal	Normal	Abnormal	Normal
1	1455.84 ± 1570.16	-134.96 ± 2799.57	1514.43 ± 484.18	13223.60 ± 922.77
2	2675.32 ± 354.80	2964.63 ± 494.86	148.39 ± 134.62	125.20 ± 114.63
3	740.69 ± 392.00	1068.92 ± 675.45	190.13 ± 194.67	165.33 ± 172.79
4	1596.88 ± 895.67	965.69 ± 1562.66	1137.10 ± 326.30	1117.53 ± 485.78
5	132.84 ± 71.41	109.89 ± 92.17	179.68 ± 143.26	174.31 ± 163.75

Table 3 Classification performance measures of various studied classifiers

Parameters	FSC	LS-SVM	SMO-SVM
CA (%)	85.3932	81.1944	81.0556
CSen (%)	88.2352	81.6667	84.1667
CSpe (%)	81.5789	84.3333	90.0000
MCC	0.7006	0.6216	0.6042

Table 4 Comparison of various studies on screening of knee-joint disorders using the VAG signals

Author	Classifier and Features	Results/Performance
Our Work	FSC, L_2 norm & LEE	Accuracy: 85.39%, MCC:0.700
[6]	LDA, WPD	Accuracy: 80%
[7]	Neural network, FF, S, kurtosis, and H entropy	AU-ROC: 0.82
[8]	Neural network, ATC, and VMS	AU-ROC: 0.9174
[38]	Neural network and FLDA, FD	AU-ROC: 1.0
[39]	SVM, SyEn, ApEn, FuzzyEn, mean, SD, RMS	AU-ROC: 0.9212, Accuracy: 0.8356

[37]. The performance is also measured in terms of Matthew's correlation coefficient (MCC) [37]. The highest CA of 85.39% was obtained using seven highly discriminative features and FSC as classifier. These seven features are: first SB of LEE, fifth SB of LEE, fifth SB of L_2 norm, second SB of LEE, first SB of L_2 norm, third SB of LEE, and second SB of L_2 norm.

We have compared the performance of our designed system with the existing systems. Table 4 gives the summary of various studies done previously for the screening of VAG signals of normal and degenerated knee-joints.

Many studies have been done in this area to obtain good classification results. Rangayyan et al. [7] used form factor (FF), skewness (S), kurtosis, and H entropy-based features and observed area under receiver operating characteristics (AU-ROC) of 0.82. In a different approach [38], they employed fisher linear discriminant analysis (FLDA), fractal dimension (FD)-based features and reported AU-ROC in range $0.92 - 0.96$ and maximum at 1.0. In [8] results were obtained by using approximate turns count (ATC), and variance of mean-squared value (VMS)-based features with a classification performance in terms of AU-ROC of 0.9172.

Other works in this area involve Wu et al. [39] who applied features based on the system entropy (SyEn), approximate entropy (ApEn), fuzzy entropy (FuzzyEn), mean, standard deviation (SD), root mean square (RMS) and classified with SVM classifier and observed an AU-ROC of 0.9212 and accuracy of 0.8356. Umapathy et al. [6] introduced another approach by means of linear discriminant analysis (LDA), wave packet decomposition (WPD), with a classification accuracy of 80%.

The presented system in this paper was able to achieve a CA of 85.39%, CSen of 88.23%, CSpe of 81.57% and MCC of 0.700. Classification using LS-SVM and SMO-SVM resulted in CA of 81.19 and 81.05%, respectively. This performance was observed by means of L_2 norm and LEE-based features classified by FSC. The results show promising performance which could be explored further. More algorithms could be devised by signal decomposition using DDDTCWT, FSC classifiers, and different feature set.

4 Conclusion

We have proposed a new automated screening system for VAG signals using DDDTCWT using FSC and SVM classifiers. The DDDTCWT decomposes VAG signals into various SBs and $L_2 norm$ and LEE-based features are extracted from these SBs. The extracted features are ranked using t-test ranking and are fed to FSC, LS-SVM, SMO-SVM classifiers. To develop a robust and resilient model, the tenfold cross-validation is employed to avoid redundancy and overfitting. So obtained model presents promising classification performance. Hence, the study also indicates further opportunities with exploring DDDTCWT in the classification of VAG signals. The proposed CAD system is noninvasive, which is a significant advantage as the best available process for disorder identification is arthroscopy which is semi-invasive and impractical for deteriorated knee-joints and to identify early symptoms.

Acknowledgements The VAG-based dataset used in this work was provided by Prof. Rangaraj M. Rangayyan, Dr. Cyril B. Frank, Dr. Gordon D. Bell, Prof. Yuan-Ting Zhang, and Prof. Sridhar Krishnan of University of Calgary, Canada. We would like to show our gratitude to them for this opportunity.

References

1. Wu, Y.: Knee Joint Vibroarthrographic Signal Processing and Analysis. Springer, Berlin (2015)
2. Laupattarakasem, W., Laopaiboon, M., Laupattarakasem, P., Sumananont, C.: Arthroscopic debridement for knee osteoarthritis. The Cochrane Library
3. Who department of chronic diseases and health promotion. http://www.who.int/chp/topics/rheumatic/en/. Accessed 19-08-2017
4. Lozano, R., Naghavi, M., Foreman, K., Lim, S., Shibuya, K., Aboyans, V., Abraham, J., Adair, T., Aggarwal, R., Ahn, S.Y., et al.: Global and regional mortality from 235 causes of death for 20 age groups in 1990 and 2010: a systematic analysis for the global burden of disease study 2010. The Lancet **380**(9859), 2095–2128 (2013)
5. U. Nations, World population to 2300, United Nations: New York, NY
6. Umapathy, K., Krishnan, S.: Modified local discriminant bases algorithm and its application in analysis of human knee joint vibration signals. IEEE Trans. Biomed. Eng. **53**(3), 517–523 (2006)
7. Rangayyan, R.M., Wu, Y.: Screening of knee-joint vibroarthrographic signals using statistical parameters and radial basis functions. Med. Biol. Eng. Comput. **46**(3), 223–232 (2008)
8. Rangayyan, R.M., Wu, Y.: Analysis of vibroarthrographic signals with features related to signal variability and radial-basis functions. Ann. Biomed. Eng. **37**(1), 156–163 (2009)
9. Sharma, M., Pachori, R.B., Acharya, U.R.: A new approach to characterize epileptic seizures using analytic time-frequency flexible wavelet transform and fractal dimension. Pattern Recognit. Lett. **94**, 172–179 (2017). https://doi.org/10.1016/j.patrec.2017.03.023
10. Ladly, K., Frank, C., Bell, G., Zhang, Y., Rangayyan, R.: The effect of external loads and cyclic loading on normal patellofemoral joint signals. Def. Sci. J. **43**(3), 201 (1993)
11. Rangayyan, R.M., Krishnan, S., Bell, G.D., Frank, C.B., Ladly, K.O.: Parametric representation and screening of knee joint vibroarthrographic signals. IEEE Trans. Biomed. Eng. **44**(11), 1068–1074 (1997)
12. Sharma, M., Dhere, A., Pachori, R.B., Acharya, U.R.: An automatic detection of focal EEG signals using new class of time-frequency localized orthogonal wavelet filter banks. Knowl. Based Syst. **118**, 217–227 (2017)
13. Sharma, M., Achuth, P.V., Pachori, R.B., Gadre, V.M.: A parametrization technique to design joint time-frequency optimized discrete-time biorthogonal wavelet bases. Signal Process. **135**, 107–120 (2017)
14. Sharma, M., Dhere, A., Pachori, R.B., Gadre, V.M.: Optimal duration-bandwidth localized antisymmetric biorthogonal wavelet filters. Signal Process. **134**, 87–99 (2017)
15. Sharma, M., Bhati, D., Pillai, S., Pachori, R.B., Gadre, V.M.: Design of time-frequency localized filter banks: Transforming non-convex problem into convex via semidefinite relaxation technique. Circuits Syst. Signal Process. **35**(10), 3716–3733 (2016)
16. Sharma, M., Gadre, V.M., Porwal, S.: An eigenfilter-based approach to the design of time-frequency localization optimized two-channel linear phase biorthogonal filter banks. Circuits Syst. Signal Process. **34**(3), 931–959 (2015)
17. Sharma, M., Pachori, R.B.: A novel approach to detect epileptic seizures using a combination of tunable-q wavelet transform and fractal dimension. J. Mech. Med. Biol. 1740003. https://doi.org/10.1142/S0219519417400036. http://www.worldscientific.com/doi/pdf/10.1142/S0219519417400036
18. Bhati, D., Sharma, M., Pachori, R.B., Gadre, V.M.: Time-frequency localized three-band biorthogonal wavelet filter bank using semidefinite relaxation and nonlinear least squares with epileptic seizure EEG signal classification. Digit. Signal Process. **62**, 259–273 (2017)
19. Bhati, D., Sharma, M., Pachori, R.B., Nair, S.S., Gadre, V.M.: Design of time-frequency optimal three-band wavelet filter banks with unit sobolev regularity using frequency domain sampling. Circuits Syst. Signal Process. **35**(12), 4501–4531 (2016)
20. Sharma, M., Kolte, R., Patwardhan, P., Gadre, V.: Time-frequency localization optimized biorthogonal wavelets. In: International Conference on Signal Processing and Communication (SPCOM), pp. 1–5 (2010)

21. Selesnick, I.W.: The double-density dual-tree dwt. IEEE Trans. Signal Process. **52**(5), 1304–1314 (2004)
22. Neumann, J., Steidl, G.: Dual-tree complex wavelet transform in the frequency domain and an application to signal classification. Int. J. Wavelets Multiresolution Inf. Process. **3**(01), 43–65 (2005)
23. Kingsbury, N.: A dual-tree complex wavelet transform with improved orthogonality and symmetry properties. In: Proceedings of the 2000 International Conference on Image Processing, vol. 2, pp. 375–378. IEEE (2000)
24. Sharma, M., Vanmali, A.V., Gadre, V.M.: Wavelets and Fractals in Earth System Sciences. CRC Press, Taylor and Francis Group (2013). Ch. Construction of Wavelets
25. Selesnick, I.W.: Hilbert transform pairs of wavelet bases. IEEE Signal Process. Lett. **8**(6), 170–173 (2001)
26. Sharma, M., Deb, D., Acharya, U.R.: A novel three-band orthogonal wavelet filter bank method for an automated identification of alcoholic eeg signals. Appl. Intell. (2017). https://doi.org/10.1007/s10489-017-1042-9
27. Han, J., Dong, F., Xu, Y.: Entropy feature extraction on flow pattern of gas/liquid two-phase flow based on cross-section measurement. J. Phys. Conf. Ser. **147**, 012041 (2009). IOP Publishing
28. Sharma, A., Amarnath, M., Kankar, P.: Feature extraction and fault severity classification in ball bearings. J. Vib. Control **22**(1), 176–192 (2016)
29. Gupta, V., Priya, T., Yadav, A.K., Pachori, R.B., Acharya, U.R.: Automated detection of focal eeg signals using features extracted from flexible analytic wavelet transform. Pattern Recognit. Lett. **94**, 180–188 (2017)
30. Kailath, T.: The divergence and Bhattacharyya distance measures in signal selection. IEEE Trans. Commun. Technol. **15**(1), 52–60 (1967)
31. Amo, Ad, Montero, J., Biging, G., Cutello, V.: Fuzzy classification systems. Eur. J. Oper. Res. **156**(2), 495–507 (2004)
32. Ishibuchi, H., Nakaskima, T.: Improving the performance of fuzzy classifier systems for pattern classification problems with continuous attributes. IEEE Trans. Ind. Electron. **46**(6), 1057–1068 (1999). https://doi.org/10.1109/41.807986
33. Suykens, J.A., Vandewalle, J.: Least squares support vector machine classifiers. Neural Process. Lett. **9**(3), 293–300 (1999)
34. Boser, B.E., Guyon, I.M., Vapnik, V.N.: A training algorithm for optimal margin classifiers. In: Proceedings of the Fifth Annual Workshop on Computational Learning Theory, pp. 144–152. ACM (1992)
35. Cristianini, N., Shawe-Taylor, J.: An Introduction to Support Vector Machines and Other Kernel-based Learning Methods. Cambridge university press, Cambridge (2000)
36. Kohavi, R., et al.: A study of cross-validation and bootstrap for accuracy estimation and model selection. In: Ijcai, pp. 1137–1145. Stanford, CA (1995)
37. Azar, A.T., El-Said, S.A.: Performance analysis of support vector machines classifiers in breast cancer mammography recognition. Neural Comput. Appl. **24**(5), 1163–1177 (2014)
38. Rangayyan, R.M., Oloumi, F., Wu, Y., Cai, S.: Fractal analysis of knee-joint vibroarthrographic signals via power spectral analysis. Biomed. Signal Process. Control **8**(1), 23–29 (2013)
39. Wu, Y., Chen, P., Luo, X., Huang, H., Liao, L., Yao, Y., Wu, M., Rangayyan, R.M.: Quantification of knee vibroarthrographic signal irregularity associated with patellofemoral joint cartilage pathology based on entropy and envelope amplitude measures. Comput. Methods Programs Biomed. **130**, 1–12 (2016)

Fault Diagnosis of Ball Bearing with WPT and Supervised Machine Learning Techniques

Ankit Darji, P. H. Darji and D. H. Pandya

Abstract In this paper, fault classification was done using RBIO 5.5 wavelet. Features were extracted at fifth level of decomposition with wavelet packet transform (WPT) where energy and kurtosis were extracted for both horizontal and vertical responses at all WPT nodes. Thus, total 400 samples were taken of defective bearing with reference to healthy bearing to minimize the experimental error. Multilayer perceptron of ANN with correlation-based feature selection has compared with sequential minimal optimization-based support vector method (SVM). Result shows that ANN with multilayer perceptron with CFS criteria has performed better than SVM for classification of ball bearing condition.

Keywords Signal processing · Fault diagnosis · MLT · WPT · ANN · SVM

1 Introduction

Ball bearings have widespread presentations from rotating machineries of process industries to household applications. Condition-based monitoring with advanced signal processing and MLT for classification of ball bearing conditions has recognized as effective soft computing technique. Vibration signature analysis is one of effective condition monitoring tools for fault diagnosis of ball bearings, and subsequent action prevents the catastrophic failure of machinery and improves operating life of process

A. Darji (✉)
C. U. Shah University, Wadhwan, Surendranagar 363030, Gujarat, India
e-mail: ankitdarjildrp@gmail.com

P. H. Darji
Mechanical Department, C. U. Shah College of Engineering & Technology, Wadhwan, Surendranagar 363030, Gujarat, India
e-mail: pranav_darji@rediffmail.com

D. H. Pandya
Mechanical Department, LDRP-ITR, Gandhinagar 382015, Gujarat, India
e-mail: veddhrumi@gmail.com

© Springer Nature Singapore Pte Ltd. 2019
M. Tanveer and R. B. Pachori (eds.), *Machine Intelligence and Signal Analysis*,
Advances in Intelligent Systems and Computing 748,
https://doi.org/10.1007/978-981-13-0923-6_25

equipments. Condition monitoring has executed by comparing vibration signature of a running machine with a healthy machine condition as reference, which would make it possible to detection of faults effectively. There have been different methods to evaluate different MLTs like supervised/unsupervised for fault diagnoses of ball bearing. Wavelets were utilized to extract feature attributes and different approaches to conclude the most suitable classifier. Samantha and Al-Balushi [1] was worked on utility of artificial intelligence for condition monitoring techniques as predictive maintenance of process industries. ANN has tried to classify healthy and defective bearing conditions with different operating speeds. Lin and Qu [2] have reviewed different techniques such as TDR spectrum, power spectral density spectrum, FFT analysis, and wavelet techniques to analyze digital vibration signals and to extract features. In time domain, signal analysis was done with statistical parameters. Effectiveness of time–frequency analysis is improved with statistical parameters of second, third, and fourth order. The evolvement of time–frequency analysis was recognized as effective data revealing techniques compared to other signal analysis techniques. Nonstationary signals can be well analyzed with wavelet transformation. In case of nonstationary data analysis, FFT has reported modulated information lack as compared to time-domain technique. Wu et al. [3] had reported successfulness of WPT and HHT as compared to usefulness of wavelet transform. Nikolaou and Antoniadis [4] had evaluated Morlet wavelet as effective mother wavelet for ball bearing fault diagnosis with WPT. Prabhakar et al. [5] and Purushotham et al. [6] have researched for stationary signal analysis of ball bearing race with DWT and justified effectiveness of DWT. SVM and PSVM have recognized as effective classifier with features extracted from time–frequency domain data Saravanan et al. [7]. Fault diagnosis of ball bearing with soft computing technique which incorporated different DSP transformations like WPT and HHT with modified feature selection criteria like radial basis function (RBF) network is projected by Lei et al. [8]. Rafiee et al. [9] have developed fault classification technique with MLP where input data were extracted from autocorrelation of CWT coefficient. Authors have also focused on selection of effective wavelet selection for a typical gearbox system. Rafiee et al. [10] have reported Db44 wavelet for effective mother wavelet for condition monitoring tool for rotating elements.

Various researchers have researched on different data mining and soft computing techniques with different classifiers based on machine learning techniques for condition monitoring. Further, various controlling parameters have researched like selection of suitable mother wavelet, selection of most effective features, number of attributes, and classifier. The conventional artificial neural networks have a few drawbacks like poor generalization capability and local solution tendency rather than global. SVM is recent supervised machine learning techniques with sequential minimal optimization (SMO) algorithm and statistical learning theory (SLT). Meyer et al. [11] has worked on SVM and its application in the area of condition monitoring of bearings with high classification accuracy deals with limited datasets. Hall [12] elucidated importance of correlation-based filter algorithm for any type of limited dataset for data mining will be analyzed with WEKA. Hall has compared filter-based algorithm with wrapper-based algorithm for feature attribute selection from wide datasets. Pandya et al. [13] developed APF-KNN algorithm with asymmetric proximity function approach with optimizes feature selection that concluded higher classification accuracy. Raw vibration signals from SKF 6205 ball bearing have processed for fast Fourier transform of intrinsic mode function, which were analyzed and reported effectively. Bearing condition could be classified more accurately with extracted statistical and acoustic features and implemented for proper data mining with or without filter. Experimental assessments were done with different bearing conditions such as healthy bearing, bearing with BPFO, BPFI, BSF, and combined defect. Pandya et al. [14] described an evaluation of artificial intelligence (AI) techniques in fault diagnosis of ball bearings. The effectiveness of ANN and SVM (SMO) were compared for five bearing conditions. Pandya et al. [15] described significant of WPT as an effective fault diagnostic tool for ball bearing for its transient signals with experimental validations.

This paper has focused on comparison of two classifiers MLP and sequential minimal optimization (SMO) latest algorithm of SVM. Wavelet transform has used to extract feature using Rbio5.5 as mother wavelet. Multilayer perceptron (MLP) of ANN with correlation-based feature selection criteria is reported as significant tool with accuracy of 99.75%.

2 Selection of Mother Wavelet for Ball Bearing

The raw signals picked up in time domain have been utilized to perform fault diag nosis. WPT has accepted substantial DSP transformation tool and reports the meaningful characteristics of time–frequency domain data. Vibration signature analysis with WPT has varying descriptive parameters. As per state of the art, in the present work, the following real wavelets from four-mother wavelet family are evaluated to classify ball bearing conditions. (1) db10 & db44—Daubechies, (2) sym2—symlets, (3) coif5—coiflet, and (4) rbio5.5—reverse biorthogonal. Result shows that rbio5.5 would be more suitable wavelet for fault classification based on maximum energy-to-Shannon entropy ratio criteria for wavelet selection. Wavelet selection methodology for vibration signal analysis was described by step-by-step procedure as follows.

- Digital signature analysis module of vibration analyzer has recorded raw vibration signals.
- According to the sampling frequency f_s, raw data have decomposed into concern level of decomposition. Characteristic defect frequency component f_c is to be recorded from the signal in Eq. 1.

$$\frac{f_s}{2^{j+1}} \leq f_c \leq \frac{f_s}{2^j} \tag{1}$$

- Now, maximum energy-to-entropy ratio has to be calculated for speed range up to 6000 rpm for all bearing conditions and lead to select the mother wavelet with maximum value.

Table 1 Selection of mother wavelet

Mother wavelet	Max. energy/Shannon entropy ratio
Symlet2	0.4477
Db44	0.2066
Coiflate5	0.3871
Rbio5.5	4.7095
Db10	1.2910

Rbio5.5 wavelet is selected as the most effective wavelet for evaluating the bearing vibration signature analysis as per Table 1.

3 Machine Learning Techniques

Catastrophic failure of operating part failures in rotating machines of process industries can cause both financial loss and individual damages. Machine learning is an approach of self-learning with provided information to system which is understood as effective application of AI. There are two learning methodologies: supervised and unsupervised. When system has defined input and concern output, then training of data is considered as supervised learning, while system process data that to be analyzed with cluster technique is known as unsupervised learning technique. Pattern recognition and classification using machine learning techniques are described here. In this paper, supervised machine learning methods, viz., artificial neural networks (MLP) and support vector machines, are studied in detail, and application of case study is presented as follows.

3.1 ANN with Multilayer Perception

ANN is self-assured big data of neurons operating at the same time to resolve a specific problem. ANN consists of a set of nodes ordered into layers and coupled with weight element called "synapses". At each node, the weighted inputs are summed, threshold, and then inputted to the activation functions to produce output for that specific node. Most of neurons in NN exchange their inputs using a scalar function called an "activation function", yielding a neuron output. Generally, used activation functions are called as "linear functions", "threshold functions", "sigmoid functions", and "bipolar sigmoid functions". The sigmoid function was chosen in the NN classifiers tested in this paper. Sigmoid function is appropriate due to the following reasons.

1. Sigmoid function in backpropagation NN will minimize the computational time.
2. Function is suitable for applications whose desired output values are between 0 and 1.

3. Activation functions should be chosen to suit the allocation of target values for output neurons.

Sigmoid function is well suited for target values that are binary [0, 1]. The sigmoid functions have additional advantages of varying boundary range which could be scaled to output responses.

In feedforward networks, neurons are organized in layers and these layers are connected in one direction and there is no loop in the network. ANN parameters could be organized as per requirement which classified it either as supervised method or unsupervised machine learning technique. In supervised learning, network neurons have trained with bounded output target and train data accordingly. Backpropagation is considered to be most effective supervised machine learning algorithm. Multilayer perception has modified version of backpropagation algorithm to train data with bounded output targets range. In backpropagation learning, the network weights are updated in the direction of the gradient of the performance function.

3.2 SVM—Updated Sequential Minimal Optimization

Support vector machines with updated sequential minimal optimization have supervised classification approach based on statistical learning theory as compared to competitive approach. Updated SMO supervised classifiers have recognized as significant tool for classification and regression in limited dataset such as fault diagnosis of ball bearings. Meyer et al. [11] have evaluated the performance of SVM with 16 other significant classifiers, using 21 training and trial data. Outcome of that research has made significant conclusion to prove the superiority of SVM classifier over and above other classifiers. In this approach, marginal area has discoursed between the two different classes and orients it to minimize the generalization error.

4 Experimental Setup and Data Preparation

In this work, experimental setup has developed in laboratory with rotor-bearing test rig, where horizontal rotor has supported with SKF 6205 bearing. DC motor with servo control has utilized to change the variable speed of shaft. A flexible coupling has connected rotor shaft weighted 2.5 kg with motor in balance condition. SKF 6205 bearings have investigated for this work. Experimental setup with data acquisition system, sensors, and instrumentation is shown in Fig. 1. Two uniaxial sensors (piezoelectric) are with sensitivity of 1.02 mv/(m/s^2). Frequency range of sensors is 1–30 kHz. Five different sets of bearings have investigated. Five different sets of bearings are healthy bearing, bearing with ORD, IRD, bearing with defect on any one ball, and bearing with combined defect which contains spall on outer race, inner race, and spall on any one ball. All the spalls have indulged in bearing with

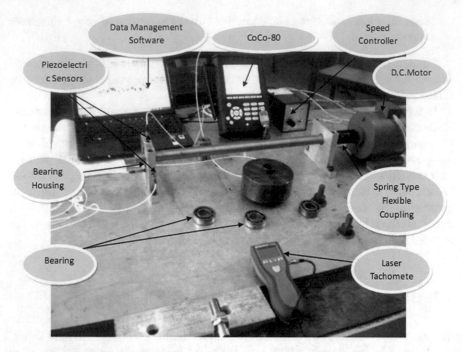

Fig. 1 Experimental setup [13]

laser engraving machining operation. The size of spall would be an average 300 ×
300 × 20 micron.

Vibration analyzer with four input channels makes crystal instrumentation that the
USA has utilized to make Coco-80 and seeded defects are shown in Fig. 2. Machine
learning techniques with experimental data have included three main steps: data
acquisition, feature extraction, and feature selection. Overview of MLT proceeds as
follows.

4.1 Extraction of Features

Feature vectors evaluated considering different statistical parameters like energy and
fourth-order kurtosis. In this paper, the energy and kurtosis of decomposed signal with
WPT have improved effectiveness of fault diagnosis and defect classification. Statistical term energy makes available the quantitative value of each decomposed wavelet
coefficient, which remarkably differentiates the signal. Mathematical expression of
energy contained in the signal is as mentioned in Eq. 2.

Fig. 2 Seeded localized defect

$$E_{x(t)} = \sum_{m=0}^{2^j-1} \int |x_j^m(t)|^2 dt \qquad (2)$$

Kurtosis is the fourth-order moment of signal that represent flatness of signal with a dimensionless statistical parameter. It means that higher kurtosis of data leads to conclude more variation in amplitude of signal as mentioned in Eq. 3.

$$K_j^m = \frac{\sum \{[x_j^m(t) - \bar{x}_j^m(t)]^4\}}{\sigma_{x_j^m(t)}^4} \qquad (3)$$

Thus, for better 2D digital signal processing, mainly two statistical terms have evaluated. Energy of signal shows strong indicator of the signal, but is not effectively identified defect at initial stage, where kurtosis shows high sensitivity to defect at early stage but has low stability. Hence, combined application of both statistical features can reflect better fault diagnosis result as compared to individual consideration of each feature.

4.2 Selection of Features

MLT algorithms robotically take out information from machine legible data. But their success is generally reliable on the data quality which may excise on. If the information is not sufficient or having nonlinear information, then machine learning algorithms may reduce accuracy and understandable outputs may fail to discover effective outcome from data mining. Feature subset selectors are algorithms that attempt to classify and minimize immaterial and superfluous information on prior to learning. Proper selection of feature may reduce hypothesis search space and enhanced performance of classifier design as WEKA. In this work, CBFS algorithm is operated using WEKA. CBFS algorithm is investigative for estimating the rank of signal subset with concern feature attributes. This experimental data considered the expediency of subset features to forecast the class tag as per concern inter-correlation among feature attribute dataset, the proposition on which the heuristic can be remarked:

Significant feature attributes data have better recognition with bearing class label, while scattered feature attributes have non-correlation with specific class label and will be wrongly classified.

$$Merit_s = \frac{kr_{cf}}{\left(k + k(k-1)r_{ff}\right)^{1/2}} \tag{4}$$

where Meritsis is the heuristic "merit" of a feature subset S containing k features, r_{cf} is the mean feature class correlation ($f \in S$), and r_{ff} is the average feature. Thus, all features have homogeneous characteristics. The numerator is representing feature attributes dataset with class label, while the denominator indicated severance intensity among the feature subsets. The experimental data with merits of all feature attributes to evaluate irrelevant features and quantified significance of relevance features are based on its predictive class. Random feature attributes have distinguished with each other and each will be highly correlated with other features undefinably. Exclusive description of algorithm is described by Pandya et al. [13].

5   Results and Discussion

Feature vectors constructed with statistical values of the energy and kurtosis of wavelet coefficients is an efficient diagnostic tool to diagnose early fault identification of ball bearing. Raw signal has decomposed to extract data with WPT up to fifth level of decomposition for signal responses in vertical and horizontal directions. Thus, two features have vector of 128 datasets. Datasets have acquired 10 sets of speed from 1000 to 6000 rpm and to minimize sigmoid error 10 times data repeated which lead to have input vector with 100 datasets. Finally, input matrix of 400 × 128 samples. Then, data were normalized in the range 0.1–0.9 to minimize the effect of spreadness of attributes. Correlation-based feature selection (CFS) algorithm was used through WEKA. CFS was estimated on overall training instant and hence tends to select a subset of feature that has low redundancy and is strongly predictive of class. Then, each feature evaluated locally with minimum redundancy and rank in ascending order. In this work, speed ranges of 1–6 K rpm generate 10 sets of datasets. Thus, each class of bearing will contain 100 input vectors and finally input matrix of 400 × 128 samples. To improve the accuracy of classifier, all statistical parameters

Table 2 Accuracy evaluation with minimum number of feature attributes

Number of attributes	ANN		SVM	
	Accuracy (%)	Time (s)	Accuracy (%)	Time (s)
14	91.25	1.32	87.5	0.17
28	**99.75**	3.16	95	0.24
42	99.75	5.62	78.75	0.36
112	99.75	36.74	**98.25**	0.42
128	99.75	48.59	98	0.48

Table 3 Comparison of % of decrement in time with optimize number of attributes

MLT	Time (s) with all 128 attributes	Optimize number of attributes	Time (s)	% Decrement in time
ANN	48.59	28	3.16	93.5
SVM	0.48	112	0.42	12.5

Table 4 Confusion matrix

ANN					SVM			
BD	IRD	ORD	CD	Classified as	BD	IRD	ORD	CD
100	0	0	0	**BD**	98	2	0	0
0	99	1	0	**IRD**	1	98	1	0
0	0	100	0	**ORD**	0	3	97	0
0	0	0	100	**CD**	0	0	0	100

like energy and kurtosis have considered with reference to the data received from a health bearing at each concern speed.

Experimental results have shown better classification rate with vertical responses as compared to horizontal responses. CBFS criteria are recommended for feature attribute selection and optimized number of attributes that have designated as tabulated in Table 2. ANN with multilayer perception has shown best classification rate with twenty-eight (28) minimum number of attributes, while SVM with 112. Efficiency of correlation-based feature selection and comparison of computational time is tabulated in Table 3.

Experimental test results have shown in two-dimensional confusion matrix with a row and column for each class with multi-class prediction. Here, actual class is represented in the row and the predicted class has shown in the column as shown in Table 4. Test result with both machine learning techniques has compared with 400 samples. ANN could be able to classify 399 samples and SVM classified 393 samples correctly. CBFS algorithm has increased accuracy of classifiers more rapidly as SVM has shown reduction of calculative time by 12.5% (SVM) and 93.5% (ANN).

6 Concluding Remarks

- Rbio5.5 is to be reported as effective real wavelet based on maximum energy-to-entropy ratio.
- ANN classifier has shown 93.5% reduction in computational time with CBFS (correlation-based feature selection) algorithm.
- SVM has shown 12.5% reduction in computational time with CBFS (correlation-based feature selection) Algorithm.
- Number of feature attributes has shown significant effect on performance of ANN classifier as compared to SVM classifier.
- Result shows that ANN with multilayer perceptron with CFS criteria performs better than SVM for fault diagnosis of ball bearing.

ANN with multilayer perceptron with CFS criteria was concluded as faster classifier to calculate in 3.16 s with a classification accuracy of 99.75%.

Acknowledgements This work is supported by LDRP Institute of Technology & Research a constituent institute of Kadi Sarva Vishwavidyalaya, Gandhinagar, Gujarat.

References

1. Samanta, B.K., Al-Balushi, R.: Artificial neural network based fault diagnostics of rolling element bearings using time-domain features. Mech. Syst. Signal Process. **17**(2), 317–328 (2003)
2. Lin, J., Qu, Liangsheng: Feature extraction based on Morlet wavelet and its application for mechanical fault diagnosis. J. Sound Vib. **234**(1), 135–148 (2000)
3. Wu, T.Y., Chen, J.C., Wang, C.C.: Characterization of gear faults in variable rotating speed using Hilbert-Huang transform and instantaneous dimensionless frequency normalization. Mech. Syst. Signal Process. **30**, 103–122 (2012)
4. Nikolaou, N.G., Antoniadis, I.A.: Demodulation of vibration signals generated by defects in rolling element bearings using complex shifted Morlet wavelets. Mech. Syst. Signal Process. **16**(4), 677–694 (2002)
5. Prabhakar, S., Mohanty, A.R., Sekhar, A.S.: Application of discrete wavelet transform for detection of ball bearing race faults. Tribol. Int. **35**(12), 793–800 (2002)
6. Purushotham, V., Narayanan, S., Suryanarayana, A.N.P.: Multi-fault diagnosis of rolling bearing elements using wavelet analysis and hidden Markov model based fault recognition. NDT E Int. **38**(8), 654–664 (2005)
7. Saravanan, N., Kumar Siddabattuni, V.N.S., Ramachandran, K.I.: A comparative study on classification of features by SVM and PSVM extracted using Morlet wavelet for fault diagnosis of spur bevel gear box. Expert Syst. Appl. **35**(3), 1351–1366 (2008)
8. Lei, Y., He, Z., Zi, Y.: Application of an intelligent classification method to mechanical fault diagnosis. Expert Sys. Appl. **36**(6), 9941–9948 (2009)
9. Rafiee, J., et al.: A novel technique for selecting mother wavelet function using an intelligent fault diagnosis system. Expert Syst. Appl. **36**(3), 4862–4875 (2009)
10. Rafiee, J., Rafiee, M.A., Tse, P.W.: Application of mother wavelet functions for automatic gear and bearing fault diagnosis. Expert Syst. Appl. **37**(6), 4568–4579 (2010)
11. Meyer, D., Friedrich, L., Kurt, H.: The support vector machine under test. Neurocomputing. **55**(1), 169–186 (2003)

12. Hall, M.A.: Correlation-based feature selection for machine learning. Ph.D. thesis. The University of Waikato (1999)
13. Pandya, D.H., Upadhyay, S.H., Harsha, S.P.: Fault diagnosis of rolling element bearing with intrinsic mode function of acoustic emission data using APF-KNN. Expert Syst. Appl. **40**(10), 4137–4145 (2013)
14. Pandya, D.H., Upadhyay, S.H., Harsha, S.P.: Fault diagnosis of rolling element bearing by using multinomial logistic regression and wavelet packet transform. Soft. Comput. **18**(2), 255–266 (2014)
15. Pandya, D.H., Upadhyay, S.H., Harsha, S.P.: Fault diagnosis of high-speed rolling element bearings using wavelet packet transform. Int. J. Signal Imaging Syst. Eng. **8**(6), 390–401 (2015)

Continuous Hindi Speech Recognition Using Kaldi ASR Based on Deep Neural Network

Prashant Upadhyaya, Sanjeev Kumar Mittal, Omar Farooq,
Yash Vardhan Varshney and Musiur Raza Abidi

Abstract Today, deep learning is one of the most reliable and technically equipped approaches for developing more accurate speech recognition model and natural language processing (NLP). In this paper, we propose Context-Dependent Deep Neural-network HMMs (CD-DNN-HMM) for large vocabulary Hindi speech using Kaldi automatic speech recognition toolkit. Experiments on AMUAV database demonstrate that CD-DNN-HMMs outperform the conventional CD-GMM-HMMs model and provide the improvement in word error rate of 3.1% over conventional triphone model.

Keywords Deep neural network (DNN) · Hidden markov model (HMM)
Speech recognition · Kaldi · Hindi language

1 Introduction

Deep learning approach has recently become the hot research area for the researcher working in the field of automatic speech recognition system. Deep learning was initially set during late 1980s that relied on backpropagation algorithm [1] and Boltzmann machine algorithm [2]. In this, feature training was performed on one or more layers of hidden units using backpropagation. However, during that period, the limitation of available hardware restricted the use of Boltzmann machine algorithms [2]. Therefore, more research was conducted for developing more efficient algorithms

P. Upadhyaya (✉) · O. Farooq · Y. V. Varshney · M. R. Abidi
Department of Electronics, Aligarh Muslim University, Aligarh 202002, Uttar Pradesh, India
e-mail: upadhyaya.prashant@gmail.com

O. Farooq
e-mail: omarfarooq70@gmail.com

S. K. Mittal
Electrical Engineering, Indian Institute of Science
Bangalore, Bengaluru 560012, Karnataka, India
e-mail: rsr.skm@gmail.com

© Springer Nature Singapore Pte Ltd. 2019
M. Tanveer and R. B. Pachori (eds.), *Machine Intelligence and Signal Analysis*,
Advances in Intelligent Systems and Computing 748,
https://doi.org/10.1007/978-981-13-0923-6_26

to train the *layered*-Boltzmann machine. One of the well-known layered-Boltzmann machines is a restricted Boltzmann machine (RBM) that contains a single hidden layer. Hinton [3] proposed the efficient algorithms for training deep layered RBM machine.

Advancements in the field of Graphics Processing Units (GPUs) hardware had made it possible to increase the depth of deep neural nets from single hidden layers to multiple hidden layers. Therefore, development of more powerful algorithms and improvement in GPUs hardware had intensively increased the use of deep neural nets. Modern GPUs are more capable of handling large database models to be trained and made it more faster in comparison to the CPUs in peak operation per second [3, 4].

Thus, advantage of using GPU is that it can minimize the computational time and can also provide the parallel programming environment that makes the algorithms run much faster. This has attracted the researcher working in the field of machine learning. Areas such as image processing [5, 6] and speech processing [7–9] had picked up and started using deep learning models to train the classifier. Experiment conducted in [10] using phone recognition based on deep neural nets showed promising results over baseline model. The same approach was reported in [7]. Dahl et al. [8] developed the first continuous large vocabulary speech recognition model based on deep neural nets. Performance obtained through deep neural nets was so good that researcher in the field of speech processing switched to develop their acoustic model based on deep neural nets. DNN became the state-of-the-art for speech recognition which provides multiple nonlinear layers for modeling the system [9, 11–16].

Previously, most of the work done in automatic speech recognition model was based on simple training and decoding of HMM-GMM model. But, today DNNs have proved to be the speedy way for most of the automatic speech recognition system. Deep Neural Networks (DNNs) along with Hidden Markov Models (HMMs) have shown the significant improvement over automatic speech recognition task [9, 11–16].

RBM, autoencoder and Deep Belief Networks (DBN) are some of the neural-based techniques which are used to improve the performance of the model system [12, 17–19]. Autoencoders are simple three-layer neural networks whose outputs are directly connected back to inputs. All of the output edges are connected back to input edges. Hidden layers are usually lesser than the input and output layers. The objective is to minimize the error at reconstruction layer and in other word is to find the most efficient compact representation or encoding for the input signal. RBM on the other hand uses stochastic (Gaussian distribution) and adjusts the weights to minimize the reconstruction error. RBMs do not use deterministic methods. In Deep Belief Networks (DBN), different layers are composed of RBMs that are stacked together to form multiple layers. Each such layer learns features from the features that are extracted from the underlying layer [12, 17–19]. However, research suggests that any of the above methods may give better performance and more likely be dependent on the dataset used for research. For this research work, RBM model is considered.

In this paper, continuous Hindi speech deep neural network recognition model based on Kaldi toolkit [20] is evaluated. Standard MFCC features are extracted from AMUAV corpus. AMUAV Hindi speech database contains 1000 phonetically balanced sentences recorded by 100 speakers consisting of 54 Hindi phones. Each

Fig. 1 The undirected graph
of an RBM with n hidden
and m visible variables

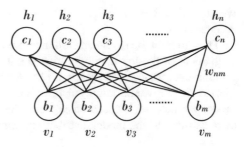

speaker in AMUAV records 10 short phonetically sentence out of which two sentences are common to each speaker. Audio was sampled at 16,000 Hz. Standard Mel-Frequency Cepstral Coefficients (MFCCs) features are extracted from the given audio files. Tri-gram language model was developed using SRILM language toolkit.

2 Restricted Boltzmann Machine

Generally, RBM can be considered as a two-state model, in which one state contains m visible units (v) to represent observable data and another state consists of n hidden units (h). Figure 1 shows the undirected graph of an RBM with n hidden and m visible variables.

Restricted Boltzmann Machine (RBM) computes the energy of every unit of visible, and hidden layers of states which are obtained as

$$E(\mathbf{v}, \mathbf{h}) = -\mathbf{b}^T\mathbf{v} - \mathbf{c}^T\mathbf{h} - \mathbf{v}^T\mathbf{W}\mathbf{h} \tag{1}$$

where $\mathbf{W}$ contains connection weights of visible/hidden layer in matrix form, b is the visible bias vector, and c is a hidden bias vector. Also, visible/hidden layers joint probability distribution given by Gibbs distribution $p(\mathbf{v}, \mathbf{h})$ [2, 3, 12], for the given configuration is defined in terms of their energy as

$$p(\mathbf{v}, \mathbf{h}) = \frac{e^{-E(\mathbf{v}, \mathbf{h})}}{Z} \tag{2}$$

where the normalization factor $Z = \sum_{\mathbf{v}, \mathbf{h}} e^{-E(\mathbf{v}, \mathbf{h})}$ is known as the partition function.

In Eq. (1), RBM energy function can be rewritten as a function dependent on at most one hidden unit, which yields

$$E(\mathrm{v}, \mathrm{h}) = -\beta(\mathbf{v}) + \sum_j \gamma_j(\mathbf{v}, h_j) \tag{3}$$

where $\beta(\mathbf{v}) = \mathbf{b}^T\mathbf{v}$ and $\gamma_j(\mathbf{v}, h_j) = -(c_j + \mathbf{v}^T\mathbf{W}_{*,j})h_j$, with $\mathbf{W}_{*,j}$ denoting the jth column of $\mathbf{W}$.

Thus, to train the restricted Boltzmann machine for an arbitrary model parameter θ, gradient of the log-likelihood of the data is computed. Thus, for a given set of training data, model parameters do the learning by finding the weight and biases (parameter) in which training dataset has high probability. Therefore, updating of model parameter θ is based on expectations of the energy gradients over the given data distribution and model distribution, computed as

$$\Delta\theta = \langle\partial E|\partial\theta_{data}\rangle - \langle\partial E|\partial\theta_{model}\rangle \tag{4}$$

3 Deep Neural Network Using Kaldi ASR Tool

Deep neural networks along with high compute capability GPU cards are now one of the core programming tools for developing model for speech recognition task. Two most important speech recognition toolkits HTK 3.5 [21] and Kaldi [22] have included DNNs for building speech recognition model. However, Kaldi which is actively maintained and contains simple documentation of DNNs script has attracted most of the researcher working in the field of speech recognition [22]. Therefore, Kaldi has become the most preferable speech recognition toolkit for building speech model using DNN.

Kaldi toolkit currently supports three codebases for DNNs [22]. First code script "nnet1" is maintained by Karel Vaseley in which training can be done using single GPU. Second code script "nnet2" is maintained by Daniel Povey which is basically based on the original script of Karel Vaseley. It allows multiple threads for programming by using multiple CPU or GPU for training. Third code script is "nnet3" which is the enhanced version of "nnet2" script of Daniel Povey.

Experiment work on DNN using Kaldi ASR toolkit has been reported in [23–29]. Povey et al. [23] perform an experiment using n-gram phone model based on sequence-discriminative training. In their work, they proposed lattice free (LF) version using Maximum Mutual Information (MMI) criterion for training of acoustic model based on neural network for five large vocabulary continuous speech recognition (LVCSR) task. Cosi [24] performed phone recognition on ArtiPhon Italian corpus [25] using Kaldi ASR toolkit. However, results obtained using DNNs do not overcome with baseline method. Cosi claim that this is due to insufficient corpus size resulted in the failure of tuning of the DNN architecture for ArtiPhon corpus.

Miao [26] developed a full-fledged DNN acoustic model based on Kaldi + PDNN model using Kaldi Switchboard 110 h setup. In their work, initial model was developed using Kaldi toolkit and DNN model was trained using PDNN. Finally, these models (GMM + PDNN) were decoded using Kaldi ASR. Presently, four DNN recipes are released using Kaldi + PDNN. PDNN is deep learning toolkit developed under Theano environment.

Another work by Chen et al. [27], based on Long Short-Term Memory (LSTM) using recurrent neural network, has reported. Using this model, robust speech recognition is performed using the integrated approach of both automatic speech recognition and speech enhancement using LSTM network. However, most of the works based on Kaldi speech recognition model are used for the English language. Here, in this paper, we propose the deep neural nets model for the Hindi language.

4 Experiment Setup

Experiment on Hindi automatic speech recognition model for Independent Speaker (IS) using Kaldi toolkit was set up under Ubuntu 16.04 LTS (64-bit Operating System). Running on dual core 4 CPUs of Intel(R) Core(TM) i5-4200U CPU @ 1.60 GHz with 4GM RAM and hosting single GPU of computing capability of 3.5, 2 GB memory running under CUDA 8.0 toolkit. Therefore, implementation of Hindi speech using DNNs is restricted to "nnet1", i.e., Karel's setup which uses pretraining and compiled using CUDA GPU (NVIDIA).

Experiments were performed on AMUAV Hindi speech database which contains 1000 phonetically balanced sentences recorded by 100 speakers using 54 Hindi phones [30]. Further, for feature-space reduction, linear discriminant analysis (LDA) along with maximum likelihood linear transform (MLLT) is applied which acts as the observation input to Gaussian mixture model (GMM)-based acoustic models. Also, for tied-state triphones, we built feature-space maximum likelihood linear regression (f-MLLR) for speaker-adapted GMM model. Further, frame-based cross-entropy along with state-level Minimum Bayes Risk (sMBR) training of Hindi database is evaluated.

Figure 2 shows the stage of DNNs for Hindi automatic speech recognition system. RBM pretraining is usually done in unsupervised stage on top of f-MLLR features. Frame cross-entropy training using mini-batch stochastic gradient descent is performed so that it correctly classifies each frame to their corresponding triphone states (i.e., PDFs). Further, sequence-training optimizing sMBR is done so that better frame accuracy with respect to reference alignment is obtained. It then performs six iterations of RBM pretraining, followed by frame cross-entropy training and sMBR sequence-discriminative training. Table 1 shows that using sequence-discriminative training, the word error rates of Hindi automatic speech recognition model were further reduced.

5 Comparative Performance

This section presents the comparative analysis of work done on Kaldi-DNN using Hindi, Arabic, English, and Italian language. Ahmed et al. [31] and Cosi [32] presented the complete recipe for building Arabic and Italian speech recognition model

Fig. 2 Training of RBM
layers for Hindi speech
model for AMUAV database

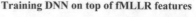

Training DNN on top of fMLLR features

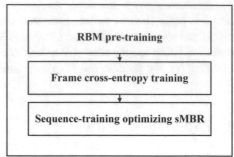

Table 1 WER performance for Monophone, Triphone, LDA + MLLT, and Karel's DNN using
MFCC features for continuous Hindi speech recognition

Features	Word error rate (%WER)
Monophone training and decoding	18.32
Deltas + Delta–Deltas training and decoding	14.73
LDA + MLLT training and decoding	12.62
DNN Karel's	
Karel's hybrid DNN (dnn4_pretrain-dbn_dnn)	12.13
System combination Karel's DNN + sMBR (dnn4_pretrain-dbn_dnn_smbr)	**11.63**

using Kaldi. Ali et al. [31] performed an experiment using broadcast news system
using 200 h Gale corpus for Broadcast Report (BR), Broadcast Conversational (BC),
and the combination of the two, i.e., BR + BC. Here, only the best results obtained for
Arabic language using Broadcast Report (BR) are presented. For Italian language,
experiment was performed using 10 h of Italian FBK ChilIt corpus [32]. For Hindi
language, experiment was performed using AMUAV corpus [30]. We also performed
the experiment on English language using TIMIT corpus [33] for same TIMIT-DNN
recipe present under Kaldi source. However, the corpus size of each language was
different, and hence comparisons were made only on the results obtained for respective language using Kaldi-DNN. MFCC features were used as a baseline for all the
four languages. Table 2 shows the comparative analysis of the work for different
languages.

From Table 2, it can easily be shown that for each case the DNN has performed
well over conventional GMM-HMM model. It can easily be observed from the table
that performance of Arabic language which was trained over 200 h of dataset has
shown significant improvement over other languages. Similarly, Italian language
which was trained over 10 h of dataset has shown better results than Hindi language.

For similar corpus size, i.e., TIMIT and AMUAV, experiment results obtained are
better for Hindi language. But improvement from GMM-HMM to DNN model was
obtained best for TIMIT. For, Hindi language, we obtained the improvement of ~1%

Table 2 Comparative analysis of DNN model developed for a particular language (Arabic, Italian, English, and Hindi) over conventional GMM-HMM (LDA + MLLT) model in terms of word error rate (%WER)

Features	Arabic language (%WER) [31]	Italian language (%WER) [32]	English language (%WER) [33]	Hindi language (%WER) [30]
Corpus set→	Gale corpus	ChiIIt corpus	TIMIT corpus	AMUAV corpus
GMM-HMM (LDA + MLLT)	22.32	12.7	24.0	12.62
Karel's Hybrid DNN	17.36	8.6	18.4	12.13
System combination Karel's DNN + sMBR	**15.81**	**8.3**	**18.1**	**11.63**

over conventional LDA + MLLT model, whereas for TIMIT it was ~6%. As the pre-training of DNN depends upon the labeled frames (phoneme-to-audio alignments) using GMM-HMM system. In case of TIMIT corpus, phoneme-to-audio alignments were done using 39 English phones, whereas for our case phoneme-to-audio alignments were done on 54 Hindi phones. Therefore, less number of phone model results in better GMM-HMM phone model alignment which can increase the performance of ASR model trained using DNN.

For our case, due to insufficient corpus size, it was not optimum approach for finding the optimum weight for updating the model parameter and tuning of the DNN layers for a given set of training data. Thus, more accurate results could be obtained with further tuning of the DNN layer thereby increasing the corpus size.

6 Conclusion and Future Work

From the results obtained, it is found that the performance of Hindi speech recognition model has shown improvement using deep neural network. Experiments were performed using two phases. In the first phase, simple DNN without sMBR was executed. In the second phase, DNN using sMBR was attempted. For the first phase, improvement in WER(%) of 6.19%, 2.6%, and 0.49% was obtained with respect to monophone, triphone, and LDA + MLLT, respectively. For the second phase, improvement in WER(%) of 6.69%, 3.1%, and 0.99% was obtained with respect to monophone, triphone, and LDA + MLLT respectively. Further, using sMBR, it had shown further improvement of 0.5% WER over pretrained DNNs.

Hence, more robust results can be obtained with further tuning of the DNN layer. We plan to continue our work by increasing the AMUAV database and developing more accurate DNN models for continuous Hindi speech. We have not used autoencoder and DBN in the present study. We think that DBN usage would improve our results and hence we definitely plan to make use of DBN in our very next experiments. As future work, we can implement our speech model using autoencoder, DBM, and compare them with RBM models.

Acknowledgements The authors would like to acknowledge Institution of Electronics and Telecommunication Engineers (IETE) for sponsoring the research fellowship during this period of research.

References

1. Rumelhart, D.E., Hinton, G.E., Williams, R.J.: Learning representations by back-propagating errors. Nature **323**, 533–536 (1986)
2. Ackley, D.H., Hinton, G.E., Sejnowski, T.J.: A learning algorithm for Boltzmann machines. Cogn. Sci. **9**, 147–169 (1985)
3. Hinton, G.E.: Training products of experts by minimizing contrastive divergence. Neural Comput. **14**, 1771–1800 (2002)
4. Raina, R., Madhavan, A., Ng, A.Y.: Large-scale deep unsupervised learning using graphics processors. In: Proceedings of 26th International Conference on Machine Learning (ICML 09), pp. 873–880 (2009)
5. Mnih, V., Hinton, G.E.: Learning to detect roads in high-resolution aerial images. Lecture Notes in Computer Science (including Subser. Lect. Notes Artif. Intell. Lect. Notes Bioinformatics) vol. 6316 LNCS, pp. 210–223 (2010)
6. Cireşan, D.C., Meier, U., Gambardella, L.M., Schmidhuber, J.: Handwritten digit recognition with a committee of deep neural nets on GPUs. Technical Report No. IDSIA-03-11. 1–8 (2011)
7. Dahl, G.E., Yu, D., Deng, L., Acero, A.: Large vocabulary continuous speech recognition with context-dependent DBN-HMMS. In: 2011 IEEE International Conference on Acoustics, Speech and Signal Processing, pp. 4688–4691 (2011)
8. Dahl, G.E., Sainath, T.N., Hinton, G.E.: Improving deep neural networks for LVCSR using rectified linear units and dropout. In: International Conference on Acoustics, Speech and Signal Processing (ICASSP), 2013, pp. 8609–8613. IEEE (2013)
9. Hinton, G., Deng, L., Yu, D., Dahl, G.E., Mohamed, A., Jaitly, N., Senior, A., Vanhoucke, V., Nguyen, P., Sainath, T.N., Kingsbury, B.: Deep neural networks for acoustic modeling in speech recognition. IEEE Signal Process. Mag. 82–97 (2012)
10. Mohamed, A., Dahl, G.E., Hinton, G.: Acoustic modeling using deep belief networks. IEEE Trans. Audio Speech Lang. Process. **20**, 14–22 (2012)
11. Hinton, G.E., Srivastava, N., Krizhevsky, A., Sutskever, I., Salakhutdinov, R.R.: Improving neural networks by preventing co-adaptation of feature detectors. In: International Conference on Acoustics, Speech and Signal Processing (ICASSP), 2013, pp. 1–18. IEEE (2012)
12. Jaitly, N., Hinton, G.: Learning a better representation of speech soundwaves using restricted boltzmann machines. In: Proceedings of IEEE International Conference on Acoustics, Speech and Signal Processing (ICASSP), vol. 1, pp. 5884–5887 (2011)
13. Deng, L., Li, J., Huang, J.T., Yao, K., Yu, D., Seide, F., Seltzer, M., Zweig, G., He, X., Williams, J., Gong, Y., Acero, A.: Recent advances in deep learning for speech research at Microsoft. In: Proceedings of IEEE International Conference on Acoustics, Speech and Signal Processing (ICASSP), pp. 8604–8608 (2013)
14. Dahl, G.E., Yu, D., Deng, L., Acero, A.: Context-dependent pre-trained deep neural networks for large-vocabulary speech recognition. IEEE Trans. Audio Speech Lang. Process. **20**, 30–42 (2012)
15. Zeiler, M.D., Ranzato, M., Monga, R., Mao, M., Yang, K., Le, Q.V., Nguyen, P., Senior, A., Vanhoucke, V., Dean, J., Hinton, G.E.: On rectified linear units for speech processing New York

University, USA Google Inc., USA University of Toronto, Canada. In: International Conference on Acoustics, Speech and Signal Processing (ICASSP), 2013, pp. 3517–3521. IEEE (2013)

16. Seide, F., Li, G., Chen, X., Yu, D.: Feature engineering in context-dependent deep neural networks for conversational speech transcription. In: Proceedings of 2011 IEEE Workshop on Automatic Speech Recognition and Understandings (ASRU 2011), pp. 24–29 (2011)

17. Gehring, J., Miao, Y., Metze, F., Waibel, A.: Extracting deep bottleneck features using stacked auto-encoders. In: International Conference on Acoustics, Speech and Signal Processing (ICASSP), pp. 3377–3381 (2013)

18. Deng, L., Seltzer, M.L., Yu, D., Acero, A., Mohamed, A.R., Hinton, G.: Binary coding of speech spectrograms using a deep auto-encoder. In: Eleventh Annual Conference of the International Speech Communication Association, pp. 1692–1695 (2010)

19. Dahl, G., Mohamed, A.R., Hinton, G.E.: Phone recognition with the mean-covariance restricted Boltzmann machine. In: Advances in Neural Information Processing Systems, pp. 469–477 (2010)

20. Povey, D., Ghoshal, A., Boulianne, G., Burget, L., Glembek, O., Goel, N., Hannemann, M., Motlicek, P., Qian, Y., Schwarz, P., Silovsky, J., Stemmer, G., Vesely, K.: The Kaldi speech recognition toolkit. In: IEEE Workshop Automatic Speech Recognition and Understanding, pp. 1–4 (2011)

21. Young, S., Gales, M., Liu, X.A., Povey, D., Woodland, P.: The HTK Book (version 3.5a). English Department, Cambridge University (2015)

22. Kaldi Home Page. www.kaldi-asr.org

23. Povey, D., Peddinti, V., Galvez, D., Ghahrmani, P., Manohar, V., Na, X., Wang, Y., Khudanpur, S.: Purely sequence-trained neural networks for {ASR} based on lattice-free {MMI}. In: Proceedings of Interspeech, pp. 2751–2755 (2016)

24. Cosi, P.: Phone recognition experiments on ArtiPhon with KALDI. In: Proceedings of Third Italian Conference on Computational Linguistics (CLiC it 2016) Fifth Evaluation Campaign of Natural Language Processing and Speech Tools for Italian 1749, 0–5 (2016)

25. Canevari, C., Badino, L., Fadiga, L.: A new Italian dataset of parallel acoustic and articulatory data. In: Proceedings of Annual Conference on International Speech Communication Association Interspeech, Jan 2015, pp. 2152–2156 (2015)

26. Miao, Y.: Kaldi+PDNN: building DNN-based ASR systems with Kaldi and PDNN. arXiv CoRR. abs/1401.6, 1–4 (2014)

27. Chen, Z., Watanabe, S., Erdogan, H., Hershey, J.R.: Speech enhancement and recognition using multi-task learning of long short-term memory recurrent neural networks. Interspeech 3274–3278 (2015)

28. Zhang, X., Trmal, J., Povey, D., Khudanpur, S.: Improving deep neural network acoustic models using generalized maxout networks. In: 2014 IEEE International Conference on Acoustics, Speech and Signal Processing, pp. 215–219 (2014)

29. Vu, N.T., Imseng, D., Povey, D., Motlicek, P., Schultz, T., Bourlard, H.: Multilingual deep neural network based acoustic modeling for rapid language adaptation. Icassp-2014, pp. 7639–7643 (2014)

30. Upadhyaya, P., Farooq, O., Abidi, M.R., Varshney, P.: Comparative study of visual feature for bimodal hindi speech recognition. Arch. Acoust. **40** (2015)

31. Ali, A., Zhang, Y., Cardinal, P., Dahak, N., Vogel, S., Glass, J.: A complete Kaldi recipe for building Arabic speech recognition systems. In: Proceedings of 2014 IEEE Workshop Spoken Language Technology (SLT 2014), pp. 525–529 (2014)

32. Cosi, P.: A KALDI-DNN-based ASR system for Italian. In: Proceedings of International Joint Conference on Neural Networks (2015)

33. Lopes, C., Perdigão, F.: Phone recognition on the TIMIT database. Speech Technol. **1**, 285–302 (2011)

Extreme Gradient Boosting with Squared Logistic Loss Function

Nonita Sharma, Anju and Akanksha Juneja

Abstract Tree boosting has empirically proven to be a highly effective and versatile approach for predictive modeling. The core argument is that tree boosting can adaptively determine the local neighborhoods of the model thereby taking the bias-variance trade-off into consideration during model fitting. Recently, a tree boosting method known as XGBoost has gained popularity by providing higher accuracy. XGBoost further introduces some improvements which allow it to deal with the bias-variance trade-off even more carefully. In this manuscript, performance accuracy of XGBoost is further enhanced by applying a loss function named squared logistics loss (SqLL). Accuracy of the proposed algorithm, i.e., XGBoost with SqLL, is evaluated using test/train method, K-fold cross-validation, and stratified cross-validation method.

Keywords Boosting · Extreme gradient boosting · Squared logistic loss

1 Introduction

In the course of recent decades, machine learning and data mining have turned out to be one of the backbones of data innovation and with that, a somewhat central, although typically hidden, part of our life. With the constantly expanding amounts of data getting to be noticeably accessible, there is justifiable reason to consider that information mining will turn out to be a significant element for technological advancements. In all aspects of life, applications of these two are utilized. Example applications include fraud detection, e-mail protection, recommender system, etc. [1].

Further, boosting is the most widely used technique in machine learning that improves the prediction accuracy of various classification models.

N. Sharma (✉) · Anju
Dr. B. R. Ambedkar National Institute of Technology Jalandhar, Jalandhar, India
e-mail: nonitasharma@nitdelhi.ac.in

A. Juneja
Jawaharlal Nehru University, New Delhi, Delhi, India

© Springer Nature Singapore Pte Ltd. 2019
M. Tanveer and R. B. Pachori (eds.), *Machine Intelligence and Signal Analysis*,
Advances in Intelligent Systems and Computing 748,
https://doi.org/10.1007/978-981-13-0923-6_27

Boosting technique is an ensemble technique created by combining various weak learners to build a strong learner with higher precision. Weak learners are those indicators that give more precision than random guessing.

However, strong learners are those classifiers that give maximum accuracy and hence coined as the base of machine learning. Boosting technique is employed when the dataset is large, and high predictive power is the vital requisite of the application. Further, it is also used to reduce the bias and variance in the prediction models. However, the technique also solves the overfitting problem for smaller dataset. Additionally, it has wide application area and applies to numerous classification techniques, viz., feature selection, feature extraction, and multi-class categorization. The applications of boosting include medical area, text classification, page ranking, business, and so on [2].

In addition, several boosting algorithms are already in place [3]. The most widely used are

1. Adaptive boosting (AdaBoost),
2. Gradient boosting,
3. Extreme Gradient boosting (XGBoost), and
4. Random forest.

Adaptive boosting also known as AdaBoost algorithm is proposed by Freund and Schapire [4]. The algorithm starts by selecting a base classification algorithm (e.g., Naïve Bayes) and repetitively enhancing its prediction accuracy by draining the inaccurately classified samples in the training dataset. Initially, AdaBoost assigns same weights to all the training samples and selects a base classifier algorithm. Further, for every iteration, the base algorithm classifies the training samples, and the weights of the inaccurate classified samples are increased. The algorithm iterates n times, repeatedly applying base classification algorithm on the training dataset with newly calculated weights. At the end, the final classification model is the weighted sum of the n classifiers [5].

Gradient boosting is an effective off-the-shelf strategy for creating accurate models for classification problems. The technique has empirically proven itself to be highly effective for a vast array of classification and regression problems. The aim of this method is to train a collection of decision trees, given the case that the training of single decision tree is known a priori. However, in this method, boosting is visualized as an optimization problem, where the objective of the technique is to minimize the loss of the classifier model by adding one weak learner at a time as done in a gradient descent. Gradient boosting is also called stage-wise additive classifier as a new weak learner is added at one time and the previously classified weak learners are left frozen, i.e., unchanged for that iteration.

Extreme Gradient Boosting (XGBoost) follows greedy approach and has demonstrated high performance and speed. It is used for a wide number of applications and it also supports outdoor memory [6]. Due to parallel computation process, it is faster than other boosting algorithms [7]. The reason behind the higher performance of

XGBoost is that it is scalable in nature. Additionally, it has the following advantages over other algorithms:

- Due to parallel processing process, it has faster performance than gradient boosting.
- It controls the overfitting problem.
- It gives better performance result on many datasets.
- Basically, it is a tree building algorithm.
- It is used for classification, regression, and ranking with custom loss functions.

Another ensemble technique that is widely used in machine learning is random forest. It is used in classification, regression, and many more prediction problems. At training time, multiple decision trees are created and the output is the mean or average prediction of each tree. The algorithm is proposed by Ho [7]. In random forest algorithm, the classifier shows low bias and high variance. Random forest follows the parallel computation process. The algorithm that is used for training and testing process is bootstrapping. However, for every iteration, data is split into number of trees using bagging process. Bagging process divides the whole dataset and creates samples. Then, classification is done on these samples using decision trees. Further, the classifier predicts the classes of samples and final class is predicted by the majority voting or it can be the simple average.

Comparing all the boosting algorithms, XGBoost and its entire variant exhibit the maximal performance among all the categories of boosting algorithms [8, 9]. On the contrary, AdaBoost displays the minimal performance lower than random forest. This can be attributed to the computation process followed in the respective boosting algorithms. XGBoost and random forest implement parallel computation process while AdaBoost implements serial computation process. Hence, the performance also displays the same realization. XGBoost and AdaBoost use the concept of weighted average, while random forest considers the simple average of weak learners. Further, in boosting, various weak learners are consolidated and give a strong learner with higher accuracy. Therefore, bias and variance are considered important parameters to measure the accuracy of these algorithms. The better algorithm is the one which provides high bias and low variance. Both XGBoost and AdaBoost depict the same. But random forest shows the opposite. Further, accuracy is also impacted by the cross-validation of error. All the four algorithms implement the cross-validation of error and hence are more accurate than single learner. To state the comparative analysis of the accuracy of all the four algorithms, accuracy of XGBoost is maximum and random forest shows the least accuracy among all. However, overfitting of data occurs due to the branches involving noisy data or outliers. It is imperative to reduce the overfitting problem to enhance the accuracy of the learners. XGBoost and random forest avoid overfitting problem, and AdaBoost does not avoid the problem completely but it is less prone to overfitting. Table 1 presents a comparative analysis of techniques used in various boosting algorithms.

Further, boosting is a representation of gradient descent algorithm for loss functions. Loss function maps a real-time event to a number representing the cost associated with that event. The goal of any optimization problem is to minimize the loss

Table 1 Comparative evaluation of boosting algorithms

Technique	Computation process	Final prediction	Training and testing algorithm	Bias	Variance
XGBoost [1]	Parallel	Weighted average	Any algorithm can be used	High	Low
AdaBoost [5]	Serial	Weighted average	Any algorithm can be used	High	Low
Random forest [7]	Parallel	Simple average	Bootstrapping	Low	High

function as much as possible [10]. Loss function is the penalty for misclassified data points in any classification problem. The main objective of estimation is to find an objective function that models its input well, i.e., it should be able to predict the values correctly. The loss function is a measure to quantify the amount of deviation between the predicted values and actual values.

Suppose that $(u_1, v_1), \ldots, (u_n, v_n)$ is observed, where u_i is the value of input space χ and v_i is the class label which takes 1 or -1 values. The collection of weak learners is denoted by $WL = (w_i(u) : \chi \rightarrow (1, -1)|i = (1, \ldots, I))$, where each learner gives class label for the input [11]. Strong learner SL is made by consolidated weak learners, where SL is given by

$$SL = \sum_{i=1}^{I} (\alpha_i w_i(u)) \tag{1}$$

The loss given by SL over sample (u, v) is $l(-vSL(u))$, where $l : R \rightarrow R$ is continuous and differentiable function except finite points. It is also called growing function; Loss function is given by

$$L(SL) = (\frac{1}{n}) \sum_{i=1}^{n} l(-v_i * SL(u_i)) \tag{2}$$

So, in simple term, a loss function is the cost of the error between the prediction $z(b)$ and the observation at the point b. Loss function is a convex function. A convex function shows that there are no local minima. In every optimization problem, the main task is to minimize the loss or cost function. It may be its objective function also. And to minimize the conditional risk, a loss function is derived that is used for outliers. Furthermore, it classifies the data points with high accuracy and it fits the outliers. Loss function always affects the accuracy. Loss function is convex in nature. Loss function gives two types of error: one is positive part error and second is negative part error. Negative part error always decreases the accuracy. Positive part truncation makes strong to any boosting algorithm [12].

In summary, XGBoost which is implemented on gradient boosting algorithm and follows the greedy approach is the best performing boosting algorithm in terms of computation performance and accuracy [1, 13]. However, the confidence of classification points achieved from the XGBoost can further be enhanced using a carefully selected loss function. Hence, in this manuscript, a squared logistics loss (SqLL) function is presented and the impact of applying SqLL on XG Boost is presented. The focus of the research work is to present an efficient boosting algorithm in terms of prediction power, which reduces the loss function of the target variables as well.

2   XGBoost with Squared Loss Function

XGBoost is tree ensemble method, i.e., a collection of Classification and Regression Trees (CART). The first step is to classify the family members into different nodes and give them the score on the corresponding node. Following steps are performed in the method:

1. Additive training: Parameters of trees are given in this. It contains information about tree structure and the node score. In this, addition of a tree is possible at anytime.
2. Complexity of model: Regularization parameter is taken as complexity of model and it controls the overfitting problem at certain points.
3. Structure score: This gives information about the best split conditions.

2.1   Objective Function

The objective function of XGBoost is given by

$$Objective = L(\varphi) + R(\varphi) \tag{3}$$

where the first term of objective function is training loss function which measures how predictive our model is and the second term is regularization term which helps us to overfitting the data.

2.2   Loss Function

In this section, we will introduce the loss function. The loss function is the measure of prediction accuracy that we define for the problem at hand [14]. The applications are interested in minimizing the expected loss, which is known as the risk. The function which minimizes the risk is known as the target function. This is the optimal prediction function we would like to obtain. The function is defined as

$$L(\varphi) = \sum_j \left[(a_j \log_{10}(1 + e^{-a_j}))^2 + \left((1 - a_j) \log_{10}\left(1 + e^{a_j}\right)\right)^2 \right] \quad (4)$$

where α_j is the target variable. SqLL holds the property of convexity and convexity means there are no local minima. The proof of SqLL convexity is given below:

$$L(\varphi) = \sum_j \left[(a_j \log_{10}(g_\sigma))^2 + \left((1 - a_j) \log_{10}(k_\sigma)\right)^2 \right] \quad (5)$$

where $g_\sigma = 1 + e^{-a_j}$ and $k_\sigma = 1 + e^{a_j}$.

To prove the function to be convex, we need to prove that above two functions are convex function.

Gradient:

$$\nabla_\sigma \left[-\log(g_\sigma(x)) \right] = (g_\sigma(x) - 1)x$$

$$= \left(\frac{-e^{\alpha^T x}}{1 + e^{\alpha^T x}} \right) x$$

$$= \nabla_\sigma \left[-\log\left(1 + e^{-\sigma^T x}\right) \right]$$

$$= \left(\frac{1}{1 + e^{\alpha^T x}} - 1 \right) x \quad (6)$$

Hessian:

$$\nabla_\sigma^2 \left[-\log(g_\sigma(x)) \right] = \nabla_\sigma (\nabla_\sigma \left[-\log(g_\sigma(x)) \right]$$

$$= \nabla_\sigma ((g_\sigma(x) - 1)x)$$

$$= g_\sigma(x)(1 - g_\sigma(x))xx^T \quad (7)$$

Next, to prove that hessian matrix is positive semi-definite:

$$\forall_s : s^T \nabla_x^2 (\log(g_\sigma(x))) = s^T [g_\sigma(x)(1 - g_\sigma(x))xx^T]s$$

$$= g_\sigma(x)(1 - g_\sigma(x))(x^8 z)^2] \geq 0 \quad (8)$$

Hence proved that $g_\sigma = 1 + e^{-a_j}$ is a convex function. Similarly, we can prove it for $k_\sigma = 1 + e^{-a_j}$. And square of these two is also convex. Additionally, their sum is also convex. So, the loss function is convex in nature.

3 Results and Discussion

3.1 Dataset

The simulations are performed on the CALTECH 101 dataset, which contains 101 categories of object pictures. Each category has almost 50 images. The size of the image is roughly 300×200 pixels. The images in various categories vary from 40 to 150.

3.2 Evaluation Parameters

Experiments are performed to evaluate the proposed algorithm in the following evaluation parameters: Accuracy using train/test dataset, k-fold cross-validation, stratified cross-validation, and gross error sensitivity.

3.3 Accuracy

Accuracy is measured as the error in the predicted results and expected results. Experiments are done to compare the accuracy of the proposed technique with the state-of-the-art techniques in the following ways.

3.3.1 With Training and Testing Dataset

Here, the dataset is split into two parts: Training and testing sets. The size of training set taken is 67%, whereas that of testing is 33%. The method is used in case of large datasets and produces performance estimates with lower bias. Table 2 shows that the proposed method outperforms other algorithms by a huge margin.

Table 2 Comparative evaluation of boosting algorithms

Boosting technique	Accuracy (%)
XGBoost with SqLL	80.02
XGBoost	77.95
AdaBoost	72.64
Random forest	73.14

Table 3 Comparison of boosting techniques using k-fold cross-validation

Boosting technique	Accuracy (%)
XGBoost with SqLL	80.02
XGBoost	77.95
AdaBoost	72.64
Random forest	73.14

Table 4 Comparison of boosting techniques using stratified cross-validation

Boosting technique	Accuracy (%)
XGBoost with SqLL	77.25
XGBoost	76.95
AdaBoost	65.61
Random forest	69.76

3.3.2 K-Fold Cross-Validation Method

This cross-validation method is used to evaluate the performance of XGBoost on the new data with k set to 3, 5, or 10. Here, the data is divided into K-folds. After evaluating the techniques on different k performance scores, the results are averaged out. This is a more reliable way to measure the performance of the algorithm (Table 3).

3.3.3 Stratified Cross-Validation

The method is particularly applicable when the categories are large and there is imbalance in the instances of each category. In this method, stratified folds are created when performing the cross-validation which in a way creates the effect of enforcing the uniform distribution of classes in each fold as the distribution is in the whole training dataset. Table 4 shows that the accuracy achieved from XGBoost with SqLL is 77.25%.

3.4 Gross Error Sensitivity (GER)

When the data is changed slightly, it is called outliers. GER is a matrix where we deliberately add outliers in the dataset and use it to compare at different portions of outliers. In our dataset, we make a comparison between different techniques after adding 0, 2, and 4% outliers [15], given in Table 5. Results show that GER in XGBoost with SQL is not impacted by the percentage of outliers in the data, in other words, the error rate remains the same. While in case of random forest, results are highly impacted by the presence of outliers.

Table 5 Gross error sensitivity of boosting techniques

Boosting technique	0%	2%	4%
XGBoost with SqLL	0.0812	0.0823	0.0826
XGBoost	0.0814	0.0846	0.0847
AdaBoost	0.895	0.9283	1.0684
Random forest	2.125	7.17	7.82

Table 6 Total loss of boosting techniques with SqLL

Boosting technique	Total loss
XGBoost with SqLL	1.577643
AdaBoost with SqLL	1.717639
Random forest with SqLL	1.764983

3.5 Techniques Comparison with Squared Logistic Loss

After applying the squared logistic loss function on different techniques, the result is very encouraging depicting that the total loss in XGBoost is minimum, while random forest depicts the maximum loss as displayed in Table 6.

4 Conclusions

In summary, we have performed both experimental and theoretical studies of the XGBoost algorithm with squared loss function. The proof of convexity of the SqLL is given and a complete description of the XGBoost algorithm with SqLL is provided. The accuracy of the proposed algorithm has been compared with state-of-the-art boosting algorithm using three established techniques, namely train and test datasets, k-fold cross-validation, and stratified cross-validation method. The results establish the efficacy of the stated approach in terms of accuracy, total loss, and gross error sensitivity.

References

1. Chen, T., Guestrin, C.: XGBoost: reliable large-scale tree boosting system. In: Proceedings of the 22nd SIGKDD Conference on Knowledge Discovery and Data Mining, San Francisco, CA, USA, 2015, pp. 13–17
2. Freund, Y., Schapire, R., Abe, N.: A short introduction to boosting. J. Jpn. Soc. Artif. Intell. **14**(771–780), 1612 (1999)
3. Zhou, Z.-H.: Ensemble Methods: Foundations and Algorithms. CRC Press (2012)
4. Schapire, R.E.: Explaining Adaboost. In: Empirical Inference, pp. 37–52. Springer (2013)

5. Culp, M., Johnson, K., Michailidis, G.: ada: an r package for stochastic boosting. J. Stat. Softw. **17**(2), 9 (2006)
6. Chen, T., Guestrin, C.: XGBoost: a scalable tree boosting system. In: Proceedings of the 22nd ACM SIGKDD International Conference on knowledge Discovery and Data Mining, pp. 785–794. ACM (2016)
7. Ho, T.K.: The random subspace method for constructing decision forests. IEEE Trans. Pattern Anal. Mach. Intell. **20**(8), 832–844 (1998)
8. Friedman, J., Hastie, T., Tibshirani, R., et al.: Additive logistic regression: a statistical view of boosting (with discussion and a rejoinder by the authors). Ann. Stat. **28**(2), 337–407 (2000)
9. Mason, L., Baxter, J., Bartlett, P.L., Frean, M.R.: Boosting algorithms as gradient descent. In: Advances in Neural Information Processing Systems, pp. 512–518 (2000)
10. Kanamori, T., Takenouchi, T., Eguchi, S., Murata, N.: The most robust loss function for boosting. In: Neural Information Processing, pp. 496–501. Springer (2004)
11. Masnadi-Shirazi, H., Vasconcelos, N.: On the design of loss functions for classification: theory, robustness to outliers, and savageboost. In: Advances in Neural Information Processing Systems, pp. 1049–1056 (2009)
12. Schapire, R.E.: The strength of weak learnability. Mach. Learn. **5**(2), 197–227 (1990)
13. Friedman, J.H.: Greedy function approximation: a gradient boosting machine. Ann. Stat. 1189–1232 (2001)
14. Zhai, S., Xia, T., Tan, M., Wang, S.: Direct 0-1 loss minimization and margin maximization with boosting. In: Advances in Neural Information Processing Systems, pp. 872–880 (2013)
15. Kearns, M., Valiant, L.: Cryptographic limitations on learning boolean formulae and finite automata. J. ACM (JACM) **41**(1), 67–95 (1994)

Distinguishing Two Different Mental States of Human Thought Using Soft Computing Approaches

Akshansh Gupta, Dhirendra Kumar, Anirban Chakraborti
and Vinod Kumar Singh

Abstract Electroencephalograph (EEG) is useful modality nowadays which is utilized to capture cognitive activities in the form of a signal representing the potential for a given period. Brain–Computer Interface (BCI) systems are one of the practical application of EEG signal. Response to mental task is a well-known type of BCI systems which augments the life of disabled persons to communicate their core needs to machines that can able to distinguish among mental states corresponding to thought responses to the EEG. The success of classification of these mental tasks depends on the pertinent set formation of features (analysis, extraction, and selection) of the EEG signals for the classification process. In the recent past, a filter-based heuristic technique, Empirical Mode Decomposition (EMD), is employed to analyze EEG signal. EMD is a mathematical technique which is suitable to analyze a nonstationary and nonlinear signal such as EEG. In this work, three-stage feature set formation from EEG signal for building classification model is suggested to distinguish different mental states. In the first stage, the signal is broken into a number of oscillatory functions through EMD algorithm. The second stage involves compact representation in terms of eight different statistics (features) obtained from each oscillatory function. It has also observed that not all features are relevant, therefore, there is need to select most relevant features from the pool of the formed features which is carried out in the third stage. Four well-known univariate feature selection algorithms are investigated in combination with EMD algorithm for forming the feature vectors for further classification. Classification is carried out with help of learning

A. Gupta (✉) · A. Chakraborti · V. Kumar Singh
School of Computational and Integrative Sciences, Jawaharlal Nehru University,
New Delhi, India
e-mail: akshanshgupta@jnu.ac.in

A. Chakraborti
e-mail: anirban@jnu.ac.in

V. Kumar Singh
e-mail: vinod.acear@gmail.com

D. Kumar
AIM & ACT Banasthali Vidyapith, Niwai, Rajasthan, India
e-mail: dhirendrakumar@banasthali.in

© Springer Nature Singapore Pte Ltd. 2019
M. Tanveer and R. B. Pachori (eds.), *Machine Intelligence and Signal Analysis*,
Advances in Intelligent Systems and Computing 748,
https://doi.org/10.1007/978-981-13-0923-6_28

the support vector machine (SVM) classification model. Experimental result on a publicly available dataset shows the superior performance of the proposed approach.

1 Introduction

The Brain–Computer Interface (BCI) is one of the regions which has sponsored up in developing techniques for assisting neurotechnologies for ailment prediction and manage motion [1, 2, 15]. BCIs are rudimentary geared toward availing, augmenting, or rehabilitating human cognitive or motor-sensory characteristic [14, 19]. To capture brain activities, EEG is one of the prevalent technology as it provides a signal with high temporal resolution in a noninvasive way [14, 15]. Mental task classification (MTC)-based BCI is one of the famed categories of BCI technology which does no longer involve any muscular activities [3], i.e., EEG responses to mental tasks.

In the literature, the EEG signals have been analyzed especially in three domains specifically temporal, spectral, and hybrid domain. In a hybrid domain, both the frequency and temporal information is utilized for analysis of the EEG signals simultaneously. Empirical mode decomposition (EMD) is this sort of heuristic hybrid approach that can examine the signal in both domains by decomposing the signal in distinctive frequency components termed as Intrinsic Mode Function (IMF) [13]. In literature, EMD has been incorporated for data analysis followed by using these decomposed signals for parametric feature vector formation for building classification model [6, 10].

In this work, final set of feature vectors for the classification process is obtained in three stages. In the first stage, the raw EEG signal is analyzed using EMD algorithms which results into a number of IMFs. A compact representation of these IMFs with the parametric feature coding has been introduced with the help of eight well-known parameters namely root mean square, variance, skewness, kurtosis, Lampleziv Complexity, central and maximum frequency, and Shannon entropy. Further to select only relevant features, four univariate feature selection methods are investigated which is the third stage of the proposed method for obtaining the final feature vectors for classification.

Outline of this article is as follows: Sect. 2 contains an overview of feature extraction and parametric feature formation. Feature selection approach is discussed in Sect. 3. In Sect. 4, a brief description of dataset and Experimental result are discussed. The conclusion is discussed in Sect. 5.

2 Feature Extraction

In this work, feature extraction from EEG signal has been carried out in two stages: First stage involves the decomposition of EEG signal from each channel into k number of intrinsic mode functions (IMFs) using Empirical Mode Decomposition (EMD) algorithm (discussed in Sect. 2.1). Later, in the second stage, these decomposed IMFs obtained from each channel were used to calculate eight parametric features. Hence, each signal can be transformed to more compact form. A brief description of EMD and parametric Feature vector construction are described in the following subsections.

2.1 Empirical Mode Decomposition (EMD)

EMD is a mathematical technique which is utilized to analyze a nonstationary and nonlinear signal. EMD assumes that a signal is composed of a series of different IMFs and decompose the signal into these continuous functions. Each IMFs have the following properties [13]:

1. Number of zero crossings and number of extrema are either equal or differ at most by one.
2. Local maxima and local minima produces the envelope whose mean value is equal to zero at a given point.

Figure 1 showed the plot of first four IMFs of an EEG segment using EMD algorithm. More details of this algorithm can be found in [13].

2.2 Parametric Feature Vector Construction

For constructing feature vector from the decomposed EEG signal, we have calculated eight parameters using moment values, complexity measure, and uncertainty values of the decomposed signal. The moments characterize the decomposed signal by certain statistical properties, a complexity measure shows repetitive nature in

Fig. 1 IMF plot obtained for a given EEG signal

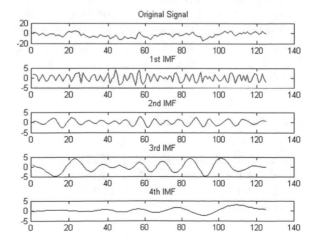

the time-series signal of decomposed signal and the uncertainty value denotes how much information contained by the signal. These parameters are root mean square, variance, skewness, kurtosis, Lampleziv Complexity, central & maximum frequency, and Shannon entropy of the signal.

3 Feature Selection

Feature selection [12, 16] is one of the approach to determine relevant features. Inspite of available rich research works on feature selection, not much work has been done in the area of mental task classification. The feature selection can be done using two methods. First method is classifier independent and relevance of the feature is measured by its inherent statistical properties, such as distance measure, correlation, etc. This approach is also known as filter method of feature selection. The second is wrapper method, where feature selection is classifier dependent and choose optimal subset of features to enhance accuracy of classifier. The wrapper based methods [16] find optimal or relevant subset of features from all possible combination of subsets of features and require classifier to evaluate the performance of the subset. Therefore, the computational cost of wrapper methods is much higher than filter methods.

Univariate (ranking) method, a filter feature selection method, is simple to compute. In these methods, a scoring function is used to measure the relevance, information content, discriminatory capability, or quality index of each feature separately without involving any classifier. Features are then ranked on the basis of their score in the order of its relevance. Many univariate feature selection methods are suggested in the literature. We have investigated four well known and commonly used univariate filter feature selection methods, namely, Fisher discriminant ratio, Pearson's correlation coefficient, Mutual information, and Wilcoxons rank sum test to determine a set of relevant features for mental task classification. A brief description of these methods is given below.

The problem under consideration has n samples of EEG signal, d features, and m distinct classes for mental task problem. Lets assume that matrix $\mathbf{X}$ represents available EEG data of dimension $n \times d$, where n is total number of samples and d represents total number of features. Here, each row $\mathbf{x}_i$ in matrix represents sample from class label c_i where $i = 1, 2, \ldots, m$ and each column $\mathbf{f}_j$ in matrix represents feature vector. Thus, the matrix $\mathbf{X}$ is represented as.

3.1 Pearson's Correlation

Pearson's correlation coefficient (CORR) [7, 18] is employed to determine linear correlation between two variables or between a variable and class label. The Pearson's correlation coefficient (CORR) of kth feature vector ($\mathbf{f}_k$) with the class label vector ($\mathbf{c}$) is given by

$$CORR\,(\mathbf{f}_k, \mathbf{c}) = \frac{cov(\mathbf{f}_k, \mathbf{c})}{\sigma_{\mathbf{f}_k}\sigma_c} = \frac{E[(\mathbf{f}_k - \mu_k)(\mathbf{c} - \bar{c})]}{\sigma_{\mathbf{f}_k}\sigma_c}, \quad for\ k = 1, 2, \ldots, d \quad (1)$$

where $\sigma_{\mathbf{f}_k}$, σ_c represent respectively the standard deviations of feature vector $\mathbf{f}_k$ and $\mathbf{c}$. $cov\,(\mathbf{f}_k, \mathbf{c})$ represents the covariance between $\mathbf{f}_k$ and $\mathbf{c}$, $\mu_k = \frac{1}{n}\sum_{i=1}^{n} X_{ik}$ and $\bar{c} = \frac{1}{n}\sum_{i=1}^{n} c_i$ are the mean of $\mathbf{f}_k$ and $\mathbf{c}$ respectively.

The value of $CORR\,(\mathbf{f}_k, \mathbf{c})$ lies between -1 and $+1$. The correlation value closer to -1 or 1, shows the stronger correlation among the prescribed variables while zero value implies no correlation between the two variables. It can measure both the degree as well as the trend of correlation. Also, it is invariant to linear transformations of underlying variables. However, the assumption of linear relationship between the variables is not always true. Also, sometimes the value of correlation coefficient may misinterpret the actual relation, as a high value does not always imply a close relationship between the two variables and it is sensitive to outliers too.

3.2 Mutual Information

Mutual information is an information theoretic-based ranking method which measures dependency between two variables. The mutual information of a feature vector $\mathbf{f}_k$ and the class vector $\mathbf{c}$ is given by [20]:

$$I\,(\mathbf{f}_k, \mathbf{c}) = \sum P\,(\mathbf{f}_k, \mathbf{c}) \log \frac{P\,(\mathbf{f}_k, \mathbf{c})}{P\,(\mathbf{f}_k)P\,(\mathbf{c})} \quad (2)$$

where $P\,(\mathbf{f}_k)$ and $P\,(\mathbf{c})$ are the marginal probability distribution functions for random variables $\mathbf{f}_k$ and $\mathbf{c}$ respectively and $P\,(\mathbf{f}_k, \mathbf{c})$ is joint probability distribution.

The maximum value of mutual information indicates the higher dependency of variable on the target class. The advantage of mutual information is that it can capture even the nonlinear relationship between the feature and the corresponding class label vector $\mathbf{c}$.

3.3 Fisher Discriminant Ratio

Fisher Discriminant Ratio (FDR) is another univariate filter feature selection method which is based on the statistical properties of the features. FDR($\mathbf{f}_k$) for kth feature for two class i and j is defined as:

$$FDR(\mathbf{f}_k) = \frac{(\mu_{i(k)} - \mu_{j(k)})^2}{\sigma_{i(k)}^2 + \sigma_{j(k)}^2} \quad (3)$$

where $\mu_{i(k)}$ and $\sigma_{i(k)}^2$ are the mean and dispersion of the data of class i, respectively, for kth feature. It takes maximum value when the square of the difference between mean of two classes for feature $\mathbf{f}_k$ is maximum and sum of variances of corresponding feature for both class i and j is minimum.

3.4 Wilcoxon's Ranksum Test

Wilcoxon Ranksum Test, proposed by [21], is a statistical test, which is carried out between data of two classes on the basis of median of the samples without assuming any probability distribution.

The statistical difference $t(\mathbf{f}_k)$ of feature $\mathbf{f}_k$ for given two classes i and j using Wilcoxon's statistics is defined as [17]:

$$t(\mathbf{f}_k) = \sum_{l=1}^{N_i} \sum_{m=1}^{N_j} DF((X_{lk} - X_{mk}) \le 0) \tag{4}$$

where N_i and N_j are the number of the samples in ith and jth class respectively, DF is the logical distinguishing function between two classes which assigns a value of 1 or 0 corresponding to true or false and X_{lk}, is the expression values of kth feature for lth sample. The value of $t(\mathbf{f}_k)$ tends from zero to $(N_i \times N_j)$. The relevance of the feature is determined as:

$$R(t(\mathbf{f}_k)) = \max(t(\mathbf{f}_k), N_i \times N_j - t(\mathbf{f}_k)) \tag{5}$$

4 Experimental Setup and Result

4.1 Dataset and Constructing Feature Vector

In order to check the effectiveness of the proposed method, experiments have been performed on a publicly available dataset[1] [15] which consists of recordings of EEG signals using six electrode channels from seven subjects with the recording protocols. Each subject was asked to perform 5 different mental tasks as namely *Baseline task relax* (B), *Letter Composing task* (L), *Non trivial Mathematical task* (M), *Visualizing Counting* (C) of numbers written on a blackboard, and *Geometric Figure Rotation* (R) task. For conducting the experiment, data from all the subjects are utilized except Subject 4; as data recorded for Subject 4 is incomplete [8].

The EEG signal corresponding to each mental task of a particular subject is formed into half-second segments which yields into 20 segments (signal) per trial per channel.

---

[1]http://www.cs.colostate.edu/eeg

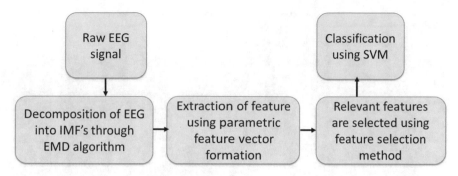

Fig. 2 Flow diagram of the proposed method

Thus, for every channel, each of 20 segments are decomposed using EMD algorithm into 4 IMFs. The eight parameters are extracted for each of these IMFs per segment per channel per trial for a given subject. A set of aforementioned eight statistical parameters is obtained for each of the six channels of the signal and these sets were concatenated to form a feature vector. Hence, the final feature vector is of 192 dimensions (four IMFs × 8 parameters × six channels) after applying the parametric feature vector formation step. As the dimension of feature vector are still high and not all features are relevant for classification so feature selection methods are utilized for selecting only relevant features for classification which results in lowering the time for building the classification model. Figure 2 shows complete pipeline for constructing the feature vector from each subject using all trial corresponding to each mental tasks labels (B, L, M, C, and R) for further classification using SVM classifier.

4.2 Results and Discussion

As discussed in the previous subsection, a set of feature vectors have been obtained corresponding to every mental task labels (B, L, M, C, and R). Binary mental task classification problem has been formulated to distinguish the different mental state of different subjects. The optimal value of SVM regularization parameters, i.e., gamma and cost, were obtained with the help of grid search algorithm. The average classification accuracy of 10 runs of 10 cross-validations has been reported. Figure 3 shows average classification accuracy of different binary combination of mental tasks averaged over all subjects corresponding to different feature selection techniques. Number of relevant features selected corresponding to given feature selection method is summarized in Fig. 4. From these figures it can be noted that incorporating the feature selection techniques will lead to better accuracy in comparison to without feature selection.

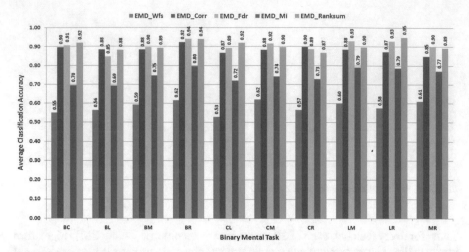

Fig. 3 Average classification accuracy for different binary mental task combinations over all subjects

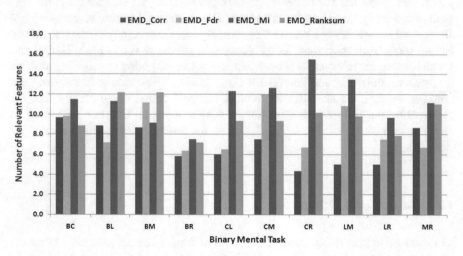

Fig. 4 Average number of relevant features corresponding to different binary mental task combinations over all subjects

The research work of Gupta and Kirar [11] used 48 features using parametric feature extraction method for the classification. However, our current approach has generated 192 features from EMD signal, which may have some irrelevant features that may hinder classification performance in terms of accuracy and time. The feature selection techniques have huge advantage in terms of dimensionality reduction and removal of irrelevant features. Hence, reduced set of relevant features increase the performance of classifiers [4]. The study compared the results of some popular feature selection techniques (Corr, MI, FDR, and Ranksum) with complete set of features

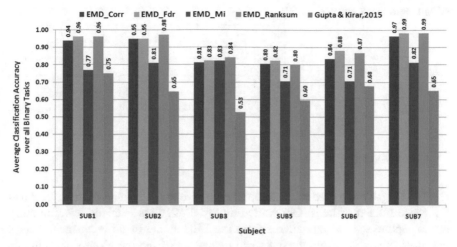

Fig. 5 Comparison of performance of various classification approaches on different subjects

(WFS) for mental task classification. The feature selection techniques had shown promising results in terms of accuracy over WFS (See Fig. 3) and the research work of Gupta & Kirar [11] (See Fig. 5). The training time of classifiers would also reduced significantly as the number of relevant features after applying the feature selection methods are much lesser than that of without feature selection (See Fig. 4). On the basis of two classification performance parameter, we observed that correlation-and FDR-based feature selection methods are well suited for mental task classification. Hence, our model can be beneficial for the differently abled persons to communicate with the machine more efficiently, i.e., quickly and accurately.

4.3 Friedman Statistical Test

For assessment of the significant difference among combinations of different feature selection methods with EMD and the research work [11], statistically, a two way [5] and nonparametric statistical test known as Friedman test [9] has been conducted. Where, H_0 was null hypothesis that assumes there is no significance difference in performance in among all approach. The H_0 was rejected at 95% of confidence level. Table 1 shows the ranking of different methods. Lowest rank for a given method shows its better performance compared to other methods. From Table 1, it can be noted that the combination of EMD feature extraction with FDR feature selection performs better in comparison to other methods.

Table 1 Friedman's ranking

Algorithms	Ranking
EMD_Corr	2.65
EMD_Fdr	1.65
EMD_Mi	4
EMD_Ranksum	1.7
Gupat & Kirar,2015	5

5 Conclusion

The EEG signals are used to capture the cognitive activities and each activity had embedded hidden patterns. Our study employed effective machine learning strategy to capture the hidden patterns from the EEG signal of different mental tasks and make prediction about the unknown mental task from the given signal. In this work, EMD algorithm is used to decomposed EEG signal into IMFs and parametric features are calculated for forming the feature vectors. Further, for selecting only relevant features, four well-known univariate feature selection techniques are investigated which reduces the dimension of feature vectors which results into reduction of time in building the classification model. The experiment has been performed on a publicly available EEG dataset which contains the responses to different mental thought regarding some task. The experimental results show the performance of the proposed approached for binary mental task classification problem is improved after incorporating the feature selection in conjunction with EMD algorithm.

Acknowledgements Authors express their gratitude to Cognitive Science Research Initiative (CSRI), DST & DBT, Govt. of India & CSIR, India for obtained research grant.

References

1. Anderson, C.W., Stolz, E.A., Shamsunder, S.: Multivariate autoregressive models for classification of spontaneous electroencephalographic signals during mental tasks. IEEE Trans. Biomed. Eng. **45**(3), 277–286 (1998)
2. Babiloni, F., Cincotti, F., Lazzarini, L., Millan, J., Mourino, J., Varsta, M., Heikkonen, J., Bianchi, L., Marciani, M.: Linear classification of low-resolution eeg patterns produced by imagined hand movements. IEEE Trans. Rehabil. Eng. **8**(2), 186–188 (2000)
3. Bashashati, A., Fatourechi, M., Ward, R.K., Birch, G.E.: A survey of signal processing algorithms in brain-computer interfaces based on electrical brain signals. J. Neural Eng. **4**(2), R32 (2007)
4. Bennasar, M., Hicks, Y., Setchi, R.: Feature selection using joint mutual information maximisation. Expert Syst. Appl. **42**(22), 8520–8532 (2015)
5. Derrac, J., García, S., Molina, D., Herrera, F.: A practical tutorial on the use of nonparametric statistical tests as a methodology for comparing evolutionary and swarm intelligence algorithms. Swarm Evol. Comput. **1**(1), 3–18 (2011)
6. Diez, P.F., Torres, A., Avila, E., Laciar, E., Mut, V.: Classification of Mental Tasks Using Different Spectral Estimation Methods. INTECH Open Access Publisher (2009)

7. Dowdy, S., Wearden, S., Chilko, D.: Statistics for Research, vol. 512. Wiley, New York (2011)
8. Faradji, F., Ward, R.K., Birch, G.E.: Plausibility assessment of a 2-state self-paced mental task-based bci using the no-control performance analysis. J. Neurosci. Methods **180**(2), 330–339 (2009)
9. Friedman, M.: The use of ranks to avoid the assumption of normality implicit in the analysis of variance. J. Am. Stat. Assoc. **32**(200), 675–701 (1937)
10. Gupta, A., Agrawal, R., Kaur, B.: Performance enhancement of mental task classification using eeg signal: a study of multivariate feature selection methods. Soft Comput. **19**(10), 2799–2812 (2015)
11. Gupta, A., Kirar, J.S.: A novel approach for extracting feature from eeg signal for mental task classification. 2015 International Conference on Computing and Network Communications (CoCoNet), pp. 829–832. IEEE (2015)
12. Guyon, I., Elisseeff, A.: An introduction to variable and feature selection. J. Mach. Learn. Res. **3**, 1157–1182 (2003)
13. Huang, N.E., Shen, Z., Long, S.R., Wu, M.C., Shih, H.H., Zheng, Q., Yen, N.C., Tung, C.C., Liu, H.H.: The empirical mode decomposition and the hilbert spectrum for nonlinear and non-stationary time series analysis. Proc. R. Soc. Lond. Ser. A: Math. Phys. Eng. Sci. (1971) **454**, 903–995 (1998)
14. Kauhanen, L., Nykopp, T., Lehtonen, J., Jylanki, P., Heikkonen, J., Rantanen, P., Alaranta, H., Sams, M.: Eeg and meg brain-computer interface for tetraplegic patients. IEEE Trans. Neural Syst. Rehabil. Eng. **14**(2), 190–193 (2006)
15. Keirn, Z.A., Aunon, J.I.: A new mode of communication between man and his surroundings. IEEE Trans. Biomed. Eng. **37**(12), 1209–1214 (1990)
16. Kohavi, R., John, G.H.: Wrappers for feature subset selection. Artif. Intell. **97**(1), 273–324 (1997)
17. Li, S., Wu, X., Tan, M.: Gene selection using hybrid particle swarm optimization and genetic algorithm. Soft Comput. **12**(11), 1039–1048 (2008)
18. Pearson, K.: Notes on the history of correlation. Biometrika, 25–45 (1920)
19. Pfurtscheller, G., Neuper, C., Schlogl, A., Lugger, K.: Separability of eeg signals recorded during right and left motor imagery using adaptive autoregressive parameters. IEEE Trans. Rehabil. Eng. **6**(3), 316–325 (1998)
20. Shannon, C.E., Weaver, W.: The Mathematical Theory of Communication (urbana, il) (1949)
21. Wilcoxon, F.: Individual comparisons by ranking methods. Biom. Bull. **1**(6), 80–83 (1945)

Automatic Attendance System Using Deep Learning Framework

Pinaki Ranjan Sarkar, Deepak Mishra
and Gorthi R. K. Sai Subhramanyam

Abstract Taking attendance in a large class is cumbersome, repetitive, and it consumes valuable class time. To avoid these problems, we propose an automatic attendance system using deep learning framework. An automatic attendance system based on the image processing consists of two steps: *face detection* and *face recognition*. Face detection and recognition are well-explored problems in computer vision domain, though they are still not solved due to large pose variations, different illumination conditions, and occlusions. In this work, we used state-of-the-art face detection model to detect the faces and a novel recognition architecture to recognize faces. The proposed face verification network is shallower than the state-of-the-art networks and it has achieved similar face recognition performance. we achieved 98.67% on LFW and 100% on classroom data. The classroom data was made by us for practical implementation of the complete network during this work.

1 Introduction

Though tremendous strides have been made in the field of face detection and recognition [3], they are still considered to be difficult problems. The difficulty comes when the captured image is in an unconstrained environment. An image in an unconstrained environment may result in different illumination, head pose, facial expressions, and occlusions. Both detection and recognition accuracy drop significantly in the presence of these variations, especially in the case of pose variations and background clutter. Classroom is a perfect example of an unconstrained environment

P. R. Sarkar (✉) · D. Mishra
Indian Institute of Space Science and Technology, Thiruvananthapuram, India
e-mail: sarkar0499pinaki@gmail.com

D. Mishra
e-mail: deepak.mishra@iist.ac.in

G. R. K. S. Subhramanyam
Indian Institute of Technology Tirupati, Tirupati, India
e-mail: rkg@iittp.ac.in

© Springer Nature Singapore Pte Ltd. 2019
M. Tanveer and R. B. Pachori (eds.), *Machine Intelligence and Signal Analysis*,
Advances in Intelligent Systems and Computing 748,
https://doi.org/10.1007/978-981-13-0923-6_29

335

where the factors like illumination, head pose, occlusions matter. In this paper, our main focus is to develop a highly accurate, implementable, light network to work as an automatic attendance system. Taking attendance in large class is cumbersome. Even though alternate attendance system schemes have provided a better accuracy, facial recognition-based framework has always been a topic of interest among academicians and researchers, mostly because face is usually the fastest means through which humans recognize a person and it can also be collected in a non-evasive manner.

Some researchers have addressed this facial recognition-based attendance management system using low-or middle-level features like principal component analysis (PCA), eigenface [2, 11, 18], discrete wavelet transform (DWT), discrete cosine transform (DCT) [10]. Features extracted in the related literatures are called low-or middle-level as they do not extract deep hierarchical representation from the high-dimensional data. Deep learning specifically Convolutional Neural Networks (CNN) have produced state-of-the-art results for unconstrained face detection and recognition tasks.

Our proposed framework has two stages: face detection and face verification. The detection of face is a crucial task as the verification result depends on the efficiency of the face detection stage. Present state-of-the-art model [6] for face detection is used in this paper to extract all the faces in highly unconstrained environment, i.e., classroom. This CNN (ResNet architecture [5])-based model has produced an average precision of 81% while prior art ranges from 29–64% in WIDER, a widely used database for face detection benchmarking. Next, the face recognition model used in this paper is inspired by the works like [12, 16, 20]. CNNs are good at learning the intricate structures with the use of backpropagation algorithm and produces high-level semantic features which lead to high accuracy in verification task.

We also used spatial transformer network (STN) [8] to align the faces which shows improvement in facial verification accuracy. The proposed recognition architecture in conjunction with STN shows improvement in LFW database, though both work on par on recognition in the considered classroom data. Details of which will be demonstrated in Sect. 5. Our contributions in this work are as follows:

- We used Deep Learning, specifically Convolutional Neural Network to develop an automatic attendance system.
- The trained model developed during training stage is of 132 MB which can be implemented easily in microprocessors or in raspberry-pi.
- We used spatial transformer network to learn alignment of the faces which leads better facial verification accuracy.

2 Databases

2.1 Labeled Faces in the Wild

Labeled Faces in the Wild (LFW) database [7] consists of 5423 unique classes with the large pose and illumination variations. A recent study [1] shows that, for a given number of images, it is better to have more people than having more images for fewer people. So, augmentation was done (9 random rotations between $[-45^0, +45^0]$) to increase our training data into 276000 training images, 92000 testing images, and 92000 validation images.

2.2 Classroom Data

A short video was taken using NIKON D5200 camera from one of the classrooms. The video was taken for 13.84 s at 50 fps and it had 692 frames. The students were told to give various facial pose during the video-shoot. To evaluate the proposed framework, 25 frames from the 692 frames are taken randomly. All of the extracted faces after the face detection are manually labeled. Out of the 25 frames, the faces from first 20 frames are taken to train the recognition network and all the face images from the remaining 5 frames are taken to test the network. The classroom which is selected for this work includes most of the challenging factors, such as extreme lighting condition, occlusion, presence of tiny faces, large head pose, etc.

3 Proposed Approach

Overall architecture of the automatic attendance system proposed in this paper is shown in Fig. 1. All the stages will be discussed in the consecutive subsections.

3.1 Face Detection

An image in an unconstrained environment like a classroom may result in different illumination, head pose, facial expressions, and occlusions. Though we have very strong face recognition models like [12, 16, 20] etc. but the face detection is still a challenging problem. Recently, Hu et al. [6] came up with a novel deep learning framework to find small and tiny faces which is found to be very effective in our classroom data. Their detector combines a novel combination of scale, context, and resolution to detect faces. Their proposed architecture resembles a region proposal network (RPN) trained for only faces as the objects. This is a binary multichannel

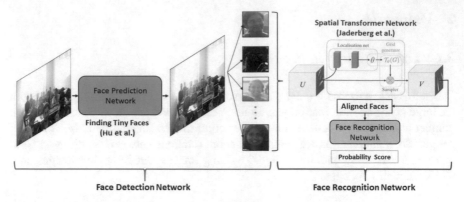

Fig. 1 Overall architecture of the automatic attendance system

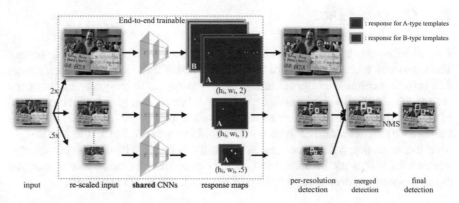

Fig. 2 Overall architecture of the detection framework. First, a coarse-level image pyramid at [0.5x, 1x, 2x] resolutions is taken as the input. At each scale, the input image is rescaled to 224×224 which is then passed through ResNet-50 [5] to extract hypercolumn features. Based on the features, response maps of the corresponding templates are predicted. Next, the detection bounding boxes are extracted and merged back in the original scale. A non-maximum suppression (NMS) is applied to get the final result. They run A-type templates which are tuned for faces with 40–140 pixels in height on the coarse image pyramid and B-type templates on 2x interpolated images to find faces less than 20 pixels in height. Please refer to [6] for more details about these two types of templates

heat-map prediction problem. The overview of the detection scheme is given in the Fig. 2. The face detection algorithm works very well for classroom images. This algorithm is very good at detecting out-of-plane rotated faces as well as tiny faces but the performance lags when in-plane rotation occurs (Fig. 3). One example from classroom database is shown in Fig. 4.

Though most of the previously developed attendance system use Viola–Jones algorithm [17] for face detection, in this paper, [6] is chosen because of its robustness to unconstrained environment. After detecting the faces from a classroom video, a cropping operation based on the bounding box location was performed and the facial

(a) Face detection result when there is controlled environment but large out of plane face variation.

(b) Face detection result when the environment is uncontrolled but faces are frontal.

Fig. 3 Results produced using [6]

(a) Face detection result when $+45^0$ in-plane rotation is introduced. 11 faces out of 14 faces were detected, 1 is falsely detected.

(b) Face detection result when -45^0 in-plane rotation is introduced. 12 faces out of 14 faces were detected, 1 is falsely detected.

Fig. 4 Results where the face detection method [6] fails

images for each frames were stored. In the next section, we discuss about the face recognition task.

3.2 Face Recognition

Deep learning paradigm has greatly influenced the development of face recognition. After the Facebook's DeepFace [16], Deep learning has proven its dominance over prior state-of-the-art. Inspired from this work a lot of deep models have been proposed [9, 12–15, 19, 20]. Deep learning has shown that automatically learnt deep features from personal identity are more effective in robust recognition than the traditional handcrafted features, such as local binary pattern (LBP), Active Appearance Model (AAM), Bayesian faces, Gaussian faces, Eigenfaces, etc.

Most of the deep learning frameworks for face recognition take aligned face images during both training and testing phase. The alignment is mainly done using geometric transformations. During the development of an automatic face recognition network, 2D and 3D transformations in facial images are experimented to see the effectiveness of alignment learning (AAM [4] and 3D Dense Face Alignment [20]) in the performance of face recognition task. To implement an automated attendance system, one can not rely on handcrafted features or any such preprocessing. It is much preferable to have a model that recognizes faces using self-alignment learning. Thus, we propose a face recognition network similar to the work [19]. In the subsequent sections, we describe our proposed network and the results on massively benchmarked database, Labeled Faces in the Wild (LFW) [7].

4 Experiments

4.1 Experiments on LFW

We used spatial transformer network [8] (STN) to learn the alignment of faces properly such that the performance of our proposed network increases. STN learns to transform optimally the out-of-plane rotated faces such that the loss function defined at the end gets minimized. The face verification network in conjunction with STN has obtained 98.67% accuracy in the LFW database. Some of the facial images after passing through trained STN is shown in Fig. 5. Later, we experimented without using the STN in our classroom data as large pose variation does not occur frequently in the classroom images.

Spatial Transformer Network: This differentiable CNN-based network [8] is very good at learning geometric transformations and it can learn alignment for out-of-plane faces by minimizing cost function through stochastic gradient descent(SGD) algorithm in conjunction with backpropagation algorithm. For more information, please refer [8].

Architecture: The overall architecture for face recognition is shown in the Fig. 1. The details of the architecture is given in the Table 1. The choice of the architecture is based on the rational that, in one classroom the detectable faces will not cross 150–200 in numbers. So, an architecture was needed which will show satisfactory recognition performance. We introduced a small and efficient facial verification architecture which can recognize faces with high accuracy. ImageNet challenge wining architectures were not used for this task as the detected face sizes will vary. In our experiment, it was observed that the size of the smallest recognizable face is around 1600 square pixels. Upsampling detected faces which are far away from the camera affects the recognition performance.

Table 1 Detailed parameters of each layer for LFW database

Spatial transformer network

Name	Filter size	Feature maps	Dropout
Pooling1	3		
Conv1	5	20	
Pooling2	3		
Conv2	3	20	
Dense			
FC1		50	
FC2		6	
Image-sampler			

Recognition Network

Name	Filter size	Feature maps	Dropout
Conv3	5	32	
ReLU			
Pooling3	2		
Conv4	5	64	
ReLU			
Pooling4	2		
Dropout			0.35
Conv5	3	96	
ReLU			
Dense			
FC3		2096	
ReLU			
FC4		2096	
ReLU			
Dropout			0.25
FC5		1000	
Softmax			

3 convolution layers, followed by subsampling layers, and nonlinear activation layers are mainly used for feature extraction. Two fully connected layers are present to mix the extracted deep features. The network is optimized with *categorical loss* and *softmax* as this is a multiclass learning problem. Evaluation of this network is done on both LFW database and classroom data. We have discussed the results in the Sect. 5.

(a)   (b)

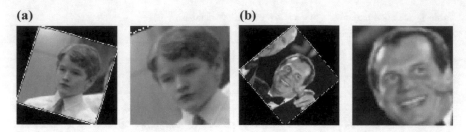

Fig. 5 Output of the STN in LFW database. Left: Input to the recognition network; Right: Output image after STN

4.2 Experiments on Classroom

In this case, mostly frontal faces or near frontal faces can be seen which can help us to exclude the STN from our proposed network. For faces larger than 200 pixels in height, tight-fitted facial images are taken while some context information is included for faces smaller than 40 pixels in height, as context is a very important factor for detecting small faces. All the processes mentioned above are done in automated way. After getting the facial images, investigation on hypertuning the ImageNet winning architectures was done in initial experiments. These models have huge number of parameters also they are tuned on 224×224 or 299×299 image sizes. The extracted faces from classroom have an average size of 120×117 and upsampling the faces in higher resolution affects the facial features. Thus, the pretrained architecture fails to achieve high accuracy during the recognition task. Our recognition model is smaller compared to the state-of-the-art architectures and due to less number of parameters in conjunction with less training images, our model does not suffer from overfitting.

Architecture: The architecture used on classroom data is slightly modified version of the Table 1. The only change is done at the final Fully Connected layers (FC5) to train and test on classroom data. Based upon the student number, the FC5 layer can be changed and fine-tuned before using the trained network.

4.3 Transfer Learning on Classroom Data

Training of the recognition network (see Table 1) is done on LFW database and this trained network is fine-tuned during training on the classroom database.

4.4 System and Training Information

For our implementation, we have used Keras with Theano backend as our simulation environment on a system that has Intels Xeon 3.07 GHz processor, 24 GB RAM, and 1 Nvidia Quadro 6000 GPU with 448 Cuda cores. The objective or loss function used in this paper is *categorical cross entropy*. Backpropagation algorithm is employed to update the weights of the CNNs. Stochastic gradient descent (SGD) algorithm is used to find local minima for this recognition cum classification problem.

5 Results and Discussion

5.1 On LFW Dataset

The training along with validation loss and accuracy plot is shown in Fig. 6. The STN network learns to align the faces such that the loss defined at the end gets minimized.

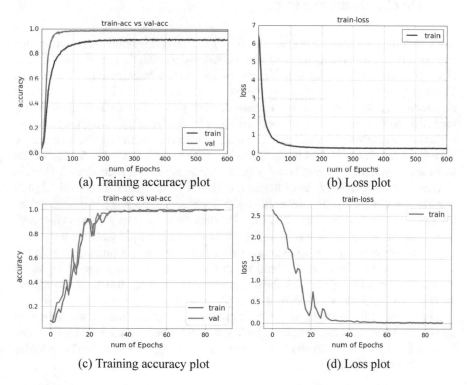

(a) Training accuracy plot (b) Loss plot

(c) Training accuracy plot (d) Loss plot

Fig. 6 Performance on different databases. **a, b** are the results in LFW database. **c, d** are fine-tuning learning curves in classroom dataset

Table 2 Comparison of face verification performances

Methods	Train set (in million)	Database	Recognition accuracy (%)
DeepFace [16]	4M	LFW	97.35
FaceNet [12]	200M	LFW	99.63
DeepID2+ [15]	0.2M	LFW	99.47
Alignment Learning [19]	0.46M	LFW	99.08
Ours	0.46M	LFW	98.67
Ours	14 images	Classroom	100

Some of the aligned output from the LFW database are given in the Fig. 5. We can see that our network has achieved near state-of-the-art performance on the clipped database in Table 2.

5.2 On Classroom Dataset

The fine-tuning learning curve (see Fig. 6c) shows that the network has multiple ups and down or non-smooth learning. One possible reason is that the network was initialized with the learned parameters of the network which was used on LFW database. There is a huge difference between the image quality between LFW database and classroom data. In LFW, there is no presence of tiny faces like we have in our classroom data so to learn from the previous network there is notifiable weight updatation which causes the initial disturbance in the learning curves. In the classroom data, we have observed that STN does not produce any significant improvement. Most of the faces are frontal so alignment is not required for an automatic attendance system.

Some of the predicted face images are shown in the Fig. 7. From this figure, one can see our recognition network works very well in presence of illumination, background clutter, pose variation, etc. The tiny face, i.e, Fig. 7b was taken with added context and large face such as Fig. 7f was taken in a tightly fitted manner.

The convergence of this network is much faster when it is initialized with the learned parameters on LFW database experiment. The following figures highlight the performance of our network during training and testing. It has taken only 60 iterations to recognize all 14 faces with 100% accuracy.

The effect of the spatial transformer network in both the databases is shown in Table 3. In LFW database, employing STN shows an improvement in recognition. Most of the detected faces will be frontal in classroom, so the effect of STN remains unnoticed in this case.

Table 3 Effect of STN on face verification performance

Method	Database	Recognition accuracy (%)
Without STN	LFW	96.28
With STN	LFW	98.67
Without STN	Classroom	100
With STN	Classroom	100

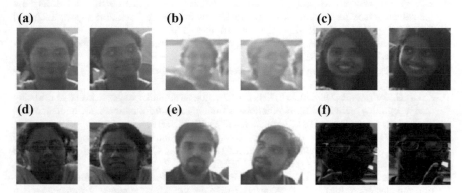

Fig. 7 Recognition output. Left side: Trained face, Right side: Test face

6 Conclusions and Future Works

We have implemented an end-to-end face recognition as well as an automatic attendance system using spatial transformer network and a small recognition network. The network has achieved 98.67% recognition accuracy in the LFW database and the trained model in this experiment was used to fine-tune in classroom data. The size of the trained model is of 132 MB, which can be easily stored in a microprocessor or raspberry-pi. The network will produce the probability scores of the present students from the captured images or videos.

In future, we want to add sentimental analysis using deep learning framework with this network. This will help to produce feedback of a particular class and it will show which topics, time, methods attract more attention of the students. Adding emotion analysis network will help the faculties to improve or modify their teaching method.

References

1. Bansal, A., Castillo, C.D., Ranjan, R., Chellappa, R.: The do's and don'ts for cnn-based face verification. CoRR **abs/1705.07426** (2017). http://arxiv.org/abs/1705.07426
2. Chintalapati, S., Raghunadh, M.: Automated attendance management system based on face recognition algorithms. In: IEEE International Conference on Computational Intelligence and Computing Research (ICCIC), pp. 1–5. IEEE (2013)

3. Hassaballah, M., Aly, S.: Face recognition: challenges, achievements and future directions. IET Comput. Vis. **9**(4), 614–626 (2015)
4. Hassner, T., Harel, S., Paz, E., Enbar, R.: Effective face frontalization in unconstrained images. Proc. IEEE Conf. Comput. Vis. Pattern Recognit. **7**, 4295–4304 (2015)
5. He, K., Zhang, X., Ren, S., Sun, J.: Deep residual learning for image recognition. Proc. IEEE Conf. Comput. Vis. Pattern Recognit. **2016**, 770–778 (2016)
6. Hu, P., Ramanan, D.: Finding tiny faces. In: 2017 IEEE Conference on Computer Vision and Pattern Recognition (CVPR), pp. 1522–1530 IEEE (2017)
7. Huang, G.B., Ramesh, M., Berg, T., Learned-Miller, E.: Labeled faces in the wild: A database for studying face recognition in unconstrained environments. Technical report, Technical Report 07–49, University of Massachusetts, Amherst (2007)
8. Jaderberg, M., Simonyan, K., Zisserman, A., et al.: Spatial transformer networks. Adv. Neural Inf. Process. Syst. 2017–2025 (2015)
9. Lu, C., Tang, X.: Surpassing human-level face verification performance on lfw with gaussianface. AAAI, 3811–3819 (2015)
10. Lukas, S., Mitra, A.R., Desanti, R.I., Krisnadi, D.: Student attendance system in classroom using face recognition technique. In: International Conference on Information and Communication Technology Convergence (ICTC), pp. 1032–1035. IEEE (2016)
11. Rekha, E., Ramaprasad, P.: An efficient automated attendance management system based on eigen face recognition. In: 7th International Conference on Cloud Computing, Data Science and Engineering-Confluence, pp. 605–608. IEEE (2017)
12. Schroff, F., Kalenichenko, D., Philbin, J.: Facenet: A unified embedding for face recognition and clustering. In: Proceedings of the IEEE Conference on Computer Vision and Pattern Recognition, pp. 815–823 (2015)
13. Sun, Y., Chen, Y., Wang, X., Tang, X.: Deep learning face representation by joint identification-verification. In: Advances in neural information processing systems, pp. 1988–1996 (2014)
14. Sun, Y., Liang, D., Wang, X., Tang, X.: Deepid3: face recognition with very deep neural networks (2015). arXiv:1502.00873
15. Sun, Y., Wang, X., Tang, X.: Deeply learned face representations are sparse, selective, and robust. Proc. IEEE Conf. Comput. Vis. Pattern Recognit. 2892–2900 (2015)
16. Taigman, Y., Yang, M., Ranzato, M., Wolf, L.: Deepface: closing the gap to human-level performance in face verification. Proc. IEEE Conf. Comput. Vis. Pattern Recognit. 1701–1708 (2014)
17. Viola, P., Jones, M.: Rapid object detection using a boosted cascade of simple features. In: Proceedings of the 2001 IEEE Computer Society Conference on Computer Vision and Pattern Recognition CVPR 2001, pp. I–I. IEEE (2001)
18. Wagh, P., Thakare, R., Chaudhari, J., Patil, S.: Attendance system based on face recognition using eigen face and pca algorithms. In: International Conference on Green Computing and Internet of Things (ICGCIoT), pp. 303–308. IEEE (2015)
19. Zhong, Y., Chen, J., Huang, B.: Towards end-to-end face recognition through alignment learning (2017). arXiv:1701.07174
20. Zhu, X., Lei, Z., Liu, X., Shi, H., Li, S.Z.: Face alignment across large poses: A 3d solution. Proc. IEEE Conf. Comput. Vis. Pattern Recognit. 146–155 (2016)

Automatic Lung Segmentation and Airway Detection Using Adaptive Morphological Operations

Anita Khanna, N. D. Londhe and S. Gupta

Abstract The respiratory system of lungs contains airway trees. The detection and segmentation of airways is a challenging job due to noise, volume effect and non-uniform intensity. We present a novel automatic method of lung segmentation and airway detection using morphological operations. Optimal thresholding combined with connected component analysis gives good results for lung segmentation. We describe a quick method of airway detection with grayscale reconstruction performed on four-connected low-pass filtered image. The results are quite satisfactory with some error due to non-uniform intensity and volume effect in the CT image.

Keywords Optimal thresholding · Connected component analysis
Four-connected filter · Grayscale reconstruction

1 Introduction

Pulmonary CT scans are used worldwide for diagnosing and treating lung diseases. Lung cancer is the common cause of death both in men and women as per the survey done by American Cancer society [1]. Early detection of the symptoms can help save mankind [2]. Detection of tumor, nodule is quite difficult due to the presence of multiple vascular structures and airways in lung parenchyma. For the purpose of investigation, lung has to be first segmented. Moreover quantification of airways can also help in diagnosis and treatment planning. Airways are the gas exchange structures in lungs and gets narrower as it penetrates deep inside the lung. Detection and segmentation of airways is a challenging task but can help in detecting pulmonary diseases.

A. Khanna (✉) · N. D. Londhe · S. Gupta
Electrical Engineering, NIT Raipur, Raipur, India
e-mail: anitadhawan0308@gmail.com

N. D. Londhe
e-mail: nlondhe.ele@nitrr.ac.in

S. Gupta
e-mail: sgupta.ele@nitrr.ac.in

© Springer Nature Singapore Pte Ltd. 2019
M. Tanveer and R. B. Pachori (eds.), *Machine Intelligence and Signal Analysis*,
Advances in Intelligent Systems and Computing 748,
https://doi.org/10.1007/978-981-13-0923-6_30

Many researchers have proposed different techniques for lung segmentation. Region Growing has been commonly used by researcher for lung segmentation. Manual selection of seed points with 3D region growing have been proposed by Kalender et al. [3]. Thresholding techniques with region growing have been suggested by Ko and Betke [4] and Lee et al. [5]. Shojaii et al. [6] performed immersion-based watershed algorithm for lung segmentation. Internal and external markers combined with gradient of the image were used with watershed algorithm for segmentation.

The different approaches, both semiautomatic and automatic, towards airway detection have been based on region growing, morphology, and neural network. Fabijanska [7] proposed a fully automated region growing technique for airway segmentation. This work includes two passes of region growing, first for coarse detection and second for refining the airways. A review on airway tree extraction, using different techniques has been done by Pu et al. [8]. A work on voxel classification as airway and non-airway voxels have been proposed by Lo et al. [9] to segment airways. In this work, airway tree was grown taking into account the leakage into parenchyma. Grayscale morphological operations were used by D. Aykac et al. [10] to detect and construct airway tree. Convolutional networks for classification of airways have been performed by Charbonnier et al. [11] with leakage detection as classification problem. Multiple segmentation helped in airway extraction avoiding leakages.

This proposed work goes in two stages: First lung segmentation is done automatically using optimal thresholding and connected component analysis. In the second stage, airways are detected from 2D CT pulmonary images using grayscale reconstruction method. The grayscale value of airways local minima is elevated which are finally thresholded.

The paper is framed as follows: Different methods used for above specified work are described in Sect. 2. Section 3 presents the results based on the techniques used in Sect. 2. Conclusion and discussion is included in Sect. 4.

2 Methods

Figure 1 gives the overview of the proposed method used for lung segmentation and airway detection. The input is 2D CT pulmonary image slices taken from VESSEL12 challenge. Different stages of work are as follows:

2.1 Lung Segmentation

Optimal thresholding: Here quick automated way of thresholding is used against manual thresholding to enhance the image. The process starts with selection of a threshold based on background pixel range at the corners and then adaptively the threshold between foreground and background creates enhanced image.

Fig. 1 Block diagram

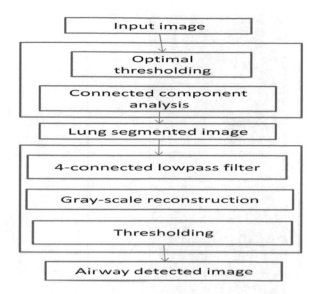

Connected component analysis: Next open and close operations are performed to get binary image clearly showing the two lungs, outer area and background. Connected component analysis of the binary image creates a label matrix for different components which helps to segment the two lungs.

2.2 Airway Detection

Four-connected low-pass filter: Four-connected neighborhood low-pass filter is a good averaging filter for noise reduction. The filter matrix is chosen based on the smoothing criteria to improve the image.

Grayscale reconstruction: Airways and vessels exhibit peaks and valleys in grayscale. For identifying potential airways in an image, grayscale closing operation is performed with structuring element B

$$P = I \bullet B = (I \oplus B) \ominus B \tag{1}$$

Structuring element B takes different sizes B_k and the process is repeated. This helps in improving the grayscale value of airway minima pixels according to their sizes (Fig. 2). Finally, grayscale reconstruction is done using the maximum pixel values among all the image outputs.

$$P_\infty = \max(P_k \ominus B_{k,} I) \tag{2}$$

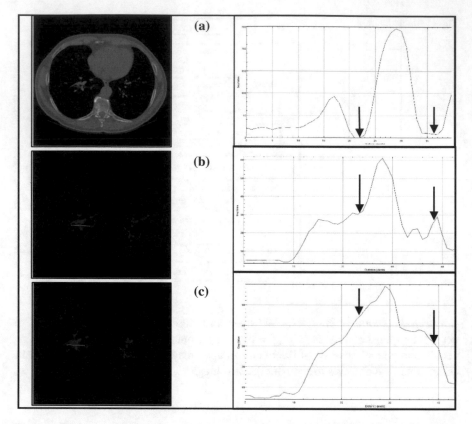

Fig. 2 **a** Original image with a line overlaid on it and its line profile **b** Improved air-way grayscale image with a line overlaid on it and its line profile **c** Final airway grayscale image with a line overlaid on it and its line profile. Arrows show the graylevels of airway being improved

This output image is eroded and subtracted from the original image to give only airways in the image.

$$K = P_\infty - I, \tag{3}$$

where K is the final image reconstructed after above mentioned operation and I is the original image.

Thresholding: In the difference image, local minima airways can be located as bright clusters but not all bright points are airways. Therefore thresholding is performed to get the final airways in the image.

3   Results

For the lung segmentation and airway detection, the data has been taken from VES-SEL12 which is in RAW form. These have been converted to the suitable form for MATLAB using Mevislab.

Lung Segmentation: The optimal thresholding operation is performed on the original image to get the threshold value and the image is thresholded. Figure 3b gives the thresholded image. Then connected component analysis is done to get the final lung segmented image as shown in Fig. 3c.

Airways Detection: For removing noise, four-connected low-pass filter is used and the results on different images are shown in Fig. 4a. Grayscale reconstruction is performed using multiple close operations and final gray-level maximum image is extracted. In Fig. 4b, improved gray-level pixel values of airways can be clearly seen.

This maximized gray-level pixel is then subtracted from the original image to highlight the airways as shown in Fig. 5.

Thresholding: This is the final step to get the airway detected image. This is achieved by thresholding. Few spurious points are also detected as airways which are due to non-uniform intensity and volume effects (that results in shading). The airways detection technique is applied on different CT pulmonary images and results (as white areas) are shown in Fig. 6. A small portion of detected airways is depicted in Fig. 7.

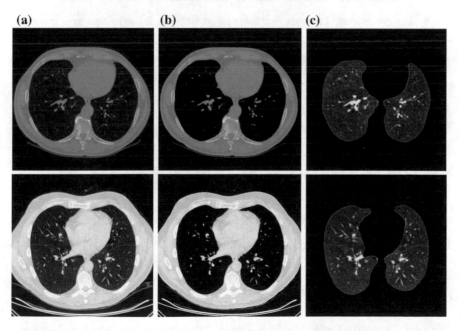

Fig. 3 **a** Original image **b** Optimally thresholded image **c** Lung segmented image

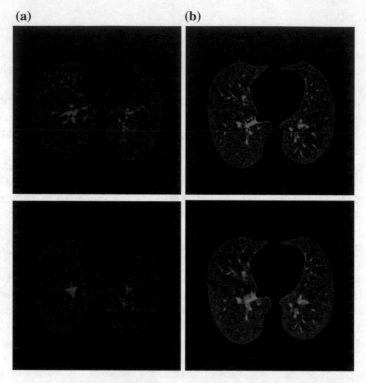

Fig. 4 Airway detection **a** Low-pass filter output image **b** Grayscale maximized image after reconstruction

Fig. 5 Results of two input images after subtraction

4   Conclusion and Discussion

We have proposed a method for lung segmentation and airway detection. Lung segmentation results are quite satisfactory which can be visualized clearly. The steps

(a) **(b)**

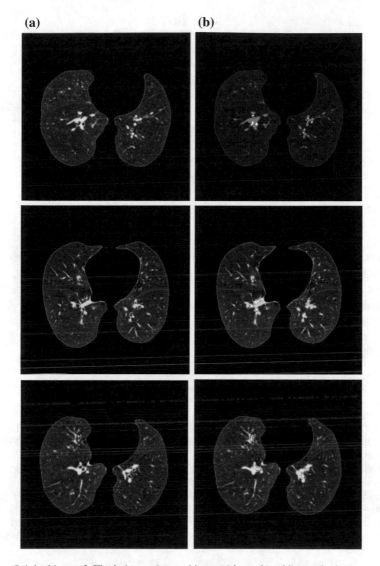

Fig. 6 **a** Original image **b** Final airway detected image (shown by white patches)

taken for airways detection have also given satisfactory results with some errors (both false positive and true negative detection) which are due to non-uniform intensity to CT images and volume effect which leads to shading. Airways are of different sizes at different places of lungs and detection of small airways is a challenging task. Moreover, there is no gold standard to validate the results so the overall results can be evaluated visually. All these operations are performed on 2D CT lung images from VESSEL12 challenge. We are working on different techniques to fulfill the

(a)   (b)

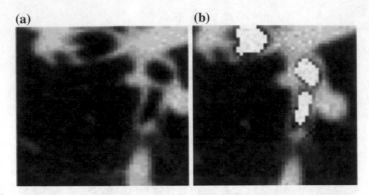

Fig. 7 **a** Portion of original image showing airways **b** Detected airways

shortcomings but the concept pursued for lung segmentation and airway detection can be appreciated In future, we will apply this technique on 3D slices of CT scans to detect and segment the airways and to construct an airway tree which will help the radiologists for further investigations.

References

1. Cancer Facts & Figures, American Cancer Society: Cancer Statistics (2012). http://www.cancer.org
2. Henchke, C.I., McCauley, D.I., Yankelevitz, D.F., Naidich, D.P., McGuinness, G., Miettinen, O.S., Libby, D.M., Pasmamntier, M.W., Koizumi, J., Altorki, N.K., Smith, J.P.: Early lung cancer action project: overall design and findings from baseline screening. Lancet **354**(9173), 99–105 (1999)
3. Kalender, W.A., Fiche, H., Bautz, W., Skalej, M.: Semiautomatic evaluation procedures for quantitative CT of the lung. J. Comput. Assist. Tomogr. **15**, 248–255 (1991)
4. Ko, J.P., Betke, M.: Chest CT: automated nodule detection and assessment of change over time—preliminary experience. Radiology **218**(1), 267–273 (2001)
5. Lee, Y., Hara, T., Fujita, H., Itoh, S., Ishigaki, T.: Automated detection of pulmonary nodules in helical CT images based on an improved template—matching technique. IEEE Trans. Med. Imaging **20**(7), 595–604 (2001)
6. Shojaii, R., Alirezaie, J., Babyn, P.: Automatic lung segmentation in CT images using watershed transform. In: ICIP 2005, Geneva, Italy, Sept 2005
7. Fabijanska, A.: Two-pass region growing algorithm for segmenting airway tree from MDCT chest scans. Computerized Med. Imaging Graph. **33**, 537–546 (2009)
8. Pu, P., Gu, S., Liu, S., Zhu, S., Wilson, D., Siegfried, J.M., Gur, D.: CT based computerized identification and analysis of human airways: a review. Med. Phys. **39**(5), 2603–2616 (2012)
9. Lo, P., Sporring, J., Ashraf, H., Pedersen, J.J.H., de Bruijne, M.: Vessel-guided airway tree segmentation: a voxel classification approach. Med. Image Anal. **14**, 527–538 (2010)
10. Aykac, D., Hoffman, E.H., Mclennan, G., Reinhardt, J.M.: Segmentation and analysis of human airway tree from three-dimensional X-ray CT images. IEEE Trans. Med. Imaging **22**(8) (2003)
11. Charbonnier, J.P., Van rikxoort, E.M., Setio, A.A.A., Prokop, C.M.S., Ginneken, B.V., Ciompi, F.: Improving airway segmentation in computed tomography using leak detection with convolutional networks. Med. Image Anal. **35**, 52–60 (2017)

ADMET Prediction of Dual PPARα/γ Agonists for Identification of Potential Anti-diabetic Agents

Neha Verma and Usha Chouhan

Abstract Since last 15 years, ex vivo experimental tools to describe ADME–Tox profiles of molecules have been exercised in initial stages of the drug development process to enhance the success percentage of discovery index and to ameliorate better candidates into drug discovery. Implementation of in silico ADMET prediction has further improved discovery support, allowing virtual screening of chemical compounds and therefore, application of ADMET property prediction at each phase of the drug development process. Recently there have been advancements in the approaches used to determine the accurateness of the prediction as well as application domain of the absorption, metabolism, distribution, excretion, and toxicity models. Developments also seen in the methods used to anticipate the physiochemical properties of leads in the initial steps of drug development. Absorption, distribution, metabolism, and excretion, ADME parameters, were calculated to investigate pharmacokinetic properties of hit compounds for the screening of new and potent anti-diabetic agents. ADME studies on a set of ligand molecules were performed by DruLiTo software. This study would permit chemists and drug-metabolism researchers to focus on compounds having maximum likelihoods of meeting the essential ADME criteria, also would add to a fall in compound attrition at late-stage.

Keywords ADMET · Pharmacokinetic properties · Anti-diabetic agents

1 Introduction

Diabetes has been a chronic malady that affects associate oversize proportion of population around the world and has assumed epidemic dimensions [1, 2]. The majority of those cases, some 90% measure of Type 2 Diabetes mellitus (T2DM) [3]. The calculable range of diabetic patients within the world was 1.712 billion (2.8%) within the year 2000 and 3.82 billion (8.3%) within the year 2013 foreseen

N. Verma (✉) · U. Chouhan
Department of Mathematics, Bioinformatics and Computer Applications, MANIT, Bhopal, India
e-mail: nehav2314@gmail.com

© Springer Nature Singapore Pte Ltd. 2019   355
M. Tanveer and R. B. Pachori (eds.), *Machine Intelligence and Signal Analysis*,
Advances in Intelligent Systems and Computing 748,
https://doi.org/10.1007/978-981-13-0923-6_31

to be 3.66 billion (4.4%) by the year 2030 [4]. Of this 0.613 billion patients were in India in the year 2011, and this range is anticipated to become 1.012 billion by the year 2030 [5]. The occurrence of T2DM is hastily increasing, notably among older, obese persons possessing associative cardiovascular (CV) risks [6]. T2DM is caused either by resistivity of insulin or by reduction in secretion of insulin [7] though there are alternative contributors. There is a gentle but rigorous decline in the beta cell function [8] with the disease progression necessitating up-gradation of the early chosen drug in addition of more, ultimately raising the use of insulin.

Thus, for the development of novel therapeutic agents for the treatment of T2DM and alternative metabolic malady, it is crucial to determine the molecular targets of the transducers markedly involved in the regulation of lipid and glucose homeostasis. Metabolic nuclear receptors (NR) molecules are found out as a particularly appealing target and govern the lipid and glucose homeostasis. Out of these receptors, members of the peroxisome proliferator-activated receptor (PPAR) family have been specially considered for more than a decade.

Issemann and Green discovered the PPARs in 1990 [9]. PPARs are representatives of the nuclear receptor-activated transcription factor superfamily composed of three subtypes as PPARα, PPARβ/δ, and PPARγ [10]. The major role of PPARs was primarily thought to be confined to glucose homeostasis, lipid metabolism, and cellular differentiation [11–14]. PPARs can be triggered by naturally occurring ligands like fatty acids or eicosanoids. Synthetic anti-diabetic agents like lipid-lowering fibrates and thiazolidinedione are found to act as agonists of PPARα and PPARγ, respectively [15–17].

The PPARα is deeply expressed in tissues that readily yield energy from the lipids, in addition to skeletal muscle and liver. PPARα controls the expression of various genes engaged in lipid uptake, homeostasis and catabolism. The PPARβ/δ has a universal expression however its role is not completely understood until now. The PPARγ is shown to vastly express itself in adipose tissues and monitor the expression of genes mediating the differentiation of adipocytes, insulin action and energy metabolism, as well as those which encodes a range of secreted adipokines [18]. From recent studies it has been found that dual agonists of PPARα/γ are credibly of much importance in medicinal chemistry. Dual PPARα/γ agonists has shown to reduce the free triglyceride (TG) plasma concentration [19], which could be related with hike in the adipocyte LPL activity and increment in b-oxidation in case of animal models with insulin resistance. Also the PPARα/γ dual agonists found to raise the plasma HDL concentration [20, 21]. Such effects on insulin sensitivity and glycaemia are proportionate to that of TZDs, a well-known PPARγ activator. These agonists are associated with an increase of sensitivity of insulin in the rodent, however this should be an aftereffect of the downturn in concentration of free fatty acid. Compelling impediment of side effects like oedema and/or obesity as observed in the substantial treatment by TZDs, is an additional interest for the co-activation of PPARα/γ.

Thus ligands simultaneously activating all the PPARs can be strong candidates in relation to drugs and can be used to treat abnormal metabolic homeostasis. PPAR receptors are activated by various metabolites and fatty acids [22]. Out of those, the

8-HETE arose notably interesting as this natural compound is a strong agonist of PPARα (EC50 = 100 nM) exhibiting a partial activity on PPARγ [23].

In silico design and development has helped prompt identification of drug-like chemical entities for varied ailments. To discern therapeutic implication of PPAR activators, here we state the absorption, distribution, metabolism, excretion and toxicity (ADMET) prediction on the series of 8-HETE compounds reported to have PPAR agonistic activity as shown in Table 1. These compounds have common scaffold and their biological assays performed under a single experiment, thus were used for the purpose of ADMET studies.

2 Materials and Methods

2.1 Dataset/Ligand Preparation

Three dimensional coordinates of 24 ligands taken from literature [23] and structures generated using chemdraw ultra software (www.cambridgesoft.com), USA. Structure files are generated with ".mol" and ".cdx" extension. The structure and LogP of the ligands considered in this study are shown in Table 1.

2.2 ADMET Prediction

The requisite for enhanced ADME–Tox throughput to completely fit the needs of discovery has led to improved and increasing curiosity in computational or in silico models which can generate predictions on the basis of chemical structures only. Absorption, distribution, metabolism and excretion, ADME parameters, were calculated to investigate pharmacokinetic properties of hit compounds. ADMET related physicochemical properties for 24 lead compounds possessing PPAR agonistic activity are predicted using Drug-Likeness Tool (DruLiTo) software [24] are shown in Table 2. DruLiTo is the tools which can be freely available package for the prediction of toxicity examined the physicochemical properties like Lipinski's rule, Veber rule, BBB rule and Quantitative Estimate of Drug-likeness (QED). Furthermore, the best PPAR agonists explained various physicochemical properties of compounds such as log P, log D, TPSA, MW, log S, HBA, HBD and nRot for the high probability of clinical success.

Table 1 Structure and biological activity of carbo- and heterocyclic analogues of 8-HETE

S. no.	Structure	Activity (nM)		S. no.	Structure	Activity (nM)	
		PPARα	PPARγ			PPAR α	PPAR γ
C1		173	642	C13		1,142	>10,000
C2		1,632	549	C14		262	1,413
C3		>10,000	>10,000	C15		723	>10,000
C4		1,543	>10,000	C16		114	617
C5		>10,000	28	C17		1,454	>10,000
C6		>10,000	>10,000	C18		1,300	1,546
C7		>10,000	>10,000	C19		1,531	1,356
C8		>10,000	>10,000	C20		>10,000	>10,000
C9		>10,000	340	C21		1,194	1,503
C10		>10,000	>10,000	C22		1,086	1,195
C11		>10,000	>10,000	C23		1,485	>10,000
C12		>10,000	>10,000	C24		1,399	1,566

3 Results and Discussion

3.1 ADMET Prediction

ADMET properties are needed to define the aggregate value to suitable the compounds for drug designing. 24 lead compounds are evaluated for drug design and development such as, molecular weight (MW), lipophilicity, i.e., calculated partition

Table 2 Drug likeliness properties of compounds

S. no.	MW	log P	alogP	HBA	HBD	TPSA	AMR	nRB	nAtom	nAcidicGroup	RC	nRigidB	nAromRing	nHB	SAlerts
C1	406.08	5.282	−0.808	1	1	29.46	58.33	11	51	0	1	12	1	2	2
[a]C2	407.07	4.492	−1.228	2	1	42.35	61.68	11	50	0	1	12	1	3	2
[a]C3	414.04	4.754	−1.02	1	1	29.46	52.6	11	48	0	1	12	1	2	3
[a]C4	415.04	3.964	−1.44	2	1	42.35	55.9	11	47	0	1	12	1	3	3
C5	334.21	5.056	−0.729	3	1	55.76	65.6	13	54	0	1	11	1	4	4
[a]C6	335.21	4.266	−1.149	4	1	68.65	68.9	13	53	0	1	11	1	5	4
C7	336.23	5.572	−2.86	3	1	55.76	58.02	14	56	0	1	10	1	4	3
[a]C8	337.23	4.782	−3.28	4	1	68.65	61.32	14	55	0	1	10	1	5	3
C9	342.18	5.088	−0.831	3	1	55.76	59.9	13	51	0	1	11	1	4	3
[a]C10	343.18	4.298	−1.252	4	1	68.65	63.2	13	50	0	1	11	1	5	3
C11	404.06	5.184	−2.599	1	1	29.46	43.04	13	49	0	1	9	1	2	2
[a]C12	405.05	4.394	−3.019	2	1	42.35	46.34	13	48	0	1	9	1	3	2
C13	456.09	7.158	−0.808	1	1	29.46	58.38	11	57	0	2	17	2	2	2
C14	457.09	6.12	−1.228	2	1	42.35	61.68	11	56	0	2	17	2	3	2
C15	464.06	6.63	−1.02	1	1	29.46	52.6	11	54	0	2	17	2	2	3
C16	465.05	5.592	−1.44	2	1	42.35	55.9	11	53	0	2	17	2	3	3
C17	384.23	6.932	−0.729	3	1	55.76	65.6	13	60	0	2	16	2	4	4
C18	385.23	5.894	−1.149	4	1	68.65	68.9	13	59	0	2	16	2	5	4
C19	386.25	7.448	−2.86	3	1	55.76	58.02	14	62	0	2	15	2	4	3
C20	387.24	6.41	−3.28	4	1	68.65	61.32	14	61	0	2	15	2	5	3
C21	392.2	6.964	−0.831	3	1	55.76	59.9	13	57	0	2	16	2	4	3
C22	393.19	5.926	−1.252	4	1	68.65	63.2	13	56	0	2	16	2	5	3
C23	380.2	6.911	−2.674	3	1	55.76	49.41	13	56	0	2	15	2	4	2
C24	381.19	5.873	−3.095	4	1	68.65	52.71	13	55	0	2	15	2	5	2

[a]Compounds qualifying all the properties

Table 3 Overall summary of ADMET property prediction

S. no.	Selected rule/filters	Number of molecule qualifying the rule	Number of molecule violated the rule
1	Lipinski's rule	8	17
2	BBB likeness rule	21	4
3	Ghose_Filter	13	12
4	Weighted QED	25	0
5	Unweighted QED	25	0
6	All selected filters	7	18

coefficient (log P), algorithm partition coefficient (alogP), hydrogen bond donors (HBD), hydrogen bond acceptor (HBA), number of rotatable bonds (nRB), topological polar surface area (TPSA), number of atom (nAtom), number of acidic group (nAG), number of rigid bond (nRigidB), number of hydrogen bond (nHB), aromatic ring number (nAR) and SAlerts (SA). The estimated values of physicochemical properties for the leads molecules are mentioned in Table 2. The MW value of the lead compounds ranges from 334 to 465, while the log P value of the lead molecules ranges from 3.96 to 7.44. The alogP value range for the lead compounds range from −0.72 to −3.28. TPSA ranges from 29.46 to 68.65 for lead compounds.

From the ADMET prediction of all the ligands, it is found that only seven ligands qualified all the filters as mentioned in summary Table 3.

Ligand C2, C3, C4, C6, C8, C10, and C12 qualified all the properties (as shown in Table 2) and can be used to screen potent ligand molecules from the chemical structure databases. Thus, this type of molecular scaffold can be exploited for the development of novel PPAR agonists. This research on lead compounds can be further modified and analyzed to design new compounds which can be used as drug molecules against Type 2 Diabetes.

4 Conclusion

In silico ADMET studies indicated the ligands which are in standard ranges. Therefore, these molecules can be used as beneficial molecule for screening large library and finding new inhibitors. Further the QSAR and molecular docking and molecular dynamic studies can be performed on the dataset for obtaining more potent molecules for the treatment of Type 2 Diabetes Mellitus.

Acknowledgements Author Neha Verma would like to acknowledge the UGC, Delhi for providing RGNFSC fellowship.

References

1. Dunstan, D.W., et al.: The rising prevalence of diabetes and impaired glucose tolerance: the Australian diabetes, obesity, and lifestyle study. Diabetes Care **25**, 829–834 (2002)
2. Rizvi, A.A.: Type 2 diabetes: epidemiologic trends, evolving pathogenic concepts, and recent changes in therapeutic approach. South. Med. J. **97**, 1008–1027 (2004)
3. Vaseem, A., Sethi, B., Kelwade, J., Ali, M.: SGLT2 inhibitor and DPP4 inhibitor co-administration in Type 2 diabetes—Are we near the "Promised Land"? Res. Rev. J. Pharmacol. Toxicol. Stud. **4**, 6–11 (2016)
4. Wild, S., et al.: Global prevalence of diabetes, estimates for the year 2000 and projections for 2030. Diabetes Care **27**, 1047–1053 (2004)
5. International Diabetes Federation 2012, 5th edn. The Global Burden
6. Carver, C.: Insulin treatment and the problem of weight gain in Type 2 diabetes. Diabetes Educ. **32**, 910–917 (2006)
7. Cavaghan, M.K., et al.: Interactions between insulin resistance and insulin secretion in the development of glucose intolerance. J. Clin. Investig. **106**, 329–333 (2000)
8. UK Prospective Diabetes Study Group: Overview of 6 years' therapy of Type II diabetes: a progressive disease. Diabetes **44**, 1249–1258 (1995)
9. Issemann, I., Green, S.: Activation of a member of the steroid hormone receptor superfamily by peroxisome proliferators. Nature **347**, 645–650 (1990)
10. Naidenow, J., Hrgovic, I., Doll, M., Jahn, T.H., Lang, V., Kleemann, J., Kippenberger, S., Kaufmann, R., Zoller, N., Meissner, M.: Peroxisome proliferator-activated receptor (PPAR) α and δ activators induce ICAM-1 expression in quiescent non stimulated endothelial cells. J. Inflamm. **13**, 27 (2016)
11. Grygiel-Górniak, B.: Peroxisome proliferator-activated receptors and their ligands: nutritional and clinical implications a review. Nutr. J. **13**, 17 (2014)
12. Hansen, M.K., Connolly, T.M.: Nuclear receptors as drug targets in obesity, dyslipidemia and atherosclerosis. Curr. Opin. Investig. Drugs **9**, 247–255 (2008)
13. Mansour, M.: The roles of peroxisome proliferator-activated receptors in the metabolic syndrome. Prog. Mol. Biol. Transl. Sci. **121**, 217–266 (2014)
14. Torra, I.P., Chinetti, G., Duval, C., Fruchart, J.C., Staels, B.: Peroxisome proliferator-activated receptors: from transcriptional control to clinical practice. Curr. Opin. Lipidol. **12**, 245–254 (2001)
15. Duan, S.Z., Usher, M.G., Mortensen, R.M.: PPARs: the vasculature, inflammation and hypertension. Curr. Opin. Nephrol. Hypertens. **18**, 128–133 (2009)
16. Gross, B., Staels, B.: PPAR agonists: multimodal drugs for the treatment of Type-2 diabetes. Best Pract. Res. Clin. Endocrinol. Metab. **21**, 687–710 (2008)
17. Monsalve, F.A., Pyarasani, R.D., Delgado-Lopez, F., Moore-Carrasco, R.: Peroxisome proliferator-activated receptor targets for the treatment of metabolic diseases. Mediat. Inflamm. 549627 (2013)
18. Lyon, C.J., Law, R.E., Hsueh, W.A.: Minireview: adiposity, inflammation, and atherogenesis. Endocrinology **144**, 2195–2200 (2003)
19. Lohray, B.B., Lohray, V.B., Bajji, A.C., Kalchar, S., Poondra, R.R., Padakanti, S., Chakrabarti, R., Vikramadithyan, R.K., Misra, P., Juluri, S., Mamidi, N.V.S.R., Rajagopalan, R.: (−)3-[4-[2-(Phenoxazin-10-yl)ethoxy]phenyl]-2-ethoxypropanoic Acid [(−)DRF 2725]: A dual PPAR agonist with potent antihyperglycemic and lipid modulating activity. J. Med. Chem. **44**, 2675–2678 (2001)
20. Duran-Sandoval, D., Thomas, A.C., Bailleul, B., Fruchart, J.C., Staels, B.: Pharmacology of PPARα, PPARγ and dual PPARα/γ agonists in clinical development. Medicine/Sciences **19**, 819–825 (2003)

21. Henke, B.R.: Peroxisome proliferator-activated receptor α/γ dual agonists for the treatment of Type 2 diabetes. J. Med. Chem. **47**, 4118–4127 (2004)

22. Forman, B.M., Chen, J., Evans, R.M.: Hypolipidemic drugs, polyunsaturated fatty acids, and eicosanoids are ligands for peroxisome proliferator-activated receptors alpha and delta. Proc. Natl. Acad. Sci. U.S.A. **94**, 4312–4317 (1997)

23. Caijo, F., et al.: Synthesis of new carbo- and heterocyclic analogues of 8-HETE and evaluation of their activity towards the PPARs. Bioorganic Med. Chem. Lett. **15**, 4421–4426 (2005)

24. DruLiTO: Drug likeness tool. http://www.niper.gov.in/pi_dev_tools/DruLiToWeb/DruLiTo_index.html

Sentiment Score Analysis for Opinion Mining

Nidhi Singh, Nonita Sharma and Akanksha Juneja

Abstract Sentiment Analysis has been widely used as a powerful tool in the era of predictive mining. However, combining sentiment analysis with social network analytics enhances the predictability power of the same. This research work attempts to provide the mining of the sentiments extracted from Twitter Social App for analysis of the current trending topic in India, i.e., Goods and Services Tax (GST) and its impact on different sectors of Indian economy. This work is carried out to gain a bigger perspective of the current sentiment based on the live reactions and opinions of the people instead of smaller, restricted polls typically done by media corporations. A variety of classifiers are implemented to get the best possible accuracy on the dataset. A novel method is proposed to analyze the sentiment of the tweets and its impact on various sectors. Further the sector trend is also analyzed through the stock market analyses and the mapping between the two is made. Furthermore, the accuracy of stated approach is compared with state of art classifiers like SVM, Naïve Bayes, and Random forest and the results demonstrate accuracy of stated approach outperformed all the other three techniques. Also, a detailed analysis is presented in this manuscript regarding the effect of GST along with time series analysis followed by gender-wise analysis.

Keywords Sentiment analysis · Goods and services tax · Classification
Text mining · Support vector machine · Naïve bayes classifier · Random forest

N. Singh (✉)
National Institute of Technology Delhi, New Delhi 110040, India
e-mail: 162211001@nitdelhi.ac.in

N. Sharma
Dr. B. R. Ambedkar National Institute of Technology Jalandhar, Jalandhar, India
e-mail: nonita@nitj.ac.in

A. Juneja
Jawaharlal Nehru University, New Delhi 110040, India
e-mail: akankshajuneja.jnu@gmail.com

© Springer Nature Singapore Pte Ltd. 2019
M. Tanveer and R. B. Pachori (eds.), *Machine Intelligence and Signal Analysis*,
Advances in Intelligent Systems and Computing 748,
https://doi.org/10.1007/978-981-13-0923-6_32

1   Introduction

Goods and Service Tax (GST) has been a trending topic on twitter since the announcement made by the Indian government for its imposition in India. It is an indirect tax or value added tax for the whole nation, which will make India one unified common market [1]. Earlier there used to be different types of taxes for state and central government (Local Body Tax (LBT), Value Added Tax (VAT), Service Tax, etc.). GST, on the other hand, is a single tax on the supply of goods and services, right from the manufacturer to the consumer. Credits of input taxes paid at each stage will be available in the subsequent stage of value addition, which makes GST essentially a tax only on value addition at each stage. The final consumer will thus bear only the GST charged by the last dealer in the supply chain, with set-off benefits at all the previous stages. Thus, the implementation of GST seems to be very promising for Indian markets since no cross utilization of credit would be permitted. Hence there is an imperative need to analyze the impact of GST on Indian economy using sentiment of common people.

Sentiment is a simple, view or opinion held or expressed by an individual or group of individuals. The process of identification or classification of sentiments in any sentence or text computationally is called sentiment analysis [2]. One of the techniques is to calculate the sentiment score of the sentence and classify it according to this. Sentiment score represents the numerical value of polarity of sentence. The sentiment polarity is a verbal representation of the sentiment. It can be "negative", "neutral", or "positive". The value less than zero represents the negative sentiment, a value greater than zero represents positive sentiment and value equal to zero represents neutral sentiment. The sentiment score can be calculated for any sentence, document, named entities, themes, and queries.

To attain a large and varied dataset of recent public opinions on the GST, we have used Twitter to gain real time access of the opinions across the country. Different classification techniques are used to better understand the dataset that is generated and then to analyze the sentiment accurately. Support Vector Machines that finds a hyperplane as decision boundary is used in this manuscript for classification of tweets. Another common classifier for text categorization i.e. Naïve Bayes which works on probabilities is also implemented. Then classification is boosted using Ensemble method. The ensemble was implemented using random forests that construct a number of decision trees for classification. The methodology involves collection of tweets, preprocessing them and then using for analysis of sentiment. For analysis purpose, we have employed different classifiers and their accuracies are tabulated in the results section.

This manuscript proceeds as follows. Related work is reviewed and discussed to provide the theoretical background and foundation for our study in Sect. 2. We then describe the data and discussion on sentiment analysis procedures in Sect. 3. In Sect. 4, we present our results about the topic. Finally, we conclude the paper by discussing study implications and suggesting future research directions.

2 Literature Review

Wilson et al. [3] has described about the method to find the contextual polarity. In contextual polarity, the words comprising a phrase are to be considered and then the sentiment is found for the complete phrase. This method has to be used for a large dataset where each sentence is complex. Complex sentence here refers to a sentence consisting of more than two phrases.

Alessia et al. [4] has described about first about the types of approaches to do sentiment classification along with their advantages and disadvantages. Then it dealt with the tools with respect to the techniques discussed. This paper also discusses about the emerging domains where sentiment analysis can be done.

Bickart and Schindler [5] discussed that social media has been exploding as a category of online discourse where people create content, share and discuss in communication network. From a business and marketing perspective, they noticed that the media landscape has dramatically changed in the recent years, with traditional now supplemented or replaced by social media. In contrast to the content provided by traditional media sources, social media content tends to be more "human being" oriented.

Yu et al. [6] discussed about the use of social media and trends that are available on social media instead of conventional media. Since the major proportion of our population uses social media, so a huge amount of information is generated on any issue. This information can be used to carry out analysis and research. This paper discussed about the sentiment analysis of social networking data for Stock Market Analysis.

Prabowo et al. [7] discussed the use of multiple classifiers in sentiment analysis. This paper combined rule-based classification, supervised learning and machine learning into a new combined method. This method was then tested on datasets and accuracy and recall was shown to be increased compared to existing methods.

Hiroshi et al. [8] have discussed about automated sentiment analysis. Various approaches have been applied to predict the sentiments of words, expressions or documents. These are Natural Language Processing (NLP) and pattern-based, machine learning algorithms, such as Naive Bayes (NB), Maximum Entropy (ME), Support Vector Machine (SVM), and unsupervised learning.

3 Methodology

3.1 Framework

The framework for this research work is presented here in this section. The process starts with twitter search API. From this API, we get keys and tokens to retrieve the tweets in real-time scenario. Then we proceed to data preparation or preprocessing. In this step the tweet is converted to text as JavaScript Object Notation (JSON)

Fig. 1 Framework of the
sentiment analysis

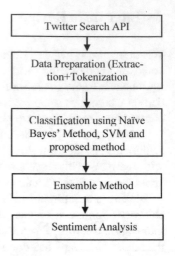

objects are returned from twitter search API. This process is called as extraction of tweets. After this, we proceed to tokenize the tweets as desired tokens. The next step is classification using different classifiers (SVM and Naïve Bayes) and the proposed method. The final step is presentation of our results of classification. The flowchart in Fig. 1 summarizes this framework.

3.2 Framework

The data used in this work is real time data in the form of tweets. It is collected using the twitter search API that allows developers to extract tweets programmatically. The tweets were collected here for the period of 30 days, i.e., from 11 July, 2017 to 15 September, 2017 using the "twitteroauth" version of public API[1] by Williams (2012). This was the period right after the launch of GST. On an average, everyday there were around 3000 tweets. The tweets were extracted on daily basis and stored in comma separated values (CSV) files. Hence the data that was under consideration comprised of more than 1,00,000 tweets. With each tweet, there comes meta data like user id, name, tweet id, text, longitude, latitude, etc. For the analysis in this work, we use text, user id and name.

Once any issue pops up on Twitter, there are numerous hashtags created from the same issue. Similarly for the GST, many hashtags were used. So we selected the best five hashtags out of all and used them as Twitter handles to collect tweets from Twitter. The most used hashtags are: GST, GSTBill, gstrollout, and GST for NewIndia [9]. The statistics of these hashtags are presented in Table 1 given below.

---

[1]https://dev.twitter.com/rest/public/search.

Table 1 Trending hashtags related to GST	Hashtag	Popularity	Correlation (%)
	#GST	59.7	100
	#GSTForNewIndia	47.6	1.80
	#GSTBill	47.1	1.80
	#gstrollout	41.1	1.80

The popularity of the hashtags is calculated based on the number of people using it and the correlation is calculated according to the topic.

The tweets are generated in three given ways: original, re-tweets and replies. Original tweets are any messages posted on twitter. Re-tweets are reposting of another person's tweet. Replies are answers to some queries or a form chat. These tweets were retrieved from the twitter as JSON objects and then converted into textual format for analysis. There are predefined functions to convert into text as shown in Fig. 2.

Once we get text from tweet, it needs to be filtered to remove unnecessary information as the nature of tweets is random and casual. In order to filter out these useless data, the Standford Natural Language Processing (SNLP group 2015) is used[2]. It is an open-source tool that gives the grammatical relations between words in a sentence as output. Then process of tokenization is carried out. Tokenization is breaking up of sentences into words, phrases, keywords, or tokens. A token is an instance of sequence of characters that are grouped together as a useful semantic unit for processing. This breaking up is as per the application requirement. These tokens are then used for the text mining and parsing purposes. There are in-built methods for tokenization in natural language processing (NLP) toolkits like Natural Language Toolkit, Apache Lucene and Solr, Apache OpenNLP, etc.

Fig. 2 Data extraction procedure

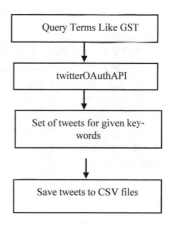

3.3 Classification

Classification is the process of categorization of new patterns with respect to the old patterns on which it is trained where class labels are already known. There are two kinds of classifiers: Lazy and Active. Lazy(e.g., Naïve Bayes) are those which do not build any model until testing starts. Active (e.g., decision trees) are those which create a model as training data is presented to them.

Support Vector Machine (SVM) is an active classifier that measures the complexity of Hypothesis based on margin that separates the plane and not number of features. The input and output pattern for the SVM is 0/1(0-negative/1-positive). Sentences or tweets as retrieved are not suitable for classification. The transformation need to be done so that the input is recognized by the classifier. For this we have already discussed the preprocessing/tokenization of tweets. There are two kinds of SVM classifiers: Linear SVM Classifier and Nonlinear SVM Classifier. In the linear classifier model that we used, we assumed that training examples are plotted in 2D space. These data points are expected to be separated by an apparent gap. It predicts a straight hyperplane dividing two classes. The primary focus while drawing the hyperplane is on maximizing the distance from hyperplane to the nearest data point of either class. The drawn hyperplane is called maximum-margin hyperplane.

Naive Bayes classifiers have the ability to classify any type of data: text, networks features, phrases, parse trees, etc. In Naïve Bayes Classifier, a set of words as bigrams or trigrams can also be used. This classifier takes a small set of positive and negative words and finds the probability of each tweet with respect to these words and then finds the conditional probability. Below is the formula for calculating the conditional probability.

$$P(X|C) = \frac{(P(C|X) * P(C))}{P(X)},\tag{1}$$

where

$P(C)$ is the probability of class C. This is known as the prior probability.

$P(X)$ is the probability of the evidence(regardless of the class).

$P(X|C)$ is the probability of the evidence given it belongs to class C.

$P(C|X)$ is the probability of the class C true and it is an evidence.

The probability that a document belongs to a class C is given by the maximum of class probability $P(C)$ multiplied by the products of the conditional probabilities of each word for that class. It is also called as maximum A Posteriori (MAP). It is given as

$$MAP(C) = max(P(X|C))\tag{2}$$

This is used to calculate the probability of tweet with respect to each class i.e. positive and negative.

Ensemble is a technique of combining two or more algorithms of similar or dissimilar types called base learners. This is done to increase the robustness. For each tweet the classification process is carried out by every classifier in the ensemble (it may also be a classifier of the same type, but trained on a different learning sample). Then the results of all classifiers are aggregated as the final ensemble classifier. This can be done in three ways: (a) Majority-based voting (b) Weighted voting (c) Rank based Methods. In the case of tie, randomly any one is chosen. Here or classification of tweets, we have used random forests as an ensemble technique. In random forest, a specified number of decision trees are constructed for classification on the partitioned datasets.

Pseudo Code:

Input: Dataset D (set of Tweets extracted from a particular handle)

Training:
 	
  1. Select "m" words from every tweet randomly.
  2. From these "m" words , calculate the splitting point "s".
  3. Break it into child nodes using split points calculated.
  4. Repeat steps 1 through 3 till threshold is reached.
  5. Build "n" trees using steps 1-4.

Testing:
  1. Select "m" words randomly and using created decision tree to predict the outcome.
  2. Find the number of votes for each class C,
  3. Find the maximum voted for class.

3.4 Proposed Method

The approach followed here is to count the positive and negative words in each domain specific tweet and assign a sentiment score. Initially, we preprocess the tweets using different tools. After this, we segregate our collected tweets on the basis of their domain using domain specific words. Then we analyze the sentiment of the tweet using domain specific positive and negative words after preprocess. The formula for sentiment score is given as

$$Sentiment\ Score = Pc - Nc, \tag{3}$$

where Pc represents positive word count and Nc represents negative word count.

This way, we can ascertain how positive or negative a tweet is.

Pseudo Code:

Input: Dataset D (set of Tweets extracted from a particular handle)

1. Create a list of domain/ sector specific words. eg(Auto, FCMG)
 2. Create a list of domain specific positive words.
 3. Create a list of domain specific negative words.
 4. Apply functions to remove punctuation symbols, extra spaces, tabs, digits and RT etc.
 5. Convert obtained text to lower case text.
 6. Filter the tweet according to the domain.
 7. Calculate the sentiment score for each domain specific tweet:
     i.   Count the number of the positive words in sentence as p.
     ii.   Count the number of the positive words in sentence as n.
     iii.   Calculate the sentiment score as: Score = p − n
 8. Categorize the tweets into 3 categories:
     i.   Positive Class where Score > 0
     ii.   Negative Class where Score < 0
     iii.   Neutral Class where Score = 0.

4 Results and Discussions

In this research, we have constructed our own dataset of about 1 lakh tweets for 30 days using the twitter search API. The hashtags used are: GST, GSTBill, gstrollout, and GSTforNewIndia. The R language was used for the analysis and classification part.

4.1 Accuracy

In this manuscript, we have used four different classification techniques: SVM, Naïve Bayes, Sentiment Score, Random forests. We have tabulated their results as follows (Table 2).

As it can be inferred from the table, the accuracy of all classifiers lie in the range of 60–70 per cent. The highest accuracy was found in proposed method, i.e.,

Table 2 Classification results

Classifier	Accuracy(%)
SVM	64.34
Naïve Bayes	58.44
Sentiment score (proposed method)	68.75
Random forests	65.21

Sentiment Score and lowest in the Naïve Bayes classifier. SVM and Random forest have performed fairly well.

4.2 National Average Sentiment

This gives generalized overview in Fig. 3 shows the attitude of twitter users towards GST. It is clearly visible that the users are neither positive nor negative about it. A neutral emotion is found for 75% people. It can be inferred from Fig. 3 that 12.1% of people are showing positive reaction to this while another 12.8% are showing negative attitude towards it.

Figure 4 shows how the people are reacting towards this bill. There are only 31% people who are originally tweeting about this. A large proportion of people around 66% are re-tweeting. This shows actual information flow and how the influences are passed from one people to another in a network. The meager 3% were the replies of the tweets.

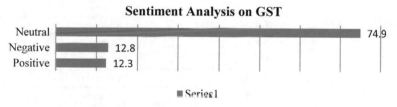

Fig. 3 Bar graph on sentiment analysis of GST

Fig. 4 Types of tweets

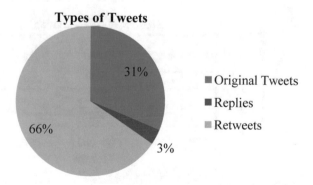

4.3   Sector-Wise Analysis

For this part we, took the data from BSE and NSE sites and compared them. We took seven top companies from each sector and analyzed their stocks after launch of GST. This would show how has GST affected them(in positive or negative manner). For the tweets collected, we applied the proposed sentiment analysis algorithm to find the sentiment. The results for the algorithm are shown in Table 3. It shows a positive sentiment for the pharmaceutical, auto, FMCG, services. A negative sentiment was found for media. The realty sector was showing a neutral sentiment.

Figure 5 shows the actual market trend of these six sectors after the launch of GST. The media, financial services, FMCG and pharma has shown increase in total turnover after the launch while realty and auto were at same.

From Table 3 and Fig. 5, the results can be validated as FMCG, services and pharmaceutical has shown a growth after the launch. The GST has affected badly to realty sector as it was found in our analysis. The media sector has shown a different behavior than expected in the analysis. The auto was expected to perform well but it gave an average turnover.

Table 3 Sector-wise sentiment analysis

Sector	Sentiment
Media	Negative
Pharma	Positive
Realty	Neutral
Auto	Positive
Fmcg	Positive
Services	Positive

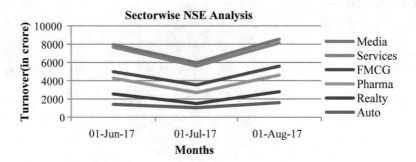

Fig. 5 NSE plots (actual)

Fig. 6 Gender-wise analysis of GST

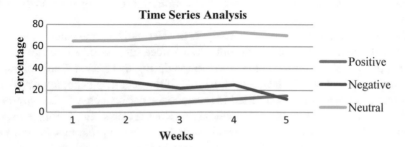

Fig. 7 Time series analysis of GST

4.4 National Male/Female Average

Twitter does not provide gender of users directly. So it is not possible to obtain gender from tweets. For getting gender from tweets, we can user the user name of the individuals and then use AI tool like NamSor to deduce gender. Figure 6 shows the gender-wise breakdown of twitter user tweeting about GST. As it can be clearly seen from the figure, only 24% of the women reacted about this. This analysis presented to us a greater participation of men on twitter.

4.5 Time Series Analysis

Trend plots were plotted for this data that show few spikes and drops in Fig. 7. This analysis of this does not show rapid changes in the sentiment of twitter user but considerate amount of change can be seen. The x-axis shows the number of weeks passed whereas the y-axis shows the percentage of sentiment. There is an increase in positive sentiment and neutral sentiment whereas a decrease is observed in negative sentiment. This shows that the citizens, over the time, started accepting this new policy and slowly all the negative elements would be vanished.

5 Conclusions

In summary, we have performed a classification of sentiment towards the launch of new taxing policy in India, i.e., GST. It was analyzed that overall the attitude of Indian population was neutral towards it. The analysis was also done gender-wise, region-wise. Apart from this, time series analysis was done. It has shown an increase in positive attitude and neutral attitude. Further, prolific users, common words and phrases, and key users were also identified. Furthermore, sector-wise analysis was also done.

As a future work, historical data analysis of GST can be done using tweets from before enforcement of the bill till the bill was enforced successfully. It can give a more a clear picture of the opinions transformations during this period. Other classifiers may also be tested on this dataset as it is large. If the accuracy is not improved, then boosting and bagging can be utilized to enhance the performance. This analysis can also be used in other domains like politics, product reviews, etc. During elections, all electoral candidates use social media for their campaign programs. So analyzing these, we can predict the results. This was done in US 2016 elections.

References

1. Dani, S.: A research paper on an impact of goods and service tax (GST) on Indian Economy. Bus Eco J **7**, 264 (2016). https://doi.org/10.4172/2151-6219.1000264
2. Liu, B.: Sentiment analysis and opinion mining. Synth. Lect. Hum. Lang. Technol. **5**(1), 1–167 (2012)
3. Wilson, T., Wiebe, J., Hoffmann, P.: Recognizing contextual polarity in phrase-level sentiment analysis. In: Proceedings of the Conference on Human Language Technology and Empirical Methods in Natural Language Processing. Association for Computational Linguistics (2005)
4. Alessia, D., et al.: Approaches, tools and applications for sentiment analysis implementation. Int. J. Comput. Appl. **125**, 3 (2015)
5. Bickart, B., Schindler, R.M.: Internet forums as influential sources of consumer information. J. Interact. Market. **15**(3), 31–40 (2001). ISSN 1094-9968. http://dx.doi.org/10.1002/dir.1014
6. Yu, Y., Duan, W., Cao, Q.: The impact of social and conventional media on firm equity value: a sentiment analysis approach. Decis. Support Syst. **55**(4), 919–926 (2013). ISSN 0167-9236. http://dx.doi.org/10.1016/j.dss.2012.12.028
7. Prabowo, R., Thelwall, M.: Sentiment analysis: a combined approach. J. Informetr. **3**(2), 143–157 (2009)
8. Hiroshi, K., Tetsuya, N., Hideo, W.: Deeper sentiment analysis using machine translation technology. In: Proceedings of the 20th International Conference on Computational Linguistics (COLING 2004), Geneva, Switzerland, 23–27 Aug 2004, pp. 494–500
9. Trending Hashtags. https://www.hashtagify.com. Accessed 15 Sept 2017

MRI Segmentation for Computer-Aided Diagnosis of Brain Tumor: A Review

Princi Soni and Vijayshri Chaurasia

Abstract Brain tumor is an uncontrolled growth of cells in the brain. Diagnosis of brain tumor is complicated and challenging task as the brain itself a complex structure and tumor have excessive variety, diversity in shape, large range in intensity and ambiguous boundaries. The validity of brain tumor segmentation is a significant issue in biomedical signal processing because it has a direct impact on surgical groundwork. Detection of brain tumor by magnetic resonance imaging (MRI) using (CAD) involves; preprocessing, segmentation, and morphological operation for analysis purpose. The magnetic resonance imaging segmentation is characterized by a high nonuniformity of both the pathology and the surrounding non-pathologic brain tissue. Computer Aided Diagnosis system can assist in the detection of suspicious brain disease as the manual segmentation is time-consuming and it reported the time-varying result. To tag tumor pixel or trace tumor area, texture and pattern remembrance, classification is performed with different algorithms. This article presents an overview of the most relevant brain tumor segmentation methods.

Keyword Brain tumor · MRI · Image segmentation · Computer-aided diagnosis

1 Introduction

The brain is soft, spongy mass of tissue. The unwanted and abnormal growth of cells inside or around the brain is a brain tumor. These unwanted cells can also damage healthy cells of the brain. This may cause inflammation; the pressure within the skull, compressing parts of conquers and brain swelling. The exact cause of brain tumor is not clear. There are two main categories of brain tumors [1].

Primary brain tumor: It starts in any part of the brain. It can be either malignant or benign.

P. Soni (✉) · V. Chaurasia
Department of Electronics and Communication, Maulana Azad National Institute of Technology, Bhopal, India
e-mail: princi100ni@gmail.com

© Springer Nature Singapore Pte Ltd. 2019 375
M. Tanveer and R. B. Pachori (eds.), *Machine Intelligence and Signal Analysis*,
Advances in Intelligent Systems and Computing 748,
https://doi.org/10.1007/978-981-13-0923-6_33

- Malignant: Nonuniform in structure and contain cancer cells.
- Benign: Uniform in structure and does not contain cancer cells.

Secondary brain tumor: It starts in any part of a body and spread to the brain. It is knowledge-based as metastases.

Meningioma is an example of the low-grade tumor that is a benign tumor and glioblastoma is a class of high-grade tumor classified as a malignant tumor. While gliomas and astrocytomas can be a low- or high-grade tumor [2].

A brain tumor is a deadly disease and CAD of brain tumor is a process that contains a number of steps for analysis. For efficient tumor detection mainly computed tomography (CT) and magnetic resonance imaging (MRI), based algorithms are used to achieve an accurate and reproducible brain tumor detection result [3]. Detection of location and size of brain tumor should be accurate and precise. Brain cancer is usually identified by imaging test that performed on MRI or/and CT scan, employ computer technology to give rise to detailed 2D pictures of the 3D brain [4]. MRI possesses good contrast resolution as compare with CT so BRAIN MRI is the better choice for most brain disorders [5]. This method is mainly images of organs with high contrast and spatial resolution. MRI technique does not use the ionizing radiation so chances of harm of healthy tissues are decreases [6]. The versatility of image capture parameter and variety in data processing tool enable its adaptation to a vast array of clinical situations. So, for structural and functional evaluations in neurology MRI based CAD techniques are used extensively [7]. For brain tumor diagnosis with MR image processing, there are three steps: preprocessing of MRI image, segmentation and analytic analysis of segmented image [8].

In rest of the paper, Sect. 2 presents a briefing about image segmentation, in Sect. 3, a survey of existing state of art MRI image segmentation techniques are presented. Section 4 contains performance analysis of existing methods followed by concluding remarks in Sect. 5.

2 Segmentation

Segmentation is prime task of MRI processing. It is a process of dividing an image into multiple parts. This parathion is made between different tumor tissues (solid tumor, edema, and necrosis) and normal brain tissues; Gray Matter (GM), White Matter (WM), and Cerebrospinal Fluid (CSF) [4]. Automatic segmentation is most accurate and swift method as an analogy with manual segmentation and semiautomatic segmentation. Although manual segmentation has done by qualified professionals for higher accuracy it is a time-consuming process and a standard deviation is much higher. This is typically used to identify objects, facts, statistics or other relevant information in digital images [9]. In this paper we will take a glance and make a view about various segmentation methods, they are mostly based on; threshold method, color-based segmentation (clustering), transform method (watershed), texture methods (texture filters). A Detailed description of some of the existing MRI segmentation methods is given in the following section.

3 Existing Methods

In 1983 J Bezdek et al. [10] gave the first method for segmentation of MRI images. The nonnegative segmentation was based on "FUZZY C MEANS" algorithm. For brain tumor segmentation, threshold method would avoid as brain tissues have a varying intensity and threshold method leads to difficulties in threshold determination. In the beginning of CAD brain tumor detection, "FUZZY C MEANS" (FCM) clustering method is the most popular method for medical image segmentation [11].

Shan Shen et al. define new extensions to FCM i.e. improved FCM (IFCM) which consider two influential factor in brain tumor segmentation; first is a distinguished between a neighboring pixel in image exact intensity and second relative locations of neighboring each pixel are clear. FCM reduced incorrect segmentation (greatly within the noise levels 7–15%) as compared with FCM, so it can prefer at higher noise levels. IFCM has reduced value of incorrect segmentation (0.026), under segmentation (0.023) and over-segmentation (0.053), but still there is a little region which cannot be classified with high accuracy due to high noise level [12].

GBM (glioblastoma multiforme) is a type of primary brain tumor and 40% of brain tumor patients of all ages suffer by this. Jason J. et al. analyses GBM and define segmentation by weighted aggregation (SWA) i.e. a method of detection. By comparing this method with conventional affinity there is 9% of improvement as it has the ability to avoid local ambiguity and null requirement of initialization in training sets. Model-based SWA is constructed on Bayes classification rule. This method classifies GBM brain tumor and accuracy increases with increment in number of datasheet. In this technique computation time can drop and it is a benefit over conventional technique [13].

As less computational time and accuracy, both can be achieved by the hybrid model in tumor detection system. Ahmed Kharrat et al. explore the idea of hybrid methods, i.e., a combination of spatial gray level dependence method (SGLDM), genetic algorithm (GA) and support vector machine (SVM) with and without wavelet transform (WT). WT is transformed that has developed for iteration of classification and for texture feature extraction; it is also used as a filter. SVM is a classifier method which shows superior performance and feasibility than neural network. SGLDM is a statistical model for extracting probability function and texture information. GA is another classifier method which maintaining acceptable classification accuracy. A hybrid approach is classified high-grade gliomas (HGG) and low-grade gliomas (LGG) tumor. For effective evaluation of tumor WT deploy which make a system more accurate and fast due to feature extraction so WT model is better than without WT model [14].

As till now, we get segmentation's root is classification so for enhancement in accuracy and efficiency some more method also comes in frame. Quratul Ain et al.

introduce discrete cosine transform (DCT); this transform is used along with Naïve Bayes classification. DCT is converted signal component to their frequency component without any distortion. Along DCT, k-means clustering used for tumor edge detection. DCT contains property of symmetric classification. Naïve Bayes segmentation applied on DCT preprocessed image. The major positive outcomes of this method are tumor region extracting is highly accurate, with low execution time and estimate root mean square error (RMSE). This system has required one time under observation training after that it can process new data itself for segmentation. Filter selection gets more flexible as in this method filter can reconstruct according to minimal RMSE [15].

Detection of position of a tumor is an essential task. N. Nandha Gopal et al. present a two technique for segmentation, which calculates the number of pixels affected by tumor cells and result has been compared with existing result. These methods are Particle Swarm Optimization (PSO) and genetic algorithm (GA) which are used as optimization method with Fuzzy algorithms. FCM to identify tumor position and pixel similarities, these are measure with an expert report. FCM classification with PSO and GA optimization technique has a higher rate of tumor detection. Execution time directly depends on abstraction layer but as hidden layer will increase there is an increment in accuracy in a reduction in error rate. The error rate for FCM with GA is 0.39 and for FCM with PSO is 0.12 so most optimize and accurate optimum technique is PSO [16].

Position and size of brain tumor detect by above algorithms but computation time should also reduce. Andac Hamamci et al. present a fast and robust segmentation technique method which deals with minimum user interaction over graph-cut and grow-cut conventional method. That method is cellular automata (CA) algorithm, i.e., tumor-cut segmentation to a graph-theoretic algorithm which has shortest path and lower standard deviation as compare with graph-cut and grow-cut. CA is perfect for the anatomic structure as it grows smooth boundaries with compare to grow-cut which produced jagged and irregular surface. CA has preferred over the graph-cut method which reports a problem of shrinking bias due to minimum cut optimization. Smoothing and contrast enhancement with low correction fraction and high sensitivity to overcome heterogeneous tumor segmentation can be attained by CA-based seeded tumor segmentation [17].

Natural smoothness to segmentation result without explicit regularization by using the contextual information can be achieved by local independent projection-based classification (LIPC) method. Meiyan Huang et al. define LIPC method which embedded in a multi-resolution framework for reducing effective cost and at the high noise level, it maintains lower error characteristics. In this method, there is no connection between training and testing datasheets. LIPC can evaluate with and without softmax regression but for high accuracy with softmax regression has preferred. LIPC method performed on both images synthetic and patient image data and it visualizes its accuracy and dice similarity parameter within a bounded time period. LIPC shows more precise classification for neighbor pixel so this response a low miss rate. This method also leads to natural smoothness to classification without direct arrange or manipulation of training data [5].

Fuzzy C means algorithms are an expeditious method for segmentation. Extension and advancement did in FCM algorithm to get regularize clustering, speed up the convergence process of an algorithm. FCM did segmentation only for GM, WM, and CSF. El-Melegy et al. present PIGFCM (prior information guide fuzzy c means algorithm) which classify GM, WM, and CSF, in addition to this BG (background) also examine and report better robustness against noise. Pre-segmentation performs and that provide prior information about tumor class. FCM is a conventional and famous technique of MRI segmentation. Extended FCM method has improved the system by increase accuracy and speed also drop a level of uncertainty under high noisy condition. PIGFCM is an advanced method which is enhances and resolute form of FCM in terms of higher dice value. Another advantage of PIG FCM is lower sensitivity to noise, intensity compensation to the nonuniform intensity and zero requirements of tuning weight factor [1].

Kamil Dimililer et al. proposed two techniques those are an amalgamation of neural network and image processing. Image with neural networks (IWNN) and image processing with a neural network (IPWNN) both methods use brain image set as input, but IWNN uses an original image whereas IPWNN uses image processing techniques on original images. The neural network is back-propagation system with a number of hidden layer and accuracy increases as a number of hidden layer increases but it increases time also. Image processing involves morphological operation such as erosion, median filtering for contrast enhancement. The image processing adoptive technique is more accurate with less no hidden layer system also for detection of abnormal tumor cells of the brain [8].

To improve classification efficiency and reduce the dimensionality of feature space, two-tier classification method raise. V. Anitha et al. put forward a two-tier classification method K-nearest neighbor (KNN) which is forwarded by self-organizing map (SOM) neural network classifier that produces a deterministic reproducible result. In KNN algorithm classification of an object is based on a majority vote of its neighbors. The resemblance between training and testing data is processed by KNN after preprocessing and segmentation by SOM. The two-tier classification system, first for eliminating noise & stripping of a skull and second for segmentation by K- means clustering algorithm is KNN. Sometimes it called adaptive pillar k-means algorithm which conquers restriction of k-means clustering algorithm [4].

Advance MRI modalities perfusion-weighted imaging (PWI), diffusion-weighted imaging (DWI), magnetic resonance spectroscopic imaging (MRSI) and conventional MRI (cMRI) are an unsupervised method for segmentation. Now combine effect of different MRI modalities is multi-parametric MRI (MP-MRI) which is used for HGG segmentation by N. Sauwen et al. As this bring hierarchical nonnegative matrix factorization (hNMF) for tumor segmentation to MP-MRI datasheets. This method has better sensitivity and the computational cost was average, relatively with computational time and efficiency then NFM [18].

Artificial neural network (ANN) and support vector machine (SVM) both utilize for classification method. Deformation feature extraction component of Brain lateral ventricles (LaVs), i.e., dynamic template image creation and 3D deformation model states by Shang-Ling Jui et al. The component consecutively performs the steps of segmentation of LaVs, 3D alignment, and transformation of LaVs to deformations feature. In LaVs feature extraction of the image with SVM and ANN methods, the extreme value that has a bright pixel of a feature is corresponding to the region of a brain tumor. This is pertinence between feature extraction and crimps from growth in a tumor. A system with LaVs feature extraction has a rise in accuracy and drop in miss rate for both using SVM (19.9–10%) and ANN (21.9–11.7%) [9].

Recently, another neural network based algorithm describes by Sérgio Pereira et al. that is a convolutional neural network (CNN). It is back-propagation multiple hidden layers with kernel weights neural network. The application of convolutional layers consists in convolving a single or an image with kernels to obtain feature maps so it links with the previous layer. A long process is done before CNN such as MRI preprocessing, patch extraction, patch preprocessing. LGG and HGG require specifying systems so to reduce complexity author suggest a CNN and tuned the intensity normalization information for each tumor grade. In designing CNN heterogeneity take place by multisite multi-scanner acquisitions of MRI images using intensity normalization. CNN resulted in segmentations with a better diagnosis of the complete tumor as well as of the intra-tumor structures, i.e., tumor inside the tumor structure. Parameter PPV, i.e., Positive predictive value which indicates an amount of false position and the true position has a higher value. It has better dice and speed as compared to the conventional neural network [19].

4 Performance Analysis

The success of any method always determines by some parameter. This parameter may be accuracy, that state of being correct or simply precise. An accurate and efficient system is a crucial requirement. High sensitivity, dice and low Root Mean Square Error (RMSE) are some more desirable parameter.

These parameters describe in terms of true and false positive and true and false negative as follows:

True positive (TP): clinical abnormality is present and classification result is positive.

True negative (TN): clinical abnormality is absent and classification result is negative. False positive (FP): clinical abnormality is absent and classification result is positive (Fig. 1).

False negative (FN): clinical abnormality is present and classification result is negative [4].

Accuracy is a degree of closeness of result of measurement and calculation of a quantity to its true or actual value.

$$Accuracy = \frac{TP + TN}{TP + TN + FP + FN} \tag{1}$$

Sensitivity is a value which represents the percentage of detection of actual value.

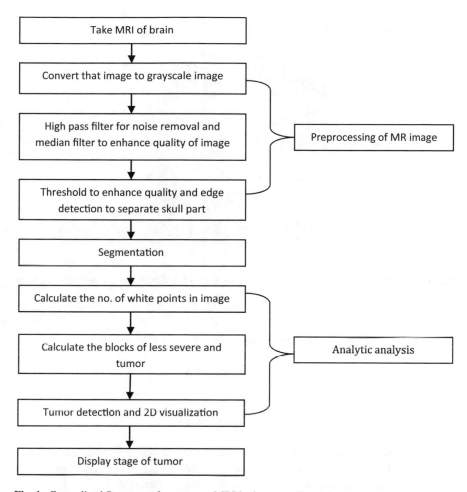

Fig. 1 Generalized Structure of a common MRI brain tumor diagnosis

$$Sensitivity = \frac{TP}{TP + FN} \tag{2}$$

Computation time is the total time required for image segmentation to attain a specified result.

RMSE has frequently used the measure of differences between values predicted by a model or an estimator and values actually observed. The term is always between 0 and 1.

$$RMSE = \sqrt{\left(\frac{\sum_{i=1}^{n}(p_i - \hat{p}_i)^2}{n}\right)} \tag{3}$$

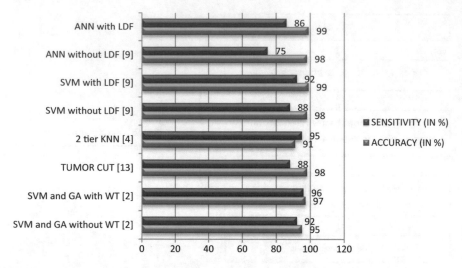

Fig. 2 Bar graph representation of performance analysis of different segmentation technique on basis of sensitivity and accuracy

Here n: number of different prediction

$$p_i : observed value$$

$$\hat{p}_i : predicted value$$

Dice is a measure of overlapping or spatial alignment between manual segmentation by expert and automatic segmentation. Its value comes in between 0 and 1 and higher value is desirable as it an indication of system efficiency.

$$Dice = \frac{2 * TP}{(FP + TP) * (TP + FN)} \qquad (4)$$

Methods have described in Sect. 3 are comparatively analyze here:

Performance of some segmentation methods is representing in terms of accuracy and sensitivity. That comparative analysis shown in Fig. 2 by bar chart and SVM with LaVs deformation feature extraction (LDF) [9] reported maximum accuracy and sensitivity than ANN with and without LDF [9], tumor cut [17] and hybrid technique using WT [14].

Another comparative analysis of some methods based on accuracy and RMSE. Preferably, the value of RMSE should be as low as possible with higher accuracy. Performance of image segmentation method is presented by bar graph in Fig. 3. As shown, DCT and Naïve Bayes classification method is reporting highest accuracy and low error.

Fig. 3 Bar graph representation of performance analysis on basis of dice and computation time

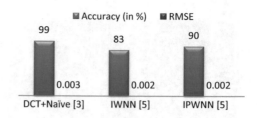

Fig. 4 Bar graph representation of performance analysis on basis of accuracy and time

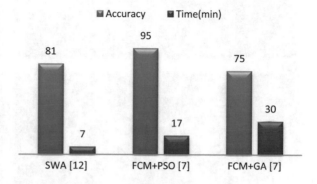

Fig. 5 Bar graph representation of performance analysis on basis of dice and computational time

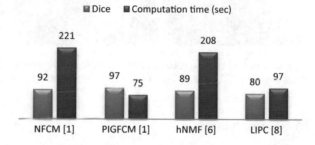

Further, the performance of some existing image segmentation methods is analogized on the premise of accuracy and execution time in Fig. 4. The accuracy of PSO with FCM [16] has reported maximum accuracy (15% more accurate than SWA) and SWA [13] algorithm has lesser execution time (2.5 times faster than PSO with FCM). So a selection of the method for brain tumor diagnosis will subject to trade-off application. It depends on specification and requirement of the application.

Figure 5 shows performance analyses of NFCM [1], PIGFCM [1], hierarchal NMF [18] and LIPC [5] technique on the basis of DICE and computational time. PIGFCM [1] is a computationally very efficient technique with leading DICE value as our appropriation. PIGFCM also achieve a specific result in minimum time with better specificity as compared with other methods.

Visualization of several MRI segmentation algorithms is shown in below figure. by our analysis and examination, PIGFCM is the best segmentation method and that can be seen in the figure also [1, 5, 18] (Fig. 6).

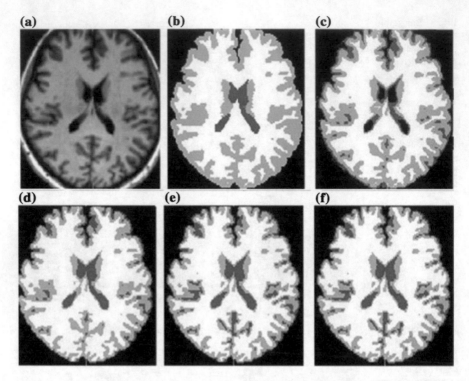

Fig. 6 Result of various methods on IBSR database. **a** A brain MRI slice from IBSR [1]. **b** Ground truth [1], **c** NFCM [1], **d** PIGFCM [1], **e** Hierarchal NMF [18], **f** LIPC [5]

5 Conclusions

MRI is very functional for analysis of soft tissues. It is robustness, accurate, sensitive, low miss rate and yield less noise. In this paper a study of knowledge-based MRI segmentation techniques have been presented with emphasis on classification technique of tissues like; fuzzy classification, neural networks, support vector machine (SVM), expectation maximization (EM) and clustering method these are base pedestal of separation of gray matter (GM), white matter (WM) and tumor. FCM based methods have an appropriate value of dice with better sensitivity. PIGFCM [1] is most efficient and robust method. PSO configuration with FCM has supreme accuracy. LIPC [5] and hNMF [18] have a low miss rate and less false over FCM based methods. Neural network based algorithm with image processing shows better efficiency as comparing only neural network based algorithm. SVM [9] has reported good accuracy and low noise effect as compare with ANN [9]. Characterization of abnormalities through MRI is still exigent and invite task because of a complex structure of brain hence segmentation of MR images is a very active area of research. Segmentation of brain tissues on medical images is not limited to brain tumor detection. It may be further extended for advanced surgical approaches.

References

1. El-Melegy, Mokhtar: Tumor segmentation in brain MRI using a fuzzy approach with class center priors. EURASIP J. Image Video Process. **2014**, 21 (2014)
2. Nabizadeh, N., Kubat, M.: Brain tumors detection and segmentation in MR images: Gabor wavelet vs. statistical features. J. Comput. Elect. Eng. **45**, 286–301 (2015). http://dx.doi.org/1 0.1016/j.compeleceng.2015.02.0070045-7906
3. Huda, S., Yearwood, J., Jelinek, H.F., Hassan, M.M., Fortino, G., Buckland, M.: A hybrid feature selection with ensemble classification for imbalanced healthcare data: a case study for brain tumor diagnosis. IEEE J. Digit. Object Identifier **4**, 9145–9155 (2017)
4. Anitha, V., Murugavalli, S.: Brain tumor classification using two-tier classifier with adaptive segmentation technique. IET Comput. Vis. **10**(1), 9–17 (2016)
5. Huang, M., Yang, W., Wu, Y., Jiang, J., Chen, W.: Brain tumor segmentation based on local independent projection-based classification. IEEE Trans. Biomed. Eng. **61**(10) (2014)
6. Bahadure, N.B., Ray, A.K., Thethi, H.P.: Image analysis for MRI based brain tumor detection and feature extraction using biologically inspired BWT and SVM. Hindawi Int. J. Biomed. Imaging, 12 (2017), Article ID 9749108. https://doi.org/10.1155/2017/9749108
7. Schad, L.R., Bluml, S., Zuna, I.: MR tissue characterization of intracranial tumors by means of texture analysis. Magn. Reson. Imaging **11**(6), 889–896 (1993)
8. Dimililer, K., İlhan, A.: Effect of image enhancement on MRI brain images with neural networks. In:12th International Conference on Application of Fuzzy Systems and Soft Computing, ICAFS 2016, August 2016, Vienna, Austria, vol. 102, pp. 39–44 (2016)
9. Jui, S.-L., Zhang, S., Xiong, W., Yu, F., Fu, M., Wang, D., Xiao, K., Hassanien, A.E.: Brain MRI tumor segmentation with 3d intracranial structure deformation features. IEEE Comput. Soc. IEEE Intell. Syst. **12** (2016)
10. Peizhuang, W.: Pattern recognition with fuzzy objective function algorithms (James C. Bezdek). SIAM Rev. **25**, 442–442 (1983). https://doi.org/10.1137/1025116
11. Bezde, J.: Pattern Recognition with Fuzzy Objective Function Algorithms. Plenum, New York (1981)
12. Shen, S., Sandham, W.: Member, IEEE, Granat, M., Sterr, A.: MRI fuzzy segmentation of brain tissue using neighborhood attraction with neural-network optimization. IEEE Trans. Inf. Technol. Biomed. **9**(3) (2005)
13. Corso, J.J., Sharon, E., Dube, S., El-Saden, S., Sinha, U., Yuille, A.: Efficient multilevel brain tumor segmentation with integrated bayesian model classification. IEEE Trans. Med. Imaging **27**(5) (2008)
14. Kharrat, A., Gasmi, K., Ben Messaoud, M., Benamrane, N., Abid, M.: A hybrid approach for automatic classification of brain mri using genetic algorithm and support vector machine. Leonardo J. Sci. (17), 71–82 (2010)
15. Ain, Q.-U., Mehmood, I., Naqi, S.M., Jaffar, M. A.: Bayesian classification using DCT features for brain tumor detection. In: Proceedings of the 14th International Conference on Knowledge-Based Intelligent Information and Engineering Systems, Heidelberg, Germany, Sept. 2010, pp. 340–349 (2010)
16. Nandha Gopal, N., Karnan, M.: Diagnose brain tumor through MRI using image processing clusteringalgorithms such as fuzzy c means along with intelligent optimization techniques. In: IEEE International Conference on Computational Intelligence and Computing Research (2010)
17. Hamamci, A., Kucuk, N., Karaman, K., Engin, K., Unal, G.: Tumor-cut: segmentation of brain tumors on contrast enhanced MR images for radiosurgery application. EEE Trans. Med. Imaging **31**(3) (2012)
18. Sauwen, N., Acou, M., Van Cauter, S., Sima, D.M., Veraart, J., Maes, F., Himmelreich, U., Achten, E., Van Huffel S.: Comparison of unsupervised classification methods for brain tumor segmentation using multi-parametric MRI. Neuro Image Clin. Sci. Direct **12**, 753–764 (2016)
19. Pereira, S., Pinto, A., Alves, V., Silva, C.A.: Brain tumor segmentation using convolutional neural networks in MRI images. IEEE Trans. Med. Imaging. **35**(5) (2016)

OFDM Based Real Time Digital Video Transmission on SDR

Rupali B. Patil⊙ and K. D. Kulat

Abstract Cognitive radio (CR) is an emergent approach used to address the upcoming spectrum crunch by increasing the spectrum efficiency by resourceful use of the existing licensed band. Also, data traffic in air has increased beyond limits. So, this ultimately increases the demand for accommodation of high data rate by the user. Transmitting multimedia demands high data rate and spectral efficiency. OFDM is a multi-carrier modulation technique which proves effective in CR. The results generated in this paper are taken from the experimentation performed on SDR-Lab using OFDM technique with the flow graphs designed in GNU Radio for real-time transmission of video. The focus of this paper is on the transmission of real-time video signal captured by webcam using OFDM modulation and simulation results generated with GNU radio along with software-defined radio (SDR). The main goal here is to provide effective solution to difficulties arising from transmission of video in CR environment and finding the best-suited modulation technique.

Keywords OFDM · Software-defined radio · GNU radio · Cognitive radio

1 Introduction

The rapid growth in wireless communication leads to deployment of new wireless devices and applications. It also increases data traffic in air and wireless radio spectrum. However, due to fixed spectrum assignment policy, a great portion of the licensed spectrum is not properly utilized which ultimately becomes bottleneck [1]. So the notion of CR is proposed to resolve the issue of spectrum efficiency. Through spectrum sensing and analysis, CR is having the competency to optimally adapt their

R. B. Patil (✉) · K. D. Kulat
Electronics & Communication Department, Visvesvaraya National Institute of Technology, Nagpur, India
e-mail: rupali1210@gmail.com

K. D. Kulat
e-mail: kdkulat@ece.vnit.ac.in

© Springer Nature Singapore Pte Ltd. 2019
M. Tanveer and R. B. Pachori (eds.), *Machine Intelligence and Signal Analysis*,
Advances in Intelligent Systems and Computing 748,
https://doi.org/10.1007/978-981-13-0923-6_34

operating parameters according to the interactions with the adjoining radio environment. CR can detect the spectrum white space, i.e., a portion of frequency band which is not being used by the primary users (licensed users) and utilize the unused spectrum [2]. CR can detect licensed user's activity through spectrum sensing and hold the transmission generated due to secondary user's transmission, thus giving the priority of the channel to primary user [3].

Also looking toward the growing demand for data traffic and applications, transmission of multimedia data like video has also increased. One such successful experimentation of GNU radio-based real-time video transmission is performed with GMSK modulation on SDR in [4]. For all this, we need to have a system which must be able to transmit the video signal which can be accommodated in the available spectrum of licensed users. But video signal needs large bandwidth for transmission. So the recent contests in wireless communication are as follows:

1. To find the effective ways of transfer of multimedia data like images, video, etc. using the same channel which is being used for the voice transmission.
2. Best-suited modulation technique which will work effectively with CR.
3. To transfer this data using unused band in licensed user spectrum, i.e., effective implementation of CR concept with spectrum sensing [5].

So this necessitates a more capacious hardware assembly to be dealt with these sorts of signal processing. Another key solution to this issue is to use the evolving technology of SDR which gives suppleness in the implementation process by replacing the hardware with software in addition to cost-effectiveness [6]. GNU radio along with SDR is a powerful development tool which performs signal processing for the SDR [7]. In addition to the processing capabilities, it has the added advantage of implementing modulation/demodulation functions along with error correcting codes. Python language with the support of C++ forms a great combination for this development tool. C++ is used for writing the code of processing blocks along with Python language which basically used to create a coordination link among these blocks [8] and also new blocks can be created. The GNU radio is used to effectively design wireless communication applications as it makes use of different operating systems efficiently. Documentation details regarding GNU radio block can easily be retrieved from the GNU radio website [9].

The hardware kits of SDR-Lab support the GNU radio software. The transmit and receive operating frequency range of SDR-Lab kits is 0.4–4 GHz [10]. These kits are as good as software programmable hardware transceiver. Looking toward the benefits of SDR-Lab kits inspired us for the use of these kits with GNU radio for the implementation of wireless communication system.

High data rate along with spectral efficiency necessitates transmission of data related to multimedia applications. OFDM is a multi-carrier modulation technique which proves effective in CR. It is being used widely in the wireless standards like DAB, DVB-T, and WiMAX IEEE 802.16. OFDM is solution to the problems that arise with high bit-rate communications, the most severe of which is time dispersion. The symbol stream of data is split into numerous lower rate streams, and these are then transmitted on subcarriers. The symbol duration is increased due to splitting of data which ultimately increases the symbol duration by the number of orthogonally overlapping subcarriers. Also, a small portion of the neighboring symbols is affected due to multipath echoes. The OFDM symbol is prolonged with cyclic prefix to remove remaining inter-symbol interference (ISI). With this technique, the dispersion effect of multipath channels due to high data rates is reduced which eventually reduces the requirement for complex equalizers. The scalability, robustness against narrowband interference, high spectral efficiency, and easy implementation are the added advantages of OFDM. CR system is assumed to operate as a secondary user in a licensed band. The CR identifies unused parts of the licensed spectrum and exploits them. The final goal is to design a system for primary (licensed) users which gives maximum throughput while keeping minimum interference [11].

OFDM-based systems to some extent tried to be realized in the literature, such as [12, 13]. These works publicized that OFDM systems are possible to be realized using SDR. However, the work presented has not been tested for transmission of real-time video signal.

In this paper, we tried to build an OFDM-based system for the transmission of real-time video on GNU radio software platform. The system is then explored with SDR-Lab kits in realistic environment. The paper is structured as follows. Section 2 introduces system overview. Section 3 considers steps for designing of OFDM transmitter and receiver. Section 4 details the results from OFDM receiver side video reception. Finally, conclusions and future scope are presented in Sect. 5.

2 System Overview

The overall system block diagram of SDR-based transmitter and receiver for OFDM is shown in Fig. 1.

The input to the SDR-based system is a real-time video captured by webcam. The input signal is then OFDM modulated. For performing, the modulation flow graph is designed using the GNU radio. The video signal captured by webcam is modulated by OFDM and transmitted wirelessly on 1.234 GHz frequency using an SDR-Lab trans-receiver device. At the receiver end, it is being received by the other SDR-Lab trans-receiver device which is further processed using GNU radio which performs the task of OFDM demodulation. The video received can be stored at the receiver side.

The use of SDR-Lab with freeware GNU radio gives highly flexible, flow graph-oriented framework.

3 Steps for Designing of OFDM Transmitter and Receiver

SDR-Lab-based project setup with laptop is shown in Fig. 2. Out of two SDRs connected to laptop one is a transmitter and other is a receiver.

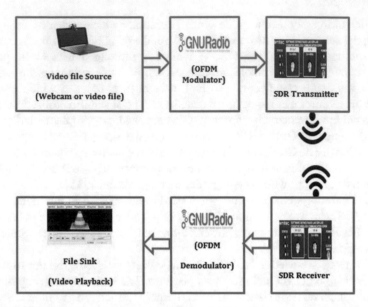

Fig. 1 Overall block diagram of SDR-based transmitter and receiver for OFDM

Fig. 2 Project setup for
OFDM transmitter and
receiver

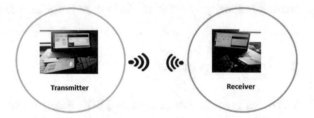

3.1 Steps for Designing of OFDM Transmitter

The flow graph of OFDM transmitter is shown in Fig. 3. Steps are given for understanding flow graph of transmitter:

Step 1   Stream of digital data is divided into packets and appends headers/tags to these packets for identification.

Step 2   This is followed by converting data into a constellation scheme. The blocks which are mainly used are *streamCRC*32, *PacketHeaderGenerator*, *ChunkstoSymbols,* and *taggedstreammux.*

Step 3   Now with the help of blocks like *OFDM carrier allocator,* channel coded bits are gathered together for mapping it to corresponding constellation points.

Step 4   Then, the data are represented in the form of complex numbers and it is ordered serially.

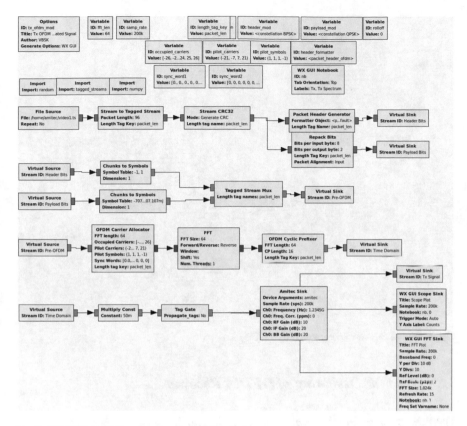

Fig. 3 OFDM transmitter flow graph (GNU radio)

Step 5   Pilot symbols which are known are then mapped and inserted with known mapping after step 4.

Step 6   A serial-to-parallel conversion is then applied along with IFFT operation on the obtained parallel complex data. The block *FFT* of GNU radio is used which is set to do IFFT operation on transmitter side.

Step 7   The output data of this operation are grouped and collected again, as per the number of requisite transmission subcarriers.

Step 8   With the help of *OFDM Cyclic Prefixer* block, insertion of cyclic prefix in every block of data is done according to the system specification and the data are multiplexed in a serial fashion. This data is then OFDM modulated and ready for transmission.

Step 9   SDR-Lab is used for the transformation of digital data in time-domain, up-convert it to transmission frequency for transmission across a wireless channel.

Step 10   Transmitted video signal can be seen with the help of sink and also various scope sinks and FFT sinks are used to see the output at various points.

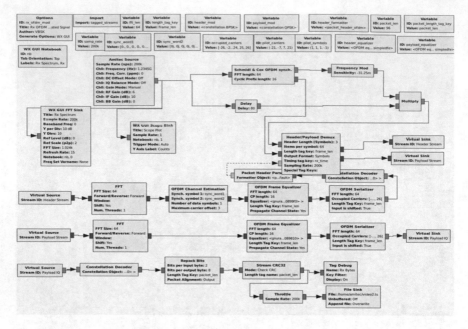

Fig. 4 OFDM receiver flow graph (GNU radio)

3.2 Steps for Designing of OFDM Receiver

The flow graph of OFDM receiver is shown in Fig. 4.

Step 1   The transmitted OFDM signal is received and is converted back to the constellation scheme and extracting the payload bits out of the received packets.

Step 2   Received signal is down-converted by receiver USRP which is finally converted to digital domain.

Step 3   Carrier frequency synchronization is performed while down-converting the received signal.

Step 4   Next step of symbol timing synchronization is achieved after this conversion.

Step 5   To demodulate, the OFDM signal *FFT block* plays a vital role. This is followed by channel estimation using the demodulated pilots.

Step 6   *OFDM Channel estimations* block gives header stream complex data which is then remapped according to the transmission constellation diagram.

Step 7   *File sink* is used to store the output.

4 Results and Discussion

An OFDM is used for real-time video signal along with SDR-Lab kits and GNU radio platform. FFT and scope plots are used to observe signals at each point. The signal is captured using webcam. This generated signal is broken into the packets, and modulated signal is then transmitted. The received signal is demodulated as per the steps given in Sect. 3. Final output seen on the receiver side will be the transmitted side video and its spectrum is shown in Figs. 5 and 6.

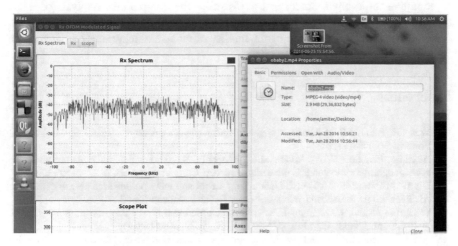

Fig. 5 Received video signal with its size and spectrum

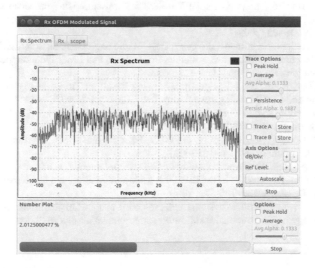

Fig. 6 Spectrum of received video

5 Conclusion and Future Scope

GNU radio-based implementation of real-time live video transmission using OFDM along with SDR-Lab devices is effectively presented in this paper. Results show the successful reception of the received video signal spectrum. The need for improved performance of this modulation technique demands the use of other advanced signal processing techniques. This can be further applied to improve the performance of the developed SDR-based OFDM system. Future work also focuses on the experimentation of various modulation techniques for improving real-time video transmission in cognitive radio environment and performing the comparative analysis of these techniques to find out the one which is more suitable to work in cognitive radio environment along with improving the quality of video and lengthens the video size.

References

1. Wang, B., Liu, K.J.R.: Advances in cognitive radio networks: A survey. IEEE J. Sel. Top. Signal Process. **5**(1), 5–23 (2011)
2. Akyildiz, I.F., Lee, W.-Y., Vuran, M.C., Mohanty, S.: Next generation/dynamic spectrum access/cognitive radio wireless networks: a survey. Comput. Netw. **50**, 2127–2159 (2006)
3. Draft: IEEE802.16h-2010 IEEE Standard for Local and metropolitan area networks Part 16: Air Interface for Broadband Wireless Access Systems Amendment 2
4. Patil, R.B., Kulat, K.D., Gandhi, A.S.: GMSK based real time video transmission on SDR platform. Int. J. Appl. Eng. Res. (IJAER) **12**(15), 5089–5093 (2017)
5. Axell, E., Leusand, G., Larsson, E.G.: Overview of spectrum sensing for cognitive radio. In: 2nd International Workshop on Cognitive Information Processing (2010)
6. Tong, Z., Arifianto, M.S., Liau, C.F.: Wireless transmission using universal software radio peripheral. In: 2009 International Conference on Space Science and Communication, Negeri Sembilan, pp. 19–23 (2009)
7. Song, W.: Configure cognitive radio using GNU Radio and USRP. In: 2009 3rd IEEE International Symposium on Microwave, Antenna, Propagation and EMC Technologies for Wireless Communications, Beijing, pp. 1123–1126 (2009)
8. Mate, A., Lee, K.-H., Lu, I-T.: Spectrum Sensing Based on Time Covariance Matrix Using GNU Radio and USRP for Cognitive Radio. IEEE (2011). ISSN:978-1-4244-9877-2
9. GNU radio getting started Guide. https://www.gnuradio.org/
10. SDR Lab getting started Guide. http://sdr-lab.com/downloads
11. Mahmoud, H.A., Yucek, T., Arslan, H.: OFDM for cognitive radio: merits and challenges. IEEE Wirel. Commun. **16**(2), 6–15 (2009)
12. Marwanto, A., et al.: Experimental study of OFDM implementation utilizing GNU Radio and USRP-SDR. In: Proceedings of the IEEE 9th Malaysia International Conference on Communications (MICC), pp. 132–135, Dec 2009
13. Tichy, M., Ulovec, K.: OFDM system implementation using a USRP unit for testing purposes. In: Proceedings of 22nd International Conference Radioelektronika 2012, Brno, pp. 1–4 (2012)

Generalized ε—Loss Function-Based Regression

Pritam Anand, Reshma Rastogi (nee Khemchandani) and Suresh Chandra

Abstract In this paper, we propose a new loss function termed as "generalized ε—loss function" to study the regression problem. Unlike the standard ε—insensitive loss function, the generalized ε—loss function penalizes even those data points which lie inside of the ε—tube so as to minimize the scatter within the tube. Also, the rate of penalization of data points lying outside of the ε—tube is much higher in comparison to the data points which lie inside of the ε—tube. Based on the proposed generalized ε—loss function, a new support vector regression model is formulated which is termed as "Penalizing ε—generalized SVR (Pen-ε—SVR)." Further, extensive numerical experiments are carried out to check the validity and efficacy of the proposed Pen-ε—SVR.

1 Introduction

Support Vector Machine (SVM) is one of the most popular machine learning algorithms (Cortes and Vapnik [1], Burges [2], Cherkassky and Mulier [3], Vapnik [4]). SVM has emerged from the research in statistical learning theory on how to regulate the trade off between the structural complexity and empirical risk. It has

P. Anand (✉) · R. Rastogi (nee Khemchandani)
Faculty of Mathematics and Computer Science, South Asian University,
New Delhi 110021, India
e-mail: ltpritamanand@gmail.com

R. Rastogi (nee Khemchandani)
e-mail: reshma.khemchandani@sau.ac.in

S. Chandra
Department of Mathematics, Indian Institute of Technology Delhi,
New Delhi 110016, India
e-mail: sureshiitdelhi@gmail.com

© Springer Nature Singapore Pte Ltd. 2019
M. Tanveer and R. B. Pachori (eds.), *Machine Intelligence and Signal Analysis*,
Advances in Intelligent Systems and Computing 748,
https://doi.org/10.1007/978-981-13-0923-6_35

outperformed the existing tools in a wide variety of applications. Some of them can be found in (Osuna et al. [5], Joachims [6], Schlkopf et al. [7], Lal TN et al. [8]).

SVM has also been extended to solve the problem of function approximation. The standard ε—Support Vector Regression (SVR) uses the ε—insensitive loss function to minimize the empirical risk. For the purpose of minimizing structural risk, it incorporates a regularization term in its optimization problem. The minimization of the ε—insensitive loss function is equivalent to finding a function $f(x) = w^T x + b$ such that most of the training points lie inside of its ε—band. The regularization term of the ε—SVR makes the regressor $f(x)$ as flat as possible. Thus, the ε—SVR finds a function $f(x) = w^T x + b$, as flat as possible, such that most of the data points lie inside of the ε—band of the $f(x)$. Data points which lie outside of the ε—tube have been penalized in the optimization problem to bring them close to $f(x)$. These data points along with the data points lying on the boundary of the ε—tube of $f(x)$ constitute "Support vectors" which decides the orientation and position of the regressor $f(x)$.

The use of ε—insensitive loss function in standard ε—SVR model makes it ignore the data points lying inside of the ε—tube. It makes the ε—SVR to avoid the over-fitting while estimating the resultant estimator $f(x)$. However, it also causes to lose some information of the training set in the estimation of the function $f(x)$.

Taking motivation from the above facts, we have proposed a new generalized loss function in this paper. The popular ε—insensitive loss function is a particular case of the proposed generalized function. The proposed loss function facilitates different rates of the penalization inside and outside of the ε—tube. It makes this loss function to use the full information of the training set without overfitting the data points.

The new loss function has been termed as "generalized ε—loss function," and is given as follows

$$L_{\tau_1,\tau_2,\varepsilon}^{Pen}(u) = \max(\tau_2|u| + (\tau_1 - \tau_2)\varepsilon, \ \tau_1|u|). \tag{1}$$

For the regression training set $T = \{(x_i, y_i), i = 1, 2, \ldots l)\}$, the above loss function can be used to measure the empirical error as follows

$$L_{\tau_1,\tau_2,\varepsilon}^{Pen}(y_i, x_i, f(x_i)) \begin{cases} = \tau_2|y_i - f(x_i)| + (\tau_1 - \tau_2)\varepsilon, \ if \ |y_i - f(x_i)| \geq \varepsilon, \\ = \tau_1|y_i - f(x_i)| \ otherwise, \end{cases} \tag{2}$$

where $\tau_2 \geq \tau_1 \geq 0$ and $\varepsilon > 0$ are parameters.

Figure 1 below shows "generalized ε—loss function" with $\tau_1 = 1$ and $\tau_2 = 4$ and $\varepsilon = 4$. The data points lying inside of the ε—tube have been penalized with the rate of τ_1 in the proposed loss function. The data points which lie outside of the ε—tube have been given the penalty $\tau_2|y_i - f(x_i)| + (\tau_1 - \tau_2)$. The proposed loss function enjoys the full use of the training set while maintaining its convexity.

The Fig. 2 below shows the "generalized ε—loss function" for different value of τ_2 and τ_1. The well-known ε—insensitive loss function is also a particular case of the "generalized ε—loss functions" with the parameter $\tau_2 = 1$ and $\tau_1 = 0$.

Fig. 1 Plot of the
$L^{Pen}_{\tau_1,\tau_2,\varepsilon}(u) = \max(\tau_2|u| + (\tau_1 - \tau_2)\varepsilon, \ \tau_1|u|)$ with
$\tau_1 = 1$ and $\tau_2 = 4$ and
$\varepsilon = 4$. For the case of
regression, $u = y - f(x)$

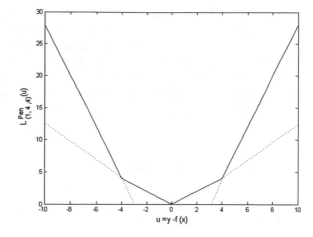

Fig. 2 Plot of the penalizing
ε—generalized loss with
different value of τ_2 and τ_1.
The proposed loss functions
reduce to popular
ε—insensitive function with
$\tau_2 = 1$ and $\tau_1 = 0$ (red line)

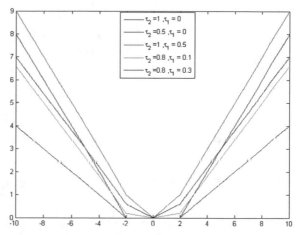

Similar to the line of standard ε—SVR model, the proposed loss function can be minimized along with the regularization term $\frac{1}{2}||w||^2$ in an optimization problem. The resultant model has been termed as "Penalizing ε—generalized SVR (Pen-ε—SVR)" model. Since the proposed loss function is convex, so the Pen-ε—SVR model can be converted into a convex programming problem which can be solved efficiently. The numerical results given in the experimental section of this paper verify that the proposed Pen-ε—SVR model owns better generalization ability than existing SVR model as the underlying generalized ε—loss function is more general than existing ε—insensitive function in nature.

We now describe notations used in the rest of this paper. All vectors will be taken as column vector unless it has not been specified. For any vector $x \in R^n$, $||x||$ will denote the l_2 norm. A vector of ones of arbitrary dimension will be denoted by e. Let (A, Y) denotes the training set, where $A = [A_1, A_2, \ldots, A_l]$ contains the l points in

$\mathbb{R}^n$ represented by l rows of the matrix A and $Y = [y_1, y_2, ..., y_l] \in \mathbb{R}^{l \times 1}$ contains the corresponding label or response value of the row of matrix A.

The rest of this paper has been organized as follows. Section 2 briefly describes standard ε—SVR models. In Sect. 3, the Penalizing ε—generalized SVR (Pen-ε—SVR)' model has been proposed for its linear and nonlinear cases. Section 4 evaluates the proposed Pen-ε-SVR model in terms of numerical results produced by the experiments carried on several artificial and UCI benchmark datasets.

2 Support Vector Regression

SVR model uses the ε-insensitive loss function to measure the empirical risks which can be given as follows

$$L_\varepsilon(y, x, f(x)) = \begin{cases} |y_i - f(x_i)| - \varepsilon, & if \ |y_i - f(x_i)| \geq \varepsilon, \\ 0 & otherwise, \end{cases} \tag{3}$$

where $\varepsilon \geq 0$ is parameter.

2.1 ε—Support Vector Regression

The standard ε-SVR minimizes

$$\frac{1}{2}||w||^2 + C \sum_{i=1}^{l} L_\varepsilon(y_i, x_i, f(x_i)),$$

which can be equivalently converted to following Quadratic Programming Problem (QPP)

$$\min_{w,b,\xi,\xi^*} \frac{1}{2}||w||^2 + C \sum_{i=1}^{l} (\xi_i + \xi_i^*)$$

subject to,
$$y_i - (A_i w + b) \leq \varepsilon + \xi_i, \ (i = 1, 2, .., l),$$
$$(A_i w + b) - y_i \leq \varepsilon + \xi_i^*, \ (i = 1, 2, .., l),$$
$$\xi_i \geq 0, \ \xi_i^* \geq 0 \ , (i = 1, 2, .., l), \tag{4}$$

where $C > 0$ is the user specified positive parameter that balances the trade off between the fitting error and the flatness of the function.

3 Penalizing ε-Generalized SVR

The Pen-ε-SVR model minimizes

$$\frac{1}{2}\|w\|^2 + C \sum_{i=1}^{l} L_{(\tau_1,\tau_2,\varepsilon)}^{Pen}(y_i, x_i, f(x_i)),$$

where $\tau_2 \geq \tau_1 \geq 0$ and $\varepsilon > 0$ are parameters.

3.1 Linear Penalizing ε-Generalized SVR

The above problem can be equivalently solved by converting the problem into quadratic programming problem by introducing the slack variables ξ_1 and ξ_2 as follow

$$\min_{(w,b,\xi_1,\xi_2)} \frac{1}{2}\|w\|^2 + Ce^T(\xi_1 + \xi_2)$$

subject to,

$$Y - (Aw + eb) \leq \frac{1}{\tau_1}\xi_1,$$

$$(Aw + eb) - Y < \frac{1}{\tau_1}\xi_2,$$

$$Y - (Aw + eb) \leq \frac{1}{\tau_2}\xi_1 + e(1 - \frac{\tau_1}{\tau_2})\varepsilon,$$

$$(Aw + eb) - Y \leq \frac{1}{\tau_2}\xi_2 + e(1 - \frac{\tau_1}{\tau_2})\varepsilon. \tag{5}$$

The above QPP reduces to the QPP (4) of standard ε-SVR model with the particular choice of parameters $\tau_2 = 1$ and $\tau_1 = 0$. Therefore, the Pen-ε-SVR can be claimed to be more general model which also includes the case of popular standard ε-SVR model.

In order to find the solution of the primal problem (5), we need to derive its corresponding Wolfe dual problem. For this, we write the Lagrangian function for the primal problem (5) as follow.

$L(w, b, \xi_1, \xi_2, \alpha_1, \alpha_2, \beta_1, \beta_2) = \frac{1}{2}\|w\|^2 + Ce^T(\xi_1 + \xi_2) + \alpha_1^T(Y - (Aw + eb) - \frac{1}{\tau_1}\xi_1) + \alpha_2^T(Aw + eb - Y - \frac{1}{\tau_1}\xi_2) + \beta_1^T(Y - (Aw + eb) - \frac{1}{\tau_2}\xi_1 - e(1 - \frac{\tau_1}{\tau_2})\varepsilon) + \beta_2^T(Aw + eb - Y - \frac{1}{\tau_2}\xi_2 - e(1 - \frac{\tau_1}{\tau_2})\varepsilon)$ where $\alpha_1 = (\alpha_1^1, \alpha_1^2, ..., \alpha_1^l)$, $\alpha_2 = (\alpha_2^1, \alpha_2^2, ..., \alpha_2^l)$, $\beta_1 = (\beta_1^1, \beta_1^2, ..., \beta_1^l)$ and $\beta_2 = (\beta_2^1, \beta_2^2, ..., \beta_2^l)$ are the Lagrangian multipliers.

The K.K.T. optimality conditions are given by

$$\frac{\partial L}{\partial w} = w - A^T(\alpha_1 - \alpha_2 + \beta_1 - \beta_2) = 0, \tag{6}$$

$$\frac{\partial L}{\partial b} = e^T(\alpha_1 - \alpha_2 + \beta_1 - \beta_2) = 0, \tag{7}$$

$$\frac{\partial L}{\partial \xi_1} = C - \frac{1}{\tau_1}\alpha_1 - \frac{1}{\tau_2}\beta_1 = 0, \tag{8}$$

$$\frac{\partial L}{\partial \xi_2} = C - \frac{1}{\tau_1}\alpha_2 - \frac{1}{\tau_2}\beta_2 = 0, \tag{9}$$

$$\alpha_1^T(Y - (Aw + eb) - \frac{1}{\tau_1}\xi_1) = 0, \tag{10}$$

$$\alpha_2^T(Aw + eb - Y - \frac{1}{\tau_1}\xi_2) = 0, \tag{11}$$

$$\beta_1^T(Y - (Aw + eb) - \frac{1}{\tau_2}\xi_1 - e(1 - \frac{\tau_1}{\tau_2})\varepsilon) = 0, \tag{12}$$

$$\beta_2^T(Aw + eb - Y - \frac{1}{\tau_2}\xi_2 - e(1 - \frac{\tau_1}{\tau_2})\varepsilon) = 0, \tag{13}$$

$$Y - (Aw + eb) \le \frac{1}{\tau_1}\xi_1, \tag{14}$$

$$(Aw + eb) - Y \le \frac{1}{\tau_1}\xi_2, \tag{15}$$

$$Y - (Aw + eb) \le \frac{1}{\tau_2}\xi_1 + e(1 - \frac{\tau_1}{\tau_2})\varepsilon, \tag{16}$$

$$(Aw + eb) - Y \le \frac{1}{\tau_2}\xi_2 + e(1 - \frac{\tau_1}{\tau_2})\varepsilon, \tag{17}$$

$$\alpha_1 \ge 0, \alpha_2 \ge 0, \beta_1 \ge 0, \beta_2 \ge 0. \tag{18}$$

Using the above KKT conditions, the Wolfe dual of the primal problem (5) can be obtained as follows

$$\min_{(\alpha_1,\alpha_2,\beta_1,\beta_2)} \frac{1}{2}(\alpha_1 - \alpha_2 + \beta_1 - \beta_2)AA^T(\alpha_1 - \alpha_2 + \beta_1 - \beta_2)$$
$$- (\alpha_1 - \alpha_2 + \beta_1 - \beta_2)^T Y + \varepsilon(1 - \frac{\tau_1}{\tau_2})(\beta_1 + \beta_2)^T e$$

subject to,

$$
\begin{aligned}
(\alpha_1 - \alpha_2 + \beta_1 - \beta_2)^T e &= 0, \\
C - \frac{1}{\tau_1}\alpha_1 - \frac{1}{\tau_2}\beta_1 &= 0, \\
C - \frac{1}{\tau_1}\alpha_2 - \frac{1}{\tau_2}\beta_2 &= 0, \\
\alpha_1, \ \alpha_2, \ \beta_1, \ \beta_2 &\ge 0.
\end{aligned}
\tag{19}
$$

After obtaining the solution of the problem (19), the value of the w can be obtained from the KKT condition (6) as follow

$$w = A^T(\alpha_1 - \alpha_2 + \beta_1 - \beta_2). \tag{20}$$

Let us define the following sets

$$S_1 = \{i : \alpha_1^i > 0, \beta_1^i > 0\},$$
$$S_2 = \{j : \alpha_2^j > 0, \beta_2^j > 0\}.$$

Using KKT conditions (10) and (12), $\forall\ i \in S_1$, we have

$$Y_i - (A_i w + b) - \tfrac{1}{\tau_1}\xi_1^i = 0, \tag{21}$$

$$Y_i - (A_i w + b) - \tfrac{1}{\tau_2}\xi_1^i - (1 - \tfrac{\tau_1}{\tau_2})\varepsilon = 0, \tag{22}$$

which gives $\xi_1^i = \tau_1\varepsilon$. Further using KKT condition (10), we can obtain

$$b = y_i - A_i w - \varepsilon \quad \forall i \in S_1 \tag{23}$$

On the similar line, $\forall\ j \in S_2$, we can obtain using KKT conditions (9) and (11)

$$b = y_j - A_j w + \varepsilon \quad \forall j \in S_2 \tag{24}$$

For each $i \in S_1$ and each $j \in S_2$, we calculate the values of the b from (23) and (24), respectively, and use their average value as the final value of the b. For the given $x \in R^n$, the estimated regressor is obtained as follows

$$f(x) = w^T x + b = (\alpha_1 - \alpha_2 + \beta_1 - \beta_2)^T A x + b \tag{25}$$

3.2 Nonlinear Penalizing ε-Generalized SVR

The nonlinear Pen-ε-SVR will seek to estimate the function.
$f(x) = w^T \phi(x) + b$, where $\phi : R^n \to \mathcal{H}$ is a nonlinear mapping.
The nonlinear Pen-ε-SVR solves following optimization problem

$$\min_{(w,b,\xi_1,\xi_2)} \frac{1}{2}\|w\|^2 + Ce^T(\xi_1 + \xi_2)$$

subject to,

$$Y - (\phi(A)w + eb) \le \frac{1}{\tau_1}\xi_1,$$

$$(\phi(A)w + eb) - Y \le \frac{1}{\tau_1}\xi_2,$$

$$Y - (\phi(A)w + eb) \le \frac{1}{\tau_2}\xi_1 + (1 - \frac{\tau_1}{\tau_2})e\varepsilon,$$

$$(\phi(A)w + eb) - Y \le \frac{1}{\tau_2}\xi_2 + (1 - \frac{\tau_1}{\tau_2})e\varepsilon. \tag{26}$$

Similar to the line of the linear Pen-ε-SVR, the corresponding Wolfe dual problem of the primal problem (26) has been obtained as follow

$$
\min_{(\gamma_1, \gamma_2, \lambda_1, \lambda_2)} \frac{1}{2}(\gamma_1 - \gamma_2 + \lambda_1 - \lambda_2)\phi(A)\phi(A)^T(\gamma_1 - \gamma_2 + \lambda_1 - \lambda_2)
$$
$$
- (\gamma_1 - \gamma_2 + \lambda_1 - \lambda_2)^T Y + \varepsilon(1 - \frac{\tau_1}{\tau_2})(\lambda_1 + \lambda_2)^T e
$$

subject to,
$$
\begin{aligned}
(\gamma_1 - \gamma_2 + \lambda_1 - \lambda_2)^T e &= 0, \\
C - \frac{1}{\tau_1}\gamma_1 - \frac{1}{\tau_2}\lambda_1 &= 0, \\
C - \frac{1}{\tau_1}\gamma_2 - \frac{1}{\tau_2}\lambda_2 &= 0, \\
\gamma_1, \ \gamma_2, \ \lambda_1, \ \lambda_2 &\geq 0.
\end{aligned} \tag{27}
$$

A positive definite kernel $K(A, A^T)$, satisfying the Mercer condition (Scholkopf B and Smola AJ [10]), is used to obtain the $\phi(A)\phi(A)^T$ without explicit knowledge of mapping ϕ. It reduces the problem (27) as follow.

$$
\min_{(\gamma_1, \gamma_2, \lambda_1, \lambda_2)} \frac{1}{2}(\gamma_1 - \gamma_2 + \lambda_1 - \lambda_2)K(A, A^T)(\gamma_1 - \gamma_2 + \lambda_1 - \lambda_2)
$$
$$
- (\gamma_1 - \gamma_2 + \lambda_1 - \lambda_2)^T Y + \varepsilon(1 - \frac{\tau_1}{\tau_2})(\lambda_1 + \lambda_2)^T e
$$

subject to,
$$
\begin{aligned}
(\gamma_1 - \gamma_2 + \lambda_1 - \lambda_2)^T e &= 0, \\
C - \frac{1}{\tau_1}\gamma_1 - \frac{1}{\tau_2}\lambda_1 &= 0, \\
C - \frac{1}{\tau_1}\gamma_2 - \frac{1}{\tau_2}\lambda_2 &= 0, \\
\gamma_1, \ \gamma_2, \ \lambda_1, \ \lambda_2 &\geq 0.
\end{aligned} \tag{28}
$$

For the given $x \in R^n$, the estimated regressors is obtained as follows

$$
\begin{aligned}
f(x) &= w^T \phi(x) + b \\
&= K(x, A^T)(\gamma_1 - \gamma_2 + \lambda_1 - \lambda_2) + b
\end{aligned} \tag{29}
$$

4 Experimental Results

We have performed a series of experiments to verify the claims made in this paper. To show the efficacy of the proposed Pen-ε-SVR, we have tested it on four artificial datasets and five UCI benchmark datasets (Blake CI and Merz CJ [11]). Further, the performance of other existing SVR methods like standard ε-SVR and L_1- norm SVR (Tanveer M [9]) has been compared with proposed Pen-ε-SVR on these datasets. It has been found that, the proposed Pen-ε-SVR owns better generalization ability than ε-SVR models.

All of the regression methods were simulated in Matlab 12.0 environment (http://in.mathworks.com/) on Intel XEON processor with 16.0 GB RAM. The L_1-Norm SVR has been solved by using the linprog function of the Matlab. The proposed Pen-ε-SVR and SVR have been solved by the interior-point convex algorithm using the quadprog function of Matlab. Throughout these experiments, we have used RBF kernel $exp(\frac{-||x-y||^2}{q})$ where q is the kernel parameter.

The optimal values of the parameters have been obtained using the exhaustive search method (Hsu CW and Lin CJ [12]) by using cross-validation. The value of the parameter ε, τ_1, and τ_2 have been searched in the set $\{0.05, 0.1, 0.2, 0.3....., 1, 1.5, 2, 2.5, 3, 3.5, 4, 4.5, 5\}$ for all relevant regression methods. The value of the parameter q of RBF kernel has been searched in the set $\{2^i, \; i = -10, -2, ..., 12\}$.

4.1 Performance Criteria

In order to evaluate the performance of the regression methods, we first introduce some commonly used evaluation criteria. Without loss of generality, let l and k be the number of the training samples and testing samples, respectively. Furthermore, for $i = 1, 2, ...k$, let $\hat{y}_i$ be the predicted value for the response value y_i and $\bar{y} = \frac{1}{k} \sum_i^k y_i$ is the average of $y_1, y_2, ..., y_k$. The definition and significance of the some evaluation criteria has been listed as follows.

  (i) SSE: Sum of squared error of testing, which is defined as SSE=$\sum_{i=1}^k (y_i - y_i')^2$. SSE represents the fitting precision.

 (ii) SST : Sum of squared deviation of testing samples, which is defined as SST = $\sum_{i=1}^k (y_i - \bar{y})^2$. SST shows the underlying variance of the testing samples.

(iii) RMSE : Root mean square of the testing error, which is defined as RMSE = $\sqrt{\frac{1}{k} \sum_{i=1}^k (y_i - y_i')^2}$.

(iv) MAE: Mean absolute error of testing, which is defined as $\frac{1}{k} \sum_{i=1}^k |(y_i - y_i')|$.

 (v) SSE/SST : SSE/SST is the ratio between the sum of the square of the testing error and sum of the square of the deviation of testing samples. In most cases, small SSE/SST means good agreement between estimations and real values.

4.2 Artificial Datasets

We have synthesized some artificial datasets to show the efficacy of the proposed method over the existing methods. To compare the noise-insensitivty of the regression methods, only training sets were polluted with different types of noises in these

artificial datasets. For the training samples (x_i, y_i) for $i = 1, 2, .., l$, following types
of datasets have been generated.

TYPE 1:-

$$y_i = \frac{sin(x_i)}{x_i} + \xi_i, \;\; \xi_i \sim U[-0.2, 0.2]$$

and x_i is from $U[-4\pi, 4\pi]$.

TYPE 2:-

$$y_i = \frac{sin(x_i)}{x_i} + \xi_i, \;\; \xi_i \sim U[-0.3, 0.3]$$

and x_i is from $U[-4\pi, 4\pi]$.

TYPE 3:-

$$y_i = \frac{sin(x_i)}{x_i} + \xi_i, \;\; \xi_i \sim N[0, 0.5]$$

and x_i is from $U[-4\pi, 4\pi]$.

TYPE 4:-

$$y_i = \frac{sin(x_i)}{x_i} + \left(0.5 - \left|\frac{x_i}{8\pi}\right|\right)\xi_i, \;\; \xi_i \sim U[-0.1, 0.1]$$

and x_i is from $U[-4\pi, 4\pi]$.

TYPE 4 datasets contain 200 training samples and 400 non-noise testing samples,
while other datasets contain 100 training samples and 500 non-noise testing samples. To avoid the biased comparison, 10 independent groups of noisy samples were
generated randomly using Matlab toolbox for all type of datasets.

Table 1 shows the performance of the Pen-ε-SVR model along with ε-SVR and
L_1-Norm SVR. It can be concluded that the proposed Pen-ε-SVR model is more
general in nature as it could find the value of the τ_1 and τ_2 which can lead to better
prediction than existing SVR methods.

4.3 UCI Datasets

For further evaluation, we have checked the performance of the proposed methods
on five UCI datasets namely, Traizines, Chwirut, Servo, Concrete Slump, and Yatch
Hydro Dyanamics which are commonly used in evaluating a regression method. For

Table 1 Results on Artificial Datasets

Dataset	Regressor	$\tau_2+\tau_1$	SSE/SST	RMSE	MAE	CPU time (s)
TYPE 1	Pen-ε-SVR	$1+0.2$	0.0110 ±0.0048	0.0333± 0.0073	0.0256 ±0.0062	0.91
	Pen-ε-SVR	$2+0.8$	0.0110 ±0.0056	0.0332± 0.0082	0.0264 ±0.0056	0.1.75
	ε-SVR		0.0115 ±0.0046	0.0344±0.0071	0.0256±0.0057	0.47
	L_1-Norm SVR		0.0163 ±0.0062	0.0408±0.0082	0.0329±0.0066	3.68
TYPE 2	Pen-ε-SVR	$0.5+0.1$	0.0231 ±0.0152	0.0472± 0.0143	0.0359 ±0.0099	1.1
	Pen-ε-SVR	$0.5+0.2$	0.0192 ±0.0127	0.0435± 0.0129	0.0339 ±0.0087	1.57
	Pen-ε-SVR	$0.5+0.3$	0.0212 ±0.0151	0.0449± 0.0159	0.0345 ±0.0087	1.73
	Pen-ε-SVR	$0.5+0.4$	0.0229 ±0.0151	0.0468± 0.0151	0.0359 ±0.0093	1.66
	ε-SVR		0.0253 ±0.0116	0.0504±0.0118	0.0387±0.0082	0.48
	L_1-Norm SVR		0.0284 ±0.0131	0.0534±0.0128	0.0425±0.0086	3.96
TYPE 3	Pen-ε-SVR	$0.5+0.1$	0.0231 ±0.0152	0.0472± 0.0143	0.0359 ±0.0099	1.13
	Pen-ε-SVR	$0.5+0.2$	0.0194 ±0.0127	0.0435± 0.0129	0.0339 ±0.0087	1.57
	Pen-ε-SVR	$0.5+0.3$	0.0212 ±0.0151	0.0449± 0.0159	0.0345 ±0.0087	1.73
	Pen-ε-SVR	$0.5+0.4$	0.0229 ±0.0151	0.0468± 0.0151	0.0359 ±0.0093	1.66
	ε-SVR		0.0253 ±0.0116	0.0504±0.0118	0.0387±0.0082	0.48
	L_1-Norm SVR		0.0284 ±0.0131	0.0534±0.0128	0.0425±0.0086	3.96
TYPE 4	Pen-ε-SVR	$0.4+0.3$	0.0130 ±0.0121	0.0348± 0.0130	0.0268 ±0.0090	3.02
	Pen-ε-SVR	$0.5+0.3$	0.0134 ±0.0129	0.0352± 0.0134	0.0264 ±0.0092	3.07
	Pen-ε-SVR	$0.6+0.2$	0.0141 ±0.0133	0.0363± 0.0136	0.0262 ±0.0091	3.22
	ε-SVR		0.0169 ±0.0142	0.0402±0.0134	0.0297±0.0095	0.82
	L_1-Norm SVR		0.0227 ±0.0153	0.0474±0.0128	0.0351±0.0100	25.92

Table 2 Results on UCI Datasets

Dataset	Regressor	$\tau_2+\tau_1$	SSE/SST	RMSE	MAE	CPU time (s)	$(p, c_1, c_3, \varepsilon)$
Traizines 186×60	Pen-ε-SVR	1.5 + 0.1	0.9254 ±0.4550	0.1337± 0.0426	0.1001 ±0.0280	9.94	$(2^6, 2^4, \_, 0.1)$
	Pen-ε-SVR	1.5 + 0.2	0.9269 ±0.4476	0.1340± 0.0424	0.1007 ±0.0290	7.01	$(2^6, 2^4, \_, 0.1)$
	Pen-ε-SVR	1.5 + 0.3	0.9259±0.4467	0.1340± 0.0424	0.1007 ±0.0291	6.70	$(2^6, 2^4, \_, 0.1)$
	ε-SVR		0.9318±0.4430	0.1347±0.0418	0.1018±0.0281	1.25	$(2^6, 2^4, \_, 0.1)$
	L_1-Norm SVR		0.9204±0.3105	0.1363±0.0400	0.1031±0.0241	17.87	$(2^6, 2^6, 2^1, 0.1)$
Chwirut 214×2	Pen-ε-SVR	1 + 0.1	0.0222 ±0.0113	3.2330± 0.8788	2.2856 ±0.5089	13.55	$(2^{-6}, 2^6, \_, 0.2)$
	Pen-ε-SVR	2.5 + 1	0.0222 ±0.0117	3.2237± 0.885	2.2778 ±0.5154	12.50	$(2^{-6}, 2^6, \_, 0.2)$
	Pen-ε-SVR	3 + 2.5	0.0222 ±0.0118	3.2263± 0.8852	2.2891 ±0.5138	13.06	$(2^{-6}, 2^6, \_, 0.2)$
	ε-SVR		0.0224±0.0120	3.2333±0.9646	2.2443±0.5523	1.75	$(2^{-6}, 2^6, \_, 0.2)$
	L_1-Norm SVR		0.0223±0.0124	3.2176 ±0.9643	2.2707±0.5373	15.11	$(2^{-6}, 2^6, 2^1, 0.1)$
Servo 167×4	Pen-ε-SVR	1 + 0.4	0.1639 ±0.1686	0.5330 ±0.4014	0.3059 ±0.1422	4.56	$(2^{-3}, 2^2, \_, 0.1)$
	Pen-ε-SVR	1 + 0.1	0.1613 ±0.1650	0.5284 ±0.4016	0.3045 ±0.1428	4.99	$(2^{-3}, 2^2, \_, 0.1)$
	Pen-ε-SVR	1 + 0.2	0.1640 ±0.1685	0.5306 ±0.4028	0.3043 ±0.1432	4.54	$(2^{-3}, 2^2, \_, 0.1)$
	ε-SVR		0.1646±0.1706	0.5381±0.4260	0.3006±0.1523	0.80	$(2^{-3}, 2^2, \_, 0.1)$
	L_1-Norm SVR		0.1785±0.1920	0.5420 ±0.4125	0.3072±0.1427	8.14	$(2^{-3}, 2^{-2}, 2^{-5}, 0.1)$
Concrete Slump 103×7	Pen-ε-SVR	1.5 + 0.1	0.0072 ±0.0052	0.5433± 0.0761	0.4141 ±0.0729	2.42	$(2^1, 2^{10}, \_, 0.1)$

(continued)

Table 2 (continued)

Dataset	Regressor	$\tau_2+\tau_1$	SSE/SST	RMSE	MAE	CPU time (s)	$(p, c_1, c_3, \varepsilon)$
	Pen-ε-SVR	2 + 1.5	0.0072 ±0.0053	0.5408± 0.0780	0.4171 ±0.0810	2.69	$(2^1, 2^{10}, \_, 0.1)$
	Pen-ε-SVR	1.5 + 1	0.0071 ±0.0052	0.5379± 0.0739	0.4127 ±0.0725	3.16	$(2^1, 2^{10}, \_, 0.1)$
	ε-SVR		0.0073±0.0049	0.5466±0.0608	0.4304±0.0664	1.03	$(2^1, 2^{10}, \_, 0.1)$
	L_1-Norm SVR		0.0073±0.0065	0.5317 ±0.1224	0.4142±0.00665	4.14	$(2^1, 2^7, 2^{-4}, 0.1)$
Yatch 308×6	Pen-ε-SVR	4 + 2.5	0.0022 ±0.0031	0.6116 ±0.3715	0.3349± 0.1019	34.19	$(2^{-2}, 2^{10}, \_, 0.2)$
	Pen-ε-SVR	4 + 3	0.0022 ±0.0031	0.5102 ±0.3733	0.3342± 0.1017	35.11	$(2^{-2}, 2^{10}, \_, 0.2)$
	Pen-ε-SVR	4 + 3.5	0.0022 ±0.0031	0.5091 ±0.3737	0.3336± 0.1010	34.06	$(2^{-2}, 2^{10}, \_, 0.2)$
	ε-SVR		0.0034±0.0008	0.8629±0.2029	0.5163±0.1057	3.68	$(2^{-2}, 2^{10}, \_, 0.1)$
	L_1-Norm SVR		0.0024±0.0025	0.5590 ±0.3165	0.3713±0.1033	44.96	$(2^{-2}, 2^{10}, 2^{-5}, 0.2)$

all the datasets, only feature vectors were normalized in the range of [0,1]. Tenfold cross-validation (Duda RO and Hart PR [13]) method has been used to report the numerical results for these datasets.

Table 2 lists the performance of the proposed Pen-ε-SVR, SVR and L_1-Norm SVR on above-mentioned UCI datasets using different evaluation criteria along with their training times in seconds. Table 2 also lists the tunned parameters of regression methods for each datasets. It can be observed that the proposed Pen-ε-SVR model could tune the value of τ_1 and τ_2 which can predict better than existing SVR models. However, the training time of Pen-ε-SVR model exceeds from the SVR in the Table 2. It is because of the fact that while moving from the optimization problem of the SVR to Pen-ε-SVR model, the two last box constraints of the SVR get converted to linear constraints in Pen-ε-SVR model.

5 Conclusion

In this paper, we have introduced a new family of loss functions which we have termed as "generalized ε-loss function." The proposed loss functions have been used to measure the empirical risk in a regression model which we have termed as "generalized ε—loss function" (Pen-ε-SVR). The proposed Pen-ε-SVR owns better generalization ability than existing SVR methods as it is more general in nature.

The Pen-ε-SVR requires lots of parameters to be tunned. The value of the τ_1 and τ_2 should be determined from the training set itself according to the scatter present in it. Future work involves finding the solution path for the Pen-ε-SVR with the value of τ_1 and τ_2. Further, the relevance and use of the proposed loss function in the case of the classification is still under investigation and can be explored in the future.

References

1. Cortes, C., Vapnik, V.: Support vector networks. Mach. Learn. **20**(3), 273–297 (1995)
2. Burges, J.C.: A tutorial on support vector machines for pattern recognition. Data Min. Knowl. Discov. **2**(2), 121–167 (1998)
3. Cherkassky, V., Mulier, F.: Learning From Data:concepts, Theory and Methods. John Wiley and Sons, New York (2007)
4. Vapnik, V.: Statistical Learning Theory, vol. 1. :Wiley, New York (1998)
5. Osuna, E., Freund, R., Girosit, F.: Training support vector machines: An application to face detection. In: Proceedings of IEEE Computer Vision and Pattern Recognition, pp. 130–136 . San Juan, Puerto Rico (1997)
6. Joachims, T.: Text categorization with support vector machines: learning with many relevant features. Eur. Conference Mach. Learn. Springer, Berlin (1998)
7. Schlkopf, B., Tsuda, K., Vert, J.P.: Kernel Methods in Computational Biology. MIT press (2004)
8. Lal, T.N., Schroder, M., Hinterberger, T., Weston, J., Bogdan, M., Birbaumer, N., Scholkopf, B.: Support vector channel selection in BCI. IEEE Trans. Biomed. Eng. **51**(6), 10031010 (2004)

9. Tanveer, M.: Linear programming twin support vector regression. Filomat, **31**(7), 2123–2142 (2017)
10. Schlkopf, B., Smola, A.J.: Learning With Kernels: Support Vector Machines, Regularization, Optimization, and Beyond. MIT press (2002)
11. Blake, C.I., Merz, C.J.: UCI repository for machine learning databases (1998). http://www.ics.uci.edu/*mlearn/MLRepository.html
12. Hsu, C.W., Lin, C.J.: A comparison of methods for multi class support vector machines. IEEE Trans. Neural Netw. **13**, 415425 (2002)
13. Duda, R.O., Hart, P.R., Stork, D.G.: Pattern Classification, 2nd edn. Wiley, USA (2001)

PICS: A Novel Technique for Video Summarization

Gagandeep Singh, Navjot Singh and Krishan Kumar

Abstract With brisk growth in video data the demand for both effective and powerful methods for video summarization is also elevated so that the users can browse the apace and comprehend a large amount of video content. Our paper highlights a novel keyframe extraction technique based on the clustering to attain video summarization (*VS*). Clustering is an unsupervised procedure and these algorithms rely on some prior assumptions to define subgroups in the given dataset. To extract the keyframes in a video, a procedure called *k-medoids clustering* is used. To find the number of optimal clusters, is a challenge here. This task can be achieved by using the cluster validation procedure, *Calinski–Harabasz index* (CH index). This procedure is based on the well-defined criteria for clustering that enable the selection of an optimal parameter value to get the best partition results for the dataset. Thus, CH index allows users a parameter independent VS approach to select the keyframes in video without bringing down the further computational cost. The quantitative and qualitative evaluation and the computational complexity are drained to compare the achievements of our proposed model with the state-of-the-art models. The experimental results on two standard datasets having various categories of videos indicate PICS model outperforms other existing models with best *F-measure*.

1 Introduction

Video is a combination of visual and speech data stored in a banausic order to reveal the events or activities. Conventionally, a video constitutes a number of scenes and

G. Singh (✉) · N. Singh · K. Kumar
Department of Computer Science and Engineering, National Institute of Technology Uttarakhand, Srinagar (Garhwal), India
e-mail: gagansidhu.cse16@nituk.ac.in

N. Singh
e-mail: navjot.singh.09@nituk.ac.in

K. Kumar
e-mail: kkberwal@nituk.ac.in

© Springer Nature Singapore Pte Ltd. 2019
M. Tanveer and R. B. Pachori (eds.), *Machine Intelligence and Signal Analysis*,
Advances in Intelligent Systems and Computing 748,
https://doi.org/10.1007/978-981-13-0923-6_36

411

each scene is having a group of shots. Further, these shots are divided into group of frames with a specific and constant frame rate. Therefore, it needs a constant amount of time to watch a video. With advancement in digital video capturing and editing techniques, there is a gain in video-based data. This rise in data results in the demand for a powerful and effective technique to capture video and better analysis of the video. Advancement in storing of digital, distribution of this content, and digital video recorders results in making the recording of the digital content procedure easy [1]. However, it will never be wise to watch the whole video content because of shortage of time. In such cases, the abstract of the videos is useful for the users instead of watching the whole of the videos. It becomes one of the critical challenges of achieving such ponderous amount of video data [2]. VS is the key solution to all such gigantic amount of digital content with easy access to summary content [3, 4]. Moreover, it is a self-sufficient video squeezing method, based on the content itself, that produce an effective and powerful summary for a video, by recognizing most constitutive and relevant content present in the video content. VS can be of two categories either be a sequence of frozen images which are also called Storyboard or moving images called Skimming [5]. Video-Storyboard and be defined as a group of stationary keyframes, which contains the important content of a video with minimal data. This class of video summaries is well explored using numerous clustering algorithms [6–11] where different clusters are formed on the basis of similarity between the frames. In other hand, the video skimming preserves the important information without losing the semantics of the video sequence [12]. In this work, we employ k-medoids [8, 10, 13] algorithm to extract the keyframe. Further, we implemented CH index [14]-based cluster validation technique to get an optimal clusters set. The CH index criterion provides the best results for clustering-based solutions [15]. The main contribution of this paper exists in these aspects:

- For reducing time for computation, we implement CH index (discussed in Sect. 2.2) technique which assists in finding a best *k-value* by choosing five videos at random instead of selecting all videos contained in a specific dataset.
- *VS* problem is transformed into clustering problem. We proposed a unified framework which is scalable for both extraction of keyframes and skimming of videos.
- Unlike to previous methods, which require number of keyframes a prior, our framework provides a parameter independent *VS* approach. Also, without incurring any additional cost of computation, the keyframes are selected effectively.

The structure for remaining paper is given as follows: Sect. 2 presents the model we proposed. Section 3 discusses some experimental results implemented using two standard datasets for videos. In Sect. 4, the conclusion and future perspectives and are given.

2 Proposed Model

The first N frames are extracted in total with size of $W \times H$ from a video **V**, here W indicates the width and H indicate height of a frame. The video employed from the datasets holds 2700 *RGB* frames on an average. All these frames are represented in three planes. To save the time of computation, all N *colored* frames of the input video (RGB form) are transformed to grayscale frames, where each frame is resized into one-dimensional vector as an input vector for the unsupervised clustering algorithm.

2.1 Keyframe Extraction

In this work, we proposed a keyframe-based extraction technique that is used to find those frames which are most likely to be centrally located in the set of frames. These selected frames are also known shots. The frames in a shot are very close to other frames in that particular group. So we used a partitioning method based on the *k-medoids*. Here, to get the set of keyframes extracted from the video *k-medoids algorithm* [8, 13] is used. To partition the given data into clusters or group is the basic objective of any clustering problem so that the data entities present in a particular cluster are having similar properties with each other than the entities present in the other clusters. Kaufman and Rousseeuw [13] propose an unsupervised clustering method having computational effort being improved and data points present in the dataset are potential candidates for the centers in the cluster. The goal function of this *k-medoids* algorithm is to divide the given data collection X into c clusters by minimizing the absolute distance among the data points and the selected centroid.

Algorithm 1: K-medoids clustering algorithm

Repeat for $l = 1, 2, \ldots, \prod_{k=1}^{n} max|h^{(l)} - h^{(l-1)}| \neq 0$

Step 1 Computing distances

$$D_{ik}^2 = (x_k - h_i)^T (x_k - h_i) \qquad , 1 \le i \le t, i \le k \le N \qquad (1)$$

Step 2 Selecting the data points with minimal distance for a cluster.

Step 3 Calculating the new centroid for the cluster

$$h_i^{(l)*} = \frac{\sum_{j=1}^{N_i} x_i}{N_i} \qquad (2)$$

Step 4 The nearest data point is then chosen to be the cluster centroid.

$$D_{ik}^{2*} = (x_k - v_i^*)^T (x_k - v_i^*) \qquad (3)$$

and

$$x_i^* = argmin_i (D_{ik}^{2*}; v_i^{(l)} = x_i^*.) \qquad (4)$$

Ending Partition matrix is then calculated

$$P = \sum_{q=1}^{k} \sum_{r=1}^{n} ||A_n^{(q)} - C_q|| \qquad (5)$$

Where, $||A_R^{(q)} - C_q||$ selected as a distance measure used to calculate the distance between a data point present in a cluster $A_r^{(q)}$ and centroid C_q of that cluster. Procedure for *k-medoids* is discussed underneath in Algorithm 1. Given X as the dataset, select from the range between $1 < t < N$ as the number of clusters. Initially, the selected clusters are randomly choosen from X.

2.2 Extracting Optimal Keyframes

In order to group similar frames together, the frames extracted from the input video have the k-medoids clustering technique applied over them, with the theory that the events extant in the video are represented by clusters. The keyframe of a particular event is represented by the centroid of that particular cluster. Thus, the centroids for $k \le N$ are:

$$A = A_1, A_2, \ldots A_k \qquad (6)$$

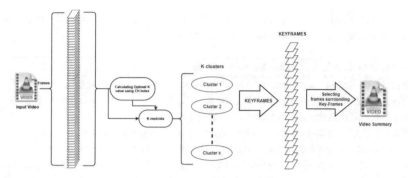

Fig. 1 Parameter independent clustering strategy PICS for video summaries

Now, obtaining the *k-value* is a summons to contest. In other words, it is a challenge to find the best possible number of clusters. Thus, we need the number of clusters a prior. To address this problem, the internal cluster evaluation method is being employed and is evaluated on the basis of intercluster and intra-cluster distances, is *CH index*. A variety of values for *k* are used to find the preeminent value of *k* using *CH index* based learning process. To achieve best *k* value, we choose the *k* value for which the *CH index* give the maximum value. The CH index [14] uses average between-cluster and within-cluster to valuate the cluster validity. The cluster index of CH index is calculated using following equation:

$$CH(K) = \frac{[trace\,\mathbf{G}/K - 1]}{[trace\,\mathbf{M}/N - K]} for\,K \in N \tag{7}$$

where **G** represents intercluster error sum of square and **M** indicates intra-cluster different between data point and cluster center of that particular cluster. The maximum index values represents the best possible cluster partitioning.

$$trace\,\mathbf{G} = \sum_{k=1}^{K} |C_k|\,||\overline{C_k} - \overline{x}||^2 \tag{8}$$

$$trace\,\mathbf{M} = \sum_{k=1}^{K} \sum_{i=1}^{N} w_{k,i}\,||x_i - \overline{C_k}||^2 \tag{9}$$

Combining an unsupervised clustering with a clustering validation technique, we propose a PICS (*Parameter Independent Clustering Strategy*) approach for extracting the keyframes from the video for VS. The basic outline for PICS approach is depicted in the Fig. 1 and as implemented in Algorithm 2.

Algorithm 2: *PICS-Parameter Independent Clustering Strategy for VS*

Step 1 Splitting video into N frames and reshape them in 1-Dimension.
Step 2 Calculating the CH index for $k = 3, 4, 5, \ldots, 15$ and choose the value for k for which the index value in maximum.
Step 3 Perform k–*mediods* (Algorithm 1) on video for chosen k–value.
Step 4 Select the k centers as the *key–frame* for each cluster or event.
Step 5 The centroid or nearby centroid frame is declared as keyframe, and the frames which have 90% similarity with keyframes in a cluster are counted in the final summary.

---

3   Experimentation and Discussion

Here several experiments are being discussed to approve the efficiency and proficiency of VS algorithm based on our suggested *PICS* strategy. Using two standard datasets, we implemented *k-medoids* clustering method. The first dataset videos of different category from Open Video Project (OVP)[1] (contain 50 videos). Each video lasts on average 110 s. The second dataset[2] of 50 videos which are having various classes like "*advertisements*," "*drawing*," "*TV–shows*," "*sports*," "*newsflash*," and "*home videos*." Our experiments are split between two different parts; the first specified part used is implemented to get value best value for k where k is the number of clusters in the video. Here, to find the value of k, we used the *CH index* validation technique. Five videos were chosen at random from a specific dataset instead of using the dataset entirely. This was done to save some computation time. Afterwards, *CH index* method was applied having a range from 3 to 15. At the end, the average of these five CH index outcomes was computed. The maximum value of *CH index* best describes finest k value. In experimental setup for both the dataset, value $k = 12$ in Table 1 and Figs. 2 and 3 are optimal.

3.1   Qualitative Analysis

VSUMM [16] official website provides the keyframe summaries which are selected by experts and used as ground-truth for both the datasets. The same website also provides the results of STIMO (VISTO approach extension) [17], Delaunay Clustering DT [18], VSUMM [16], and Video Project storyboard OVP [19].We consider the mean value of the five randomly selected ground-truth result for keyframes for

---

[1] " https://open-video.org/".

[2] " https://sites.google.com/site/vsummsite/download".

calculation of quantitative metrics. Fig. 4 depicts the keyframe results for video 5th of 1st dataset. From Fig. 4 we observe that

- Proposed model gives very good results as most of the keyframes extracted matches the ground-truth.
- Certainly proposed approach outperforms state-of-the-art techniques (Fig. 5).

Table 1 CH index values for different k values on 1st dataset and 2nd dataset

k-value	Dataset 1					Average	Dataset 2					Average
	vid1	vid2	vid3	vid4	vid5		vid1	vid2	vid3	vid4	vid5	
3	648	1416	1189	1108	338	0940.4	537	1308	438	1240	1146	0933.8
4	663	1434	1122	1128	317	0932.8	552	1323	406	1231	1139	0930.2
5	658	1754	1180	1036	332	0992.0	759	1865	224	1197	1027	1014.4
6	614	1740	1336	1101	336	1025.4	503	1588	380	1363	1109	0988.6
7	623	1832	1503	1122	310	1078.0	531	1855	340	1530	1132	1077.6
8	627	1657	1808	1102	314	1101.6	716	1431	414	1854	1168	1116.6
9	700	1879	1871	1153	330	1186.6	811	1233	430	1641	1223	1067.6
10	126	1754	2078	1185	334	1215.4	635	1635	434	2077	1145	1185.2
11	729	1563	2114	1235	337	1195.6	638	1452	437	2643	1269	1287.8
12	736	1512	2586	1259	324	**1283.4**	847	1782	424	2594	1296	**1388.6**
13	708	1496	2043	1223	344	1162.8	808	1686	446	2154	1231	1265.0
14	699	1483	2032	1218	333	1153.0	787	1574	323	2043	1228	1191.0
15	683	1473	2019	1203	334	1143.4	697	1478	316	2008	1208	1141.4

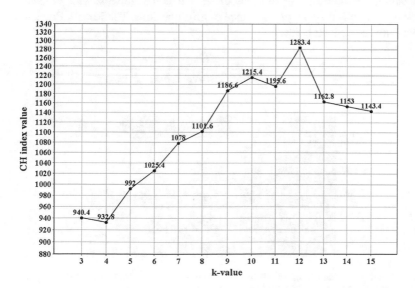

Fig. 2 Selecting maximum CH index value to get an optimal k-value on 1st dataset

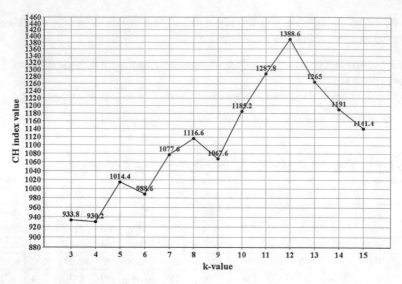

Fig. 3 Selecting maximum CH index value to get an optimal k-value on 2nd dataset

3.2 Quantitative Results

A thorough investigation has been done for extracting keyframes to achieve ideal VS but no standard has been established yet to quantify the performance. Different approaches for keyframe extraction and ground-truth summaries are compared to get quantitative results. Summaries provided by different methods are equated with user summaries for assessing different metrics, which includes Recall, Precision, and F-measure, to quantify the performance of all algorithms. The Recall, Precision, and F-measure are defined below:

$$Precision = \frac{True\ Positive}{True\ Positive + False\ Positive} \tag{10}$$

Fig. 4 Results for video number 5th from 1st dataset, top to bottom: User summary, DT [18], STIMO [17], OVP [19], VSUMM1 [16], VSUMM2 [16], and PICS Model

Fig. 5 Results for video number 49th from 2nd dataset, top to bottom: User summary, VSUMM1 [16], VSUMM2 [16] and PICS Model

Table 2 Performance of PICS with existing models on 1st dataset

Algorithm	Precision (%)	Recall (%)	F-measure (%)
DT [18]	47.0	55.0	48.5
STIMO [17]	39.0	65.0	48.8
OVP [19]	43.0	64.0	51.4
VSUMM1 [16]	42.0	**77.0**	54.4
VSUMM2 [16]	48.0	63.0	54.5
AVS [8]	65.8	60.7	63.1
PICS model	**66.0**	61.0	**64.0**

Table 3 Performance of PICS with existing models on 2nd dataset

Algorithm	Precision (%)	Recall (%)	F-measure (%)
VSUMM1 [16]	38.0	**72.0**	49.7
VSUMM2 [16]	44.0	54.0	48.5
AVS [8]	53.0	49.7	50.3
PICS model	**58.0**	46.0	51.0

$$Recall = \frac{True\ Positive}{True\ Positive + False\ Negative} \tag{11}$$

$$F - measure(L_\beta) = \frac{(1 + \beta)^2 \times Precision \times Recall}{\beta^2 \times Precision + Recall} \tag{12}$$

VS is a method of shrinking the original video's length. So, a large number of algorithms work with minimal Recall value and greater Precision value. If the summary of the video contains every frame of the input video then it will result in 100% recall, which is quite inefficient. Greater Precision signifies the algorithm as effective in removing the irrelevant frames. Therefore, In VS Recall is not the primary concern. Moreover, some algorithms choose Precision while others use Recall as their quality measure. So, F-measure is better to use as a quality measure.

Table 4 Comparison of computation time

Approach	Frame rate [frame per Sec]	Total time [S]
STIMO [17]	25	32.8
DT [18]	30	92.5
VSUMM [16]	30	71.8
OVP [19]	30	63.9
PICS model	30	68

Tables 2 and 3 depicts distinctions between our work and existing procedures. Important observations are:

- The proposed clustering-based VS approach delivers improved attainments comparing with other methods.
- In Both the datasets, our proposed model is having better Precision value.
- Some algorithms choose Precision while others use Recall as their quality measure. So, F-measure is better to use as a quality measure. The F-measure value for PICS model both the datasets is better (maximum) as compared to other models.

3.3 Computational Complexity

The videos from the Open Video Project OVP [19] are of size 352×240. For our experiments, we used 3.1 GHz Dual core system, for video of an average of 2000 frames which is approx. 60 s of duration. Our model takes 13.2 s to estimate the prior k-value for the clustering model. The average computation time per video is equated between PICS technique and some other models is presented in Table 4.

4 Conclusion

In our work, we came up with Parameter Independent Clustering Strategy for extracting keyframes for Video Summarization to generate abstract and concise video summaries also called event summarization (ES). In order to ensure that the summary has a minimum number of frames, users access to large amounts of digital content is facilitated by ES in a powerful and robust way and this is based on keyframes extraction. The clustering-based method groups the frames of the video into best cluster set. The centroid of each cluster known as cluster head can be represented as a keyframe for the summary. Calinski–Harabasz Index is the technique used for finding the optimal set of clusters. After experimentations, the results on two standard video datasets shows the proposed model outperforms the existing techniques with better *F-measure*. In future, we will work with *multiview* videos to implement more precise summarization techniques for different applications.

References

1. Singh, N., et al.: Performance enhancement of salient object detection using superpixel based Gaussian mixture model. MTAP, 1–19 (2017)
2. Kumar, K., et al.: Event BAGGING: a novel event summarization approach in multi-view surveillance videos. In: IEEE IESC'17 (2017)
3. Gao L, et al.,: Learning in high-dimensional multimedia data: the state of the art. Multimed. Syst. 1–11 (2017)
4. Kumar, K., et al.: Eratosthenes sieve based key-frame extraction technique for event summarization in videos. MTAP, 1–22 (2017)
5. Truong, B.T., Venkatesh, S.: Video abstraction: a systematic review and classification. ACM Trans. Multimed. Comput. Commun. Appl. 3(1, Article 3), 37 (2007). https://doi.org/10.1145/1198302.1198305
6. Vermaak, J., Perez, P., Gangnet, M., Blake, A.: Rapid summarization and browsing of video sequences. In: British machine vision conference, pp 1–10 (2002)
7. Zhuang, Y., Rui, Y., Huang, T.S., Mehrotra, S.: Adaptive key frame extraction using unsupervised clustering. In: Proceedings of the International Conference on Image Processing, vol. 1, pp 866–870. IEEE (1998)
8. Kumar, K., et al.: Equal partition based clustering approach for event summarization in videos. In: The 12th IEEE SITIS'16, pp. 119–126 (2016)
9. Hadi, Y., Essannouni, F., Thami, R.O.H.: Unsupervised clustering by k-medoids for video summarization. In: ISCCSP'06 (2006)
10. Hadi, Y., Essannouni, F., Thami, R.O.H. (2006). Video summarization by k-medoid clustering. In: Proceedings of the 2006 ACM Symposium on Applied Computing, pp. 1400–1401. ACM. https://doi.org/10.1145/1141277.1141601
11. Anirudh, R., Masroor, A., Turaga, P.: Diversity promoting online sampling for streaming video summarization. In: 2016 IEEE International Conference on Image Processing (ICIP), pp. 3329–3333. IEEE (2016)
12. Kumar, K., Shrimankar, D.D.: F-DES: fast and deep event summarization. IEEE TMM (2017). https://doi.org/10.1109/TMM.2017.2741423.
13. Kaufman, L., Rousseeuw, P.J.: Clustering by means of Medoids. In: Dodge, Y. (ed.) Statistical Data Analysis Based on the L1 - Norm and Related Methods, pp. 405–416. North-Holland, New York (1987)
14. Calinski, T., Harabasz, J.: A dendrite method for cluster analysis. Commun. Stat. 3(1), 1–27 (1974)
15. Van Craenendonck, T., Blockeel, H.: Using internal validity measures to compare clustering algorithms. In: Benelearn 2015 Poster presentations (online), pp. 1–8 (2015)
16. de Avila, S.E.F., Lopes, A.P.B., et al.: Vsumm: a mechanism designed to produce static video summaries and a novel evaluation method. Pattern Recognit. Lett. 32(1), 56–68 (2011)
17. Furini, M., Geraci, F., Montangero, M., Pellegrini, M.: Stimo: still and moving video storyboard for the web scenario. Multimed. Tools Appl. 46(1), 47–69 (2010)
18. Mundur, P., Rao, Y., Yesha, Y.: Keyframe-based video summarization using Delaunay clustering. Int. J. Digit. Libr. 6(2), 219–232 (2006)
19. Video open project storyboard (2016). https://open-video.org/results.php?size=extralarge

Computationally Efficient ANN Model for Small-Scale Problems

Shikhar Sharma, Shiv Naresh Shivhare, Navjot Singh and Krishan Kumar

Abstract In this current age of digital photography, the digital information is expanding exponentially. The use of such information in fields like research, automation, etc. has experienced a rise over the last decade. Also, employing machines to automate any task has been performed since forever. This leads to extensive use of the machine in solving the task of understanding the digital information called computer vision. Machine learning has always played an eminent role in various computer vision challenges. But, with the emergence of deep learning, machines are now outperforming humans. This has led to exaggerate the use of such deep learning techniques like convolutional neural network (CNN) in almost every machine vision task. In this paper, a new technique is proposed that could be used in place of CNN for solving elementary computer vision problems. The work uses the ability of the spatial transformer networks (STN) to effectively extract the spatial information from an input. Such an information is invariant and could be used as input to more plain neural networks like artificial neural network (ANN) without performance being compromised.

Keywords Spatial transformer networks · Artificial neural networks · CNN
Deep learning

S. Sharma (✉) · S. N. Shivhare · N. Singh · K. Kumar
Department of Computer Science and Engineering, National Institute
of Technology, Uttarakhand, Srinagar (Garhwal), India
e-mail: shikhar01.cse14@nituk.ac.in

S. N. Shivhare
e-mail: shiv.naresh@nituk.ac.in

N. Singh
e-mail: navjot.singh.09@nituk.ac.in

K. Kumar
e-mail: kkberwal@nituk.ac.in

© Springer Nature Singapore Pte Ltd. 2019
M. Tanveer and R. B. Pachori (eds.), *Machine Intelligence and Signal Analysis*,
Advances in Intelligent Systems and Computing 748,
https://doi.org/10.1007/978-981-13-0923-6_37

423

1 Introduction

The human mind is one of the most powerful, complex biological machines that perform trillions of computation in a second. The ability of the machine to perform complex tasks like human mind might become true one day, but many state-of-the-art-models already explored architecture similar to that of the human brain. Artificial neural network is one of such model that tries to mimic the functionalities of the human brain to some extent. The flow of information through the network changes as the model learns based on the inputs and outputs fed to it. These find nonlinear patterns residing in the input data. Such complex relationships between input and output are the model using nonlinear statistical analysis. Learning patterns just by observing datasets make them a very good approximation tool which can be reasoned as one of the main assets of ANN. Moreover, in contrast to traditional state-of-the-art models like support vector machine (SVM), they utilize the exhaustive information regarding the dataset rather than just considering spatial data in proximity to support vectors [1].

The concept of ANN came into existence by the work of Lecun et al. [2] in 1988. Thereafter, the relevancy of these networks in every possible pattern recognition tasks made them the prominent piece of work. The applications of ANN varies from regression [3, 4], classification [5, 6], verification [7], and recognition [8]. Initially, ANN tend to have only 3 layers, *input layer*, *hidden layer*, and *output layer*. Recently, with the availability of dedicated computing tools like GPU, the number of hidden layers in ANN experienced a rapid escalation as shown in Fig. 1.

The major detriment of ANN is that they rely way too much on the pose, structure, shape, texture, color like properties of an object for recognition. Any variation incurred in any of the above-mentioned features may cause misclassification.

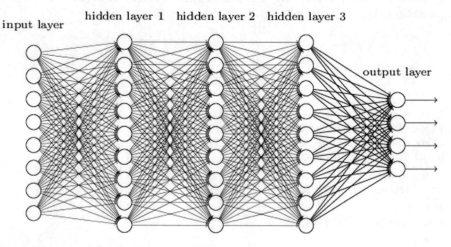

Fig. 1 Illustration of ANN with many #hidden layers

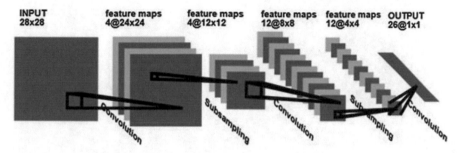

Fig. 2 Architecture of typical CNN [21]

Various attempts like *PCA*, *LDA*, and *data augmentation* have been proposed over last decade for making an ANN stable to such varying data. However, the results obtained are not up-and-coming as expected. This demerit of ANN is removed by more powerful and stable network known as convolutional neural networks. Convolutional neural networks, scalable, fast, efficient, and end-to-end deep learning framework that has pushed the boundaries of computer vision forward exceptionally. The list of its achievements varies from regression [9], classification [10], segmentation [11, 12], localization [13–17], and many more. Recognition despite deformation in shape and texture or varying object pose could be the reasons by which it manages to pull all of its tricks. A simple CNN uses convolution layers, ReLU layers followed by subsampling layers repeatedly as shown in Fig. 2, thus, forming a rich and more robust understanding of complex patterns residing inside input.

CNN is best in generalizing patterns as the increased number of layer helps to recognize all features between classification and input information. Moreover, greater depth of network helps to learn complex patterns. For example, to recognize objects in images, the first layer learns to understand patterns in edges, the second layer combines that pattern of edges to form motifs, the next layer learns to combine motifs to attain patterns in parts, and the final layer learns to recognize objects from the parts identified in the previous layer. However, these networks are slower in learning patterns in contrast to ANN. Also, the data required by CNN is way too much than data required by its predecessor counterpart ANN. Spatial transformer networks are learnable modules that perform spatial transforming of objects in input and could be used with any neural network configuration [18]. Moreover, this spatial transforming can also vary for each input sample, so an appropriate behavior is learned while training without any extra aid. So, these networks can be thought as similar to convolution layers in CNN. Instead of fixed reception on input as done by CNN, they perform more dynamic and adaptive reception using spatial transformation as shown in the Fig. 3.

The use of CNN in last few years has increased exponentially. This has led to overexploitation of CNN, simpler problems are solved using state-of-the-art-models while they could be solved using simple networks. The proposed work tries to explore the capabilities of STN with ANN in common computer vision problems.

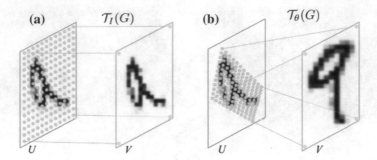

Fig. 3 Spatial transformation by STN [18]

This approach could replace the use of CNN in simple computer vision tasks. Further, an evaluation of both CNN and STN infused ANN is performed on different datasets and a detailed comparison between CNN and ANN is presented. The main contributions of proposed work are outlined below.

- The use of STN along with ANN to address common computer vision problems. Thus, proving that STN infused ANN can replace CNN in such cases.
- The training time required by ANN with STN is far less than time requirement of CNN. Moreover, the data samples needed for appropriate training of ANN is very less as compared to CNN.
- The evaluation of both ANN and CNN shows that ANN could perform comparably to CNN while using lesser training time and data.

The rest of the paper is structured as follows: Sect. 2 begins with some background on deep learning and then presents the proposed models in this paper, Sect. 3 presents the empirical trials, and Sect. 4 presents the conclusions and future work.

2 Proposed Work

In this section, we describe how ANN, CNN, and STN. First, Sect. 2.1 describes the CNN and its implementation and Sect. 2.2 discusses STN and its inclusion in ANN. The two methodologies, i.e., spatial transformer network (STN) and convolutional spatial transformer network (STCN) that are implemented in the work are shown in Fig. 4.

2.1 Deep Learning Using CNN

The concept of deep learning came into existence with the work of Hinton et al. [19, 20]. As the term suggests, it has a number of layers than traditional models.

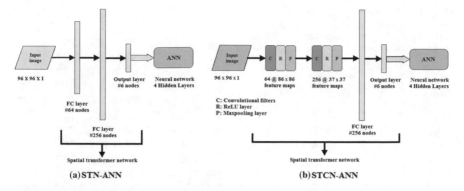

Fig. 4 Spatial transformation networks as implemented in work

They develop a hierarchy in feature recognition, i.e., first layer act as an abstraction of features which are explored deeply by following layers and classification task is performed by the last layer. For example, to recognize objects in images, the first layer learns understanding complex patterns in edges, the second layer combines these patterns to form motifs, the next layer learns to combine motifs to attain patterns in parts, and the final layer learns to recognize objects from the parts identified in the previous layer [21].

The inputs from various datasets are first range normalized (0, 1), then fed into the network for training and testing. Since each dataset has input having different dimensions, each sample in each dataset is resized to 96 × 96. The various architecture of CNN is implemented in proposed work. The configuration of these models is presented in Fig. 5. The convolution operation is widely used in the field of image processing. The convolution layers in CNN also work on the same principle. The convolution of an input x with kernel k is computed by Eq. (1), where x is an image in the input layer or a feature map in the subsequent layers. The convolution kernel, k is a square matrix having dimension specified by the user. The number of feature maps is a hyperparameter that is determined experimentally. For a 2D kernel, the convolution can be written as:

$$(x * k)_{ijm} = \sum_{p,q=0}^{s-1} (x_{i+p,j+q}) \times (k_{s-p,s-q})$$ (1)

In our work, several CNN configuration is used and compared with each other. The kernel size defines the receptive field of the hidden neurons in feature maps. It acts as the filter for searching specific pattern in the input image. The stride defines the movement of kernel across the input image. Lesser the stride more accurate the feature maps regarding the patterns. In all the CNN models presented, we took the stride of 1 in all the kernel dimensions. We used ReLU (Rectified Linear Unit) as activation function [22], which enhances the learning process of the network. for the input x the output of ReLU is given by

Layer (type)	Output Shape	Param #
dense_1 (Dense)	(None, 96, 9216)	893952
dense_2 (Dense)	(None, 96, 64)	589888
dropout_1 (Dropout)	(None, 96, 64)	0
dense_3 (Dense)	(None, 96, 256)	16640
dropout_2 (Dropout)	(None, 96, 256)	0
dense_4 (Dense)	(None, 96, 512)	131584
dropout_3 (Dropout)	(None, 96, 512)	0
dense_5 (Dense)	(None, 96, 1024)	525312
dropout_4 (Dropout)	(None, 96, 1024)	0
dense_6 (Dense)	(None, 96, 6)	6150

Total params: 2,163,526
Trainable params: 2,163,526
Non-trainable params: 0

(a) Artificial neural network

Layer (type)	Output Shape	Param #
conv2d_1 (Conv2D)	(None, 96, 96, 64)	7808
max_pooling2d_1 (MaxPooling2	(None, 48, 48, 64)	0
conv2d_2 (Conv2D)	(None, 48, 48, 128)	663680
max_pooling2d_2 (MaxPooling2	(None, 24, 24, 128)	0
conv2d_3 (Conv2D)	(None, 24, 24, 256)	819456
max_pooling2d_3 (MaxPooling2	(None, 12, 12, 256)	0
dropout_1 (Dropout)	(None, 12, 12, 256)	0
conv2d_4 (Conv2D)	(None, 12, 12, 512)	1180160
max_pooling2d_4 (MaxPooling2	(None, 6, 6, 512)	0
flatten_1 (Flatten)	(None, 18432)	0
dropout_2 (Dropout)	(None, 18432)	0
dense_1 (Dense)	(None, 512)	9437696
dropout_3 (Dropout)	(None, 512)	0
dense_2 (Dense)	(None, 50)	25650

Total params: 12,134,450
Trainable params: 12,134,450
Non-trainable params: 0

(b) CNN with 4 layers

Layer (type)	Output Shape	Param #
conv2d_1 (Conv2D)	(None, 96, 96, 64)	7808
max_pooling2d_1 (MaxPooling2	(None, 48, 48, 64)	0
conv2d_2 (Conv2D)	(None, 48, 48, 128)	663680
max_pooling2d_2 (MaxPooling2	(None, 24, 24, 128)	0
conv2d_3 (Conv2D)	(None, 24, 24, 256)	819456
max_pooling2d_3 (MaxPooling2	(None, 12, 12, 256)	0
dropout_1 (Dropout)	(None, 12, 12, 256)	0
conv2d_4 (Conv2D)	(None, 12, 12, 512)	1180160
max_pooling2d_4 (MaxPooling2	(None, 6, 6, 512)	0
conv2d_5 (Conv2D)	(None, 6, 6, 128)	589952
dropout_2 (Dropout)	(None, 6, 6, 128)	0
flatten_1 (Flatten)	(None, 4608)	0
dropout_3 (Dropout)	(None, 4608)	0
dense_1 (Dense)	(None, 512)	2359808
dropout_4 (Dropout)	(None, 512)	0
dense_2 (Dense)	(None, 100)	51300

Total params: 5,672,164
Trainable params: 5,672,164
Non-trainable params: 0

(c) CNN with 5 layers

Layer (type)	Output Shape	Param #
conv2d_1 (Conv2D)	(None, 96, 96, 20)	200
conv2d_2 (Conv2D)	(None, 96, 96, 45)	8145
conv2d_3 (Conv2D)	(None, 96, 96, 60)	24360
dropout_1 (Dropout)	(None, 96, 96, 60)	0
conv2d_4 (Conv2D)	(None, 96, 96, 75)	40575
max_pooling2d_1 (MaxPooling2	(None, 48, 48, 75)	0
conv2d_5 (Conv2D)	(None, 48, 48, 100)	67600
dropout_2 (Dropout)	(None, 48, 48, 100)	0
conv2d_6 (Conv2D)	(None, 48, 48, 125)	112625
dropout_3 (Dropout)	(None, 48, 48, 125)	0
max_pooling2d_2 (MaxPooling2	(None, 24, 24, 125)	0
conv2d_7 (Conv2D)	(None, 24, 24, 175)	197050
dropout_4 (Dropout)	(None, 24, 24, 175)	0
conv2d_8 (Conv2D)	(None, 24, 24, 256)	403456
max_pooling2d_3 (MaxPooling2	(None, 12, 12, 256)	0
conv2d_9 (Conv2D)	(None, 12, 12, 275)	633875
dropout_5 (Dropout)	(None, 12, 12, 275)	0
conv2d_10 (Conv2D)	(None, 12, 12, 400)	990400
max_pooling2d_4 (MaxPooling2	(None, 6, 6, 400)	0
flatten_1 (Flatten)	(None, 14400)	0
dropout_6 (Dropout)	(None, 14400)	0
dense_1 (Dense)	(None, 1024)	14746624
dropout_7 (Dropout)	(None, 1024)	0
dense_2 (Dense)	(None, 512)	524800
dropout_8 (Dropout)	(None, 512)	0
dense_3 (Dense)	(None, 100)	51300

Total params: 17,801,010
Trainable params: 17,801,010
Non-trainable params: 0

(d) CNN with 10 layers

Fig. 5 Various neural network architecture presented

$$f(x) = max(0, x) \tag{2}$$

A smooth approximation to ReLU is the analytic function also called soft plus function is given by

$$f(x) = ln(1 + e^x) \tag{3}$$

To avoid exploding gradient problem, we employed dropout layers having the ratio to be 0.5 and 0.25 [23]. Dropout layers neglect the input from some neurons in previous layers. This avoids the exploding as well as the vanishing of the gradient. Moreover, this also avoids the overfitting of the network while training, promising higher accuracy on test data. A pooling operation is applied to reduce the impact of translations and reduces the number of trainable parameters that would be needed. All the layers discussed above collectively act as single convolution layer. Moreover, we also changed kernel size in next layer for better training and testing. A fully connected layer could be understood as the feedforward neural network or simple ANN. The feature maps obtained after both convolution layers act as input to fully connected layers. The last layer or softmax layer transforms the output of various neurons into probability.

With the passage of time many robust and fast training algorithms have been presented by many authors. Kingma et al. [24] proposed a new training algorithm called Adam optimization basically in neural networks for speeding up the learning process. They used the concept of second-order moments and their correction in training. We used Adam optimization technique for backpropagating the error. It has many benefits over traditional Stochastic gradient descent method (SGD). It removes major drawback of SGD viz. slow training. The various parameters in Adam optimization are stepsize (α), exponential decay rates for moment estimation (β_1, β_2), objective function $f(\theta)$, and moment vectors (m_0, v_0). The gradient w.r.t $f(\theta)$ at any time instance is

$$g_t = \nabla_\theta f_t(\theta_{t-1}) \tag{4}$$

The moment estimates are updated using equations given below.

$$
\begin{aligned}
m_t &= \beta_1 \cdot m_{t-1} + (1 - \beta_1) \cdot g_t, \\
v_t &= \beta_2 \cdot v_{t-1} + (1 - \beta_2) \cdot g_t^2
\end{aligned}
\tag{5}
$$

These moments are then used to obtain corrected moments as shown by Eq.

$$
\begin{aligned}
m_t &= \frac{m_t}{1 - \beta_1^t}, \\
v_t &= \frac{v_t}{1 - \beta_2^t}
\end{aligned}
\tag{6}
$$

These moments are then used to update the parameters used in network,

$$\theta_t = \theta_{t-1} - \frac{\alpha \cdot m_t}{\sqrt{v_t} + \epsilon} \tag{7}$$

Where the ϵ is set to 10^{-8} by default. We used categorical cross-entropy as our objective function.

$$f(\theta) = -\frac{1}{n} \sum_{i=1}^{n} \sum_{j=1}^{m} y_{ij} log(p_{ij}) \tag{8}$$

Where n is number of sample and m is number of categories. The traditional mean squared error (MSE) emphasis more on the incorrect outputs, so it is slightly ineffective if used along with softmax layer. For the sake of comparison, the adam optimization discussed above is used to train all the neural networks implemented in the work.

2.2 Spatial Transforming Artificial Neural Network (ST-ANN)

This section describes how spatial transforming is achieved by STN. It deploys transformation depending upon the input map in the forward pass and results in the single output. If the number of channels in the input is more than one, each channel is subjected to the same transformation. According to Max et al. [18], each transformation undergoes 3 different operations as shown in Fig. 6.

Localization net: the input to be transformed is fed into it, after application of operations by various hidden layers it generates the parameters required for transformation. This net can be configured to work as any network like CNN or fully connected network. The last layer in that network is regression layer for generating transformation.

Grid generator: this layer performs the task of generating bounding box around the region of interest. For a multichannels input, such boxes are generated around each channel. The sampler takes the generated grid as input along with the original map to perform the spatial transformation. The location where the kernel is applied is given by coordinate in $T_0(G)$ which results in transformation. The above-mentioned STN is used with multilayer neural network having 4 hidden layers. Two different STN are implemented to contrast the effectiveness of convolutional filters in STN, one with simple feedforward neural network (STN) and other using the convolutional neural

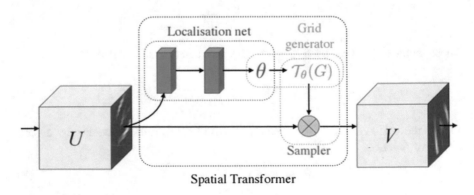

Fig. 6 Architecture of typical STN [18]

Table 1 Configuration of STN

Layer type	Parameters
Fully connected	#neurons: 6
Dropout	Ratio : 0.5
ReLU	
Fully connected	#neurons: 256
Dropout	Ratio : 0.25
ReLU	
Fully connected	#neurons: 64
Input	96 × 96 gray scale image

Table 2 Configuration of STCN

Layer type	Parameters
Fully connected	#neurons: 6
Dropout	Ratio : 0.5
ReLU	
Fully connected	#neurons: 256
Dropout	Ratio : 0.25
Maxpooling	kernel : 2 × 2, stride : 1
ReLU	
Convolution	#filters: 128, kernel : 7 × 7, stride : 1
Maxpooling	kernel : 2 × 2, stride : 1
ReLU	
Convolution	#filters: 64, kernel : 11 × 11, stride : 1
Input	96 × 96 grayscale image

network in ST layer (STCN). The architecture of different ST layers are shown in Tables 1 and 2. The STN-ANN model consists of 4 hidden layers in ANN architecture while 2 hidden layers in ST layer. The last layer in STN-ANN is softmax layer for classification purpose. The number of neurons in softmax layer is same as the number of classes in different datasets. While STCN-ANN model is same as STN-ANN but having convolutional filters in ST layers.

3 Experiment and Results

The architecture is implemented using the Python library Keras for deep learning based on the CUDA as well CuDNN on standard dual-core computer having NVIDIA GeForce GTX 660 GPU (2GB memory and 960 CUDA cores). To enable comparison among different networks, each network is trained and tested using the same amount of data using same training algorithm. As mentioned earlier, 3 different CNN configuration is implemented for comparison along with STN-ANN, STCN-ANN, and simple ANN. The different datasets used for experimentation are: *MNIST Handwritten digits dataset (MNIST),*[1] *Cifar-10 dataset,*[2] *ASL fingerspelling dataset (ASL),*[3] *Cifar-100 dataset,*[4] *Facial expression research group dataset (FERG-DB),*[5] and *Georgia Tech face database (GTF-DB).*[6]

The different datasets obtained are split into two parts: Training (70%) and Testing (30%) datasets. The models are evaluated using standard metrics. The above process of training and testing is repeated for 20 times (experimental value) and the values obtained are averaged and compared for each model. The results are evaluated by computing the recall and precision measures for dataset and comparing the results with different models. The experiments were run until the network converged. Table 3 shows the results for different models for different datasets based on precision, recall, and F-measure values given by Eqs. 9, 10, and 11, respectively.

$$P = \frac{TP}{TP + FP} \tag{9}$$

$$R = \frac{TP}{TP + FN} \tag{10}$$

$$F_\beta = \frac{(1 + \beta^2) \times P \times R}{\beta^2 \times P + R} \tag{11}$$

where, P denotes precision, R denotes recall, and $Fthe_\beta$ denotes F-measure and assumed that precision and recall both have equal priorities, $\beta = 1$. The time requirement of ST-ANN model is also very less in compared to other models. The use of Adam optimization in training the network instead of traditional Stochastic Gradient Descent (SGD) approach has reduced the training time drastically in case of each model. Also, the use of GPU for computation also aids in saving training time. So, the computation *time per epoch (seconds)* of STN-ANN model is less as compared to other models but not at the cost of accuracy as discussed in the last column of Table 3.

---

[1] http://yann.lecun.com/exdb/mnist/.

[2] https://www.cs.toronto.edu/~kriz/cifar.html.

[3] http://empslocal.ex.ac.uk/people/staff/np331/index.php?section=FingerSpellingDataset.

[4] https://www.cs.toronto.edu/~kriz/cifar.html.

[5] http://grail.cs.washington.edu/projects/deepexpr/ferg-db.html.

[6] http://www.anefian.com/research/.

Table 3 Comparison of precision, recall, and F-measure

Model	Dataset	#params (M)	Precision (%)	Recall (%)	F-measure (%)	Time/ep (s)
ANN	MNIST	2.16	67.1	67.8	67.4	3.11
	Cifar-10	2.16	62.1	62.3	62.2	3.12
	ASL	2.18	63.1	63.9	63.5	3.10
	Cifar-100	2.25	51.2	52.8	51.9	3.23
	FERG	2.16	64.9	66.1	65.5	3.18
	GTF	2.20	71.4	73.2	72.3	3.09
CNN-4	MNIST	12.11	97.5	98.7	98.1	20.1
	Cifar-10	12.11	78.1	79.1	78.6	20.3
	ASL	12.12	99.7	99.9	99.8	20.4
	Cifar-100	12.16	67.1	69.7	67.9	20.7
	FERG	12.13	87.1	88.9	88.0	20.5
	GTF	12.09	84.9	84.7	84.8	20.0
CNN-5	MNIST	5.60	94.5	95.7	94.1	5.11
	Cifar-10	5.61	82.1	83.1	82.6	5.11
	ASL	5.62	99.7	99.9	99.8	5.12
	Cifar-100	5.66	77.1	78.7	77.9	5.15
	FERG	5.63	92.1	92.9	92.5	5.13
	GTF	5.59	93.9	94.7	93.8	5.10
CNN-10	MNIST	17.6	96.5	97.7	97.1	28.11
	Cifar-10	17.6	86.1	87.1	86.6	28.11
	ASL	17.7	99.7	99.9	99.8	28.12
	Cifar-100	17.8	79.1	80.7	79.9	28.15
	FERG	17.8	95.1	95.9	95.5	28.13
	GTF	17.5	94.9	96.7	95.8	28.10
STN-ANN	MNIST	3.66	96.1	97.1	96.5	3.11
	Cifar-10	3.66	85.1	85.7	85.4	3.11
	ASL	3.68	96.7	97.9	97.3	3.12
	Cifar-100	3.75	76.1	78.5	77.3	3.15
	FERG	3.71	94.1	94.9	94.5	3.13
	GTF	3.66	93.9	94.7	94.2	3.10
STCN-ANN	MNIST	21.66	98.5	99.7	99.1	38.11
	Cifar-10	21.66	92.1	93.1	92.6	38.11
	ASL	21.68	99.7	99.9	99.8	38.12
	Cifar-100	21.75	81.1	82.7	81.9	38.15
	FERG	21.71	97.1	97.9	97.5	38.13
	GTF	21.65	96.9	98.7	97.8	38.10

4 Conclusion and Future Work

The various neural networks mentioned in the work were implemented and compared with each other on the basis of precision, recall, F-measure, and time. The major drawback of a simple neural network that they rely exceptionally upon the orientation of training data. Preexisting techniques for resolving this issue requires a lot of time and data. The proposed methodology tries to resolve the mentioned challenge with minimum data and time requirements. So, the performance behavior of neural networks was observed to change in presence of additional spatial transformer layers with a small increase in computational complexity. Moreover, results proved that such smaller and time efficient models can be used in simpler computer vision problems without comprising the accuracy.

References

1. Vapnik, V., et al.: Support vector machine. Mach. Learn. **20**(3), 273–297 (1995)
2. Lecun, Y., Galland, C.C., Hinton, G.E.: GEMINI: Gradient estimation through matrix inversion after noise injection. In: NIPS, pp. 141–148 (1988)
3. Palanisamy, P., et al.: Prediction of tool wear using regression and ANN models in end-milling operation. Int. J. Adv. Manuf. Technol. **37**(1), 29–41 (2008)
4. Shirsath, et al.: A comparative study of daily pan evaporation estimation using ANN, regression and climate based models. Water Res. Manag. **24**(8), 1571–1581 (2010)
5. Khan, Javed, et al.: Classification and diagnostic prediction of cancers using gene expression profiling and artificial neural networks. Nat. Med. **7**(6), 673 (2001)
6. Louis, David, N., et al.: The 2007 WHO classification of tumours of the central nervous system. Acta Neuropathol. **114**(2), 97–109 (2007)
7. Al-Shoshan, et al.: Handwritten signature verification using image invariants and dynamic features. In: International Conference on Computer Graphics, Imaging and Visualisation. IEEE (2006)
8. Parra, et al.: Automated brain data segmentation and pattern recognition using ANN. In: Computational Intelligence, Robotics and Autonomous Systems (2003)
9. Szegedy, C., et al.: Going Deeper with Convolutions. CVPR (2015)
10. Schroff, F., Kalenichenko, D., Philbin, J., et al.: Facenet: A unified embedding for face recognition and clustering (2015). arXiv:1503.03832
11. Long, J., et al.: Fully convolutional networks for semantic segmentation. In: CVPR (2015)
12. Kumar, K., et al.: F-DES: fast and deep event summarization. IEEE TMM. https://doi.org/10.1109/TMM.2017.2741423
13. Jaderberg, M., et al.: Synthetic data and artificial neural networks for natural scene text recognition. NIPS DLW (2014)
14. Gkioxari, G., et al.: Contextual action recognition with r-cnn (2015). arXiv:1505.01197
15. Simonyan, K., et al.: Very deep convolutional networks for large-scale image recognition. ICLR (2015)
16. Karen, et al.: Two-stream convolutional networks for action recognition in videos. In: NIPS, pp. 568–576 (2014)
17. Tompson, J.J., et al.: Joint training of a convolutional network and a graphical model for human pose estimation. In: NIPS, pp. 1799–1807 (2014)
18. Jaderberg, M., et al.: Spatial transformer networks. Adv. Neural Inf. Process. Syst. (2015)
19. Hinton, G., Osindero, S., Teh, Y.: A fast learning algorithm for deep belief nets. Neural Comput. **18**, 1527–1554 (2005)

20. Hinton, G.E., Salakhutdinov, R.R.: Reducing the dimensionality of data with neural networks. Science **313**, 504–507 (2006)
21. Lecun, Y., Bengio, Y., Lhinton, G.: Deep learning. Nature **521**, 436–444 (2015)
22. Zeiler, M.D., Ranzato, M., Monga, R., Mao, M., Yang, K., Le, Q.V., Nguyen, P., Senior, A., Vanhoucke, V., Dean, J., Hinton, G.E.: On Rectified Linear Units for Speech Processing. Proc. ICASSP (2013)
23. Srivastava, N., Hinton, G.E., Krizhevsky, A., Sutskever, I., Salakhutdinov, R.R.: Dropout: a simple way to prevent neural networks from overfitting. J. Mach. Learn. Res. **15**, 1929–1958 (2014)
24. Kingma, D., Jimmy Ba, : Adam: A method for stochastic optimization (2014). arXiv:1412.6980

Investigating the Influence of Prior Expectation in Face Pareidolia using Spatial Pattern

Kasturi Barik, Rhiannon Jones, Joydeep Bhattacharya and Goutam Saha

Abstract The perception of an external stimulus is not just stimulus-dependent but is also influenced by the ongoing brain activity prior to the presentation of stimulus. In this work, we directly tested whether spontaneous electroencephalogram (EEG) signal in prestimulus period could predict perceptual outcome in face pareidolia (visualizing face in noise images) on a trial-by-trial basis using machine learning framework. Participants were presented with only noise images but with the prior information that some faces would be hidden in these images while their electrical brain activities were recorded; participants reported their perceptual decision, face or no-face, on each trial. Using features based on the Regularized Common Spatial Patterns (RCSP) in a machine learning classifier, we demonstrated that prestimulus brain activities could discriminate face and no-face perception with an accuracy of 73.15%. The channels corresponding to the maximal coefficients of spatial pattern vectors may be the channels most correlated with the task-specific sources, i.e., frontal and parieto-occipital regions activate for 'face' and 'no-face' imagery class, respectively. These findings suggest a mechanism of how prior expectations in the prestimulus period may affect post-stimulus decision-making.

K. Barik (✉) · G. Saha
Department of Electronics & Electrical Communication Engineering,
Indian Institute of Technology Kharagpur, Kharagpur, India
e-mail: kasturibarik@iitkgp.ac.in

G. Saha
e-mail: gsaha@ece.iitkgp.ernet.in

R. Jones
Department of Psychology, University of Winchester, Winchester, UK
e-mail: rhiannon.jones@winchester.ac.uk

J. Bhattacharya
Department of Psychology, Goldsmiths University of London, London, UK
e-mail: j.bhattacharya@gold.ac.uk

© Springer Nature Singapore Pte Ltd. 2019   437
M. Tanveer and R. B. Pachori (eds.), *Machine Intelligence and Signal Analysis*,
Advances in Intelligent Systems and Computing 748,
https://doi.org/10.1007/978-981-13-0923-6_38

Keywords EEG · Prior Expectation · Face Pareidolia · Single-trial Classification · Spatial Pattern · Artificial Neural Network

1 Introduction

There is growing evidence that the ongoing brain activity is not meaningless, rather carries a functional significance that largely determines how an incoming stimulus will be processed [24]. In this framework, perception is understood as a process of inference, whereby sensory inputs are combined with prior knowledge [26], i.e., the integration of bottom-up sensory inputs and top-down prior expectations. To date, there has been no satisfactory functional explanation of the predictive role of prestimulus brain states. Although the role of prestimulus neural activity is unclear, it is found that perception is not entirely determined by the visual inputs, but it is influenced by individual's expectations [13]. One extreme example of how expectation primes our perception is pareidolia, which refers to the perception of concrete images, such as letters, animals, or faces—in random or undefined stimuli. Specifically, "Face Pareidolia" is a psychological tendency to see faces in random stimuli. Among all forms of pareidolia, face pareidolia is more explored: individuals have reported seeing a face in the clouds [8], Jesus in toast [17], or the Virgin Mary in a tortilla [7]. We employed face pareidolia as an extreme example of the extent to which anticipation can influence our perception. Face pareidolia indicates how the visual system is strongly predisposed to perceive faces, due to the societal importance of faces and our highly developed ability to process them.

In this study, participants were presented with only noise images but with the prior information that some faces would be hidden in these images while their electrical brain activities were recorded; participants reported their perceptual decision, face or no-face, on each trial. Since, the participants were instructed that a face was present in some of the trials, therefore, they expected to visualize it in the upcoming white noise images. Since this expectation is in the prestimulus period, it is referred to as prior expectation. This study offered a novel understanding of the usefulness of machine learning techniques in decoding mental states from prior brain states where each trial was categorized as one of the two classes, face class or no-face class, depending on participants response on trial-by-trial basis.

Multichannel EEG signals have now been popularly used in a wide variety of applications, such as neurological disorder detection or prediction, mental task classification, emotion recognition, cognitive psychology prediction. In this study, we use multichannel EEG in a machine learning framework to predict the perceptual outcomes from the participants' prestimulus activity. For feature extraction, we consider the Common Spatial Pattern (CSP) method [14], which has the ability to identify discriminating brain activity patterns between the two classes. The spatial filters employed in CSP are extremely useful in improving the signal-to-noise ratio as well as obtaining the distinguishing spatial patterns. CSP discriminates each EEG trial based on the covariance of signals at the electrode sites in multichannel EEG. These patterns can also be depicted unambiguously in terms of source geometry and thus can have functional significance. The CSP aims at learning spatial filters that maximize the variance of EEG signals from one class while minimizing their variance from the other class. However, CSP algorithm is built in sample-based covariance matrix estimation. Therefore, its performance is limited when the number of available training samples is small that can lead to overfitting. Thus, the Regularized Common Spatial Patterns (RCSP) is particularly effective in the small-sample setting [20]. We compared the classification performance of CSP and RCSP at individual participant level. Additionally, we explored whether any specific brain oscillation plays a crucial role in predicting the perceptual decision.

2 Materials and Methods

2.1 Participants and Stimuli

Seven healthy human adults (six females, age 23.43 ± 4.20 years) gave written informed consent and participated in this study. The experimental protocol was approved by the Local Ethics Committee at Goldsmiths. Visual noise stimuli were created using Adobe Photoshop V.9®. A total of 402 images were used, which all differed slightly, but were made to the same specifications. These were rectangular images on a black background, with monochromatic noise and a 100% gaussian distribution, and had a gaussian blur with 1 pixel radius.

2.2 Procedure

The experiment was composed of six blocks, each separated by 2 min breaks. Each block contained 67 trials. In each trial, a central fixation cross was presented for 1000 ms, followed by the visual noise stimulus presented centrally, for 350 ms. A screen then appeared asking participants whether they had seen a face, to which participants responded with an appropriate button press to indicate their response. Stimulus presentation and responses were controlled by the E-prime® (Psychology Software Tools, Inc., USA). Before beginning the task, participants were informed that faces had been hidden in some of the images; however, only noise images were used throughout.

2.3 Data Acquisition and Preprocessing

EEG signals were acquired using 64 active electrodes placed according to the international 10–10 system of electrode placement. The vertical and horizontal eye movements were recorded by placing additional electrodes above and below right eye and at the outer canthus of each eye, respectively. The sampling rate was 512 Hz.

The EEG data was algebraically re-referenced to the average of two earlobes. We applied notch filter at 50 Hz to reduce any powerline interferences. Blink-related artifacts were corrected using Independent Component Analysis (ICA). Further, any epochs-containing large artifacts were rejected based on the visual inspection. In this study, as we focused on investigating the predictive power of the prestimulus brain responses, we epoched our data from 1000 ms before the presentation of an image to stimulus onset. The experimental paradigm and the epoch formation are shown in Fig. 1.

Each trial was categorized as one of the two classes, *face* class or *no-face* class, depending on the participant's response. The number of trials in each class for an individual participant is listed in Table 1. The EEG data were preprocessed and analyzed by Matlab-based toolboxes, EEGLAB [6] and FieldTrip [22], and by custom-made Matlab scripts.

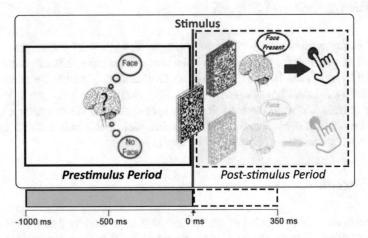

Fig. 1 Experimental paradigm: Here an example of an epoch (−1000 to +350 ms) is presented. In this study, we focused on the 1000 ms time period (shown in gray) before the stimulus onset

Table 1 Number of trials of each subject

Subject	No. of trials present in face class	No. of trials present in No-face class
1	67	193
2	68	226
3	116	212
4	104	187
5	90	116
6	116	216
7	159	170

2.4 Feature Extraction

In brain signal studies, the time–frequency analysis is widely used, where a large number of features are generated. To handle a large number of features, probability score feature selection is generally used. However, there are a few drawbacks of the probability score feature selection with respect to spatial pattern analysis [14]. Hence, in this study, we have chosen spatial filter feature extraction procedure. Spatial filters are created by defining weights for each electrode such that the weighted sum of activity at all electrodes helps to highlight aspects of the signal that are present but difficult to isolate in the spatially unfiltered data. In conventional feature selection methods, the statistical scores are computed at different electrode sites assuming the EEG processes at these sites are independent. As the spatial covariance of the EEG is not taken into account and because of the multitude of electrode sites involved, the calculated statistical scores may indicate differences that are due to chance alone. These spurious results, when interpolated, can appear to be well defined on a significance-probability map and thus lead to misleading conclusions about the differences between the populations. Spatial pattern presents a method of discriminating EEGs which is based on the covariance between the potential variations at the electrode sites in a multichannel EEG.

2.4.1 CSP Algorithm

The common spatial pattern (CSP) algorithm is frequently employed to extract the most discriminating information from EEG signals. CSP method was first suggested for classification of multichannel EEG during imagined hand movements by [23]. The CSP aims at learning spatial filters that maximize the variance of EEG signals from one class while minimizing their variance from the other class. The main idea is to use a linear transform to project the multichannel EEG data into low-dimensional spatial subspace. This transformation can maximize the variance of two-class signal matrices. CSP is a spatial filtering method that seeks projections with the most differing power/variance ratios in the feature space. The projections are calculated by a simultaneous diagonalization of the covariance matrices of two classes. Usually, only the first few most discriminatory filters are needed for classification. Figure 2 is an example of CSP filtering in 2D. Spatial filter maps the samples in Fig. 2a to those in Fig. 2b; the strong correlation between the original two axes is removed and both distributions are simultaneously decorrelated.

2.4.2 RCSP Algorithm

In spite of its popularity and efficacy, CSP is also known to be highly sensitive to noise and to severely overfit with small training set. A small number of training samples tend to result in a biased estimation of eigenvalues [20]. In this case, the

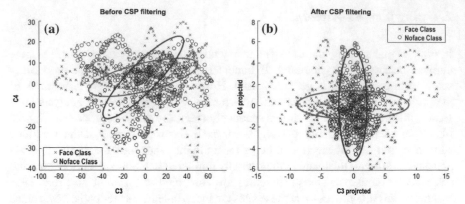

Fig. 2 Example of CSP filtering in 2D: Signal activation is scatter plotted for channel C3 and C4 where different source activities/samples for face epoch vs. no-face epoch are shown. Two sets of samples marked by blue crosses (Face) and red circles (No-face) are scatter plotted from two Gaussian distributions. In **a**, the distribution of samples before filtering is shown. Two ellipses show the estimated covariances. In **b**, the distribution of samples after the filtering is shown. Note that both the classes are uncorrelated at the same time; the horizontal (vertical) axis gives the largest variance in the blue (red) class and the smallest variance in the red (blue) class, respectively

estimated parameters can be highly unstable, giving rise to high variance. In order to address these drawbacks, prior information are added into the CSP learning process, under the form of regularization terms [12, 20]. Regularized CSP (RCSP) algorithm is used to regularize the covariance matrix estimation in CSP extraction and uses two regularization parameters. The first regularization parameter (γ) controls the shrinkage of the sample-based covariance matrix estimation toward a scaled identity matrix to account for the bias due to the limited number of samples. The second regularization parameter (β) controls the shrinkage of a subject-specific covariance matrix toward a more generic covariance matrix to lower the estimation variance. This generic matrix represents a given prior on how the covariance matrix for the mental state considered should be. RCSP has been shown to outperform classical CSP in terms of classification performance [19, 20].

Based on the regularization technique introduced in [19], the regularized average spatial covariance matrix for each class is calculated as

$$\widehat{\overline{R}}_{Class}(\gamma, \beta) = (1 - \gamma)\,\widehat{\overline{R}}_{Class}(\beta) + \frac{\gamma}{N_{Ch}}\mathbf{tr}\left[\widehat{\overline{R}}_{Class}(\beta)\right].\mathbf{I} \tag{1}$$

where γ ($0 \leq \gamma \leq 1$) and β ($0 \leq \beta \leq 1$) are two regularization parameters, $\mathbf{tr}$ represents *trace* of a matrix and $\mathbf{I}$ is an identity matrix of size $N_{Ch} \times N_{Ch}$, and $\widehat{\overline{R}}_{Class}(\beta)$ is defined as following:

$$\widehat{\overline{R}}_{Class}(\beta) = \frac{(1 - \beta)\,.R_{Class} + \beta.\widehat{R}_{Class}}{(1 - \beta)\,.K + \beta.\widehat{K}} \tag{2}$$

R_{Class} and $\widehat{R}_{Class}$ are the sum of the sample covariance matrices for K training trials and a set of $\widehat{K}$ generic training trials, *class* can be "face" (R_F) or "no-fac" (R_{NF}). The $\widehat{R}_{Class}$ is introduced to reduce the variance in the covariance matrix estimation and it tends to produce more stable results. To generate $\widehat{R}_{Class}$, we adopt the idea of generic learning introduced for one-training-sample face recognition [28]. For a subject whose EEG signals are to be classified, the training process employs the corresponding EEG trials collected for other subjects in the regularization term $\widehat{R}_{Class}$. Here, we followed a heuristic procedure [4] to automatically select the parameters that are required for regularized spatial pattern analysis.

The procedures in the classical CSP method are followed to get the RCSP algorithm. The composite spatial covariance of RCSP can be factorized as

$$\widehat{R}\left(\gamma, \beta\right) = \widehat{R}_F\left(\gamma, \beta\right) + \widehat{R}_{NF}\left(\gamma, \beta\right) = \widehat{U}_0 \widehat{\Sigma} \widehat{U}_0^T \tag{3}$$

where $\widehat{U}_0$ is the eigenvector matrix and $\widehat{\Sigma}$ is the diagonal matrix of corresponding descending ordered eigenvalues. The whitening transformation matrix

$$\widehat{P} = \widehat{\Sigma}^{-\frac{1}{2}} \widehat{U}_0^T \tag{4}$$

transform the average covariance matrices as S_F and S_{NF}, which share common eigenvectors and the sum of the corresponding eigenvalues for the two matrices will always be one,

$$S_F = \widehat{U} \Lambda_F \widehat{U}^T \quad S_{NF} = \widehat{U} \Lambda_{NF} \widehat{U}^T \quad \Lambda_F + \Lambda_{NF} = \mathbf{I} \tag{5}$$

The eigenvectors with the largest eigenvalues for S_F have the smallest eigenvalues for S_{NF} and vice versa; and the full projection matrix is formed as

$$\widehat{W}_0 = \widehat{U}^T \widehat{P} \tag{6}$$

where $\widehat{W}_0 \in \mathbb{R}^{D \times N_{Ch}}$. For feature extraction, a trial $X_{(Class, k=1)} \in \mathbb{R}^{N_{Ch} \times N_T}$ is first projected as

$$\widehat{Z} = \widehat{W}_0 X \tag{7}$$

Then, the feature vector $\mathfrak{X}_0$ is formed from the variance of the rows of $\widehat{Z}$ as

$$\mathbf{x}_{0_d} = log\left(\frac{var\left(\widehat{z}_d\right)}{\sum_{d=1}^{D} var\left(\widehat{z}_d\right)}\right) \tag{8}$$

where $\mathbf{x}_{0_d}$ is the dth component of $\mathfrak{X}_0$, $\widehat{z}_d$ is the dth row of $\widehat{Z}$ and $var\left(\widehat{z}_d\right)$ is the variance of the vector $\widehat{z}_d$. For the most discriminative patterns, only the first and last m columns of $\mathfrak{X}_0$ are kept to form feature vector $\mathfrak{X}$. As we follow supervised learning, the class labels are already given according to the participants perceptual decision, face or no-face, on each trial. In our experimental framework only feature extraction (using CSP or RCSP) and classification process is included. Because here the feature dimension is low, feature selection is not required. Figure 3a explains the spatial pattern feature extraction processes along with dimensions using block diagram.

2.5 Single-Trial Classification

As stated earlier, we had two classes of trials depending on the participant's responses: *Face* and *No-face*. Our classifier, based on the prestimulus EEG data, aimed to categorize each trial to one of these two classes. We considered personalized average model (PAM) where trials of individual participants were handled independently for studying participant-dependent characteristics [2]. The number of trials in the *No-face* class was much higher than that in the *Face* class (Table 1). To overcome the class imbalance, we used random downsampling approach [11, 15, 18]. In this method, the majority class was randomly downsampled to equate the number of minority and majority class samples, ensuring the balance between two classes. Here, 66 trials were used from each class. Since this method used only a subset of majority class samples, the data was randomized 25 times to minimize selection bias. Figure 3b shows the block diagram of the detailed classification process. In this study, we performed subject-dependent classification where models for each subject were trained separately.

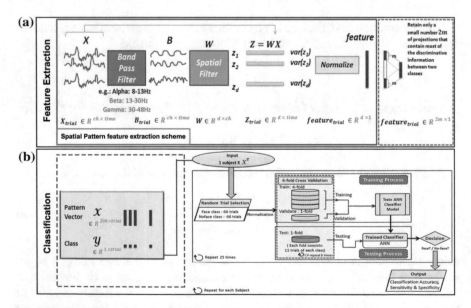

Fig. 3 Block diagram of feature extraction **a** and classification process **b** for personalized average model

We used Artificial Neural Network (ANN) [3, 9] as a classifier with six-fold cross-validation (CV). The three-layered feedforward back-propagation ANN consisted of an input layer, a hidden layer of 10 neurons and an output layer with two neurons representing the two classes. Prior to classification, the feature vectors were normalized between 0 and 1. To prevent the overfitting of the ANN classifier, early stopping of training using an inner validation set was done. The six-fold CV was performed with different randomly selected dataset of a participant to address data imbalance. To increase reliability, this procedure was performed 25 times, and the final classification accuracy was averaged across these 25 runs. We evaluated average classification accuracy, standard deviation, sensitivity, and specificity of the classifier. Sensitivity and specificity are statistical measures to evaluate the class-wise performance of a classifier. Here, the sensitivity or the true positive rate refers to the accuracy of classifying face trials to *face* class, i.e., the percentage of face trials that were correctly identified as *face* class, and specificity or the true negative rate refers to the proportion of no-face trials that were correctly identified as the *no-face* class.

3   Results

3.1   Comparison of CSP and RCSP Analysis

The classical measure for a spatial pattern algorithm is the most discriminative subset of filters, which is based on the eigenvalues [4]. In order to compare the classification performance of CSP and RCSP at the individual participant level, first, we have to find the most discriminative filters of each spatial pattern algorithm. Selection procedure of the most discriminative subset of the spatial filters is described in [4]. By following [4], we found $m = 6$ is the most discriminant filter for CSP method, where as $m = 3$ is the most discriminant filter for RCSP method in this experiment. In this analysis, the spatial pattern features are extracted from overall frequency band range, i.e., 0.1–48 Hz, and the whole prestimulus 1 *sec* time-window is considered. The comparison of the performances of the spatial filters is shown in Table 2. It represents the results of the performance (in %) of each spatial filters for their corresponding most discriminant filters for the frequency range of 0.1 to 48 Hz. It is found that RCSP yields better performance than CSP. Classification accuracy, sensitivity, and specificity of CSP are around chance level. Regularized spatial pattern features yield $73.15 \pm 1.78\%$ classification accuracy. An average improvement of 23.06% is achieved through RCSP method than CSP algorithm. As a sample-based covariance matrix estimation is the fundamental root of the classical CSP, in our study, it can lead to overfitting for a small number of training samples. RCSP overcomes this problem and hence performs better.

Table 2 Comparison of the performance (in %) of CSP and RCSP filters

| Results of CSP (m = 6) [All frequency: 0.1–48 Hz] | | | | Results of RCSP (m = 3) [All frequency: 0.1–48 Hz] | | | |
Subject	Accuracy ± Std.	Sensitivity	Specificity	Subject	Accuracy ± Std.	Sensitivity	Specificity
1	52.84 ± 3.67	52.27	53.40	1	71.40 ± 6.81	72.80	70.00
2	48.75 ± 3.28	41.74	55.75	2	76.25 ± 6.52	76.74	75.76
3	49.92 ± 3.63	48.63	51.21	3	71.70 ± 7.76	71.44	71.97
4	50.41 ± 4.70	51.51	49.31	4	74.09 ± 7.32	73.86	74.32
5	48.82 ± 4.97	47.80	49.84	5	73.14 ± 6.66	75.37	70.91
6	50.11 ± 5.44	45.60	54.62	6	73.98 ± 8.52	73.71	74.24
7	49.84 ± 4.20	54.69	45.00	7	71.52 ± 6.84	70.68	72.35
PAM	50.09 ± 1.36	48.89	51.30	PAM	73.15 ± 1.78	73.51	72.79

3.2 Frequency Band-Wise Spatial Pattern Analysis

As RCSP performs better than CSP, we further investigate frequency band-specific regularized spatial pattern in classical EEG frequency bands: delta (0.1–4 Hz), theta (4–8 Hz), alpha (8–13 Hz), beta (13–30 Hz), and gamma (30–48 Hz). For each frequency band, the most discriminative subset of filters are selected using the subset selection algorithm [4].

Here, the number of filters (for delta m = 6, i.e., for face class 6th filter and for no-face class 59th filter) represents the number of pairs of the RCSP filters. The frequency band-wise classification performance of RCSP filters is shown in Table 3. Here, we have found that for PAM model, alpha band yields best performance of 73.78 ± 1.98 % among other bands. To interpret the spatial pattern we investigated the topographical maps of PAM model for both face and no-face classes. The electrode weights ($\widehat{W}_0$) corresponding to most discriminant filters obtained with RCSP algorithm on each band are illustrated on Fig. 4a. The channels corresponding to the maximal coefficients of spatial pattern vectors are termed as optimal channels. The thick dot '•' of topoplots indicates the selected optimal channel. In delta band, O1 electrode has the maximum variance for *face* class relative to *no-face* class and the actual direction is aligned with the electrode direction, and similarly for no-face class, CP4 captures the maximum variance of delta band data averaged over all trials relative to the face class. These two channels are most correlated with the task-specific sources in delta band. Similarly, for each EEG frequency band, the channels that are most correlated with the task-specific sources are highlighted in the Fig. 4a. The barplot of Fig. 4b represents frequency band-wise classification accuracy that reflects the results of Table 3. It clearly shows that alpha frequency band plays a strong role in predicting the effect of prior expectation. In Fig. 4c, the most discriminant filters over all bands, for all the subjects are highlighted with thick black dot '•' in the scalp. It is clearly seen that FC5 and PO8 are most discriminant electrodes for "face" and

Table 3 Average classification performances (in %) for each frequency band with corresponding number of most discriminant filters (m)

Accuracy ± standard deviation					
Band:	Delta	Theta	Alpha	Beta	Gamma
m:	6	4	5	6	5
Sub1	69.13 ± 5.38	68.75 ± 4.39	71.59 ± 6.85	72.68 ± 4.61	73.71 ± 5.48
Sub2	68.14 ± 5.48	68.14 ± 5.31	74.85 ± 6.11	72.05 ± 5.92	71.82 ± 7.17
Sub3	72.57 ± 5.54	70.76 ± 5.68	72.27 ± 6.18	68.64 ± 7.96	70.79 ± 5.22
Sub4	72.08 ± 4.37	69.88 ± 5.23	76.63 ± 7.35	75.07 ± 5.17	75.68 ± 5.82
Sub5	72.46 ± 4.39	74.81 ± 5.67	75.00 ± 5.52	72.76 ± 6.46	74.13 ± 6.46
Sub6	73.37 ± 6.72	74.51 ± 5.85	74.62 ± 6.01	77.80 ± 5.19	76.09 ± 7.32
Sub7	69.73 ± 4.99	72.16 ± 7.34	71.52 ± 5.76	71.52 ± 5.43	72.31 ± 7.22
PAM	**71.07 ± 2.03**	**71.28 ± 2.65**	**73.78 ± 1.98**	**72.93 ± 2.87**	**73.50 ± 1.98**

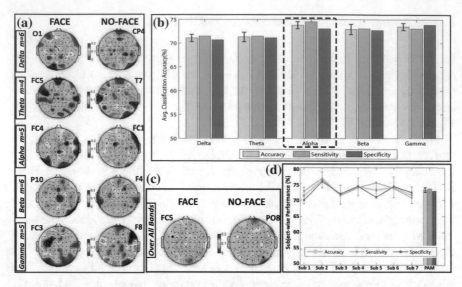

Fig. 4 **a** The topographical maps of *face* and *no-face* classes for the PAM model are shown. The color of the topoplots represents the electrode weights corresponding most discriminant filters obtained with RCSP algorithm on each band. **b** The barplot represents frequency band-wise classification accuracy, sensitivity, and specificity (in %) of RCSP filters for PAM model. Error bars indicate standard error of mean (SEM). **c** The topoplots highlight the most discriminant filters obtained over all bands for all the subjects. **d** The subject-wise and personalized average model performance for overall frequency band is shown

"no-face" classes, respectively. The subject-wise and personalized average model performance for overall frequency band (0.1–48 Hz) is shown in Fig. 4d. For PAM model, the classification accuracy was 73.15 ± 1.78 % for overall frequency band.

4 Discussion

Using features based on the RCSP in a machine learning classifier, we demonstrated that prestimulus brain oscillations could systematically predict post-stimulus perceptual decision in a face pareidolia task on a trial-by-trial basis.

This study inspected the causal impact of prior expectation before the stimulus onset on the post-stimulus perception in face pareidolia. Participants were presented with noise images, but prior information on the faces being hidden in these images led to the participants reporting seeing faces on many trials, in spite of the fact that no faces were hidden in the images. We demonstrated that it was possible to capture regularized spatial features of large scale ongoing brain activities prior to the presentation of stimuli that could reliably predict the participants' responses, face, or no-face, on trial-by-trial basis. Our classifier model produced a mean accuracy of around 73.15% that was substantially above the chance level of 50%. This finding is consistent with a growing body of literature establishing the existence of neural signals that predetermine perceptual decisions [5, 10, 25]. It is known that any decision made in the post-stimulus period is not entirely dependent on the stimulus alone; instead, it relies on several top-down processes including expectations, prior knowledge and goals, formed in the prestimulus period [26]. Although our findings purely represent functional brain activities, future research should explore the potential contribution of individual differences at the structural brain level.

It is found that, for prior expectation in face pareidolia, most active electric fields at the scalp are frontal and parieto-occipital regions. The channels corresponding to the maximal coefficients of spatial pattern vectors may be the channels most correlated with the task-specific sources, i.e, frontal and parieto-occipital regions activate for "face" and "no-face" imagery class, respectively.

Ongoing oscillations in the alpha frequency range play a strong role in predicting the effect of prior expectation. Different frequency bands are related to various cognitive and perceptual processes [27]. In our study, we found that the alpha band prestimulus oscillations were more critically involved with the prediction of future decision. Along with the spatial pattern features of alpha band, gamma band RCSP features also yield the best classification performance and capture the prior influence as well. This result was in line with other studies demonstrating the causal role of alpha oscillations in the prestimulus period in shaping post-stimulus task processing. For example, the strength of prestimulus alpha power was associated with detecting near-threshold stimuli [1, 16]. In a recent work, it has been found that low-frequency alpha oscillations can serve as a mechanism to carry and test prior expectation about stimuli [21]. Our results extend these studies by demonstrating that the regularized spatial pattern features in the alpha band could be captured at the single-trial level that possess significant discrimination ability to influence future decisions.

5 Conclusion

Using a pattern classification approach for spatial pattern of large scale EEG signals, we found that it is indeed feasible to predict the perceptual decision considerably higher than chance level based on the prestimulus activity alone. Prediction was performed using machine learning framework on single-trial basis. Further, the perceptual decision information was specifically coded in the prestimulus alpha oscillations and in the regularized spatial pattern features.

References

1. Babiloni, C., Vecchio, F., Bultrini, A., Romani, G.L., Rossini, P.M.: Pre- and poststimulus alpha rhythms are related to conscious visual perception: a high-resolution EEG study. Cereb. Cortex 16(12), 1690–1700 (2006)

2. Bhushan, V., Saha, G., Lindsen, J., Shimojo, S., Bhattacharya, J.: How we choose one over another: predicting trial-by-trial preference decision. PloS One **7**(8), e43351 (2012)
3. Bishop, C.M.: Neural Networks for Pattern Recognition. Oxford University Press, Oxford (1995)
4. Blankertz, B., Tomioka, R., Lemm, S., Kawanabe, M., Muller, K.R.: Optimizing spatial filters for robust EEG single-trial analysis. IEEE Signal Process. Mag. **25**(1), 41–56 (2008)
5. Bode, S., Sewell, D.K., Lilburn, S., Forte, J.D., Smith, P.L., Stahl, J.: Predicting perceptual decision biases from early brain activity. J. Neurosci. **32**(36), 12488–12498 (2012)
6. Delorme, A., Makeig, S.: EEGLAB: an open source toolbox for analysis of single-trial EEG dynamics including independent component analysis. J. Neurosci. Methods **134**(1), 9–21 (2004)
7. Fitts, A.: Review of the holy tortilla and a pot of beans. J. Caribb. Lit. **7**(1), 197 (2011)
8. Guthrie, S.: Faces in the Clouds. Oxford University Press, Oxford (2015)
9. Haykin, S., Network, N.: A comprehensive foundation. Neural Netw. **2** (2004)
10. Hsieh, P.J., Colas, J., Kanwisher, N.: Pre-stimulus pattern of activity in the fusiform face area predicts face percepts during binocular rivalry. Neuropsychologia **50**(4), 522–529 (2012)
11. Japkowicz, N., Stephen, S.: The class imbalance problem: a systematic study. Intell. Data Anal. **6**(5), 429–449 (2002)
12. Kang, H., Nam, Y., Choi, S.: Composite common spatial pattern for subject-to-subject transfer. IEEE Signal Process. Lett. **16**(8), 683–686 (2009)
13. Kok, P., Brouwer, G.J., van Gerven, M.A., de Lange, F.P.: Prior expectations bias sensory representations in visual cortex. The J. Neurosci. **33**(41), 16275–16284 (2013)
14. Koles, Z.J., Lazar, M.S., Zhou, S.Z.: Spatial patterns underlying population differences in the background EEG. Brain Topogr. **2**(4), 275–284 (1990)
15. Kubat, M., Matwin, S., et al.: Addressing the curse of imbalanced training sets: one-sided selection. In: ICML. vol. 97, pp. 179–186. Nashville, USA (1997)
16. Linkenkaer Hansen, K., Nikulin, V.V., Palva, S., Ilmoniemi, R.J., Palva, J.M.: Prestimulus oscillations enhance psychophysical performance in humans. J. Neurosci. **24**(45), 10186–10190 (2004)
17. Liu, J., Li, J., Feng, L., Li, L., Tian, J., Lee, K.: Seeing Jesus in toast: neural and behavioral correlates of face pareidolia. Cortex **53**, 60–77 (2014)
18. Liu, Y., Chawla, N.V., Harper, M.P., Shriberg, E., Stolcke, A.: A study in machine learning from imbalanced data for sentence boundary detection in speech. Comput. Speech Lang. **20**(4), 468–494 (2006)
19. Lotte, F., Guan, C.: Regularizing common spatial patterns to improve BCI designs: unified theory and new algorithms. IEEE Trans. Biomed. Eng. **58**(2), 355–362 (2011)
20. Lu, H., Plataniotis, K.N., Venetsanopoulos, A.N.: Regularized common spatial patterns with generic learning for EEG signal classification. In: Engineering in Medicine and Biology Society, 2009. EMBC 2009. Annual International Conference of the IEEE. pp. 6599–6602. IEEE (2009)
21. Mayer, A., Schwiedrzik, C.M., Wibral, M., Singer, W., Melloni, L.: Expecting to see a letter: alpha oscillations as carriers of top-down sensory predictions. Cereb. Cortex **26**(7), 3146–3160 (2016)
22. Oostenveld, R., Fries, P., Maris, E., Schoffelen, J.M.: FieldTrip: open source software for advanced analysis of MEG, EEG, and invasive electrophysiological data. In: Computational Intelligence and Neuroscience 2011 (2010)
23. Ramoser, H., Muller-Gerking, J., Pfurtscheller, G.: Optimal spatial filtering of single trial EEG during imagined hand movement. IEEE Trans. Rehabil. Eng. **8**(4), 441–446 (2000)
24. Sadaghiani, S., Hesselmann, G., Friston, K.J., Kleinschmidt, A.: The relation of ongoing brain activity, evoked neural responses, and cognition. Front. Syst. Neurosci. **4**, 20 (2010)
25. Schölvinck, M.L., Friston, K.J., Rees, G.: The influence of spontaneous activity on stimulus processing in primary visual cortex. Neuroimage **59**(3), 2700–2708 (2012)

26. Summerfield, C., de Lange, F.P.: Expectation in perceptual decision making: neural and computational mechanisms. Nat. Rev. Neurosci. **15**(11), 745–756 (2014)
27. Von Stein, A., Sarnthein, J.: Different frequencies for different scales of cortical integration: from local gamma to long range alpha/theta synchronization. Int. J. Psychophysiol. **38**(3), 301–313 (2000)
28. Wang, J., Plataniotis, K.N., Lu, J., Venetsanopoulos, A.N.: On solving the face recognition problem with one training sample per subject. Pattern Recognit. **39**(9), 1746–1762 (2006)

Key-Lectures: Keyframes Extraction in Video Lectures

Krishan Kumar, Deepti D. Shrimankar and Navjot Singh

Abstract In this multimedia era, the education system is going to adopt the video technologies, i.e., video lectures, e-class room, virtual classroom, etc. In order to manage the content of the audiovisual lectures, we require a huge storage space and more time to access. Such content may not be accessed in real time. In this work, we propose a novel key frame extraction technique to summarize the video lectures so that a reader can get the critical information in real time. The qualitative, as well as quantitative measurement, is done for comparing the performances of our proposed model and state-of-the-art models. Experimental results on two benchmark datasets with various duration of videos indicate that our key-lecture technique outperforms the existing previous models with the best *F-measure* and *Recall*.

1 Introduction

Due to advancement in the video technologies, the video content is growing around the clock. Handling such volume of content become a challenge for the implementation of the real-time applications, such as video surveillance, educational purposes, video Lectures, sports highlights, etc. [1].

We have watched that YouTube and other multimedia sources are pushing the bounds of video consuming during last decay. With a growing the size of the video

K. Kumar (✉) · N. Singh
National Institute of Technology Uttarakhand, Srinagar (Garhwal), India
e-mail: kkberwal@nituk.ac.in

N. Singh
e-mail: navjot.singh.09@gmail.com

D. D. Shrimankar · K. Kumar
VNIT, Nagpur, India
e-mail: dshrimankar@cse.vnit.ac.in

© Springer Nature Singapore Pte Ltd. 2019
M. Tanveer and R. B. Pachori (eds.), *Machine Intelligence and Signal Analysis*,
Advances in Intelligent Systems and Computing 748,
https://doi.org/10.1007/978-981-13-0923-6_39

collection, a technology is required to efficiently and effectively access through the video with entertaining the important data and semantic of the video [2]. The user might not have always adequate time to access/watch/browse the entire video data or the integral video content might not be the interest or important for the user. In other research work, a novel summarization model for video known as VISCOM is proposed which generate the video synopsis based on the most representative forms and draw the video frames through color co-occurrence matrices [3]. In such scenarios, the user might just want to access the compact view of the video content summary, besides watching or browsing the entire video content [4–6].

In the educational system, the video lectures are one of the important e-resources for the readers. As video lectures, the content challenge for more of a readers time every day, one possible deficiency is a video lecture summarization system. The lecture slides are an example of a video Lectures summary. Still, not everyone has the time to edit/filter their videos for a succinct and concise version. The salient contributions of the study are as follows:

- A fast and effective technique using online clustering for generating a video summary.
- This is first video Lectures summarization technique which focuses on the visual data [7].
- It is aimed especially at low-quality media real-time applications, specifically for conference, classroom, and NTPEL Lecture video.

2 Problem Formulation

The purpose of video Lectures summarization is to opt a set of frames that attains the maximum the importance of video data under a given circumstance, let's say, browsing/accessing time. For example, let say a video Lecture with N number of frames requires t_N for browsing. The authors would like to summarize these N frames into a set of $S(<N)$ frames that can be browsed in $t_S(<t_N)$. In order to aim this target, there are numerous of the most interesting frames have to be elected without worrying about the overall semantic/narrative of the lecture. In our study, we notice that the important pages should carry the following interesting characteristics:

- Adequate content to be worth accessing/browsing.
- Unique video content.

The first characteristic will choose a page containing texts which are useful to cover the maximum content of a frame or comprises the figures, or/and tables to assisting in understanding of the video contents/Lectures. The second one lets up the duplicate pages, including animations. We utilized the image processing as well as text processing for analyzing these characteristics.

2.1 Video Lectures Summarization Through Clustering

Foremost, we extract total N frames with size $W \times H$ from a video V Lecture, where H and W represent the height and breadth of a frame, respectively. On an average, the video is used for the datasets comprises 30,000 RGB frames. In video Lectures, the variation in the visual content of the successive frames may be considerable, however, the computation in processing these frames needs processing in all the three planes of frames. In order to reduce the computation cost, all N *colored* frames of a video/Lecture are reshaped into grayscale frames. Then, each shape is resized into the one-dimensional vector.

We applied the interframe difference clustering approach to extract visual features from a video Lecture. The interframe difference approach estimates the difference/changes between two consecutive frames (as evidenced in the Fig. 1). In order to estimate the difference score $DS_{i,i-1}$ using Euclidean Distance between current frame f_i and the previous frame f_{i-1}, we used the binarized subtracted image as follows:

$$DS_{i,i-1} = \|f_i - f_{i-1}\|_2 = \left(\sum_{k=1}^{n} |e_{k,i} - e_{k,i-1}| \right)^{\frac{1}{2}} \tag{1}$$

Where $e_{k,i}$ is kth element of f_i. We observed that when a Professor/Lecturer is writing the content, then the content will be increased frame by frame and when entire Blackboard or Whiteboard is full then, the content should be erased from the board. Before erasing the content, the full content frame should be kept open in order to capture the full visual capacity of the speech. Using this proposed approach, such frames are extracted and saved as the keyframes of the Lectures and we call this technique as Key-Lectures as implemented in the Algorithm 1.

Fig. 1 Example interframe difference

Algorithm 1 *keyframes* extraction from video Lecture

1: **procedure** *Key-Lect*($\{f_1, f_2, , f_N\}$ INPUT FROM VIDEO **V**)
2:     Initially Cluster_set=∅, KF(Keyframes)=∅, $Cluster_j = ∅$, where j=1;
3:     **for** frame $f_{i=1\ to\ N} \in V$ **do**
4:         Compute the difference score $DS_{i,i-1}$ between two consecutive frames
            using Euclidean distance $DS_{i,i-1} = (\sum_{k=1}^{n} |e_{k,i} - e_{k,i-1}|)^{\frac{1}{2}}$, where $e_{k,i}$
            is kth element of f_i.
5:         **if** $DS_{i,i-1} == 0\ ||\ DS_{i,i-1} > 0$ **then**
6:             Add the current frame f_i to current cluster $Cluster_j$.
7:         **else**
8:             Select the previous frame f_{i-1} as keyframe and add to $KF = KF \cup f_{i-1}$.
9:             Cluster_set= Cluster_set $\cup$ $Cluster_j$. Update $j = j + 1$.
10:             Create new $Cluster_j$, add current frame f_i to newly cluster $Cluster_j$.
11:         **end if**
12:     **end for**
13:     **return** KF <- final *keyframes* from video Lecture **V** after removal of the
        keyframes in which speaker is present.
14: **end procedure**

3   Experiments and Results

We discuss both qualitative and quantitative discussion on the proposed system. We selected total *101* video from two course video datasets (*Cloud Computing*[1]*:CC (64 Lectures), Operating System*[2]*:OS (37 Lectures)*) from the NTPEL Project.

In order to evaluate our model, we define the human selected Ground truth summaries. Both datasets are employed for several experiments and assessments which are done on a desktop PC with *i7* processor and *8* GB of DDR4-RAM.

3.1   Qualitative Analysis

We prepare the human selected ground truth summaries to assess the state-of-the-art-models and the proposed model. In the ground truth, there are 16 slides select as the final output, and Para Free [4] and EVS [5] summarization model generate 13 and 14 summarized slides in their final output, respectively, while our proposed key-lecture summarization model extract the 15 slides. As Fig. 2 depicts that our key-lecture proposed technique is brought forth the maximum number of keyframes according to ground truth. Hence, our proposed model quantitatively works better than the state-of-the-art models.

---

[1]http://nptel.ac.in/courses/106106129/
[2]http://nptel.ac.in/courses/106106144/

Ground
Truth

Para
Free [1]

EVS [2]

Proposed
Approach

Fig. 2 *keyframes* of 2nd Lecture of *CC* dataset arranged in temporal order

3.2 Quantitative Analysis

The objective to achieve the best video summary, the process of *keyframes* extraction
has been widely exploited. Nevertheless, in that respect is no existing the best technique to judge their operations. To evaluate the tone of each summarization model,
summaries of video generated by many approaches are equated to the human selected
ground summaries which depend on the three evaluation metrics Precision, Recall,
and F-measure which consists often as defined in Eqs. 2, 3, 4 respectively:

$$Precision = \frac{True\ Positive}{True\ Positive + False\ Positive} \tag{2}$$

$$Recall = \frac{True\ Positive}{True\ Positive + True\ Negative} \tag{3}$$

$$F - measure(F_\beta) = \frac{(1 + \beta^2) \times Precision \times Recall}{\beta^2.Precision + Recall} \tag{4}$$

When $\beta = 1$ it means F_1 become the Harmonic mean between *Recall* and *Precision*
and known as balanced *F-score* due to *Recall* and *Precision* are equal weighted.
Hence, maximum value of $F_1 - measure$ shows more accurate model. We compared
our extracted *keyframes* with the ground user summaries with threshold parameter
$\delta = \frac{H \times W}{10}$, where H (height of a form), W (the width of a human body). Tables 1 and
2 show the quantitative assessment of our technique with state-of-the-art on both the

Table 1 A comparison of our Key-Lecture model with with state-of-the-art

Dataset	Model	Precision (%)	Recall (%)	F-measure (%)
CC	Para Free [4]	**98.3**	71.2	82.6
	EVS [5]	96.1	78.7	86.5
	Proposed approach	88.7	**90.0**	**89.3**

Table 2 A comparison of our Key-Lecture proposed model with with state-of-the-art

Dataset	Model	Precision (%)	Recall (%)	F-measure (%)
OS	Para Free [4]	95.7	76.7	84.9
	EVS [5]	**97.9**	77.4	86.4
	Proposed approach	92.1	**87.2**	**89.6**

datasets, where the best assessment results are demonstrated in the bold font. Here, we watched that

- *Recall* of our key-lecture model on both the datasets is better in comparison to the state-of-the-art-models.
- Low precision indicates that less number of frames have been removed. Still, in case of high precision, greater number of frames will be taken out, which may lead to removal of some active frames unnecessary from the final summary.
- There are some of the techniques perform better in the context of *Precision* while the others may be in the context of *Recall*. Hence, *F-measure* which is the harmonic mean of both *Recall* and *Precision* and which is better measurement for summarization. *F-measure* of our key-lectures model on both benchmark datasets is also the maximum among all the previous existing models, which represents better performance of the proposed keyframes summarization model.

3.3   Computational Complexity

In our tryouts, the frame video/Lectures size (from NPTEL which is openly available) is *352 × 240* and approximately 30,000 frames (i.e., 1500 s). We employ the online clustering approach in order to save computational cost. The proposed method holds only two frames at a time in order to save the runtime memory space which makes the faster execution. The total taken by our approach is about 9430 s for 101 number videos, i.e., 97.37 s per video. Hence, the proposed model is almost 50% faster than existing models [4, 5].

The proposed technique can extract the highlights of a video Lectures on an average 1.5 min for a video Lecture of about 30 min. Thus, it can fulfill the requirement of the practical application in real time, such as video Lectures, Slides summarization,

Table 3 Computational cost comparison

Model	Sampling Rate [frame per Sec]	Average Time per video [s]
Para free [4]	25	202.25
EVS [5]	30	180.00
Proposed approach	25	**97.37**

video surveillance, sports highlights, etc. The average computation time per video is equated with the nominated key-lecture model and the state-of-the-art models in Table 3.

Conclusion and Future Perspectives

This study highlights the Key-Lectures technique to study and sum up the video Lectures with a clustering as unsupervised learning. The primary target was to detect the highlights with the minimum storage as easily as accessing time of the videos Lectures. The observational results on both benchmark datasets with various duration of Lecture videos demonstrate that our Key-Lectures technique outperforms the state-of-the-art models with the best *F-measure* and *Recall*.

Since the summarized videos, events streamed to the host are usually processed within a minimal amount of time, in order to analysis the video Lectures content, we may apply this technique for real-time applications. Presently, the primary application of this study can be counted below the project NDL of India or for other multimedia-based educational project.

References

1. Meng, J., Wang, H., Yuan, J., Tan, Y.P.: From keyframes to key objects: video summarization by representative object proposal selection. In: Proceedings of the IEEE Conference on Computer Vision and Pattern Recognition, pp. 1039–1048 (2016)
2. Sun, M., Farhadi, A., Taskar, B., Seitz, S.: Summarizing unconstrained videos using salient montages. IEEE Trans. Pattern Anal. Mach. Intell. **39**(11), 2256–2269 (2017)
3. Cirne, M.V.M., Pedrini, H.: VISCOM: a robust video summarization approach using color co-occurrence matrices. Multimed. Tools Appl. 1–19 (2017)
4. Mishra, D.K., Singh, N.: Parameter free clustering approach for event summarization in videos. In: Proceedings of International Conference on Computer Vision and Image Processing, pp. 389–397. Springer, Singapore (2017)
5. Kumar, K., Shrimankar, D.D., Singh, N.: Equal partition based clustering approach for event summarization in videos. In: 12th International Conference on Signal-Image Technology and Internet-Based Systems (SITIS), pp. 119–126. IEEE (2016)
6. Kumar, K., Shrimankar, D.D.: F-DES: fast and deep event summarization. IEEE Trans. Multimed. (2017). https://doi.org/10.1109/TMM.2017.2741423
7. Cong, Y., Yuan, J., Luo, J.: Towards scalable summarization of consumer videos via sparse dictionary selection. IEEE Trans. Multimed. **14**(1), 66–75 (2012)

A Novel Saliency Measure Using Entropy and Rule of Thirds

Priyanka Bhatt and Navjot Singh

Abstract Human intelligence can easily identify the visually attractive object, i.e., salient object with high accuracy in real time. It is an issue of concern to design an efficient computational model which can imitate human behavior such that the model attains better detection accuracy and takes less computation time. Until now, many models have been designed which are either better regarding detection accuracy or computation time but not both. This paper aims to realize a model that takes less computational time and at the same time attains higher detection accuracy. In this work, we propose a novel saliency detection model via the efficient use of entropy and boost the performance of saliency detection by employing the concept of the rule of thirds. The paper compares the performance of the proposed model with eighteen existing models on six publicly available datasets. With regard to precision, recall, and F-measure on all the six datasets, experimental results indicate better performance of the proposed model. Less computation time required by the proposed model in comparison to many state-of-the-art models.

Keywords Salient object detection · Entropy · Rule of thirds · Saliency map

1 Introduction

Saliency of an object or thing can be defined as the quality of being prominent, conspicuous, and noticeable that makes it stand out from its surroundings or background. The process of detection and segmentation of salient object in natural scenes

P. Bhatt (✉) · N. Singh
Department of Computer Science and Engineering, NIT Uttarakhand,
Srinagar (Garhwal), India
e-mail: erpriyankabhatt@gmail.com

N. Singh
e-mail: navjot.singh.09@gmail.com

© Springer Nature Singapore Pte Ltd. 2019
M. Tanveer and R. B. Pachori (eds.), *Machine Intelligence and Signal Analysis*,
Advances in Intelligent Systems and Computing 748,
https://doi.org/10.1007/978-981-13-0923-6_40

461

or image is known as salient object detection [1]. There are various applications of salient object detection, such as object detection and recognition, image and video compression [2], video summarization, remote sensing [3], photo retargeting [4], automatic cropping and recentering [5], image quality assessment, advertising a design [2], image segmentation, content-based image retrieval [6, 7], image editing and manipulating, object discovery, and human–robot interaction. There are two categories of visual saliency [2]-(a) Bottom-up saliency—attracting regions of the image are content dependent, and are not dependent on the behavior or the experience relative to the human (b) Top-down saliency—In this, the observer has an information in his mind about the regions that could attract his attention (he already knows the subject of the content of the image). Most of the research works focused on the bottom-up approach of visual attention. There are three categories of saliency model: objectness proposals [1, 9], salient object detection [1], and fixation prediction [1, 8]. In objectness proposals, a small number (few hundreds or thousands) of proposals are generated to cover all the objects in an image. Salient object detection aims to accurately detect where the salient object should be. In fixation prediction, there is a prediction of where human eye look at when a certain image shown for few seconds. Salient object detection involves two steps—(i) To detect the most salient object and (ii) To segment the accurate boundary of that object. One of the earliest and well-known computational models who generated the first wave of interest in visual saliency was given by Itti et al. [8]. This model uses the concept of feature integration theory to combine color, intensity, and orientation feature to produce a final saliency map. Liu et al. (LIU) [10] who generated the second wave of interest, defines saliency detection as a binary segmentation problem. Later, Bruce and Tsotsos (AIM) [11] used information maximization concept to modeled visual saliency. Harel et al. (GBVS) [12] proposed a saliency model which includes two steps. The first step includes the formation of activation map and the second step includes the normalization of the activation map. Hou and Zhang (SR) [13] extracted the spectral residual of an image in spectral domain by analyzing the log-spectrum of an input image. Zhang et al. (SUN) [14] based on the Bayesian framework, determined the salient object by computing the probability of the potential target at each location in the image. Achanta et al. (FT) [15] presented a frequency tuned saliency model that evaluates center-surround contrast using color and luminance features. Achanta and Susstrunk (ASS) [16] computed the saliency map based on the idea of maximum symmetric surround difference. Goferman et al. (GOF) [17] design a saliency model whose aim is to detect not only the dominant objects but also the parts of their surroundings that convey the context. Shen and Wu (SHEN) [18] proposed a saliency model using the idea of a low-rank matrix and sparse noise. Vikram et al. (VIKRAM) [19] estimates the local saliency by extracting a set of random rectangles from the whole image. Imamoglu et al. (WT) [20] generated a saliency model by using the concept of a wavelet transform. Xie et al. (BSLM) [21] proposed a saliency model which is based on the Bayesian approach and uses the idea of low-level and mid-level cues. Jiang et al. (AMC) [22] enhanced the saliency detection by using the concept of absorbing Markov chain. Zhang (COSAL) [23] proposed a unified co-saliency detection framework by calculating the co-saliency score using principled

Bayesian formulation. Singh et al. [24] use the constrained particle swarm optimization technique to combine features for salient object detection. Liu et al. (STREE) [25] presented a saliency tree as novel saliency detection framework. Zhu et al. (Zhu) [26] proposed a new segment-based saliency detection method which consists of a new representation of superpixels by multivariate normal distributions. Singh and Agarwal (SA) [27] suggested a model which involves the modification of the Liu et al. model and combination of Kullback–Leibler divergence and Manhattan distance is used to determine center-surround histogram difference. Qin et al. (BSCA) [28] proposed a novel saliency model by using the idea of single and multilayer cellular automata. Singh et al. (ROTC) [29] used the concept of Davies–Bouldin index and proposed a model that based on the fusion of the rule of thirds and image center. Singh et al. (CH) [30] introduced a salient object detection model by using the convex hull approach. Singh et al. (SP-GMM) [31] proposed an enhanced saliency detection model using super pixel-based Gaussian mixture model. The above-mentioned state-of-the-art models are either better regarding of detection accuracy or computation time but not both. Most of these previous saliency detection models have considered that the salient object lies in the center location of the image, but it is not necessary that the salient object always lies on the center location of the image. It may happen that the object of interest may lie in the other region of the image. In this research work, we attempted to realize a saliency detection model by using the concept of entropy and then combine the resultant saliency map with the saliency map generated by the rule of thirds. Our work gives importance to other regions of the image as well apart from the image center, by using the concept of the rule of thirds. To examine the efficiency of the proposed model, experiments are done using six publicly accessible image datasets. Here, some quality metrics, such as precision, recall, F-measure, and computation time, are used to evaluate the performance of the proposed model and compared the proposed model with the 18 existing models. The paper further is summarized as follows: Sect. 2 involves the description of the proposed model, Sect. 3 contains experimental setup and results, Sect. 4 includes the conclusion and future work.

2 Proposed Model

The proposed model utilizes the abstract idea of entropy and rule of thirds to compute saliency which is discussed in detail underneath:

2.1 Entropy

Saliency is related to the local image complexity, i.e., Shannon entropy. Entropy is the measure of information content, the higher the entropy, more the information content. The information tells us about what class the instance belongs to. The entropy can

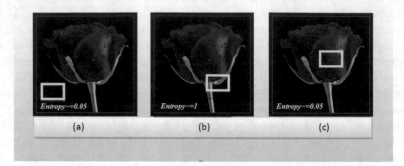

Fig. 1 **a** Region covered by window is the part of the background, **b** Window lies partially in the object and partially in the background, **c** Region covered by window is the part of the object

be defined as follows: Entropy $= \sum p_i log p_i$, where p_i is the probability that an instance belongs to class i. In this work, we have taken a rectangular box of window size 5×5. We have moved this window all over the pixels in an image and found the entropy of each bounding pixel and is computed as

$$E(j) = - \sum_{i \in N(j)} p_i log p_i \qquad (1)$$

where $N(j)$ is the 5×5 neighborhood of the jth pixel. As we know, entropy value of any homogeneous and structured region is closer to zero (zero if the region has the same color), while the heterogeneous and unstructured region must have higher entropy value (higher if the region has the different colors).
We have observed three cases:

- **First case**: When the region covered by window is the part of the background and have similar color values. In this case, the entropy value will be closer to 0.
- **Second case**: When the window lies partially in the object and partially in the background, i.e., the region covered by the window is having sharp changes. In this case, entropy has some higher value.
- **Third case**: When the region covered by window is the part of the object and have similar color values, then the entropy value, in this case, will also be closer to 0 (Fig. 1).

2.2 Rule of Thirds

Rule of Thirds (ROT) is a widely used principle in photography which is used to create aesthetics in a photograph. In ROT the image is segmented into nine equal parts as shown in Fig. 2. The four points highlighted in Fig. 2 are considered important as the object of interest is placed on these four points to increase the aesthetics. Many existing models considered center prior to detecting salient objects. They have given more importance to the objects which are placed near the center of the image and less importance to the ones near the corners of the image. But there may be images in which object of interest is placed near the image corner. So, we have given more weightage to the corners of the image (Fig. 3).

Let r be the set of four points such that $r=\{r_k\}_{k=0}^3$. The closer the object to these points higher is its saliency. The position prior weight $R(j)$ of the jth pixel with spatial location P_j, using the rule of thirds is computed as:

Fig. 2 ROT pixels position

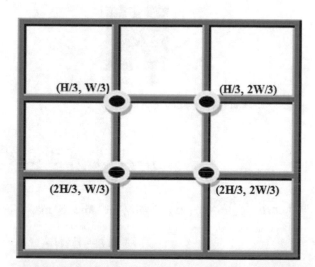

Fig. 3 ROT-based position prior

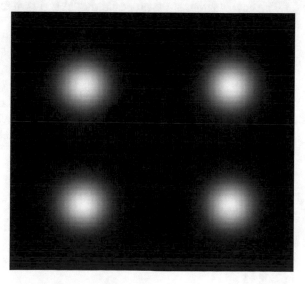

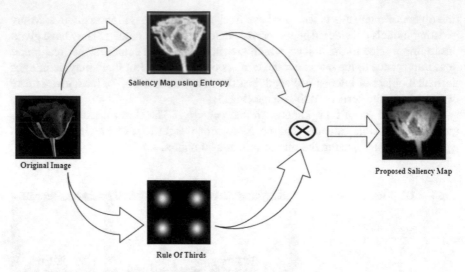

Fig. 4 The Outline of the proposed model

$$R(j) = \frac{1}{4}\sum_{k=0}^{3} exp\left(\frac{-|P_j - r_k|^2}{2\sigma^2}\right) \qquad (2)$$

where $\sigma = \frac{1}{6}min(W, H)$. Finally, the saliency map is computed as

$$SM(j) = E(j).R(j) \qquad (3)$$

The flow diagram of the proposed model is shown in Fig. 4.

3 Experimentation and Results

The qualitative and quantitative evaluation is carried out to determine the performance of the proposed model and compared with the existing models. All the experiments are done using Windows 8 environment over Intel(R) Xeon(R) processor with a speed of 2.27 GHz and 4 GB RAM.

3.1 Salient Object Database

The performance of the proposed model and eighteen other state-of-the-art models is evaluated using the six publicly available datasets whose details can be observed in Table 1.

Table 1 Datasets used for performance analysis

SNO	Dataset	Image	Object
1	MSRA-B[1]	5000	1
2	ASD[2]	1000	1
3	SAA-GT[3]	5000	1
4	SOD[4]	300	3
5	SED1[5]	100	1
6	SED2[6]	100	2

[1]http://www.research.microsoft.com/enus/um/people/jiansun/salientobject/salient_object.htm
[2]http://ivrgwww.epfl.ch/supplementary_material/RK_CVPR09/GroundTruth/binarymasks.zip
[3]http://elderlab.yorku.ca/~vida/SOD/index.html
[4]http://www.wisdom.weizmann.ac.il/~vision/Seg_Evaluation_DB
[5]http://www.wisdom.weizmann.ac.il/~vision/Seg_Evaluation_DB

3.2 Qualitative Evaluation

In Fig. 5. we can observe the qualitative assessment of the proposed model and other 18 state-of-the-art models. We have selected some images from the test dataset which contain objects that are different in size, shape, type, and position.

In Fig. 4, it is observed that our proposed model produces a better saliency map concerning the other saliency maps of existing 18 models.

3.3 Quantitative Evaluation

The quantitative assessment of the proposed model and eighteen other state-of-the-art models is carried out with regard to precision, recall, F-measure, and computation time. By taking the ground truth G and the detection result R, precision, recall, F-measure are computed as:

$$precision = \frac{TP}{TP + FP} \tag{4}$$

$$Recall = \frac{TP}{TP + FN} \tag{5}$$

$$F_\beta = \frac{(1 + \beta^2) precision \times Recall}{\beta^2 \times precision + Recall} \tag{6}$$

where TP=$\sum_{G(x,y)=1} R(x, y)$, FP=$\sum_{G(x,y)=0} R(x, y)$, FN=$\sum_{R(x,y)=0} G(x, y)$ and considered $\beta = 1$ as we have given equal importance to both precision and recall.

In Fig. 6, the quantitative comparison of the proposed model and other 18 state-of-the-art models based on the precision, recall, and F-measure on all six publically

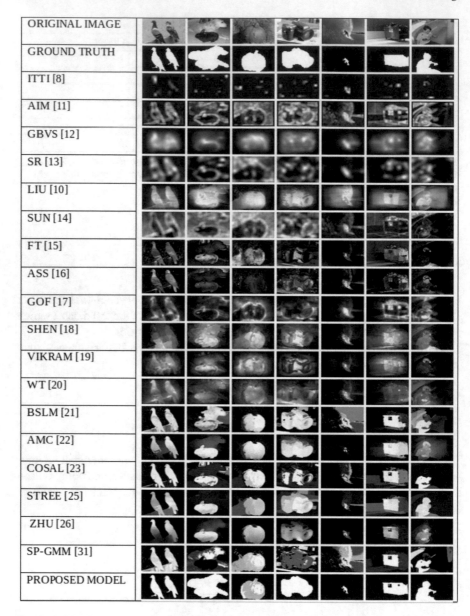

ORIGINAL IMAGE							
GROUND TRUTH							
ITTI [8]							
AIM [11]							
GBVS [12]							
SR [13]							
LIU [10]							
SUN [14]							
FT [15]							
ASS [16]							
GOF [17]							
SHEN [18]							
VIKRAM [19]							
WT [20]							
BSLM [21]							
AMC [22]							
COSAL [23]							
STREE [25]							
ZHU [26]							
SP-GMM [31]							
PROPOSED MODEL							

Fig. 5 Qualitative comparison of the proposed model with existing 18 models

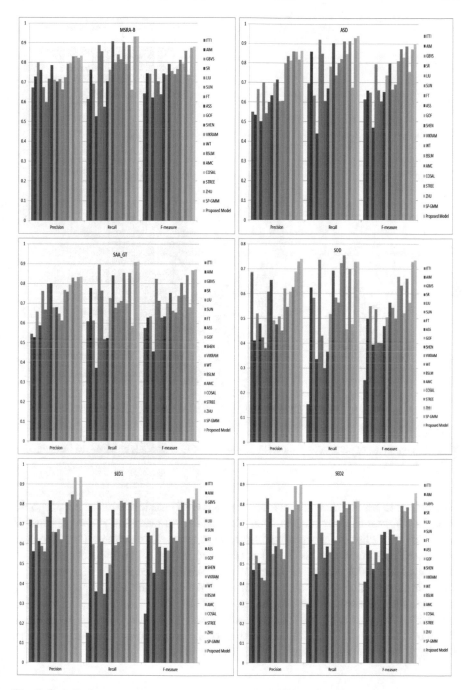

Fig. 6 Quantitative comparison on six datasets based on the Precision, Recall, and F-measure

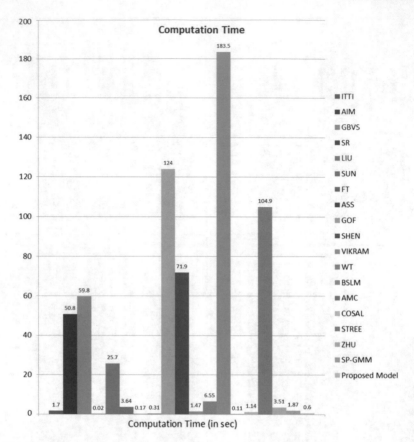

Fig. 7 Comparison of the proposed model and 18 state-of-the-art models based on the Computation time

available datasets are shown. It can be seen that the Precision value of the proposed model is highest as compared to other 18 models. As precision is the measure of exactness so higher the value of precision more exact the results will be. Recall is the measure of completeness, and it is observed in Fig. 6 that the Recall value of the proposed model is highest among the 18 existing models. The other metrics, i.e., F-measure which is the harmonic mean of precision and recall is also used for evaluating the performance. It can be seen (Fig. 6) that the value of F-measure is also highest for the proposed model as compared to other 18 existing models. In the Fig. 7 it can be observed that the computation time of the proposed model nearly equals to 0.6 that is very less as compared to other 18 state-of-the-art models.

4 Conclusion and Future Work

In this work, we have proposed a novel saliency detection model via the effective use of entropy and the rule of thirds. The model works well for the images where the object of interest is placed away from the image center. Regarding precision, recall, and F-measure on six publicly available datasets, experimental results demonstrate the better performance of the proposed model in comparison to eighteen existing state-of-the-art methods. The proposed model also requires lesser computation time. There are few challenges in detecting salient objects which may be dealt in our future work. These include noise in images, articulation, poor illumination, background clutter, etc. More complicated datasets may be used in our experiments which may contain images with only one or multiple salient objects. Research work may also be extended over datasets containing images without any salient object.

References

1. Borji, A., Cheng, M.M., Jiang, H., Li, J.: Salient object detection: a survey (2014). arXiv preprint arXiv:1411.5878
2. Itti, L.: Models of bottom-up and top-down visual attention. Doctoral Dissertation, California Institute of Technology (2000)
3. Li, Z., Itti, L.: Saliency and gist features for target detection in satellite images. IEEE Trans. Image Process. **20**(7), 2017–2029 (2011)
4. Chen, L.Q., Xie, X., Fan, X., Ma, W.Y., Zhang, H.J., Zhou, H.Q.: A visual attention model for adapting images on small displays. Multimed. syst. **9**(4), 353–364 (2003)
5. Santella, A., Agrawala, M., DeCarlo, D., Salesin, D., Cohen, M.: Gaze-based interaction for semi-automatic photo cropping. In: Proceedings of the SIGCHI Conference on Human Factors in Computing Systems, pp. 771–780. ACM, USA (2006)
6. Amit, Y.: 2D target detection and recognition, models, algorithms and networks (2002)
7. Gonzalez, R.C., Woods R.E.: Digital image processing (2002)
8. Itti, L., Koch, C., Niebur, E.: A model of saliency-based visual attention for rapid scene analysis. IEEE Trans. Pattern Anal. Mach. Intell. **20**(11), 1254–1259 (1998)
9. Alexe, B., Deselaers, T., Ferrari, V.: Measuring the objectness of image windows. IEEE Trans. Pattern Anal. Mach. Intell. **34**(11), 2189–2202 (2012)
10. Liu, T., Yuan, Z., Sun, J., Wang, J., Zheng, N., Tang, X., Shum, H.Y.: Learning to detect a salient object. IEEE Trans. Pattern Anal. Mach. Intell. **33**(2), 353–367 (2011)
11. Bruce, N., Tsotsos, J.: Saliency based on information maximization. In: Advances in Neural Information Processing Systems, pp. 155–162 (2006)
12. Harel, J., Koch, C., Perona, P.: Graph-based visual saliency. In Advances in Neural Information Processing Systems, pp. 545–552 (2007)
13. Hou, X., Zhang, L.: Saliency detection: a spectral residual approach. In: IEEE Conference on Computer Vision and Pattern Recognition (CVPR'07), pp. 1–8. IEEE, USA (2007)
14. Zhang, L., Tong, M.H., Marks, T.K., Shan, H., Cottrell, G.W.: SUN: a Bayesian framework for saliency using natural statistics. J. Vis. **8**(7), 32–32 (2008)
15. Achanta, R., Hemami, S., Estrada, F., Susstrunk, S.: Frequency-tuned salient region detection. In: IEEE Conference on Computer Vision and Pattern Recognition (CVPR 2009). pp. 1597–1604. IEEE, USA (2009)
16. Achanta, R., Süsstrunk, S.: Saliency detection using maximum symmetric surround. In: 2010 17th IEEE International Conference on Image Processing (ICIP), pp. 2653–2656. IEEE, USA (2010)

17. Goferman, S., Zelnik-Manor, L., Tal, A.: Context-aware saliency detection. IEEE Trans. Pattern Anal. Mach. Intell. **34**(10), 1915–1926 (2012)
18. Shen, X., Wu, Y.: A unified approach to salient object detection via low rank matrix recovery. In: 2012 IEEE Conference on Computer Vision and Pattern Recognition (CVPR), pp. 853-860. IEEE, USA (2012)
19. Vikram, T.N., Tscherepanow, M., Wrede, B.: A saliency map based on sampling an image into random rectangular regions of interest. Pattern Recognit. **45**(9), 3114–3124 (2012)
20. Imamoglu, N., Lin, W., Fang, Y.: A saliency detection model using low-level features based on wavelet transform. IEEE Trans. Multimed. **15**(1), 96–105 (2013)
21. Xie, Y., Lu, H., Yang, M.H.: Bayesian saliency via low and mid level cues. IEEE Trans. Image Process. **22**(5), 1689–1698 (2013)
22. Jiang, B., Zhang, L., Lu, H., Yang, C., Yang, M. H.: Saliency detection via absorbing markov chain. In: Proceedings of the IEEE International Conference on Computer Vision, pp. 1665–1672 (2013)
23. Fu, H., Cao, X., Tu, Z.: Cluster-based co-saliency detection. IEEE Trans. Image Process. **22**(10), 3766–3778 (2013)
24. Singh, N., Arya, R., Agrawal, R.K.: A novel approach to combine features for salient object detection using constrained particle swarm optimization. Pattern Recognit. **47**(4), 1731–1739 (2014)
25. Liu, Z., Zou, W., Le Meur, O.: Saliency tree: a novel saliency detection framework. IEEE Trans. Image Process. **23**(5), 1937–1952 (2014)
26. Zhu, L., Klein, D.A., Frintrop, S., Cao, Z., Cremers, A.B.: A multisize superpixel approach for salient object detection based on multivariate normal distribution estimation. IEEE Trans. Image Process. **23**(12), 5094–5107 (2014)
27. Singh, N., Agrawal, R.K.: Combination of Kullback-Leibler divergence and Manhattan distance measures to detect salient objects. Signal Image Video Process. **9**(2), 427–435 (2015)
28. Qin, Y., Lu, H., Xu, Y., Wang, H.: Saliency detection via cellular automata. In: Proceedings of the IEEE Conference on Computer Vision and Pattern Recognition, pp. 110–119 (2015)
29. Singh, N., Arya, R., Agrawal, R.K.: A novel position prior using fusion of rule of thirds and image center for salient object detection. Multimed. Tools Appl. **76**(8), 10521–10538 (2017)
30. Singh, N., Arya, R., Agrawal, R.K.: A convex hull approach in conjunction with Gaussian mixture model for salient object detection. Digit. Signal Process. **55**, 22–31 (2016)
31. Singh, N., Arya, R., Agrawal, R. K.: Performance enhancement of salient object detection using superpixel based Gaussian mixture model. Multimed. Tools Appl. pp. 1–19 (2017)

An Automated Alcoholism Detection Using Orthogonal Wavelet Filter Bank

Sunny Shah, Manish Sharma, Dipankar Deb and Ram Bilas Pachori

Abstract Alcohol misuse is a common social issue related to the central nervous system. Electroencephalogram (EEG) signals are used to depict electrical activities of the brain. In the proposed study, a new computer-aided diagnosis (CAD) has been developed to recognize alcoholic and normal EEG patterns, accurately. In this paper, we present an automatic system for the classification of normal and alcoholic EEG signals using orthogonal wavelet filter bank (OWFB). First, we derive sub-bands (SBs) of EEG signals. Then, we compute logarithms of the energies (LEs) of the SBs. The LEs are employed as the discriminating features for the separation of alcoholic and normal EEG signals. A supervised machine learning algorithm called K nearest neighbor (KNN) has been employed to classify normal and alcoholic patterns. The proposed model has yielded very good classification results. We have achieved a classification accuracy (CA) of 94.20% with tenfold cross-validation (CV).

Keywords Alcoholism · Electroencephalogram (EEG) · Ensemble subspace
KNN · Feature · CAD

S. Shah (✉) · M. Sharma · D. Deb
Institute of Infrastructure Technology Research and Management,
Ahmedabad 380026, Gujarat, India
e-mail: shahsunny1416@gmail.com

M. Sharma
e-mail: manishsharma.iitb@gmail.com

D. Deb
e-mail: dipankardeb@iitram.ac.in

R. B. Pachori
Discipline of Electrical Engineering, Indian Institute
of Technology Indore, Indore 453552, India
e-mail: pachori@iiti.ac.in

© Springer Nature Singapore Pte Ltd. 2019   473
M. Tanveer and R. B. Pachori (eds.), *Machine Intelligence and Signal Analysis*,
Advances in Intelligent Systems and Computing 748,
https://doi.org/10.1007/978-981-13-0923-6_41

1   Introduction

Drinking too much alcohol leads to accidents and dependence, but can also cause damaged organs, weakened immune system and cancers. In our brain, two types of the signals are present: one is electrical and other is chemical. Neurotransmitter is a chemical transmitter which transmits a message between neurons and is responsible for our feelings, body response and mood [4]. In normal condition, the chemical signal is balanced. When a person drinks alcohol, there is an imbalance in chemical signal and our response slows down. Due to variation in the chemical signal, there is also a change in the electrical signal which is recorded using scalp placed on the head of the person. Electroencephalogram (EEG) signal is the measurement of the instantaneous electrical action of the brain over a small period of time [23]. It can depict the status of the entire body, and so it inspires the use of digital signal processing approach. Electrical signal (in μV) is generated due to current flow within the tip of dendrites and axons, and also between different dendrites.

Someone working in alcoholism detection has to use digital signal processing and machine learning tools to detect if the given signal is alcoholic or normal. Recorded data of brain (EEG) is used in digital signal processing wherein data is divided into sub-bands (SBs) for signal reconstruction. EEG signals provide the observable display of waveform and are applied to the computer-aided digital signal processing techniques to characterize using machine learning techniques.

Realistic automatic EEG signal analysis in a medical professional is important in the detection of neurobiological diseases. By using computer-aided diagnosis we can prevent the possibility of misjudging the information. However, automatic classification of recorded signals is a difficult task as the morphological and temporal characteristics of recorded signals demonstrate significant variations for distinct patients under different temporal and physical conditions. For accurate analysis of recorded signals, high sampling rate and larger quantization levels are required.

For discrimination of measured EEG signals, different methods have been proposed. Acharya et al. [13, 17, 18] used nonlinear features like largest Lyapunov exponent (LLE) and approximate entropy to separate the two signals and for classification, support vector machine (SVM) was used [1]. Faust et al. have used wavelet packet transform to decompose signals into SBs for energy measures [8]. Recently, Patidar et al. [11] have suggested a novel alcoholism scrutiny theory using tunable-Q wavelet transform (TQWT) wherein correntropy-dependent feature extraction is given to least square SVM (LS-SVM) classifier for data classification.

In our work, two-band wavelet filter banks are used to decompose EEG signals into four SBs [3, 14, 16]. The t-test is used to isolate each feature. Then, the logarithm of energies (LEs) of particular SBs are calculated, and the computed feature vectors of LEs are utilized by classifier to separate out normal and alcoholic EEG signals. The tenfold cross-validation (CV) is utilized and the classification performance is expressed in terms of classification accuracy (CA), classification sensitivity (CS), and classification specificity (CSF) [14].

2 Materials and Method

The operational flow of the proposed scheme is shown in Fig. 1. The initial step of filtering the raw EEG data is known as preprocessing which removes the undesired noise from the recorded EEG signals for noise-free EEG data that is decomposed into different SBs using orthogonal wavelet filter bank. The LEs of each SB are then computed and used as isolating features. Student's t-test [15] is used for feature ranking. The classification has been performed using K nearest neighbor (KNN) and least square support vector machine (LS-SVM).

2.1 EEG Data Collection and Noise Removal

Current study based on EEG dataset which was collected from [2]. Preference is given to alcoholism in the informational data file which was gathered by an outsized research of EEG signals. For data collection, 64 electrodes were used and 120 trials were undertaken. A nose electrode is used to ground the subjects and the electrodes were placed on the scalp of the subjects.

The sampling frequency of signals is 256 Hz. During sampling, the responses with high movements of the body and eyes are discarded for better results. Then for alcoholic and nonalcoholic subjects, 30 EEG recordings were acquired. The collected data files are then divided into four sections where every segment contains

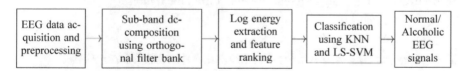

Fig. 1 The proposed algorithm

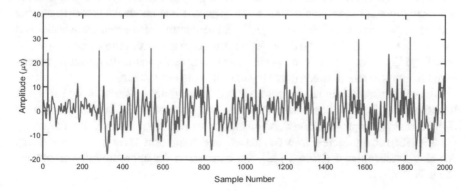

Fig. 2 Alcoholic EEG signal segment

Table 1 Details of the dataset used in this study

Group	Total segments	Sampling frequency (Hz)	Total samples
Normal	120	256	2048
Alcoholic	120	256	2048

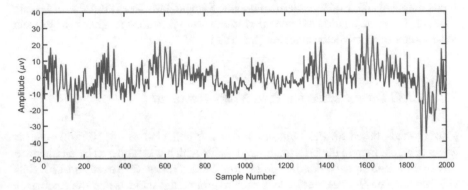

Fig. 3 Normal EEG signal segment

2,048 samples. In this manner, for alcoholic and normal subjects 120 information segments were acquired. Table 1 shows outline of dataset used in this study. Figures 2 and 3 demonstrate sample data segment for alcoholics and nonalcoholic subjects, respectively, and they are different as evident from the figures.

2.2 Design of Daubechies Orthogonal Wavelet Filter Bank

Wavelet change acts as a numerical magnifier that zooms into little scales to uncover minimally divided occasions in time and zooms out into extensive scales to display waveform patterns in a global sense [19]. Determination of appropriate wavelet and the total number of decomposition levels are needed to examine discrete wavelet transform [5]. Selection of decomposition levels depends on the dominant frequency components wherein desired frequency components are available.

Since the recorded signals posses no helpful recurrence segments over 30 Hz, the decomposition level is chosen to be 4. The EEG signals were decomposed into details d_1 to d_4 and last approximation a_4. Considering F_s representing sampling frequency ($= 256$ Hz) for the alcoholic task dataset, and transform level $n = 4$, and using wavelet: Daubechies 4 ("$db4$") [22], the expression for approximation frequency range is

$$\left[0, \frac{F_s}{2^{(n+1)}}\right], \tag{1}$$

Table 2 Frequency bands for Daubechies 4 filter decomposition with $F_s = 256\,\text{Hz}$

Signal decomposition	Frequency bands (Hz)
d_1	64–128
d_2	32–64
d_3	16–32
d_4	8–16
a_4	0–8

and the frequency range for details are expressed as

$$\left[\frac{F_s}{2^{(n+1)}}, \frac{F_s}{2^n} \right]. \tag{2}$$

Using these expressions, we have different frequency bands as given in Table 2.

2.3 Extracting and Ranking the Features

Feature extraction process should be quick and trustworthy to reduce the time taken for diagnosis. After getting appropriate SB, the next task is to calculate the LE of all four SBs, as shown in Figs. 4 and 5.

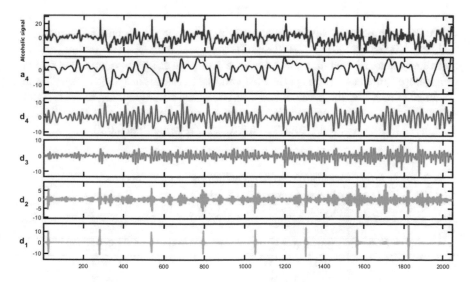

Fig. 4 Approximate and detailed SBs of EEG signal from alcoholic subject

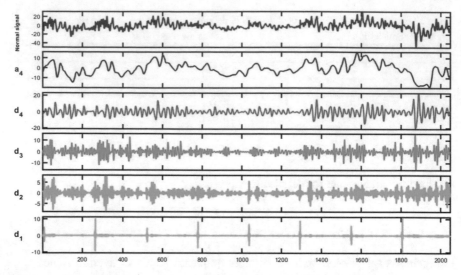

Fig. 5 Approximate and detailed SBs of EEG signal taken from normal subject

EEG signals were decomposed into details d_1 to d_4 and last approximation a_4 for both normal and alcoholic signals. To classify a group of signals, the obtained LEs features were used to separate those signals, as follows:

$$x_{LE} = \ln\left(\sum_{n\varepsilon\mathbb{Z}} |x(n)|^2\right), \qquad (3)$$

where $x(n)$ represent wavelet time series of a particular SBs of an EEG signal.

Every LE feature has statistical significance, and the statistical parameters like mean and standard deviation are calculated using t-test and p-values. For ranking of a feature, student's t-test is performed and these features are ranked based on the results obtained, and the p-values are also computed corresponding to each feature using the t-test. Lower p-value indicates better separation of features. From Table 3, we can see that except d_1 all four SBs have lesser p-value than 0.0001.

Table 3 Range (mean $\pm$ standard deviation) of normal and alcoholic features

SBs	Alcoholic feature	Normal feature	p-values
a_4	10.5309 ± 0.7459	11.8638 ± 0.4362	5.9466e−22
d_4	8.5939 ± 0.5826	9.4915 ± 0.3111	26697e−08
d_3	7.3214 ± 0.5242	8.4046 ± 0.2837	1.5215e−18
d_2	7.0745 ± 0.5315	8.1975 ± 0.2951	3.7374e−37
d_1	7.0755 ± 0.5468	8.2958 ± 0.4256	0.2875

2.4 Classification Using Ensemble Subspace KNN and LS-SVM

The KNN is a nonparametric method for data classification which compares the training and testing data while estimating feature values usually calculated using standard Euclidean distance. This method uses the available cases so as to classify new cases based on a measure of similarity. Majority vote of k neighbors is used for classification and assignment to the class most common among those neighbors is done. The value of k refers to the number of nearest values to be considered before the output class is decided. Given x_i and x_j, where k varies from 1 to p, the Euclidean distance is given by

$$p(x_i, x_j) = \sqrt{\sum_{k=1}^{p} (x_{ik} - x_{jk})^2}. \tag{4}$$

Ensemble classifier refers to a group of individual classifiers cooperatively trained on dataset in a supervised classification problem wherein learning from instances, patterns/examples takes place, and each instance is associated with a label/class. Depending on the pattern distribution, it is possible that not all the patterns are learned well by an individual classifier [12].

Therefore, by training a group of classifiers on the same problem, we can achieve higher accuracy through ensemble classifiers. The accuracy for the KNN is high as compared to the other classifiers [24] and so an ensemble KNN classifier is preferred.

The LS-SVM [20, 25] is used to solve the linear system problem in a binary classification task by finding support vectors by optimize the given function. LS-SVM is a variant of SVM with a slightly altered objective function given by

$$J(w, b, e) = \frac{1}{2} w^T w + \gamma \frac{1}{2} \sum_{i=1}^{N} e_i^2, \tag{5}$$

where the objective function is to be minimized so as to find the hyperplane expressed by w, for separating the data into two parts. The above formulation is an implicit consequence of a least square regression problem. We have calculated the performance using LS-SVM.

Performance evaluation: The classification performance measurement has been performed from the following parameters [10]:

$$CA = \frac{TP + TN}{TP + FN + FP + TN} \tag{6}$$

$$CS = \frac{TP}{TP + FN}, \quad CSF = \frac{TN}{FP + TN}, \tag{7}$$

where TN (true negative): number of normal subjects classified as normal, TP (true positive): number of alcoholic subjects classified as alcoholic, FN (false negative): number of normal subjects classified as alcoholic, and FP (false positive): number of alcoholic subjects classified as normal.

3 Results and Discussion

Using student's t-test, we obtain five LE features shown in Table 3 which are important because four LEs have p-value < 0.0001. The box blot of the five LEs is shown in Fig. 6 which clearly indicates separation (nonoverlapping) of normal and alcoholic EEG signals. The classification results are shown in Table 4. For classification of normal and alcoholic LE features, LS-SVM and KNN classifier are used, and the classification accuracy and specificity of KNN are greater.

In Table 5, we compare the proposed CAD system with the existing alcoholism identification systems. For EEG signal analysis a couple of researchers [6, 9, 21] have proposed linear model techniques, but they have not used any classification algorithms, and in such cases classification performance is shown as not applicable (NA) [6, 9, 21].

Table 4 Classification performance using KNN and LS-SVM

Classifier	CA (%)	CS (%)	CSF (%)
KNN	94.2	92.5	95.8
LS-SVM	92.08	93.33	90.83

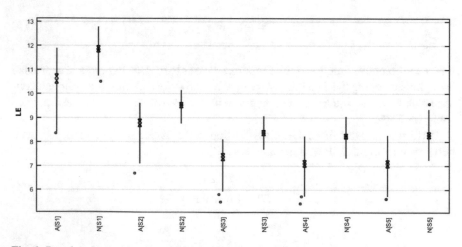

Fig. 6 Boxplots for five features: A[Si], N[Si] denote ith SB of alcoholic and normal signals

Table 5 Comparison of the classification performance with the existing methods

Authors	Features	Classification technique	Classification performance
Ehlers et al. [6]	CD, discriminant analysis	No classifier	NA
Kannathal et al. [9]	CD, LLE, H, entropy	No classifier	NA
Faust et al. [7]	PSD	ANN and SVM	0.822 (AUC-ROC)
Tcheslavski and Gonen [21]	Parametric spectral, statistical analysis coherence measure	No classifier	NA
Acharya et al. [1]	APPENT, Sampen, LLE, HOS	SVM	91.3% (CA)
Our work	Daubechies orthogonal wavelet filter bank	KNN	94.2% (CA)

CD = Correlation dimension, LLE = Largest Lyapunov exponent, H = Hurst's exponent, PSD = Power spectral density, APPENT = Approximate entropy, Sampen = Sample entropy, HOS = Higher order spectra, ANN = Artificial neural network, AUC = Area under the curve, ROC = Receiver operating characteristics

Acharya et al. [1] have taken different nonlinear features and applied to LS-SVM classifiers. and a maximum classification accuracy of 91.3% is achieved among all the compared methods. Also, computation cost of nonlinear features is higher than the computational cost of LE feature used by us in this study.

For good and sturdy performance, the data is divided into 10 subsets. For calculation of accuracy any nine subsets are taken as training and remaining one is used as testing set. The above procedure is repeated ten times, each time with an alternate testing subset. On every iteration classification, the accuracy is calculated, and the average of all iterations is obtained. We can see that the designed algorithm of KNN classifier with tenfold cross-validation gives promising classification performance with CA of 94.2%.

4 Conclusion

It is challenging to diagnose a disease like alcoholism because it requires more attention and patience to analyze the recorded EEG signal. If due to some reason the medical expert fails to detect the signal variations then in some cases it may lead to serious disease also. Due to this reason, automated diagnosis system which is also financially savvy and nonintrusive system is used. Henceforth, it can be effortlessly utilized in every hospital centers and remote towns. The proposed computer supported framework empowers medical experts to identify alcoholism in a case who does not indicate signs of an alcohol abuse. Also, road accidents can be reduced if an EEG-based suitable system is developed to catch a driver who drinks alcohol and drives.

References

1. Acharya, U.R., Vinitha Sree, S., Chattopadhyay, S., Suri, J.S.: Automated diagnosis of normal and alcohlic EEG signals. Int. J. Neural Syst. **22**(03), 1250011 (2012)
2. Begleiter, H.: https://archive.ics.uci.edu/ml/datasets/eeg+database (2018)
3. Bhattacharyya, A., Sharma, M., Pachori, R.B., Sircar, P., Acharya, U.R.: A novel approach for automated detection of focal EEG signals using empirical wavelet transform. Neural Comput. Appl. (2016)
4. Charles, H.: Hundred Questions and Answers about Alcoholism. Jones and Bartlett Publishers, Burlington (2007)
5. Daubechies, I.: Ten Lectures on Wavelets. CBMS-NSF Regional Conference Series in Applied Mathematics. Society for Industrial and Applied Mathematics (1992)
6. Ehlers, C.L., Havstad, J., Prichard, D., Theiler, J.: Low doses of ethanol reduce evidence for nonlinear structure in brain activity. J. Neurosci. **18**(18), 7474–7486 (1998)
7. Faust, O., Acharya, R., Allen, A., Lin, C.: Analysis of EEG signals during epileptic and alcoholic states using AR modeling techniques. IRBM **29**(1), 44–52 (2008)
8. Faust, O., Yu, W., Kadri, N.A.: Computer-based identification of normal and alcoholic EEG signals using wavelet packets and energy measures. J. Mech. Med. Biol. **13**(03), 1350033 (2013)
9. Kannathal, N., Choo, M.L., Acharya, U.R., Sadasivan, P.: Entropies for detection of epilepsy in EEG. Comput. Methods Progr. Biomed. **80**(3), 187–194 (2005)
10. Mumtaz, W., Vuong, P.L., Xia, L., Malik, A.S., Rashid, R.B.A.: Automatic diagnosis of alcohol use disorder using EEG features. Knowl. Based Syst. **105**, 48–59 (2016)
11. Patidar, S., Pachori, R.B., Upadhyay, A., Acharya, U.R.: An integrated alcoholic index using tunable-Q wavelet transform based features extracted from EEG signals for diagnosis of alcoholism. Appl. Soft Comput. **50**, 71–78 (2017)
12. Rahman, A., Tasnim, S.: Ensemble classifiers and their applications: a review. Int. J. Comput. Trends Technol. **10**(1), 31–35 (2014)
13. Sharma, M., Pachori, R.B.: A novel approach to detect epileptic seizures using a combination of tunable-q wavelet transform and fractal dimension. J. Mech. Med. Biol. **17**(07), 1740003 (2017)
14. Sharma, M., Deb, D., Acharya, U.R.: A novel three-band orthogonal wavelet filter bank method for an automated identification of alcoholic EEG signals. Appl. Intell. (2017)
15. Sharma, M., Dhere, A., Pachori, R.B., Acharya, U.R.: An automatic detection of focal EEG signals using new class of time–frequency localized orthogonal wavelet filter banks. Knowl. Based Syst. **118**, 217–227 (2017)
16. Sharma, M., Pachori, R.B., Acharya, U.R.: A new approach to characterize epileptic seizures using analytic time-frequency flexible wavelet transform and fractal dimension. Pattern Recognit. Lett. **94**, 172–179 (2017)
17. Singh, P., Pachori, R.B.: Classification of focal and nonfocal EEG signals using features derived from Fourier-based rhythms. J. Mech. Med. Biol. **17**(07), 1740002 (2017)
18. Singh, P., Joshi, S.D., Patney, R.K., Saha, K.: Fourier-based feature extraction for classification of EEG signals using EEG rhythms. Circuits Syst. Signal Process. **35**(10), 3700–3715 (2016)
19. Subasi, A.: EEG signal classification using wavelet feature extraction and a mixture of expert model. Expert Syst. Appl. **32**(4), 1084–1093 (2007)
20. Suykens, J., Vandewalle, J.: Least squares support vector machine classifiers. Neural Process. Lett. **9**(3), 293–300 (1999)
21. Tcheslavski, G.V., Gonen, F.F.: Alcoholism-related alterations in spectrum, coherence, and phase synchrony of topical electroencephalogram. Comput. Biol. Med. **42**(4), 394–401 (2012)
22. Tolić, M., Jović, F.: Classification of wavelet transformed EEG signals with neural network for imagined mental and motor tasks (2013)
23. Übeyli, E.D.: Statistics over features: EEG signals analysis. Comput. Biol. Med. **39**(8), 733–741 (2009)

24. Umale, C., Vaidya, A., Shirude, S., Raut, A.: Feature extraction techniques and classification algorithms for EEG signals to detect human stress - a review. Int. J. Comput. Appl. Technol. Res. **5**(1), 8–14 (2016)
25. Vapnik, V.N.: The Nature of Statistical Learning Theory. Springer, New York (2000)

An Efficient Brain Tumor Detection and Segmentation in MRI Using Parameter-Free Clustering

Shiv Naresh Shivhare, Shikhar Sharma and Navjot Singh

Abstract Automation in detecting and segmenting brain tumor is the need of the era in order to diagnose human brain magnetic resonance images (MRIs) and required for better treatment planning as compared to the manual process. Manual diagnosis of brain tumor MRI is a time-consuming process and often depends on the expertise of the clinician or radiologist which may lead to a chance of human error. However, automatic brain tumor detection has been a complex task in medical image analysis due to unknown, unstructured nature of abnormality and a huge variability in shape, location, and characteristics of different sub-compartments of the tumor. In this paper, we propose a fully automatic model for brain tumor detection based on parameter-free clustering algorithm and morphological dilation and hole-filling operations. The method is applied to an axial slice of the T1c modality of BRATS 2015 training dataset. In our experiments, we segmented the tumor from the contrast-enhanced T1-weighted image and compared the results with the available ground truth. Results of tumor segmentation achieved of 75% of the Dice similarity coefficient (DSC) for a tumor core region when compared to the ground truth.

S. N. Shivhare (✉) · S. Sharma · N. Singh
National Institute of Technology, Uttarakhand, Srinagar (Garhwal), India
e-mail: shiv.naresh@nituk.ac.in

S. Sharma
e-mail: shikhar01.cse14@nituk.ac.in

N. Singh
e-mail: navjot.singh.09@nituk.ac.in

© Springer Nature Singapore Pte Ltd. 2019
M. Tanveer and R. B. Pachori (eds.), *Machine Intelligence and Signal Analysis*,
Advances in Intelligent Systems and Computing 748,
https://doi.org/10.1007/978-981-13-0923-6_42

1   Introduction

Brain image analysis in the field of medical imaging has been a popular choice of researchers in the last decade. As an emerging area of research, it bridges the gap between technological and medical fields and facilitates doctors and clinicians by enabling automation in the task of brain image analysis. Based on the intensity of tumor and its cancerous property, brain tumors are commonly divided into two parts: benign tumor (noncancerous) and malignant tumor (cancerous). However, World Health Organization (WHO) categorized brain tumors into four grades: grade I, II (benign and low-grade tumors) and grade III, IV (for malignant and high-grade tumors) [1].

According to American Brain Tumor Association, nearly 80,000 patients are expected to diagnose in the year 2017 and out of which approximately 32% cases are of malignant brain tumor [2]. Magnetic resonance imaging (MRI) is the most widely used and effective technique to diagnose brain tumors. The most commonly used MRI sequences or modalities, i.e., T1-weighted image (T1), T1-weighted image with contrast enhancement (T1c), T2-weighted image (T2) and T2-weighted image with fluid-attenuated inversion recovery (FLAIR) are acquired by the radiologist by applying variations in repetition time (TR) and time to echo (TE) [3]. Figure 1 shows brain images of a glioma patient in different MRI modalities. These modalities are quite useful for delineation and segmentation of brain lesion and its substructures due to specific characteristics and histological functions of MRI modalities.

Glioblastoma multiform (GBM) is the most common malignant brain tumor type which is very severe with a poor survival rate. Recently in 2016, Dupont et al. [6] have given an extensive survey on image segmentation methods applied to glioblastoma tumor type. Dupont et al. [6] divided segmentation methods into four categories: region-based approach, threshold-based approach, classification- and clustering-based approach, and model-based approach. Multiple regions are created from an image based on pixel properties and similarity measures in region-based approach [7]. Threshold-based approach segment a given image based on a predefined threshold, e.g., a specific intensity value is used to classify the whole image into two parts.

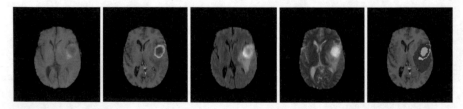

Fig. 1   Axial slice of HGG patient1 from BRATS 2015 training dataset [4]. From *left* to *right*: T1-weighted image (T1), contrast-enhanced T1-weighted image (T1c), T2-weighted image with fluid-attenuated inversion recovery (FLAIR), T2-weighted image (T2), and tumor subparts displayed in colors using BraTumIA [5]: edema (blue), enhancing tumor (red), non-enhancing solid core (sky blue), and necrotic core (green)

Thresholding can be either global or adaptive based on the number of thresholds applied [8]. Classification and clustering-based approaches are the most common for tumor segmentation task. Feature extraction of the brain image voxel (volumetric pixel in case of 3D MRI) is an important step in this approach. Feature characteristics are extracted through intensity and texture information of the set of voxels [9]. In the last, model-based approaches have been widely used by many researchers in recent years for brain tumor segmentation.

A model-based approach is incorporated in semi-automated methods by developing deformable models for a specific atomic structure using some prior information and features of brain tissues such as shape, location, and orientation. Parametric and geometric deformable models are two major variants of models-based approaches [10]. Enormous work has been done in the last decade and all the methods discussed above are applied by various researchers for segmentation purpose in the field of medical image analysis. Few of the latest research articles work is discussed underneath and comparative analysis is shown in Table 1.

Stefan Bauer et al. [11] proposed an automatic method for brain tissue segmentation, which combines support vector machine classification using multispectral intensities and textures with subsequent hierarchical regularization based on conditional random fields. Hamamci et al. [12] proposed a segmentation method named Tumor-Cut for radiosurgery applications on contrast-enhanced MR images. The proposed method requires the user to draw the maximum diameter of the tumor on input MRI images. Jainy Sachdeva et al. [13] proposed a model-based approach and performed several experiments to select regions of interest, measure the efficacy of the framework, to remove bias, and to test the robustness of the framework based on active contour model.

Havaei et al. [14] proposed a semi-automated method and transformed the segmentation problem into a classification problem and a brain tumor is segmented by training and classifying within that same brain only. In this framework, a subset of a voxel is selected by the user that belongs to each tissue type within a single brain instead of having many different brain scans for each case to train on with the observers truth. Moreover, Menze et al. [15] proposed a generative probabilistic model in multidimensional MR images using Gaussian mixtures and brain atlas. The author introduced an expectation-minimization (EM) segmentation method to estimate the label map for a new image. They proposed an estimation algorithm with closed-form EM update equations in order to extract a latent atlas prior distribution and the tumor posterior distributions jointly from the image data.

Model-based approach is further implemented by many authors in which Song et al. [16] proposed a level set-based model by classifying the pixels of MR images into three classes, tumor, edema, and healthy tissue. While Pratondo et al. [17] proposed a segmentation method that joins machine learning algorithms and region-based active contour model. We have also gone through some of the novel and recent research works like the one proposed by Shoeb A. Bandey et al. [18] who developed a semi-automated segmentation method based on active contour model to obtain the region of interest with the help of a texture-based image feature map. Nabizadeh et al. [19] proposed an automated framework for brain tumor segmentation using brain

Table 1 Comparative study of the state of the art. [HGG—High-Grade Glioma, DSC—Dice Similarity Coefficient, JSC—Jaccard Similarity Coefficient]

Model	Modality	Method	Tumor	Accuracy		Time
				Complete	Core	
Bauer et al. (2011)	T1, T1c, T2, FLAIR	Hierarchical SVM and CRF	Glioma	0.77(DSC)	0.54(DSC)	<2 min
Hamamci et al. (2012)	T1c	Tumor-Cut algorithm	HGG	0.89(DSC)	-	1–16 min
Sachdeva et al. (2013)	T1, T1c, T2, FLAIR	Content-based active contour model	HGG	85.5%	-	<2 min
Havaei et al. (2015)	T1, T2, FLAIR	Kernelized-SVM, kNN and random forests	HGG	0.86(DSC)	0.71(DSC)	<35 s
Menze et al. (2016)	T1, T1c, T2, FLAIR	Probabilistic generative model	HGG,	0.78(DSC)	0.58(DSC)	-
Song et al. (2017)	T1, T1c, T2, FLAIR	Region-based level set method + K-means	HGG	0.8(DSC)	0.76(DSC)	-
Pratondo et al. (2017)	MRI	Region-based active contour model		0.9(DSC)	-	<6 min
Bandey et al. (2017)	MRI	Active contour model	HGG	0.92(JSC)	-	21 s
Nabizadeh et al. (2017)	T1, T2, FLAIR	Texture-based and contour-based algorithm	HGG	93%	-	-
Usman et al. (2017)	T1, T1c, T2, FLAIR	Wavelets and machine learning	HGG	0.88(DSC)	0.75(DSC)	<2 min
Kaya et al. (2017)	T1	PCA-based clustering with FCM & K-means	-	-	-	-

image texture features and active contour model. Usman et al. [20] proposed a framework for brain tumor segmentation using wavelet and machine learning concepts. They extracted intensity, intensity differences, local neighborhood, and wavelet texture from the preprocessed multimodality brain images. Moreover, Kaya et al. [21] proposed a method for brain tumor segmentation using K-means and fuzzy C-means clustering based on different variations of principal component analysis (PCA) for dimensionality reduction in T1-weighted MR modality.

Segmentation of abnormality in medical images has been always a difficult task due to irregular shape, size, and orientation of the lesion. Automation of such task is still challenging and therefore no such tool available which is universally accepted for the diagnosis of brain MRI. However, classification and clustering-based approaches are found successful and efficient in the process of automated brain tumor segmentation. In this paper, we propose a fully automated framework for brain tumor segmentation using parameter-free clustering approach and mathematical morphological operations. The rest of the paper is organized as follows. Section 2 explains the proposed model and the methodology adopted. The proposed model framework is explained with the help of Fig. 2 and Algorithm 1 which explains the overall step-wise procedure. Section 3 illustrates the dataset selected and the results obtained and discuss the performance measure. Later we showed a comparative analysis of results obtained with the existing techniques. Finally, the conclusion and future works are presented in Sect. 4.

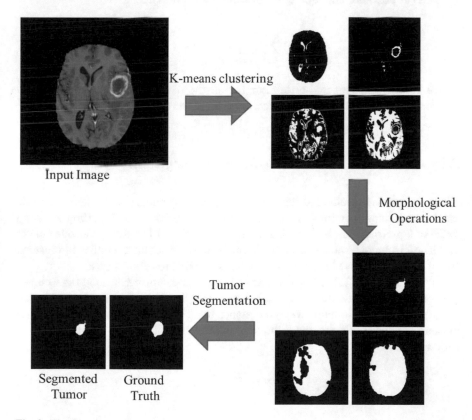

Fig. 2 The flowchart of the proposed model

2   Proposed Model

Time and accuracy are two important measures in the field of medical image analysis. It has been observed that techniques which obtain good results lag in terms of computation time and vice versa. Automation of medical image segmentation needed a robust and efficient algorithm in terms of both detection accuracy and computation time. In this path, we have made an attempt to develop a framework using clustering and morphological operations. For this purpose, we have used parameter-free k-means clustering in association with dilation and hole filling morphological operations. The flowchart of the proposed model is shown in Fig. 2.

Objective of k-means algorithm is such that similarity of data points inside a cluster should be as much as possible, whereas the partitioned clusters should be dissimilar to one another. The squared-error function is used to make the result such that intracluster correspondence is as high as possible and intercluster correspondence is as less as possible. The squared-error function is given as

$$E = \sum_{i=1}^{k} \sum_{p \in C_i} |p - m_i{}^2| \tag{1}$$

where E represents squared-error sum for all the data points in input pattern, p represents a data point, and m_i represents mean of the cluster C_i.

2.1   Cluster Validation with Silhouette Coefficient

Clustering with K-means algorithm always requires the number of clusters (k) while execution as input parameter as discussed in the Algorithm 1. Inputting a wrong value of k may lead to obtaining bad results in the form of clusters. Various cluster validation techniques are popular for identifying or selecting a number of clusters. In our experiments, we have used the Silhouette index for this purpose.

Silhouette index [22] for a particular data point is computed the steps as follows: (1) computing average distance to all data points that belong to the same cluster, p_i; (2) computing minimum average distance to all data points that belong to other clusters, q_i; and (3) finally, the Silhouette index of the given data point is given as follows:

$$S_i = \frac{q_i - p_i}{Max(p_i, q_i)} \tag{2}$$

where $S_i \in [-1, 1]$ and higher the value of S_i better the results. The proposed model for the tumor segmentation of brain MRI is depicted in Fig. 2. The proposed model consists of three major steps: (1) parameter-free k-means clustering applied in contrast-enhanced T1-weighted image to partition the image into different clusters based on pixel intensities, (2) applying dilation and hole filling morphological operations on the clustered image obtained from step 1, (3) tumor detection and comparison with the ground truth. The stepwise algorithm of the proposed model is shown in Algorithm 1.

Algorithm 1: Brain Tumor Segmentation

Input: T1c modality of the MRI
Clustering:
  1. Partitioning the input image into different clusters
  2. Extracting the tumor core (enhancing tumor + necrotic core) using
K-means clustering.
Morphological Operations:
  1. Dilation operation on clustered image
  2. Hole filling operation on the dilated image
Tumor Segmentation: Final segmented tumor region
Similarity with Ground Truth: Calculate the Dice similarity coefficient
(DSC) with the available ground truth
Output: Segmented tumor

---

3 Experiments and Results

3.1 Data Selection

We performed our experiments on BRATS 2015 training dataset which is the skull stripped dataset of glioma patients acquired from different medical institutions and freely available online [4, 23]. The BRATS benchmark dataset is the collection of 3D MRI sequences acquired with the magnetic strength of 1.5T and 3T in four modalities, T1 (T1-weighted image), T2 (T2-weighted image), T1c (T1-weighted with the contrast-enhanced agent (Gad), and FLAIR (T2-weighted image with fluid-attenuated inversion recovery). The entire dataset is divided into two groups, high-grade glioma images, and low-grade glioma images. In our experiments, we have used only T1c modality 3D images of the high-grade tumor (HGG) patients which are acquired with 1 mm isotropic voxel size. All the images are with their ground truth in which the tumor is segmented in four subparts including edema, enhancing tumor, non-enhancing tumor, and necrotic core. However, we have segmented only tumor core which includes enhancing tumor, non-enhancing tumor, and necrotic core but excludes edema.

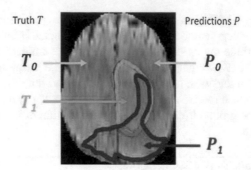

Fig. 3 Predicted tumor core and ground truth for computation of Dice similarity coefficient. where T_1: ground truth for tumor region, T_0: region of brain excluding T_1, P_1: predicted tumor region, and P_0: Predicted region excluding P_1. Adapted from [23]

3.2 Performance Measure

We have used the popular similarity index, Dice similarity coefficient (DSC) for performance measurement of the proposed model for brain tumor segmentation. We obtained a binary map for the predicted tumor region through proposed model, $P \in 0, 1$ with the ground truth, $T \in 0, 1$.

Dice similarity coefficient also known as similarity index, computes the overlapped region between the predicted area and the ground truth area as shown is Fig. 3. Dice similarity coefficient is defined as follows:

$$DSC(P, T) = \frac{|P_1 \wedge T_1|}{(|P_1| + |T_1|)/2} \tag{3}$$

3.3 Results

Results obtained in terms of the Dice similarity coefficient (DSC) and compared to other state-of-the-art methods on the BRATS 2015 challenge training dataset. We have used high-grade glioma (HGG) images for our experiments. Images obtained during various steps of the proposed model are shown in Fig. 4 which depicts an overall procedure of the proposed model from taking the contrast-enhanced T1-weighted image as input to the final computation of tumor core as a prediction of the model.

Our proposed model obtained 75.04% accuracy in terms of Dice similarity coefficient (DSC) when segmenting tumor core when compared with the ground truth. A comparison of the accuracy of various models is shown is Table 2 below which shows that the performance of the proposed method is better than many existing methods.

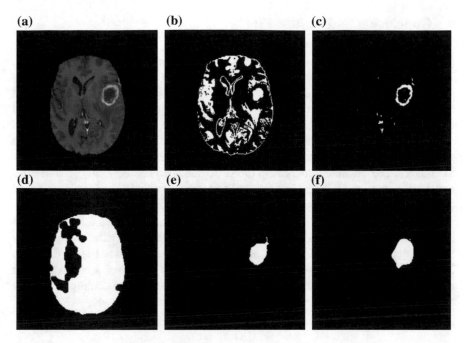

Fig. 4 Results obtained using proposed model for brain tumor segmentation. **a** Contrast-enhanced T1-weighted image slice as input, **b** one of the intermediate clustered image, **c** another intermediate clustered image, **d** one of the intermediate image obtained after dilation and hole filling morphological operations, **e** segmented tumor region, and **f** ground truth

Table 2 Comparison with the existing state-of-the-art models of Dice similarity coefficient of high-grade glioma while segmenting tumor core

Sr. No.	Model	Tumor core (HGG)
1	Bauer et al.(2011) [11]	0.54
2	Zhao et al. (2013) [24]	0.7
3	Reza et al. (2013) [25]	0.72
4	Meier et al. (2013) [26]	0.73
5	Menze et al. (2016) [15]	0.58
6	Pei et al. (2017) [27]	0.64
7	Sauwen et al. (2017) [28]	0.72
8	Proposed model	0.75

4   Conclusion and Future Work

In this work, we proposed an efficient model for lesion detection in brain MR images. The proposed method comprised of two main modules: (1) parameter-free clustering approach which is used to partition input image into different clustered images, and (2) morphological dilation followed by hole filling operations on previously obtained clustered images which lead to obtain segmented tumor core region. The proposed model was applied to the publicly available BRATS 2015 challenge training dataset which consists of a sufficient number of glioma (high-grade and low-grade) patients 3D images. The obtained binary map tumor region is compared to the ground truth and 75.06% accuracy is found in terms of Dice similarity coefficient, which is better than the accuracy of many state-of-the-art methods as shown in Table 2.

In the future, we will use other techniques for segmenting abnormality in medical images discussed in the Introduction section. Since brain tumor patients are increasing day by day, the medical industry needs an efficient automation tool to diagnose brain MRI and provide a better treatment for it. In our future work, we will try to enhance the performance of the system incorporating machine learning techniques and deformable models.

References

1. Louis, D.N., Ohgaki, H., Wiestler, O.D., Cavenee, W.K., Burger, P.C., Jouvet, A., Kleihues, P.: The 2007 WHO classification of tumours of the central nervous system. Acta Neuropathol. **114**(2), 97–109 (2007)
2. Brain tumor statistics. American brain tumor association (2017). http://www.abta.org/about-us/news/braintumorstatistics
3. Drevelegas, A., Papanikolaou, N.: Imaging modalities in brain tumors. Imaging of Brain Tumors with Histological Correlations, pp. 13–33. Springer, Berlin (2011)
4. Kistler, M., et al.: The virtual skeleton database: an open access repository for biomedical research and collaboration (2017). https://ww.smir.ch/BRATS/Start2015
5. Porz, N., Bauer, S., Pica, A., Schucht, P., Beck, J., Verma, R.K., Wiest, R.: Multi-modal glioblastoma segmentation: man versus machine. PloS One **9**(5), e96873 (2014)
6. Dupont, C., Betrouni, N., Reyns, N., Vermandel, M.: On image segmentation methods applied to glioblastoma: state of art and new trends. IRBM **37**(3), 131–143 (2016)
7. Wong, K.P.: Medical image segmentation: methods and applications in functional imaging. Handbook of Biomedical Image Analysis, pp. 111–182. Springer, US
8. Masters, B.R., Gonzalez, R.C., Woods, R.: Digital image processing. J. Biomed. Opt. **14**(2), 029901 (2009)
9. Gordillo, N., Montseny, E., Sobrevilla, P.: State of the art survey on MRI brain tumor segmentation. Magn. Reson. Imaging **31**(8), 1426–1438 (2013)
10. Yao, J.: Image processing in tumor imaging. New techniques in oncologic imaging, pp. 79–102 (2006)
11. Bauer, S., Nolte, L.P., Reyes, M.: Fully automatic egmentation of brain tumor images using support vector machine classification in combination with hierarchical conditional random field regularization. In: International Conference on Medical Image Computing and Computer-Assisted Intervention, pp. 354–361. Springer, Berlin (2011)

12. Hamamci, A., Kucuk, N., Karaman, K., Engin, K., Unal, G.: Tumor-cut: segmentation of brain tumors on contrast enhanced MR images for radiosurgery applications. IEEE Trans. Med. Imaging **31**(3), 790–804 (2012)
13. Sachdeva, J., Kumar, V., Gupta, I., Khandelwal, N., Ahuja, C.K.: Segmentation, feature extraction, and multiclass brain tumor classification. J. Digit. Imaging **26**(6), 1141–1150 (2013)
14. Havaei, M., Larochelle, H., Poulin, P., Jodoin, P.M.: Within-brain classification for brain tumor segmentation. Int. J. Comput. Assist. Radiol. Surg. **11**(5), 777–788 (2016)
15. Menze, B.H., Van Leemput, K., Lashkari, D., Riklin-Raviv, T., Geremia, E., Alberts, E., Ayache, N.: A generative probabilistic model and discriminative extensions for brain lesion segmentation with application to tumor and stroke. IEEE Trans. Med. Imaging **35**(4), 933–946 (2016)
16. Song, Y., Ji, Z., Sun, Q., Zheng, Y.: A novel brain tumor segmentation from multi-modality MRI via a level-set-based model. J. Signal Process. Syst. **87**(2), 249–257 (2017)
17. Pratondo, A., Chui, C.K., Ong, S.H.: Integrating machine learning with region-based active contour models in medical image segmentation. J. Vis. Commun. Image Represent. **43**, 1–9 (2017)
18. Banday, S.A., Mir, A.H.: Statistical textural feature and deformable model based brain tumor segmentation and volume estimation. Multimed. Tools Appl. **76**(3), 3809–3828 (2017)
19. Nabizadeh, N., Kubat, M.: Automatic tumor segmentation in single-spectral MRI using a texture-based and contour-based algorithm. Expert Syst. Appl. **77**, 1–10 (2017)
20. Usman, K., Rajpoot, K.: Brain tumor classification from multi-modality MRI using wavelets and machine learning. Pattern Anal. Appl. 1–11 (2017)
21. Kaya, I.E., Pehlivanl, A.Ç., Sekizkardeş, E.G., Ibrikci, T.: PCA based clustering for brain tumor segmentation of T1w MRI images. Comput. Methods Progr. Biomed. **140**, 19–28 (2017)
22. Rousseeuw, P.J.: Silhouettes: a graphical aid to the interpretation and validation of cluster analysis. J. Comput. Appl. Math. **20**, 53–65 (1987)
23. Menze, B.H., Jakab, A., Bauer, S., Kalpathy-Cramer, J., Farahani, K., Kirby, J., Lanczi, L.: The multimodal brain tumor image segmentation benchmark (BRATS). IEEE Trans. Med. Imaging **34**(10), 1993–2024 (2015)
24. Wu, W., Chen, A.Y., Zhao, L., Corso, J.J.: Brain tumor detection and segmentation in a CRF (conditional random fields) framework with pixel-pairwise affinity and superpixel-level features. Int. J. Comput. Assist. Radiol. Surg. **9**(2), 241–253 (2014)
25. Reza, S., Iftekharuddin, K.: Multi-class abnormal brain tissue segmentation using texture features. In: Proceedings of NCIMICCAI BRATS, vol. 1, pp. 38–42 (2013)
26. Meier, R., Bauer, S., Slotboom, J., Wiest, R., Reyes, M.: Appearance-and context-sensitive features for brain tumor segmentation. In: Proceedings of MICCAI BRATS Challenge, 020-026 (2014)
27. Pei, L., Reza, S.M., Li, W., Davatzikos, C., Iftekharuddin, K.M.: Improved brain tumor segmentation by utilizing tumor growth model in longitudinal brain MRI. In: SPIE Medical Imaging (pp. 101342L–101342L). International Society for Optics and Photonics (2017)
28. Sauwen, N., Acou, M., Sima, D.M., Veraart, J., Maes, F., Himmelreich, U., Van Huffel, S.: Semi-automated brain tumor segmentation on multi-parametric MRI using regularized non-negative matrix factorization. BMC Med. Imaging **17**(1), 29 (2017)

Analysis of Breathy, Emergency and Pathological Stress Classes

Amit Abhishek, Suman Deb and Samarendra Dandapat

Abstract Recently, man–machine interaction based on speech recognition has taken an increasing interest in the field of speech processing. The need for machine to understand the human stress levels in a speaker-independent manner, to prioritize the situation, has grown rapidly. A number of databases have been used for stressed speech recognition. Majority of the databases contain styled emotions and Lombard speech. No studies have been reported on stressed speech considering other stress conditions like emergency, breathy, workload, sleep deprivation and pathological condition. In this work, a new stressed speech database is recorded by considering emergency, breathy and pathological conditions. The database is validated with statistical analysis using two features, mel-frequency cepstral coefficient (MFCC) and Fourier parameter (FP). The results show that these recorded stress classes are effectively characterized by the features. A fivefold cross-validation is carried out to assess how the statistical analysis results are independent of the dataset. Support vector machine (SVM) is used to classify different stress classes.

Keywords Emotion · Stress · Breathy · Emergency · Pathological

1 Introduction

Stress recognition is the process of identifying the stress class of the person from his/her speech. The characteristics of speech signal changes under stress conditions. Due to this, the performance of machine is affected in case of human–machine

A. Abhishek (✉) · S. Deb · S. Dandapat
Electronics and Electrical Engineering, Indian Institute of Technology,
Guwahati 781039, Assam, India
e-mail: amit.abhishek@iitg.ernet.in

S. Deb
e-mail: suman.2013@iitg.ernet.in

S. Dandapat
e-mail: samaren@iitg.ernet.in

© Springer Nature Singapore Pte Ltd. 2019   497
M. Tanveer and R. B. Pachori (eds.), *Machine Intelligence and Signal Analysis*,
Advances in Intelligent Systems and Computing 748,
https://doi.org/10.1007/978-981-13-0923-6_43

interaction. The causes of this stressed speech can be due to emergency conditions, fatigue/physical environmental factor, pathological condition (disease), sleep deprivation, perceived threat, glottal abnormalities, workload and noisy environments (Lombard effect) [12, 17].

Stress recognition plays a significant role in speech recognition [2]. Accuracy of speech recognition under stressed environment needs careful consideration of retraining the model in simulated stressed environment. This simulated stress conditions improve the stress classification in a speaker-dependent manner, but reduce drastically in speaker-independent scenarios due to the large variation of the stress parameter among people. This is due to the fact that the stress causes change in the vocal tract and the breathing pattern.

Stress classification not only increase the speech recognition rate but also can be applied to (*i*) prioritize the emergency situation [19], (*ii*) medical situation [19], (*iii*) analysis of breathing pattern of sports person, (*iv*) assess the quality of customer satisfaction in telecommunication industry and (*v*) forensic analysis of the caller by the law enforcement [24]. Earlier studies have explored the prospect of the using continuous features such as pitch, formant and energy-related features for stress cues [5, 10]. Speech spectral features like MFCC, LPC and LPCC are also proven good speech stress descriptor [13, 25]. But the studies in [1, 8, 9] have found that MFCC, LPCC outperforms the LPC in detecting speech under stress. The non-linear Teager energy operator (TEO) based features are also proved to be efficient in stress speech detection [32].

So for an efficient recognition of the speech stress levels, the appropriate features must be extracted and the further processed to detect stress. It has been demonstrated that the voice quality features [14, 21, 22, 30] are related to speech stress levels. The first approach to determining the voice quality parameter is to removing the filtering effect of the vocal tract [25] and measuring the glottal wave parameters. Second approach is to directly extracting the quality parameter from speech signal, represented by jitter and shimmer. A set of harmony features, first proposed by Yang and Lugger [22], which came from the well-known psychoacoustic harmony perception in music theory, for automatic emotion recognition. According to music theory, the harmony structure of chord is mainly responsible for producing a positive or negative impression on listeners. Acoustic interpretation explains that the unique quality (tone) of each instrument is due to the unique structure of a harmonic sequence. To detect this perceptual content of the speech a set a harmonic sequence is taken named Fourier parameter and its first-order (Δ) and second-order differences ($\Delta\Delta$). The organization of the paper is summarized as follows. The detail explanation of database creation is given in Sect. 2. Feature extraction and analysis is carried out in Sect. 3. The results of the proposed work are explained in Sect. 4. Finally, we conclude the work in Sect. 5.

2 Database Description

2.1 Existing Database

In literature, a number of stressed speech databases have been investigated. EMODB database [4] is one of the most popular databases for stress/emotion classification is recorded for ten German sentences and ten speakers participated for data recording. The database contains seven emotions as anger, anxiety, boredom, disgust, happiness, neutral and sadness. The speech under simulated and actual stress (SUSAS) database has been used extensively [16, 32]. The SUSAS database is partitioned into five domains: (*i*) talking style, (*ii*) single tracking task or speech produced in noise (Lombard effect), (*iii*) dual tracking task, (*iv*) actual speech under stress and (*v*) psychiatric analysis data. Lombard speech was created by adding 85 dB SPL pink noise through headphones. In [26], two databases were used: one is the English language database, and the other one is the Telugu language (an Indian language) database. Both the databases contain three stress classes, anger, happiness and compassion. Shukla et al. created a simulated stressed speech database [27]. The database is of Hindi language (an Indian language). The database consists of four stress conditions, angry, sad, Lombard and neutral. Fifteen non-professional speakers participated in the recording. A total of 35 isolated keywords were selected for the data recordings. Deb and Dandapat created a stressed speech database, which contains three stress classes, out-of-breath speech, low out-of-breath speech and normal speech [11].

2.2 Limitations of Existing Databases

Majority of the databases in the literatures have investigated stress as an analogy to emotion. But for man–machine interaction to prioritize the situation, analysis of different classes of stress speech is necessary. Again, to best of our knowledge, no database has recorded a separate class for emergency. In this paper, we have classified stress into four classes as urgency, breathy, pathological and normal. A new database is created for these four stress classes and an evaluation of the database is done using the existing methods.

2.3 IITG-Stress Database

Database creation is the most important part of any speech processing algorithm. For a speaker-independent stress recognition system, a database with varied speaking style and culture plays a significant role. In this paper, we have created a new database named IITG-Stress database. The database is created with the help of 17 non-professional speakers (3 female, 14 male), from different parts of the country

Table 1 Recoding details of IITG-Stress database

Number of speaker	17 (3 female, 14 male)
Number of emotions	4
Number of sentences	5
Number of utterances	340
Sampling frequency (F_s)	11025 Hz

Table 2 Sentences used for recording of IITG-Stress database

S.N.	Sentences
1	The fire is spreading
2	Give me some water
3	The storm is coming
4	Call the ambulance
5	Hurry up, there is an accident on the highway

(India). The speakers are research scholars of Indian Institute of Technology Guwahati (India), and are knowledgeable and familiar with primary speech processing methodologies. All the speakers fall in the age group of 23–30 years. Five sentences for each of the four stress level were recorded from each speaker. Table 1 and Table 2 represents the recording details and recorded sentences, respectively.

Recording has been done in a controlled studio environment. Breathy utterances have been recorded by the speaker after some physical exercise for 5–10 min.

3 Features Extraction and Analysis

In this paper, the database is validated using two features, MFCC and FP. The detailed discussion is given below.

3.1 Fourier Parameter (FP)

Speech signal is a combination of the different sinusoidal signal vibrating at different resonance frequencies. It can be modelled as the passing of the glottal impulses through a time-varying vocal tract filter [23]. Fourier transformation of a signal decomposes the signal to its fundamental and harmonic frequencies. So using Fourier transformation as an analytical tool we will decompose the speech signal having M frames.

So speech signal can be modelled as the combination of the harmonics (1):

$$s(n) = \sum_{k=1}^{K} A_k^m(n) cos\left(2\pi \frac{f_k^m}{F_s} n + \phi_k^m\right) \tag{1}$$

where $A_k^m(n)$ and ϕ_k^m is the amplitude and phase of the kth harmonics, m is the index of the frame, F_s is the sampling frequency, K is the total number of speech harmonic components.

Discreet Fourier transform (DFT) deals with representing $s(n)$ with samples of its spectrum $S(w)$. As $S(w)$ is periodic in 2π radians, DFT is established from its sampled Fourier transform in fundamental range $0 \leq w \leq 2\pi$.

$$A(k) = \sum_{n=0}^{N-1} s(n) e^{-j\frac{2\pi}{N} nk} \qquad k = 0, 1, 2, \ldots N-1 \tag{2}$$

Harmonics include magnitude and phase.

3.2 Mel-Frequency Cepstral Coefficient

Mel-frequency cepstral coefficients (MFCC) represents the psychoacoustic nature [3, 18, 30] of human ear which is sensitive to the logarithmic value of audio frequency.

13-dimensional MFCC features are extracted from the frames along with its delta (Δ) and delta–delta ($\Delta\Delta$). From Fig. 1 it can be inferred that there is a significant variation across frames of different stress utterances. In Sect. 4, we will validate the effectiveness of using MFCC feature.

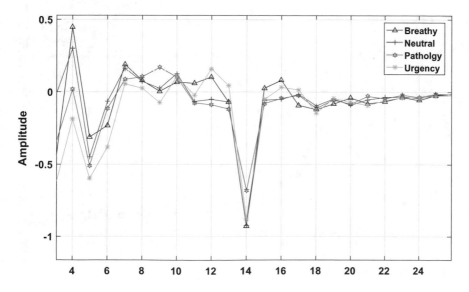

Fig. 1 Averaged MFCC feature

3.3 Features Analysis

For every frame, Fourier parameter is estimated using Fourier analysis. As shown in (1) and (2), $A_k^m(n)$ is the magnitude of kth Fourier parameter of mth frame. Figure 2 shows 60 Fourier parameter averaged over frame for individual stress. The harmonic amplitude has large variation only at lower harmonics and smoother in lower harmonics. The peak of the stress utterances are at the lower harmonics. For example, the peak of urgency stress is at A_{12}. These variations may be used to classify the stress from speech more effectively.

Figure 3 represents the FP averaged over similar stress sentences for $k = 9$, A_9^m. It shows averaged harmonic coefficient A_9 over the frames for various stress utterances. It is evident that the amplitude of the harmonic magnitudes is different for different stress. It can be seen that the averaged A_9 for breathy is higher among all stress levels whereas normal has the lowest.

3.4 Features Extraction

The database created contains 340 utterances, 85 utterances from each stress class. Fourier transform is done along frames and 120 Fourier parameters are taken as the stress features. The delta (Δ) and delta–delta ($\Delta\Delta$) differences of the Fourier parameters are also derived. The mean, maximum, minimum, median and standard

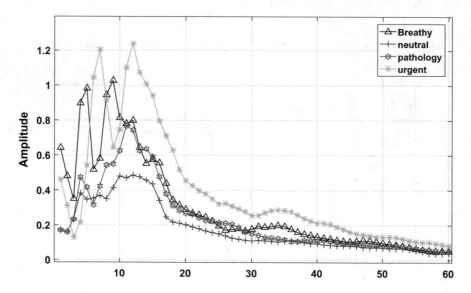

Fig. 2 Harmonic coefficients

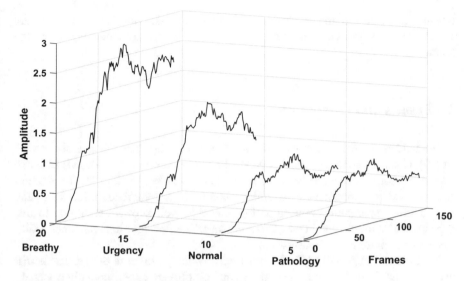

Fig. 3 Averaged harmonic magnitudes for $k = 9$

deviation [2, 7, 15, 22] of the Fourier parameters are calculated and appended, constructing a 1785 feature vector.

Similarly, the MFCC features are derived along frames so as the delta (Δ) and delta–delta ($\Delta\Delta$) MFCC. The mean, maximum, minimum, median and standard deviation of the MFCC, delta and delta–delta MFCC are calculated, making it a 195 feature vector.

So a combined 1980 features are extracted from each utterance.

3.5 Features Normalization

Feature normalization is a method to standardize the range of the features [20]. The range of the features extracted varies widely. As a result, the distance between two features varies broadly. Therefore, the range of all features should be normalized so that each feature contributes approximately proportionately to the final distance. The goal is to eliminate speaker and recording variability while keeping the effectiveness of stress discrimination. Generally, two type of normalization is used for feature scaling, zeros-mean-unit-variance, min-max normalization. Here we have used min-max normalization given by following equation:

$$labelequ3 X = \frac{X - min(X)}{max(X) - min(X)} \tag{3}$$

where X is the feature matrix of n × 1980 dimension, and min(X) and max(X) are 1 × 1980 dimension. The minimum and maximum values are calculated along the feature.

4 Results and Discussion

In the field of speech, a number of classifiers including artificial neural network (ANN), Gaussian mixture model (GMM), hidden Markov model (HMM) [29] and support vector machine (SVM) have been studied. SVM [18, 28, 31] has demonstrated good performance on several problems of pattern recognition including speech stress recognition and outperformed other well-known classifiers. In this paper, we have used SVM in two different ways. The first solution was 'one-vs-all'. The second solution was 'one-vs-one'.

We have used LIBSVM [6] which is freely available, for ease of use. The kernel used is 'radial basis function', gamma (g) and cost (c) are experimentally selected.

4.1 Result Analysis Using One-vs-All Multi-Class Classification

Figure 4 represents results of 'one-vs-all' with different feature set FP, MFCC and combination of both. Using FP features we have achieved an accuracy of 72.4% for urgency, 57% for breathy, 54.2% for pathology, 60.6% for neutral, with an average accuracy of 61.05%. While using MFCC features we got an average accuracy of 55.35%. We have tested the result with the combined feature set and achieved an accuracy of 73.6% for urgency, 57.2% for breathy, 54.2% for pathology, 51.2% for neutral with an average accuracy of 59%. It can be observed that with the combined feature set, the result has been improved for all of the stress levels except neutral condition.

4.2 Result Analysis Using One-vs-One Multi-Class Classification

'One-vs-one' classification is carried out on the database and the confusion matrices of the fivefold cross-validation are given in Tables 3, 4, 5, 6 and 7.

The total dataset is separated into training set with 80% utterance and test set with 20% utterance. A fivefold validation has been done and the confusion matrix is given below.

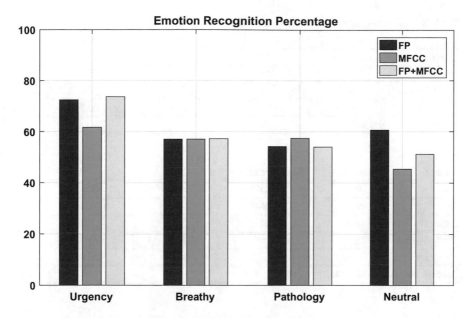

Fig. 4 Comparison with different features

Table 3 Confusion matrix (%) of classification result for first fold

	Breathy	Neutral	Pathological	Urgency
Breathy	*58.8*	41.2	0	0
Neutral	11.7	*70.7*	17.6	0
Pathological	0	47	*41.2*	11.8
Urgency	0	0	0	*100*

Average accuracy = *69.17%*

Table 4 Confusion matrix (%) of classification result for second fold

	Breathy	Neutral	Pathological	Urgency
Breathy	*64.7*	0	0	35.3
Neutral	47	*41.2*	11.8	0
Pathological	11.8	11.8	*64.6*	11.8
Urgency	0	0	0	*100*

Average accuracy = *64.7%*

For first set of data, an accuracy of 69.17% has been achieved. Similarly, 64.7, 58, 54.42 and 44.17% results have been achieved with different sets of data for fivefold validation. Total average accuracy was 57.2%.

Table 5 Confusion matrix (%) of classification result for third fold

	Breathy	Neutral	Pathological	Urgency
Breathy	*53*	17.6	0	29.4
Neutral	23.5	*58.8*	5.9	11.8
Pathological	17.6	11.8	*53*	17.6
Urgency	5.9	11.8	11.8	*70.5*

Average accuracy = *58%*

Table 6 Confusion matrix (%) of classification result for fourth fold

	Breathy	Neutral	Pathological	Urgency
Breathy	*41.3*	17.6	17.6	23.5
Neutral	5.9	*76.5*	11.8	5.8
Pathological	23.5	23.6	*29.4*	23.5
Urgency	17.6	11.8	0	*70.6*

Average accuracy = *54.41%*

Table 7 Confusion matrix (%) of classification result for fifth fold

	Breathy	Neutral	Pathological	Urgency
Breathy	*17.6*	70.6	11.8	0
Neutral	0	*94.1*	5.9	0
Pathological	0	53	*47*	0
Urgency	41.2	35.3	5.9	*17.6*

Average accuracy = *44.17%*

5 Conclusion

In previous studies, different emotional databases are used for stress evaluation. In this paper, we developed a new database (IITG-Stress database) for different stress classes as emergency, breathy and pathological conditions and neutral. IITG-Stress database is evaluated with the Fourier parameter and MFCC. It is observed that the Fourier parameters and MFCC are different for different stress classes.

Furthermore, both FP and MFCC features were evaluated for speaker-independent stress speech recognition by using SVM. The study showed that FP features are effective in characterizing and recognizing different stress classes, breathy, urgent and pathological, in speech signals than MFCC. Moreover, it is possible to improve the performance of stress recognition by combining both FP and MFCC features. These results establish that the new IITG-Stress database is helpful for speaker-independent speech stress recognition.

Acknowledgements The author would like to thank all the speakers participated in the data recordings for IITG-Stress database.

References

1. Atal, B.S.: Effectiveness of linear prediction characteristics of the speech wave for automatic speaker identification and verification. J. Acoust. Soc. Am. **55**(6), 1304–1312 (1974)
2. Ayadi, M.E., Kamel, M.S., Karray, F.: Survey on speech emotion recognition: features; classification schemes; and databases. Pattern Recogn. **44**(3), 572–587 (2011)
3. Bou-Ghazale, S.E., Hansen, J.: A comparative study of traditional and newly proposed features for recognition of speech under stress. IEEE Trans. Speech Audio Process. **8**(4), 429–442 (2000)
4. Burkhardt, F., Paeschke, A., Rolfes, M., Sendlmeier, W., Weiss, B.: A database of German emotional speech. In: Proceedings of Interspeech, Lissabon pp. 1517–1520 (2005)
5. Busso, C., Lee, S., Narayanan, S.: Analysis of emotionally salient aspects of fundamental frequency for emotion detection. IEEE Trans. Audio Speech; Lang. Process **17**(4), 582–596 (2009)
6. Chang, C.C., Lin, C.J.: Libsvm: a library for support vector machines. CM Trans. Intel. Syst. Technol. (TIST) **2**(3), 27 (2011)
7. Clavel, C., Vasilescu, I., Devillers, L., Richard, G., Ehrette, T.: Feartype emotion recognition for future audio-based surveillance systems. Speech Commun. **50**, 487–503 (2008)
8. Davis, S., Mermelstein, P.: Comparison of parametric representations for monosyllabic word recognition in continuously spoken sentences. IEEE Trans. Acoust. Speech Signal Process. **28**(4), 357–366 (1980)
9. Deb, S., Dandapat, S.: Emotion classification using residual sinusoidal peak amplitude. In: 2016 International Conference on Signal Processing and Communications (SPCOM), pp. 1–5, June 2016
10. Deb, S., Dandapat, S.: Classification of speech under stress using harmonic peak to energy ratio. Comput. Electr. Eng. **55**, 12–23 (2016)
11. Deb, S., Dandapat, S.: Fourier model based features for analysis and classification of out-of-breath speech. Speech Commun. **90**, 1–14 (2017)
12. Deb, S., Dandapat, S.: A novel breathiness feature for analysis and classification of speech under stress. In: 2015 Twenty First National Conference on Communications (NCC), pp. 1–5. IEEE (2015)
13. Deb, S., Dandapat, S.: Emotion classification using segmentation of vowel-like and non-vowel-like regions. IEEE Trans. Affect. Comput. (2017)
14. Gobl, C., Chasaide, A.N.: The role of voice quality in communicating emotion mood and attitude. Speech Commun. **40**, 189–212 (2003)
15. Grimm, M., Kroschel, K., Mower, E., Narayanan, S.: Primitives based evaluation and estimation of emotions in speech. Speech Commun. **49**, 787–800 (2007)
16. Hansen, J.H.: Analysis and compensation of speech under stress and noise for environmental robustness in speech recognition. Speech Commun. **20**(1), 151–173 (1996). Speech under Stress
17. Hansen, J.H., Patil, S.: Speech under stress: Analysis, modeling and recognition. In: Speaker Classification I, pp. 108–137. Springer (2007)
18. Kamaruddina, N., Wahabb, A., Quek, C.: Cultural dependency analysis for understanding speech emotion. Expert Syst. Appl. **39**, 5115–5133 (2012)
19. Kotti, M., Paterno, F.: Speaker-independent emotion recognition exploiting a psychologically-inspired binary cascade classification schema. Int. J. Speech Technol. **15**, 131–150 (2012)
20. Kustner, O., Tato, R., Kemp, T., Meffert, B.: Towards real life applications in emotion recognition. In: Proceedings of the Conference on Affective Dialogue Systems, pp. 25–35 (2004)

21. Li, X., Tao, J., Johnson, M.T., Soltis, J., Savage, A., Leong, K.M., Newman, J.D.: Stress and emotion classification using jitter and shimmer features. In: Proceedings of the IEEE International Conference on Acoustics Speech and Signal Processing, vol. 4, pp. IV–1081–IV–1084 (2007)

22. Lugger, M., Yang, B.: Combining classifiers with diverse feature sets for robust speaker independent emotion recognition. In: Proceedings of the 17th European Signal Processing Conference, pp. 1225–1229 (2009)

23. McAulay, R.J., Quatieri, T.F.: Speech analysis/synthesis based on a sinusoidal representation. IEEE Trans. Acoust. Speech Signal Process. **34**(4), 744–754 (1986)

24. Ntalampiras, S., Potamitis, I., Fakotakis, N.: An adaptive framework for acoustic monitoring of potential hazards. EURASIP J. Audio Speech Music Process. **2009**(13), 1–15 (2009)

25. Rabiner, L., Schafer, R.: Digital Processing of Speech Signals, 1st ed. Prentice Hall, Upper Saddle River, New Jersey 07458, USA (1978)

26. Ramamohan, S., Dandapat, S.: Sinusoidal model-based analysis and classification of stressed speech. IEEE Trans. Audio Speech Lang. Process. **14**(3), 737–746 (2006)

27. Shukla, S., Dandapat, S., Prasanna, S.R.: Spectral slope based analysis and classification of stressed speech. Int. J. Speech Technol. **14**(3), 245–258 (2011)

28. Vapnik, V.N.: Statistical Learning Theory. Wiley, New York, NY, USA (1998)

29. Wagner, J., Vogt, T., Andre, E.: A systematic comparison of different hmm designs for emotion recognition from acted and spontaneous speech. In: Proceedings of the 2nd International Conference on Affective Computing and Intelligent Interaction, vol. 4738, pp. 114–125 (2007)

30. Yang, B., Lugger, M.: Emotion recognition from speech signals using new harmony features. Signal Process. **90**, 1415–1423 (2010)

31. You, M.Y., Chen, C., Bu, J.J., Liu, J., Tao, J.H.: Emotion recognition from noisy speech. In: Proceedings of the IEEE International Conference on Multimedia Expo, pp. 1653–1656, Jul 2006

32. Zhou, G., Hansen, J.H.L., Kaiser, J.F.: Nonlinear feature based classification of speech under stress. IEEE Trans. Speech Audio Process. **9**(3), 201–216 (2001)

An Empirical Investigation of Discretization Techniques on the Classification of Protein–Protein Interaction

Dilip Singh Sisodia and Maheep Singh

Abstract Protein–protein interaction is a biological process, which plays a vital role in the functioning of the metabolic process inside the organism. More than 80% of protein does not perform function alone but performs in combination. Some non-identified protein can be identified with their interaction with a protein whose function is already known. Protein–protein interactions (PPI) and Protein–protein non-interactions (PPNI) display different levels of growth rate, and the number of PPI is significantly greater than that of PPNI. This significant difference in the number of PPI and PPNI increases the cost of constructing a balanced data set. In this paper, the effect of various discretization techniques including Ameva, Class-Attribute Inter-Dependence Maximization (CAIM), Chi-merge, and Fu sinter is investigated with different classification techniques. The CAIM Discretization with SVM has a significant impact on the result as compared to normal SVM using 10-fold cross-validation. Experiments are performed on E. coli and H. Sapiens protein datasets, and we achieved excellent results with accuracies 92.8% and 93.8% on average in CAIM Discretization using SVM classifier, with AUC values of 80.7% and 82.1% respectively.

Keywords Discretization · Metabolic · SVM · C4.5 · Protein–protein interaction

1 Introduction

The participation of "Proteins" in different biological processes results in cellular functions by the interrelation of other proteins in order to function it properly. PPI is the most vital thing inside a cell. It is important for both functional and structural

D. S. Sisodia (✉)
National Institute of Technology Raipur, Raipur, India
e-mail: dssisodia.cs@nitrr.ac.in

M. Singh
Guru Ghasidas Central University, Bilaspur, India
e-mail: maheeps99@gmail.com

© Springer Nature Singapore Pte Ltd. 2019
M. Tanveer and R. B. Pachori (eds.), *Machine Intelligence and Signal Analysis*,
Advances in Intelligent Systems and Computing 748,
https://doi.org/10.1007/978-981-13-0923-6_44

aspects, i.e., Muscle contraction, cell signaling, cellular transport, and biochemical pathways functions are possible due to protein–protein interaction. Knowledge of PPI can give valuable intuitions about the mechanism of the functions of genes in a living cell.

The interaction between protein–protein is done by chemical signaling method. This interaction is helpful in recovering from diseases. The effect of PPI interaction is the damages of protein and change the kinetic equity of enzymes [1]. There are two types of PPI interaction transient, and stable, which are categorized by interaction for a short duration and a long duration such as stable interaction interacts for a long duration, and transient interaction interacts for a short duration of time. The important problem is that the PPI. Prediction is not well equitable classification problem. Hence, the finding of PPI has been the core issue in biological systems and gnomonical functions. In the recent past, several experimental approaches were proposed to detect PPI. However, there are some drawbacks in existing experimental methods, like low AUC values, high cost and time-consuming. Therefore, computational investigation of PPI is considered one of the most important in biological datasets.

This paper presents a method for predicting PPI with discretization. The purpose of this study is the use of discretization technique with different classifiers. Discretization is the process of categorizing the continuous variables, values, and attributes into discrete values. We use many discretization techniques like Ameva, CAIM, Fusinter, Chi-merge. Ameva generates a minimum number of discrete intervals based on chi-square method [2]. It only works in supervised learning. CAIM is used to boost the class and attribute interrelationship to provide a minimum number of discrete intervals [3]. Chi-merge uses the χ^2 statistical method to convert the numerical attributes into discrete attributes [4]. Fusinter is an absolute dominant method which is used for discretizing the numerical values [5]. In machine learning, classification comes under supervised learning in which it predicts the new observation belongs to which category, by training data set observation. The algorithm to implement the classification is called as classifiers. In this study, we use two classifiers; they are SVM and C4.5. SVM (Support vector machine) is a selective approach to separate the class label through a hyperplane. C4.5 (Decision tree): It has a tree-like structure and predicts whether the happening condition is right or wrong [6]. This is applicable for both supervised and unsupervised learning. Datasets used in the problems may contain many features.

2 Related Work

In the starting phase of PPI research, many efficient experimental techniques were applied, like, Mass Spectrometry [7], yeast two-hybrid system [8]. The method proposed by Subhadip Basu et al. [9] known as fuzzy SVM, with 10-fold cross-validation, which performs better as compared to classical SVM. On the test datasets, their average AUC for 10-fold cross-validation using fuzzy SVM was 76.59% and 80.17% for H. Sapiens and E. coli respectively. V. Srinivasa Rao et al. proposed a

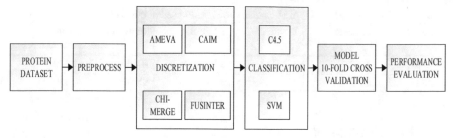

Fig. 1 Block diagram of proposed method

method [10] for PPI, which has three approaches, namely "in vitro", "in vivo" and "in silico". Again, each approach has different techniques. (1) The "in vitro" method was performed outside of the living organism. (2) The "in vivo" method was performed within a living organism. (3) The "in silico" method was performed via a computer, also called as a computer simulation. The method proposed by Wang et al. [11] to predict PPI, which is the combination of Discrete cosine transform (DCT) and Ensemble Rotation forest (RF) algorithm. First, protein sequences were converted into the position-specific scoring matrix (PSSM) and then the feature is drawn out with the help of DCT technique, and at last ensemble, rotation forest algorithm was applied. According to Du et al. [12], the ensemble learning method was used for PPI sites with weighted feature descriptor (EL-WFD). The weighted feature descriptor was used to find the distance between protein residues. Guo et al. [13] proposed a method based on hexagon structure similarity, which identifies residues on interfaces from an input protein with both sequence and 3D structure information. According to Zhou and Shan [14], PPI sites were predicted from a neural network with sequence profiles of neighboring residues and solvent exposure as input. The main strong point of the network predictor lies in the fact that neighbor lists and solvent exposure are comparatively insensitive to structural changes accompanying complex formation.

3 Material and Methodology

In this section description of the methodology used to perform the present work is given. First of all, preprocessing is applied to protein dataset [15, 16] and after preprocessing dataset is divided into test and train set. Different discretization technique is used on the preprocessed dataset. Different classification algorithms are trained using the training dataset. The test dataset is used for testing and do performance evaluation with respect to different parameters (Fig. 1).

Table 1 Statistics of PPI dataset of H. Sapiens and E. coli used in this paper

Datasets	Instances	Target	Feature	Interacting	Non-interacting
H. Sapiens	7338	2	336	1010	6228
E. coli	8259	2	336	1220	7039

3.1 Protein Dataset

For the experimental work, the core subset which contained 7338 pairs is employed for H. Sapiens, which has 336 features from which 1010 are interacting, and 6228 are not-interacting, and in E. coli, there are 8259 pairs which have 336 features from which 1220 are interacting, and 7039 are non-interacting given in Table 1. Dataset is taken from [17], and both datasets have binary feature.

3.2 Preprocessing

Design of Dataset. Before applying any algorithm, we must make data set according to the algorithm. Let p_x and p_y be a protein interaction pair, which can be characterized by the amino acid arrangements $w_1, w_2, \ldots, w_n$ and $v_1, v_2, \ldots, v_m$ respectively. Where $w_k, v_l \in \{A, R, N, D, L, K, M, F, C, Q, E, G, H, I, P, S, T, W, Y, V\} \forall k = w_k, v_l \in \{A, R, N, D, L, K, M, F, C, Q, E, G, H, I, P, S, T, W, Y, V\} \forall k = 1$ to n and $\forall l = 1$ to m.

We have to calculate the distance (w_k, v_l) in order to find out the interatomic distance between p_x and p_y. If the distance is less than $3.5°$ A [18], then the corresponding residue pair (w_k, v_l) which belongs to the protein (pair p_x and p_y) is said to be interacting. Otherwise, they are said to be non-interacting. The sequence of proteins is theoretically divided into sub sequences having multiple overlapping sections, where each section consists of 21 amino acids.

For each local sub-sequence pair from proteins p_x and p_y, we have considered all residues from $w_1, w_2, \ldots, w_n$ and $v_1, v_2, \ldots, v_m$, where, the section contains 21 amino acids respectively, and also verified whether any of the residue pairs has Distance $(w_k, v_l) < 3.5°$A. Regarding the threshold of the interatomic distance, we followed many relevant works. After extracting dominant residue pair to be interacting one, we marked the residues of the pair of subsequences which are obtained from p_x and p_y as positive. We then take out HQI-8 features [17, 19] for 42 residues (two proteins each containing 21 amino acids), resulting in a feature vector of 428 * 336, representing the positive training case. The shifting of overlapping subsequences was done as a hypothetical sliding window, in order to analyze further interactions. The vector feature is taken from [9].

3.3 Description of Discretization Technique

Several supervised machine learning algorithms need a discrete feature space. In machine learning applications, feature representation is an important aspect. This paper used discretization to predict PPI. The process of discretization method involves in converting continuous values into discrete values by incorporating intervals where attribute values can be put rather than singleton values [20]. There are various techniques, but in this paper, we use only four discretization techniques which predicted PPI well.

Ameva [2]. To find a discretization scheme with a globally ideal value is a major problem. The first method, which starts with a single interval [2], functions in a top-down technique, where two new intervals are formed by dividing one of the existing interval, which produces the optimal value. This method performs the discretization task at minimal computational cost. Therefore, it may be applied to continuous attributes which are having a large number of exclusive values and then locating the local maximum values of the Ameva Criterion. The second method is that, when the number of labels is small, we can apply a genetic algorithm for this type of problems which calculates global maximum values of the Ameva criterion. Both approaches find the less number of intervals.

CAIM [3]. The main aim of CAIM algorithms is to minimize the discrete interval numbers while reducing the loss of inter-dependency between class attribute [3]. This process makes use of class attribute inter-dependency information as the criterion for the ideal discretization. The ideal discretization structure can be obtained by searching the entire space of all possible discretization structures to find one having the largest value of the CAIM criterion. That searching of the structure is highly time-consuming. Therefore, the CAIM Discretization method uses a greedy approach, which searches for the local maximum value of the criterion as the optimal value of CAIM discretization.

Chi-merge. The Chi-merge algorithm first initializes and then bottom-up merging process is applied until a termination condition is met [4]. It is initialized by first sorting the training samples according to their value for the attributes being discretized and then making the initial discretization, in which each sample is put into its own interval. The interval merging process comprised of two steps and repeated continuously. First, we calculate the value of χ^2 for each pair of adjacent intervals; the second step is to combine the pair of adjacent intervals with the smallest value of χ^2. The merging process continues until all pairs have χ^2 values beyond the parameter of the χ^2 threshold.

Fusinter. This algorithm uses the same approach as the chi-merge method. Fusinter is a bottom-up algorithm to search the partition which improves the measures. The main feature of Fusinter is based on measure sensitive to sample size and used in SIPINA induction graph construction technique. The main advantage of Fusinter is to avoid very thin partitioning due to its specific properties.

3.4 Description of Classification Models

Classification is the technique to predict, which category the new observation belongs to, based on the training samples, which is used by the system during learning. It also identifies group membership. Classification is a type of supervised learning. The algorithm used for classification is known as a classifier. In classification, the target variable is in categorical or discrete form. Suppose $(x_1, y_1), \ldots, (x_n, y_n)$ is a set of observation then choose a function $f(x) = y$, where y is a finite set. x_i represents data points and y_i represents class label. There are various types of classifiers. In this paper, we used only two classifiers, which is efficient for our datasets.

Decision Tree (C 4.5). C4.5 is a learning algorithm developed by Quinlan [21], which is used to generate a decision tree. It is an extension of the ID3 algorithm. The tree, which is generated by C4.5 is used for classification problems. It is also known as a statistical classifier.

Algorithm 1. C 4.5 for constructing TREE

```
Input: Dataset (S) of attributes
T = {}
if dataset S is not fulfilling the optimal evaluation
criteria, then
    return
endif
for all attribute A belong to S
  calculate information gain of A
end for
A_best=Attribute with highest information gain
Assign A_best as the root node
S_p= bring out sub-dataset from S, on the basis of A_best
for all S_p do
    T_p= C 4.5 (Sp)
    Assign T_p as the corresponding node of the tree
end for
```

It builds the decision tree from the set of training samples by using the concept of information gain and entropy gain. C4.5 finds the best attribute at each node of the decision tree of the data that splits the set of samples into subset enriched in one class or the other. The criteria of splitting the attributes are done by the normalized information gain, and the attribute with the highest information gain is chosen to make the decision. The C4.5 then repeats the process for the subtrees. The C4.5 algorithm has some limitations, like, when all the samples in the list belong to the same class, then it normally creates leaf node for the decision tree. If none of the attributes provide any information gain, in that case, C4.5 make a decision node higher up the tree using the excepted value of the class. If instances of any previously unseen class encountered again, then C4.5 decides node higher up the tree using the excepted value.

Support Vector Machine(SVM). This comes under supervised learning, which represents the examples of sample points in space, and a hyperplane divides that sample points of other categories, and the gap is as large as possible [22]. SVM is concurrently minimized the empirical classification error and maximizes the geometric margin. For a given dataset $(A_1, B_1), \ldots, (A_N, B_N)$, where, is a multi-dimensional vector, and $B_N = \{-1, 1\}$, is the label of the given dataset and N is the number of instances, the training data can be viewed by dividing through a hyper plane (Eq. 1):

$$W.A + c = 0 \tag{1}$$

where, c is a scalar, and W is p-dimensional Vector. Adding the offset value (c) permits us to raise the margin. In the absence of c, the hyperplane is enforced to pass through the origin. As we are interested in SVM and the parallel hyperplanes, the Equation can be defined as parallel hyperplanes (Eqs. 2 and 3)

$$W.A + c = 1 \tag{2}$$

$$W.A + c = -1 \tag{3}$$

If the training data set is linearly distinguishable, then we can select these hyperplanes so that there are no points between them and then try to maximize their distance. The distance between the hyperplane is 2/IWI. So, we want to Minimize IWI. We need to certify that for all "i", either (Eqs. 4 or 5)

$$W.Ai - c \geq 1 \tag{4}$$

or

$$W.Ai - c \leq 1 \tag{5}$$

and can be written as (Eq. 6),

$$Bi(W.Ai - c) \geq 1, 1 \leq i \leq N \tag{6}$$

SVM needs more time to evaluate the dataset when the number of samples is very large.

3.5 Performance Measure

This paper used 10-fold cross-validation technique to evaluate the predictions of the used models by using the complete performance factors, i.e., Accuracy (Eq. 7), Precision (Eq. 8) Sensitivity (recall) (Eq. 9), FPR (Eq. 10), AUC (Eq. 11). Precision is the number of pertinent instances among the retrieved instances, while recall or

Table 2 Performance of simple SVM and C4.5

	Classifier	Precision (%)	Sensitivity (%)	Accuracy (%)	AUC (%)
H. Sapiens	C4.5	58.6	69.2	88.5	78.4
	SVM	66.6	61.4	90.15	78.1
E. coli	C4.5	62.1	65.6	89.1	79.3
	SVM	71.1	61.1	90.6	78.4

sensitivity is the number of pertinent instances that have been retrieved instances. Following are the different evaluation [23] criteria.

$$Accuracy = \frac{TP + TN}{Total\ no\ of\ Samples} \tag{7}$$

$$Precision = \frac{TP}{TP + FP} \tag{8}$$

$$Sensitivity\ (Recall) = \frac{TP}{TP + FN} \tag{9}$$

$$FRR = \frac{FP}{TP + FP} \tag{10}$$

$$AUC = 0.5 + \frac{TPR - FPR}{2} \tag{11}$$

where TP denotes true positive, i.e., the number of examples classified as positives which are actually positive, FP denotes False positive, i.e., the number of examples are classified as positive which are actually negative, TN denotes true negative, i.e., the number of examples are classified as negative which are actually negative, and FN denotes false negative, i.e., the number of examples classified as negative which are actually positive. The Receiver Operating Characteristic (ROC) curves are generated to calculate the prediction performance. The curve is plotted using the false positive rate against the true positive rate (or sensitivity), which can also be calculated as (1-specificity) at several threshold values. AUC (Area under the curve) is calculated either by the area under the ROC curve or by above AUC formula.

4   Results and Discussion

First, experiments are performed using simple SVM and C4.5 algorithm on the protein datasets given in Table 1. The summaries of results are shown in Table 2.

We calculate the evaluation parameters from Eqs. (7–9) and (11) respectively.

Table 2 shows the results of C4.5 [20] and SVM [24] with poly kernel [25] algorithm and find AUC of 78.4 and 78.1% for H. Sapiens data sets and AUC of

Table 3 Performance of different methods using H. Sapiens data set

Discretization	Classifier	Sensitivity (%)	Precision (%)	Accuracy (%)	AUC (%)
Ameva	C4.5	95.3	92.6	89.5	74.4
	Svm	99.8	92.7	93.1	75.7
CAIM	C4.5	93.6	93.1	88.6	75.6
	Svm	98.3	94.6	93.8	82.1
Chi-Merge	C4.5	94.04	93.3	89.1	76.4
	Svm	98.3	94.4	93.5	81.1
Fusinter	C4.5	97.8	92.08	90.9	72.9
	Svm	100	86.05	86	50

Table 4 Performance of different methods using E. coli dataset

Discretization	Classifier	Sensitivity (%)	Precision (%)	Accuracy (%)	AUC (%)
Ameva	C4.5	97.5	94	92.6	80.3
	SVM	97.4	93.9	90.3	79.7
CAIM	C4.5	94.3	93.9	89.7	77.6
	SVM	97.7	94.1	93.8	80.7
Chi-Merge	C4.5	95.5	94.2	87.1	75.6
	SVM	92.3	93.7	88.8	76.8
Fusinter	C4.5	94.7	92.8	86.5	78.9
	SVM	99.6	88.5	84.1	62.1

79.3 and 78.4% for E. coil which is not up to the mark. These results are obtained by SVM algorithm. So, we applied other techniques.

However, the obtained result is not satisfactory. Therefore, discretization techniques are used on the datasets H. Sapiens and E. coli DataSet. Discretization is used to discretize the continuous values into discrete values. Then we apply 10-fold cross-validation and at last using SVM with parameters values (kernel = Poly, c = 100, eps = 0.001, degree = 1, gamma = 0.01, nu = 0.1, p = 1.0, shrinking = 1) and find AUC given in Table 3 and Table 4 respectively for H. Sapiens and E. coli.

4.1 Evaluation of Proposed Method

To evaluate the performance of the proposed method, we also perform the proposed method on the H. Sapiens dataset and E. coli data set. When we are using the proposed method to predict this dataset, we obtained the average AUC of 82.1% and 80.7% respectively. The prediction results are shown in Tables 3 and 4.

The accuracies from Table 3 indicate CAIM Discretization along with SVM model is effective for predicting PPI. The standard deviation of these values is less, which indicates that the method is efficient and robust. Generally, the pair of proteins having high in similarity is more likely to interact with each other, and the similarity among protein sequences depends on their deviation time and substitution rates. Hence, the proposed method makes use of evolutionary information and predicts the interaction among proteins.

4.2 Comparison with Other Methods

Prediction of PPI had performed by several machine learning techniques. We compared the proposed technique with the Fuzzy Support Vector Machine technique for the H. Sapiens dataset and E. coli dataset. Majority of the methods were performed with standard average values which are lower than the proposed method. Many methods of computations had been proposed to predict PPI. We have used the H. Sapiens dataset and E. coli dataset to compare the CAIM Discretization model by implementing the SVM algorithm with the other methods. Table 3 shows the average results of predictions on H. Sapiens dataset.

It can be seen that the AUC achieved by these seven methods are in between 50 and 81.1%. The accuracy, precision, sensitivity, and AUC of the methods mentioned are lower than the proposed method. Table 4 shows the performance of the predictions on the E. coli dataset using seven different methods. The AUC obtained by these seven methods were between 62.1% and 80.3%, which was also lower than the proposed method shown in Table 6. We use SVM with parameters values of (kernel $=$ RBF, $c = 100$, eps $= 0.001$, degree $= 1$, gamma $= 0.01$, nu $= 0.1$, p $= 1.0$, shrinking $= 1$) and C4.5 algorithm with parameters values (pruned $=$ true, confidence $= 0.25$, instances per leaf $= 2$).

From Tables 3 and 4, we can see that CAIM with SVM algorithm gives an excellent result as compared to other techniques. SVM is the slow process when the number of the dataset is large, but it is the most reliable and robust algorithm [25]. The same results can also be seen from Figs. 2 and 3, which is a graphical representation of the table data.

4.3 Comparison Between SVM and Fuzzy SVM

The outcome expressed that the SVM with discretization performs well in PPI prediction on both the H. Sapiens dataset and E. coli data set. The comparison between proposed and previous method of H. Sapiens and E. coli dataset is given in Table 5 and Table 6 respectively.

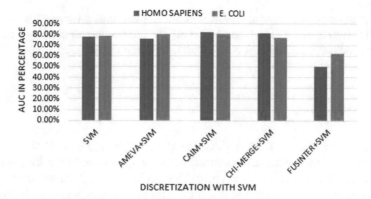

Fig. 2 Comparison of different discretization methods using SVM in H. Sapiens, E. coli

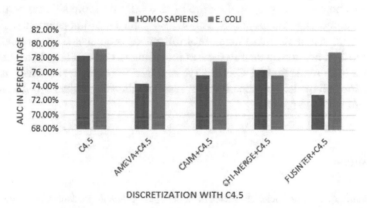

Fig. 3 Comparison of different discretization methods using C4.5 in H. Sapiens and E. coli

Table 5 Performance comparison between proposed and previous method of H. Sapiens

Method	Precision (%)	Sensitivity (%)	Accuracy (%)
Discretization+SVM	94.6	98.3	82.1
Fuzzy SVM [9]	56.8	59.1	76.5

Table 6 Performance comparison between proposed and previous method of E. coli

Method	Precision (%)	Sensitivity (%)	Accuracy (%)
Discretization+SVM	94.4	97.7	80.7
Fuzzy SVM [9]	82.1	62.8	80.1

The comparison between SVM with discretization and fuzzy SVM [9] is given in above table, and it is found that proposed method shows a better enhancement of AUC than traditional methods.

The comparison between the SVM with discretization and fuzzy SVM on E. coli dataset is given in Table 6, and the AUC is 80.7% and 80.1% respectively which tells that SVM with discretization is better than traditional methods.

5   Conclusion

We developed a computational model that the combination of evolutionary information of protein embedding, discretization, and Support vector machine classifier. The main enhancements come from robust SVM classifier with the use of the CAIM Discretization. These results prove that the CAIM Discretization in combination with the SVM can improve the accuracy and AUC of the prediction. So CAIM is very effective technique. The proposed method has a little bit more significant than the previous methods, and the proposed method also reduce the dimensionality. The proposed method can be useful in many bioinformatics problems. This method, using 10-fold cross-validation on the PPI dataset including E. coli and H. sapiens data sets, achieves a high prediction AUC of 80.7% and 82.1% respectively, which is better than previous methods. The experimental results indicate that the proposed method is more efficient and robust.

References

1. Scientific, T.F.: Thermo Scientific Pierce Protein Assay Technical Handbook. Thermo Scientific (2009)
2. Gonzalez-Abril, L., Cuberos, F.J., Velasco, F., Ortega, J.A.: Ameva: An autonomous discretization algorithm. Expert Syst. Appl. **36**, 5327–5332 (2009)
3. Kurgan, L.A., Cios, K.J.: CAIM discretization algorithm. IEEE Trans. Knowl. Data Eng. **16**, 145–153 (2004)
4. Kerber, R.: Chimerge: Discretization of numeric attributes. In: Proceedings of the Tenth National Conference on Artificial Intelligence, pp. 123–128 (1992)
5. Zighed, D.A., Rabaséda, S., Rakotomalala, R.: FUSINTER: a method for discretization of continuous attributes. Int. J. Uncertain. Fuzziness Knowl.-Based Syst. **6**, 307–326 (1998)
6. Chauhan, H., Chauhan, A.: Implementation of decision tree algorithm c4. 5. Int. J. Sci. Res. Publ. **3** (2013)
7. Ho, Y., Gruhler, A., Heilbut, A., Bader, G.D., Moore, L., Adams, S.-L., Millar, A., Taylor, P., Bennett, K., Boutilier, K.: others: Systematic identification of protein complexes in Saccharomyces cerevisiae by mass spectrometry. Nature **415**, 180–183 (2002)
8. Fields, S., Song, O.: A novel genetic system to detect protein-protein interactions. Nature **340**, 245–246 (1989)
9. Sriwastava, B.K., Basu, S., Maulik, U.: Protein???Protein interaction site prediction in Homo sapiens and E. coli using an interaction-affinity based membership function in fuzzy SVM. J. Biosci. **40**, 809–818 (2015)
10. Rao, V.S., Srinivas, K., Sujini, G.N., Kumar, G.N.: Protein-protein interaction detection: methods and analysis. Int. J. Proteomics **2014** (2014)
11. Wang, L., You, Z.-H., Xia, S.-X., Liu, F., Chen, X., Yan, X., Zhou, Y.: Advancing the prediction accuracy of protein-protein interactions by utilizing evolutionary information from the position-specific scoring matrix and ensemble classifier. J. Theor. Biol. **418**, 105–110 (2017)

12. Du, X., Sun, S., Hu, C., Li, X., Xia, J.: Prediction of protein-protein interaction sites by means of ensemble learning and weighted feature descriptor. J. Biol. Res. **23**, 10 (2016)
13. Guo, F., Ding, Y., Li, S.C., Shen, C., Wang, L.: Protein-protein interface prediction based on hexagon structure similarity. Comput. Biol. Chem. **63**, 83–88 (2016)
14. Zhou, H.-X., Shan, Y.: Prediction of protein interaction sites from sequence profile and residue neighbor list. Proteins Struct. Funct. Bioinform. **44**, 336–343 (2001)
15. Berman, H.M., Westbrook, J., Feng, Z., Gilliland, G., Bhat, T.N., Weissig, H., Shindyalov, I.N., Bourne, P.E.: The protein data bank. In: 1999-International Tables for Crystallography Volume F: Crystallography of biological macromolecules, pp. 675–684 (2006)
16. Salwinski, L., Miller, C.S., Smith, A.J., Pettit, F.K., Bowie, J.U., Eisenberg, D.: The database of interacting proteins: 2004 update. Nucl. Acids Res. **32**, D449–451 (2004)
17. Sriwastava, B.K., Basu, S., Maulik, U.: Predicting protein-protein interaction sites with a novel membership based fuzzy SVM classifier. IEEE/ACM Trans. Comput. Biol. Bioinforma. **12**, 1394–1404 (2015)
18. Singh, R., Park, D., Xu, J., Hosur, R., Berger, B.: Struct2Net: a web service to predict protein-protein interactions using a structure-based approach. Nucl. Acids Res. **38**, W508–W515 (2010)
19. Saha, I., Maulik, U., Bandyopadhyay, S., Plewczynski, D.: Fuzzy clustering of physicochemical and biochemical properties of amino acids. Amino Acids **43**, 583–594 (2012)
20. Dougherty, J., Kohavi, R., Sahami, M.: others: Supervised and unsupervised discretization of continuous features. In: Machine Learning: Proceedings of the Twelfth International Conference, pp. 194–202 (1995)
21. Quinlan, J.R.: C4. 5: Programs for Machine Learning. Elsevier (2014)
22. Hsu, C.-W., Chang, C.-C., Lin, C.-J.: Others: A Practical Guide to Support Vector Classification (2003)
23. Huang, J., Ling, C.X.: Using AUC and accuracy in evaluating learning algorithms. IEEE Trans. Knowl. Data Eng. **17**, 299–310 (2005)
24. Vishwanathan, S.V.M., Murty, M.N.: SSVM: a simple SVM algorithm. In: Proceedings of the 2002 International Joint Conference on Neural Networks, 2002. IJCNN'02, pp. 2393–2398 (2002)
25. Markowetz, F.: Classification by support vector machines. Pract. DNA Microarray Anal. (2003)

A Comparative Performance of Classification Algorithms in Predicting Alcohol Consumption Among Secondary School Students

Dilip Singh Sisodia, Reenu Agrawal and Deepti Sisodia

Abstract The increased consumption of alcohol among secondary school students has been a matter of concern these days. Alcoholism not only affects individual's decision-making ability but also have a negative effect on academic performance. The early prediction of a student consuming alcohol can be helpful in preventing them from such risks and failures. This paper evaluates classification algorithms for prediction of certain risks of secondary school student due to alcohol consumption. The classification algorithms considered here are three individual classifiers including Naïve Bayes Classifier, Random Tree, Simple Logistic and three ensemble classifiers: Random Forest, Bagging, and Adaboost. The dataset is taken from the UCI repository. The performance of these algorithms is evaluated using standard evaluation metrics such as Accuracy, Precision, Recall and F-Measure. The results suggested that Simple Logistic and Random Forest performed better than the other classifiers.

Keywords Alcohol consumption · Classifiers · Performance measures
Prediction · Ensemble learners

1 Introduction

There are lots of risks associated with drinking alcohol underage. It not only affects the body-brain, heart, liver, and pancreas but also leads to cardiovascular diseases and car accidents which are considered as one of the major risk factors of alcoholism. Excessive drinking of alcohol increases the risk of developing different types of cancers, including cancers of the Mouth, Esophagus, Throat, Liver, and Breast [1]. Alcohol consumption affects the decision-making ability, and tends to get involved in violence and harm others. The brain is on developing stage during teenage years of adolescents. Consumption of alcohol at this stage is the major risk factor. It shows

D. S. Sisodia (✉) · R. Agrawal · D. Sisodia
National Institute of Technology Raipur, Raipur, India
e-mail: dssisodia.cs@nitrr.ac.in

© Springer Nature Singapore Pte Ltd. 2019   523
M. Tanveer and R. B. Pachori (eds.), *Machine Intelligence and Signal Analysis*,
Advances in Intelligent Systems and Computing 748,
https://doi.org/10.1007/978-981-13-0923-6_45

adverse effects on one's physical and mental learning ability due to lack of maturity during their school years.

Those children who involve in alcohol consumption by age 13 lacks in less self-control. This can lead to aggression and fights. Alcohol consumption increases the risk of being involved in violence which could lead to the criminal record like threatening or damaging in a nonphysical way [2]. Their natural tendency to experiment and take risks is also increased.

In this paper, individual and ensemble classification algorithms such as Naive Bayes classification (NBC), random tree, simple logistic, random forest, bagging and Adaboost have been considered for comparing their performance on the student alcohol consumption data. The dataset is taken from the UCI repository which was collected from two public schools from the Alentejo region of Portugal during the 2005–2006 year. For comparison of selected classification algorithms, 10-fold cross-validation methods are used in this work study.

The rest of the paper is organized as follows: Section 2 briefly describes the related previous works. Section 3 describes the classification algorithms used for the purpose of classification. Section 4 describes the evaluation metrics used for the evaluation of classification algorithms. Section 5 reports the results obtained in our experiments. Section 6 concludes our work along with insight into the future work.

2   Related Work

In this section already reported work on prediction and detection of alcohol consumption is discussed brief Pagnotta et al. [3] used decision tree and random forest to predict secondary school student alcohol consumption and achieved 92% accuracy for the same. They have also described the correlation between alcohol usages, sex and study time of the students. To reduce the failures in prediction and to improve the prediction of the performance of the students, Cortez and Silva [4] used four classification models and three different feature selection methods in their experiment. Excessive intake of Alcohol is the serious risk factor regarding health problems among teenagers in various states. Hence, Bi et al. [5] has proposed a method which is a combination of cluster analysis, and feature selection where on the basis of average daily drinking behavior drinking patterns are identified using cluster analysis and risk factors which are associated with each pattern are identified by feature selection.

Sharma et al. [6] has proposed a new machine learning and signal processing-based automated system to detect epileptic episodes accurately. The proposed algorithm employs a promising time-frequency tool called Tunable-Q Wavelet Transform (TQWT) to decompose EEG signals into various Sub-Bands (SBs). The method achieved 100%, the highest classification accuracy as well as the largest area under ROC curve (AUC) for all classes.

Sharma and Pachori [7] has designed an automated system to classify alcoholic and normal EEG signals using a recently designed duration-bandwidth product (DBP), optimized three-band orthogonal wavelet filter bank (TBOWFB) and log-energy (LE). Using the 10-fold cross-validation strategy, the result obtained gives the classification accuracy (CA) of 97.08%.

3 Classification Algorithms

The purpose of supervised learning is to include a classification for categorical response values to separate the data into specific classes. Classification is a two-step process. Model is created in its first step using classification algorithm. In second step model is trained, and performance and accuracy is measured. The main goal of classification is to classify data into different classes according to constraints. It is used to predict the target class by analyzing the training dataset. To determine each target class, classification uses training data set to find proper boundaries. Once the task is done, it makes predictions of the response values. This process is known as classification [8].

3.1 Individual Classifiers

Naive Bayes, random tree, and simple logistic classifier are the individual learners or classifiers which are employed in this work.

Naïve Bayes Classifier. Naive Bayes is a probabilistic algorithm based on Bayes' Theorem to predict the category of a sample. By probabilistic it means that for a given sample, the probability of each category is calculated and provides the category with the highest one as an output. On the basis of prior knowledge probability of an event can be calculated. It is a conditional probability model: given a problem instance to be classified represented by a vector $F = f1, f2, \ldots, fn$ representing some n features; it calculates this instance probabilities $P(C|f1, f2, \ldots, fn)$ for each possible classes C, which is represented in the Eq. 1. As stated earlier, each feature f_i is conditionally independent of every other feature f_j [9].

$$P(C|f1, f2, \ldots, fn) = \frac{1}{Z} P(C) \prod_{i=1}^{n} P(fi|C) \qquad (1)$$

Naïve Bayes classifier uses Bayes theorem to calculate the conditional probability. Bayes theorem calculates the posterior probability using Eq. 2.

$$P(A|B) = \frac{P(B|A) * P(A)}{P(B)}, \qquad (2)$$

where,

- P (A) is the prior probability.
- P (B | A) is known as likelihood.

Random Tree Classifier: At each node, random tree classifier constructs a tree that considers K randomly chosen attributes. No pruning is performed by the Random tree. Based on back fitting it allows estimation of class probabilities or target mean in the regression case. When using this method, two important parameters are chosen namely: the height h of the random tree and the number N of base classifiers. Training efficiency and minimal memory requirements are the advantages of using a random tree classifier. The algorithm uses only one pass over the data to create a random decision tree [10].

Simple Logistic Classifier: For building linear logistic regression models, this kind of classifier is used. To analyze a dataset, a statistical method like logistic regression is used where to determine an outcome more than one independent variable is used.

The dichotomous variable is used to measure the outcome in which there are only two possible outcomes. The dependent variable is binary or dichotomous in logistic regression, i.e., it only contains data coded as 1 when TRUE, success, pregnant, etc. or 0 when FALSE, failure, non-pregnant, etc. Finding the best model is the goal of logistic regression to describe the relationship between the dichotomous characteristic of interest and a set of independent variables. Logistic regression generates the coefficients and its standard errors and significance levels of a formula to predict a logit transformation of the probability of the presence of the characteristic of interest [11].

3.2 Ensemble Classifier

The main idea of ensemble classifier is to construct set of models that are cooperatively trained on data set in a supervised classification problem and classify new data points using the weighted vote of their predictions.

Random Forest Classifier: Random Forest is an ensemble classifier or an ensemble learning method for classification that consists of many decision trees and outputs the class that is the mean prediction of individual trees [12]. The term was first proposed by Tin Kam Ho of Bell Labs in 1995 [13]. It generates a forest of the decision tree using lots of tree classifier. Every time algorithm selects a random partition and performs the classifier. In this work, the classifier performance is improved after applying smote to the imbalanced dataset [14].

Bagging Classifier: It is one of the ensemble method based on bootstrap sampling that generates various versions of a predictor and then use them to get an aggregated predictor. The aggregated predictor is constructed by averaging the versions when predicting the numerical result and does a voting when the class is predicted. It is a bootstrap ensemble method where multiple versions are formed by making bootstrap replicates of the learning set and then using them as new learning sets [15].

Breiman [16] showed that Bagging is effective on unstable learning algorithms where small changes in the training set can lead to large changes in resulting predictions. Neural networks and decision trees are examples of unstable learning algo-

Table 1 Confusion matrix

		Predicted class	
		Yes	No
Actual class	Yes	True positive (TP)	True negative (TN)
	No	False positive (FP)	False negative (FN)

rithms. Bagging which is known as bootstrap aggregating is one of the machines learning an ensemble algorithm which not only improves the stability and accuracy but also it reduces the variance of various machine learning algorithms. Bagging is considered as an effective technique as it increases the accuracy of the single model by using multiple copies of it which are trained on different sets of data.

Adaboost Classifier: Adaptive Boosting also known as Adaboost is one of popular machine learning ensemble algorithm. Via an iterative process, it improves the simple boosting algorithm. In order to reduce the classifier's error, it generates such classifier who proves to be better comparatively random guessing.

The main focus is given to those patterns which are harder to classify using the concept of 'pseudo losses.' The amount of focus is quantified by employing a weight in multi-label which is assigned to every pattern in the training set. In this method, the same weight is assigned to all the patterns initially. Every time during the iterations weights of correctly classified instances is decreased while the weights of all misclassified instances are increased. By creating as many classifiers and performing a number of additional iterations. As a result, the weak learner is forced to focus on the difficult instances of the training set [17]. Furthermore, every individual classifier is assigned by weight where it measures the overall accuracy of the classifier.

Adaboosting follows an approach where based on the probabilities of examples it selects set of examples and by the probability weight the error of each example for that example [18].

4 Evaluation Metrics

Some standard sort of evaluation metrics must be used in order to evaluate the classification algorithms and their prediction. Measuring the performance is the way to evaluate a solution to the problem [19]. In this experiment, four measures such as accuracy, precision, recall, and F-measure are used to evaluate the outcome of the results. Table 1 represents the confusion metrics which is used to calculate these measures.

True Positives. (TP) are the examples whose are correctly classified as positive by the classifier.

True Negative. (TN) are the examples which are correctly classified as negative by the classifier.

False Positive. (FP) are the examples which are incorrectly classified as true by the classifiers.

False Negative. (FN) are the examples which are incorrectly classified as false by the classifiers.

Accuracy (A). Accuracy determines the accuracy of the algorithm in predicting instances. Equation (3) is used to find the accuracy of the classifier.

$$A = \frac{TP + TN}{Total\,no\,of\,samples} \tag{3}$$

Precision (P). Classifier's correctness/accuracy is measured by Precision. If precision is high, then false positive is less, and if precision is low, then false positive is high. Equation (4) is used to evaluate the precision of a classifier.

$$P = \frac{TP}{TP + FP} \tag{4}$$

Recall (R). To measures the classifier's completeness or sensitivity, recall is used. If the recall is high, then the false negative is less, and lower recall means a false negative. Precision is often decreased when a recall is improved. As the sample space increases, it becomes hard to be precise [20]. Equation (5) is used to evaluate recall.

$$R = \frac{TP}{TP + FN} \tag{5}$$

F-Measure (F). F-Measure is the combination of precision and recall, which is the weighted average of precision and recall. It is represented in Eq. (6).

$$F = 2 * \frac{P * R}{P + R} \tag{6}$$

5   Experimental Verification

To evaluate the classifiers using the evaluation metrics discussed earlier we used some standard data sets available using Weka tool [21]. The description of the datasets is written below which is followed by the experimental results.

Table 2 Performance metrics of individual classifiers

Classifiers	Accuracy	Precision	Recall	F-measure
Naïve Bayes	0.868	0.868	0.868	0.868
Random tree	0.8702	0.871	0.87	0.87
Simple logistic	0.8702	0.871	0.87	0.87

5.1 Data Set

The dataset is taken from the UCI repository. From the University of Minho, Portugal, Paulo Cortez and Alice Silva gathered the data. The data was collected during the 2005–2006 year from two public schools from the Alentejo region of Portugal. The attributes in the dataset consisted of grades, social, demographic and school-related variables [4].

Two separate files student-mat.csv (with 395 examples) and student-Portuguese.csv (with 649 records) was provided in the UCI. The 382 instances of which were common in both hence were integrated into one dataset with 662 instances in total. The description of 33 attributes in the dataset is listed in Table 1.

The dataset was imbalanced consisting of 595 instances of majority class and only 67 instances of a minority class. So, to avoid imbalance in the dataset, we have used Synthetic Minority Oversampling Technique (SMOTE). The total instances after applying SMOTE were 1131 which included 595 instances of a student not consuming alcohol and 536 instances of student consuming alcohol.

5.2 Results

Using 10-fold cross-validation technique the above data set is used, and the classifier used. The following results have been obtained by taking the average value for all the classes (i.e., A and B).

Table 2 indicates that both random tree and simple logistic classifier performing equally well than naïve Bayes classifier. The same result can also be seen from Fig. 1.

Similarly, Table 3 represents the different performance values obtained by applying different ensemble classifier techniques. It shows that random forest is performing better than the other two classifiers in ensemble classifiers. The same result can also be seen from Fig. 2. As compared with the individual classifiers, Random forest also outperforms all other classifiers in our experiment.

The RoC (Receiver Operating Characteristic) area for all the classifier is represented in Table 4. Same is also being represented in Fig. 3. From the above results, it can be conclude that random forest outperforms all other classifiers for the used dataset.

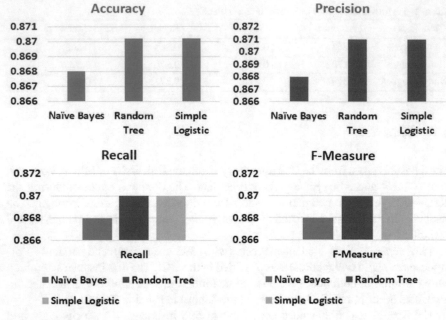

Fig. 1 Performance of individual classifiers

Table 3 Performance metrics of ensemble classifiers

Classifiers	Accuracy	Precision	Recall	F-measure
Random forest	93.5455	0.937	0.935	**0.935**
Adaboost	85.7648	0.865	0.858	0.858
Bagging	90.2741	0.903	0.903	0.903

Table 4 RoC area of all classifiers

Classifiers	RoC values
Naïve Bayes	0.935
Random tree	0.871
Simple logistics	0.936
Random forest	**0.981**
Adaboost	0.931
Bagging	0.961

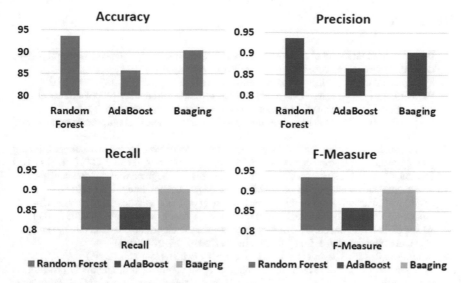

Fig. 2 Performance of ensemble classifiers

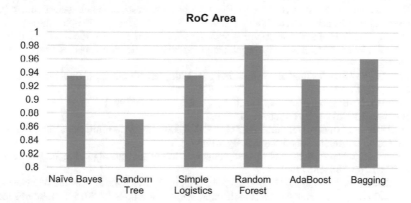

Fig. 3 RoC area of classifiers

6 Conclusion

In this work, popular machine learning classification algorithms are used for evaluating their classification performance regarding Accuracy, Precision, Recall, and F-Measure in classifying Student Alcohol Consumption dataset. With the selected dataset, Simple Logistic as an individual classifier and Random Forest in ensemble classifier gives better results with all the feature set combinations in the balanced dataset. More research can be possible if any additional features will be added to the dataset. The main purpose is to prevent the students from being misguided. Hence, it will be helpful for a smooth academic environment.

References

1. Bateman, M.: Does alcohol cause breast cancer. http://www.drinkaware.co.uk/alcohol-and-yo u/health/does-alcohol-cause-breast-cancer (2011). Accessed 11, 2011
2. Spear, L.P.: Alcohol's effects on adolescents. Alcohol Res. Health **26**, 287–291 (2002)
3. Pagnotta, F., Amran, H.M.: Using data mining to predict secondary school student alcohol consumption. Department of Computer Science, University of Camerino (2016)
4. Cortez, P., Silva, A.M.G.: Using data mining to predict secondary school student performance (2008)
5. Bi, J., Sun, J., Wu, Y., Tennen, H., Armeli, S.: A machine learning approach to college drinking prediction and risk factor identification. ACM Trans. Intell. Syst. Technol. (TIST) **4**, 72 (2013)
6. Sharma, M., Deb, D., Acharya, U.R.: A novel three-band orthogonal wavelet filter bank method for an automated identification of alcoholic EEG signals. Appl. Intell. 1–11 (2017)
7. Sharma, M., Pachori, R.: A novel approach to detect epileptic seizures using a combination of tunable-Q wavelet transform and fractal dimension. J. Mech. Med. Biol. 1740003 (2017)
8. AL-Nabi, D.L.A., Ahmed, S.S.: Survey on classification algorithms for data mining: comparison and evaluation. Int. J. Comput. Eng. Intell. Syst. **4**, 18–27 (2013)
9. Murphy, K.P.: Naive bayes classifiers. University of British Columbia (2006)
10. Jagannathan, G., Pillaipakkamnatt, K., Wright, R.N.: A practical differentially private random decision tree classifier. In: IEEE International Conference on Data Mining Workshops, ICDMW'09, pp. 114–121 (2009)
11. Feng, J., Xu, H., Mannor, S., Yan, S.: Robust logistic regression and classification. In: Proceedings of Advances in Neural Information Processing Systems, pp. 253–261 (2014)
12. Gäš, B.: Analysis of a random forests model. J. Mach. Learn. Res. **13**, 1063–1095 (2012)
13. Liaw, A., Wiener, M., et al.: Classification and regression by randomForest. R News **2**, 18–22 (2002)
14. Loh, W.-Y.: Classification and regression trees. Wiley Interdiscip. Rev. Data Mining Knowl. Discov. **1**, 14–23 (2011)
15. Breiman, L.: Bagging predictors. Mach. Learn. **24**, 123–140 (1996)
16. Breiman, L.: Out-of-bag estimation (1996)
17. Falaki, H.: AdaBoost algorithm. Startrinity, 202 (2009). http://startrinity.com/VideoRecogniti on/Resources/Adaboost/boosting%20algorithm
18. Freund, Y., Schapire, R.E., et al.: Experiments with a new boosting algorithm. In: Proceedings of ICML, pp. 148–156 (1996)
19. Eberhardinger, B., Anders, G., Seebach, H., Siefert, F., Reif, W.: A research overview and evaluation of performance metrics for self-organization algorithms. In: 2015 IEEE International Conference on Self-Adaptive and Self-Organizing Systems Workshops (SASOW), pp. 122–127 (2015)
20. Tiwari, M., Jha, M.B., Yadav, O.: Performance analysis of data mining algorithms in Weka. IOSR J. Comput. Eng. **6**, 32–41 (2012)
21. Witten, I.H., Frank, E., Hall, M.A., Pal, C.J.: Data Mining: Practical Machine Learning Tools and Techniques. Morgan Kaufmann (2016)

Agglomerative Similarity Measure Based Automated Clustering of Scholarly Articles

Dilip Singh Sisodia, Manjula Choudhary, Tummala Vandana and Rishi Rai

Abstract The flooding of online scholarly articles necessitates the automated organization of documents according to their most descriptive attributes. In this paper, an agglomerative similarity measure based on common features associated with research articles, such as number of references, authors, citations, and contents are used for automated clustering of scholarly articles. The agglomerative similarity matrix is based on a combination of citation matrix, author matrix, and the content matrix for feature vector representation. The experiments are performed on agglomerative feature vector derived from wiki20 dataset using different unsupervised learning algorithms such as K-Means, K-medoids, and Fuzzy C-means. The clustering result obtained with modified feature vector is compared to the existing bag of words model using separation and cohesion as performance metrics. The Dunn's index is used for finding the optimal number of clusters.

Keywords Similarity matrix · K-means · K-medoids · Fuzzy c-mean · TF-IDF

1 Introduction

There is a rapid growth in electronic information such as online newspapers, journals, conference proceedings, websites, E-mails, etc. Using all this electronic information for controlling, indexing, or searching is not feasible for search engines. So, automatic document clustering is an important issue. By using document clustering methods, we can provide insight into data distribution, or we can preprocess data for other applications [1]. Document clustering is an automatic clustering operation of text documents so that similar or related documents are presented in the same cluster, and dissimilar or unrelated documents are presented in different clusters [2]. Clustering is an unsupervised learning method. So, there is no need for a training set while applying the clustering algorithms. It just uses the input data to find regularities in

D. S. Sisodia (✉) · M. Choudhary · T. Vandana · R. Rai
National Institute of Technology Raipur, Raipur, India
e-mail: dssisodia.cs@nitrr.ac.in

© Springer Nature Singapore Pte Ltd. 2019
M. Tanveer and R. B. Pachori (eds.), *Machine Intelligence and Signal Analysis*,
Advances in Intelligent Systems and Computing 748,
https://doi.org/10.1007/978-981-13-0923-6_46

the cluster. The goal of clustering is to maximize intra-cluster similarities among the documents while minimizing inter-cluster similarities.

In this paper, the aim is to form clusters of documents by using proposed feature vector and make a comparison with existing Bag of words model method. Research papers contain metadata like index, citation and author's name, which are associated with them. Clustering of Research papers can be done either by using word frequency matrix, citation matrix, author matrix or a combination of all these matrices. If two papers cite the same research papers or contain same author the probability of the papers being in the same cluster increases.

The rest of the research paper is organized as follows: Sect. 2 briefly describes the related previous works. Section 3 describes the methodology used for the purpose of clustering. Section 4 describes the performance indices used for the evaluation of clustering algorithms. Section 4 reports the results obtained in our experiments. Section 6 concludes our work along with insight into the future work.

2 Related Work

A knowledge-transfer analysis model for any technology field was proposed in 2012 [3]. In this model, patent data with backward citations to non-patent literature and forward citations by later patents were analyzed. Another approach for clustering scientific documents proposed by Aljaber et al. [4] is based on the utilization of citation contexts. Where a citation context is essentially the text surrounding the reference markers used to refer to other scientific works. Papers [5–7] presented a novel method to cluster research papers into hierarchical, overlapping clusters using the topic as similarity measure. Nakazawa et al. [6] also used citation network along with the topic for the clustering of the research papers.

A paper or document can satisfy the property of more than one clusters, to find the accurate cluster fuzzy c-mean is used which is described in [8–10]. Term frequency and inverse document frequency is used for document representation [7, 10–13]. A document clustering algorithm that represents documents as a time series of words was proposed in [14]. In this paper, timed series representation of the document was used for the computation of the distance between documents.

3 Methodology

Text documents are taken as input, then all the documents are initially preprocessed and then feature matrices of the documents are obtained using similarity matrix based on word frequency matrix, citation matrix, and author matrix. After this, various clustering methods are applied to the feature vector obtained. Description of Steps adopted for KDD (knowledge discovery from data) process are described as follows.

3.1 Document Preprocessing

Document preprocessing is the process of representing the document in term of indices. The primary objective of document preprocessing is to represent the documents in such a manner that it is efficient to store and retrieve the extracted information in the system. Document preprocessing includes the following stages:

Tokenization. Tokenization is the process of converting the character sequence or sentences into meaningful elements called tokens. Tokens can be in the form of words, symbols, or phrases which are grouped and used in the further processing. In the English language, generally, white spaces are used as delimiters in tokenization. In our project, white spaces have been used as delimiters.

Stop Word Removal. Stop words like articles, preposition, etc., do not give any significance to the documents and it also makes the documents heavier and less important. So, for analysis, all these words should be removed. Stop words are not measured as keywords in text mining applications [15].

Stemming. Stemming method is used to find out the root/stem of a word. For example, the words know, knows, knowing, known, all can be stemmed to the word "KNOW." The purpose of this method is to remove various suffixes, to reduce the number of words, to have exactly matching stems and to save memory space and time. The stemming process is done using various algorithms. Most popularly used algorithm is "M.F. Porters Algorithm" [16].

3.2 Document Representation

In document processing and information retrieval systems, a key process is document representation. It is used to extract the relevant documents from the vast collection of documents. It is essential to transform the full-text version of the documents to vector form. Such transformed document describes the contents of the original documents based on the constituent terms called index terms. These terms are used in indexing, the relevant ranking of the keywords for optimized search results, information filtering, and information retrieval. The vector space model, also called vector model, is the popular algebraic model to represent textual documents as vectors. Using the vector space model, documents are represented using the term frequency (TF)—inverse document frequency (IDF) or TF- IDF weighting scheme.

3.3 Feature Matrix

Here similarity matrix [17] is used as feature matrix, which is the combination of word frequency matrix, citation matrix, and author matrix.

Word Frequency Matrix. If two research papers contain common words the probability of two papers belonging to same cluster increases. To measure that, word frequency matrix is used. Word frequency matrix is obtained using TF-IDF. Equation 1 represents the TF-IDF value of index i for research paper j.

$$\text{value}_{i,j} = \begin{cases} 0 \text{ if } j \text{ doesn't contain the word } i \\ \dfrac{1 * \log\left(\frac{k}{m}\right)}{\text{number of words in } j} \end{cases} \tag{1}$$

Here k represents the total number of the research paper and m, the number of articles where the term i appears.

Citation Matrix. If two research papers cite a common research paper the probability of two papers belonging to the same cluster increases. To measure that, we introduce a citation matrix as a feature vector represented by Eq. 2.

$$\text{value}_{i,j} = \begin{cases} 0 \text{ if } i \text{ is not referenced by } j \\ \dfrac{1 * \log\left(\frac{k}{m}\right)}{\text{number of references in } j} \end{cases} \tag{2}$$

Here, k represents a total number of research papers and m, number of research papers where the reference i appears.

Author Matrix. If two research papers are written by the same author the probability of both the papers being similar increases. To measure that, we introduced an author matrix as the feature vector. We took m * k matrix where m represents the total number of authors in all reference papers, and k represents the total number of research papers. We assign a value using the method term frequency-inverse document frequency, but here we take each author as a single term if the specific author writes that research paper.

Thus, we introduce an author matrix as a feature vector, which is obtained using the Eq. 2.

$$\text{value}_{i,j} = \begin{cases} 0 \text{ if } i \text{ is not an author of } j \\ \dfrac{1 * \log\left(\frac{k}{l}\right)}{\text{number of authors in } j} \end{cases} \tag{3}$$

Here, k represents the total number of the research paper and l the number of research papers where one of the authors is i.

3.4 Proposed Model

Using the above-said measures, we have followed the steps as represented in Fig. 1. First, we load all the articles which will be followed by the steps as given in Sect. 3.3.

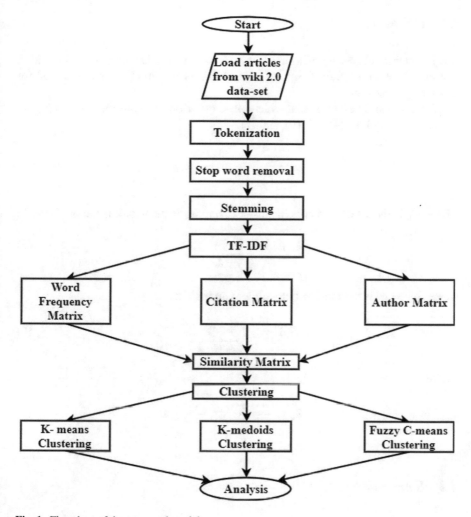

Fig. 1 Flowchart of the proposed model

Clustering will then be applied to the database after finding the similarity matrix and then each document will be categorized using three different clustering techniques.

4 Performance Index

The following are the performance indices used in the experiment [18].

4.1 Cohesion

It is an internal clustering measure. It measures how similar the objects are within clusters. Here we perform a Square Sum of Errors within clusters (SSE) [17] for measuring Cohesion.

SSE is the simplest and most widely used criterion measure for clustering. It is calculated using Eqs. (4), (5), (6), and (7)

$$SSE = \sum_{k=1}^{K} \sum_{\forall x_i \in C_k} \|x_i - \mu_k\|^2, \tag{4}$$

where, C_k is the set of instances in cluster k, and μ_k is the vector mean of cluster k.

$$\mu_k = \sum_{\frac{1}{N_k} \forall x_i \in c_k} x_{i,j}, \tag{5}$$

where, $N_k = c_k$ is the number belonging to an instance.

$$SSE = \frac{1}{2} \sum_{x_i, y_j \in c_k} N_k \overline{S_k}, \tag{6}$$

where,

$$S_k = \frac{1}{N_k^2} \sum_{\forall x_i \in C_k} \|x_i - x_j^2\| \tag{7}$$

4.2 Separation

Separation [19] is the dissimilarity between two clusters. If the separation between the clusters is maximum, then the clustering is considered better. This can be done by using Eq. 8.

$$S = \sum_i \sum_j \|x_i - x_j^2\|, \tag{8}$$

where, x_i, x_j does not belong to the same cluster.

4.3 Dunn's Index

Dunn Index [20] identifies the number of clusters which are well separated and compact. It combines dissimilarity between the clusters and their diameters (Intra-cluster distance) to find the required number of clusters.

$$DI = \frac{\min_{1 \le i < j \le m} d(C_i, C_j)}{\max_{1 \le k \le m} d(i, j)_k},$$ (9)

where, "m" is the number of clusters, $d(C_i, C_j) =$ distance between clusters, $d(i, j)_k$ = diameter of cluster "k".

The diameter of a cluster can be calculated in many possible ways; it can be the distance between the two farthest points within the cluster, or the mean of pairwise distance between all data points within a cluster, or the mean of the distances of all data points in a cluster from the centroid. Higher the value of Dunn's Index the better classified the dataset is. The major drawback is that the index is sensitive to noise data. Calculating Dunn's index is also time-consuming.

5 Results and Discussion

Table 1 shows the values of Dunn Index for varying number of clusters. Dunn Index is used to find the optimal number of clusters. From the table, it can be inferred that the optimal number of clusters for K-means, K-medoids, and fuzzy c-means [21] when using citation and author matrix along with word frequency matrix are 9, 11, and 10 respectively. Similarly, when only word frequency matrix is used, the optimal number of clusters for K-means, K-medoids, and fuzzy C-means remains same.

It is to be noted that, for an optimal number of clusters the values for Dunn's index is always higher when citation and author matrix are also considered, Indicating better performance when citation and author matrix are also being considered.

The graphs following the table are the results of internal clustering measures, i.e., cohesion and separation with and without considering author and citation matrix performed on wiki20 dataset [22, 23]. Directory documents contain 20 computer science articles in text format.

Figures 2, 4 and 6 show the comparison between cohesion and number of clusters. It measures how closely related the objects in a cluster are. Here blue line represents cohesion for k-mean, k-medoids, and fuzzy C-means without author and citation matrix. Similarly, red line represents cohesion for k-mean, k-medoids, and fuzzy C-means with author and citation matrix.

Figures 3, 5 and 7 are the graphs between separation and number of clusters. Here blue line represents separation for k-means, k-medoids, and fuzzy C-means without author and citation matrix whereas red line represents separation for k-means, k-medoids, and fuzzy C-means with author and citation matrix.

Table 1 Dunn's index value for k-means, k-medoids and fuzzy c-means clustering with and without using citation and author matrix

No. of clusters	Dunn's index with citation and author matrix			Dunn's index without citation and author matrix		
	K-means	K-medoids	Fuzzy C-means	K-means	K-medoids	Fuzzy C-means
2	0.5	0.43	0.5	0.6	0.3	0.4
3	0.76	0.82	0.8134	0.705	0.8172	0.8059
4	0.9	0.871	0.873	0.844	0.877	0.861
5	0.943	0.9375	0.938	0.9293	0.917	0.904
6	0.947	0.9777	0.967	0.9328	0.971	0.953
7	0.9692	0.9993	0.975	0.9591	0.967	0.9624
8	0.9834	0.996	0.9799	0.9677	0.959	0.967
9	**0.9995**	0.982	0.9834	**0.9922**	0.967	0.985
10	0.9964	0.996	**0.998**	0.9891	0.974	**0.9976**
11	0.9781	**0.999**	0.9833	0.9805	**0.997**	0.9833
12	0.971	0.984	0.9833	0.975	0.989	0.979
13	0.973	0.9716	0.974	0.9485	0.962	0.965
14	0.965	0.9716	0.9734	0.952	0.962	0.9621
15	0.942	0.9426	0.962	0.9396	0.942	0.958
16	0.942	0.9426	0.961	0.93	0.942	0.946
17	0.91	0.9316	0.936	0.9096	0.9396	0.916
18	0.82	0.816	0.887	0.827	0.87	0.8634
19	0.6	0.6	0.72	0.5	0.5	0.64

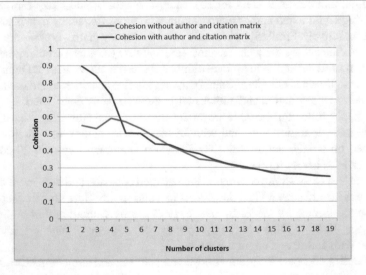

Fig. 2 Cohesion for k-means algorithm

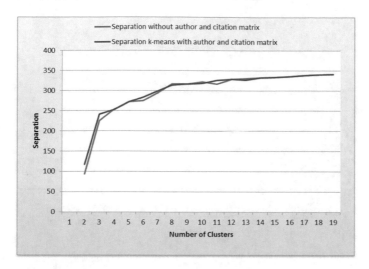

Fig. 3 Separation for k-means

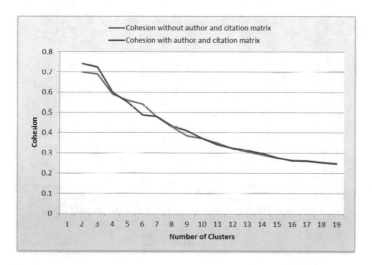

Fig. 4 Cohesion of k-medoids

For some cases in Cohesion and Separation, when using citation and author matrix along with word frequency matrix, the performance is better as compared to the one in which citation and author matrix are not considered. In other cases, it is the other way around.

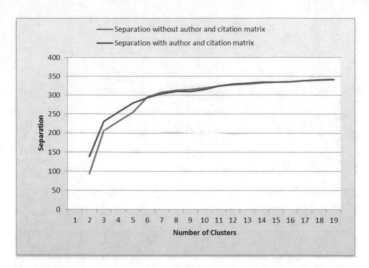

Fig. 5 Separation for k-medoids

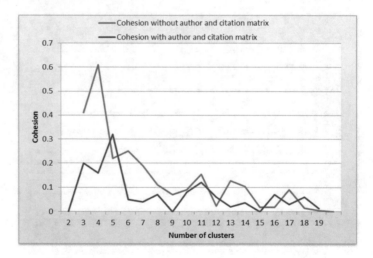

Fig. 6 Cohesion for fuzzy c-means

6 Conclusion and Future Work

In this paper, an agglomerative similarity measure based on common features associated with research articles is evaluated. The feature vector containing citation and
author matrix along with word frequency matrix outperformed the feature vector
containing only words frequency matrix in almost all the cases when using Dunn's
index as a performance measure. But we are unable to conclude the same with cohesion and separation performance indices since it is not following a particular pattern.

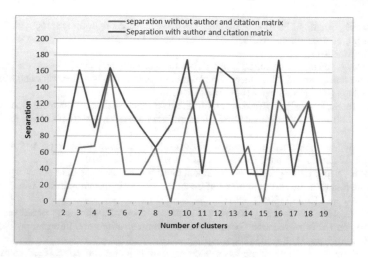

Fig. 7 Separation for fuzzy c-means

The Dunn indices were also used to find an optimal number of clusters, and it is 9 in the case of K-Means, 11 in the case of K-Medoids and 10 in the case of Fuzzy C-Means. The results in case of cohesion and separation cannot be concluded. Hence, it is needed that these should be verified with a large dataset and with different algorithms to get more accurate results. Here we used K-medoids clustering algorithm because, in K-means, the mean can be affected due to the presence of outliers, to decrease that we used K medoids. Fuzzy C-Means was also used since in research papers, there could be a possibility that a research can belong to more than one field. In future work, the number of citations of each paper may be included to access the quality of a research paper.

References

Han, J., Pei, J., Kamber, M.: Data Mining: Concepts and Techniques. Elsevier (2011)

Kaufman, L., Rousseeuw, P.J.: Clustering large applications (Program CLARA). In: Finding Groups in Data: An Introduction to Cluster Analysis, pp. 126–163 (2008)

Wang, X., Zhao, Y., Liu, R., Zhang, J.: Knowledge-transfer analysis based on co-citation clustering. Scientometrics **97**, 859–869 (2013)

Aljaber, B., Stokes, N., Bailey, J., Pei, J.: Document clustering of scientific texts using citation contexts. Inf. Retr. **13**, 101–131 (2010)

Sun, X.: Textual document clustering using topic models. In: 10th International Conference on Semantics, Knowledge and Grids (SKG), pp. 1–4 (2014)

Nakazawa, R., Itoh, T., Saito, T.: A visualization of research papers based on the topics and citation network. In: 19th International Conference on Information Visualisation (iV), pp. 283–289 (2015)

Shubankar, K., Singh, A., Pudi, V.: A frequent keyword-set based algorithm for topic modeling and clustering of research papers. In: 3rd Conference on Data Mining and Optimization (DMO), pp. 96–102 (2011)

Gao, T., Du, J., Wang, S., Chen, L.: Topic detection for emergency events based on FCM document clustering. In: 3rd IEEE International Conference on Broadband Network and Multimedia Technology (IC-BNMT), pp. 1181–1185 (2010)

Kummamuru, K., Dhawale, A., Krishnapuram, R.: Fuzzy co-clustering of documents and keywords. In: The 12th IEEE International Conference onFuzzy Systems, 2003(FUZZ'03), pp. 772–777 (2003)

Win, T.T., Mon, L.: Document clustering by fuzzy c-mean algorithm. In: 2nd International Conference on Advanced Computer Control (ICACC), pp. 239–242 (2010)

Mishra, R.K., Saini, K., Bagri, S.: Text document clustering on the basis of inter passage approach by using K-means. In: International Conference on Computing, Communication & Automation (ICCCA), pp. 110–113 (2015)

Chang, H.-C., Hsu, C.-C., Deng, Y.-W.: Unsupervised document clustering based on keyword clusters. In: IEEE International Symposium on Communications and Information Technology (ISCIT 2004), pp. 1198–1203 (2004)

Chim, H., Deng, X.: Efficient phrase-based document similarity for clustering. IEEE Trans. Knowl. Data Eng. **20**, 1217–1229 (2008)

Matei, L.S., Trăuşan-Matu, Ş.: Document clustering based on time series. In: 19th International Conference on System Theory, Control and Computing (ICSTCC 2015), pp. 128–133 (2015)

Porter, M.F.: An algorithm for suffix stripping. Program **14**, 130–137 (1980)

Ramasubramanian, C., Ramya, R.: Effective pre-processing activities in text mining using improved porter's stemming algorithm. Int. J. Adv. Res. Comput. Commun. Eng. **2**, 4536–4538 (2013)

Sisodia, D.S., Verma, S., Vyas, O.P.: A discounted fuzzy relational clustering of web users' using intuitive augmented sessions dissimilarity metric. IEEE Access. **4**, 6883–6893 (2016)

Sisodia, D.S., Verma, S., Vyas, O.P.: Augmented intuitive dissimilarity metric for clustering of Web user sessions. J. Inf. Sci. **43**, 480–491 (2016)

Ben-Gal, I.: Outlier detection. Data Mining and Knowledge Discovery Handbook, pp.131–146 (2005)

Dunn, J.C.: A Fuzzy Relative of the ISODATA Process and its use in Detecting Compact Well-Separated Clusters (1973)

Sisodia, D.S., Verma, S., Vyas, O.P.: Performance evaluation of an augmented session dissimilarity matrix of web user sessions using relational fuzzy C-means clustering. Int. J. Appl. Eng. Res. **11**, 6497–6503 (2016)

Medelyan, O., Witten, I.H., Milne, D.: Topic indexing with Wikipedia. In: Proceedings of the AAAI WikiAI workshop, pp. 19–24 (2008)

Medelyan, O.: Human-Competitive Automatic Topic Indexing (2009)

Performance Evaluation of Large Data Clustering Techniques on Web Robot Session Data

Dilip Singh Sisodia, Rahul Borkar and Hari Shrawgi

Abstract Web robots are scripts that automatically surf the Web's server structure to locate and index information. These robots are sometimes used maliciously to create a myriad of problems in the functioning of servers. Such automated programs are difficult to trace and triangulate as they mask their identities. A weblog file which comprises of server requests can be used for identifying these robots by using clustering techniques. These log files contain a massive amount of data, and large data clustering algorithms are used to partition the requests into robotic sessions or human sessions. In this paper, a study is conducted, comparing the primary large clustering techniques. For clustering of the HTTP requests, we implemented BIRCH—Balanced Iterative Reducing and Clustering using Hierarchy (Hierarchical clustering technique), DBSCAN—Density-Based Spatial Clustering of Applications with Noise (Density-based clustering technique) and CLIQUE—Clustering in Quest (Grid-based method) using open-source ELKI & JBIRCH java packages. The performances of the three algorithms are compared using internal validating measures -Dunn's Index, DB Index, and Average Silhouette Index. As a result of the study, we found the optimal number of clusters to be four that produces the best validation measures.

Keywords Clustering · BIRCH · DBSCAN · CLIQUE · Clustering feature
Web robots · Web server logs · Web sessions

1 Introduction

Websites are frequently footprinted by automated software agents known as web robots or crawlers that commit various acts including productive activities like indexing, but also malicious activities such as spamming or intruding [1]. Harmful actions of web robots lead to different damaging effects over the internet like DDOS attacks on reputed websites.

D. S. Sisodia (✉) · R. Borkar · H. Shrawgi
National Institute of Technology Raipur, Raipur, India
e-mail: dssisodia.cs@nitrr.ac.in

© Springer Nature Singapore Pte Ltd. 2019
M. Tanveer and R. B. Pachori (eds.), *Machine Intelligence and Signal Analysis*,
Advances in Intelligent Systems and Computing 748,
https://doi.org/10.1007/978-981-13-0923-6_47

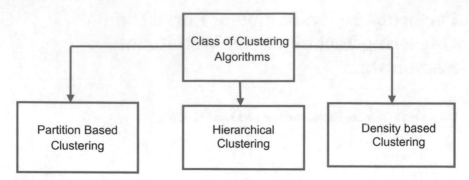

Fig. 1 Taxonomy of clustering techniques

To deal with such situations, it is essential for the servers to distinguish web robots from actual users. This will help the companies in deriving improved results related to advertising analytics and interactions with real customers; it will assist in optimizing the various Machine Learning algorithms that are deployed today. Also, it will help the administrators in analyzing robot activities going on in the websites and affecting its performance.

The information of all the HTTP sessions is stored in a log file which is automatically created and managed by the web server. The log file contains a significant amount of data, which is useful in identifying the users accessing the network [2]. This data includes IP addressing information, the URI of requests, request method and status, protocols used, timestamps and the size of the file transferred. All of this incisive information can help in identifying robotic sessions. Classification into user or robot sessions is straightforward if the session data in the log file is labeled [3]. But there is a dearth of such labeled data. In the absence of marked data, clustering is a robust method which helps in categorizing the data into smaller groups which might be easier to work with. Since the data associated with these log files is massive; it cannot be processed in raw form as it can turn out to be quite tedious. Therefore, pre-processing of the data is required. Structuring the data into groups or clusters helps in better analysis of data. This process of grouping data points together is called Clustering. In a cluster, objects are grouped on the basis of similarity. Clustering is also used in several fields such as pattern recognition [4], genetic science [5], artificial intelligence and much more. Clustering can be divided into the following major categories. Hierarchy Based [6], Density-Based [7] and Grid-based clustering [8].

Numerous algorithms have been introduced in each category of clustering, and the use of these techniques depends on the type of data being used and its application. Figure 1 represents the basic taxonomy of these techniques. Some of the algorithms are very efficient, but only when the size of input data is small (e.g., K-Means [9], K-Medoids [10], Agglomerative Clustering [11] etc.), whereas some of them are suitable for application on large-scale data (BIRCH [6], DBSCAN [7], CLIQUE [12], etc.). In this paper, we are concerned with algorithms that can be applied to large-scale data, and we would be focusing only on those.

The rest of this paper consists of four more sections. This section is followed by a review of related work. Section 3 describes the methodology followed in the study and explanations of three clustering algorithms are presented. In Sect. 4, the dataset used is specified, and the experimental results are compared based on specific evaluation measures. Lastly, the study is concluded in Sect. 5.

2 Related Work

Clustering of web session data is generally categorized as either supervised or unsupervised. It is clear that most of the literature concerns itself with supervised methods [13–17]; better data and improved results are the attributed reasons for this observation.

Two of the significant unsupervised techniques used are a self-organizing map (SOM) based on Artificial neural networks (ANN) and the modified adaptive resonance theory 2 (Modified ART2). These are used for classifying malicious and non-malicious users/robots [18].

A recent study conducted by Ferrari and Castro [19] presents a new meta-learning system which provides another dimension for identifying robotic sessions.

3 Methodology

The outline of this study is shown in Fig. 2. The bulk of data samples available is fed into one of the clustering algorithms. The selection of clustering algorithm is based on proximity measures. Data points are grouped together by these algorithms if they resemble one another. This resemblance is checked in accordance with the similarity measures, which can be either the Euclidian distance, Minkowski distance or the Manhattan distance [20].

Different criteria and starting points lead to different clustering algorithms [21]. A common framework is to classify clustering algorithms as hierarchical clustering and density-based clustering [9, 21, 22].

As stated earlier, not all algorithms are applicable to large-scale data. Some of them are either time or memory inefficient, while some produce poor quality clusters when a large dataset (many instances, attributes or both) is supplied to them. To handle extensive data, as present in web server log files, specially designed large

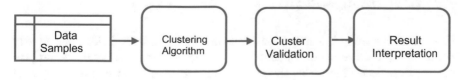

Fig. 2 General clustering process

data clustering algorithms are required. One such special-purpose algorithm was selected from each of the three categories of clustering techniques, namely Hierarchical, Density-based and Grid-based to cluster our data. Following three large data clustering algorithms were chosen.

3.1 BIRCH Algorithm

This class contains algorithms that create a nested sequence of partitions in the datasets at different levels. The clusters are represented by distinctive tree-like patterns called Dendrograms. The algorithms may function in two ways—agglomerative and divisive. Zhang et al. [6] contributed BIRCH (Balanced Iterative Reducing and Clustering using Hierarchy) clustering technique which cluster input data points incrementally as well as dynamically. Being a hierarchical algorithm, the first phase of the algorithm constructs a CF (Clustering Feature) Tree, which serves as a mechanism for clustering. It is a height-balanced tree, created dynamically as new data points are inserted in the CF tree which is similar to the B + tree. The non-leaf nodes store the CF entries for its child node, while the leaf nodes consist of the subclusters. Each CF entries consist of three parameters {N, Linear_sum, Square_sum}.

3.2 DBSCAN Algorithm

This algorithm finds dense regions in data space, i.e., the region segregated by low-density points at the boundaries, and identifies them as clusters. Daszykowski and Walczak [23] proposed DBSCAN (Density-Based Spatial Clustering with Noise) clustering algorithm which overcame the limitations of partition based and hierarchical clustering algorithms. DBSCAN can discover arbitrarily shaped clusters which other algorithms were unable to discover. For the experiments, synthetic sample database, as well as SEQUOIA 2000 benchmark data [24], was used. The data points are clustered according to two necessary global parameters—MinPts and Eps. Eps is the Epsilon Neighborhood of a point where Epsilon is the radius of the circle considering the point as the center. MinPts is the minimum number of points surrounding a point within the epsilon neighborhood.

3.3 CLIQUE Algorithm

Grid-based clustering is a unique technique in which dataset is divided into equally partitioned cells. The major benefit of Grid-based clustering is that it has fast processing time depending on the size of the grid instead of the size of data. Agrawal et al. [12] devised a clustering technique CLIQUE that can cluster high dimensional

Table 1 Label distribution in original dataset

Type of Sessions (as labeled in dataset)	# of Sessions
Total # of Sessions	334579
Human (or Browser)	304488
Web Robot (S/w Agents)	18206
Probable Web Robot (S/W Agents)	3994
Probable Human (or Browser)	7891

data. The clusters generated are same irrespective of the orders in which input data is fed into the algorithm. Experiments were performed in this study with CLIQUE algorithm using the data from synthetic data generator of [25]. We used the algorithm present in Algorithm 3 to test the performance of CLIQUE technique. The performance of the experiment is measured in the regards of efficiency and accuracy.

The advantage of CLIQUE algorithm is that the clustering complexity depends on the number of populated grid cells instead of a number of data points. Also, it is insensitive to the input order of the data. As the size of input is increased, it also scales linearly with respect to the change and is also efficient when the number of dimensions increases.

4 Experimental Results

In order to validate the algorithms mentioned above, we have used the dataset described in Table 1.

To perform an accurate and precise evaluation of the discussed techniques, a dataset containing a multitude of instances was required. For this purpose, the dataset presented in [26] was used. It comprises of the web server log of Computer Science Department, the University of Minnesota which logs at lacs of HTTP sessions. The temporal attributes calculated are total time, standard deviation time, average time. The width and depth attributes are calculated by the graph represented by HTML requests. There were 26 derived features in the dataset along with 334579 instances.

Since this study concerns itself with evaluation on large datasets, there was a requirement to choose evaluation measures [27] which can scale to large data and provide us with a clear comparison between the discussed techniques. Thus, the results obtained by clustering were validated using internal measures—Silhouette index [28], Dunn's index [29] and Davies-Bouldin (DB) index [30]. Silhouette index was chosen as it is a measure independent of the diameter of the cluster and thus suits our needs perfectly. Dunn's and DB Index when used in combination are useful in finding the optimal number of clusters in any data. A different number of clusters viz. 4, 8, 12, 16 clusters were used, and analysis of internal measures was performed to find out the number of clusters that optimizes the performance of web robots [26]

Table 2 Values of measures for different number of clusters formed using BIRCH

Number of clusters	Avg. Silhouette index	Dunn's index	Davies-Bouldin index	Time taken (s)
3	0.154	13.089	57.132	4.531
4	**0.208**	16.342	45.567	4.354
6	0.004	11.24	60.243	4.657
8	0.002	11.11	62.354	4.335

Table 3 Values of measures for different number of clusters formed using DBSCAN

Number of clusters	Avg. Silhouette index	Dunn's index	Davies-Bouldin index	Time taken (s)
3	0.598	13.505	54.286	3890.43
4	0.475	15.265	**50.143**	3982.22
6	0.018	10.487	60.576	4023.5
8	0.002	9.298	59.966	4678.12

Table 4 Values of measures for different number of clusters formed using CLIQUE

Number of clusters	Avg. Silhouette index	Dunn's index	Davies-Bouldin index	Time taken (s)
3	0.238	16.957	56.056	11.576
4	**0.455**	**18.154**	**52.096**	14.089
6	0.102	13.445	59.364	12.785
8	0.098	15.333	64.857	10.079

dataset. Finally, we evaluated which algorithm gives a better result for large-scale clustering data with respect to time complexity and cluster quality.

The values of various internal measures for clusters formed using all algorithms are mentioned in the tables.

It is known that the value of Dunn's index should be high and Davies–Bouldin index should be low for best clustering quality. Thus, it can be inferred from Tables 2, 3 and 4 that keeping the number of clusters as 4 will provide the best clustering quality.

Figure 3 compares the performance of the three techniques over the number of clusters. As mentioned in [29, 30], the more the value of avg. Silhouette index is closer to 1.0 the better-clustered points it represents. So, it is clearly seen from Fig. 3 that the web robot's data, when clustered into four clusters (by adjusting the values of input parameters to clustering algorithms), corresponds to better-clustered points. Also, CLIQUE and DBSCAN produced better values of internal measures (mentioned in Tables 1, 2 and 3), but DBSCAN took 1000 times more time to run. So, CLIQUE emerges to be the better algorithm in terms of time complexity and producing better quality clusters.

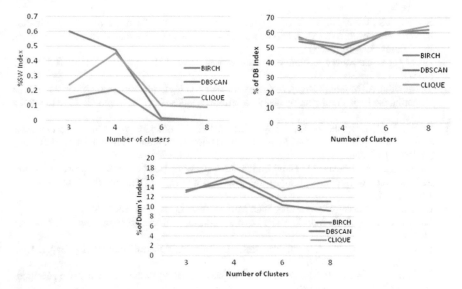

Fig. 3 Value measures for various indexes

5   Conclusion

In this paper, we evaluated three large data clustering techniques viz. CLIQUE, BIRCH, DBSCAN on Web robot's dataset. Each of these belongs to a different category of clustering—hierarchical, grid, and density-based. The output of clustered Web Robots was evaluated using Internal Validation measures, i.e., Avg. Silhouette Index, Dunn's Index, and DB Index. For web robot's data, the optimal clusters number came out to be 4, as when clustered into four clusters, it produces the maximum value for Silhouette index (0.475) and Dunn's index (18.154), also producing minimum value for Davies–Bouldin Index (45.567). The DBSCAN algorithm is about 1000 times slower in terms of time complexity than other two algorithms, and BIRCH produced a significantly lower value corresponding to Silhouette Index (0.204). CLIQUE outperformed both BIRCH and DBSCAN by having a lower time complexity as well as producing a better value corresponding to Silhouette Index (0.455). So, CLIQUE algorithm was more significantly better regarding both time complexity and clustering quality.

References

1. Sun, Y., Zhuang, Z., Giles, C.L.: A large-scale study of robots.txt. In: Proceedings of the 16th International Conference on World Wide Web, pp. 1123–1124. ACM (2007)
2. Sisodia, D.S., Verma, S., Vyas, O.: A comparative analysis of browsing behavior of human visitors and automatic software agents. Am. J. Syst. Software. **3**, 31–35 (2015)

3. Sisodia, D.S., Verma, S., Vyas, O.: Agglomerative approach for identification and elimination of web robots from web server logs to extract knowledge about actual visitors. J. Data Anal. Inf. Process. **3**, 1–10 (2015)
4. Hu, M.K.: visual pattern recognition by moment invariant. IRE Trans. Inf. Theory **8**, 179–187 (1962)
5. Thalamuthu, A., Mukhopadhyay, I., Zheng, X., Tseng, G.C.: Evaluation and comparison of gene clustering methods in microarray analysis. Bioinformatics (Oxford, England) **22**, 2405–2412 (2006)
6. Zhang, T., Ramakrishnan, R., Livny, M.: BIRCH: an efficient data clustering databases method for very large. In: ACM SIGMOD International Conference on Management of Data, pp. 103–114 (1996)
7. Zabihi, M.: Vafaei Jahan, M., Hamidzadeh, J.: A density based clustering approach to distinguish between web robot and human requests to a web server. The ISC Int. J. Inf. Secur. **6**, 77–89 (2014)
8. Berkhin, P.: A Survey of Clustering Data Mining Techniques. Springer (2006)
9. Jain, A.K.: Data clustering: 50 years beyond K-means. Pattern Recogn. Lett. **31**, 651–666 (2010)
10. Park, H.-S., Jun, C.-H.: A simple and fast algorithm for K-medoids clustering. Expert Syst. Appl. **36**, 3336–3341 (2009)
11. El-Hamdouchi, A., Willett, P.: Comparison of Hierarchical Agglomerative Clustering Methods for Document Retrieval (1989)
12. Agrawal, R.: Automatic subspace clustering of high dimensional data for data mining applications. US Patent No 6,003,029 (1999)
13. Tan, P.N., Kumar, V.: Discovery of web robot sessions based on their navigational patterns (2002)
14. Dikaiakos, M.D., Stassopoulou, A., Papageorgiou, L.: An investigation of web crawler behavior: characterization and metrics. Comput. Commun. **28**, 880–897 (2005)
15. Stassopoulou, A., Dikaiakos, M.: Web robot detection: a probabilistic reasoning approach. Comput. Netw. (2009)
16. Doran, D., Gokhale, S.S.S.: Web robot detection techniques: overview and limitations. Data Min. Knowl. Discov. **22**, 183–210 (2011)
17. Stevanovic, D., An, A., Vlajic, N.: Feature evaluation for web crawler detection with data mining techniques. Expert Syst. Appl. **39**, 8707–8717 (2012)
18. Stevanovic, D., Vlajic, N., An, A.: Detection of malicious and non-malicious website visitors using unsupervised neural network learning. Appl. Soft Comput. J. **13**, 698–708 (2013)
19. Ferrari, D.G., De Castro, L.N.: Clustering algorithm selection by meta-learning systems: a new distance-based problem characterization and ranking combination methods. Inf. Sci. **301**, 181–194 (2015)
20. Kouser, K., Sunita, A.: A comparative study of K Means algorithm by different distance measures. Int. J. Innov. Res. Comput. Commun. Eng. **1** (2013)
21. Xu, R., Wunsch, D.: Survey of clustering algorithms. IEEE Trans. Neural Netw. **16**, 645–678 (2005)
22. Äyrämö, S., Kärkkäinen, T.: Introduction to partitioning-based clustering methods with a robust example. Reports of the Department of Mathematical Information Technology Series C, Software Engineering and Computational Intelligence 1/2006 (2006)
23. Daszykowski, M., Walczak, B.: Density-based clustering methods. Compr. Chemom. **2**, 635–654 (2010)
24. Stonebraker, M., Frew, J., Gardels, K., Meredith, J.: The Sequoia 2000 storage benchmark. In: ACM SIGMOD Record, pp. 2–11 (1993)
25. Zait, M., Messatfa, H.: A comparative study of clustering methods. Future Gener. Comput. Syst. **13**, 149–159 (1997)
26. Tan, P.-N., Kumar, V.: Discovery of web robot sessions based on their navigational patterns. In: Intelligent Technologies for Information Analysis, pp. 193–222. Springer, Berlin, Heidelberg (2004)

27. Sisodia, D.S., Verma, S., Vyas, O.: Augmented intuitive dissimilarity metric clustering of web user sessions. J. Inf. Sci. **43**, 480–491 (2016)
28. Kaufman, L., Rousseeuw, P.: Finding Groups in Data: An Introduction to Cluster Analysis. John Wiley & Sons (1990)
29. Dunn, J.C.: Well-separated clusters and optimal fuzzy partitions. J. Cybern. **4**, 95–104 (1974)
30. Davies, D.L., Bouldin, D.W.: A cluster separation measure. IEEE Trans. Pattern Anal. Mach. Intell. 224–227 (1979)

Keystroke Rhythm Analysis Based on Dynamics of Fingertips

Suraj, Parthana Sarma, Amit Kumar Yadav, Amit Kumar Yadav
and Shovan Barma

Abstract The proposed work presents an analysis of rhythmic patterns based on the dynamics of the fingertips observed during keystroke events on a traditional computer keyboard. In this work, a detailed analysis of pressure and acceleration applied by the user has been taken into consideration, in contrast to the earlier works which have focused primarily on the timing variations of the keys. In this purpose, two different types of sensors pressure and accelerometer were embedded on a traditional keyboard. Groupings of numerical digits and special keys were used to design different kinds of tasks to acquire sensor and timing variations of each keystroke event. In total, six subjects (two females and four males) were asked to provide the data to validate the proposed idea. Subsequently, two types of analysis, interpersonal and intrapersonal keystroke rhythm, have been carried out. The analysis results show uniquely identifiable keystroke sensor and timing variations for different users. However, individual users almost maintained his/her keystroke pressure and acceleration variation irrespective of the type of tasks. Such results demonstrate that keystroke rhythm based on pressure and acceleration variations of the fingertips can be used as a behavioural feature for developing more sophisticated biometric systems for intrusion detection.

Suraj (✉) · P. Sarma · A. K. Yadav · A. K. Yadav · S. Barma
Department of Electronics and Communication Engineering,
Indian Institute of Information Technology Guwahati, Guwahati, Assam, India
e-mail: hrishabhsuraj52@gmail.com

P. Sarma
e-mail: parthana.sarma28@gmail.com

A. K. Yadav
e-mail: aysanny@gmail.com

A. K. Yadav
e-mail: yadavamit23797@gmail.com

S. Barma
e-mail: shovan.barma@gmail.com

© Springer Nature Singapore Pte Ltd. 2019
M. Tanveer and R. B. Pachori (eds.), *Machine Intelligence and Signal Analysis*,
Advances in Intelligent Systems and Computing 748,
https://doi.org/10.1007/978-981-13-0923-6_48

Keywords Keystroke dynamics · Behavioural biometrics · Fingertip dynamics
MEMS sensors

1 Introduction

In this digital era, the passwords play a very crucial role in our everyday life including interpersonal communication (email, social media conversations, etc.), financial
transactions (banking, shopping etc.) and cloud based data storage. Even though
several high-level security measures like one-time passwords (OTP) based verification, secret question-answer, etc., have been incorporated in the existing biometric
systems, still hackers have been able to find a way to access personal data and
accounts, etc. Certainly, creating and remembering long complex passwords and
changing passwords frequently are not considered as a good solution. Therefore,
several biometric-based approaches like iris recognition [1], face recognition [2] and
fingerprint recognition [3] have been considered to overcome the drawbacks. Nevertheless, such methods can provide a better safeguard than the existing password-based
systems. Deployment of these sophisticated technologies is a bit expensive to find
its place in a common man's lifestyle. Thus, incorporation of behavioural features
such as keystroke dynamics in the existing systems could be a reasonable solution.

The keystroke dynamics is a method of intrusion detection by incorporating the
unique rhythmic patterns exhibited by users while typing on the keyboard. Using
this technique, people can be identified by their typing rhythms [4] similar to the
identification of a user by his handwriting. In this regard, most of the works conducted
are features made up of timing variations of the user during the keystroke.

As per literature, in the broad sense, keystroke verification techniques can be
divided into two major classes—static and continuous keystroke verification [5]. The
static keystroke verification approach analyzes and authenticates the user based upon
keyboard dynamics at discrete instants of time like the log-in process, etc., whereas
continuous keystroke verification-based approaches are more concerned with real-
time keystroke dynamics. For the timing-based keystroke rhythm analysis, release
and hold time as discriminative features are taken into account [6]. In literature, in
1977, Forsen et al. [7] investigated whether the authentication of users can be done
by analyzing the way in which the user types. Later, Peacock et al. [8] proposed new
metrics named cost to a user to enroll (CUE) and cost to a user to authenticate (CUA)
for measuring the usability of keystroke systems. The CUE measures the number of
keystrokes a user must submit to the system before being enrolled as a valid user,
whereas the CUA measures the number of keystrokes a user must submit to the system
each time he/she authenticates. Gaines et al. [9] distinguished an individual user by
analyzing his/her keystroke timing variation based on typing performed over a long
passage. Their results suggest the existence of typing consistency whenever the user
is allowed to perform the similar keystroke events. However, one of the limitations
of timing-based keystroke system is the system clock resolution which effects the
intrusion detection accuracy of biometric systems. In this concern, Kurihara et al.

[10] proposed a system by embedding pressure sensors inside the keys and achieved the satisfactory result; however, they did not consider the system time information. Nevertheless, the reliability and performances of the system could be improved by introducing extra sensors [11] along with pressure sensors in addition to the timing features obtained during keystroke events [12].

Hence, motivated by need of such a biometric system, we have proposed a new system to analyze the field of keystroke dynamics by combining the hold, release timing variation features along with the sensor variations features (pressure and acceleration). A traditional keyboard embedded with sensors has been designed for this purpose. To observe the variation in the keystroke dynamics among different users under different scenarios, four distinct tasks were designed [13]. The task includes typing phone numbers, typing random numbers, and typing digits over logical tasks and arithmetic calculations. The keystroke timing and sensor data were acquired from six subjects, and consequently, interpersonal and intrapersonal keystroke rhythm analysis were performed.

The rest of the paper is structured as follows. Section 2 illustrates the system overview which describes the keyboard layout with pressure and accelerometer sensors. Section 3 describes the methodology followed by Sect. 4, which summarizes the experimental procedures. The results are summarized in Sect. 5. Finally, Sect. 6 draws the conclusion and future work of the proposed research.

2 System Overview

An overview of the proposed system is shown in Fig. 1. It includes major blocks like data acquisition, feature extraction, and data analysis. During the data acquisition phase, two different types of data have been collected for any particular keystroke event by a user: (a) *Timing data*: timing features obtained from the operating system's clock and (b) *Sensor data*: sensor features (acceleration and pressure variation) obtained during the same keystroke events as of timing data. During the feature extraction phase, four features were selected—hold time, release time [14] from timing data and pressure, acceleration variations from sensor data variation during keystroke. Further, the data analysis phase involved analyzing the collected features from the users. The keyboard keys used and their manual notations used for designing the keyboard are listed in Table 1. The ideology behind incorporating the use of 12 keys for building the keyboard involves the fact that most common input layout across devices is the numeric keypad layout with digits 0-9 and symbols. The data from three modalities (the system clock, the force-sensitive resistor, and the accelerometer) were logged using a USB 2.0 data cable and Arduino Mega 256 boards, respectively. The pressure and acceleration variations during keystroke events were logged using the 12 analog input pins of two Arduino Mega 256 boards connected to the laptops through a USB cable. The timing variations of the same keystroke event were simultaneously logged using the USB data cable of a traditional desktop computer keyboard.

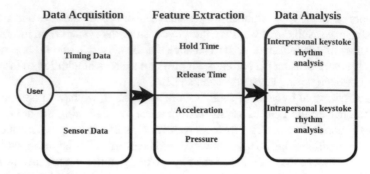

Fig. 1 System block diagram

Table 1 Keyboard key notation

Keyboard key	Q	W	E	R	T	Y	U	I	O	P	[	]
Manual notation	0	1	2	3	4	5	6	7	8	9	*	#

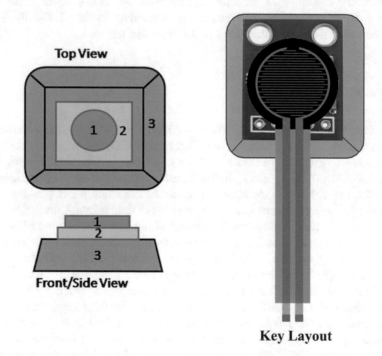

Fig. 2 Key layout

Figure 2 shows the layout of the key with accelerometer stacked in between the key's outer surface and force-sensitive resistor. The Force-sensitive resistor was kept exposed to the surface where fingers interact directly during the keystroke events to ensure high precision pressure readings. In Fig. 2, the key, the accelerometer, and the force-sensitive resistor are labeled as 3, 2, 1, respectively.

3 Methodology

3.1 Data Acquisition

The experiments have been conducted by considering six subjects (two female and four male) of the age group of 18–23 years. Prior data acquisition, all the subjects were briefed about the intention of the task and the use of their keystroke behavioural data. The data acquisition took place in two sessions of three trials each. As afore-mentioned, two kinds of data including timing variations and sensors-data variations during keystroke events have been acquired simultaneously. The data acquisition was performed within a laboratory environment. During each session, the subjects were asked to sit calmly on a traditional chair and perform the keystroke events as per the instructed tasks. Consequently, the subjects were instructed how to perform their typing events by the instructor. In this proposed work, we have considered the numerical and special keys as mentioned in Table 1 to conduct the experiment. Hence, the subjects were instructed to use those keys only. The entire data of tim-ing, pressure, and acceleration variations of the keystroke was collected using the proposed keyboard architecture which incorporated the use of USB 2.0 to log the timing variations and UART to log the pressure and acceleration variations. The data acquisition was performed using 64-bit Windows-10 platform systems with AMD A4-5000 processor (1500 MHz, 4 Cores, 4 GB RAM). There were four different sets of tasks performed by the user over auditory and visual modality. The tasks were based on different reasoning and logical skills of the users. The numbers in the tasks were presented to the user either by the visual, auditory modalities or using both. In case of visual-based tasks, the users were asked to type the sequence of 12 digits presented to them on a sheet of paper, whereas the similar 12-digit sequence narrated by a prerecorded audio was asked to be typed in the auditory mode.

The description of each of the four tasks performed by the users are as follows:

(1) *Typing phone numbers (Task 1)*: It is assumed that the user remembers his/her own phone number and can type their phone number rapidly. Therefore, the users were asked to type their own 12-digit cell phone number (including country code, eg. 91 for India).
(2) *Typing random numbers (Task 2_v and Task 2_a)*: It was performed in two modes—visual and auditory. In Task 2_v, the users were asked to type a 12-digit random number presented before them in the group of two, three, or four digits/special keys as shown in Fig. 3. Such grouping of numbers aids to analyze

Fig. 3 Grouping of digits

32 42 72 82 42 61

431 316 821 129

2342 6212 1832

Table 2 Description of notations

Notation	Description
T_hold	It denotes the net hold timing of a key during keystroke event
T_release	It denotes the net release timing of a key during keystroke event
Pressure	It denotes the net pressure exerted on a key during keystroke event
Acceleration	It denotes the net acceleration of a key during keystroke event

the keystroke variations of a user while a user remembers a long digit sequence. In Task2_a, the users were asked to complete a similar 12-digit sequence narrated to them by a pre-recorded audio clip.

(3) *Typing digits in a logical task (Task 3_v and Task 3_a):* It was performed in two modes—visual and auditory. In this case, the user was asked to complete and type a random arithmetic progression up to six/seven terms which assumed user capable of guessing the common difference after some terms in the sequence. This task was also carried out in both visual and auditory modalities.

(4) *Typing digits after an arithmetic calculation (Task 4):* In this case, the user was asked to type the two summands of four digits along with their four-digit sum comprising altogether the 12 digits, after performing the arithmetic task.

The final dataset for all six users was logged by performing the above mentioned tasks. The features in the dataset are timing, pressure and acceleration variations of the keystroke. The exact features are described in Table 2.

4 Experiment

Once the data were logged, the features were extracted from the dataset and consequently significant changes were profiled as a dataset. The experiments were conducted in two phases: first one being the features from the timing data of keystroke of several users, and the second one being the features from the pressure and acceleration variations. The timing data was collected using the pyHook library [15] of windows platform which returned a unique eight-digit time stamp for each keystroke which comes straight from the "time" member of the Win32 EVENTMSG struct. The pressure and acceleration variation of sensors were measured in Volts by mapping the analog readings obtained from the microcontrollers. Moreover, as no standard formula existed to map the analog readings from Arduino into the S.I units of pressure and acceleration, the sensor variations were profiled in Volts only.

5 Results and Discussions

The results are presented on the basis of analysis of keystroke timing and sensor variations. Table 3 illustrates the mean timing and sensor variations of each of the six users subjected to perform all the tasks. From the table, it can be observed that the data variations for a particular set of input are almost consistent for the six users. Therefore, the designed system is suitable for further analysis of the keystroke dynamics of users.

In interpersonal keystroke rhythm analysis, the timing and sensor data variations of all six users were analyzed. Whereas, intrapersonal keystroke rhythm analysis of users was aimed to analyze the keystroke timing sensor variation of individual user while performing all the instructed tasks.

5.1 Interpersonal Keystroke Rhythm Analysis

The graphs describing the sensor variation (acceleration and pressure) of all users during keystroke events in one of the tasks are shown in Figs. 4 and 5 respectively. Whereas Figs. 6 and 7 show the plots describing timing variation (hold time and release time) during keystroke events in the same task.

The sensor data variations (pressure and sensor) during keystroke events in Task 1 are represented in the plots by Volts whereas the timing data variations (hold and release time) during the keystroke events in Task 1 are represented in the plots by

Table 3 Mean variation of sensor and timing data of all users across tasks

Users	Accel.(in mV)	P (in mV)	T hold (in eight-digit time stamp)	T release (in eight-digit time stamp)
1	1602	2403	9.61×10^7	9.162×10^7
2	1309.2	1689	4.985×10^7	4.985×10^7
3	852	1082	1.524×10^7	1.524×10^7
4	1311.1	1646.8	4.959×10^7	4.959×10^7
5	1277	1560.6	4.816×10^7	4.815×10^7
6	1428	1754.2	5.033×10^7	5.033×10^7

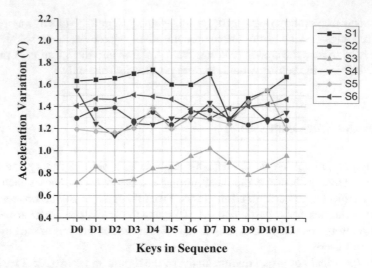

Fig. 4 Acceleration variations of six users while performing Task 1

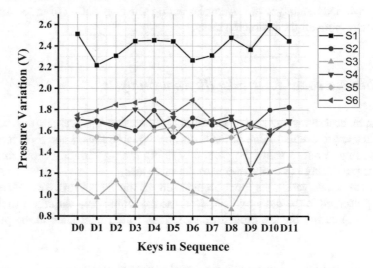

Fig. 5 Pressure variations of six users while performing Task 1

eight digit number denoting the time stamps returned by the PyHook library. The notations S1-S6 in the legend section of each plot describes the Users (U1 to U6). The horizontal axis of each plot containing labels D0-D11 represents the digits in the 12-digit sequence representing the cellphone number of the user used in Task 1. The vertical axis in Figs. 4, 5, 6 and 7 represent the variation of acceleration, pressure, release time, and hold time of six users obtained by performing Task 1. However, it can be observed from Fig. 4, it can be observed that each user (U1-U6) has its own characteristics keystroke rhythm pattern described by the acceleration variation

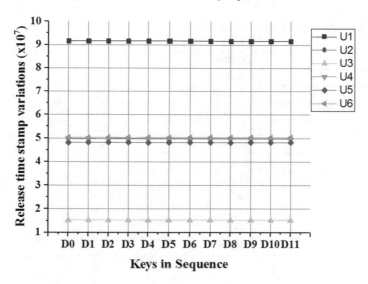

Fig. 6 Release timing variations of six users while performing Task 1

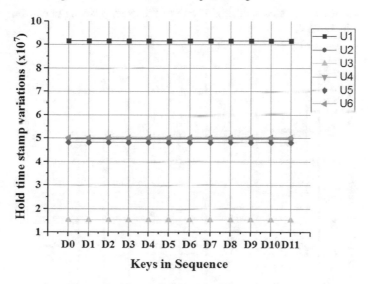

Fig. 7 Hold timing variations of six users while performing Task 1

obtained while typing his/her phone number during Task 1. The same inference can be deduced from Fig. 5 which shows users U1-U6 displaying unique keystroke rhythm pattern described by pressure variations. However in Figs. 6 and 7, the release and hold time variations of the user were found to be relatively to be significantly noticed in the eight-digit time stamps. Hence, the keystroke rhythm pattern of the users described by the release and hold key timings appear to be uniform. The small

Table 4 Hold and release time variations of a user

Digit	T_hold	T_release
D0	91609859	91608421
D1	91610062	91610000
D2	91610937	91610203
D3	91611171	91611093
D4	91611656	91611312
D5	91611875	91611765
D6	91612265	91612031
D7	91612750	91612390
D8	91612953	91612906
D9	91613203	91613093
D10	91613437	91613343
D11	91613781	91613609

variations in hold and release timing of one of the users (User 1) performing Task 1 are shown in Table 4.

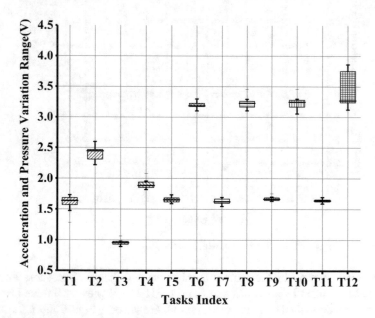

Fig. 8 Boxplot of pressure and acceleration variations

5.2   Intrapersonal Keystroke Rhythm Analysis

The timing and sensor data variation for a single user while performing the four tasks have been observed in this sub-section. The analysis of the sensor and timing variations of one of six users while performing keystroke events in all the tasks is done by plotting boxplots. The boxplot describing the acceleration and pressure sensor variation range of a user U1 while performing all the Tasks after two sessions are shown in Fig. 8. Similarly, the boxplot describing the hold time and release time variation range of a user U1 while performing all the tasks after two sessions is shown in Fig. 9. The horizontal axis of the plot in Fig. 8 represents labels denoted by T1, T2, …, T12 where T1, T3, T5, …, T11 represent tasks Task 1, Task 2_v, Task 2_a, Task 3_v, Task3_a, and Task 4 for keystroke events involving acceleration variations whereas T2, T4, T6, …, T12 represent Task 1, Task 2_v, Task 2_a, Task 3_v, Task3_a, and Task 4 for keystroke events involving pressure sensor variations. However, the vertical axis of the plot in Fig. 9 denotes the range of acceleration and pressure variations in Volts. Similarly, the horizontal axis in the same plot represents markers denoted by T1 to T12 where T1, T3, T5, …, T11 represent tasks Task 1, Task 2_v, Task 2_a, Task 3_v, Task3_a, and Task 4 for keystroke events involving hold time variations whereas T2, T4, T6, …, T12 represent Task 1, Task 2_v, Task 2_a, Task 3_v, Task3_a, and Task 4 for keystroke events involving release time variations. Consequently, the vertical axis of the plot denotes the range of hold and release time variations in eight digit numbers denoting time stamps returned by the PyHook library. The boxplot shown in Fig. 8 shows that User 1 has almost uniform pressure

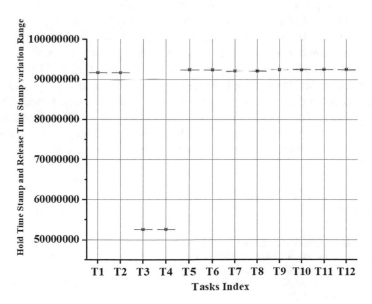

Fig. 9 Boxplot of hold and release timing variations

and acceleration variation for keystroke events performed during performing Task 1, Task 2_a, Task 3_v, Task3_a and Task 4. However, an inconsistency in sensor variations is observed when the user was asked to perform the Task 2_v. Similar observations were noted from the boxplot of Fig. 9 which shows an invariance in the hold and release time of user while performing Task2_v. However, as the user shows relatively small sensor and timing variations across different domains of instructed tasks, it can be inferred from the sensor and timing profile of the users that irrespective of the tasks given to a user, the sensor and timing variations of his/her keystroke events remains almost constant.

6 Conclusions and Future Work

This paper presents an analysis of keystroke dynamics by designing a traditional keyboard embedded with sensors(pressure and acceleration). The analysis of the collected timing and sensor features was carried out to study interpersonal and intrapersonal keystroke rhythm variations. The observations obtained after the interpersonal keystroke rhythm analysis of six users for the task shows that users had their unique keystroke rhythm patterns obtained from timing and sensor variations. Furthermore, in case of the intrapersonal keystroke rhythm analysis, the results showed that keystroke rhythms reflected by the user were almost task independent. Thus, embedding pressure and acceleration sensor variations in the keystroke dynamics can be used as an accurate behavioural feature.

Although the proposed novel keyboard has been found to be a better behavioural feature for analyzing keystroke rhythms among individuals, there are a few more works yet to be taken into consideration as future work. One of which is incorporating more number of users with wide range of age groups in data acquisition phase. Moreover, as the current work has been done on the numerical keypad layout, extending the sensor embedding keyboard layout to the whole keyboard needs to be done. Further, performances evaluation of the fused sensor and timing keystroke variations will be conducted by incorporating machine learning methods. Besides all these, the newer paradigm will be included to incorporate more human behaviors during typing to understand the keystroke dynamics in a better way.

References

1. De Marsico, M., Galdi, C., Nappi, M., Riccio, D.: Firme: face and iris recognition for mobile engagement. Image Vis. Comput. **32**(12), 1161–1172 (2014)
2. Schroff, F., Kalenichenko, D., Philbin, J.: Facenet: a unified embedding for face recognition and clustering. In: Proceedings of the IEEE Conference on Computer Vision and Pattern Recognition, pp. 815–823 (2015)
3. Labati, R.D., Genovese, A., Piuri, V., Scotti, F.: Toward unconstrained fingerprint recognition: a fully touchless 3-d system based on two views on the move. IEEE Trans. Syst. Man Cybern.

Syst. **46**(2), 202–219 (2016)
4. Zheng, N., Bai, K., Huang, H., Wang, H.: You are how you touch: user verification on smartphones via tapping behaviors. In: IEEE 22nd International Conference on Network Protocols (ICNP 2014), pp. 221–232. IEEE, USA (2014)
5. Monrose, F., Rubin, A.D.: Keystroke dynamics as a biometric for authentication. Future Gener. Comput. Syst. **16**(4), 351–359 (2000)
6. Sim, T., Janakiraman, R.: Are digraphs good for free-text keystroke dynamics? In: IEEE Conference on Computer Vision and Pattern Recognition (CVPR'07), pp. 1–6. IEEE, USA (2007)
7. Forsen, G.E., Nelson, M.R., Staron Jr., R.J.: Personal attributes authentication techniques. Technical report, Pattern Analysis and Recognition Corp. Rome (1977)
8. Peacock, A., Ke, X., Wilkerson, M.: Typing patterns: a key to user identification. IEEE Secur. Priv. **2**(5), 40–47 (2004)
9. Gaines, R.S., Lisowski, W., Press, S.J., Shapiro, N.: Authentication by keystroke timing: some preliminary results. Technical Report, Rand Corp. Santa Monica (1980)
10. Nonaka, H., Kurihara, M.: Sensing pressure for authentication system using keystroke dynamics. Int. J. Comput. Intell. **1**(1), 19–22 (2004)
11. Giuffrida, C., Majdanik, K., Conti, M., Bos, H.: I sensed it was you: authenticating mobile users with sensor-enhanced keystroke dynamics. In: International Conference on Detection of Intrusions and Malware, and Vulnerability Assessment. pp. 92–111, Springer, Berlin (2014)
12. Saevanee, H., Bhattarakosol, P.: Authenticating user using keystroke dynamics and finger pressure. In: 2009 6th IEEE Consumer Communications and Networking Conference (CCNC), pp. 1–2. IEEE, USA (2009)
13. Conti, M., Zachia-Zlatea, I., Crispo, B.: Mind how you answer me!: transparently authenticating the user of a smartphone when answering or placing a call. In: Proceedings of the 6th ACM Symposium on Information, Computer and Communications Security, pp. 249–259. ACM, USA (2011)
14. Kambourakis, G., Damopoulos, D., Papamartzivanos, D., Pavlidakis, E.: Introducing touchstroke: keystroke-based authentication system for smartphones. Secur. Commun. Netw. **9**(6), 542–554 (2016)
15. Rehim, R.: Effective python penetration testing. Packt Publishing Ltd, Birmingham (2016)

A Fuzzy Universum Support Vector Machine Based on Information Entropy

B. Richhariya and M. Tanveer

Abstract Universum-based support vector machines (USVMs) are known to give better generalization performance than standard SVM methods by incorporating prior information about the data. In datasets involving noise and outliers, this universum-based scheme is not so effective because the generated universum data points do not lie in between the two classes. In this paper, we propose a fuzzy universum support vector machine (FUSVM) by introducing the weights to the universum data points based on their information entropy. Since there is no standard approach of selecting the universum, our information entropy based approach is helpful in giving less weight to the outlier universum points and thus gives prior information about the data in an appropriate manner. In addition, we also propose a fuzzy-based approach for universum twin support vector machine named as fuzzy universum twin support vector machine (FUTSVM). Experimental results on several benchmark datasets indicate that, comparing to SVM, USVM, TWSVM and UTSVM our proposed FUSVM and FUTSVM have shown better generalization performance.

Keywords Universum · Fuzzy membership · Information entropy
K-nearest neighbour (KNN)

1 Introduction

In the recent years, support vector machine (SVM) [1, 2] has been widely used for classification problems [3–5], pattern recognition [6], speaker verification [7] and intrusion detection [8]. Due to its low VC (Vapnik—Chervonenkis) dimension, it gives good generalization performance even with high-dimensional data. In contrast

B. Richhariya (✉) · M. Tanveer
Discipline of Mathematics, Indian Institute of Technology Indore,
Simrol, Indore 453552, India
e-mail: brichhariya@iiti.ac.in

M. Tanveer
e-mail: mtanveer@iiti.ac.in

© Springer Nature Singapore Pte Ltd. 2019　　　　　　　　　　　　　　569
M. Tanveer and R. B. Pachori (eds.), *Machine Intelligence and Signal Analysis*,
Advances in Intelligent Systems and Computing 748,
https://doi.org/10.1007/978-981-13-0923-6_49

to artificial neural network (ANN) which suffers from local minima, the convex optimization problem in SVM gives a global optimal solution. Further, to reduce the computational cost of SVM, Jayadeva and Chandra [9] proposed a twin support vector machine (TWSVM) where two hyperplanes are constructed instead of one as in the case of standard SVM. The two hyperplanes in TWSVM are closer to their own class and as far as possible from the other class. Experimental results show that TWSVM can effectively improve the generalization performance over SVM. Due to its better generalization ability, researchers have proposed various variants of TWSVM [10–17]. Some improvements on SVM are proposed in [18, 19].

Weston et al. [20] proposed universum-based support vector machine (USVM) to include prior information about the distribution of data in SVM. In USVM, the universum data points do not belong to any of the classes and the procedure for selecting universum points is dependent on the data. Researchers have used their own techniques for the selection of universum data. Bai and Cherkassky [21] have used random averaging for facial images and took the average of the faces as the universum. An In-between-universum (IBU) approach is proposed in [22] for classification of images. Cherkassky et al. [23] discussed the practical conditions for the appropriate selection of universum data. Recently, a universum twin support vector machine (UTSVM) is proposed by Qi et al. [24] where random averaging of data is used as universum. A least squares twin support vector machine (LSTSVM) for universum data is proposed in Xu et al. [25] where random averaging scheme is used.

For classification of noisy data, fuzzy-based approaches are used for SVM [26–31]. In 2016, Fan et al. [32] used entropy based fuzzy membership for support vector machine (EFSVM) in case of class imbalance problem. This entropy-based approach is used to give higher fuzzy membership to the data points which lie at the boundary of the two classes. Motivated by this concept, we have proposed entropy-based fuzzy membership in our proposed FUSVM and FUTSVM. In our entropy-based fuzzy approach, the universum points are assigned fuzzy membership based on their uncertainty of belonging to any one class. So the universum points are calculated using the random averaging scheme and then higher membership is assigned to those universum points which lie in between the two classes using their entropy values. The universum points which are in between the two classes have higher entropy values as compared to the universum points lying nearer to one of the classes. By the use of a fuzzy-based approach for the universum data, the effect of outlier universum points is reduced which leads to higher generalization performance.

The paper is organized as follows: Sect. 2 gives a review on the work related to universum support vector machine (USVM) and universum twin support vector machine (UTSVM). Section 3 describes the details of the proposed FUSVM and FUTSVM algorithms. Numerical experiments are performed on well known real-world benchmark datasets and their results are compared with SVM, USVM, TWSVM and UTSVM in Sect. 4. Finally, we conclude our work in Sect. 5.

2 Related Work

2.1 Universsum Support Vector Machine (USVM)

In USVM [20], the universum data is incorporated by using an $\varepsilon-$insensitive loss function. The objective function of USVM is written as

$$\min_{w,\,b,\xi,\,\eta} \frac{1}{2}||w||^2 + C\left(\sum_{i=1}^{m} \xi_i\right) + C_u \sum_{j=1}^{2|u|} (\eta_j)$$

$$\text{subject to } y_i(\varphi(x_i)^t w + b) \geq 1 - \xi_i, \tag{1}$$

$$y_j(\varphi(x_j)^t w + b) \geq -\varepsilon - \eta_j,$$

$$\xi_i \geq 0, \eta_j \geq 0, \forall i = 1, \ldots, m, \forall j = 1, \ldots, 2|u|,$$

where C and C_u are penalty parameters, $\varphi : R^n \to R^p$ where $p > n$ is the mapping to higher dimension, ε is the tolerance value, m is the total number of samples and u is the number of universum data points.

For solving the optimization problem we treat the set of universum points as belonging to both the classes. We use the universum set twice and assign the target values as +1 to one set and −1 to the other.

The dual formulation of USVM is written as

$$\min_{\lambda} - \sum_{i=1}^{m+2|u|} \rho_i \lambda_i + \frac{1}{2} \sum_{i=1}^{m+2|u|} \sum_{j=1}^{m+2|u|} \lambda_i \lambda_j K(x_i, x_j) y_i^t y_j$$

$$\text{subject to } \quad 0 \leq \lambda_i \leq C, \forall i = 1, \ldots, m$$

$$\rho_i = 1, \forall i = 1, \ldots, m$$

$$0 \leq \lambda_i \leq C_u, \forall i = m+1, \ldots, m+2|u|$$

$$\rho_i = -\varepsilon, \forall i = m+1, \ldots, m+2|u|$$

$$\text{and} \quad \sum_{i=1}^{m+2|u|} \lambda_i y_i = 0,$$

where $K(x_i, x_j) = \varphi(x_i)^t \varphi(x_j)$ is the kernel function and λ is the vector containing the Lagrange multipliers.

The classifier is written as

$$f(x) = sign\left(\sum_{i=1}^{m+2|u|} \lambda_i y_i K(x_i, x) + b\right). \tag{2}$$

2.2   Universum Twin Support Vector Machine (UTSVM)

In UTSVM [24], the universum points are used as constraints in two optimization problems instead of one. This helps in reducing the computation cost of USVM. In nonlinear case, the objective functions of UTSVM are written as

$$\min_{w_1, b_1, \xi, \psi} \frac{1}{2}||K(X_1, D^t)w_1 + e_1b_1||^2 + C_1e_2^t\xi + C_ue_u^t\psi$$

$$\text{subject to } -(K(X_2,D^t)w_1 + e_2b_1) + \xi \geq e_2, \xi \geq 0 \tag{3}$$

$$(K(U, D^t)w_1 + e_2b_1) + \psi \geq (-1 + \varepsilon)e_u, \psi \geq 0$$

$$\min_{w_2, b_2, \eta, \psi^*} \frac{1}{2}||K(X_2, D^t)w_2 + e_2b_2||^2 + C_2e_1^t\eta + C_ue_u^t\psi^*$$

$$\text{subject to } (K(X_1, D^t)w_2 + e_1b_2) + \eta \geq e_1, \eta \geq 0 \tag{4}$$

$$-(K(U, D^t)w_2 + e_ub_1) + \psi^* \geq (-1 + \varepsilon)e_u, \psi^* \geq 0$$

where ξ, η, ψ, ψ^* are the slack variables; C_1, C_2 and C_u are penalty parameters; $D = [X_1; X_2]$; e_1, e_2, e_u are vectors of suitable dimension having all values as 1s and $K(x^t, D^t) = (k(x, x_1), \ldots, k(x, x_m))$ is a row vector in R^m space.

The Wolfe dual of UTSVM is written as

$$\max_{\alpha_1, \mu_1} e_2^t\alpha_1 - \frac{1}{2}(\alpha_1^t T - \mu_1^t O)(S^t S)^{-1}(T\alpha_1^t - O\mu_1^t) + (\varepsilon - 1)e_u^t\mu_1$$

$$\text{subject to } 0 \leq \alpha_1 \leq C_1, 0 \leq \mu_1 \leq C_u, \tag{5}$$

$$\max_{\alpha_2, \mu_2} e_1^t\alpha_2 - \frac{1}{2}(\alpha_2^t S - \mu_2^t O)(T^t T)^{-1}(S^t\alpha_2 - O^t\mu_2) + (\varepsilon - 1)e_u^t\mu_2$$

$$\text{subject to } 0 \leq \alpha_2 \leq C_2, 0 \leq \mu_2 \leq C_u, \tag{6}$$

where $S = [K(X_1, D^t) \ e_1], T = [K(X_2, D^t) \ e_2]$ and $O = [K(U, D^t) \ e_u]$.

The non-linear hyperplanes $K(x^t, D^t)w_1 + b_1 = 0$ and $K(x^t, D^t)w_2 + b_2 = 0$ are obtained by using the following Eqs. (7) and (8),

$$\begin{bmatrix} w_1 \\ b_1 \end{bmatrix} = -(S^t S + \delta I)^{-1}(T^t\alpha_1 - O^t\mu_1), \tag{7}$$

$$\begin{bmatrix} w_2 \\ b_2 \end{bmatrix} = (T^t T + \delta I)^{-1}(S^t\alpha_2 - O^t\mu_2). \tag{8}$$

A new data point $x \in R^n$ is assigned to a class 'i' on the basis of its distance from the two hyperplanes using the following formula:

$$class \ i = \min|K(x^t, D^t)w_i + b_i| \ \text{for } i = 1, 2. \tag{9}$$

3 Proposed Approach

3.1 Fuzzy Universum Support Vector Machine (FUSVM)

To reduce the effect of noise and outlier data points, we propose a fuzzy-based approach for USVM. The objective function of FUSVM with the corresponding constraints is written as

$$\min_{w,\,b,\,\xi,\,\eta} \frac{1}{2}||w||^2 + C\left(\sum_{i=1}^{m}\xi_i\right) + C_u\sum_{j=1}^{2|u|}f_j(\eta_j)$$

$$\text{subject to} \quad y_i(\varphi(x_i)^t w + b) \geq 1 - \xi_i, \tag{10}$$
$$y_j(\varphi(x_j)^t w + b) \geq -\varepsilon - \eta_j,$$
$$\xi_i \geq 0, \eta_j \geq 0, \forall i = 1,\ldots,m, \forall j = 1,\ldots,2|u|$$

where C and C_u are penalty parameters, $\varphi : R^n \to R^p$ where $p > n$ is the mapping to higher dimension, ε is the tolerance value, f_j is the fuzzy membership value, m is the total number of samples and u is the number of universum data points.

The Lagrangian of the objective function (10) is given as

$$L = \frac{1}{2}||w||^2 + C\left(\sum_{i=1}^{m}\xi_i\right) + C_u\left(\sum_{j=1}^{2|u|}f_j\eta_j\right) - \sum_{i=1}^{m}\lambda_i(y_i(\varphi(x_i)^t w + b) - 1 + \xi_i) - \sum_{i=1}^{m}\psi_i\xi_i$$

$$- \sum_{j=1}^{2|u|}\alpha_j(y_j(\varphi(x_j)^t w + b) + \varepsilon + \eta_j) - \sum_{j=1}^{2|u|}\beta_j\eta_j, \tag{11}$$

where λ, ψ, α and β are the Lagrange multipliers.

For solving the optimization problem we treat the universum points as belonging to both the classes and assign the target values accordingly.

The dual formulation of (11) after applying the Karush–Kuhn–Tucker (KKT) necessary and sufficient conditions is written as

$$\min_{\lambda} - \sum_{i=1}^{m+2|u|}\rho_i\lambda_i + \frac{1}{2}\sum_{i=1}^{m+2|u|}\sum_{j=1}^{m+2|u|}\lambda_i\lambda_j K(x_i, x_j)y_i^t y_j$$

$$\text{subject to} \quad 0 \leq \lambda_i \leq C, \forall i = 1,\ldots,m$$
$$\rho_i = 1, \forall i = 1,\ldots,m$$
$$0 \leq \lambda_i \leq f_i C_u, \forall i = m+1,\ldots,m+2|u|$$
$$\rho_i = -\varepsilon, \forall i = m+1,\ldots,m+2|u|$$

$$\text{and} \quad \sum_{i=1}^{m+2|u|}\lambda_i y_i = 0,$$

where $K(x_i, x_j) = \varphi(x_i)^t\varphi(x_j)$ is the kernel function and λ is the vector containing the Lagrange multipliers.

For any data point $x \in R^n$, the classifier is written as

$$f(x) = sign\left(\sum_{i=1}^{m+2|u|} \lambda_i y_i K(x_i, x) + b\right).$$ (12)

3.2 Fuzzy Universum Twin Support Vector Machine (FUTSVM)

By introducing weights to the universum data points based on their information entropy, the non-linear FUTSVM comprises of the following pair of minimization problems:

$$\min_{w_1, b_1, \xi, \psi} \frac{1}{2}||K(X_1, D^t)w_1 + e_1 b_1||^2 + C_1 e_2^t \xi + C_u f_u^t \psi$$

$$\text{subject to } -(K(X_2, D^t)w_1 + e_2 b_1) + \xi \geq e_2, \xi \geq 0$$ (13)

$$(K(U, D^t)w_1 + e_2 b_1) + \psi \geq (-1 + \varepsilon)e_u, \psi \geq 0$$

$$\min_{w_2, b_2, \eta, \psi^*} \frac{1}{2}||K(X_2, D^t)w_2 + e_2 b_2||^2 + C_2 e_1^t \eta + C_u f_u^t \psi^*$$

$$\text{subject to } (K(X_1, D^t)w_2 + e_1 b_2) + \eta \geq e_1, \eta \geq 0$$ (14)

$$-(K(U, D^t)w_2 + e_u b_1) + \psi^* \geq (-1 + \varepsilon)e_u, \psi^* \geq 0,$$

where ξ, η, ψ, ψ^* are the slack variables; C_1, C_2 and C_u are penalty parameters; $D = [X_1; X_2]$; e_1, e_2 are vectors of suitable dimension having all values as 1s, f_u is the vector containing the fuzzy membership values and $K(x^t, D^t) = (k(x, x_1), \ldots, k(x, x_m))$ is a row vector in R^m space.

In comparison to UTSVM, the fuzzy-based approach of FUTSVM is helpful in reducing the effect of outliers in the universum data which is due to the random averaging scheme of selecting the universum. In this manner, appropriate membership value can be given to the universum points based on their information entropy. This approach reduces the effect of noise on the universum and results in better generalization performance.

The Lagrangians of problems (13) and (14) are written as

$$L_1 = \frac{1}{2}||K(X_1, D^t)w_1 + e_1 b_1||^2 + C_1 e_2^t \xi + C_u f_u^t \psi$$
$$+ \alpha_1^t((K(X_2, D^t)w_1 + e_2 b_1) - \xi + e_2)$$
$$- \beta_1^t \xi - \mu_1^t((K(U, D^t)w_1 + e_2 b_1) + \psi + (1 - \varepsilon)e_u) - \gamma_1^t \psi$$ (15)

$$L_2 = \frac{1}{2}\|K(X_2,D^t)w_2 + e_2b_2\|^2 + C_2e_1^t\eta + C_uf_u^t\psi^*$$
$$+ \alpha_2^t((-K(X_1,D^t)w_2 - e_1b_2) - \eta + e_1)$$
$$- \beta_2^t\eta + \mu_2^t((K(U,D^t)w_2 + e_1b_2) - \psi^* - (1-\varepsilon)e_u) - \gamma_2^t\psi^* \quad (16)$$

where $\alpha_1 = (\alpha_{11}, \ldots, \alpha_{1q})^t$, $\beta_1 = (\beta_{11}, \ldots, \beta_{1q})^t$, $\mu_1 = (\mu_{11}, \ldots, \mu_{1r})^t$, $\gamma_1 = (\gamma_{11}, \ldots, \gamma_{1r})^t$, $\alpha_2 = (\alpha_{21}, \ldots, \alpha_{2p})^t$, $\beta_2 = (\beta_{21}, \ldots, \beta_{2p})^t$, $\mu_2 = (\mu_{21}, \ldots, \mu_{2r})^t$ and $\gamma_2 = (\gamma_{21}, \ldots, \gamma_{2r})^t$ are the Lagrange multipliers where p, q and r are the number of data points in 'class 1', 'class 2' and universum respectively.

The Wolfe duals of Eqs. (15) and (16) are obtained by applying the Karush–Kuhn–Tucker (KKT) necessary and sufficient conditions as

$$\max_{\alpha_1, \mu_1} \ e_2^t\alpha_1 - \frac{1}{2}(\alpha_1^tT - \mu_1^tO)(S^tS)^{-1}(T\alpha_1 - O\mu_1^t) + (\varepsilon - 1)e_u^t\mu_1 \quad (17)$$

subject to $0 \leq \alpha_1 \leq C_1, 0 \leq \mu_1 \leq f_uC_u$,

$$\max_{\alpha_2, \mu_2} \ e_1^t\alpha_2 - \frac{1}{2}(\alpha_2^tS - \mu_2^tO)(T^tT)^{-1}(S^t\alpha_2 - O^t\mu_2) + (\varepsilon - 1)e_u^t\mu_2 \quad (18)$$

subject to $0 \leq \alpha_2 \leq C_2, 0 \leq \mu_2 \leq f_uC_u$,

where $S = [K(X_1, D^t) \ e_1]$, $T = [K(X_2, D^t) \ e_2]$ and $O = [K(U, D^t) \ e_u]$.

The non-linear hyperplanes $K(x^t, D^t)w_1 + b_1 = 0$ and $K(x^t, D^t)w_2 + b_2 = 0$ are obtained by using the following Eqs. (19) and (20):

$$\begin{bmatrix} w_1 \\ b_1 \end{bmatrix} = -(S^tS)^{-1}(T^t\alpha_1 - O^t\mu_1), \quad (19)$$

$$\begin{bmatrix} w_2 \\ b_2 \end{bmatrix} = (T^tT)^{-1}(S^t\alpha_2 - O^t\mu_2). \quad (20)$$

To avoid the ill-conditioning in the calculation of inverse $(S^tS)^{-1}$ and $(T^tT)^{-1}$, we add a regularization term δI to the matrices in (19) and (20) as $(S^tS + \delta I)^{-1}$ and $(T^tT + \delta I)^{-1}$ to make them positive definite where δ is a small positive value.

For a data point $x \in R^n$, it is assigned to a class $'i'$ on the basis of the following formula

$$class \ i = \min|K(x^t, D^t)w_i + b_i| \quad for \ i = 1, 2. \quad (21)$$

3.3 Calculation of Fuzzy Membership

Motivated by Fan et al. [32], the fuzzy membership is calculated based on the information entropy of the universum data points. The fuzzy calculation based on information entropy helps to identify those points which are having highest uncertainty of belonging to one of the two classes. In this work, we have used the random averaging method for selecting the universum data. The fuzzy membership value for the universum data points is calculated as per the following:

i. Calculate the information entropy of the universum data points based on a K-nearest neighbour (KNN) approach using Euclidean distance. The probability value of the universum data point is calculated based on the class label of its neighbours.

ii. Assign the universum data points to 10 subsets in decreasing order of entropy values [32]. The formula for information entropy is as follows:

$$E = -p_+ \ln(p_+) - p_- \ln(p_-),$$

where p_+ is the probability of belonging to the positive class, p_- is the probability of belonging to the negative class and 'ln' is the natural logarithm.

The fuzzy membership value is calculated as

$$f = 1 - \delta(n - 1),$$

where $\delta = 0.05$ is the fuzzy parameter and $n = 10$ is the number of subsets.

The proposed approach of assigning fuzzy membership based on information entropy gives more weight to those universum points which are lying in between the two classes. In case of universum data with outliers, this fuzzy-based approach is useful in order to reduce the effect of outliers on the SVM classifier.

4 Numerical Experiments

In this section, we have performed numerical experiments to check the effectiveness of the proposed methods and compared with other baseline methods. The datasets are taken from KEEL imbalanced datasets [33] and UCI repository [34] for binary classification. The computations are performed on a PC installed with Windows 10 OS of 64 bit, 3.60 GHz Intel® core™ i7-4790 processor having 32 GB of RAM under MATLAB R2008a environment. To solve the quadratic programming problems, we have used MOSEK optimization toolbox which is taken from http://www.mosek.com. Gaussian kernel is used for the classifier which is defined as $k(a, b) = \exp\left(-\frac{1}{2\sigma^2}\|a - b\|^2\right)$ where vector $a, b \in R^m$ and σ is the kernel parameter. We have used 5-fold cross validation technique for selecting the optimum parameters.

For FUSVM and FUTSVM, the value of K for KNN is chosen as $K = 5$. For FUSVM, USVM, FUTSVM and TWSVM the value of ε is chosen from $\{0.1, 0.3, 0.5, 0.6\}$. For all the methods, the value of σ is calculated as per the following formula [35],

Table 1 Performance comparison of proposed FUSVM with SVM and USVM

Dataset (Train size, Test size)	SVM (C, σ) time (s)	USVM $(C = C_u, \sigma, \varepsilon)$ time (s)	FUSVM $(C = C_u, \sigma, \varepsilon)$ time (s)
German (400 × 24, 600 × 24)	76 (10^1, 6.80536) 0.3027	**77.5** (10^0, 6.80536, 0.1) 3.7708	**77.5** (10^0, 6.80536, 0.3) 3.872
Cleveland (150 × 13, 147 × 13)	80.95 (10^1, 5.26173) 0.0365	81.63 (10^0, 5.26173, 0.1) 0.5113	**82.99** (10^0, 5.26173, 0.3) 0.5453
Ionosphere (150 × 33, 201 × 33)	**89.55** (10^1, 4.38631) 0.0377	87.06 (10^3, 4.38631, 0.1) 0.5274	89.05 (10^0, 4.38631, 0.1) 0.546
Transfusion (350 × 4, 398 × 4)	82.41 (10^2, 2077.88) 0.192	82.66 (10^5, 2077.88, 0.3) 2.8532	**84.17** (10^5, 2077.88, 0.6) 2.916
Cmc (500 × 9, 973 × 9)	74.2 (10^3, 13.4139) 0.4126	69.37 (10^1, 13.4139, 0.3) 5.779	**74.51** (10^3, 13.4139, 0.1) 6.0853
Heart-stat (180 × 13, 90 × 13)	77.78 (10^2, 85.982) 0.053	**81.11** (10^2, 85.982, 0.5) 0.7478	80 (10^2, 85.982, 0.5) 0.7644
Monk3 (250 × 7, 304 × 7)	83.22 (10^5, 163.314) 0.1036	**84.21** (10^5, 163.314, 0.1) 1.4406	83.88 (10^5, 163.314, 0.1) 1.5038
Ndc1k (400 × 32, 700 × 32)	89.29 (10^2, 571.157) 0.269	93.29 (10^3, 571.157, 0.3) 3.865	**93.71** (10^3, 571.157, 0.3) 3.9502
Pima-Indians (350 × 8, 418 × 8)	**80.38** (10^3, 2.22928) 0.1966	77.51 (10^4, 2.22928, 0.1) 2.8562	78.71 (10^3, 2.22928, 0.5) 2.9653
Wdbc (250 × 30, 319 × 30)	94.98 (10^4, 944.407) 0.111	95.61 (10^5, 944.407, 0.3) 1.4658	**95.92** (10^5, 944.407, 0.6) 1.5308
Yeast1 (600 × 8, 2368 × 8)	75.34 (10^3, 0.41735) 0.6244	**76.39** (10^0, 0.41735, 0.1) 8.5181	75.97 (10^0, 0.41735, 0.1) 8.7554
Ripley (200 × 2, 1050 × 2)	90.48 (10^0, 0.766998) 0.066	89.81 (10^0, 0.766998, 0.6) 0.923	**91.05** (10^4, 0.766998, 0.6) 0.9578
Yeast3 (500 × 8, 984 × 8)	94.31 (10^0, 0.411172) 0.4341	**95.43** (10^0, 0.411172, 0.1) 5.9856	95.33 (10^0, 0.411172, 0.1) 6.1587
Monk2 (150 × 7, 451 × 7)	**67.63** (10^-5, 151.849) 0.0748	58.76 (10^1, 151.849, 0.3) 0.5508	**67.63** (10^0, 151.849, 0.5) 0.5779

$$\sigma = \frac{1}{N^2} \sum_{i,j=1}^{N} ||x_i - x_j||^2$$

To reduce the computational cost of the parameter selection, we set $C = C_1 = C_2 = C_u$ and chosen from the set $\{10^{-5}, \ldots, 10^5\}$. For USVM, FUSVM, UTSVM and FUTSVM, the universum is calculated using random averaging scheme and the size of the universum is taken as 30% of the size of training data. The results are shown in Table 1 for the proposed FUSVM in comparison with SVM and USVM for prediction accuracy and training time. The corresponding rank comparison for FUSVM with SVM and USVM is shown in Table 2. The results for the comparison of FUTSVM with TWSVM and UTSVM are shown Table 3 for accuracy and training

Table 2 Average rank comparison of proposed FUSVM with SVM and USVM

Dataset (Train size, Test size)	SVM	USVM	FUSVM
German (400 × 24, 600 × 24)	3	1.5	1.5
Cleveland (150 × 13, 147 × 13)	3	2	1
Ionosphere (150 × 33, 201 × 33)	1	3	2
Transfusion (350 × 4, 398 × 4)	3	2	1
Cmc (500 × 9, 973 × 9)	2	3	1
Heart-stat (180 × 13, 90 × 13)	3	1	2
Monk3 (250 × 7, 304 × 7)	3	1	2
Ndc1k (400 × 32, 700 × 32)	3	2	1
Pima-Indians (350 × 8, 418 × 8)	1	3	2
Wdbc (250 × 30, 319 × 30)	3	2	1
Yeast1 (600 × 8, 2368 × 8)	3	1	2
Ripley (200 × 2, 1050 × 2)	2	3	1
Yeast3 (500 × 8, 984 × 8)	3	1	2
Monk2 (150 × 7, 451 × 7)	1.5	3	1.5
Average rank	2.4643	2.0357	1.5

Table 3 Performance comparison of proposed FUTSVM with TWSVM and UTSVM

Dataset (Train size, Test size)	TWSVM ($C_1 = C_2, \sigma$) time (s)	UTSVM ($C_1 = C_2 = C_u, \sigma, \varepsilon$) time (s)	FUTSVM ($C_1 = C_2 = C_u, \sigma, \varepsilon$) time (s)
German (400 × 24, 600 × 24)	**72** (10^−4, 6.80536) 0.0911	**72** (10^−4, 6.80536, 0.1) 0.131	**72** (10^−4, 6.80536, 0.1) 0.2365
Bupa or Liver-disorders (240 × 6, 105 × 6)	70.48 (10^−1, 66.1988) 0.0183	69.52 (10^1, 66.1988, 0.6) 0.0357	**71.43** (10^1, 66.1988, 0.6) 0.068
Cleveland (150 × 13, 147 × 13)	75.51 (10^−5, 5.26173) 0.009	75.51 (10^−2, 5.26173, 0.5) 0.0148	**76.19** (10^−2, 5.26173, 0.6) 0.0274
Ionosphere (150 × 33, 201 × 33)	**92.54** (10^−2, 4.38631) 0.0094	91.54 (10^−2, 4.38631, 0.5) 0.0145	91.54 (10^−2, 4.38631, 0.6) 0.0288
Transfusion (350 × 4, 398 × 4)	82.16 (10^−5, 2077.88) 0.0387	82.41 (10^0, 2077.88, 0.3) 0.0712	**82.66** (10^0, 2077.88, 0.3) 0.1417
Heart-stat (180 × 13, 90 × 13)	**81.11** (10^−1, 85.982) 0.0121	77.78 (10^0, 85.982, 0.5) 0.0183	80 (10^0, 85.982, 0.6) 0.0376
Monk3 (250 × 7, 304 × 7)	57.89 (10^−5, 163.314) 0.018	68.42 (10^3, 163.314, 0.3) 0.0388	**78.29** (10^4, 163.314, 0.3) 0.0782
Ndc1k (400 × 32, 700 × 32)	90.14 (10^−1, 571.157) 0.0559	93.14 (10^0, 571.157, 0.5) 0.103	**93.43** (10^0, 571.157, 0.6) 0.1971

(continued)

Table 3 (continued)

Dataset (Train size, Test size)	TWSVM ($C_1 = C_2, \sigma$) time (s)	UTSVM ($C_1 = C_2 = C_u, \sigma, \varepsilon$) time (s)	FUTSVM ($C_1 = C_2 = C_u, \sigma, \varepsilon$) time (s)
Pima-Indians (350×8, 418×8)	75.12 ($10\hat{\ }0$, 2.22928) 0.0374	**77.27** ($10\hat{\ }1$, 2.22928, 0.6) 0.0724	**77.27** ($10\hat{\ }1$, 2.22928, 0.5) 0.1474
Wdbc (250×30, 319×30)	**95.92** ($10\hat{\ }1$, 944.407) 0.0212	91.85 ($10\hat{\ }-1$, 944.407, 0.6) 0.0362	**95.92** ($10\hat{\ }0$, 944.407, 0.1) 0.0768
Vehicle2 (400×18, 446×18)	**98.43** ($10\hat{\ }0$, 269.333) 0.0579	97.76 ($10\hat{\ }1$, 269.333, 0.1) 0.1151	97.98 ($10\hat{\ }1$, 269.333, 0.1) 0.21
Yeast1 (600×8, 2368×8)	**75.84** ($10\hat{\ }0$, 0.41735) 0.1632	75.3 ($10\hat{\ }-1$, 0.41735, 0.5) 0.2959	75.3 ($10\hat{\ }-1$, 0.41735, 0.5) 0.4888
Yeast3 (500×8, 984×8)	94.41 ($10\hat{\ }-2$, 0.411172) 0.1349	93.09 ($10\hat{\ }0$, 0.411172, 0.1) 0.247	**94.61** ($10\hat{\ }-3$, 0.411172, 0.6) 0.3805
Breast Cancer Wisconsin (350×9, 333×9)	98.5 ($10\hat{\ }-4$, 12.5292) 0.0774	**98.8** ($10\hat{\ }-3$, 12.5292, 0.1) 0.1154	**98.8** ($10\hat{\ }-3$, 12.5292, 0.6) 0.1935

Table 4 Average rank comparison of proposed FUTSVM with TWSVM and UTSVM

Dataset (Train size, Test size)	TWSVM	UTSVM	FUTSVM
German (400×24, 600×24)	2	2	2
Bupa or Liver-disorders (240×6, 105×6)	2	3	1
Cleveland (150×13, 147×13)	2.5	2.5	1
Ionosphere (150×33, 201×33)	1	2.5	2.5
Transfusion (350×4, 398×4)	3	2	1
Heart-stat (180×13, 90×13)	1	3	2
Monk3 (250×7, 304×7)	3	2	1
Ndc1k (400×32, 700×32)	3	2	1
Pima-Indians (350×8, 418×8)	3	1.5	1.5
Wdbc (250×30, 319×30)	1.5	3	1.5
Vehicle2 (400×18, 446×18)	1	3	2
Yeast1 (600×8, 2368×8)	1	2.5	2.5
Yeast3 (500×8, 984×8)	2	3	1
Breast Cancer Wisconsin (350×9, 333×9)	3	1.5	1.5
Average Rank	2.0714	2.3929	1.5357

time and the corresponding rank comparison is shown in Table 4. From Tables 2 and 4, it is evident that our proposed FUTSVM and FUTSVM are giving better generalization performance in comparison to other baseline algorithms. Our proposed methods FUSVM and FUTSVM are taking some additional computational time due to the fuzzy calculation which can be traded for the generalization ability.

Figure 1 shows the insensitivity analysis for FUSVM and FUTSVM. One can observe from Fig. 1 that FUSVM gives better accuracy for large values of C and FUTSVM shows better generalization performance for lesser values of C.

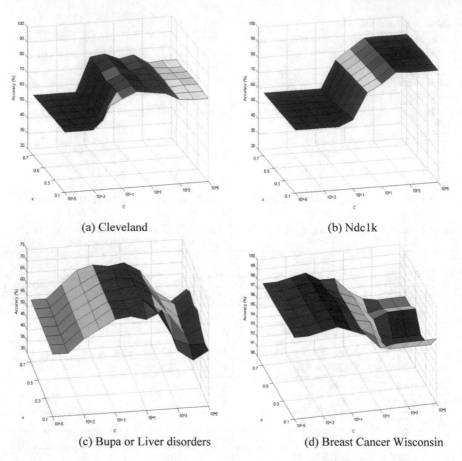

(a) Cleveland   (b) Ndc1k

(c) Bupa or Liver disorders   (d) Breast Cancer Wisconsin

Fig. 1 Insensitivity performance for classification of FUSVM is shown in (**a**) and (**b**) and for FUTSVM in (**c**) and (**d**) to the user specified parameters (C, ε) on real world datasets using Gaussian kernel

5   Conclusions and Future Work

In this paper, we propose fuzzy-based approach for USVM and UTSVM algorithms which is useful in the classification of data with noise and outliers. Our proposed FUSVM and FUTSVM have shown better generalization performance for most of the datasets. This fuzzy-based approach for universum helps in giving prior information to the data in an effective manner. The use of information entropy of the universum points is helpful in giving optimum membership values to the universum data points. Thus, we anticipate that the proposed FUSVM and FUTSVM could be considered as new baselines for the researchers owing to their superior performance. The source codes will be available from authors' homepages.

Our fuzzy-based approach for universum support vector machine can be extended to multiclass classification problems and other variants of USVM and will be addressed in future.

Acknowledgements This work was supported by Science and Engineering Research Board (SERB) as Early Career Research Award grant no. ECR/2017/000053 and Department of Science and Technology as Ramanujan fellowship grant no. SB/S2/RJN-001/2016. We gratefully acknowledge the Indian Institute of Technology Indore for providing facilities and support.

References

1. Vapnik, V.N.: Statistical Learning Theory. Wiley, New York (1998)
2. Vapnik, V.N.: The Nature of Statistical Learning Theory, 2nd edn. Springer, New York (2000)
3. Osuna, E., Freund, R., Girosi, F.: Training support vector machines: an application to face detection. In: Proceedings of the IEEE Computer Society Conference on Computer Vision and Pattern Recognition, pp. 130–136 (1997)
4. Phillips, P.J.: Support vector machines applied to face recognition, In: Proceedings of the 1998 Conference on Advances in Neural Information Processing Systems, vol. 11, pp. 803–809 (1998)
5. Michel, P., El Kaliouby R.: Real time facial expression recognition in video using support vector machines. In: Proceedings of the 5th International Conference on Multimodal Interfaces, pp. 258–264 (2003). ISBN 1-58113-621-8
6. Borovikov, E.: An evaluation of support vector machines as a pattern recognition tool. University of Maryland at College Park (2005). http://www.umiacs.umd.edu/users/yab/SVMForPatternRecognition/report.pdf
7. Schmidt, M. Gish, H.: Speaker identification via support vector classifiers. In: Proceedings of the IEEE International Conference on Acoustics, Speech, and Signal Processing. ICASSP-96, Atlanta, GA, vol. 1, pp. 105–108 (1996)
8. Khan, L., Awad, M., Thuraisingham, B.: A new intrusion detection system using support vector machines and hierarchical clustering. VLDB J. **16**, 507–521 (2007)
9. Jayadeva, K.R., Chandra, S.: Twin support vector machines for pattern classification. In: IEEE Transactions on Pattern Analysis and Machine Intelligence (TPAMI), vol. 29, pp. 905–910 (2007)
10. Kumar, M.A., Gopal, M.: Least squares twin support vector machines for pattern classification. Expert Syst. Appl. **36**(4), 7535–7543 (2009)
11. Shao, Y.H., Zhang, C.H., Wang, X.B., Deng, N.Y.: Improvements on twin support vector machines. IEEE Trans. Neural Netw. **22**(6), 962–968 (2011)
12. Tanveer, M., Khan, M.A., Ho, S.S.: Robust energy-based least squares twin support vector machines. Appl. Intell. **45**(1), 174–186 (2016)
13. Tanveer, M.: Robust and sparse linear programming twin support vector machines. Cogn. Comput. **7**(1), 137–149 (2015)
14. Tanveer, M.: Application of smoothing techniques for linear programming twin support vector machines. Knowl. Inf. Syst. **45**(1), 191–214 (2015)
15. Tanveer, M., Shubham, K., Aldhaifallah, M., Ho, S.S.: An efficient regularized K-nearest neighbor based weighted twin support vector regression. Knowl. Based Syst. **94**, 70–87 (2016)
16. Chen, W.J., Shao, Y.H., Li, C.N., Deng, N.Y.: MLTSVM: a novel twin support vector machine to multi-label learning. Pattern Recogn. **52**, 61–74 (2016)
17. Khemchandani, R., Goyal, K., Chandra, S.: TWSVR: regression via twin support vector machine. Neural Netw. **74**, 14–21 (2016)
18. Tanveer, M., Mangal, M., Ahmad, I., Shao, Y.H.: One norm linear programming support vector regression. Neurocomputing **173**, 1508–1518 (2016)

19. Khemchandani, R., Chandra, S.: Knowledge based proximal support vector machines. Eur. J. Oper. Res. **195**(3), 914–923 (2009)
20. Weston, J., Collobert, R., Sinz, F., Bottou, L., Vapnik, V.: Inference with the universum. In: Proceedings of the 23rd International Conference on Machine Learning, pp. 1009–1016. ACM, 25 Jun 2006
21. Bai, X., Cherkassky V.: Gender classification of human faces using inference through contradictions. In: Proceedings of the IEEE International Joint Conference (IEEE World Congress on Computational Intelligence) on Neural Networks. IJCNN 2008, pp. 746–750, 1 Jun 2008
22. Chen, S., Zhang, C.: Selecting informative universum sample for semi-supervised learning. In: IJCAI, vol. 6, pp. 1016–1021, 11 Jul 2009
23. Cherkassky, V., Dhar, S., Dai, W.: Practical conditions for effectiveness of the universum learning. IEEE Trans. Neural Netw. **22**(8), 1241–1255 (2011)
24. Qi, Z., Tian, Y., Shi Y.: Twin support vector machine with universum data. Neural Netw. **36**, 112–119 (2012)
25. Xu, Y., Chen M., Li, G.: Least squares twin support vector machine with universum data for classification. Int. J. Syst. Sci. **47**(15), 3637–3645 (2016)
26. Lin, C.F., Wang, S.D.: Fuzzy support vector machines. IEEE Trans. Neural Netw. **13**(2), 464–471 (2002)
27. Batuwita, R., Palade, V.: FSVM-CIL: fuzzy support vector machines for class imbalance learning. IEEE Trans. Fuzzy Syst. **18**(3), 558–571 (2010)
28. Balasundaram, S., Tanveer, M.: On proximal bilateral-weighted fuzzy support vector machine classifiers. Int. J. Adv. Intel. Paradig. **4**(3–4), 199–210 (2012)
29. Tian, D.Z., Peng G.B., Ha, M.H.: Fuzzy support vector machine based on non-equilibrium data. International Conference on Machine Learning and Cybernetics, Xi'an, China, pp. 15–17 (2012)
30. Wang, Y., Wang, S., Lai, K.K.: A new fuzzy support vector machine to evaluate credit risk. IEEE Trans. Fuzzy Syst. **13**(6), 820–831 (2005)
31. Chaudhuri, A., De, K.: Fuzzy support vector machine for bankruptcy prediction. Appl. Soft Comput. **11**(2), 2472–2486 (2010)
32. Fan, Q., Wang, Z., Li, D., Gao, D., Zha, H.: Entropy-based fuzzy support vector machine for imbalanced datasets. Knowl. Based Syst. **115**, 87–99 (2017)
33. Alcalá-Fdez, J., Fernandez, A., Luengo, J., Derrac, J., García, S., Sánchez, L., Herrera, F.: KEEL Data-mining software tool: data set repository, integration of algorithms and experimental analysis framework. J. Multiple-valued Log. Soft Comput. **17**, 255–287 (2011)
34. Murphy, P.M., Aha, D.W.: UCI repository of machine learning databases, University of California, Irvine (1992). http://www.ics.uci.edu/~mlearn
35. Tsang, I., Kocsor, A., Kwok, J.: Efficient kernel feature extraction for massive data sets. In: International Conference on Knowledge Discovery and Data Mining (2006)

Automated Identification System for Focal EEG Signals Using Fractal Dimension of FAWT-Based Sub-bands Signals

M. Dalal, M. Tanveer and Ram Bilas Pachori

Abstract The classification of focal and non-focal electroencephalogram (EEG) signals for diagnosis of epilepsy at an early stage is one of the most difficult problems. There have been many attempts to develop automated detection algorithms to assist clinical research for presurgical analysis of epilepsy. In this paper, a novel approach for studying EEG signals has been proposed using flexible analytic wavelet transform (FAWT) which is a nonstationary signal processing technique. In this study, EEG signals are decomposed into the desired number of sub-bands (SBs). Fractal dimension (FD) is used as a feature and then computed it for all SB signals which are obtained from FAWT. The significant features obtained from the Kruskal–Wallis statistical test and are classified using robust energy-based least square twin support vector machine (RELS-TSVM). In order to show the effectiveness of the proposed method for classification of focal (F) and non-focal (NF) EEG signals, publicly available database termed as Bern-Barcelona EEG dataset is used for the study.

1 Introduction

Epilepsy is a brain disorder which is characterized by recurring seizures and these seizures happen due to an excessive electrical discharge in a group of brain cells [1]. The EEG signals are greatly used in examining the electrical activity of our brain. Biomedical signals contain a great deal of information which can be processed by

M. Dalal (✉) · M. Tanveer
Discipline of Mathematics, Indian Institute of Technology Indore,
Simrol, Indore 453552, India
e-mail: mamta.dalal7@gmail.com; phd1601241003@iiti.ac.in

M. Tanveer
e-mail: mtanveer@iiti.ac.in

R. B. Pachori
Discipline of Electrical Engineering, Indian Institute of Technology Indore,
Simrol, Indore 453552, India
e-mail: pachori@iiti.ac.in

© Springer Nature Singapore Pte Ltd. 2019   583
M. Tanveer and R. B. Pachori (eds.), *Machine Intelligence and Signal Analysis*,
Advances in Intelligent Systems and Computing 748,
https://doi.org/10.1007/978-981-13-0923-6_50

advanced signal processing techniques. Many methods based on the signal processing techniques for performing automated analysis and detection of EEG signals in
order to classify focal (F) and non-focal (NF) EEG signals have been proposed
[1–10]. Focal EEG signals are recorded from the areas of brain, where first ictal
EEG signals are detected, while NF EEG signals are recorded from brain areas that
are not involved at seizure onset [11]. Studying F and NF characteristics may lead
to help in localizing the area of focal epilepsy which eventually will be useful in
diagnosing process at early stage. Still, in many countries like India, doctors do not
use signal processing techniques and detection of seizures are performed by an experienced neurophysiologists which is obviously time-consuming and an error-prone
approach. So automated detection with the help of signal processing techniques and
various machine learning tools has become much more organized and less time taking
procedure.

A lot of the literature can be easily found regarding the characterization of EEG
signals using different signal processing techniques [1, 4, 5, 12]. Various features
have also been studied for different domain of EEG signals [2, 4, 9, 13, 14]. Even
though there are so many tools and techniques to study EEG signals but still a lot of
research work is needed in order to do better classification of F and NF EEG signals
with higher accuracy. Recently, authors have analyzed seizures activities in EEG
signals using analytic time–frequency flexible wavelet transform and used fractal
dimension (FD) as a feature [15]. EEG dataset from University of Bonn, Germany
[16] is used to study various EEG signals like as normal, seizure, seizure-free,s and
classified using least square support vector machine (LS-SVM).

In our work, we decompose EEG signals using flexible analytic wavelet transform
(FAWT) approach [17]. We use FD as a feature for classification of F and NF EEG
signals. The FD is determined using the method presented in [18] and significant
features are calculated using the Kruskal–Wallis statistical test [1]. Many researchers
preferred LS-SVM for classification as compared to standard support vector machine
(SVM), since it solves set of linear equations instead of quadratic programming [19].
A new nonparallel plane classifier, called twin SVM (TSVM) has been proposed with
the advantage of solving a pair of quadratic programming problems (QPPs) with the
distribution of data points in such a way that data points in one class gives the
constraints of the other QPP and vice versa [20]. This results in reducing the size of
QPPs and makes this algorithm more suitable. Many variants of TSVM have been
proposed [21–24] due to its strong generalization ability. Among all the variants of
TSVM, least square twin support vector machine (LS-TSVM) is the most widely
used. LS-TSVM is more efficient as it replaces the convex QPPs with a convex
linear system using the squared loss function and hence improves computation of
the algorithm [25–29]. There have been many variants of LS-TSVM in the literature
[24, 28–30]. We use robust energy-based LS-TSVM (RELS-TSVM) as a classifier
in this work [31], which is basically a modified approach of energy-based LS-TSVM
(ELS-TSVM) [32]. RELS-TSVM is introduced in such a way that it reduces the
overfitting problem along with minimizing the effect of noise and outliers.

The rest of the paper is organized as follows: Sect. 2 describes the details of the
database which we used in this work. Signal decomposition technique and features

selection procedure have been explained in Sects. 2.2 and 2.3. Section 3 describes the details of RELS-TSVM algorithm. The experimental results and conclusions are given in Sects. 4 and 5.

2 Methodology

2.1 Dataset Used

We have implemented the proposed method on publicly available database, Bern-Barcelona EEG database [33]. This EEG dataset is obtained from epilepsy patients who went through long-range intracranial EEG at the Neurology Department of the Bern University [11]. This is a bivariate EEG dataset which contains two columns, x and y, respectively. The total number of samples in each dataset are 10240, with a sampling rate of 512 Hz. We have used 50 signals for both F and NF bivariate EEG signals. The differencing operation is used as a preprocessing step on bivariate EEG signals [1, 3]. The length of signals considered for this study are 10240, 5120, 2560, and 1280 samples. We have used first time-series, second time-series, average time-series, and differenced time-series EEG signals for this study. The block diagram of our proposed methodology is shown in Fig. 1.

2.2 FAWT: Overview

In order to study various classes of EEG signals, many wavelet-based methods have been explored which allow to analyze the nonstationary nature of EEG signals [7, 9, 34, 35]. The FAWT is a novel method which was proposed by Bayram [17]. This transform method helps in adjusting various parameters in order to study time–

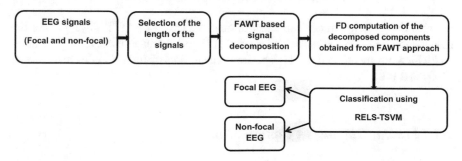

Fig. 1 Block diagram of the proposed methodology for classification of F EEG signals

frequency localization of the signals and has been studied for the analysis of various biomedical signals [36–39].

The sampling rates in both the low-pass and high-pass channels can be adjusted arbitrary using FAWT approach which help in flexible partitioning of time–frequency covering, and hence leads to in adjusting dilation and translation factors easily. The iterative filter bank used in FAWT consists of one low-pass and two high-pass channels. The high-pass channel analyzes positive frequencies while the other analyze negative frequencies. The wavelet SB signals are reconstructed for each signal by putting the coefficients 0 in all channels except one desired channel. The filters used in this approach are specified directly in the frequency domain. Let $F(\omega)$ and $H(\omega)$ are the frequency response of analytic wavelet function in FAWT.

Mathematically, $F(\omega)$ and $H(\omega)$ are written as follows [17]:

$$F(\omega) = \begin{cases} \sqrt{ab}, & |\omega| \leq \omega_a, \\ \sqrt{ab}\theta((\omega_s - \omega_a)^{-1}(\omega - \omega_a)), & \omega_a \leq \omega \leq \omega_s, \\ \sqrt{ab}\theta((\omega_s - \omega_a)^{-1}(\pi - \omega + \omega_a)), & -\omega_s \leq \omega \leq -\omega_a, \\ 0, & |\omega| \geq \omega_s, \end{cases} \quad (1)$$

where a and b are filter's parameters and

$$\omega_a = (1 - \beta)\pi/a + \varepsilon/a, \ \omega_s = \pi/b.$$

The $H(\omega)$ can be written as [17]:

$$H(\omega) = \begin{cases} \sqrt{gh}\theta((\omega_1 - \omega_0)^{-1}(\pi - \omega - \omega_0)), & \omega_0 \leq \omega \leq \omega_1, \\ \sqrt{gh}, & \omega_1 \leq \omega \leq \omega_2 \\ \sqrt{gh}\theta((\omega_3 - \omega_2)^{-1}(\omega - \omega_2)), & \omega_2 \leq \omega \leq \omega_3, \\ 0, & \omega \in [0, \omega_0) \cup (\omega_2, 2\pi), \end{cases} \quad (2)$$

where g and h are the parameters of the above-mentioned filter and

$$\omega_0 = (1 - \beta)(\pi + \varepsilon)/g, \omega_1 = a\pi/(bg), \omega_2 = (\pi - \varepsilon)/g, \omega_3 = (\pi + \varepsilon)/g.$$

It states that β helps in determining the quality factor of FAWT. R and d are the redundancy and dilation factors of FAWT [17] and we have used FAWT parameters which are mentioned in [1]. The implementation of FAWT codes are done in MATLAB [40].

2.3 Fractal Dimension Computation

In this work, we have used FD as a single feature to classify the F and NF EEG signals. The EEG signals are the outcome of a highly complex nonlinear system. FD is used to quantify the complexity. It has been shown that FD is a promising method for transient detection [41]. The FD of a time-series can be computed by several different techniques [42–44]. Different algorithms have been examined for the measurement of FD to correctly estimate the dimension over brief intervals [45]. We have used Higuchi's algorithm [18] which is based on the measure of the mean length of the curve $L(p)$ by using a segment of p samples as a unit of measure. The algorithm constructs p new time-series; each of them is defined as:

Table 1 FD (*mean* ± *std*) values for all SBs of differenced time-series EEG signals

All SB	10240 samples		5120 samples		2560 samples		1280 samples	
	F	NF	F	NF	F	NF	F	NF
1	1.98 ± 0.01	1.98 ± 0.01	1.98 ± 0.01	1.97 ± 0.01	1.99 ± 0.01	1.98 ± 0.01	1.99 ± 0.03	2.01 ± 0.03
2	1.98 ± 0.01	1.98 ± 0.01	1.98 ± 0.01	1.98 ± 0.01	1.99 ± 0.01	1.98 ± 0.01	1.99 ± 0.03	2.01 ± 0.03
3	1.98 ± 0.01	1.99 ± 0.01	1.98 ± 0.01	1.98 ± 0.01	1.98 ± 0.01	1.98 ± 0.01	1.99 ± 0.05	2.00 ± 0.05
4	1.97 ± 0.01	1.98 ± 0.01	1.96 ± 0.01	1.97 ± 0.01	1.96 ± 0.01	1.97 ± 0.01	1.97 ± 0.01	1.97 ± 0.01
5	1.96 ± 0.01	1.97 ± 0.01	1.95 ± 0.01	1.96 ± 0.01	1.94 ± 0.01	1.95 ± 0.01	1.97 ± 0.05	1.97 ± 0.05
6	1.95 ± 0.02	1.96 ± 0.01	1.93 ± 0.01	1.94 ± 0.01	1.92 ± 0.02	1.93 ± 0.02	1.90 ± 0.04	1.91 ± 0.04
7	1.93 ± 0.02	1.95 ± 0.02	1.91 ± 0.02	1.92 ± 0.02	1.90 ± 0.01	1.91 ± 0.01	1.82 ± 0.20	1.82 ± 0.20
8	1.92 ± 0.02	1.93 ± 0.02	1.89 ± 0.02	1.90 ± 0.02	1.86 ± 0.03	1.87 ± 0.03	1.72 ± 0.26	1.72 ± 0.26
9	1.90 ± 0.03	1.90 ± 0.02	1.86 ± 0.04	1.86 ± 0.04	1.78 ± 0.17	1.78 ± 0.16	1.59 ± 0.30	1.60 ± 0.30
10	1.87 ± 0.03	1.88 ± 0.03	1.83 ± 0.03	1.84 ± 0.03	1.69 ± 0.23	1.69 ± 0.22	1.47 ± 0.32	1.46 ± 0.03
11	1.84 ± 0.06	1.84 ± 0.05	1.76 ± 0.13	1.76 ± 0.13	1.59 ± 0.27	1.59 ± 0.27	1.30 ± 0.25	1.29 ± 0.25
12	1.82 ± 0.04	1.82 ± 0.04	1.69 ± 0.20	1.70 ± 0.21	1.47 ± 0.30	1.47 ± 0.30	1.15 ± 0.12	1.15 ± 0.12
13	1.77 ± 0.09	1.77 ± 0.83	1.60 ± 0.26	1.60 ± 0.26	1.33 ± 0.27	1.33 ± 0.27	1.08 ± 0.06	1.08 ± 0.06
14	1.70 ± 0.19	1.70 ± 0.18	1.49 ± 0.30	1.49 ± 0.30	1.17 ± 0.14	1.17 ± 0.14	1.04 ± 0.03	1.04 ± 0.03
15	1.61 ± 0.23	1.61 ± 0.22	1.38 ± 0.29	1.39 ± 0.29	1.10 ± 0.07	1.10 ± 0.07	1.03 ± 0.02	1.03 ± 0.02
16	1.30 ± 0.17	1.28 ± 0.16	1.12 ± 0.09	1.11 ± 0.08	1.03 ± 0.02	1.03 ± 0.02	1.01 ± 0.01	1.01 ± 0.01

Table 2 FD (*mean ± std*) values for all SBs of first time-series EEG signals

All SB	10240 samples		5120 samples		2560 samples		1280 samples	
	F	NF	F	NF	F	NF	F	NF
1	1.98 ± 0.01	1.98 ± 0.01	1.98 ± 0.01	1.98 ± 0.01	1.99 ± 0.01	1.98 ± 0.01	2.01 ± 0.03	2.00 ± 0.02
2	1.99 ± 0.01	1.99 ± 0.01	1.98 ± 0.01	1.99 ± 0.01	1.99 ± 0.01	1.99 ± 0.01	2.01 ± 0.03	2.01 ± 0.02
3	1.98 ± 0.01	1.99 ± 0.01	1.98 ± 0.01	1.98 ± 0.01	1.98 ± 0.01	1.98 ± 0.01	2.00 ± 0.04	2.00 ± 0.05
4	1.97 ± 0.01	1.97 ± 0.01	1.96 ± 0.01	1.97 ± 0.01	1.96 ± 0.01	1.97 ± 0.01	1.97 ± 0.01	1.97 ± 0.01
5	1.96 ± 0.01	1.97 ± 0.01	1.95 ± 0.01	1.96 ± 0.01	1.95 ± 0.01	1.95 ± 0.01	1.97 ± 0.05	1.97 ± 0.05
6	1.95 ± 0.02	1.96 ± 0.01	1.94 ± 0.01	1.94 ± 0.01	1.92 ± 0.02	1.92 ± 0.02	1.91 ± 0.04	1.91 ± 0.04
7	1.94 ± 0.02	1.95 ± 0.02	1.91 ± 0.02	1.92 ± 0.02	1.90 ± 0.01	1.91 ± 0.01	1.82 ± 0.19	1.82 ± 0.19
8	1.92 ± 0.02	1.93 ± 0.02	1.90 ± 0.02	1.90 ± 0.02	1.86 ± 0.03	1.86 ± 0.03	1.72 ± 0.26	1.72 ± 0.26
9	1.90 ± 0.03	1.91 ± 0.02	1.86 ± 0.04	1.86 ± 0.04	1.78 ± 0.17	1.78 ± 0.17	1.59 ± 0.30	1.59 ± 0.30
10	1.87 ± 0.03	1.88 ± 0.03	1.83 ± 0.03	1.84 ± 0.04	1.69 ± 0.23	1.69 ± 0.23	1.47 ± 0.32	1.47 ± 0.32
11	1.84 ± 0.06	1.84 ± 0.06	1.76 ± 0.13	1.76 ± 0.13	1.58 ± 0.27	1.59 ± 0.27	1.29 ± 0.25	1.30 ± 0.25
12	1.82 ± 0.04	1.82 ± 0.04	1.69 ± 0.21	1.70 ± 0.20	1.47 ± 0.30	1.47 ± 0.30	1.15 ± 0.12	1.15 ± 0.12
13	1.77 ± 0.09	1.77 ± 0.09	1.60 ± 0.26	1.60 ± 0.26	1.33 ± 0.27	1.33 ± 0.28	1.08 ± 0.06	1.09 ± 0.06
14	1.70 ± 0.19	1.70 ± 0.18	1.49 ± 0.28	1.49 ± 0.28	1.17 ± 0.14	1.17 ± 0.14	1.04 ± 0.03	1.04 ± 0.04
15	1.61 ± 0.23	1.61 ± 0.23	1.38 ± 0.29	1.38 ± 0.29	1.09 ± 0.07	1.09 ± 0.07	1.03 ± 0.02	1.03 ± 0.02
16	1.30 ± 0.17	1.29 ± 0.16	1.12 ± 0.08	1.11 ± 0.08	1.03 ± 0.02	1.03 ± 0.02	1.01 ± 0.01	1.01 ± 0.01

$$X_m^p : X(m), X(m+1p), \ldots, X\left(m + \left[\frac{N-m}{p}\right]p\right), m = 1, 2, \ldots, p.$$

Then the length,

Table 3 FD (*mean ± std*) values for all SBs of second time-series EEG signals

All SB	10240 samples		5120 samples		2560 samples		1280 samples	
	F	NF	F	NF	F	NF	F	NF
1	1.98 ± 0.01	1.98 ± 0.01	1.98 ± 0.01	1.98 ± 0.01	1.99 ± 0.01	1.98 ± 0.01	2.01 ± 0.03	2.00 ± 0.03
2	1.99 ± 0.01	1.99 ± 0.01	1.98 ± 0.01	1.99 ± 0.01	1.99 ± 0.01	1.99 ± 0.01	2.01 ± 0.03	2.01 ± 0.03
3	1.98 ± 0.01	1.99 ± 0.01	1.98 ± 0.01	1.98 ± 0.01	1.98 ± 0.01	1.98 ± 0.01	2.00 ± 0.04	2.01 ± 0.05
4	1.97 ± 0.01	1.98 ± 0.01	1.96 ± 0.01	1.97 ± 0.01	1.96 ± 0.01	1.97 ± 0.01	1.97 ± 0.01	1.97 ± 0.01
5	1.96 ± 0.01	1.97 ± 0.01	1.95 ± 0.01	1.96 ± 0.01	1.95 ± 0.01	1.95 ± 0.01	1.97 ± 0.05	1.97 ± 0.05
6	1.95 ± 0.01	1.96 ± 0.01	1.94 ± 0.01	1.94 ± 0.01	1.92 ± 0.02	1.93 ± 0.02	1.91 ± 0.04	1.91 ± 0.04
7	1.94 ± 0.02	1.95 ± 0.02	1.91 ± 0.02	1.92 ± 0.02	1.90 ± 0.01	1.91 ± 0.01	1.82 ± 0.20	1.82 ± 0.20
8	1.92 ± .02	1.93 ± 0.02	1.89 ± 0.02	1.90 ± 0.02	1.86 ± 0.03	1.86 ± 0.04	1.72 ± 0.26	1.72 ± 0.26
9	1.90 ± 0.03	1.90 ± 0.03	1.86 ± 0.04	1.86 ± 0.04	1.78 ± 0.17	1.78 ± 0.17	1.59 ± 0.30	1.59 ± 0.30
10	1.88 ± 0.03	1.88 ± 0.03	1.83 ± 0.03	1.84 ± 0.04	1.69 ± 0.23	1.69 ± 0.23	1.47 ± 0.32	1.47 ± 0.32
11	1.84 ± 0.06	1.84 ± 0.06	1.76 ± 0.13	1.76 ± 0.13	1.59 ± 0.27	1.59 ± 0.27	1.29 ± 0.24	1.30 ± 0.25
12	1.82 ± 0.04	1.82 ± 0.04	1.69 + 0.21	1.69 ± 0.21	1.47 ± 0.30	1.47 ⊥ 0.30	1.15 ± 0.12	1.13 ± 0.12
13	1.77 ± 0.09	1.77 ± 0.08	1.60 ± 0.26	1.60 ± 0.25	1.33 ± 0.27	1.33 ± 0.27	1.08 ± 0.06	1.09 ± 0.06
14	1.70 ± 0.19	1.70 ± 0.18	1.49 ± 0.28	1.49 ± 0.28	1.17 ± 0.14	1.17 ± 0.14	1.04 ± 0.03	1.05 ± 0.03
15	1.61 ± 0.23	1.61 ± 0.23	1.38 ± 0.29	1.39 ± 0.29	1.09 ± 0.07	1.09 ± 0.07	1.03 ± 0.02	1.03 ± 0.02
16	1.30 ± 0.17	1.28 ± 0.16	1.12 ± 0.09	1.11 ± 0.08	1.03 ± 0.02	1.03 ± 0.02	1.01 ± 0.01	1.01 ± 0.01

$$l_m(p) = \left(\sum_{i=1,\,\mathrm{int}((N-m)/p)} |X(m+ip) - X(m+(i-1)p)| \right)$$
$$\times \left((n-1) / \left[\frac{N-m}{p} \right] p \right),$$

$$(3)$$

$$L_m(p) = l_m(p)p^{-1},$$

$$(4)$$

Table 4 FD (*mean ± std*) values for all SBs of average time-series EEG signals

All SB	10240 samples		5120 samples		2560 samples		1280 samples	
	F	NF	F	NF	F	NF	F	NF
1	1.98 ± 0.01	1.98 ± 0.01	1.98 ± 0.01	1.98 ± 0.01	1.99 ± 0.01	1.98 ± 0.01	2.01 ± 0.03	2.00 ± 0.03
2	1.99 ± 0.01	1.99 ± 0.01	1.98 ± 0.01	1.99 ± 0.01	1.99 ± 0.01	1.99 ± 0.01	2.01 ± 0.03	2.01 ± 0.03
3	1.98 ± 0.01	1.99 ± 0.01	1.98 ± 0.01	1.98 ± 0.01	1.98 ± 0.01	1.98 ± 0.01	2.00 ± 0.05	2.00 ± 0.05
4	1.97 ± 0.01	1.98 ± 0.01	1.96 ± 0.01	1.97 ± 0.01	1.96 ± 0.01	1.97 ± 0.01	1.97 ± 0.01	1.97 ± 0.01
5	1.96 ± 0.01	1.97 ± 0.01	1.95 ± 0.01	1.96 ± 0.01	1.95 ± 0.01	1.95 ± 0.01	1.97 ± 0.05	1.97 ± 0.05
6	1.95 ± 0.02	1.96 ± 0.01	1.94 ± 0.01	1.94 ± 0.01	1.92 ± 0.02	1.92 ± 0.02	1.91 ± 0.04	1.91 ± 0.04
7	1.94 ± 0.02	1.94 ± 0.02	1.91 ± 0.02	1.92 ± 0.02	1.90 ± 0.01	1.91 ± 0.01	1.82 ± 0.20	1.82 ± 0.20
8	1.92 ± 0.02	1.93 ± 0.02	1.89 ± 0.02	1.90 ± 0.02	1.86 ± 0.03	1.86 ± 0.03	1.72 ± 0.26	1.72 ± 0.26
9	1.89 ± 0.03	1.90 ± 0.03	1.86 ± 0.04	1.86 ± 0.04	1.78 ± 0.17	1.78 ± 0.17	1.60 ± 0.30	1.60 ± 0.30
10	1.87 ± 0.03	1.88 ± 0.03	1.83 ± 0.03	1.84 ± 0.04	1.69 ± 0.23	1.69 ± 0.23	1.47 ± 0.32	1.47 ± 0.32
11	1.83 ± 0.06	1.84 ± 0.06	1.76 ± 0.13	1.76 ± 0.13	1.58 ± 0.27	1.59 ± 0.27	1.29 ± 0.24	1.30 ± 0.25
12	1.82 ± 0.04	1.82 ± 0.04	1.69 ± 0.20	1.69 ± 0.20	1.47 ± 0.30	1.47 ± 0.30	1.15 ± 0.12	1.15 ± 0.12
13	1.77 ± 0.09	1.77 ± 0.08	1.60 ± 0.26	1.60 ± 0.25	1.33 ± 0.27	1.33 ± 0.27	1.08 ± 0.06	1.09 ± 0.06
14	1.69 ± 0.19	1.69 ± 0.18	1.49 ± 0.28	1.49 ± 0.28	1.17 ± 0.14	1.17 ± 0.14	1.04 ± 0.03	1.05 ± 0.03
15	1.61 ± 0.23	1.61 ± 0.23	1.38 ± 0.29	1.39 ± 0.29	1.09 ± 0.07	1.09 ± 0.07	1.03 ± 0.02	1.03 ± 0.02
16	1.31 ± 0.17	1.28 ± 0.16	1.12 ± 0.09	1.11 ± 0.08	1.03 ± 0.02	1.03 ± 0.02	1.01 ± 0.01	1.01 ± 0.01

where the term $(N - 1)/([(N - m)/p])p$ is a normalization factor, N is the total number of samples, m is the initial time, and p is the interval time. The value of p is taken as the average of various interval time according to the samples size which we used in analysis of EEG SBs. We calculate FD for individual SB and for the reconstructed signals, hence, we have total 17 discriminating features. The mean and standard deviation (std) of FD corresponding to each SB is given in Tables 1, 2, 3, and 4.

3 RELS-TSVM-based Classification

Recently, an energy-based model of LS-TSVM (ELS-TSVM) has been proposed by Nasiri et al. [32]. ELS-TSVM satisfy only empirical risk minimization principle and the matrices appear in their formulations are always positive semidefinite. To overcome these drawbacks, Tanveer et al. [31] proposed a robust energy-based LS-TSVM method called RELS-TSVM for classification problems. RELS-TSVM implements the structural risk minimization (SRM) principle which embodies the narrow of statistical Learning theory. Also, RELS-TSVM algorithm maximizes the margin with a positive definite matrix formulation. Similar to ELS-TSVM, RELS-TSVM solves two systems of linear equations instead of solving two QPPs. We consider that all the data points which belong to class +1 are denoted by a matrix $U \in R^{m_1 \times n}$ while the data points which belong to class -1 are denoted by matrix $V \in R^{m_2 \times n}$.

The following kernel-generated surfaces are used in RELS-TSVM algorithm [31]:

$$K(x^t, T^t)w_1 + b_1 = 0 \, and \, K(x^t, T^t)w_2 + b_2 = 0.$$

where $T = [U; V]$ and K is an appropriately chosen kernel.

The pair of two QPPs for nonlinear RELS-TSVM can be expressed as follows:

$$\min_{(w_1, b_1) \in R^{m+1}} \frac{1}{2}\|K(U, T^t)w_1 + eb_1\|^2 + \frac{c_1}{2}\xi_1^t\xi_1 + \frac{c_3}{2}\left\|\begin{bmatrix}w_1\\b_1\end{bmatrix}\right\|^2$$

$$\text{s.t.} \quad -(K(V, T^t)w_1 + eb_1) + \xi_1 = E_1, \tag{5}$$

$$\min_{(w_2, b_2) \in R^{m+1}} \frac{1}{2}\|K(V, T^t)w_2 + eb_2\|^2 + \frac{c_2}{2}\xi_2^t\xi_2 + \frac{c_4}{2}\left\|\begin{bmatrix}w_2\\b_2\end{bmatrix}\right\|^2$$

$$\text{s.t.} \quad (K(U, T^t)w_2 + eb_2) + \xi_2 = E_2, \tag{6}$$

Here, $K(U, T^t)$ and $K(V, T^t)$ are kernel matrices with sizes $m_1 \times m$ and $m_2 \times m$, respectively, and $m = m_1 + m_2, c_1, c_2, c_3, c_4$ are positive parameters and E_1, E_2 are energy parameters of the hyperplane. The solution of above QPPs are given by the following equations:

$$z_1 = -(c_1 A^t A + B^t B + c_3 I)^{-1} c_1 A^t E_1, \tag{7}$$

$$z_2 = (c_2 B^t B + A^t A + c_4 I)^{-1} c_2 B^t E_2. \tag{8}$$

where $A = [K(V, T^t)e]$ and $B = [K(U, T^t)e]$. One can observe that the matrices $c_1 A^t A + B^t B + c_3 I$ and $c_2 B^t B + A^t A + c_4 I$ are positive definite and it makes RELS-TSVM algorithm more robust and stable. The decision function is defined as follows:

$$f(x_i) = \begin{cases} +1, & \text{if} \left| \frac{K(x_i, T^t)w_1 + eb_1}{K(x_i, T^t)w_2 + eb_2} \right| \leq 1, \\ -1, & \text{if} \left| \frac{K(x_i, T^t)w_1 + eb_1}{K(x_i, T^t)w_2 + eb_2} \right| > 1, \end{cases}$$

where $|.|$ implies the absolute value.

4 Experimental Results and Discussion

In this work, the automated classification of F and NF EEG signals have been carried out using RELS-TSVM. We have decomposed EEG signals using FAWT approach into fifteen SBs and one approximate band. We calculate FD for each SB of both F and NF EEG signals. Initially, we have the size of feature matrix 100×17 but only significant features are considered for the classification purpose, which we computed using the Kruskal–Wallis statistical test [1, 46, 47]. The less p value implies more discrimination ability of features. Size of significant feature matrices for various time-series are different. The mean and std of FD for each SB for different time-series of F and NF EEG signals are given in Tables 1, 2, 3, and 4. Significant features are used as an input for RELS-TSVM classifier.

Figure 2 shows the classification accuracy of different time-series of EEG database for various samples. The value of other parameters like c_1, c_2, c_3, c_4, and E_1, E_2 are selected from the set $\{2^k (|k = -5, -3, \ldots, 3, 5)\}$ and $\{0.6, 0.7, 0.8, 0.9, 1\}$, respectively. We set $c_1 = c_2$, $c_3 = c_4$, and $E_1 = E_2$ in order to reduce the computational cost [48].

Automated classification of EEG signals with various transform methods have been carried out [6–9]. In [6], mean frequency and root mean-square bandwidth features are extracted by using discrete Fourier transform to study rhythms of EEG signals and their envelopes. The obtained classification accuracy with LS-SVM classifier in their study is 89.7%, 89.52% for 50 and 750 pairs of EEG signals, respec-

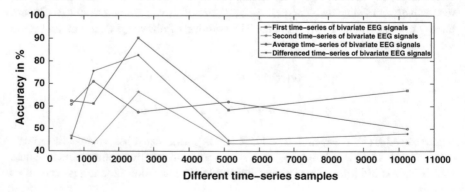

Fig. 2 Plots of accuracies for different time-series with various sample lengths

tively. An orthogonal wavelet filter banks approach coupled with entropy features and LS-SVM classifier has been studied for the classification of F and NF of EEG signals and achieved maximum accuracy of 94.25% [9]. An automated identification of F and NF with tunable-Q wavelet transform along with nonlinear features like k-nearest neighbor entropy estimator, centered correntropy, fuzzy entropy, bispectral entropies, permutation entropy, sample entropy, FD, and largest Lyapunov exponent have been studied and obtained accuracy of 94.06% using LS-SVM [7]. In [13], 50 and 750 pairs of F and NF EEG signals have been studied using delay permutation entropy along with SVM classifier. Classification accuracy of 84% and 75% are obtained, respectively. The authors examined 50 sets of each F and NF EEG signals using features namely, average sample entropy and average variance of instantaneous frequencies of intrinsic mode functions (IMFs) of signals and found the accuracy of 85% using LS-SVM [2]. In [5], authors analyzed 50 set of signals with the features extracted from IMFs of F and NF EEG signals. Various entropies like Renyi entropy, sample entropy, approximate entropy, average Shannon, and phase entropies along with LS-SVM have been used for classification and achieved an accuracy of 87%. In [4], 50 pairs of F and NF EEG signals have been studied with features extracted from discrete wavelet transform. Several ranking methods for features selection have been mentioned and the average classification accuracy of 84% is obtained using different classifiers like k-nearest neighbor, probabilistic neural networks, and LS-SVM.

In this work, average accuracy is computed using FD feature of each reconstructed SB signals for different samples with various time-series EEG signals. For each different time-series of EEG signals we got different size of significant feature matrices. The accuracy is calculated for each feature matrix separately. We used tenfold cross-validation [1, 5, 7] to obtain the optimal parameters by using RELS-TSVM as a classifier and the percentage of training and testing data is 70 and 30, respectively. The maximum average accuracy 90.2% is obtained for the classification of F and NF EEG signals.

5 Conclusions and Future Work

In this study, we have proposed a novel approach in the automated detection of F and NF EEG signals. We examined the performance using RELS-TSVM classifier along with FD as a feature. Different time-series with different samples of EEG signals are used in order to get better classification of F and NF EEG signals. The proposed method in this work can be studied for automated classification of other biomedical signals like electrocardiogram (ECG) corresponding to normal and abnormal heart activities, magnetic resonance imaging (MRI), and magnetoencephalogram (MEG) in order to examined various brain disorders.

In future, our proposed methodology can be studied on an entire dataset and more optimal filter parameters can be obtained. Accuracy can be improved by using different and more suitable classifiers. This methodology can also be used to study normal and abnormal EEG signals corresponding to Alzheimer's disease and autism.

Acknowledgements This work was supported by Science and Engineering Research Board (SERB) as Early Career Research Award grant no. ECR/2017/000053. We are thankful to National Board for Higher Mathematics (NBHM) for fellowship to Mamta Dalal. We gratefully acknowledge the Indian Institute of Technology Indore for providing facilities and support.

References

1. Gupta, V., Priya, T., Yadav, A.K., Pachori, R.B., Acharya, U.R.: Automated detection of focal EEG signals using features extracted from flexible analytics wavelet transform. Pattern Recogn. Lett. **94**, 180–188 (2017)
2. Sharma, R., Pachori, R.B., Gautam, S.: Empirical mode decomposition based classification of focal and non-focal EEG signals. In: International Conference on Medical Biometrics, Shenzhen, China, pp. 135–140 (2014)
3. Das, A.B., Bhuiyan, M.I.H.: Discrimination and classification of focal and non-focal EEG signals using entropy based features in the EMD-DWT domain. Biomed. Signal Process. Control **29**, 11–21 (2016)
4. Sharma, R., Pachori, R.B., Acharya, U.R.: An integrated index for the identification of focal electroencephalogram signals using discrete wavelet transform and entropy measures. Entropy **17**, 5218–5240 (2015)
5. Sharma, R., Pachori, R.B., Acharya, U.R.: Application of entropy measures on intrinsic mode function for the automated identification of focal electroencephalogram signals. Entropy **17**, 669–691 (2015)
6. Singh, P., Pachori, R.B.: Classification of focal and non focal EEG signals using features derived from fourier-based rhythms. J. Mech. Med. Biol. **17**(4), 2017
7. Sharma, R., Kumar, M., Pachori, R.B., Acharya, U.R.: Decision support system for focal EEG signals using tunable-Q wavelet transform. J. Comput. Sci. **20**, 52–60 (2017)
8. Bhattacharyya, A., Pachori, R.B., Acharya, U.R.: Tunable-Q wavelet transform based multivariate sub- band fuzzy entropy with application to focal EEG signal analysis. Entropy **99**, 114 (2017)
9. Sharma, M., Dhere, A., Pachori, R.B., Acharya, U.R.: An automatic detection of focal EEG signals using new class of time-frequency localized orthogonal wavelet filter banks. Knowl.-Based Syst. **118**, 217–227 (2017)
10. Sharma, R., Pachori, R.B.: Automated classification of focal and non-focal EEG signals based on bivariate empirical mode decomposition. In: Kolekar, M.H., Kumar, V. (eds.) Biomedical Signal and Image Processing in Patient Care, IGI Global, Book Chapter (2017)
11. Andrzejak, R.G., Schindler, K., Rummel, C.: Nonrandomness, nonlinear dependence and nonstationarity of electroencephalographic recording from epilepsy patients. Phys. Rev. E **86**, 046206 (2012)
12. Joshi, V., Pachori, R.B., Vijesh, A.: Classification of ictal and seizure-free EEG signals using fractional linear prediction. Biomed. Signal Process. Control **9**, 1–5 (2014)
13. Zhu, G., Li, Y., Wen, P.P., Wang, S., Xi, M.: Epileptogenic focus detection in intracranial EEG based on delay permutation entropy. In: International Symposium on Computational Models for Life Sciences, Sydney, Australia, vol. 1559, pp. 31–36 (2013)
14. Bhattacharyya, A., Sharma, M., Pachori, R.B., Acharya, U.R.: A novel approach for automated detection of focal EEG signals using empirical wavelet transform. Neural Comput. Appl. (2016)
15. Sharma, M., Pachori, R.B., Acharya, U.R.: A new approach to characterize epileptic seizures using analytic time-frequency flexible waveform transform and fractal dimension. Pattern Recogn. Lett. **94**, 172–179 (2017)
16. Andrzejak, R.G., Lehnertz, K., Mormann, F., Rieke, C., David, P., Elger, C.E.: Indication of nonlinear deterministic and finite-dimensional structures in time series of brain electrical activity: dependence on recording region and brain state. Phys. Rev. E **64**(6), 061907 (2001)

17. Bayram, I.: An analytic wavelet transform with a flexible time-frequency covering. IEEE Trans. Signal Process. **61**, 1131–1142 (2013)
18. Higuchi, T.: Approcah to an irregular time series on the basis of the fractal theory. Physica D **31**, 277–283 (1988)
19. Suykens, J.A.K., Vandewalle, J.: Least squares support vector machine classifiers. Neural Process. Netw. **9**(3), 293–300 (1999)
20. Jayadeva, Khemchandani, R., Chandra, S.: Twin support vector machines for pattern classification. IEEE Trans. Pattern Anal. Mach. Intell. **29**(5), 905–910 (2007)
21. Tanveer, M.: Application of smoothing techniques for linear programming twin support vector machines. Knowl. Inf. Syst. **45**, 191–214 (2015)
22. Tanveer, M., Shubham, K., Aldhaifallah, M., Ho, S.S.: An efficient regularized k-nearest neighbor based weighted twin support vector regression. Knowl.-Based Syst. **94**, 70–87 (2016)
23. Tanveer, M.: Robust and sparse linear programming twin support vector machines. Cogn. Comput. **7**(1), 137–149 (2015)
24. Khemchandani, R., Saigal, P., Chandra, S.: Angle-based twin support vector machine. Ann. Oper. Res. 1–31 (2017)
25. Arun Kumar, M., Gopal, M.: Least squares twin support vector machines for pattern classification. Expert Syst. Appl. **36**, 7535–7543 (2009)
26. Shao, Y.H., Zhang, C.H., Wang, X.B., Deng, N.Y.: Improvements on twin support vector machines. IEEE Trans. Neural Netw. **22**(6), 962–968 (2011)
27. Shao, Y.H., Chen, W.J., Wang, Z., Li, C.N., Deng, N.Y.: Weighted linear loss twin support vector machine for large scale classification. Knowl.-Based Syst. **73**, 276–288 (2014)
28. Zhang, Z., Zhen, L., Deng, N.Y.: Sparse least square twin support vector machine with adaptive norm. Appl. Intell. **41**(4), 1097–1107 (2014)
29. Ye, Q., Zhao, C., Ye, N.: Least square twin support vector machine classification via maximum one-class within class variance. Optim. Methods Softw. **27**(1), 53–69 (2012)
30. Kumar, M.A., Khemchandani, R., Gopal, M., Chandra, S.: Knowledge based least squares twin support vector machines. Inf. Sci. **180**, 4606–4618 (2010)
31. Tanveer, M., Khan, M.A., Ho, S.S.: Robust energy-based least squares twin support vector machines. Appl. Intell. **45**(1), 174–186 (2016)
32. Nasiri, J.A., Charkari, N.M., Mozafari, K.: Energy-based model of least squares twin support vector machines for human action recognition. Signal Process. **104**, 248–257 (2014)
33. Nonlinear time series analysis, The Bern-barcelona EEG database (2013)
34. Subasi, A.: Automatic detection of epileptic seizure using dynamic fuzzy neural networks. Expert Syst. Appl. **31**(2), 320–328 (2006)
35. Guo, L., Rivero, D., Dorado, J., Rabunal, J.R., Pazos, A.: Automatic epileptic seizure detection in EEGs based on line length feature and artificial neural networks. J. Neurosci. Methods **191**(1), 101–109 (2010)
36. Kumar, M., Pachori, R.B., Acharya, U.R.: Use of accumulated entropies for automated detection of congestive heart failure in flexible analytic wavelet transform framework based on short-term HRV signals. Entropy **19**(3), 01–21 (2017)
37. Kumar, M., Pachori, R.B., Acharya, U.R.: Automated diagnosis of myocardial infarction ECG signals using sample entropy in flexible analytic wavelet transform framework. Entropy **19**(9), 01–14 (2017)
38. Kumar, M., Pachori, R.B., Acharya, U.R.: Characterization of coronary artery disease using flexible analytic wavelet transform applied on ECG signals. Biomed. Signal Process. Control **31**, 301–308 (2017)
39. Kumar, M., Pachori, R.B., Acharya, U.R.: An efficient automated technique for CAD diagnosis using flexible analytic wavelet transform and entropy features extracted from HRV signals. Expert Syst. Appl. **63**, 165–172 (2016)
40. Istanbul Technical University: An analytic wavelet transform with a flexible time-frequency covering. http://web.itu.edu.tr/ibayram/AnDwt
41. Arle, J.E., Simon, R.H.: An application of fractal dimension to the detection of transients in the electroencephalogram. Electroencephalogr. Clin. Neurophysiol. **75**, 296–305 (1990)

42. Pickover, C.A., Khorasani, A.L.: Fractal characterization of speech waveform graphs. Comput. Graph. **10**, 51–61 (1986)
43. Katz, M.J.: Fractals and the analysis of waveforms. Comput. Biol. Med. **18**, 145–156 (1988)
44. Bullmore, E.T., Brammer, M.J., Bourlon, P., Alarcon, G., Polkey, C.E., Elwes, R., Binnie, C.D.: Fractal analysis of electroencephalographic signals intra cerebrally recorded during 35 epileptic seizures: evaluation of a new method for synoptic visualisation of ictal events. Electroencephalogr. Clin. Neurophysiol. **91**(5), 337–345 (1994)
45. Accardo, A., Affinito, M., Carrozzi, M., Bouquet, F.: Use of the fractal dimension for the analysis of electroencephalogram time series. Biol. Cybern. **77**(5), 339–350 (1997)
46. McKight, P.E., Najab, J.: Kruskal-Wallis Test. Corsini Encyclopedia of Psychology (2010)
47. Sharma, R., Pachori, R.B.: Classification of epileptic seizures in EEG signals based on phase space representation of intrinsic mode functions. Expert Syst. Appl. **42**(3), 1106–1117 (2015)
48. Hsu, C.W., Lin, C.J.: A comparison of methods for multi-class support vector machines. IEEE Trans. Neural Netw. **13**, 415–425 (2002)

Automated CAD Identification System Using Time–Frequency Representation Based on Eigenvalue Decomposition of ECG Signals

Rishi Raj Sharma, Mohit Kumar and Ram Bilas Pachori

Abstract Coronary artery disease (CAD) is a condition where coronary arteries become narrow due to the deposition of plaque inside them. It may result in heart failure and heart attack which are life-threatening conditions. Therefore, human life can be saved by detection of CAD at an early stage. Electrocardiogram (ECG) signals can be used to detect CAD. Manual inspection of ECG recordings is not reliable as the accuracy of the outcome depends on the skills and experience of clinicians. Therefore, an automated detection method for CAD based on a time–frequency representation (TFR) known as improved eigenvalue decomposition of Hankel matrix and Hilbert transform (IEVDHM-HT) using ECG beats is proposed in the present work. Time–frequency flux (TFF), coefficient of variation (COV), and energy concentration measure (ECM) are computed from the TFR matrix and fed to the random forest classifier. The proposed method has yielded 93.77% classification accuracy.

Keywords CAD · Eigenvalue decomposition · Hankel matrix · Time–frequency representation

1 Introduction

Around the globe, almost one-third of all the deaths are the result of the cardiovascular diseases (CVDs). These diseases are responsible for the death of nearly 17 million people every year globally [34]. CVDs are also found to be a major cause of death in all over India. Nearly 24.8% of all the deaths in India are due to the CVDs [28].

R. R. Sharma (✉) · M. Kumar · R. B. Pachori
Discipline of Electrical Engineering, Indian Institute of Technology Indore,
Indore 453552, India
e-mail: phd1501102018@iiti.ac.in

M. Kumar
e-mail: phd1401202005@iiti.ac.in

R. B. Pachori
e-mail: pachori@iiti.ac.in

© Springer Nature Singapore Pte Ltd. 2019
M. Tanveer and R. B. Pachori (eds.), *Machine Intelligence and Signal Analysis*,
Advances in Intelligent Systems and Computing 748,
https://doi.org/10.1007/978-981-13-0923-6_51

597

Coronary artery disease (CAD) is a major death contributor among all types of heart diseases [34]. It is a condition in which flow of blood toward the heart muscles is interrupted due to the deposition of the waxy substances (plaque) inside the coronary arteries [25]. This affects the strength of the heart muscles and weakens the pumping ability of the heart [25]. With the passing of the time, it may result to heart failure. Hence, early-stage diagnosis of CAD is a necessary step.

Exercise stress test (EST) is commonly used to diagnose CAD. A possibility of cardiac arrest is there with the EST [2, 29]. Another problem with the EST test is that the targetted heart rate may not be achievable by some of the CAD patients which are required for the accurate analysis [1]. Cardiac catheterization and coronary angiography are also used by the physicians for the diagnosis of CAD [25]. These techniques require the presence of expert clinicians due to their invasive nature [10, 25]. Electrocardiogram (ECG) recording is a noninvasive technique and can also be used to diagnose the CAD patients [2, 21]. Manual detection of the changes in ECG signals may not provide an accurate diagnosis as sometimes these signals do not indicate a significant difference between CAD and normal subjects [10]. To overcome these difficulties, we have proposed a computer-aided methodology to diagnose the CAD patients based on the ECG signals in this work.

Various methods for the automated diagnosis of CAD patients are suggested in the literature using heart rate variability (HRV) signals and ECG signals. Time and frequency domain parameters extracted from HRV signals are explored and their values are found higher for normal subjects [6]. Nonlinear features, such as cor-rentropy [27], k-nearest neighbor entropy estimator [20], higher order spectrum [1], fuzzy entropy [20], and block entropy [19] are extracted from HRV signals and showed their effectiveness in CAD detection. In [10], discrete wavelet transform (DWT) and principal component analysis (PCA)-based method applied on HRV signals has also shown good classification accuracy for separating the CAD and normal classes. A method based on fuzzy, probabilistic, and combined uncertainty models is proposed for the CAD detection using stress ECG signals in [4]. In [3], various non-linear features extracted from resting ECG signals are found effective for the CAD detection. In [33], empirical mode decomposition (EMD)-based features showed good discrimination for the two classes.

The eigenvalue decomposition of Hankel martix (EVDHM)-based multicomponent nonstationary signal decomposition method has been presented in [18]. It is an iterative approach which decomposes a multicomponent nonstationary signal by selecting eigenvalue pairs of Hankel matrix. The analysis of nonstationary signals in time–frequency domain has been given in [32] using EVDHM and Hilbert transform. The EVDHM method has been used for the analysis of speech signals [16, 17]. Moreover, a time–frequency representation (TFR) method has been recently presented which is based on the improved EVDHM (IEVDHM) and Hilbert transform (IEVDHM-HT) [31]. We have explored this method for CAD diagnosis in the work.

Our aim is to propose an automated and noninvasive technique based on ECG beats for CAD diagnosis. ECG beats are effectively used for the separation of the CAD and normal subjects in [2, 21]. To achieve the objective, first, beats are extracted from the ECG signals of both normal and CAD class. Thereafter, TFR has been formed for all the beats using IEVDHM-HT method. We have studied parameters for the TFR of CAD and normal beats. These parameters are namely, time–frequency flux (TFF), coefficient of variation (COV), and energy concentration measure (ECM).

The remaining of the paper is arranged as follows: Sect. 2 provides dataset description, preprocessing, IEVDHM-HT method, features used, and classifier. The achieved results are shown and discussed in Sect. 3. The work is concluded in Sect. 4.

2 Methodology

The methodology for CAD diagnosis using IEVDHM-HT method has been represented in Fig. 1. First of all, baseline wander and noise has been removed from ECG signal. Thereafter, segmentation of ECG beats has been performed. TFR has been obtained for all the beats using IEVDHM-HT method. Further, features are extracted from TFR plane. Finally, classification has been performed. All these steps and ECG dataset information have been explained in this section.

2.1 ECG Dataset

In this work, we have used the ECG signals available at St. Petersburg Institute of Cardiological Technics 12-lead Arrhythmia Database for seven CAD subjects [11] and Fantasia open-access database for 40 normal subjects [11, 15]. The duration of the normal and CAD ECG signals are 2 h and 0.5 h, respectively. The normal and CAD ECG signals are acquired at the sampling rate of 250 and 257 samples per second, respectively.

2.2 Preprocessing and Beat Segmentation

First, normal ECG signals are upsampled to sampling rate of 257 samples per second. The Daubechies 6 (db6) wavelet is utilized for eliminating the noise and baseline

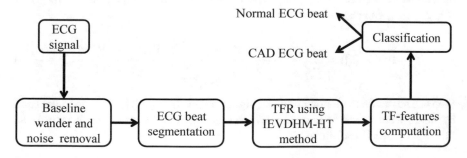

Fig. 1 Block diagram for CAD identification using proposed method

wander exist in the ECG signals [23]. Thereafter, R-peak is tracked by employing the Pan-Tompkin's method [26]. ECG beats are segmented by considering the 104 samples after R peak and 64 samples before the R peak. Hence, each ECG beat consists of 169 samples [2]. In this work, we have used 44426 ECG beats of both classes (CAD and normal).

2.3 Improved Eigenvalue Decomposition of Hankel Matrix and Hilbert Transform (IEVDHM-HT)

The TFR method based on IEVDHM and Hilbert transform is explained in steps as follows:

Step 1: A multicomponent nonstationary signal $y[s]$ of length S is used for preparing a Hankel matrix H of size $A \times A$, where $S = 2A - 1$ [18]. The matrix H is characterized as follows:

$$H = \begin{bmatrix} y[1] & y[2] & \cdots & y[A] \\ y[2] & y[3] & \cdots & y[A+1] \\ \vdots & \vdots & \vdots & \vdots \\ y[A] & y[A+1] & \cdots & y[2A-1] \end{bmatrix} \tag{1}$$

Step 2: Determine eigenvalue matrix λ_H and eigenvector matrix V_H for H, where λ_H is a diagonal matrix and its non-diagonal elements are zero. The significant eigenvalues can be obtained by using modified significant threshold point (MSTP), which states that sum of magnitude of highest eigenvalue pairs should be greater than or equal to the 95% of sum of magnitude of all the eigenvalues [31].

Step 3: Each eigenvalue pair is used to obtain the decomposed component. The jth decomposed component can be obtained using the mathematical expression given as [31]:

$$x_j[s] = \Lambda_j \, \text{fun}(V_j[s]) + \Lambda_{A-j+1} \, \text{fun}(V_{A-j+1}[s]) \tag{2}$$

where $x_j[s]$ is the jth decomposed component, Λ_j is the jth eigenvalue, and mathematical expression for 'fun$(V_j[s])$' operation applied on jth eigenvector $V_j[s]$ for $s = 1, 2, \ldots, A$, can be explained as:

$$\text{fun}(V_j[s]) = \sum_{k=1}^{A} \frac{V_j[s-k].V_j[k]}{k} + \sum_{k=A+1}^{2A-1} \frac{V_j[s-k].V_j[k]}{(2A-k)} \tag{3}$$

Step 4: The modified monocomponent signal criteria (MMSC) [31] has to be satisfied by all the decomposed components, which states two conditions of the intrinsic mode functions [13, 31] as follows:

Condition 1: The number of zero-crossings and the sum of local maxima and local minima should differ by at most one.

Condition 2: The average of both upper envelope and lower envelope should be zero.

Step 5: Check the energy of components which do not satisfy the MMSC. If sum of energy of these components is not more than 5% energy of original signal $y[s]$, then process can be terminated. Otherwise, the components which do not satisfy the MMSC are allowed to repeat the process from Step 1. We have applied the process up to three iterations in this paper.

Step 6: To obtain the decomposed components, we have merged the components which have overlapping of at least 3 dB bandwidth.

Step 7: Apply Hilbert transform [14] on all the decomposed components to find out their instantaneous frequency and instantaneous amplitude. A complete TFR has been obtained by using instantaneous frequency and square of instantaneous amplitude.

The parameters used for feature extraction are explained in the next section.

2.4 Extracted Features

The features are extracted from TFR plots formed for CAD and normal ECG beats. We have used three parameters and obtained from the TFR plane. These parameters are given below.

2.4.1 Time–Frequency Flux:

The rate of change of energy of signal can be measured in terms of TFF in time–frequency plane. Mathematical expression of TFF of time–frequency distribution $\rho[p, q]$, can be given as [8]:

$$\text{TFF} = \sum_{p=1}^{P-r} \sum_{q=1}^{Q-s} |\rho[p + r, q + s] - \rho[p, q]| \tag{4}$$

where, r and s are predefined values which are correlated with the rate of change of the signal energy. We have considered the values r and s as 1 [8].

2.4.2 Coefficient of Variation:

The COV can be explained as the ratio of the standard deviation of the time–frequency plane ($\sigma(p, q)$) to the mean of the time–frequency plane ($\mu(p, q)$). It can be expressed as [8]:

$$\text{COV}(p, q) = \frac{\sigma(p, q)}{\mu(p, q)} \tag{5}$$

2.4.3 Energy Concentration Measure

The ECM indicates the distribution of energy over time–frequency plane. It is mathematically expressed as [7]:

$$\text{ECM} = \left(\sum_{p=1}^{P} \sum_{q=1}^{Q} |\rho[p, q]|^{0.5} \right)^2 \tag{6}$$

2.5 Classification

All the features extracted from TFR are given to classifier. The classification of ECG beat for CAD and normal datasets has been performed using random forest (RF) classifier. The RF contains a large number of decision trees. Every tree makes its own decision about the class and a weight is assigned to each tree. The class label is finalized based on the collective decisions of all the classification trees. More details of RF classifier are provided in [9]. Recently, RF classifier is used in the classification

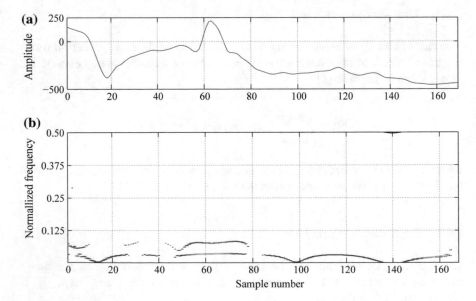

Fig. 2 Plot of **a** ECG beat of CAD subject and **b** its TFR using IEVDHM-HT method

of physiological signals [22, 30]. We have used Waikato environment for knowledge analysis (WEKA) toolbox [12] for the implementation of RF classifier in the present work. Previously, WEKA has been used for seizure EEG detection in [5].

3   Results and Discussion

In this work, ECG signals are segmented into the beats. The beats of the two classes are represented in TF-plane using IEVDHM-HT method. The typical CAD and normal ECG beats and their TFR are shown in Figs. 2 and 3, respectively.

The features TFF, COV, and ECM are computed from the decomposed components. The standard deviation is denoted by sd and mean value is denoted by μ for these features and provided in Table 1. The Kruskal–Wallis statistical test [24] has been employed to know the significance of the features in terms of p-values which are also provided in Table 1. From Table 1, it can be observed that all the features have significant discrimination ability as the p-values are very less ($p < 0.05$). The values of TFF and ECM are higher for normal subjects, while COV shows higher values for CAD subjects.

The computed μ values of ECM for the normal class are higher than CAD class. It indicates that the energy of normal ECG beats is higher than that of CAD ECG beats. The energy of signal correlates to the signal amplitude, which may show that amplitude of normal ECG beats is higher as compared to the CAD ECG beats. This can also be observed in Figs. 2 and 3. The Box plots for TFF, COV, and ECM features

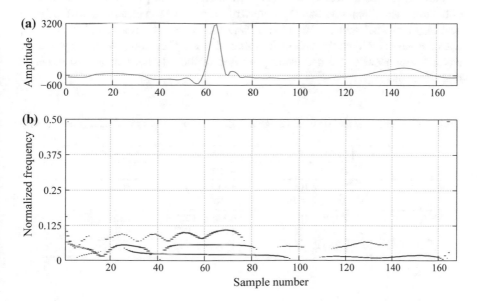

Fig. 3 Plot of **a** ECG beat of normal subject and **b** its TFR using IEVDHM-HT method

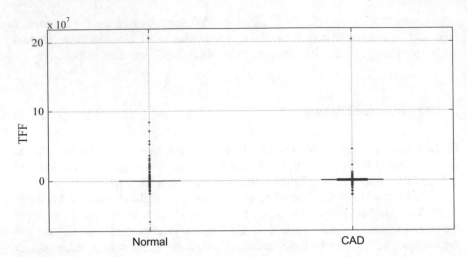

Fig. 4 Box plot for TFF feature for normal and CAD ECG beats

are depicted in Figs. 4, 5, and 6, respectively. The method proposed in [2], also used ECG beats for CAD detection and computed up to 69 features. In [21], authors have decomposed ECG beats using flexible analytic wavelet transform (FAWT) technique and computed five features to detect CAD disease. We are using only three features for CAD classification in the proposed method. Thus, the feature dimension for CAD detection in our method is lesser as compared to methods given in [2, 21].

Further, the features are applied to the RF classifier. The plot between the number of trees and classification accuracy is shown in Fig. 7. We can observe from Fig. 7 that accuracy varies between 93.74 and 93.77%. Results show that classification accuracy does not vary significantly with the number of trees. Therefore, we can say that the classification results obtained using proposed method do not depend too much on tree selection.

Table 1 The mean (μ) and standard deviation (sd) values of computed TFR features with their p-values

Features	CAD class	Normal class	p-values
	$\mu \pm$ sd	$\mu \pm$ sd	
TFF	1.077×10^5 $\pm 1.372 \times 10^6$	$4.588 \times 10^5 \pm 1.93$ $\times 10^6$	0
COV	7.155×10^{-9} $\pm 1.71 \times 10^{-9}$	$6.835 \times 10^{-9} \pm$ 1.588×10^{-9}	6.93×10^{-191}
ECM	8.107 ± 1.451	10.207 ± 0.640	0

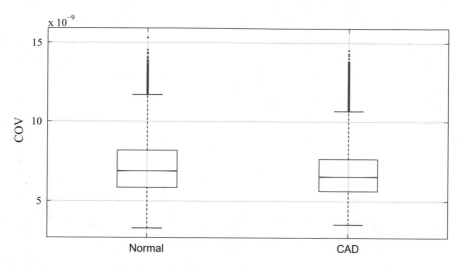

Fig. 5 Box plot for COV feature for normal and CAD ECG beats

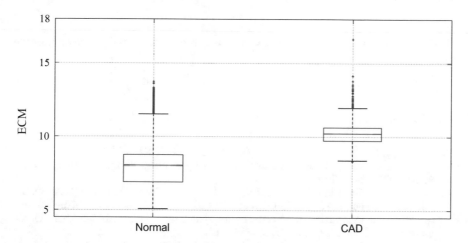

Fig. 6 Box plot for ECM feature for normal and CAD ECG beats

4   Conclusion

We have proposed a novel method to diagnose the CAD automatically using the ECG beats. The TFR for all ECG beats are obtained using IEVDHM-HT method, and thereafter, three parameters are computed from TFR plane. The obtained features achieved 93.77% accuracy of classification when used with the RF classifier. In the proposed method, we have used only three features for achieving this accuracy which may be good for fast detection of CAD disease. This method can also be used for

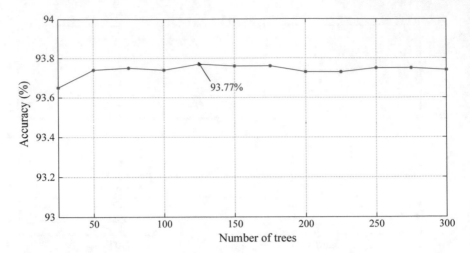

Fig. 7 Plot of accuracy versus the number of trees used in RF classifier for normal and CAD ECG beat dataset

the diagnosis of other cardiac diseases such as myocardial infarction and congestive heart failure.

References

1. Acharya, U.R., Faust, O., Sree, V., Swapna, G., Martis, R.J., Kadri, N.A., Suri, J.S.: Linear and nonlinear analysis of normal and CAD-affected heart rate signals. Comput. Methods Prog. Biomed. **113**, 55–68 (2014)
2. Acharya, U.R., Sudarshan, V.K., Koh, J.E.W., Martis, R.J., Hong, T.J., Lih, O.S., Adam, M., Hagiwara, Y., Mukiah, M.R.K., Poo, C.K., Chua, C.K., San, T.R.: Application of higher-order spectra for the characterization of coronary artery disease using electrocardiogram signals. Biomed. Signal Process. Control **31**, 31–43 (2017)
3. Antanavičius, K., Bastys, A., Blužas, J., Gargasas, L., Kaminskienė, S., Urbonavičienė, G., Vainoras, A.: Nonlinear dynamics analysis of electrocardiograms for detection of coronary artery disease. Comput. Methods Prog. Biomed. **92**, 198–204 (2008)
4. Arafat, S., Dohrmann, M., Skubic, M.: Classification of coronary artery disease stress ECGs using uncertainty modeling. In: ICSC Congress on Computational Intelligence Methods and Applications, pp. 1–4 (2005)
5. Bhattacharyya, A., Pachori, R.B.: A multivariate approach for patient specific EEG seizure detection using empirical wavelet transform. IEEE Trans. Biomed. Eng. **64**, 2003–2015 (2017)
6. Bigger, J.T., Fleiss, J.L., Steinman, R.C., Rolnitzky, L.M., Schneider, W.J., Stein, P.K.: RR variability in healthy, middle-aged persons compared with patients with chronic coronary heart disease or recent acute myocardial infarction. Circulation **91**, 1936–1943 (1995)
7. Boashash, B., Azemi, G., O'Toole, J.M.: Time-frequency processing of nonstationary signals: advanced TFD design to aid diagnosis with highlights from medical applications. IEEE Signal Process. Mag. **30**, 108–119 (2013)
8. Boashash, B., Khan, N.A., Ben-Jabeur, T.: Time-frequency features for pattern recognition using high-resolution TFDs: a tutorial review. Digit. Signal Process. **40**, 1–30 (2015)

9. Breiman, L.: Random forests. Mach. Learn. **45**, 5–32 (2001)
10. Giri, D., Acharya, U.R., Martis, R.J., Sree, S.V., Lim VI, T.C., T.A., Suri, J.S.: Automated diagnosis of coronary artery disease affected patients using LDA, PCA, ICA and discrete wavelet transform. Knowl.-Based Syst. **37**, 274–282 (2013)
11. Goldberger, A.L., Amaral, L.A.N., Glass, L., Hausdorff, J.M., Ivanov, P.C., Mark, R.G., Mietus, J.E., Moody, G.B., Peng, C.K., Stanley, H.E.: Physiobank, physiotoolkit, and physionet components of a new research resource for complex physiologic signals. Circulation **101**, e215–e220 (2000)
12. Hall, M., Frank, E., Holmes, G., Pfahringer, B., Reutemann, P., Witten, I.H.: The WEKA data mining software: an update. SIGKDD Explor. **11**, 10–18 (2009)
13. Huang, N.E., Shen, Z., Long, S.R., Wu, M.C., Shih, H.H., Zheng, Q., Yen, N.C., Tung, C.C., Liu, H.H.: The empirical mode decomposition and the Hilbert spectrum for nonlinear and non-stationary time series analysis. Proc. R. Soc. Lond. A: Math. Phys. Eng. Sci. **454**, 903–995 (1998)
14. Huang, N.E.: Hilbert-Huang Transform and Its Applications, vol. 16. World Scientific (2014)
15. Iyengar, N., Peng, C., Morin, R., Goldberger, A.L., Lipsitz, L.A.: Age-related alterations in the fractal scaling of cardiac interbeat interval dynamics. Am. J. Physiol.-Regul. Integr. Comp. Physiol. **271**, 1078–1084 (1996)
16. Jain, P., Pachori, R.B.: GCI identification from voiced speech using the eigen value decomposition of Hankel matrix. In: 2013 8th International Symposium on Image and Signal Processing and Analysis (ISPA), pp. 371–376 (2013)
17. Jain, P., Pachori, R.B.: Event-based method for instantaneous fundamental frequency estimation from voiced speech based on eigenvalue decomposition of the Hankel matrix. IEEE/ACM Trans. Audio Speech Lang. Process. **22**, 1467–1482 (2014)
18. Jain, P., Pachori, R.B.: An iterative approach for decomposition of multi-component non-stationary signals based on eigenvalue decomposition of the Hankel matrix. J. Franklin Inst. **352**, 4017–4044 (2015)
19. Karamanos, K., Nikolopoulos, S., Hizanidis, K., Manis, G., Alexandridi, A., Nikolakeas, S.: Block entropy analysis of heart rate variability signals. Int. J. Bifurc. Chaos **16**, 2093–2101 (2006)
20. Kumar, M., Pachori, R.B., Acharya, U.R.: An efficient automated technique for CAD diagnosis using flexible analytic wavelet transform and entropy features extracted from HRV signals. Expert Syst. Appl. **63**, 165–172 (2016)
21. Kumar, M., Pachori, R.B., Acharya, U.R.: Characterization of coronary artery disease using flexible analytic wavelet transform applied on ECG signals. Biomed. Signal Process. Control **31**, 301–308 (2017)
22. Kumar, M., Pachori, R.B., Acharya, U.: Automated diagnosis of myocardial infarction ECG Signals using sample entropy in flexible analytic wavelet transform framework. Entropy **19**, 488 (2017)
23. Martis, R.J., Acharya, U.R., Min, L.C.: ECG beat classification using PCA, LDA, ICA and discrete wavelet transform. Biomed. Signal Process. Control **8**, 437–448 (2013)
24. McKight, P.E., Najab, J.: Kruskal-Wallis Test. Corsini Encyclopedia of Psychology (2010)
25. What is coronary heart disease? (2015). http://www.nhlbi.nih.gov/health/health-topics/topics/cad/. Accessed 01 Apr 2016
26. Pan, J., Tompkins, W.J.: A real-time QRS detection algorithm. IEEE Trans. Biomed. Eng. **32**, 230–236 (1985)
27. Patidar, S., Pachori, R.B., Acharya, U.R.: Automated diagnosis of coronary artery disease using tunable-Q wavelet transform applied on heart rate signals. Knowl.-Based Syst. **82**, 1–10 (2015)
28. Prabhakaran, D., Jeemon, P., Roy, A.: Cardiovascular diseases in India current. Circulation **133**, 1605–1620 (2016)
29. San Roman, J., Vilacosta, I., Castillo, J., Rollan, M., Hernandez, M., Peral, V., Garcimartin, I., de la, Torre, M.d.M., Fernández-Avilés, F.: Selection of the optimal stress test for the diagnosis of coronary artery disease. Heart **80**, 370–376 (1998)

30. Sharma, R., Pachori, R.B., Upadhyay, A.: Automatic sleep stages classification based on iterative filtering of electroencephalogram signals. Neural Comput. Appl. **28**, 2959–2978 (2017)
31. Sharma, R.R., Pachori, R.B.: Time-frequency representation using IEVDHM-HT with application to classification of epileptic EEG signals. IET Sci. Meas. Technol. **12**(1), 72–82 (2018)
32. Sharma, R.R., Pachori, R.B.: A new method for non-stationary signal analysis using eigenvalue decomposition of the Hankel matrix and Hilbert transform. In: Fourth International Conference on Signal Processing and Integrated Networks (SPIN 2017), Noida India, pp. 484–488, Feb, 2017
33. Sood, S., Kumar, M., Pachori, R.B., Acharya, U.R.: Application of empirical mode decomposition-based features for analysis of normal and CAD heart rate signals. J. Mech. Med. Biol. **16**, 1640002 (2016)
34. Wong, N.D.: Epidemiological studies of CHD and the evolution of preventive cardiology. Nat. Rev. Cardiol. **11**, 276–289 (2014)

Optical Imaging with Signal Processing for Non-invasive Diagnosis in Gastric Cancer: Nonlinear Optical Microscopy Modalities

Shyam Singh and Hem Chandra Jha

Abstract Gastric cancer or stomach cancer has high incident rate and the leading cause of mortality worldwide. GC is usually undetected and asymptotic till the advanced chronic stages of its progression. Despite much advancement of technologies, still diagnosis is poor. This makes GC a fatal chronic disease. Over two decades of advancement in nonlinear optical (NLO) microscopy, it has become a powerful tool for laser-based imaging of tissue. Each of NLO modality is sensitive for specific molecule or structure. This may be useful for the understanding of the complex biological system in cancer detection. Here, we will discuss label-free, non-invasive endoscopy-based methods for morphological imaging by combining coherent anti-Stokes Raman scattering (CARS), Two-photon excited fluorescence (TPEF) and second-harmonic generation (SHG) methods of NLO microscopy.

Keywords Nonlinear optical microscopy · Gastric cancer · Early diagnosis
Coherent anti-Stokes raman scattering · Two-photon excited fluorescence
Second-harmonic generation

1 Introduction

Gastric cancer (GC) is the fifth most common cancer and third leading cause of cancer-associated mortality worldwide [1]. This contributes 6.8% of cancer cases and 8.8% of the death of total cancer [1]. Since many years endoscopy is a gold standard method for diagnosis of abnormalities in gastric mucosa [2], whereas, early detection is still not convincing [3]. High mortality reasons with GC are due to treatment of the patients in the latter stages of cancer on the basis of advanced symptoms [4].

S. Singh · H. C. Jha (✉)
Centre for Biosciences and Biomedical Engineering,
Indian Institute of Technology Indore, Indore 453552, India
e-mail: hemcjha@iiti.acin

S. Singh
e-mail: phd1501171013@iiti.ac.in

© Springer Nature Singapore Pte Ltd. 2019
M. Tanveer and R. B. Pachori (eds.), *Machine Intelligence and Signal Analysis*,
Advances in Intelligent Systems and Computing 748,
https://doi.org/10.1007/978-981-13-0923-6_52

609

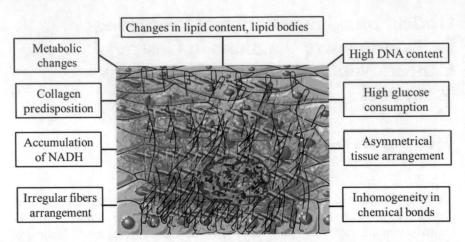

Fig. 1 The biophysical and biochemical microenvironment of a tumor: Biophysical parameters are asymmetric and predisposition of ECM including collagen and elastin and well as collagen alignment angle differences in tissue. Biochemical parameters include changes in lipid content, high DNA and protein presence, higher glucose consumption and changes in the cellular level of NADH

Importantly, the sensitivity of endoscopy does not seem to be so much efficient to detect malignant tissue at the beginning of gastric carcinogenesis [3]. Studies suggested by Schmidt and Maconi et al. [5, 6] that these symptoms are more in benign dyspepsia and nearly undetectable in the early stages of cancers. An invasive detection method of combining positron emission tomography (PET) with computed tomography (CT) is widely utilizing which represent cancer on the basis of histology, size, and tumor location [7]. However, many times PET/CT methods are unable to detect and represent the false results [8]. PET/CT also has some limitation on the basis of early staging index of GC and clinical diagnosis [8]. Various other methods are being used for imaging include X-ray, CT, MRI, and OCT [9]. However, low spatial resolution capabilities of all these methods make cancer indistinguishable mostly at an early stage. These all, in turn, hampers therapeutics outcomes and overall survivals of GC patients [10]. Tumor microenvironments have many biological characteristics including an increased amount of DNA, protein, collagen, elastin, NADH, FAD, lipids, irregular shape, the unsymmetrical appearance of the extracellular matrix, presence of uneven tissue organization, etc. [11] (Fig. 1). NLO microscopy enables molecules and chemical properties based imaging tools which have a major advantage over just imaging of lump, cyst, or other gastric abnormality by focusing on a single property or as organized mass [12].

2　Multimodal Nonlinear Optical (NLO) Microscopy

In NLO microscopy, nonlinear interactions occur between ultrafast laser light of wavelength nearly (~680–1080) nm with biological matter to generate images [13]. This nonlinear interaction occurs with the specific intrinsic biomolecule and allows the image formation without any exogenous stains. This ultimately makes this method most suitable for in vivo application [13]. However, NLO microscopy can also be used with fluorescence molecules for imaging [14]. This microscopy comprises various methods modality including second and third-harmonic generation (SHG and THG), two-photon excited fluorescence (TPEF), and coherent Raman scattering (CRS) either in the form of Stimulated Raman Scattering (SRS) or Coherent Anti-Stokes Raman Scattering (CARS) [15]. Although each modality can be operated separately for imaging purpose, advancement in ultrafast lasers system allows a combination of various NLO modalities together into a unified multimodal NLO form in a single microscopic architecture [16].

2.1　*Coherent Anti-stokes Raman Scattering (CARS) Microscopy*

Coherent anti-Stokes Raman scattering (CARS) microscopy produces images based on the vibrational signatures of molecules [17]. CARS methods are label free and molecular specific information can be obtained from tissue or cells [18]. Every bond in the universe has a unique rate of frequencies and can be identified particularly. Biomolecules are rich in C-H, O-H, N-H and other covalent chemical bonds [19]. On the basis of oscillatory motions and vibrational frequencies, CARS produces results about the chemical structure of biological tissues [19]. High accumulation of lipid content has been observed in many cancers and tumors including GC and also associated with the aggressiveness of many types of cancers [20]. CARS microscopy may detect single lipid molecules and may be useful in visualizing total lipid content, lipid bodies, saturated and non-saturated lipids. CARS can represent information of types of lipid in tumor growth and metastasis, movements and trafficking of lipid molecules, and lipid metabolic patterns [21]. CARS also detect the density of DNA and protein present in tissue [22].

2.2　*Two-Photon Excited Fluorescence (TPEF) Microscopy*

TPEF is a two-photon based imaging techniques in which the tissue molecule absorbs two photons instead of one for excitation and emission of a higher energy single photon than either of the incident photons [23]. This method offers three-dimensional (3-D) resolution images with high contrast [24]. TPEF method allows deeper pen-

Table 1 NLO microscopy modalities mechanism and information

Methods	Mechanics	Information	References
CARS	Vibrational level of molecule, extracellular morphology	Chemical information (Lipid, DNA)	[19, 20, 27]
TPEF	Electronic state, chemical bonds,	Auto-fluorescence of biological molecules (NADH, FAD etc.)	[23, 24, 26]
SHG	Nonlinear asymmetrical property of tissue	Non centrosymetrical molecules (Collagen, elastin)	[27–29]

CARS Coherent anti-Stokes Raman Scattering; *TPEF* Two-photon excited fluorescence; *SHG* Second-harmonic generation

etration of biological tissue as it utilizes near-infrared excitation wavelength [25]. TPEF is now widely being accepted in biological studies and imaging of nicoti-namide adenine dinucleotide (NADH) and oxidized flavin adenine dinucleotide to determine cellular metabolism in tissue in terms of radiometric redox [24]. These molecules have auto-fluorescence and upon excitation provides a metabolic index of the tissues about optical redox ratio (ORR) [26]. TPEF produces real-time data without hampering viable cells functioning or cellular metabolism which may very useful to estimate conclusion in living organelles or tissue. TPEF techniques can also measure oxygen consumption of distal tissue with metabolic rate, differentiation of cells, higher rate of proliferation, and uneven distribution of metabolomics in tissue [23] (Table 1).

2.3 Second-Harmonic Generation (SHG) Microscopy

SHG is also a two-photon method while in this process, both excitation photons are transformed to one emission photon with energy precisely equal to the sum of the incident photons unlike the TPEF, i.e., both excitation photons degenerate [28]. In SHG process, scattering of photons occurs with no energy loss, and two photons inter-act simultaneously in a noncentrosymmetric fashion. SHG microscopy is a sensitive method for biomolecules in tissue and organism [28]. SHG use a coherent beam which enables detection on the basis of amount, direction and polarization properties of bio-logical tissue just by the scatter concentration. In case of cancer, extracellular matrix (ECM) especially collagen deposition occurs in an unorganized manner hence SHG microscopy can be a powerful tool for structure-based detection method [29]. Emit-ted signals from SHG can be monitored using two-photon microscopy (TPM) which collected and analyzed with TPEF signal to represent structural information. The

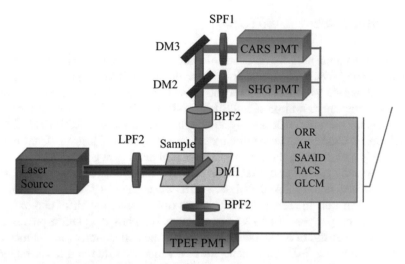

Fig. 2 Schematic of multimodal nonlinear optical microscope system: SPF, short-pass filter; BPF, band-pass filter; LPF, long pass filter; PMT, photomultiplier tube; DM, deformable mirror: TPEF, Two-photon excitation fluorescence; SHG, second-harmonic generation; CARS, coherent anti-stokes Raman scattering; ORR, optical redox ratio; SAAID, second harmonic to auto-fluorescence ageing index of dermis; TACS, tumor associated collagen signature; AR, aspect ratio; GLCM, gray level co-occurrence matrix analysis

structural pattern at molecular organization level about collagen and ECM in in vivo or ex vivo tumor tissue provides valuable insight of the cancer progression [27].

3 Analysis of the NLO Modalities Parameters by Combining CARS, SHG and PTEF in Tumorigenesis

Simultaneous acquisition of images by combining NLO modalities is still tricky [30]. However, this can be done using the properties of the coherent signal, many technical difficulties still needed to resolve [31]. Multimodal NLO microscopies are now effectively utilizing in tissues and small animals whereas depth penetration and use in human is still challenging and needed to explore [30] (Fig. 2). Various parameters are developed for the precise information about the tissue of interest on the basis of chemical, biological and biophysical properties which are described below.

3.1 Optical Redox Ratio (ORR)

ORR is based on the auto-fluorescence of endogenous amount of nicotinamide adenine dinucleotide (NADH) and flavin adenine dinucleotide (FAD) present in the cell types. Both are metabolic co-factors and have a crucial role in ATP production in oxidative phosphorylation process [26]. In mitochondria, electron transport chain process occurs after the citric acid cycle in which oxidation of NADH to NAD^+ and $FADH_2$ to FAD taken place via complex I and II system. Electrons from NADH and $FADH_2$ pass through the electron transport chain to oxygen, which is reduced to water. As only NADH and FAD molecule have auto-fluorescence property, measurement of the ratio of FAD/(FAD+NADH) gives the ORR value [23, 26, 32]. ORR ratio interprets the relative variation rate of glucose catabolism and oxidative phosphorylation process in tissue [23]. In oxidative phosphorylation process, the fluorescence intensity of NADH decreases due to the conversion of non-fluorescent molecule NAD^+ and FAD fluorescence intensity increases due to the conversion of the non-fluorescent molecule, $FADH_2$ to FAD [33]. Decreasing in ORR value represents high glucose consumption of with the low ATP generation. In tumors, oxygen availability is reduced, which requires increased glucose catabolism, leading to an increase in total NADH fluorescence thus a decrease in the ORR [34]. The ORR strongly can track metabolic changes in tumor tissue in vitro and in vivo. However, till date, no investigation has been finding with optical redox ratio and metastatic cancer progression [34].

3.2 Second-Harmonic to Auto-fluorescence Ageing Index of Dermis (SAAID)

SAAID represented the proportion of collagen and elastin present in the tissue [35]. As tumorigenesis progresses, collagen deposition increases in the stroma [36]. Stroma is primarily composed of elastin and collagen [36]. This parameter is obtained when both SHG and TPEF are used together. SHG enable detection of collagen in stroma while PTEF emits an auto-fluorescent signal from elastin. The SAAID is calculated by the ratio of (ISGH − ITPEF)/(ISHG + ITPEF). I represent the intensity of signals from NLO modalities SHG and TPEF respectively [37]. More collagen present at the tumor-stromal interface gives the corresponding high value of SAAID [37].

3.3 Tumor-Associated Collagen Signature (TACS)

TACS represent the collagen fibers orientation at the tumor-stromal boundary [38]. This parameter is obtained when SHG microscopy used. Currently, total three subtypes of TACS are characterized which fit the different stages of tumorigenesis and are highly reproducible [38]. TACS subtype1 (TACS-1) outcome represent of dense collagen surrounding a small tumor which state early stages of tumorigenesis. TACS subtype 2 (TACS-2) outcome represent the middle stage of tumorigenesis and the parallel collagen fibers arrangement near the tumor boundary at an angle nearly 0°. Finely, TACS subtype 3 (TACS-3) SHG outcome represents the vertical arrangement of collagen fibers at the tumor-epithelial interface with angle almost 90°, at the tumorigenic stage when a tumor becomes invasive [39]. Angle calculation between collagen fiber and tumor boundary identify the normal epithelial zone and tumor zone. Tissue area having some angle (0°–90°) are also categorized as abnormal appearance zone [39].

3.4 Aspect Ratio (AR)

It represents the aspect ratio (AR) of the short and long axis of eclipse using SHG modality and characterizes the alignment of fibers in tissue [40]. Many studies suggested this as a proven method to assign the degree of alignment of fibers in tissue [41]. If the fibers are well aligned, the FFT value using the SHG modality will be high and FFT plot would be more ellipsoid shape along the orthogonal path. Perfectly aligned fibers will have ellipsoid look like a line along the long axis. AR form FFT plot for non-aligned and randomly arranged fibers will tend to ellipsoid to circle shape [42]. A tissue with high AR ratio closer to one would be more isotropic whereas if AR approaching to zero, tissue will be more anisotropic [42] (Table 2).

3.5 Gray Level Co-occurrence Matrix (GLCM) Analysis

GLCM analysis represents geometrical collagen arrangement in tissue using SHG microscopy [12, 37]. This analysis method classifies the tissues based collagen arrangement. This gives information of the formed image on the basis of the spatial relationship between pixel brightness and by counting the number of occurrence of gray level at a specified pixel distance adjacent to other gray level and divided by the total count numbers [43, 44]. Various methods are available for analysis of GLCM that are classified as statistical methods, contrast methods and orderliness methods [43, 44]. Statistical methods rely on the statistical analysis of pixel values, as the value determine homogeneous in a pattern of tissue. The pixel value in repetitive form correlates homogeneous arrangement of collagen, while discrete value correlates non-uniformity of tissue in that area. Thus statistical method characterizes the pattern of collagen in tissue [45]. Similarly, contrast method also provides the information about homogeneity on the basis of pixel contrast. It gives quantitative data about the homogeneity of the medium [45]. Fibrillar structures and organization of tissue are determined by orderliness method in which GLCM analysis quanti-

Table 2 Methods for the analysis of nonlinear image in NLO modalities outputs (CARS, TPEF and SHG)

Modality	Target molecules in tumor	Parameters	Modality
ORR (optical redox ratio)	Glucose catabolism versus oxidative phosphorylation	FAD/(FAD + NADH)	[23, 26, 32, 33]
TACS (Tumor associated collagen signature)	Collagen	TACS 1, 2 and 3	[38, 39]
SAAID (second-harmonic to auto-fluorescence ageing index of dermis)	Elastin and collagen	(ISGH − ITPEF)/(ISHG + ITPEF) I = intensity	[35, 37]
FFT (Fourier transform)	Fibers alignment	Aspect ratio (AR)	[40–42]
GLCM (gray level co-occurrence matrix) analysis	Quantitative analysis of nuclear size and collagen, homogeneity in tissue	Pixel brightness values	[43–46]

tatively provide information about the mutual orientation of collagen bundles [12, 37]. GLCM can also measure inverse difference moment (IDM), energy, inertia and entropy in tissue samples [46].

4 Future Direction

Early detection of GC is an urgent need to minimize the high mortality with GC. NLO modalities are emerging as valuable tools for early diagnosis in cancer research. These techniques provide nondestructive, high-resolution imaging of tumor tissue in which different tissue structure and complex biomolecule interaction, as well as the biological system can be visualized and analyzed. Combination of CARS, SHG, and TPEF further enable more precise and composite visualization of biochemicals and structures in affected tissue. This provides more accurate information for therapeutic purposes. However, this is still very challenging to obtain images using NLO microscopy in human. Combination of CARS, SHG, and TPEF are frequently being used however, the utilization in human for diagnosis is in developing process. Till date, the penetration capability of NLO modalities is a major setback. However, combining these NLO modalities would be a valuable platform for the advancement of the ability to non-invasive diagnosis of GC in the near future.

Acknowledgements We are thankful to the Centre for Biosciences and Biomedical Engineering, Indian Institute of Technology Indore. We are also thankful to the Council of Scientific and Industrial Research grant no 37(1693)/17/EMR-II and Department of Science and Technology as Ramanujan fellowship grant no SB/S2/RJN-132/20/5. We appreciate our lab colleagues for insightful discussions and advice.

Conflict of Interest Authors have no conflicts of interest to disclose.

References

1. Ferlay, J., Soerjomataram, I., Dikshit, R., et al.: Cancer incidence and mortality worldwide: sources, methods and major patterns in GLOBOCAN 2012. Int. J. Cancer **136**, E359–E386 (2015). https://doi.org/10.1002/ijc.29210
2. Schneeweiss, S.: Sensitivity analysis of the diagnostic value of endoscopies in cross-sectional studies in the absence of a gold standard. Int. J. Technol. Assess. Health Care **16**, 834–841 (2000)
3. Yoon, H., Kim, N.: Diagnosis and management of high risk group for gastric cancer. Gut Liver **9**, 5–17 (2015). https://doi.org/10.5009/gnl14118
4. Rahman, R., Asombang, A.W., Ibdah, J.A.: Characteristics of gastric cancer in Asia. World J. Gastroenterol. **20**, 4483–4490 (2014). https://doi.org/10.3748/wjg.v20.i16.4483
5. Schmidt, N., Peitz, U., Lippert, H., Malfertheiner, P.: Missing gastric cancer in dyspepsia. Aliment. Pharmacol. Ther. **21**, 813–820 (2005). https://doi.org/10.1111/j.1365-2036.2005.02425.x
6. Maconi, G., Manes, G., Porro, G.-B.: Role of symptoms in diagnosis and outcome of gastric cancer. World J. Gastroenterol. **14**, 1149–1155 (2008). https://doi.org/10.3748/wjg.14.1149
7. Kim, S.J., Cho, Y.S., Moon, S.H., et al.: Primary Tumor 18F-FDG avidity affects the performance of 18F-FDG PET/CT for detecting Gastric Cancer recurrence. J. Nucl. Med. **57**, 544–550 (2016). https://doi.org/10.2967/jnumed.115.163295
8. Schöder, H., Gönen, M.: Screening for cancer with PET and PET/CT: potential and limitations. J. Nucl. Med. **48**, 4–18 (2007)
9. Hallinan, J.T.P.D., Venkatesh, S.K.: Gastric carcinoma: imaging diagnosis, staging and assessment of treatment response. Cancer Imaging **13**, 212–227 (2013). https://doi.org/10.1102/1470-7330.2013.0023
10. Bentley-Hibbert, S., Schwartz, L.: Use of Imaging for GI Cancers. J. Clin. Oncol. **33**, 1729–1736 (2015). https://doi.org/10.1200/JCO.2014.60.2847
11. Balkwill, F.R., Capasso, M., Hagemann, T.: The tumor microenvironment at a glance. J. Cell Sci. **125** (2013)
12. Adur, J., Carvalho, H.F., Cesar, C.L., Casco, V.H.: Nonlinear optical microscopy signal processing strategies in cancer. Cancer Inform **13**, 67–76 (2014). https://doi.org/10.4137/CIN.S12419
13. Kobat, D., Durst, M.E., Nishimura, N., et al.: Deep tissue multiphoton microscopy using longer wavelength excitation. Opt. Express **17**, 13354–13364 (2009). https://doi.org/10.1364/OE.17.013354
14. De Kumar, A., Goswami, D.: Towards controlling molecular motions in fluorescence microscopy and optical trapping: a spatiotemporal approach. Int. Rev. Phys. Chem. **30**, 275–299 (2011). https://doi.org/10.1080/0144235X.2011.603237
15. Huff, T.B., Shi, Y., Fu, Y., et al.: Multimodal nonlinear optical microscopy and applications to central nervous system imaging. IEEE J. Sel. Top. Quantum Electron. **14**, 4–9 (2008). https://doi.org/10.1109/JSTQE.2007.913419

16. Streets, A.M., Li, A., Chen, T., Huang, Y.: Imaging without fluorescence: nonlinear optical microscopy for quantitative cellular imaging. https://doi.org/10.1021/ac5013706

17. Duncan, M.D., Reintjes, J., Manuccia, T.J.: Scanning coherent anti-stokes Raman microscope. Opt. Lett. **7**, 350 (1982). https://doi.org/10.1364/OL.7.000350

18. Adur, J., Carvalho, H.F., Cesar, C.L., Casco, V.H.: Nonlinear microscopy techniques: principles and biomedical applications. Microsc. Anal. (2016). https://doi.org/10.5772/63451

19. Cheng, Ji-Xin, X.: Coherent anti-stokes Raman scattering microscopy: instrumentation, theory, and applications (2003). https://doi.org/10.1021/jp035693v

20. Luo, X., Cheng, C., Tan, Z., et al.: Emerging roles of lipid metabolism in cancer metastasis. Mol. Cancer **16**, 76 (2017). https://doi.org/10.1186/s12943-017-0646-3

21. Potma, E.O., Xie, X.S.: Detection of single lipid bilayers with coherent anti-stokes Raman scattering (CARS) microscopy. J. Raman Spectrosc. **34**, 642–650 (2003). https://doi.org/10.1002/jrs.1045

22. Légaré, F., Evans, C.L., Ganikhanov, F., et al.: Towards CARS endoscopy. Opt. Express **14**, 4427 (2006). https://doi.org/10.1364/OE.14.004427

23. Perry, S.W., Burke, R.M., Brown, E.B.: Two-photon and second harmonic microscopy in clinical and translational cancer research. Ann. Biomed. Eng. **40**, 277–291 (2012). https://doi.org/10.1007/s10439-012-0512-9

24. Benninger, R.K.P., Piston, D.W.: Two-photon excitation microscopy for the study of living cells and tissues. Curr. Protoc. cell Biol. Chapter 4: Unit **4**(11), 1–24 (2013). https://doi.org/10.1002/0471143030.cb0411s59

25. Denk, W., Strickler, J.H., Webb, W.W.: Two-photon laser scanning fluorescence microscopy. Science **248**, 273–276 (1990). https://doi.org/10.1126/science.2321027

26. Hou, J., Wright, H.J., Chan, N., et al.: Correlating two-photon excited fluorescence imaging of breast cancer cellular redox state with seahorse flux analysis of normalized cellular oxygen consumption. https://doi.org/10.1117/1.jbo.21.6.060503

27. Lee, J.W., Kim, E.Y., Yoo, H.M., et al.: Changes of lipid profiles after radical gastrectomy in patients with gastric cancer. Lipids Health Dis. **14**, 21 (2015). https://doi.org/10.1186/s12944-015-0018-1

28. Campagnola, P.J., Dong, C.-Y.: Second harmonic generation microscopy: principles and applications to disease diagnosis. Laser Photon. Rev. **5**, 13–26 (2011). https://doi.org/10.1002/lpor.200910024

29. Campagnola, P.J., Loew, L.M.: Second-harmonic imaging microscopy for visualizing biomolecular arrays in cells, tissues and organisms. Nat. Biotechnol. **21**, 1356–1360 (2003). https://doi.org/10.1038/nbt894

30. Samim, M., Sandkuijl, D., Tretyakov, I., et al.: Differential polarization nonlinear optical microscopy with adaptive optics controlled multiplexed beams. Int. J. Mol. Sci. **14**, 18520–18534 (2013). https://doi.org/10.3390/ijms140918520

31. So, P.T.C., Yew, E.Y.S., Rowlands, C.: High-throughput nonlinear optical microscopy. Biophys. J. **105**, 2641–2654 (2013). https://doi.org/10.1016/j.bpj.2013.08.051

32. Hung, Y.P., Albeck, J.G., Tantama, M., Yellen, G.: Imaging cytosolic NADH-NAD(+) redox state with a genetically encoded fluorescent biosensor. Cell Metab. **14**, 545–554 (2011). https://doi.org/10.1016/j.cmet.2011.08.012

33. Skala, M.C., Riching, K.M., Gendron-Fitzpatrick, A., et al.: In vivo multiphoton microscopy of NADH and FAD redox states, fluorescence lifetimes, and cellular morphology in precancerous epithelia. Proc. Natl. Acad. Sci. USA **104**, 19494–19499 (2007). https://doi.org/10.1073/pnas.0708425104

34. Skala, M., Ramanujam, N.: Multiphoton redox ratio imaging for metabolic monitoring in vivo. Methods Mol. Biol. **594**, 155–162 (2010). https://doi.org/10.1007/978-1-60761-411-1_11

35. Pittet, J.-C., Freis, O., Vazquez-Duchêne, M.-D., et al.: Evaluation of elastin/collagen content in human dermis in-vivo by multiphoton tomography—variation with depth and correlation with aging. Cosmetics **1**, 211–221 (2014). https://doi.org/10.3390/cosmetics1030211

36. Barcus, C.E., O'Leary, K.A., Brockman, J.L., et al.: Elevated collagen-I augments tumor progressive signals, intravasation and metastasis of prolactin-induced estrogen receptor alpha

positive mammary tumor cells. Breast Cancer Res. **19**, 9 (2017). https://doi.org/10.1186/s130 58-017-0801-1

37. Cicchi, R., Kapsokalyvas, D., De Giorgi, V., et al.: Scoring of collagen organization in healthy and diseased human dermis by multiphoton microscopy. J. Biophoton. **3**, 34–43 (2009). https://doi.org/10.1002/jbio.200910062
38. Case, A., Brisson, B.K., Durham, A.C., et al.: Identification of prognostic collagen signatures and potential therapeutic stromal targets in canine mammary gland carcinoma. PLoS ONE **12**, e0180448 (2017). https://doi.org/10.1371/journal.pone.0180448
39. Provenzano, P.P., Inman, D.R., Eliceiri, K.W., et al.: Collagen density promotes mammary tumor initiation and progression. https://doi.org/10.1186/1741-7015-6-11
40. Wang, B.-L., Wang, R., Liu, R.J., et al.: Origin of shape resonance in second-harmonic generation from metallic nanohole arrays. https://doi.org/10.1038/srep02358
41. Ambekar, R., Lau, T.-Y., Walsh, M., et al.: Quantifying collagen structure in breast biopsies using second-harmonic generation imaging. Biomed. Opt. Express. **3**, 2021–2035 (2012). https://doi.org/10.1364/BOE.3.002021
42. Adur, J., Zeitoune, A., Sanchez Salas, K., et al.: Epithelial ovarian cancer diagnosis of second-harmonic generation images: a semiautomatic collagen fibers quantification protocol. Cancer Inform. (2017). https://doi.org/10.1177/1176935117690162
43. Hu, W., Li, H., Wang, C., et al.: Characterization of collagen fibers by means of texture analysis of second harmonic generation images using orientation-dependent gray level co-occurrence matrix method. J. Biomed. Opt. **17**, 26007 (2012). https://doi.org/10.1117/1.JBO.17.2.026007
44. Mostaço-Guidolin, L.B., Ko, A.C.-T., Wang, F., et al.: Collagen morphology and texture analysis: from statistics to classification. https://doi.org/10.1038/srep02190
45. Wu, P.-C., Hsieh, T.-Y., Tsai, Z.-U., Liu, T.-M.: In vivo quantification of the structural changes of collagens in a melanoma microenvironment with second and third harmonic generation microscopy. https://doi.org/10.1038/srep08879
46. Mohanaiah, P., Sathyanarayana, P., Gurukumar, L.: Image texture feature extraction using GLCM approach. Int. J. Sci. Res. Publ. **3**, 2250–3153 (2013)

Reduce the Risk of Dementia; Early Diagnosis of Alzheimer's Disease

Shweta Jakhmola and Hem Chandra Jha

Abstract Alzheimer's Disease (AD) is the most prevalent form of dementia. Till date, no cure for AD has been reported. Multiple therapies have been developed to reduce the progression of the disease. The physiopathology of AD starts before the disease appears. This interval thrills intervention for early diagnosis of AD. However, mechanisms that lead to AD are not revealed yet. Hence, analysis of certain predisposing factors needed to explore. These includes assessment of biomarkers through, genetics, cerebrospinal fluid analysis, imaging techniques, and electrophysiological methods. Early diagnosis would allow maximum benefit of the treatment to the affected people and prepare individuals well in advance to prevent this disorder. Further, it would reduce the mortality rates, economic and social burden caused due to AD. Although, the search for new qualified biomarkers and refinements in the health care techniques are urgently needed. Nevertheless, it would prove to be an important indicator of the disorder.

Keywords Alzheimers disease · Mild cognitive impairment · Biomarkers
Cerebrospinal fluid · Neurodegeneration

1 Introduction

Dr. Alois Alzheimer, a German psychiatrist, and neuropathologist was the first to describe Alzheimer disease (AD) [1]. In his 1906 conference lecture and a subsequent 1907 article, Auguste D case was described by him. Auguste D was a 51-year-old lady with a 'strange cerebral cortex disease'. The lady presented with language impairment with progressive memory deterioration, disorientation, unusual

S. Jakhmola · H. C. Jha (✉)
Centre for Biosciences and Biomedical Engineering, Indian Institute of Technology Indore,
Indore 453552, India
e-mail: hemcjha@iiti.ac.in

S. Jakhmola
e-mail: phd1601271004@iiti.ac.in

© Springer Nature Singapore Pte Ltd. 2019
M. Tanveer and R. B. Pachori (eds.), *Machine Intelligence and Signal Analysis*,
Advances in Intelligent Systems and Computing 748,
https://doi.org/10.1007/978-981-13-0923-6_53

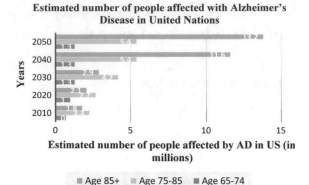

Estimated number of people affected with Alzheimer's Disease in United Nations

Fig. 1 Number of people estimated to be affected by AD by 2050 in the US. According to the study by Hebert et al. done in 2000, there would be a continuous rise in number of people with AD. It is estimated that individuals in the age group of 85+ years would be affected the most (13.2 million people) followed by persons in the age group of 75–85 years (5.4 million people) by 2050

behavioral symptoms (paranoia, hallucinations, delusions), and further psychological impairment [1]. Notably, most of the clinical observations and pathological findings that Alzheimer described at his time continue to remain central to the understanding of AD today [1]. AD damages the neurons, and the abnormal neuronal network affect processes underlying memory, learning, and other cognitive activities. It is believed that the primary cause of AD is the enhanced production or amassing of sticky beta-amyloid (Aβ) and tau protein, in the brain that leads to nerve cell death [2]. Moreover, overstimulation of receptors for excitatory neurotransmitter glutamate could contribute to neurodegeneration in AD [3]. Many studies have suggested that the changes associated with AD start early before the symptoms appear [4, 5]. A group study by Flicker et al. suggested that 70% of people with language disability, not associated with age, develop dementia within a period of 4 years [6]. Initially, the damages are replenished by the brain; this capacity gradually diminishes with time. This decline results in cognitive impairments.

With gradually increasing incidents of AD, there is a urgent need for an early diagnosis [7]. AD is one of the major issues of developing countries like India. According to a report India has third largest population of dementia patients, 4.1 million [8]. AD is reported to constitute 60% of total dementia cases. Additionally, a study suggests that South India is more prone to dementia than other regions [9].

According to a study by Hebert Le et al., it is estimated that over 13.8 million people would be affected by AD [7] (Fig. 1). Also, since the medical treatment of AD is only effective in initial stages and for patients having mild cognitive impairment (MCI), there appears the need for early diagnosis. A number of imaging techniques exist that help in detecting early AD (Table 1) [10]. Apart from these imaging techniques concentrations of Aβ42 are also taken into account as they appear from abnormal β-amyloid metabolism [11]. Moreover, detection of high levels of hyperphosphorylated Tau signifies neurodegeneration [11].

Table 1 Various imaging techniques available for MCI brain analysis

Imaging techniques	Subtype of the imaging techniques	Interpretations of the results obtained after analysis
(1) Structural imaging	(1.1) MRI (Magnetic Resonance Imaging)	Analyses reduced hippocampal volume and temporal lobe atrophy, representing tissue loss and neurodegeneration
	(1.2) CT (Computed Tomography)	Analyses medial temporal lobe and hippocampal atrophy
(2) Functional imaging	(2.1) FDG-PET (^{18}F-fluorodeoxyglucose-PET)	Measures reduced glucose utilization: hypometabolism and neurodegeneration of posterior cingulate-precuneus and temporoparietal cortex
	(2.2) fMRI (functional MRI)	Displays the pattern in glucose metabolism
(3) Molecular imaging	(3.1) PET-PIB (^{11}C-PiB and fluorinated tracers for amyloid-PET)	Analyses accumulation of β-amyloid in the cortical region
	(3.2) SPECT (Single-photon emission computed tomography)	Determines lower Cerebral blood flow (CBF) representing cognitive decline

Causes

What causes AD is still not fully revealed, however, this is certainly due to a number of interlinked series of events in the brain over an extended period of time. Likely causes includes genetic, infections, environmental and lifestyle factors. Out of these most likely causes, some can serve as early detection markers.

2 Early Detection Markers

2.1 Genetic Susceptibility

2.1.1 Possession of APOE4

Apolipoprotein (apo) E4 is linked to increased risk of late-onset (>60 years) sporadic and familial AD. ApoE has three isoforms, apoE2, apoE3, and apoE4. The isoforms differ by single amino acid substitution in humans. The risk for AD is 50% in homozygous APOE4 individuals, and for APOE3 and APOE4 heterozygotes are 20–30% [12]. In the brain, the primary source of ApoE is astrocytes. The expression of apoE4 is increased during aging in astrocytes [13].

After neuronal degeneration, ApoE takes up the lipids generated and redistributes it to the lipid requiring cells for membrane repair, remyelination, or proliferation. ApoE promotes Aβ deposition by depressing Aβ clearance. Due to stress response, ApoE4 degrades to neurotoxic fragments to a much greater extent than its other isoforms. These particles disintegrate the cytoskeleton and also hamper the mitochondrial functions. Glutamate receptor functions and synaptic plasticity are also regulated by ApoE [14]. Notably, the GABAergic interneurons are inclined more to apoE4 fragment toxicity, resulting in learning and memory impairments [15]. Decreased somatostatin and GABA levels are observed in the brain and CSF of apoE4 carriers [15]. Impairments of GABAergic interneurons may contribute to AD pathogenesis. More than 20 genetic loci have been found to be associated with increased AD risk. These genes are the ones involved in the lipid cholesterol metabolic and endosomal vesicle recycling pathway, immune system regulation, and inflammatory responses [16]. These polymorphisms contribute to some extent to the risk for AD. Repressor element 1-silencing transcription factor (REST) gene controls cell survival and is usually lost from the nucleus in mild cognitive impaired (MCI) brains [17]. Chromatin immunoprecipitation with deep sequencing (ChIP-Seq) studies have demonstrated that REST gene suppresses the genes responsible for AD pathology. REST protects neuronal cells from Aβ toxicity and oxidative damages [17]. Activation status of REST may be a potential signature of neurodegeneration. Genome analysis techniques can be useful in identification of such mutations.

2.1.2   Formation of Aβ plaques

Several studies suggest that the synaptic functions are depressed by Aβ oligomers, and it also influences various signaling pathways [18]. It interferes with the neuronal activities and stimulates the glial cells to release neurotoxic mediators. Mutations in genes—amyloid precursor protein (APP), presenilin (PS)-2 and PS-1 promote the formation of Aβ aggregates in the brain, which is responsible for early-onset (<60 years) autosomal dominant AD [19]. APP gene resides on chromosome 21. APP processing is affected by mutations thereby producing altered Aβ oligomers. These are the main constituent of amyloid plaques. Chances of early-onset dementia with pathological hallmarks of AD in the brains are higher in Down's syndrome (trisomy of chromosome 21) patients [20]. Alone the duplication of APP gene leads to early-onset AD. PS1 and PS2 encode for the g-secretase protein catalytic subunit that processes Aβ peptides. Aβ induced toxicity include disruption of intracellular calcium homeostasis, excitatory effects, i.e., change in the activity of the neurotransmitter receptors and related signaling molecules and impairments of axonal transport and mitochondrial functions [21]. Further, a significant reduction in glutamine-glutamate levels is observed in mild cognitive impaired brains [22]. A number of other genes have also been identified that affect and influence the occurrence of late-onset AD, including CLU, PICALM, BIN1, HLA-DRB1/5, SORL1, GAB2, EPHA1, ABCA7, MS4A4/MS4A6E, CD2AP, CR1, and CD33 [16]. However, the negligible effect is seen when compared to apoE4 in AD pathogenesis.

2.1.3 Tau Pathologies

Neurofibrillary tangles (NFTs), present intracellularly, are made up of primarily of aggregates of abnormally modified tau proteins [23]. These modifications include excessive phosphorylation and acetylation [24]. ApoE4 enhances accumulation of phosphorylated tau protein in the neuronal soma and dendrites. These fragments in the cytosol cause mitochondrial impairment and tau pathology [25]. NFTs become enriched in dendritic spines where it interferes with neurotransmission. Aβ oligomers incite the postsynaptic enrichment of the tau protein through a pathway involving the members of the microtubule affinity-regulating kinase (MARK) family [26], Several mutations in presenilins and APP follow in Aβ accumulation with amyloid plaques and NFTs [20]. However, the formation of amyloid plaques nor AD results due to mutations in the gene that encodes tau, MAPT. Mutations in MAPT instead cause frontotemporal lobar degeneration.

2.1.4 Accumulation of α-Synuclein

In many cases of AD, higher concentrations of RNA binding protein TDP-43 and presynaptic protein, α-synuclein is seen in the brain [27]. However, the altered forms of these proteins rather lead to other neurodegenerative disorders. It has been studied that in vivo and in vitro the amounts of misfolded α-synuclein is enhanced by Aβ oligomers [28]. Also, the TDP-43 protein is increased due to ApoE4 fragments in the neurons.

Early diagnosis of Alzheimer's disease: contribution of structural neuroimaging
MRI recognizes certain markers in the brain that comprise a part of early AD diagnosis. These markers include the detection of MCI, by identifying hippocampal and entorhinal cortex atrophy, declined temporal neocortex volume [29] and amygdala atrophy [30]. Although, the data collection by multiple studies has drawbacks, like sample size, variation in use of MRI technique. Still, these studies coincide with postmortem results that represent AD progression [2].

2.2 Mild Cognitive Impairment (MCI)

MCI is an early phase of dementia. It accounts for a condition in which the individual shows mild thinking disabilities without affecting the day to day functioning [2]. It is well documented that people possessing MCI are more prone to AD when compared to people without MCI [31]. Diagnosis of AD with revised guidelines establishes the identification of MCI as early detection [32]. However, people with MCI do not end up in AD always, MCI may regress to normal or may become stable.

2.3 Morphological Analysis—Hippocampal and Medial Temporal lobe Atrophy by MRI/CT

Alzheimer's greatly affects memory by affecting the hippocampus and the nearby structures [29]. People with MCI display histopathological hippocampal changes displaying features already similar to AD brain. 70% of these people end up having AD in a time frame of 4 years [11].

Medial Temporal lobe Atrophy and hippocampal atrophy occurs early in Alzheimer's disease which is of clinical importance in probable AD diagnosis. A study reported 14% reduction in hippocampal volume for MCI group when compared to healthy individuals [11]. In early stages to conversion to AD, atrophy was identified in the precuneus. Further, it gradually expands to the posterior parietal and frontal regions [33]. Posterior parts are more vulnerable than the anterior parts. A reliable marker for AD can, therefore, be early atrophy in posterior cortical regions. AD-related damage includes reduction volume of cornu ammonis by 40%, entorhinal cortex by 60%, and subicular complex by 40% [11]. CT [34] and MRI focuses on these damaged areas. It can be analyzed and quantified easily using plain MRI films [35]. CT images display widened hippocampal fissures [34]. However, temporal lobe imaging is hampered because of bone hardening. MRI stands better when imaging the temporal lobe than CT [35]. However, hippocampal changes are also seen in the case of normal healthy individuals.

2.4 CSF Biomarkers—Amyloid β_{1-42} ($A\beta_{42}$), Phosphorylated Tau (P-tau181) and Total Tau (T-tau)

A direct relation occurs between amyloid deposition and hypometabolism in the posterior cortical regions. Hypometabolism is associated with increasing temporoparietal and retrosplenial cortex atrophy which was identified using [18F]-2-fluoro-2-deoxy-D-glucose positron emission tomography (FDG-PET) [36] and fMRI [37].

The $A\beta$ deposition was analyzed using amyloid-imaging tracer [11C] Pittsburgh Compound-B (PIB) [38]. In vivo analysis of amyloid plaques by PIB correlate with atrophy in AD [4]. Amyloid β_{1-42} ($A\beta_{42}$), total tau (T-tau), and phosphorylated tau (P-tau181) are demonstrated in the CSF of people with MCI (Fig. 2) [39]. Results of the survey revealed that 42% of the people advanced to AD, 21% to other forms of dementia, while 41% remained stable [39]. The study proved that the risk for AD increases greatly in MCI people with concentrations of $A\beta_{42}$ and T-tau at the baseline (hazard ratio 17.7, p < 0.0001), in the experiment [39]. This method of detection of $A\beta_{42}$, and total tau (T-tau) successfully predicted probable AD. One model for studying the disorder in mice is 3xTg-AD mice. These mice display cognitive dysfunction, synaptic plasticity, tau pathology, plaque development similar to AD [40]. Accumulation of $A\beta$ plaques and tau in amygdala and hippocampus relate to early signs and symptoms of AD [30]. International criteria include hippocampal atrophy

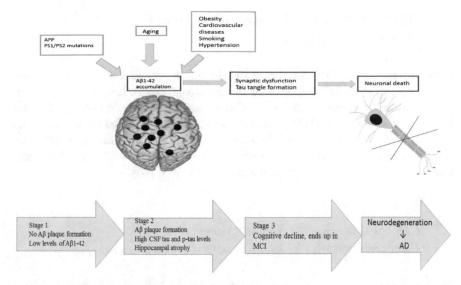

Fig. 2 Factors affecting the accumulation of Aβ1-42 and stages leading to neurodegenerative AD. Multiple factors contribute to Aβ accumulation ranging from cardiovascular disorders, hypertension, obesity to genetics like mutation in APP, PS1/PS2 gene. Aβ plaque formation begins much before the actual symptoms of AD, the levels of phosphorylated tau also increase along with hippocampal atropy. These changes are possible to be diagnosed by MRI, PET and CT. Subsequent MCI begins with more profound symptoms of AD. Finally neuronal damage is observed. At this stage the diseased person becomes completely reliant on others for care and help to perform even the basic tasks of life

along with positive CSF biomarker as a signature for AD pathology. Further confirmation is obtained from MRI, FDG-PET, and amyloid-PET [41]. Myoinositol (mIns) is another biomarker which detected by 1 H-MRS. This is a proposed biomarker in early detection of Alzheimer's disease [42].

2.5 Cerebral Blood Flow (CBF) Analysis

Cerebral blood flow (CBF) can be used in the prognostic element in cognitive decline measured by arterial spin labeling (ASL) [43]. Synaptic failures are signified by reduced CBF; this further reduces throughout AD [44]. Single-photon emission computed tomography (SPECT), determines the lower baseline CBF [45]. Analysis of posterior CBF, which is also related to glucose metabolism of the brain, gives information about AD progression [44].

2.6   Diffusion-Weighted Imaging (DWI) and DTI Diffusion Tensor Imaging (DTI)

DWI measures water diffusion based on Brownian motion. Quantitatively water diffusion is expressed as ADC (apparent diffusion coefficient). The variation in signal obtained from restricted movement is a characteristic of cerebral abscess [46]. Loss in volume of white matter (WM), a sign of early neurodegeneration can be quantified by DTI. WM integrity loss and injury are some of the information obtained from DTI measures. DTI reports of AD patients demonstrate enhanced radial, axial and mean diffusivity and lower fractional anisotropy in WM [46].

2.7   Electrophysiological Technique—Electro Encephalograph (EEG), Event-Related Potential (ERP)

Abnormalities in posterior regions can be diagnosed by EEG which is also associated with AD progression. Event-related potentials (ERPs) can be used as an early diagnostic tool in AD [47]. In the absence of such specific biomarker for predementia, this method allows better chances for early AD detection. This electrophysiological technique described by Hansenne et al., measures the alterations in brain functions due to the motor and cognitive tasks [48]. ERPs are obtained from responses of individuals like answering, perception, decision-making, also considering attention, language and memory process. Two most extensively recorded ERP include P3 and N2. P3 is a positive wave preceded by N2 which is a negative wave. P3 emanates from parietal area followed by a stimulus with a lag of 300 ms. P3 is further subcategorized in P3a and P3b. P3a is associated with attention whereas P3b to memory [49]. N2 reflects the attentive responses to a stimulus. These ERPs are altered in case of AD.

2.8   Memory Impairment as an Early Sign for Probable AD

Studies suggest that there is a correlation between the default cerebral activity in early adulthood and the risk for AD [50]. Cerebral functioning of specific brain areas predispose these regions to changes observed in Alzheimer's. These changes include atrophy, the formation of amyloid plaques and metabolic dysfunction. Supporting the notion is the study by Buckner et al. that shows memory being a prominent component of the default mode is impaired early in AD [50]. These default mechanisms may trigger cascade responsible for AD progression.

2.9 Other

A well-established reason for dementia is an improper supply of oxygen to the brain. The enhanced risk of cardiovascular problems, obesity [51], hypertension [52], diabetes [53], result in elevated risk for AD (Fig. 2). Contrary to these facts keeping the heart healthy may reduce the risk of probable dementia, AD.

3 Conclusion

Individuals with genetic risk may exhibit the symptoms of AD much aforetime than actual disease appearance. Numerous cognitive and genetic polymorphism studies demonstrated that the alterations in young adults could signify the development of AD. Early phases of AD may affect medial temporal lobe and result in tau pathology. Mostly affected regions include the posterior cortical area. With time the volume reduction extends to lateral and medial temporal lobe. Likewise, dilated transverse fissures. Further, Aβ deposition is correlated to posterior cortical atrophy. Moreover, patterns in glucose metabolism strongly overlap with atrophy. According to the National Institute on Aging and the Alzheimer's Association (NIA-AA), the presence of Aβ detected by amyloid-PET or CSF in addition to medial temporal atrophy or increased amounts of CSF tau and p-tau detect probable AD.

Moreover, AD enhances the burden on society and reduces the productive workforce, as the affected person completely becomes reliant on others for care and daily tasks. Hence, early diagnosis of the disorder is critical.

Acknowledgements This project was supported by Council of Scientific and Industrial Research grant no 37(1693)/17/EMR-II and Department of Science and Technology as Ramanujan fellowship grant no SB/S2/RJN-132/20/5. We are thankful to Ministry of Human Resource and Development for fellowship to Shweta Jakhmola. We also thank Indian Institute of Technology Indore, for providing necessary resources. The funding organization has not played any role in study design, the decision to publish, or the preparation of the manuscript. We appreciate our lab colleagues for insightful discussions and advices.

References

1. Maurer, K., Volk, S., Gerbaldo, H., et al.: Auguste D and Alzheimer's disease, vol. 349, pp. 1546–1549. Lancet, London, England (1997)
2. Alzheimer, Association, science staff, alz org: 2017 Alzheimer's disease facts and figures (2017)
3. Bell, K.F.S., de Kort, G.J.L., Steggerda, S., Shigemoto, R., Ribeiro-da-Silva, A., Cuello, A.C.: Structural involvement of the glutamatergic presynaptic boutons in a transgenic mouse model expressing early onset amyloid pathology. Neurosci. Lett. **353**, 143–147 (2003)

4. Jack, C.R., Petersen, R.C., Xu, Y., O'Brien, P.C., Smith, G.E., Ivnik, R.J., Boeve, B.F., Tangalos, E.G., Kokmen, E.: Rates of hippocampal atrophy correlate with change in clinical status in aging and AD. Neurology 55, 484–489 (2000)

5. Villemagne, V.L., Burnham, S., Bourgeat, P., Brown, B., Ellis, K.A., Salvado, O., Szoeke, C., Macaulay, S.L., Martins, R., Maruff, P., Ames, D., Rowe, C.C., Masters, C.L.: Amyloid β deposition, neurodegeneration, and cognitive decline in sporadic Alzheimer's disease: a prospective cohort study. Lancet Neurol. 12, 357–367 (2013)

6. Flicker, C., Ferris, S.H., Crook, T., Bartus, R.T.: Implications of memory and language dysfunction in the naming deficit of senile dementia. Brain Lang. 31, 187–200 (1987)

7. Hebert, L.E., Weuve, J., Scherr, P.A., Evans, D.A.: Alzheimer disease in the United States (2010–2050) estimated using the 2010 census. Neurology 80, 1778–1783 (2013)

8. Chakraborty, P.: Mercury exposure and Alzheimer's disease in India—an imminent threat? Sci. Total Environ. 589, 232–235 (2017)

9. Mathuranath, P.S., George, A., Ranjith, N., Justus, S., Kumar, M.S., Menon, R., Sarma, P.S., Verghese, J.: Incidence of Alzheimer's disease in India: a 10 years follow-up study. Neurol. India 60, 625–630 (2012)

10. Chetelat, G. aë., Baron, J.-C.: Early diagnosis of alzheimer's disease: contribution of structural neuroimaging. Neuroimage 18, 525–541 (2003)

11. de Leon, M.J., Convit, A., DeSanti, S., Bobinski, M., George, A.E., Wisniewski, H.M., Rusinek, H., Carroll, R., Louis, L.A.: Saint: contribution of structural neuroimaging to the early diagnosis of Alzheimers disease. Int. Psychogeriatr. 9, 183–190 (1997)

12. Genin, E., Hannequin, D., Wallon, D., Sleegers, K., Hiltunen, M., Combarros, O., Bullido, M.J., Engelborghs, S., De Deyn, P., Berr, C., Pasquier, F., Dubois, B., Tognoni, G., Fiévet, N., Brouwers, N., Bettens, K., Arosio, B., Coto, E., Del Zompo, M., Mateo, I., Epelbaum, J., Frank-Garcia, A., Helisalmi, S., Porcellini, E., Pilotto, A., Forti, P., Ferri, R., Scarpini, E., Siciliano, G., Solfrizzi, V., Sorbi, S., Spalletta, G., Valdivieso, F., Vepsäläinen, S., Alvarez, V., Bosco, P., Mancuso, M., Panza, F., Nacmias, B., Bossù, P., Hanon, O., Piccardi, P., Annoni, G., Seripa, D., Galimberti, D., Licastro, F., Soininen, H., Dartigues, J.-F., Kamboh, M.I., Van Broeckhoven, C., Lambert, J.C., Amouyel, P., Campion, D.: APOE and Alzheimer disease: a major gene with semi-dominant inheritance. Mol. Psychiatry. 16, 903–907 (2011)

13. Vuletic, S., Peskind, E.R., Marcovina, S.M., Quinn, J.F., Cheung, M.C., Kennedy, H., Kaye, J.A., Jin, L.-W., Albers, J.J.: Reduced CSF PLTP activity in Alzheimer's disease and other neurologic diseases; PLTP induces ApoE secretion in primary human astrocytes in vitro. J. Neurosci. Res. 80, 406–413 (2005)

14. Chen, Y., Durakoglugil, M.S., Xian, X., Herz, J.: ApoE4 reduces glutamate receptor function and synaptic plasticity by selectively impairing ApoE receptor recycling. Proc. Natl. Acad. Sci. USA 107, 12011–12016 (2010)

15. Li, Y., Sun, H., Chen, Z., Xu, H., Bu, G., Zheng, H.: Implications of GABAergic neurotransmission in Alzheimer's disease. Front. Aging Neurosci. 8, 31 (2016)

16. Guerreiro, R., Hardy, J.: Genetics of Alzheimer's disease. Neurotherapeutics 11, 732–737 (2014)

17. Lu, T., Aron, L., Zullo, J., Pan, Y., Kim, H., Chen, Y., Yang, T.-H., Kim, H.-M., Drake, D., Liu, X.S., Bennett, D.A., Colaiácovo, M.P., Yankner, B.A.: REST and stress resistance in ageing and Alzheimer's disease. Nature 507, 448–454 (2014)

18. Hoe, H.-S., Harris, D.C., Rebeck, G.W.: Multiple pathways of apolipoprotein E signaling in primary neurons. J. Neurochem. 93, 145–155 (2005)

19. Shen, J., Kelleher, R.J.: The presenilin hypothesis of Alzheimer's disease: evidence for a loss-of-function pathogenic mechanism. Proc. Natl. Acad. Sci. USA 104, 403–409 (2007)

20. Millan Sanchez, M., Heyn, S.N., Das, D., Moghadam, S., Martin, K.J., Salehi, A., et al.: Neurobiological elements of cognitive dysfunction in down syndrome: exploring the role of app. Biol. Psychiatry. 71, 403–409 (2012)

21. Huang, Y., Mucke, L.: Alzheimer mechanisms and therapeutic strategies. Cell 148, 1204–1222 (2012)

22. Huang, D., Liu, D., Yin, J., Qian, T., Shrestha, S., Ni, H.: Glutamate-glutamine and GABA in brain of normal aged and patients with cognitive impairment. Eur. Radiol. **27**, 2698–2705 (2017)
23. Zhang, Y., Hong, X., Zhang, J., Wang, J.-Z., Liu, G.: Role of Microtubule-associated protein tau Phosphorylation in Alzheimer's disease. J Huazhong Univ. Sci. Technol. [Med Sci] **37**
24. Morris, M., Maeda, S., Vossel, K., Mucke, L.: The many faces of tau. Neuron **70**, 410–426 (2011)
25. Silva, D.F.F., Esteves, A.R., Oliveira, C.R., Cardoso, S.M.: Mitochondria: the common upstream driver of amyloid-β and tau pathology in Alzheimers disease. Curr. Alzheimer Res. **8**, 563–572 (2011)
26. Zempel, H., Thies, E., Mandelkow, E., Mandelkow, E.-M.: A oligomers cause localized $Ca2^+$ elevation, missorting of endogenous tau into dendrites, tau phosphorylation, and destruction of microtubules and spines. J. Neurosci. **30**, 11938–11950 (2010)
27. Gitcho, M.A., Baloh, R.H., Chakraverty, S., Mayo, K., Norton, J.B., Levitch, D., Hatanpaa, K.J., White, C.L., Bigio, E.H., Caselli, R., Baker, M., Al-Lozi, M.T., Morris, J.C., Pestronk, A., Rademakers, R., Goate, A.M., Cairns, N.J.: TDP-43 A315T mutation in familial motor neuron disease. Ann. Neurol. **63**, 535–538 (2008)
28. Masliah, E., Rockenstein, E., Veinbergs, I., Sagara, Y., Mallory, M., Hashimoto, M., Mucke, L.: Amyloid peptides enhance—synuclein accumulation and neuronal deficits in a transgenic mouse model linking Alzheimer's disease and Parkinson's disease. Proc. Natl. Acad. Sci. **98**, 12245–12250 (2001)
29. Frisoni, G.B., Laakso, M.P., Beltramello, A., Geroldi, C., Bianchetti, A., Soininen, H., Trabucchi, M.: Hippocampal and entorhinal cortex atrophy in frontotemporal dementia and Alzheimer's disease. Neurology **52**, 91–100 (1999)
30. Lehericy, S., Baulac, M., Chiras, J., Pierot, L., Martin, N., Pillon, B., Deweer, B., Dubois, B., Marsault, C.: Amygdalohippocampal MR Volume Measurements in the Early Stages of Alzheimer Disease
31. Kantarci, K., Weigand, S.D., Przybelski, S.A., Shiung, M.M., Whitwell, J.L., Negash, S., Knopman, D.S., Boeve, B.F., O'Brien, P.C., Petersen, R.C., Jack, C.R.: Risk of dementia in MCI: combined effect of cerebrovascular disease, volumetric MRI, and 1H MRS. Neurology. **72**, 1519–1525 (2009)
32. Sperling, R.A., Aisen, P.S., Beckett, L.A., Bennett, D.A., Craft, S., Fagan, A.M., Iwatsubo, T., Jack, C.R., Kaye, J., Montine, T.J., Park, D.C., Reiman, E.M., Rowe, C.C., Siemers, E., Stern, Y., Yaffe, K., Carrillo, M.C., Thies, B., Morrison-Bogorad, M., Wagster, M.V., Phelps, C.H.: Toward defining the preclinical stages of Alzheimer's disease: recommendations from the National Institute on Aging-Alzheimer's Association workgroups on diagnostic guidelines for Alzheimer's disease. Alzheimer's Dement. **7**, 280–292 (2011)
33. Scahill, R.I., Schott, J.M., Stevens, J.M., Rossor, M.N., Fox, N.C.: Mapping the evolution of regional atrophy in Alzheimer's disease: unbiased analysis of fluid-registered serial MRI. Proc. Natl. Acad. Sci. U. S. A **99**, 4703–4707 (2002)
34. De Leon, M., George, A., Stylopoulos, L., Smith, G., Miller, D.: Early marker for Alzheimer's disease: the atropic hippocampus. Lancet **334**, 672–673 (1989)
35. Scheltens, P., Leys, D., Barkhof, F., Huglo, D., Weinstein, H.C., Vermersch, P., Kuiper, M., Steinling, M., Wolters, E.C., Valk, J.: Atrophy of medial temporal lobes on MRI in "probable" Alzheimer's disease and normal ageing: diagnostic value and neuropsychological correlates. J. Neurol. Neurosurg. Psychiatry **55**, 967–972 (1992)
36. Herholz, K.: FDG PET and differential diagnosis of dementia. Alzheimer Dis. Assoc. Disord. **9**, 6–16 (1995)
37. McKiernan, K.A., Kaufman, J.N., Kucera-Thompson, J., Binder, J.R.: A parametric manipulation of factors affecting task-induced deactivation in functional neuroimaging. J. Cogn. Neurosci. **15**, 394–408 (2003)
38. Jack, C.R., Lowe, V.J., Weigand, S.D., Wiste, H.J., Senjem, M.L., Knopman, D.S., Shiung, M.M., Gunter, J.L., Boeve, B.F., Kemp, B.J., Weiner, M., Petersen, R.C.: Serial PIB and MRI in normal, mild cognitive impairment and Alzheimer's disease: implications for sequence of pathological events in Alzheimer's disease. Brain **132**, 1355–1365 (2009)

39. Hansson, O., Zetterberg, H., Buchhave, P., Londos, E., Blennow, K., Minthon, L.: Association between CSF biomarkers and incipient Alzheimer's disease in patients with mild cognitive impairment: a follow-up study. Lancet Neurol. **5**, 228–234 (2006)

40. Billings, L.M., Oddo, S., Green, K.N.: Intraneuronal A^N_L causes the onset of early Alzheimer's disease-related cognitive deficits in transgenic mice. Neuron **45**, 675–688 (2005)

41. Frisoni, G.B., Boccardi, M., Barkhof, F., Blennow, K., Cappa, S., Chiotis, K., Démonet, J.-F., Garibotto, V., Giannakopoulos, P., Gietl, A., Hansson, O., Herholz, K., Jack, C.R., Nobili, F., Nordberg, A., Snyder, H.M., Ten Kate, M., Varrone, A., Albanese, E., Becker, S., Bossuyt, P., Carrillo, M.C., Cerami, C., Dubois, B., Gallo, V., Giacobini, E., Gold, G., Hurst, S., Lönneborg, A., Lovblad, K.-O., Mattsson, N., Molinuevo, J.-L., Monsch, A.U., Mosimann, U., Padovani, A., Picco, A., Porteri, C., Ratib, O., Saint-Aubert, L., Scerri, C., Scheltens, P., Schott, J.M., Sonni, I., Teipel, S., Vineis, P., Visser, P.J., Yasui, Y., Winblad, B.: Strategic roadmap for an early diagnosis of Alzheimer's disease based on biomarkers. Lancet Neurol. **16**, 661–676 (2017)

42. Liang, S., Huang, J., Liu, W., Jin, H., Li, L., Zhang, X., Nie, B., Lin, R., Tao, J., Zhao, S., Shan, B., Chen, L.: Magnetic resonance spectroscopy analysis of neurochemical changes in the atrophic hippocampus of APP/PS1 transgenic mice. Behav. Brain Res. **335**, 26–31 (2017)

43. Fällmar, D., Haller, S., Lilja, J., Danfors, T., Kilander, L., Tolboom, N., Egger, K., Kellner, E., Croon, P.M., Verfaillie, S.C.J., van Berckel, B.N.M., Ossenkoppele, R., Barkhof, F., Larsson, E.-M.: Arterial spin labeling-based Z-maps have high specificity and positive predictive value for neurodegenerative dementia compared to FDG-PET. Eur. Radiol. **27**, 4237–4246 (2017)

44. Musiek, E.S., Chen, Y., Korczykowski, M., Saboury, B., Martinez, P.M., Reddin, J.S., Alavi, A., Kimberg, D.Y., Wolk, D.A., Julin, P., Newberg, A.B., Arnold, S.E., Detre, J.A.: Direct comparison of fluorodeoxyglucose positron emission tomography and arterial spin labeling magnetic resonance imaging in Alzheimer's disease. Alzheimer's Dement. **8**, 51–59 (2012)

45. Hanyu, H., Sato, T., Hirao, K., Kanetaka, H., Iwamoto, T., Koizumi, K.: The progression of cognitive deterioration and regional cerebral blood flow patterns in Alzheimer's disease: a longitudinal SPECT study. J. Neurol. Sci. **290**, 96–101 (2010)

46. Hespel, A.-M., Cole, R.C.: Advances in high-field MRI. Vet. Clin. North Am. Small Animal, Pract (2017)

47. Bennys, K., Portet, F., Touchon, J., Rondouin, G.: Diagnostic value of event-related evoked potentials N200 and P300 subcomponents in early diagnosis of Alzheimer's disease and mild cognitive impairment. J. Clin. Neurophysiol. **24**, 405–412 (2007)

48. Hansenne, M.: The p300 cognitive event-related potential. I. Theoretical and psychobiologic perspectives. Neurophysiol. Clin. **30**, 191–210 (2000)

49. Polich, J., Criado, J.R.: Neuropsychology and neuropharmacology of P3a and P3b. Int. J. Psychophysiol. **60**, 172–185 (2006)

50. Buckner, R.L., Snyder, A.Z., Shannon, B.J., LaRossa, G., Sachs, R., Fotenos, A.F., Sheline, Y.I., Klunk, W.E., Mathis, C.A., Morris, J.C., Mintun, M.A.: Molecular, structural, and functional characterization of Alzheimer's disease: evidence for a relationship between default activity, amyloid, and memory. J. Neurosci. **25** (2005)

51. Loef, M., Walach, H.: Midlife obesity and dementia: meta-analysis and adjusted forecast of dementia prevalence in the united states and china. Obesity **21**, E51–E55 (2013)

52. Debette, S., Seshadri, S., Beiser, A., Au, R., Himali, J.J., Palumbo, C., Wolf, P.A., DeCarli, C.: Midlife vascular risk factor exposure accelerates structural brain aging and cognitive decline. Neurology **77**, 461–468 (2011)

53. Gudala, K., Bansal, D., Schifano, F., Bhansali, A.: Diabetes mellitus and risk of dementia: a meta-analysis of prospective observational studies. J. Diabetes Investig. **4**, 640–650 (2013)

Diagnosis of Tumorigenesis and Cancer

Charu Sonkar and Hem Chandra Jha

Abstract Detection of cancer is crucial as higher mortality due to cancer became a social burden on a day-to-day life. Still, early detection of cancer remains the utmost importance to increase survival rate and decrease severe economic loss. This study is intended to give updated information about cancer, detection, and its epidemiology. These techniques are used for cancer determination along with their limitations and advantages. Several new techniques are also discussed with its applications to give a brief overview of recent techniques.

1 Introduction

1.1 Cancer and Tumorigenesis

Cancer is a disease where cells abnormally proliferate and divide uncontrollably as well as has a property of invading other cells [1]. As defined clinically, it is a "set of disease characterized by uncontrolled cell growth and leading to metastasis." At least 80% of all human cancer is a contribution of environmental carcinogens and is also affected by chemical carcinogen.

This kind of exposure leads to damage which may result in phenotypic changes, as well as mutation. The body can react to this mutation in three ways (a) cells can repair it, (b) cells could die, and (c) cell can have a permanent mutation which can lead to cancer.

C. Sonkar · H. C. Jha (✉)
Centre for Biosciences and Biomedical Engineering, Indian Institute of Technology Indore, Indore 453552, Madhya Pradesh, India
e-mail: hemcjha@iiti.ac.in

C. Sonkar
e-mail: charusonkar@gmail.com

© Springer Nature Singapore Pte Ltd. 2019
M. Tanveer and R. B. Pachori (eds.), *Machine Intelligence and Signal Analysis*,
Advances in Intelligent Systems and Computing 748,
https://doi.org/10.1007/978-981-13-0923-6_54

633

1.2   Neoplasm, Tumor, and Cancer

Due to variable use of definitions, there is always misperception between terms neoplasm, tumour, and cancer. Neoplasm is normal cells with the property of hyperproliferation and high metastasis rate [2]. Pathology of many neoplasms is described by Bichat (1771–1802) and suggested that formation of cancer is "accidental formation" of tissue, which is formed like any other cells. However, later it was found that cancer does have hyperproliferation with minimal internal and external inhibitory signals.

The body has several cell regulatory mechanisms and disruption in this regulation may cause dysregulation of the cell cycle. In cancer, the cell can grow continuously even if not required it may form abnormal masses of tissue known as a tumor [1]. This tumor can be localized and metastasized. There are two types of tumor, namely benign tumor which is localized, and malignant tumor which can metastasize mostly forming cancer.

Cancer usually occurs in somatic cells; however, they can have familial content further, it can produce recessive disorder in offsprings like Xeroderma pigmentosa. Cancer is generally the accumulation of mutations, especially in proto-oncogenes and tumor suppressor genes. 95% of cancers are carcinoma and they mostly appear in epithelial cells.

2   Epidemiology

Epidemiology is the study of determinants of health-related groups and their distribution among a specific population. Unlike clinical medicine, it also focuses on those people who do not have the disease rather than targeting on only the diseased person.

In 2012, 14 million new cases were associated with cancer and about 8.8 million died in the world. Hence, it became a leading cause of mortality in the world [3]. Many times, the cause of cancer can be because of diet, age, pollution, and low income. Here, we are presenting a list of different geographical locations and causes of death due to cancer (Table 1).

3   Diagnostic Techniques

There are several technologies used to detect various constituents of the cell. Cell morphological changes and different characteristics are important to explore. Here are some current technologies that are being used: (a) Immunohistochemistry (b) Flow cytometry (c) Amperometric (d) Optical detection.

Table 1 Different geographical locations and causative agents of cancer

S. no	Name	Causes
1	North America	High rate of smoking
2	South America	High rate of smoking
3	Uruguay	High rate of smoking
4	Columbia	Dietary factors
5	Venezuela	Dietary factors
6	China	Low income
7	Russia	Low income
8	India	Tobacco, age, air and water pollution, dietary
9	Europe	Smoking, aging

Table 2 Describes uses, preservation and duration of preservation done by paraffin imbedding, frozen tissue and snap freeze tissue

S. no.		Paraffin embedding	Frozen tissues	Snap freeze tissue
1	Uses	DNA/RNA PCR amplification	Cell cycle analysis	Post translational modification
2	Preservation	Tissue morphology	Enzyme, function	Enzyme, function
3	Time	Multiple years	One year	One year

4 Immunohistochemistry

This technique is used to identify tissue components and search antigens of cells through specific antibodies which can be visualized by staining. These antigens can be amino acids or infectious agents and can be specific populations. This technique is termed as immunocytochemistry when used for cellular preparation. Here tissue slices are used. In this technique, tissues can be taken from liver, spleen, brain or any other organ and it includes four steps which are sample preparation, sectioning, mounting, and immunohistostaining. Three techniques are used for preservation which is paraffin embedding, frozen, snap freeze technique and their uses and duration of time till when they can be preserved is given in Table 2. In paraffin embedding, perfusion through formaldehyde can be done to remove blood antigen that can interfere with the staining. Further, it is dehydrated by a slowly increasing concentration of organic solvent like alcohol or acetone which is proceeded by immersion of tissues in paraffin. For preservation through freezing, tissues can be immersed in liquid nitrogen and tissues can also be snap frozen by immersing tissues in dry ice. These tissues can be sectioned by the microtome (paraffin embedding) or cryostat (frozen/snap freeze tissues) in thin slices, which can be then mounted on a glass slide coated with adhesive like poly L-lysine, etc., which leaves amino acid that crosslinks with the tissues. Paraffin-embedded tissues are dried, however, for frozen and snap

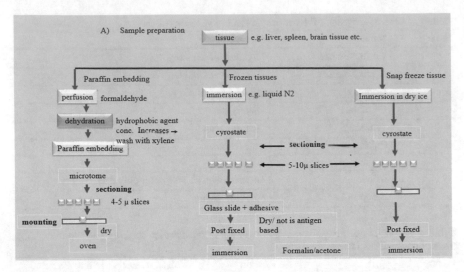

Fig. 1 Sample preparation through paraffin embedding, freezing, and snap freeze technique. This sectioning through microtome and cryostat technique and mounting by drying along with post-fixation method is shown

freeze drying it is based on the type of antigen looked for and then they are fixed by immersion with formalin/acetone (Fig. 1). Immunohistostaining can be performed by deparaffinization of tissue which is not required in snap freeze and frozen tissues followed by blocking of unwanted antigen through serum and permeabilization for exposing the desired antigen. Further, exposing them with primary (1°) antibody that can be detected by fluorescent and microscopy (through electrical signal), however for the chromogenic detection secondary antibody is given which acts as substrate produces color when reacted with 1° antibody that acts as an enzyme (Fig. 2).

Nakane heralded took immunohistochemistry beyond fluorescence microscopy by his innovation of using enzymes as marked antibodies because of which it was possible to visualize the reaction to optical microscopy [4].

Application: It is a simple technique, it can identify the origin of cells and its function inside the cells if specific antibodies are used. Identification of cellular secretion and their structures [5], diagnosis of neoplasia [6], their subtyping [7], and characterization of the primary site of malignant neoplasia [8]. A therapeutic indication of some diseases [9] and can distinguish between the benign and malignant type of particular cell [10].

This technique was considered as "brown revolution" because of its contribution and application of histopathology laboratory.

Limitations: Simple technique, however, require proficient handling and observation skills to interpret the results. Acquisition of antigen retrieval techniques is required.

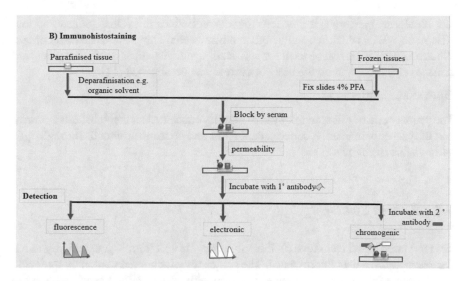

Fig. 2 Immunohistostaining of tissues and its visualization through fluorescence, electronic, and chromogenic techniques are described

5 Flow Cytometry

It allows measurement of individual cells up to the range of 1–150 μm size. This provides information of particles in fluid through optical properties.

Application: This technique along with biochemical selection can do simultaneous measurements of various properties of individual cells, hence can do a multiparameter analysis of single cells. Individual cells can be sorted on the basis of fluorescence of light scatter. It has great potential of drug screening. This can potentially monitor selectivity of drugs and stimuli in the intracellular pathway.

Limitation: Sample, when expanded to the range of thousands of antibodies, reagent becomes highly expensive and have high sample acquisition throughput. Plate-based assays can suffer from variable staining [10].

Amperometric Technique:

This is a family of electrochemical methods when instrumentally controlled sensing electrode is given potential and current is recorded as the analytical signal.

Amperometric detection: This is usually consisting of amperometric biosensors which can be used for the detection of cancer cells. Glucose oxidase enzyme is used in cancer cells which consume more glucose than normal cells. This enzyme is considered as a potential enzyme for cancer detection. This technique uses three electrodes, namely working electrode, counter electrode and reference electrode respectively. Through oxidation and reduction reaction between working and counter electrodes current is produced. First, glucose oxidase is dropped cast on working electrode then cells to be analyzed are added. If cells are normal cells they produce high current and if cells are cancerous it produces low current. Since it has lost contact and low RBCs and WBCs which is usually high in normal cells. This current can be recorded. Electrocom software can be used for verification and visualization of cancerous and normal cells and graph on the basis of current can be obtained [11].

Optical Modalities:

The primary aim of this technology is to identify cancer. Also, their absolute removal and the proper demarcation between cancerous and noncancerous cells through optically visible technique.

5.1 Raman Scattering

This is a laser-based technique. In this vibration, the spectrum is generated by much narrower features than fluorescence. The frequency of scattered photon by molecule has the same frequency as of incident photon with only fraction of light having the lower frequency than incident photon leading to inelastic scattering known as Raman effect. Raman spectrum is obtained by the difference between incident and dispersed photon which is equivalent to the vibration of scattering molecule for every molecule.

Applications: This is useful in the detection of molecular class in a biological sample. This is highly sensitive and has a high level of multiplexing.

Limitation: If molecules are very close and there is rough or uneven gold or silver, it provides inaccurate and several times increased Raman scattering [12].

5.2 Fluorescence Imaging

This technique can identify morphological as well as physiological changes. This also has the ability to visualize the specimen deeper in the tissue. Fluorophores are aromatic compounds that reemit light on excitation. Native fluorophores used are tryptophan, elastin, NADH, etc. They have a spectral range from ultraviolet to visible light, which cannot penetrate deeper into the tissues since it has high absorption. This can be overcome by using far-red to near-far-red range which has low absorption and high penetration range from 1650 to 1100 nm. As a tumor detection vehicle fluorescent dye, ligand, metallic nanoparticle and semiconductor dots can be used to identify and locate cancerous cells by identifying overexpressed receptors present on them. Selection of ligands depends on overexpression of receptors on cancerous cells like somatostatin receptor (SSTR), bombesin and folic acid receptors. SSTR is overexpressed in prostate cancers and human adenomas like non-endocrine tumors, somatotrophs.

Applications: This can be used for in vivo imaging, provide fast and precise means to identify the location of cancer. There is the possibility of making library so that bioactive molecules can be identified.

Limitations: This cannot identify considerably deep within tissue and research for the potential tumor-specific delivery system is still in progress [13]; here different photoemission rates can create background noise [14].

5.3 Quantum Dots (QD)

This technique uses minuscule-sized particles known as quantum that release fluorescence when exposed to light. However, due to their different and unstable photoemission rates, it has lots of background noise [11]. Therefore, recently these quantum dots are attached to the etchant which quenches surplus quantum dots and only tumor-specific quantum dots gives fluorescence. Etchant has silver present in them and quantum dots have an attachment to zinc. When zinc is replaced by silver quenching of fluorescence occurs, in this way it clears the metal ions released by QD. QD can cross the membranes to reach tumor cells. This etchant is not capable which also increases its specificity. This technique has been tested on mice having human gastric, prostate and breast cancer.

Applications: This technique was found to give fivefold increased visualization when compared to other techniques.

Limitations: This has been tested on mice, however not a single human study reported till date [14].

5.4 Early Diagnosis

Cancer is one of the leading causes of death. This is mostly recognized at later stages where cure remains a no longer option. Hence, early detection becomes of prime importance for limiting disease progression and increasing survival benefits. This would not only save valuable time while helping reduce economic burden. Cancer can respond well to the treatment when given at early stages. Here are some technologies that can be useful in early detection.

Techniques Using RNA-aptamer Includes:

RNA-aptamer:

Aptamers are RNAs and DNAs' oligonucleotides which have high affinity to specific ligands and they are originated from in vitro assortment experiments (systematic evolution of ligands by exponential enrichment known as SELEX). This may start from random sequence libraries, optimize the nucleic acids for its specificity and high affinity to particular ligands [15].

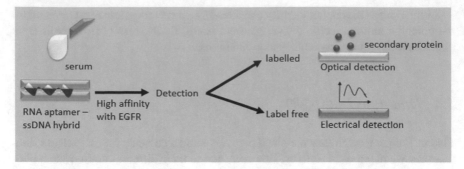

Fig. 3 Here is the diagrammatic representation of electrical detection through nanogap break junctions. Here silicon dioxide RNA–DNA hybrid is present which has the affinity for EGFR present in serum. This is detected by a label-free or labeled method

5.5 Electrical Detection Through Nanogap Break Junction

Epidermal growth factor receptor (EGFR) is a cell surface protein which is overexpressed in cancerous cells and its concentration is increased in cancer patients. EGFR is considered as the important biomarker and is useful for the detection of presence and onset of disease. Also, the specific phase of disease and monitoring changes in homeostasis. In this technique, pair of nanogold electrodes with nanometer separation made through the convergence of focused ion beam scratching and electromigration. In between the surface of gold nanoelectrodes, silicon dioxide surface is present where single-stranded DNA molecule is immobilized and hybridized with anti-EGFR RNA-aptamer to recognize EGFR protein in serum through optical detection. If EGFR protein is present, then increase in orders of magnitude in direct current which was measured and optical detection was evaluated for the characterization of surface function [16]. This technique can be useful in lung cancer, ovarian cancer, and bladder cancer (Fig. 3).

5.6 Enhanced Cancer Isolation by Nanotextured Polymer Substrate

Nanotextured polydimethylsiloxane (PDMS) is used to mimic basement membrane to enhance affinity toward cancer cells including blood. This affinity is further enhanced by fabricating it with the micro-reactive-ion etching (micro-RIE) technique by using immobilized aptamers against cell membrane. This overexpresses EGFR, hence providing better control over surface properties. This technique is used for the detection of circulating tumor cells and is found useful in human glioblastoma disease. This PDMS can be prepared by using oxygen and carbon tetrafluoride.

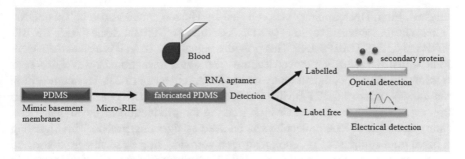

Fig. 4 Diagrammatic representation of cancer isolation by nanotextured polymer substrate

Micro-RIE nanotextured PDMS and increased surface area for immobilized anti-EGFR aptamer which results in the increased capture of tumor cells [17].

6 Detection

Cancer detection can be performed by labeled and label-free technique. In labeled detection, secondary molecule like organic, inorganic and radioactive molecule fluorescently tagged while label free is direct detection of proteins by electrical signal detection (Fig. 4).

6.1 Cancer Detection by Nanowire Sensor Assays

Nanowire sensor wire assay is a technique that can simultaneously detect many markers in a label-free manner rapidly. Previously, silicon nanowire and carbon tubes were used for the detection of bonded and unbonded protein in aqueous solution. However, this technique was not very sensitive and cannot detect multiple markers. Therefore, to improve sensitivity, discrete nanowires and surface receptors were incorporated in the single device. This device can, detect multiple protein cancer markers with high sensitivity. This can also do efficient quantitative detection of the undiluted sample at femtomolar concentrations. Serum could be used as the sample. It is used for detection of carcinoembryonic antigen, mucin-1, and prostate-specific antigen (PSA) [18].

Chemiluminescence-based ATP Detection in Cancer Cells:

Chemiluminescence resonance energy transfer (CRET) is a technique where the nonradiative transfer of energy from chemiluminescent donor to acceptor molecule occurs without excitation in a source. The chemiluminescent donor gets oxidized. In this technique magnetic nanoparticles (MNP) are used in which DNA aptamer is attached. First, DNA aptamer was selected by Huizenga and Szostak. This cDNA linker is further attached to either two DNA or three DNA to fabricate CRET-BMBP-MNP and CRET-TMBP-MNP. This provides signal, i.e., if ATP is absent then MNP along with their cDNA is removed leading to no signal formation. However, if ATP is present then there is competition between cDNA linkers and ATP. This competition leads to separation of CRET-BMBP and CRET-TMBP from MNP as a result aptamer structure switching occurs and which leads to the production of the signal. Here, binary and triplex DNA molecules can be used as signaling probes. This signaling is based on a complex of luminol, hydrogen peroxide, horse radish peroxidase and fluorescence resonance energy transfer. CRET takes place from luminol to fluorescein which is oxidized by H_2O_2 and catalyzed by horse radish peroxidase.

Application: It provides enhanced sensitivity toward ATP. This is low cost, fast, precise, and has simple manipulation (25).

7 Conclusion

This study gives information about tumorigenesis and cancer with information on diagnostic tools and their advantages and limitations. This is also intended to give a brief information regarding recent techniques and their usability. This study is also intended to give information about the sample being used and their preservation techniques.

Acknowledgements This project was supported by Council of Scientific and Industrial Research grant no 37(1693)/17/EMR-II and Department of Science and Technology as Ramanujan Fellowship grant no SB/S2/RJN-132/20/5. We are thankful to the Ministry of Human Resource and Development and University Grant Commission for the fellowship to Charu Sonkar. The funding organization has not played any role in study design, the decision to publish, or the preparation of the manuscript. We appreciate our lab colleagues for insightful discussions and advice.

References

1. Dollinger, M., Ko, A.H., Rosenbaum, E.H., et al.: Understanding cancer (2008). Everyone's Guide to Cancer Therapy: How Cancer is Diagnosed, Treated and Managed Day to Day, vol. 1, 5th edn, pp. 3–16. Andrews Mc Meel Publishing, Kansas City
2. La Vecchia, C., Negri, E., Gentile, A., Franceschi, S.: Family history and the risk of stomach and colorectal cancer. Cancer **70**(1), 50–55 (1992). https://doi.org/10.1002/1097-0142(19920701)70:1<50::aid-cncr2820700109>3.0.co;2-i
3. Ferlay, J., Soerjomataram, I., Dikshit, R., Eser, S., Mathers, C., Rebelo, M., Bray, F.: Cancer incidence and mortality worldwide: sources, methods and major patterns in GLOBOCAN 2012. Int. J. Cancer **136**(5) (2015)
4. Avrameas, S., Uriel, J.: Method of antigen and antibody labelling with enzymes and its immunodiffusion application. Comptes rendus hebdomadaires des seances de l'Academie des sciences. Serie D: Sciences naturelles **262**(24), 2543–2545 (1966)
5. Hsu, S.M., Raine, L., Fanger, H.: Use of avidin-biotin-peroxidase complex (ABC) in immunoperoxidase techniques: a comparison between ABC and unlabeled antibody

(PAP) procedures. J. Histochem. Cytochem.c **29**, 577–80 (1981). https://doi.org/10.1177/2 9.4.6166661

6. Leong, A.S., Wright, J.: The contribution of immunohistochemical staining in tumour diagnosis. Histopathology **11**, 305–1295 (1987). https://doi.org/10.1111/j.1365-2559.1987.t b01874.x

7. Taylor, C.R., Cote, R.J.: Immunomicroscopy: a diagnostic tool for the surgical pathologist (No. 19). WB Saunders Company (2006)

8. Bodey, B.: The significance of immunohistochemistry in the diagnosis and therapy of neo-plasms. Expert Opin. Biol. Therapy **2**(4), 371–393 (2002). https://doi.org/10.1517/14712598. 2.4.371

9. Werner, M., Chott, A., Fabiano, A., Battifora, H.: Effect of formalin tissue fixation and pro-cessing on immunohistochemistry. Am. J. Surg. Pathol. **24**(7), 1016–1019 (2000)

10. Krutzik, P.O., Nolan, G.P.: Fluorescent cell barcoding in flow cytometry allows high-throughput drug screening and signaling profiling. Nat. Methods (New York) **3**(5), 361–368 (2006). https://doi.org/10.1038/nmeth872

11. Lim, S.J., Zahid, M.U., Le, P., Ma, L., Entenberg, D., Harney, A.S., Smith, A.M.: Brightness-equalized quantum dots. Nat. Commun. **6**, 8210 (2015)

12. Sha, M.Y., Xu, H., Penn, S.G., Cromer, R.: SERS nanoparticles: a new optical detection modal-ity for cancer diagnosis. (2007). https://doi.org/10.2217/17435889.2.5.725

13. Pu, Y., Achilefu, S., Alfano, R.R.: Cancer Detection/Fluorescence Imaging: 'Smart beacons' target cancer tumors

14. Liu, X., Braun, G.B., Qin, M., Ruoslahti, E., Sugahara, K.N.: In vivo cation exchange in quantum dots for tumor-specific imaging. Nat. Commun. **8** (2017). https://doi.org/10.1038/s4 1467-017-00153-y

15. Hermann, T., Patel, D.J.: Adaptive recognition by nucleic acid aptamers. Science **287**(5454), 820–825 (2000)

16. Ilyas, A., Asghar, W., Allen, P.B., Duhon, H., Ellington, A.D., Iqbal, S.M.: Electrical detection of cancer biomarker using aptamers with nanogap break-junctions. Nanotechnology **23**(27), 275502 (2012)

17. Islam, M., Sajid, A., Mahmood, M.A.I., Bellah, M.M., Allen, P.B., Kim, Y.T., Iqbal, S.M.: Nanotextured polymer substrates show enhanced cancer cell isolation and cell culture. Nan-otechnology **26**(22), 225101 (2015)

18. Zheng, G., Patolsky, F., Cui, Y., Wang, W.U., Lieber, C.M.: Multiplexed electrical detection of cancer markers with nanowire sensor arrays. Nat. Biotechnol. **23**(10), 1294 (2005)

Image Denoising using Tight-Frame Dual-Tree Complex Wavelet Transform

Shrishail S. Gajbhar and Manjunath V. Joshi

Abstract In this paper, we propose a new approach to design the 1D biorthogonal filters of dual-tree complex wavelet transform (DTCWT) in order to have almost tight-frame characteristics. The proposed approach involves use of triplet halfband filter bank (THFB) and optimization of free variables obtained using factorization of generalized halfband polynomial (GHBP) to design the filters of two trees of DTCWT. The wavelet functions associated with these trees exhibit better analyticity in terms of qualitative and quantitative measures. Transform-based image denoising using the proposed filters shows comparable performance to the best performing orthogonal wavelet filters.

1 Introduction

In recent years, dual-tree complex wavelet transform (DTCWT) has been established as one of the important transform-domain processing tools in a wide range of multimedia applications [1, 3, 6, 12]. Unlike discrete wavelet transform (DWT), it offers better directionality, near-shift-invariance, and phase information with limited redundancy. In practice, DTCWT is implemented using two branches of DWT referred to as primal and dual tree and outputs of these are considered as the real and imaginary parts of the complex coefficient representation of an input signal. With the use of orthogonal/biorthogonal finite impulse response (FIR) filters in these trees,

S. S. Gajbhar (✉) · M. V. Joshi
DA-IICT, Gandhinagar, Gujarat, India
e-mail: shrishail_gajbhar@daiict.ac.in

M. V. Joshi
e-mail: mv_joshi@daiict.ac.in

S. S. Gajbhar
WIT, Solapur, Maharashtra, India

© Springer Nature Singapore Pte Ltd. 2019
M. Tanveer and R. B. Pachori (eds.), *Machine Intelligence and Signal Analysis*,
Advances in Intelligent Systems and Computing 748,
https://doi.org/10.1007/978-981-13-0923-6_55

the transform is approximately *analytic* with a redundancy factor of just 2^m for an input of m-dimensional (m-D) signal. The idea for constructing DTCWT was first proposed by Nick Kingsbury [7, 9] and subsequently developed by Selesnick in [14, 15]. We refer to [16] as an excellent tutorial paper on various aspects of DTCWT.

Although, DTCWT output representation is complex valued, real-valued filter coefficients are used in the construction and no complex arithmetic is required which is very much advantageous. However, design of such filters is quite challenging [18], since the filter coefficients need to satisfy various constraints. Selesnick [14] was the first researcher to arrive at certain conditions that must be satisfied by the DTCWT filters in order to have desired *analyticity* property. He showed that if the wavelet functions associated with the two trees of DTCWT are Hilbert transform pairs, the transform is completely analytic and shift-invariant. Since, obtaining perfect ana-lyticity is difficult using compactly supported filters, approximate analyticity and near-shift-invariance can be achieved using FIR orthogonal/biorthogonal wavelet filters [15]. In order to have these properties, filters must satisfy *perfect reconstruc-tion* (PR), *vanishing moment* (VM), and *half-sample delay* (HSD) constraints as minimum requirements. HSD condition plays the role of coupling between two trees of DTCWT to have approximate Hilbert transform relationship.

In this paper, we only consider the design of biorthogonal FIR filters. In [7, 9], Kingsbury proposed the use of odd/even filter design approach to obtain biorthogo-nal wavelet filters (BWFs) based DTCWT construction. In [15], Selesnick devised *common factor* technique capable of designing biorthogonal filters with prescribed number of vanishing moments (VMs). Most of the biorthogonal filter design tech-niques proposed in the literature are variants of the above two approaches and few works in this context can be found in [17, 21].

In this paper, we propose a new approach to design biorthogonal wavelet filters of DTCWT using odd/even filter settings. The motivation for the proposed approach is to design biorthogonal filters having near-orthogonal filter response characteristics that can provide a near-tight-frame DTCWT, which is a l^2-norm (energy) preserving transform. By *tight-frame* transform, we mean an orthogonal transform with certain amount of redundancy [4]. Such transforms have improved noise decorrelation and are useful in image denoising application. We also require the resulting wavelet func-tions to have exact symmetry which improves the directional selectivity of DTCWT [15]. This condition makes the image features such as edges to be handled without the unwanted oscillatory behavior. In addition, we also desire filters giving Hilbert pairs of wavelets with improved analyticity for better shift-invariance.

In order to obtain these properties, we make use of an optimization in the proposed approach. We propose to use odd-length biorthogonal filters having near-orthogonal frequency response in the first tree and accordingly design the second tree filters to have nearly tight-frame transform. Our approach provides an effective way to handle the filter response characteristics of these filters which is obtained by using uncon-strained optimization of free variables. Also, the proposed approach can be modified to design the orthogonal filters by designing a halfband polynomial satisfying certain nonnegativity constraints followed by spectral factorization.

The paper is organized as follows. In Sect. 2, we give the background to understand the DTCWT basics and briefly describe the odd/even technique. In Sect. 3, the proposed approach is described while Sect. 4 details the design examples of the proposed approach along with their qualitative and quantitative measures. In Sect. 5, we discuss image denoising application using the designed filters. Section 6 concludes the paper.

2 Background Review

Figure 1 shows the core structure of the DTCWT. It has two trees consisting of 2-channel filter banks that use 1D biorthogonal wavelet filters.

In Tree-1, the filters $\tilde{h}_0(n)$ and $\tilde{h}_1(n)$ represent the analysis of low-pass and high-pass filters, respectively. Similarly, the $h_0(n)$ and $h_1(n)$ represent the same on the synthesis side referred to as synthesis low-pass and high-pass filters. They are related to each other as follows:

$$\tilde{h}_1(n) = -(-1)^n h_0(n), \quad 0 \leq n \leq N - 1$$
$$h_1(n) = (-1)^n \tilde{h}_0(n), \quad 0 \leq n \leq \tilde{N} - 1. \tag{1}$$

Here, $\tilde{N}$ and N represent lengths of the filters $\tilde{h}_0(n)$ and $h_0(n)$, respectively. Similar relations hold good in the case of filters in Tree-2.

Let, $\phi_h(t)$ and $\phi_g(t)$ be the synthesis scaling functions of Tree-1 and Tree-2, respectively, and $\psi_h(t)$ and $\psi_g(t)$ be their corresponding wavelet functions related through the two-scale equations associated with them. In a similar way, one can define analysis wavelet functions $\tilde{\psi}_h(t)$ and $\tilde{\psi}_g(t)$. In order to have approximate analyticity of DTCWT, we require that $\psi_g(t) \approx \mathcal{H}\{\psi_h(t)\}$ and $\tilde{\psi}_g(t) \approx \mathcal{H}\{\tilde{\psi}_h(t)\}$ [14, 15] representing Hilbert transform pairs criteria. This indicates that the synthesis and analysis wavelet functions of Tree-2 are *approximately* Hilbert transforms of Tree-1 wavelet functions. In Fourier domain, these relations are given as

Fig. 1 Two trees of 2-channel filter banks used in DTCWT construction

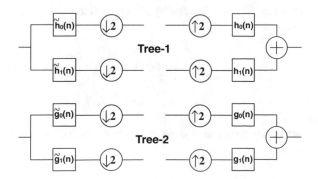

$$\Psi_g(\omega) \approx \begin{cases} -j\Psi_h(\omega), & \omega > 0 \\ j\Psi_h(\omega), & \omega < 0. \end{cases} \tag{2}$$

Similar expressions exist for $\tilde{\Psi}_g(\omega)$. Here, $\Psi_h(\omega), \Psi_g(\omega), \tilde{\Psi}_h(\omega)$ and $\tilde{\Psi}_g(\omega)$ represent Fourier transforms of $\psi_h(t), \psi_g(t), \tilde{\psi}_h(t)$ and $\tilde{\psi}_g(t)$, respectively. Since, wavelet functions depend on the scaling functions which in turn depend on the low-pass filters associated with that scaling function, the problem of designing the Hilbert transform pairs of wavelet bases reduces to designing the low-pass filters that satisfy $g_0(n) \approx h_0(n - 0.5)$ which is known as *half-sample delay* (HSD) constraint [16]. In Fourier domain, this can be expressed as

$$G_0(\omega) \approx e^{-j\frac{\omega}{2}} H_0(\omega), \tag{3}$$

where, $G_0(\omega)$ and $H_0(\omega)$ are Fourier transforms of $g_0(n)$ and $h_0(n)$, respectively. One can design these filters by approximating the magnitude and phase responses as

$$|G_0(\omega)| = |H_0(\omega)| \tag{4}$$

$$\angle G_0(\omega) = -\frac{\omega}{2} + \angle H_0(\omega). \tag{5}$$

Due to the nature of the Eq. (3), one of the two conditions given in Eqs. (4) and (5) is satisfied *exactly* or both are approximated. Odd–even design approaches satisfy the phase condition exactly while approximate the magnitude condition. A general concept of odd–even design approach is discussed in the next subsection.

2.1 Odd–Even Filter Design Approach

In this approach, well-known symmetric odd-length filters such as CDF 9/7 are used in the first tree. The filters in the second tree are chosen as symmetric even-length filters which are designed such that their amplitude responses are matched with corresponding odd length filters in Tree-1. Let the frequency response of odd-length low-pass filters is denoted by $X_0^h(\omega)$, which can be represented in terms of amplitude response as [10]:

$$X_0^h(\omega) = x_0^h(0) + \sum_{n \neq 0} x_0^h(n) cos(\omega n) \tag{6}$$

$$= F_{0,R}^h(\omega). \tag{7}$$

Similarly, response of even-length low-pass filters denoted using $X_0^g(\omega)$ can be written as

$$X_0^g(\omega) = e^{\frac{-j\omega}{2}} F_{0,R}^g(\omega), \quad \text{where} \tag{8}$$

$$F_{0,R}^g(\omega) = \sum_n 2x_0^g(n) \cos((n - \frac{1}{2})\omega). \tag{9}$$

Here, $F_{0,R}^h(\omega)$ and $F_{0,R}^g(\omega)$ represent real-valued functions of ω representing *amplitude responses* of odd- and even-length low-pass filters, respectively. In Eq. (8), one may see the presence of inherent half-sample delay due to factor $e^{\frac{-j\omega}{2}}$. Assuming the same signs for $F_{0,R}^h(\omega)$ and $F_{0,R}^g(\omega)$, we get $\angle X_0^g(\omega) = -\frac{\omega}{2} + \angle X_0^h(\omega)$, i.e., phase condition in Eq. (5) is perfectly satisfied. The problem thus reduces to approximating the amplitude responses as

$$F_{0,R}^h(\omega) \approx F_{0,R}^g(\omega). \tag{10}$$

Approximation given in the Eq. (10) is realized in practise by minimizing the mean squared error function of the form

$$E = \int_0^\pi (F_{0,R}^h(\omega) - F_{0,R}^g(\omega))^2 d\omega. \tag{11}$$

3   Proposed Approach

In the proposed approach, we use factorization of generalized halfband polynomial (GHBP) [11] to obtain the DTCWT filters of the Tree-2 as shown in Fig. 1. In contrast to [11] where the aim is to improve the frequency selectivity, our aim is to tailor the overall frequency response characteristics of the Tree-2 filters to that of Tree-1 filters to obtain the almost tight-frame transform.

3.1   Proposed Odd–Even Technique

We use two stages in the design procedure. In the first stage, we design Tree-2 filters by using GHBP and unconstrained optimization approach. In the second stage referred as *shape-parameter optimization*, we modify the Tree-1 filters in order to improve the analytic quality of the designed filters. The first stage has two parts: (1) selection of Tree-1 filters and (2) design of Tree-2 filters using GHBP optimization explained as follows:

   Selection of Tree-1 filters: we obtain the Tree-1 low-pass filters $\tilde{h}_0(n)$ and $h_0(n)$ by choosing them from a class of triplet halfband filter bank (THFB) proposed in [2]. The filter coefficients are chosen as:

$$\tilde{H}_0(z) = \frac{1+p}{2} + \left(\frac{1+p}{2}\right) z T_1(z^2) \left(\frac{1 - pzT_0(z^2)}{1+p}\right) \tag{12}$$

$$H_0(z) = \left(\frac{1 + pzT_0(z^2)}{1+p}\right) + \left(\frac{1-p}{1+p}\right) z T_2(z^2) H_0(-z). \tag{13}$$

The method uses three halfband filter kernels $T_0(z^2)$, $T_1(z^2)$ and $T_2(z^2)$ to take care of perfect reconstruction. Here, p represents shape parameter and it gives flexibility to tailor the frequency response characteristics of the designed filters. Using different values of $p \in [0, 1]$, the response value at $\omega = \frac{\pi}{2}$ for both the filters can be easily fixed. One may obtain biorthogonal filters having near-orthogonal response characteristics for $p = \sqrt{2} - 1$ which gives us $|H(\omega)|$ at $\omega = \frac{\pi}{2}$ same as in the case of orthogonal filters. Here, N_0, N_1 and N_2 are the lengths of the three halfband polynomials $T_0(z^2)$, $T_1(z^2)$ and $T_2(z^2)$, respectively. The vanishing moments for $\tilde{h}_0(n)$ and $h_0(n)$ are $2 * \min(N_0, N_1)$ and $2 * \min(N_0, N_1, N_2)$, respectively, where $*$ denotes multiplication operation.

Designing Tree-2 filters using GHBP optimization: We restrict the Tree-2 low-pass filters i.e., $\tilde{g}_0(n)$ and $g_0(n)$ to be symmetric and even-length having their magnitude responses *matched* approximately to the magnitude responses of corresponding Tree-1 filters ($\tilde{h}_0(n)$ and $h_0(n)$). In order to do this, we use factorization of GHBP with order D as given in Eq. (14)

$$P^D(z) = a_0 + a_2 z^{-2} + \cdots + a_{(D/2)-1} z^{-(D/2)-1} + z^{-(D/2)}$$
$$+ a_{(D/2)-1} z^{-(D/2)+1} + \cdots + a_0 z^{-D}, \tag{14}$$

where, polynomial order D is chosen as

$$D = 2 * (M - 1) + 4 * n_f. \tag{15}$$

Here, $M = \tilde{K} + K$ with $\tilde{K}$ and K representing the number of VMs of $\tilde{g}_0(n)$ and $g_0(n)$, respectively, while, n_f represents the number of free variables used in the unconstraned optimization. $\tilde{K}$ and K must be odd and they are chosen as:

$$\tilde{K} = 2 * \min(N_0, N_1) - 1$$
$$K = 2 * \min(N_0, N_1, N_2) - 1. \tag{16}$$

The final product filter $P(z)$ in this case is as given in Eq. (17).

$$P^D(z) = (1 + z^{-1})^M R(z) = (1 + z^{-1})^{\tilde{K}+K} R(z) \tag{17}$$

Here, $R(z)$ is a remainder polynomial expressed in terms of n_f number of free variables. After factorizing $R(z)$ into two symmetric polynomials $R_1(z)$ and $R_2(z)$, the low-pass filters $\tilde{G}_0(z)$ and $G_0(z)$ can be expressed as

$$\tilde{G}_0(z) = (1 + z^{-1})^{\tilde{K}} R_1(z) \tag{18}$$

$$G_0(z) = (1 + z^{-1})^K R_2(z). \tag{19}$$

Now, the expressions for $\tilde{G}_0(\omega)$ and $G_0(\omega)$ to be used in the objective function can be written as

$$\tilde{G}_0(\omega) = (1 + z^{-1})^{\tilde{K}} R_1(z)|_{z=e^{j\omega}} \tag{20}$$

$$G_0(\omega) = (1 + z^{-1})^K R_2(z)|_{z=e^{j\omega}}. \tag{21}$$

We then use the MATLAB optimization function *fminunc* to minimize the following objective function with respect to the chosen number of free variables (n_f) and obtain the filters $\tilde{g}_0(n)$ and $g_0(n)$.

$$F_{obj} = \int_0^\pi \left(\left| \tilde{H}_0(\omega) \right| - \left| \tilde{G}_0(\omega) \right| \right)^2 d\omega + \int_0^\pi (|H_0(\omega)| - |G_0(\omega)|)^2 d\omega, \tag{22}$$

Here, the remainder polynomial $R(z)$ is factorized into symmetric factors $R_1(z)$ and $R_2(z)$ while performing the unconstrained minimization of the above objective function. With the optimized values of the free variables, we obtain symmetric even-length linear- phase biorthogonal FIR filters $\tilde{g}_0(n)$ and $g_0(n)$ to be used in the Tree-2.

In order to have improved shift-invariance, we use second stage where we modify the Tree-1 filters using the Tree-2 filters obtained in first stage by further decreasing the error of magnitude approximation. Here, we modify the Tree-1 filters such that their magnitude responses better match to that of Tree-2 filters. We make use of Eqs. (12) and (13) where shape parameter p is subjected to unconstrained optimization as follows.

Shape-parameter optimization: We minimize the error E given in Eq. (11) by *matching* the amplitude responses of the Tree-1 filters to their corresponding Tree-2 filters obtained at the first stage output. Here, we make use of one degree of freedom associated with the Tree-1 filters $\tilde{h}_0(n)$ and $h_0(n)$ in the form of shape parameter p. We use it for minimizing the magnitude error between filters of the two trees. The optimized shape parameter p for the Tree-1 filters is obtained by minimizing the following objective function

$$F_{obj} = \int_0^\pi \left(\left| \tilde{G}_0(\omega) \right| - \left| \tilde{H}_0(\omega) \right| \right)^2 d\omega + \int_0^\pi (|G_0(\omega)| - |H_0(\omega)|)^2 d\omega, \tag{23}$$

where $\tilde{G}_0(\omega)$ and $G_0(\omega)$ are the outputs of the first stage, i.e., the even-length filters $\tilde{g}_0(n)$ and $g_0(n)$. The expressions for $\tilde{H}_0(\omega)$ and $H_0(\omega)$ are from Eqs. (12) and (13), respectively, where z is replaced by $e^{j\omega}$. During the optimization, for a specific value of p, the objective function value is evaluated and optimization is continued until F_{obj} is minimized. Note that in this case $\tilde{G}_0(\omega)$ and $G_0(\omega)$ are fixed. Once the optimal value of p is known output filters $\tilde{h}_0(n)$ and $h_0(n)$ can be easily obtained using Eqs. (12) and (13). The optimized filters of the first and second stage are then considered as the final DTCWT filters. Due to second stage optimization, analyticity of the associated wavelets is improved due to better *matching* between the magnitude responses given in Eq. (10).

4   Design Example

In this section, we give in detail a design example using the proposed approach. The error measuring *analyticity* is quantified using two measures E_1 and E_2 given in [19]. Both measures have ideal value of 0. We express quantitative measures in terms of E_1_A, E_2_A, E_1_S, E_2_S, and E_1_Avg, E_2_Avg, where A, S and Avg represent analysis, synthesis, and average, respectively.

For better understanding, we analyze the outputs of the two stages. In the first stage, Tree-1 filters $\tilde{h}_0(n)$ and $h_0(n)$ are obtained using Eqs. (12) and (13), respectively with chosen parameters as $N_0 = N_1 = N_2 = 2$ and $p = \sqrt{2} - 1$ for the design specifications used as $\tilde{K} = K = 4$. The lengths of resulting filters are 13 and 19, respectively. To design the Tree-2 filters $\tilde{g}_0(n)$ and $g_0(n)$ using GHBP optimization, we choose $n_f = 1$ i.e., one free variable. GHBP order D and number of vanishing moments are set automatically using Eqs. (15) and (16), respectively. Coefficients of the remainder polynomial $R(z)$ are then obtained using Eqs. (14) and (17). During the unconstrained optimization, remainder polynomial $R(z)$ is factorized into $R_1(z)$ and $R_2(z)$. For the optimized value of the free variable $a_0 = -9.2448 \times 10^{-04}$, F_{obj} in Eq. (22) is minimum and Tree-2 filters $\tilde{g}_0(n)$ and $g_0(n)$ of length 10 and 6, respectively, are obtained. The coefficients for the same are as given in Table 1.

In second or shape-parameter optimization stage, optimized Tree-1 filters are obtained for better analyticity. During the optimization, different combinations of lengths of halfband filters are considered to obtain the optimum filter coefficients. For $N_0 = N_1 = 2$, $N_2 = 1$ and $p = 0.4327$ combination, objective function in Eq. (23)

Table 1 Coefficients of the filters $\tilde{g}_0(n)$ and $g_0(n)$ using the proposed technique

n	1	2	3	4	5	6	7	8	9	10
$\tilde{g}_0(n)$	0.0076	0.0080	−0.0641	0.0602	0.4883	0.4883	0.0602	−0.0641	0.0080	0.0076
$g_0(n)$	0.0610	−0.0640	−0.4970	−0.4970	−0.0640	0.0610	−	−	−	−

Table 2 Coefficients of $\tilde{h}_0(n)$ and $h_0(n)$ after shape-parameter optimization stage

n	1	2	3	4	5	6	7	8	9	10
$\tilde{h}_0(n)$	−0.0008	0.0000	0.0152	−0.0312	−0.0532	0.2812	0.5777	0.2812	−0.0532	−0.0312
$h_0(n)$	−0.0002	0.0000	0.0028	0.0062	−0.0264	−0.0495	0.2737	0.5866	0.2737	−0.0495
n	11	12	13	14	15					
$\tilde{h}_0(n)$	0.0152	0.0000	−0.0008	−	−					
$h_0(n)$	−0.0264	0.0062	0.0028	0.0000	−0.0002					

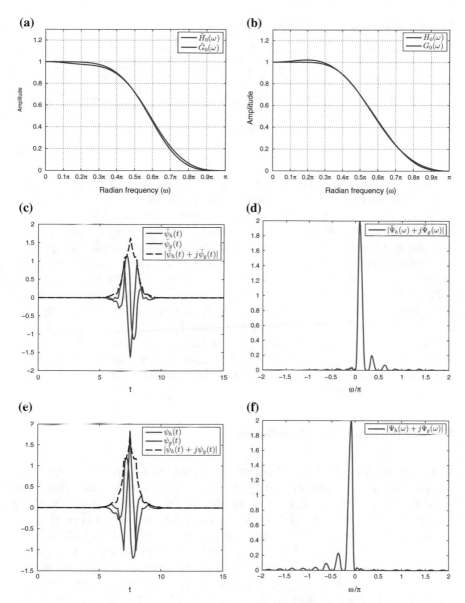

Fig. 2 Plots for the design example **a** magnitude response comparison between analysis of low-pass filters of Tree-1 and Tree-2 i.e., $|\tilde{H}_0(\omega)|$ and $|\tilde{G}_0(\omega)|$ **b** magnitude response comparison between synthesis low-pass filters of Tree-1 and Tree-2 i.e., $|H_0(\omega)|$ and $|G_0(\omega)|$. **c** Analysis wavelet functions $\tilde{\psi}_h(t)$, $\tilde{\psi}_g(t)$ and $|\tilde{\psi}_h(t) + j\tilde{\psi}_g(t)|$ **d** Magnitude frequency spectrum for $|\tilde{\Psi}_h(\omega) + j\tilde{\Psi}_g(\omega)|$ **e** Synthesis wavelet functions $\psi_h(t)$, $\psi_g(t)$ and $|\psi_h(t) + j\psi_g(t)|$ **f** Magnitude frequency spectrum for $|\Psi_h(\omega) + j\Psi_g(\omega)|$

Table 3 Analyticity quality comparison with state-of-the-art methods

	E_1_A	E_2_A	E_1_S	E_2_S	E_1_Avg	E_2_Avg
Proposed	0.0198	**0.0004**	**0.0184**	**0.0004**	**0.0191**	**0.0004**
Table I of [21]	0.0222	0.0007	0.0328	0.0008	0.0275	0.0007
Table III of [17]	0.0202	0.0008	0.0242	0.0004	0.0222	0.0006

was *minimum*. Optimized Tree-1 filters $\tilde{h}_0(n)$ and $h_0(n)$ are of lengths 13 and 15, respectively. Their filter coefficients are given in Table 2. Figures 2a and b show the magnitude response comparison between the analysis of low-pass filters of the newly optimized Tree-1 and the proposed Tree-2 filters. One can observe that magnitude response approximation is much better. Figure 2c and e show analysis and synthesis of wavelet functions. Figure 2d and f show plots to analyze the analyticity where one can observe that the frequency contents for $\omega/\pi < 0$ in Fig. 2d and $\omega/\pi > 0$ in Fig. 2f are negligible and hence showing near-analyticity. Quantitative evaluation of the designed filters for their analytic quality is tabulated in Table 3. We have also compared results of the proposed design example with the examples given in state-of-the-art odd–even techniques [17, 21]. One can see that, wavelet functions obtained using the filters designed using the proposed approach have better analyticity when compared to these methods which use constrained optimization and are computationally expensive.

5 Image Denoising Application

In this section, we show the performance of the proposed filter set for the image denoising application. Coefficients of the analysis and synthesis of low-pass filters of the chosen set for Tree-1 and Tree-2 are given in Tables 2 and 1, respectively. We use these filters to obtain the 2D DTCWT by using the construction given [16]. For comparing the image denoising performance, we have used the MATLAB software provided by Ivan W. Selesnick on his website [13]. We have compared our results with their 2D DTCWT given for the best performing filters namely 6-tap orthogonal Q-shift filters of [8]. Additive white Gaussian noise (AWGN) of standard deviation σ is added to the original image in order to test the performance on noisy images. We have used *hard thresholding* method [5] with a threshold value of $T = 2\sigma$ as used in the DTCWT software mentioned above. Figure 3 shows hard thresholding results on a part of Barbara image containing oriented texture for AWGN of standard deviation $\sigma = 30$. It can be observed that the 2D DTCWT obtained using the proposed filter set performs comparably to that of best-known orthogonal 6-tap Q-shift filters of [8] in terms of Peak Signal-to-Noise Ratio (PSNR) value while slight improvement is observed in case of recent image quality indicator values namely Structural Similarity Index Measure (SSIM) [20] and Feature Similarity Index Measure (FSIM) [22]. Both

(a) (b) (c) (d)

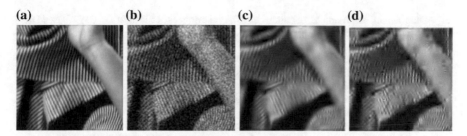

Fig. 3 Image denoising using 2-D DTCWT **a** original image **b** noisy image with $\sigma = 30$, PSNR = 18.72 dB. Denoising using **c** 6-tap orthogonal Q-shift filters of [8], PSNR = 23.35 dB, SSIM [20] = 0.6555 and FSIM [22] = 0.7944. **d** proposed biorthogonal wavelet filters given in the design example, PSNR = 23.23 dB, SSIM [20] = 0.6637 and FSIM [22] = 0.8243

SSIM and FSIM have ideal value of 1. Denoising output shown in Fig. 3c for Q-shift filters have slightly better visual performance when compared to the output for proposed filter set shown in Fig. 3d. However, directional features are better captured using the proposed filters (see the textural features visible in the bottom right hand corner) due to better directional selectivity of their 2D dual-tree directional wavelets. Better visual performance of Q-shift filters may be reasoned for their shorter lengths which help in suppressing the *ringing* artifacts near the image edges.

6 Conclusion

In this paper, we have proposed a new approach to design a set of biorthogonal wavelet filters of DTCWT having near-orthogonal filter response characteristics which are useful to get tight-frame transform. The proposed approach is based on optimization of free variables obtained through factorization of generalized halfband polynomial. Use of unconstrained optimization makes the approach simple and computationally effective. Associated wavelets of the filters obtained using the proposed approach have better analytic properties leading to improved shift-invariance. Image denoising performance using simple thresholding applied on the multiscale six directional subbands of the 2D DTCWT constructed using the proposed set of filters shows better directional selectivity and comparable performance as compared to one of the best performing orthogonal DTCWT filter set.

References

1. Anantrasirichai, N., Achim, A., Kingsbury, N.G., Bull, D.R.: Atmospheric turbulence mitigation using complex wavelet-based fusion. IEEE Trans. Image Process. **22**(6), 2398–2408 (2013)
2. Ansari, R., Kim, C.W., Dedovic, M.: Structure and design of two-channel filter banks derived from a triplet of halfband filters. IEEE Trans. Circuits Syst. II Analog. Digit. Signal Process. **46**(12), 1487–1496 (1999)

3. Asikuzzaman, M., Alam, M.J., Lambert, A.J., Pickering, M.R.: Robust DT-CWT based DIBR 3D video watermarking using chrominance embedding. IEEE Trans. Multimed. **18**(9), 1733–1748 (2016)
4. Burrus, C.S.: Bases, orthogonal bases, biorthogonal bases, frames, tight frames, and unconditional bases. https://cnx.org/contents/Es1rEfS5@4/Bases-Orthogonal-Bases-Biortho. Accessed 24 April 2017
5. Donoho, D.L., Johnstone, I.M.: Ideal spatial adaptation by wavelet shrinkage. Biometrica **81**(3), 425–455 (1994). Aug
6. Fierro, M., Ha, H.G., Ha, Y.H.: Noise reduction based on partial-reference, dual-tree complex wavelet transform shrinkage. IEEE Trans. Image Process. **22**(5), 1859–1872 (2013)
7. Kingsbury, N.: Image processing with complex wavelets. Philos. Trans. R. Soc. Lond. A Math. Phys. Eng. Sci. **357**(1760), 2543–2560 (1999)
8. Kingsbury, N.: A dual-tree complex wavelet transform with improved orthogonality and symmetry properties. In: ICIP. vol. 2, pp. 375–378. IEEE, USA (2000)
9. Kingsbury, N.: Complex wavelets for shift invariant analysis and filtering of signals. Appl. Comput. Harmon. Anal. **10**(3), 234–253 (2001)
10. Oppenheim, A.V., Schafer, R.W.: Discrete-Time Signal Processing. Pearson Higher Education (2010)
11. Patil, B.D., Patwardhan, P.G., Gadre, V.M.: On the design of FIR wavelet filter banks using factorization of a halfband polynomial. IEEE Signal Process. Lett. **15**, 485–488 (2008)
12. Rabbani, H., Gazor, S.: Video denoising in three-dimensional complex wavelet domain using a doubly stochastic modelling. IET Image Process. **6**(9), 1262–1274 (2012)
13. Selesnick, I.W. http://eeweb.poly.edu/iselesni/WaveletSoftware/. Accessed 08 April 2014
14. Selesnick, I.W.: Hilbert transform pairs of wavelet bases. IEEE Signal Process. Lett. **8**(6), 170–173 (2001)
15. Selesnick, I.W.: The design of approximate Hilbert transform pairs of wavelet bases. IEEE Trans. Signal Process. **50**(5), 1144–1152 (2002)
16. Selesnick, I.W., Baraniuk, R.G., Kingsbury, N.C.: The dual-tree complex wavelet transform. IEEE Signal Process. Mag. **22**(6), 123–151 (2005)
17. Shi, H., Luo, S.: A new scheme for the design of Hilbert transform pairs of biorthogonal wavelet bases. EURASIP J. Adv. Signal Process. **2010**, 98 (2010)
18. Tay, D.B.H.: Designing Hilbert-pair of wavelets: recent progress and future trends. In: 6th International Conference on Information Communication and Signal Process. pp. 1–5. IEEE, USA (2007)
19. Tay, D.B., Kingsbury, N.G., Palaniswami, M.: Orthonormal Hilbert-pair of wavelets with (almost) maximum vanishing moments. IEEE Signal Process. Lett. **13**(9), 533–536 (2006)
20. Wang, Z., Bovik, A.C., Sheikh, H.R., Simoncelli, E.P.: Image quality assessment: From error visibility to structural similarity. IEEE Trans. Image Process. **13**(4), 600–612 (2004)
21. Yu, R., Ozkaramanli, H.: Hilbert transform pairs of biorthogonal wavelet bases. IEEE Trans. Signal Process. **54**(6), 2119–2125 (2006)
22. Zhang, L., Zhang, L., Mou, X., Zhang, D.: FSIM: a feature similarity index for image quality assessment. IEEE Trans. Image Process. **20**(8), 2378–2386 (2011)

On Construction of Multi-class Binary Neural Network Using Fuzzy Inter-cluster Overlap for Face Recognition

Neha Bharill, Om Prakash Patel, Aruna Tiwari and Megha Mantri

Abstract In this paper, we propose a Novel Fuzzy-based Constructive Binary Neural Network (NF-CBNN) learning algorithm for multi-class classification. Our method draws a basic idea from Expand and Truncate Learning (ETL), which is a neural network learning algorithm. The proposed method works on the basis of unique core selection, and it guarantees to improve the classification performance by handling overlapping issues among data of various classes by using inter-cluster overlap. To demonstrate the efficacy of NF-CBNN, we tested it on the ORL face data set. The experimental results show that generalization accuracy achieved by NF-CBNN is much higher as compared to the BLTA classifier.

1 Introduction

Neural networks have been successfully applied to problems in pattern classification, function approximation, fault tolerance, medical diagnosis. Usually, real-world problems involve data having multiple classes. Data belongs to multiple classes aims at generating a map from the input data to the corresponding desired output, for a given training set. This kind of mapping is called a classifier. Then, the designed classifier is used to predict class labels of new input instances. One of the most popular neural

N. Bharill (✉) · O. P. Patel · A. Tiwari
Department of Computer Science and Engineering,
Indian Institute of Technology Indore, Indore, India
e-mail: phd12120103@iiti.ac.in

O. P. Patel
e-mail: phd1301201003@iiti.ac.in

A. Tiwari
e-mail: artiwari@iiti.ac.in

M. Mantri
Department of Networth the Finance Club of Indian
Institute of Management Bangalore, Bengaluru, India
e-mail: meghamantri.iiti@gmail.com

© Springer Nature Singapore Pte Ltd. 2019
M. Tanveer and R. B. Pachori (eds.), *Machine Intelligence and Signal Analysis*,
Advances in Intelligent Systems and Computing 748,
https://doi.org/10.1007/978-981-13-0923-6_56

657

network learning algorithms is Backpropagation Learning (BPL) algorithm, which requires an extremely high number of iterations to obtain even a simple binary-to-binary mapping. Also, in the BPL algorithm, finding the number of neurons required in the hidden layer to solve a given problem is a major issue. It has been seen that Stone–Weierstrass's theorem does not give the practical guideline in finding the required number of neurons [1]. Gray and Michel [2] reported Boolean-like training algorithm (BLTA). Its learning speed is vastly increased over the BPL. But the disadvantage is the formation of the four-layer network, and too many neurons must be used for the generation of binary-to-binary mapping.

In 1995 [4], ETL is proposed by Kim Park which initiates learning by selecting a core vertex from the available input data and then form a set of separating hyperplanes based on a geometrical analysis of the training inputs. But the selection of core vertex in ETL is not unique in the process of finding separating hyperplanes. Therefore, the number of separating hyperplanes for a given problem can vary depending upon the selection of the core vertex and the order of adding the training data during learning. Also, it works for two classes of problems. In 2003, Yi Xu and Chaudhari [3] proposed mETL. It uses geometric concepts of ETL [4] for classifying the given patterns into multiple categories. But they did not handle the overlapping issues of data belonging to multiple classes. If the data points belong to the multiple classes then overlapping between the data points occurs because the data does not belong to a particular cluster or neuron. It happens because the data sets may contain the noisy data or due to the effect of outliers. The overlap measure quantifies the degree of overlap between the neurons occurred due to the inclusion of data points within the neuron by computing an inter-cluster overlap [5]. Therefore, the inter-cluster overlap is an important issue that needs to be addressed in case of multi-class classification. In 2015, Wang et al. [6] proposed a feedforward kernel neural networks (FKNN) deep architecture for multi-class classification but the architecture is highly complex and computation intensive for multi-class classification and cannot handle overlapping issue.

In this paper, we proposed a Novel Fuzzy-based Constructive Binary Neural Network (NF-CBNN) learning algorithm for learning multi-class classification problems, which form a three-layer network structure and works on the unique core selection strategy. It also guarantees convergence by grouping all the samples for a given core selection. Further, it can handle the inter-cluster overlap between the neurons which occurs while grouping the data points belonging to multiple classes. Moreover, it eliminates the drawback of ETL [4] in which a number of trials with multiple core selections are required to get the convergence. In addition to it, the proposed learning algorithm NF-CBNN trains the network with single core selection which guaranteed convergence by handling learning of samples that are lying in the overlapped regions of multiple classes. Due to the single core selection, NF-CBNN eliminates processing overheads in hidden layer learning.

The remainder of the paper is organized as follows: In Sect. 2, we present the proposed methodology along with the mathematical formulations. In Sect. 3, we apply the proposed technique on face recognition data set. Experimental results demonstrate that the proposed method performs more accurate face recognition by drastically improving the generalization accuracy and also identifying the variety of faces in ORL face data sets. Finally, concluding remarks are presented in Sect. 4.

2 Proposed Model Description

In this section, the overall concept of Novel Fuzzy-based Constructive Binary Neural Network (NF-CBNN) method is presented. Section A is presented with the general idea of the model. Section B is presented with the preprocessing of images corresponding to faces so that they can be applied as the input to the neural network. In Sect. 1, we present the working of the hidden layer. Finally, section D is presented with the formation of the output layer.

2.1 Overview of the Proposed Model

The Novel Fuzzy-based Constructive Binary Neural Network is proposed which is based on the concept of Expand and Truncate Learning (ETL) [4]. It forms three-layer network structure consisting of the Input layer, Hidden layer, and the Output layer. The proposed approach works for multi-class classification problem, therefore, we need to classify the given set of variables to the multiple classes. Therefore, the input variables are partitioned into multiple groups such as $\{G_1, G_2, \ldots, G_m\}$ based on their outputs. For the multi-class classification problem, all the input variables belong to their respective groups and based on their output will be trained similarly as in case of two-class classification problem. Therefore, at a time the variables present in one group, i.e., G_1 are considered as a true variable concerning the other variables present in the remaining groups, i.e., $\{G_2, \ldots G_m\}$ are considered as false variables.

The proposed approach works by selecting a single core variable for finding all the neurons in the hidden layer because the selection of core affects the number of neurons in the hidden layer. The core variable is selected using the Euclidean distance measure because it works faster than other means of determining correlation. Once, the core is selected then to learn all the variables present in the group G_1, we find the separating hyperplanes, i.e., the hidden layer neurons based on the concept of geometric learning [4]. But, if some of the variables present in group G_1 are still left unlearned within the presently hidden layer neurons, then we compute the fuzzy inter-cluster overlap [5] measure of such variables belonging to group G_1. In this way, we can learn all the remaining variables. If Inter-cluster overlap measure is less than the desired threshold then, that variable is learned within the existing neurons otherwise, the new hyperplane, i.e., new neuron at the hidden layer is required which is generated using single core corresponds to the unlearned variable present in group G_1. Therefore, in this way, all the variables present in group G_1 are learned. Thus, it guarantees convergence by finding the hidden layer neuron corresponding to all the variables present in group G_1 with single core selection.

After training all the variables present in group G_1, these variables are regarded as "don't care" variables concerning the variables present in rest of the groups. All the variables present in group G_2 will be taken as the true variables and rest of the variables present in other groups such as $\{G_3, \ldots, G_m\}$ will be treated as the false variables. Similarly, all the input variables present in group G_2 are trained similarly as of group G_1. During training the input variables of group G_2, it may require more separating hyperplanes, i.e., hidden layer neurons in addition to the number of neurons found corresponding to group G_1 which will always separate the input variables present in group G_2. The same procedure is repeated for training all the groups till last group G_m until all the variables present in their respective groups get trained.

2.2 Preliminaries

Suppose, the set S contains the M input images denoted by $S = \{M_1, M_2, \ldots, M_n\}$ belongs to the multiple classes. Each input image is represented by a $m \times n$ matrix. First, the features are extracted corresponding to each input images. Therefore, the PCA algorithm is used for extracting the features [7]. The feature vectors are generated using the PCA algorithm varies in the range of [0–255]. Once, the feature vectors are extracted using the PCA algorithm, then these feature vectors are quantized by dividing the complete range of available values in three bins which varies in the range of [0–2]. Therefore, the first, second and third bin are given values 0, 1 and 2. The quantization is done to reduce the size of total number of bits of the feature vector. Once, the quantized feature vector is obtained for each image, it is directly converted into the binary form using normal decimal to binary conversion. Once, the quantized feature vector is converted into the binary form, it is fed as the input to the input layer of proposed NF-CBNN approach. Figure 1 describe the flowchart including all the steps involved in the preprocessing phase.

The learning of the network starts with the selection of the core variable. The process of selecting the core variable is discussed as follows: Suppose, the set $S = \{M_1, M_2, \ldots, M_n\}$ contain of all the input variables which belong to the multiple class where each input variable consist of l-$bits$. Then, input variables which belong to the same class are considered as true variables while other input variables are considered as false variables. Let set $S_1 = \{M_1, \ldots, M_c, \ldots, M_p\}$ represent the set of true input variables taken from set S and one variable M_c is selected from set S_1 as a core variable. The formulations involve in the selection of core variable from set S_1 is discussed next.

Randomly, choose one true input variable M_c from set S_1 and assign $v = M_c$ and then, compute the Euclidean Distance(ED) between the variable v from the remaining M_{p-1} variables present in set S_1, is given as

Fig. 1 Flowchart for
preprocessing of Image into
Binary format as input to
Binary Neural Network

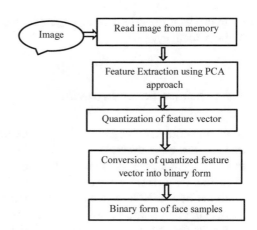

$$(ED)_j = \sqrt{\sum_{j=1}^{p-1}(M_j - v)^2} \tag{1}$$

Compute the Membership Degree (MD) corresponding to all the M_{p-1} true variables by taking the fractional distance of each true variable from variable v.

$$(MD)_j = \frac{(\frac{1}{ED_j})^{\frac{1}{m-1}}}{\sum_{j=1}^{p-1}(\frac{1}{ED_j})^{\frac{1}{m-1}}} \tag{2}$$

Check that Membership Degree (MD) obtained in Eq. 2 corresponding each true variable present in set S_1 should be equal to 1 where $j = \{1, 2, \ldots, p-1\}$.

$$MD_j = 1; \forall j \tag{3}$$

Update the value of the core variable using the Membership Degree obtained in Eq. 2.

$$C_v = \frac{\sum_{j=1}^{p-1}[(MD)_j]^2 M_j}{\sum_{j=1}^{p-1}[(MD)_j]^2} \tag{4}$$

Assign the value of the updated core variable to the initial core variable v.

$$v = C_v \tag{5}$$

In each iteration, the value of the core variable is updated using Eq. 4. Thus, the algorithm terminates when change between the two successive iterations in the value of core variable is less than the predefined threshold $\varepsilon = 0.001$ [8]. After finalizing

C_v in this way, the value of core variable is subsequently used in finding the neurons in the hidden layer.

2.3 Hidden Layer Learning

Once, the core variable is found using the above discussed formulations then we find the separating hyperplanes, i.e., the hidden layer neurons corresponding to each group. Suppose, set S_1 contains all the true variables present in set S, a core variable C_v is found from set S_1 using the above-stated formulations. Then Hamming distance of all the true variables present in set S_1 from a core variable C_v is computed. The number of true variables whose Hamming distance from the core variable is less than d then the generated hyperplane in Eq. 6 will always separate the true variables from the rest of the variables.

$$w_1 x_1 + w_2 x_2 + \cdots + w_n x_n - T = 0, \tag{6}$$

$$w_i = 1 \qquad if \quad C_v^i = 1$$
$$w_i = -1 \qquad if \quad C_v^i = 0$$

$$T = \sum_{p=1}^{n} w_p C_v^k - (d - 1)$$

where C_v^i indicates the ith bit of the core variable C_v, T indicates the threshold.

All these true variables which are linearly separable from the rest of the variables are included in the TRUE list. If the true variables are remaining which are not included in the TRUE list, then the hyperplane is expanded so that the remaining true variables can be included in the TRUE list. Therefore, the Hamming distance of all the remaining true variables with the variables present in the TRUE list is computed. The first nearest true variable whose hamming distance from the variables present in the TRUE list is minimum is chosen as a trial variable and included in the TRUE list. After including the trial variable, the hyperplane is expanded using Eq. 7.

$$(2HC_1 - HC_0)x_1 \cdots + (2HC_i - HC_0)x_2 \ldots$$
$$+ (2HC_n - HC_0)x_n - T = 0 \tag{7}$$

where, HC_0 is the total number of elements present in the TRUE list including the trial variable, HC_i is calculated as follows:

$$HC_i = \sum_{p=1}^{HC_0} M_p^i \qquad (8)$$

If $f_{min} > f_{max}$ then there exists a separating hyperplane after including the trial variable in the TRUE list as defined in Eq. 7 where,

$f_{min} = \min(\sum_{i=1}^{n}(2HC_i - HC_0)M_t)$ among all the variables present in the TRUE list

$f_{max} = \max(\sum_{i=1}^{n}(2HC_i - HC_0)M_r)$ among all the rest of the variables which are not present in the TRUE list.

The threshold T in Eq. 7 is computed using the equation defined as follows:

$$T = \left\lceil \frac{f_{min} + f_{max}}{2} \right\rceil \qquad (9)$$

If $f_{min} \le f_{max}$, then there does not exist a separating hyperplane. Then, the trial variable is removed from the TRUE list. Similarly, for the other true variables, the procedure is repeated similarly, i.e., another true variable is selected using the same criteria and test is performed to check whether the new trail variable can be included in the TRUE list or not. This procedure is repeated until no more true variables can be included in the TRUE list. However, if all the true variables are not included in the TRUE list, then more than one neuron is required at the hidden layer. Therefore, the other neurons are found by converting the true variables which are not present in the TRUE list into the false variables. Similarly, the false variable which is present in set S but not in set S_1 is converted into true variable. This is the temporary conversion of the desired output of each variable. The conversion is performed to find the separating hyperplanes. After conversion, as soon as all the variables are learned, then the proposed method converges by finding the separating hyperplanes. On the contrary, if some of the false variables are still left unlearned along with the true variables even after conversion. The reason why these variables are still left unlearned because adding them to any particular neuron might lead to the overlapping. Thus, we introduced the concept of inter-cluster overlap [5] based on fuzzy set theory [9], for which the new separating hyperplane corresponding to the unlearned samples can only be found out by membership degree. The discussion and formulation for the same are carried out in the subsequent section.

Further, the training of remaining instances is done on the following basis. Instead of choosing multiple core variables for training all variables as discussed in [4], we train the neural network architecture for the remaining variables (unlearned true and false variables) with the help of fuzzy inter-cluster overlap proposed in [5]. Suppose true variables are still left unlearned, we take those unlearned true variables as a trail variable and compute its membership degree (which determines the belongingness of true variable with the neurons) with an odd number of neurons which are generated earlier. The membership degree of the true variable with the odd number of neurons is computed only after updating the center of odd neurons using Eq. 8. Similarly, for the unlearned false variables, we compute the membership degree with an even

number of neurons only after updating the center of even neurons using Eq. 8. The
formulation for computing the membership degree is defined as follows:

$$\mu_i(x_j) = \frac{\|x_j - HC_i\|^{\frac{2}{m-1}}}{\sum_{k=1}^{n} \|x_j - HC_k\|^{\frac{2}{m-1}}}; \forall i, j \qquad (10)$$

where x_j represents the jth false or the true variables, HC_i represents the center of
ith neurons, m is the weighting component.

Once, the cluster membership degree of all the unlearned variables with there
respective neurons are computed then we compute the inter-cluster overlap measure.
The formulations used for computing the inter-cluster overlap measure is defined as
follows:

$$R(x_j, N_l, N_r) = \begin{cases} \delta(x_j), & if\left(Dom_{\min}(x_j) > 0 \ \& \right. \\ & \left. Dom_{\max}(x_j) < 1\right) \\ 0.0, & Otherwise \end{cases} \qquad (11)$$

where

$$Dom_{\min}(x_j) = \min(\mu_{F_l}(x_j), \mu_{F_r}(x_j)) \qquad (12)$$

$$Dom_{\max}(x_j) = \max(\mu_{N_l}(x_j), \mu_{N_r}(x_j)) \qquad (13)$$

Conditions:

1. If data point x_j is highly vague, i.e., if maximum degree of membership
   ($Dom_{\max}$) is less than or equal to 0.5 then, assign degree of overlap equal to
   1.0.

   (a) $Dom_{\max}(x_j) \leq 0.5$ then, $\delta(x_j) = 1.0$.

2. Conversely, if the data point x_j is not vague, i.e., if maximum degree of membership ($Dom_{\max}$) is greater than 0.5 and less than 1.0 then, assign degree of
   overlap between 0.1 to 0.9

   (a) $0.5 < Dom_{\max}(x_j) \leq 1.0$ then, $\delta(x_j) \in [0.1, \ldots 0.9]$.

3. Otherwise, if the data point x_j is clearly classified to particular cluster then,
   assign degree of overlap equal to 0.

   (a) If $Dom_{\max}(x_j) = 1$ then, $\delta(x_j) = 0$.

where,

$Dom_{\max}(x_j)$ is the maximum degree of membership of trail variable x_j.
$Dom_{\min}(x_j)$ is the minimum degree of membership of trail variable x_j.

The inter-cluster overlap measure computes the overlap due to the inclusion of
true or false variable x_j as a trial variable within the respective neurons which is
formally represented by $R(x_j, N_l, N_r)$ in Eq. 11. Each trail variable x_j is assigned
a degree of overlap based on the belongingness of trail variable to the respective

neuron denoted by $\delta(x_j)$. If the degree of overlap $\delta(x_j)$ achieved by the trial variable is less than 1, then we can learn the trial variable within the existing neuron to which its $\mu_i(x_j)$, i.e., the membership degree of trail variable x_j concerning the neurons N_l and N_r are higher. However, if the degree of overlap $\delta(x_j)$ achieved by the trial variable is equal to the defined threshold 1.0, then the trial variable cannot be included within the existing neurons. This is because the inclusion of trial variable in the existing neuron will result in vagueness then, the new separating hyperplane is found out using Eq. 6. If the learned variable is a false variable then it is added to the FALSE list but, if the learned variable is true variable, then it is added to the TRUE list. Similarly, the process is repeated for training all the unlearned false and true variables. Thus, the algorithm converges by ensuring that all the true and false variables are learned in their respective neurons. The same process works for different groups $\{G_1, G_2, \ldots, G_m\}$ belongs to different classes.

2.4 Formation of Output Layer

Once all the neurons in the hidden layer are found using the single core variable, then another problem in case of multi-class classification problem is that we need to classify the given set of variables to the multiple classes. For this purpose, we required multiple output neuron in the output layer so that the input variables can be classified to the multiple classes. As discussed earlier that the input variables are partitioned into multiple groups such as $\{G_1, G_2, \ldots, G_m\}$ based on their outputs. Since only one output neuron will be fired corresponding to each group. As soon as, all the required neurons in hidden layer are found, the number of hidden layer neuron found corresponding to group G_1 will only be connected to the Ist output neuron. However, the number of hidden layer neuron found corresponding to group G_1 and G_2 will be connected to the IInd output neuron. Similarly, the mth output neuron is connected to the hidden layer neurons of only up to G_m. Therefore, it is inferred that proposed approach requires the less number of neurons in the hidden layer as compared to the multi-class architecture proposed in [3]. However, the weights and threshold of the output neurons are set as follows: The weight of a link from the odd-numbered hidden neuron to the output neuron is set to 1. Similarly, the weight of a link from the even-numbered hidden neuron to the output neuron is set to -1. The threshold of output neuron is set to 1 if the total number of hidden neuron layer generated by the particular group is odd. Otherwise, the threshold of output neuron is set to 0, if the total number of hidden neuron layer generated by the particular group is even. Thus, the proposed approach will always produce the correct output corresponding to the all the input variables. Figure 2 depicts the proposed three layers neural network architecture for the multi-class classification problem.

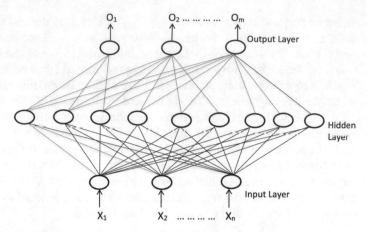

Fig. 2 Neural Network Architecture for the Multi-class Classification

3 Experiments and Analysis

In this section, the proposed NF-CBNN classifier for the multi-class problem is used for face recognition, and its performance is evaluated on the ORL face database. This section is presented with the description of face database, experimental setup, experimental results and discussion.

3.1 ORL Database Description

Experiments were performed on the ORL face database [10–12] as shown in Fig. 3, which consist a set of faces taken by the Olivetti Research Laboratory in Cambridge, U.K.[3] between April 1992 and 1994. The database contains 10 different images of each of the 40 distinct subjects numbered from 1 to 40. For some persons, the images were taken at different times by varying light with different facial expressions (open/closed eyes, smiling/not smiling) and facial details (glasses/no glasses).

3.2 Experimental Setup

The experimentation is carried out on Intel(R) Xeon(R) $E5 - 1607$ Workstation PC with 64 GB of memory and running with a processing speed of 3.0 GHz on Windows 7 Professional operating system. All codes are written in the MATLAB computing environment. To evaluate the effectiveness of proposed approach, we use stratified tenfold cross-validation method. The images of 40 distinct subjects are divided into ten subsets in which each subset is partitioned into two halves: one

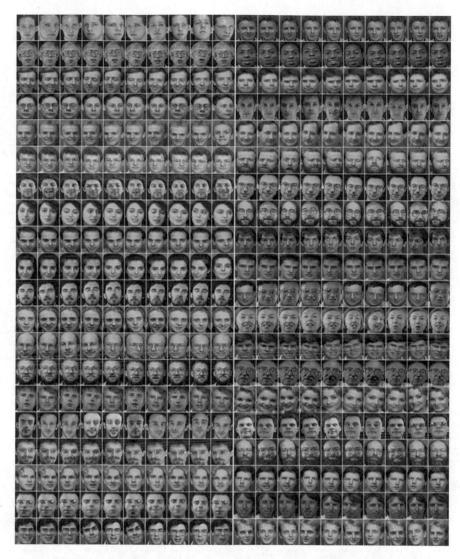

Fig. 3 The ORL face database. There are 10 sample images of each of 40 distinct persons

half is used for training contain 70% images of each subject and the other half is used for testing contain 30% images of each subject. Thus, in all 280 images are used for training and 120 images are used for testing. Then, ten rounds of training and testing runs are conducted repeatedly and independently. Results produced are reported subsequently.

Table 1 Results obtained with NF-CBNN classifier and BLTA classifier on face database. Comparative results are reported across tenfold in terms of three parameters are mean, standard deviations (Std. Dev.), classification accuracy

Set of images	NF-CBNN classifier			BLTA classifier		
	Mean	Std. Dev.	Classification accuracy	Mean	Std. Dev.	Classification accuracy
Image1	0.1000	0.0000	100.0 ± 00.00	0.4666	0.1721	46.66 ± 17.21
Image2	0.1000	0.0000	100.0 ± 00.00	0.1333	0.1721	13.33 ± 17.21
Image3	0.1000	0.0000	100.0 ± 00.00	0.3333	0.2721	33.33 ± 27.21
Image4	0.1000	0.0000	100.0 ± 00.00	0.2333	0.2249	23.33 ± 22.49
Image5	0.1000	0.0000	100.0 ± 00.00	0.1999	0.2811	19.99 ± 28.10
Image6	0.1000	0.0000	100.0 ± 00.00	0.2333	0.2744	23.33 ± 27.44
Image7	0.5666	0.2250	56.66 ± 22.50	0.3333	0.2222	33.33 ± 22.22
Image8	0.1000	0.0000	100.0 ± 00.00	0.1666	0.2357	16.66 ± 23.57
Image9	0.1000	0.0000	100.0 ± 00.00	0.2333	0.2249	23.33 ± 22.49
Image10	0.1000	0.0000	100.0 ± 00.00	0.2999	0.2918	29.99 ± 29.18
Image11	0.1000	0.0000	100.0 ± 00.00	0.2666	0.2629	26.66 ± 26.29
Image12	0.3999	0.3063	39.99 ± 30.63	0.0666	0.1405	06.66 ± 14.05
Image13	0.1000	0.0000	100.0 ± 00.00	0.1999	0.2331	19.99 ± 23.30
Image14	0.5666	0.2250	56.66 ± 22.50	0.1666	0.2357	16.66 ± 23.57
Image15	0.4666	0.2811	46.66 ± 28.11	0.0666	0.1405	14.05 ± 6.660
Image16	0.1000	0.0000	100.0 ± 00.00	0.1666	0.2367	23.57 ± 16.66
Image17	0.1000	0.0000	100.0 ± 00.00	0.1333	0.2331	23.30 ± 13.33
Image18	0.1000	0.0000	100.0 ± 00.00	0.0999	0.1609	16.10 ± 9.990
Image19	0.7666	0.1611	76.66 ± 16.11	0.0333	0.1054	10.54 ± 3.330
Image20	0.1000	0.0000	100.0 ± 00.00	0.2666	0.2108	21.08 ± 26.66
Image21	0.0000	0.0000	00.00 ± 00.00	0.0000	0.0000	00.00 ± 00.00
Image22	0.0000	0.0000	00.00 ± 00.00	0.0000	0.0000	00.00 ± 00.00
Image23	0.1000	0.0000	100.0 ± 00.00	0.0333	0.1054	03.33 ± 10.54
Image24	0.6333	0.3668	63.33 ± 36.68	0.0666	0.2108	06.66 ± 21.08
Image25	0.0000	0.0000	00.00 ± 00.00	0.0000	0.0000	00.00 ± 00.00
Image26	0.0000	0.0000	00.00 ± 00.00	0.0000	0.0000	00.00 ± 00.00
Image27	0.0000	0.0000	00.00 ± 00.00	0.0000	0.0000	00.00 ± 00.00
Image28	0.0000	0.0000	00.00 ± 00.00	0.0000	0.0000	00.00 ± 00.00
Image29	0.0000	0.0000	00.00 ± 00.00	0.0000	0.0000	00.00 ± 00.00
Image30	0.0000	0.0000	00.00 ± 00.00	0.0000	0.0000	00.00 ± 00.00
Image31	0.0000	0.0000	00.00 ± 00.00	0.0000	0.0000	00.00 ± 00.00
Image32	0.1000	0.0000	100.0 ± 00.00	0.1666	0.2357	16.66 ± 23.57
Image33	0.0000	0.0000	00.00 ± 00.00	0.0000	0.0000	00.00 ± 00.00
Image34	0.0000	0.0000	00.00 ± 00.00	0.0000	0.0000	00.00 ± 00.00
Image35	0.6999	0.2918	69.99 ± 29.19	0.0666	0.1405	06.66 ± 14.05
Image36	0.0000	0.0000	00.00 ± 00.00	0.0000	0.0000	00.00 ± 00.00
Image37	0.0000	0.0000	00.00 ± 00.00	0.0000	0.0000	00.00 ± 00.00
Image38	0.0000	0.0000	00.00 ± 00.00	0.0000	0.0000	00.00 ± 00.00
Image39	0.0333	0.1054	03.33 ± 10.54	0.0000	0.0000	00.00 ± 00.00
Image40	0.1000	0.0000	100.0 ± 00.00	0.0999	0.1609	09.99 ± 16.10
Overall	0.1483	0.2148	65.00 ± 33.86	0.1133	0.1228	58.25 ± 23.31

3.3 Experimental Results and Discussion

In this section, experimental results are reported to investigate the effectiveness of proposed NF-CBNN classifier for multi-class face recognition problem in comparison with another classifier works on a multi-class problem known as BTLA [2]. In Table 1 the results of the NF-CBNN classifier in comparison with BLTA classifier across stratified tenfold cross-validation is reported in terms of three parameters; Mean, Standard Deviation, Classification Accuracy. It can be inferred from the Table 1, that the maximum classification accuracy achieved by the proposed NF-CBNN classifier 100 ± 00.00 on images sets are $Image1$-$Image6$, $Image8$-$Image11$, $Image13$, $Image16$-$Image18$, $Image20$, $Image23$, $Image32$ and $Image40$. However, for the other images sets the classification accuracy varies in the range from 00.00 ± 00.00 to 100 ± 00.00. On the contrary, the BLTA classifier achieves 46.66 ± 00.00 as the maximum classification accuracy only on the images set $Image1$. However, it achieves the minimum classification accuracy as 00.00 ± 00.00 on images sets are $Image21$-$Image22$, $Images25$-$Image31$, $Image33$-$Image34$ and $Image36$-$Image39$ whereas on the other sets of images the classification accuracy varies in the range from 00.00 ± 00.00 to 46.66 ± 00.00. Thus, we can observe that the classification accuracy achieved by the proposed NF-CBNN classifier is comparatively much higher than the classification accuracy achieved by the BLTA classifier on stratified tenfold cross-validation. Hence, the above observation leads to the realization that NF-CBNN classifier outperforms over the BLTA classifier.

4 Conclusion

In this paper, we proposed a Novel Fuzzy-based Constructive Binary Neural Network (NF-CBNN) learning algorithm for solving multi-class classification problem. The proposed approach works on the basic idea taken from the Expand and Truncate Learning (ETL) [4]. Thus, it guarantees convergence with unique core selection. The proposed approach converges by learning all the samples including the samples which are confusing in a sense belonging to more than one classes. This issue is handled by using the inter-cluster overlap measure [5]. Thus, it ensures improvement in the classification accuracy. To validate the efficacy of the proposed NF-CBNN classifier, we perform the experimentation on face data set [10] using 10 fold cross-validation approach. The performance of NF-CBNN classifier is judged in comparison with the BLTA classifier [2] using same data set. It is observed that classification accuracy achieved by NF-CBNN classifier is improved greatly as compared to BLTA classifier because the proposed approach handles the confusing samples (samples with belongs to various classes) using the fuzzy inter-cluster overlap measure [5]. Also, NF-CBNN classifier achieves higher accuracy with 3 layer networks structure in comparison with BLTA classifier which forms four-layer network structure. From the results reported in Table 1, it is obvious that BLTA classifier is unable to classify many images sets properly and thus it achieves very poor classification accuracy. However, the proposed NF-CBNN classifier can classify more image sets. Reported results are verifying that the identification of images with our method is very high in comparison with BLTA classifier and achieves significant improvement in the classification accuracy. In addition to that, our method claims that if more variations of the same image samples have been taken, then there is an improvement of membership degree. Thus, it helps in improving the learning of unclassified samples more appropriately.

References

1. Cotter, N.E.: The stone-weierstrass theorem and its application to neural networks. IEEE Trans. Neural Netw./a Publ. IEEE Neural Netw. Counc. **1**(4), 290–295 (1989)
2. Gray, D.L., Michel, A.N.: A training algorithm for binary feedforward neural networks. IEEE Trans. Neural Netw. **3**(2), 176–194 (1992)
3. Xu, Y., Chaudhari, N.: Application of binary neural networks for classification. In: 2003 International Conference on Machine Learning and Cybernetics, pp. 1343–1348. IEEE (2003)
4. Kim, J.H., Park, S.K.: The geometrical learning of binary neural networks. IEEE Trans. Neural Netw. **6**(1), 237–247 (1995)
5. Bharill, N., Tiwari, A.: Enhanced cluster validity index for the evaluation of optimal number of clusters for fuzzy c-means algorithm. In: IEEE World Congress on Computational Intelligence. International Conference on Fuzzy Systems(FUZZ-IEEE), pp. 1526–1533. IEEE, Beijing, China (2014)
6. Wang, S., Jiang, Y., Chung, F.L., Qian, P.: Feedforward kernel neural networks, generalized least learning machine, and its deep learning with application to image classification. Appl. Soft Comput. **37**, 125–141 (2015)
7. Pentland, A.: Looking at people: sensing for ubiquitous and wearable computing. IEEE Trans. Pattern Anal. Mach. Intell. **22**(1), 107–119 (2000)
8. Setnes, M., Babuska, R.: Fuzzy relational classifier trained by fuzzy clustering. IEEE Trans. Syst. Man Cybern. Part B: Cybern. **29**(5), 619–625 (1999)
9. Kim, D.W., Lee, K.H., Lee, D.: On cluster validity index for estimation of the optimal number of fuzzy clusters. Pattern Recogn. **37**(10), 2009–2025 (2004)
10. Pigeon, S., Vandendorpe, L.: The m2vts multimodal face database (release 1.00). In: Audio-and Video-Based Biometric Person Authentication, pp. 403–409. springer (1997)
11. Angadi, S.A., Kagawade, V.C.: A robust face recognition approach through symbolic modeling of polar FFT features. Pattern Recogn. **71**, 235–248 (2017)
12. Li, H., Suen, C.Y.: Robust face recognition based on dynamic rank representation. Pattern Recogn. **60**, 13–24 (2016)

Electromyogram Signal Analysis Using Eigenvalue Decomposition of the Hankel Matrix

Rishi Raj Sharma, Pratishtha Chandra and Ram Bilas Pachori

Abstract The identification of neuromuscular abnormalities can be performed using electromyogram (EMG) signals. In this paper, we have presented a method for the analysis of amyotrophic lateral sclerosis (ALS) and normal EMG signals. The motor unit action potentials (MUAPs) have been extracted from EMG signals. The proposed method is based on improved eigenvalue decomposition of the Hankel matrix (IEVDHM). Two significant decomposed components obtained from IEVDHM, are considered for analysis purpose. These components are obtained on the basis of higher energy of components. Correntropy (CORR) and cross-information potential (CIP) are computed for two components. Thereafter, statistical analysis has been performed using the Kruskal–Wallis statistical test. We have observed that the IEVDHM method is able to provide the components, which can distinguish the ALS and normal EMG signals using CORR and CIP parameters.

Keywords Amyotrophic lateral sclerosis · Correntropy · Eigenvalue decomposition · Electromyogram · Hankel matrix

R. R. Sharma (✉) · P. Chandra · R. B. Pachori
Discipline of Electrical Engineering, Indian Institute of Technology Indore,
Indore 453552, India
e-mail: phd1501102018@iiti.ac.in

P. Chandra
e-mail: mt1602102012@iiti.ac.in

R. B. Pachori
e-mail: pachori@iiti.ac.in

© Springer Nature Singapore Pte Ltd. 2019
M. Tanveer and R. B. Pachori (eds.), *Machine Intelligence and Signal Analysis*,
Advances in Intelligent Systems and Computing 748,
https://doi.org/10.1007/978-981-13-0923-6_57

1 Introduction

The neuromuscular activation associated to the contraction potential of the skeletal muscles can be seen in the form of electrical activity. These electrical activities are recorded to study the action of muscles and known as electromyogram (EMG) [29]. The EMG signals are characterized as nonstationary in nature and affected by the structural and functional characteristics of muscles. The status of muscles and their functioning are also studied by the information contained in the EMG signals. Aforesaid information recorded in EMG signals have several applications in the field of medical, rehabilitation, sports science, and ergonomics [13]. Muscle action is controlled by electrical signals produced from the human brain. Electrical impulses transferred by brain follow a path formed by motor neurons. The disordered motor neurons may not provide proper path to electrical impulses from brain which is termed as disease [13]. The recorded EMG signals can be used to study these neuromuscular diseases such as myopathy and amyotrophic lateral sclerosis (ALS) [19, 21, 30].

The ALS is a promptly growing disease, which may lead to death of patient [19]. Symptoms of ALS are respiratory failure, atrophy, and weakness. Respiratory failure due to ALS may give rise to death of the victim, which is normally after 3–5 years from the commencement of disease [19]. Generally, middle-to-old-aged persons are affected by ALS [19]. Other age groups of persons may also be affected by it. The diagnosis of ALS disease using visual scanning of EMG signals is very time consuming and may be inaccurate. Moreover, it is performed by trained professionals and clinicians. Therefore, an automated method which can provide accurate results is required for detection of ALS EMG signals.

The motor unit action potentials (MUAPs) [22] play an important role in diagnosis of disease and used by neurophysiologists. The structure of MUAPs may become different for abnormal muscles and can be used for identification of disease [5]. Manual inspection of characteristics of MUAPs to find abnormalities also needs experienced neurophysiologist. The complicated structures of MUAPs make manual assessment difficult for disease detection. Therefore, a quantitative analysis of MUAPs is necessary to identify the abnormalities in the structure of MUAPs.

Several methodologies have been developed for detection of ALS disease [4, 20, 23]. In [23], a methodology has been presented for features extraction to distinguish abnormal EMG signals. Time-domain and frequency-domain-based features have been studied for ALS disease diagnosis in [4]. The wavelet transform-based EMG signal analysis has been proposed to characterize them in terms of singularity in [1]. The empirical mode decomposition (EMD) method has been applied on EMG signals directly for feature extraction [20]. The wavelet packet transform has also been used for EMG signal analysis [3]. Classification of neuromuscular disease can be performed by applying features directly on nonoverlapping frames obtained from the EMG signals [6]. A considered frame contains several MUAPs. In MUAP- based methods, features are applied on each MUAP for further analysis [22]. A method based on tunable-Q wavelet transform has been proposed for abnormal EMG signal detection, in which MUAPs are decomposed in to its components and thereafter features are computed from decomposed components and directly from MUAPs [12].

A signal decomposition method based on eigenvalue decomposition of Hankel matrix (EVDHM) has been proposed in [11]. The nonstationary signal analysis has been performed using combination of EVDHM and Hilbert transform [26]. A method for time–frequency representation (TFR) using improved eigenvalue decomposition of Hankel matrix (IEVDHM) method is recently proposed in [25]. The EVDHM-based method has also been applied for voiced speech signal analysis [9, 10]. In this work, we have applied IEVDHM method for the analysis of EMG signals. The MUAP signals are decomposed using IEVDHM method. Thereafter, two parameters namely correntropy which is denoted by CORR and cross- information potential which is denoted by CIP [28] have been computed from the decomposed components. In this work, our main aim is to study these two parameters to discriminate the normal and ALS EMG signals using decomposed components. The Kruskal–Wallis statistical test [18] has been performed on the obtained features for determining statistical significant difference.

2 Data Used

The EMG dataset has been obtained from EMGLABs which is publicly available online [17]. Concentric needle electrodes which have leading of area $0.07 \, mm^2$ are used for signal recording [21]. A 16 bit analog-to-digital converter has been used to digitize the signals and sampling rate of this data is 23437.5 Hz. A bandpass filter has been used to filter the signal in the frequency range from 2 Hz to 10 kHz. The data consists of three distinct groups of people which are ALS, myopathy, and normal. We are using ALS and normal groups. The ALS group has 8 persons including 4 males and 4 females between age 35 and 67 years. Five out of eight persons of this group expired after few years of the disease disclosure. Persons of normal group are healthy. Neither they are suffering from neuromuscular disease presently, nor they have history of any kind of disease related to the neuromuscular disorder. Normal group has 10 persons out of which 4 are females and 6 are males between age 21 and 37 years.

3 Methodology

Methodology proposed for EMG signal analysis is shown in Fig. 1. First of all, EMG signals are segmented into several MUAPs. The IEVDHM method has been applied to decompose the MUAPs of EMG signals. The components corresponding to highest energy are obtained. Thereafter, CORR and CIP have been computed for the selected decomposed components. The statistical analysis has been performed

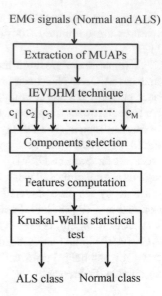

Fig. 1 Block diagram of the proposed methodology based on IEVDHM for EMG signal analysis

using the Kruskal–Wallis statistical test. All these steps of proposed methodology are explained below.

3.1 MUAP Extraction

We have extracted MUAPs from EMG signals. The task of MUAPs extraction has been performed in two steps. First, segmentation has been accomplished. Thereafter, clustering has been done. Both the steps are explained below:

Segmentation has been done by dividing the EMG signal into several time intervals, which hold the MUAPs. There may be one MUAP or several superimposed MUAPs in one segment. A window with 5.6 s duration is applied on the EMG signal [22]. If the variance of the signal presented within the window is more than a threshold value, that part of signal is considered as MUAP. The variance is computed as follows [22]:

$$\text{var}(q) = \frac{1}{N-1} \sum_{i=-k}^{k} s^2[q+i] - \left(\frac{1}{N-1} \sum_{i=-k}^{k} s[q+i] \right)^2 \tag{1}$$

Where $\text{var}(q)$ represents the variance at qth sample of signal $s[q]$ calculated in $\pm k$ sample range and N represents the segment length. The threshold value is set by applying amplitude density function on the normalized variance of signal.

After performing segmentation, the segments which look similar are brought together to form clusters. There may be several segments in a group. In the process of clustering, segments are made equal by padding zeros in the short segments if they are not equal. Cluster which contains minimum five segments is known as potential class (PCL). Each PCL is represented by the template, which signifies the active MUAP of the EMG signal [22].

3.2 IEVDHM Method

Consider a multi-component nonstationary signal $s(n)$ of length $2N - 1$. A Hankel matrix H_M has been formed using signal $s(n)$. The size of Hankel matrix H_M is M. The matrix H_M can be represented as [11]:

$$H_M = \begin{bmatrix} s(1) & s(2) & .. & s(N) \\ s(2) & s(3) & .. & s(N+1) \\ . & . & .. & . \\ . & . & .. & . \\ s(N) & s(N+1) & .. & s(2N-1) \end{bmatrix} \tag{2}$$

The relation between eigenvalue matrix λ_s and eigenvector matrix V_s of H_M, can be represented as follows:

$$H_M = V_s \lambda_s V'_s \tag{3}$$

Where, the eigenvalue matrix λ_s of H_M holds nonzero entries in its diagonal and can be represented as:

$$\lambda_s = \begin{bmatrix} \lambda_1 & 0 & .. & 0 \\ 0 & \lambda_2 & .. & 0 \\ . & . & ... & . \\ 0 & . & .. & 0 \\ 0 & 0 & .. & \lambda_M \end{bmatrix} \tag{4}$$

We have applied the modified significant threshold point (MSTP) [25] for obtaining significant eigenvalue pairs. The MSTP states that the sum of magnitude of significant eigenvalue pairs must be more than 95% of sum of magnitude of all the eigenvalues.

Each eigenvalue pair has been used to obtain the decomposed component using the mathematical expression. The kth decomposed component has been presented as [25]

$$s_k(n) = \lambda_k \, \text{fn}(V_k(n)) + \lambda_{M-k+1} \, \text{fn}(V_{M-k+1}(n)) \tag{5}$$

Where the kth decomposed component is $s_k(n)$ and an operation 'fn$(V_k(n))$' applied on kth eigenvector $V_k(n)$ for $n = 1, 2, \dots, M$, can be described as

$$\text{fn}(V_k(n)) = \sum_{p=1}^{M} \frac{V_k(n-p)V_k(p)}{p} + \sum_{p=M+1}^{2M-1} \frac{V_k(n-p)V_k(p)}{(2M-p)} \qquad (6)$$

The obtained decomposed components are required to satisfy the modified mono-component signal criteria (MMSC) [25]. The MMSC has two conditions which are given as follows [8, 25]:

Condition 1: It states that the number of zero-crossings and sum of local maxima and local minima should be different by at most one.
Condition 2: It states that the average of both upper envelope and lower envelope should be zero.

The components which do not satisfy the MMSC conditions are advised to follow the above-mentioned process again. If the energy of the components which does not satisfy the MMSC is less than 5% energy of main signal, process can be terminated.

The obtained decomposed components are merged to each other if they have overlapping of 3 dB bandwidth or more. Finally, the components obtained after merging are the decomposed components of nonstationary multi-component signal.

In the present study, we have used up to three iterations for decomposing MUAPs of EMG signals. A MUAP extracted from EMG signal taken from ALS class is depicted in Fig. 2a and its decomposed components using IEVDHM method are depicted in Fig. 2b. Similarly, in Fig. 3a, a MUAP extracted from normal EMG signal has been presented and in Fig. 3b, its decomposed components have been shown. We have used two decomposed components corresponding to higher energies. Features extraction has been explained next in this section.

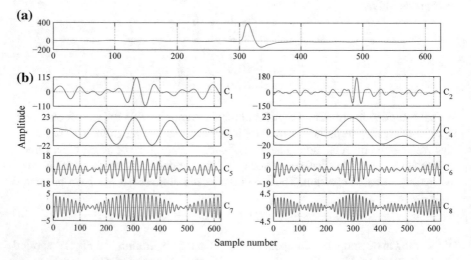

Fig. 2 Plot of **a** MUAP of EMG signal recorded from a patient suffering from ALS disease; **b** its decomposed components using IEVDHM method

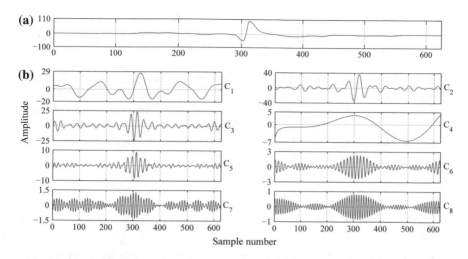

Fig. 3 Plot of **a** MUAP of EMG signal recorded from a healthy person; **b** its decomposed components using IEVDHM method

3.3 Features Extraction

It is noticed that time-domain and frequency-domain features are applied directly on the EMG signal to find the discrimination between abnormal and normal EMG signals [4]. Features applied on the decomposed components of EMG signal are also useful for classification. In the proposed method, we have studied CORR and CIP for the two components of a MUAP signal which have highest energy between all the decomposed components. The CORR and CIP have been explained below.

3.3.1 Correntropy and Cross-Information Potential:

The CORR represents the similarity of signal with its delayed samples. The CIP measures the correlation between two signals. The ITL toolbox taken for CORR and CIP computation is available at http://www.sohanseth.com/Home/codes. In this toolbox, incomplete Cholesky decomposition is applied for the computation of CIP with Gaussian kernel.

The CORR for signal S can be estimated as [16]

$$\text{CORR} = \frac{1}{P} \Sigma_{i=1}^{P} \frac{1}{\sqrt{2\pi}\sigma} \left(e^{\frac{|S-S_i|^2}{2\sigma^2}} \right) \tag{7}$$

Similarity within the probability density functions of two components is measured by CIP [28], which can be given as [27]

$$\text{CIP}(S_1, S_2) = \frac{1}{P} \sum_{a=1}^{P} \sum_{b=1}^{P} k(S_{1_a} - S_{2_b}) \tag{8}$$

Where $k(S_{1_a} - S_{2_b})$ is the kernel function. The ath sample of data set S_1 is S_{1_a} and bth sample of data set S_2 is S_{2_b}. There are totally P samples and the kernel parameter, σ is taken 1 [15]. In this work, Gaussian kernel is used with kernel size 1 for obtaining discriminatory features. Previously, CORR has been studied for classification of focal electroencephalogram (EEG) signal [7, 24] and CIP has been used for coronary artery disease characterization in [15].

3.4 Statistical Analysis

Statistical analysis is performed for finding the nature of data which provides the relation of data with the population. We have applied the Kruskal–Wallis statistical test which is a nonparametric test [18]. In this method, both class data is arranged in single series and a rank is assigned to all the data from both group. We have applied the Kruskal–Wallis statistical test [18] for getting the statistical significance (p<0.05) of features in different oscillatory levels of the MUAPs extracted from EMG signals. The Kruskal–Wallis statistical test has also been studied for EEG signal analysis in [2] and for coronary artery disease detection in [14].

4 Results and Discussion

We have extracted MUAPs from EMG signal. Thereafter, MUAPs are decomposed using IEVDHM method. An MUAP signal of ALS class and its decomposed components are depicted in Fig. 2. Similarly, in Fig. 3 a normal EMG signal and its decomposed components have been shown.

We have obtained two components of each MUAP signal which have highest energies. Thereafter, CORR and CIP parameters have been computed. The features CORR computed for first component (CORR_1), CORR computed for second component (CORR_2), and CIP have been considered for analysis. The Kruskal–Wallis statistical test [18] has been performed for getting the statistical significance ($p < 0.05$) of all features in different oscillatory levels of the MUAPs extracted from EMG signals. The mean and standard deviation (std) of features have been presented in Table 1. It is noticed that the mean values of CORR feature of ALS class are higher than mean values of the features of normal class. Moreover, the mean values of CIP feature of normal class are higher than that of ALS class. The p-values for these features are also given in Table 1, which are less than 0.05. This shows the discriminatory behavior of the proposed method for EMG signals. The box plots for CORR_1,

Table 1 Mean and standard values of features with p-value for ALS and normal class MUAPs extracted from EMG signals

Features	ALS (mean ± std)	Normal (mean ± std)	p-value
CORR$_1$	0.179 ± 0.019	0.160 ± 0.024	8.98×10^{-128}
CORR$_2$	0.168 ± 0.026	0.141 ± 0.032	1.77×10^{-84}
CIP	0.0192 ± 0.014	0.045 ± 0.024	1.69×10^{-135}

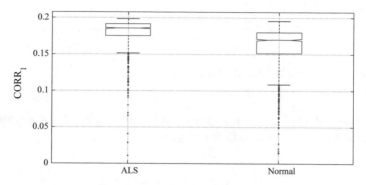

Fig. 4 Box plot of CORR$_1$ feature for ALS and normal MUAPs

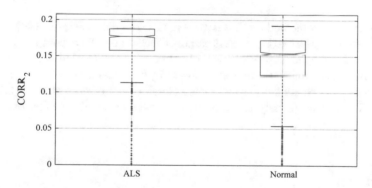

Fig. 5 Box plot of CORR$_2$ feature for ALS and normal MUAPs

CORR$_2$, and CIP are shown in Figs. 4, 5, and 6. All the features have high mean values for normal class MUAPs as compared to the ALS class MUAPs.

In the methodology proposed in [12], total 12 parameters are applied on MUAPs for EMG signal classification. Eight time domain-based parameters are applied on the MUAPs directly and four entropy-based parameters are applied on the decomposed components. In [20], parameters namely bandwidth, centre tendency measurement, mean of first derivative of instantaneous frequency are applied on the decomposed components. Our aim was to check CORR and CIP parameters of the components decomposed by using IEVDHM method for discrimination between normal and ALS EMG signals using MUAPs. This method provides the decomposed components in

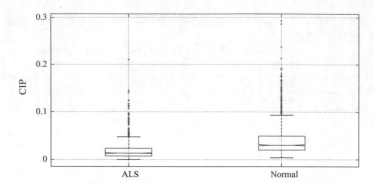

Fig. 6 Box plot of CIP feature for ALS and normal MUAPs

the decreasing order of energy it helps us to select the required two components for our proposed analysis based on their energy values.

5 Conclusion

In this work, a new method for the analysis of EMG signals using IEVDHM method has been studied. The two parameters namely CORR and CIP are computed from the decomposed components. The p-values obtained using the Kruskal–Wallis statistical test implies that extracted features are statistically significant. The presented features in this paper can be studied for automated classification of normal and ALS EMG signals. The proposed analysis technique can also be studied for other biomedical signals.

References

1. Abel, E.W., Meng, H., Forster, A., Holder, D.: Singularity characteristics of needle EMG IP signals. IEEE Trans. Biomed. Eng. **53**, 219–225 (2006)
2. Bhattacharyya, A., Pachori, R.B., Acharya, U.R.: Tunable-Q wavelet transform based multivariate sub-band fuzzy entropy with application to focal EEG signal analysis. Entropy **19**, 99 (2017)
3. Englehart, K., Hudgin, B., Parker, P.A.: A wavelet-based continuous classification scheme for multifunction myoelectric control. IEEE Trans. Biomed. Eng. **48**, 302–311 (2001)
4. Fattah, S.A., Iqbal, M.A., Jumana, M.A., Doulah, A.S.U.: Identifying the motor neuron disease in EMG signal using time and frequency domain features with comparison. Signal Image Process. **3**, 99–114 (2012)
5. Fuglsang-Frederiksen, A.: The utility of interference pattern analysis. Muscle Nerve **23**, 18–36 (2000)
6. Güler, N.F., Koçer, S.: Classification of EMG signals using PCA and FFT. J. Med. Syst. **29**, 241–250 (2005)

7. Gupta, V., Priya, T., Yadav, A.K., Pachori, R.B., Acharya, U.R.: Automated detection of focal EEG signals using features extracted from flexible analytic wavelet transform. Pattern Recognit. Lett. **94**, 180–188 (2017)
8. Huang, N.E., Shen, Z., Long, S.R., Wu, M.C., Shih, H.H., Zheng, Q., Yen, N.C., Tung, C.C., Liu, H.H.: The empirical mode decomposition and the Hilbert spectrum for nonlinear and non-stationary time series analysis. Proc. R. Soc. Lond. A Math. Phys. Eng. Sci. **454**, 903–995 (1998)
9. Jain, P., Pachori, R.B.: GCI identification from voiced speech using the eigen value decomposition of Hankel matrix. In: 8th International Symposium on Image and Signal Processing and Analysis, pp. 371–376 (2013)
10. Jain, P., Pachori, R.B.: Event-based method for instantaneous fundamental frequency estimation from voiced speech based on eigenvalue decomposition of the Hankel matrix. IEEE/ACM Trans. Audio Speech Lang. Process. **22**, 1467–1482 (2014)
11. Jain, P., Pachori, R.B.: An iterative approach for decomposition of multi-component non-stationary signals based on eigenvalue decomposition of the Hankel matrix. J. Frankl. Inst. **352**, 4017–4044 (2015)
12. Joshi, D., Tripathi, A., Sharma, R., Pachori, R.B.: Computer aided detection of abnormal EMG signals based on tunable-Q wavelet transform. In: Fourth International Conference on Signal Processing and Integrated Networks (SPIN 2017), Noida, India, pp. 544–549 (2017)
13. Ko, K.D., Kim, D., El-ghazawi, T., Morizono, H.: Predicting the severity of motor neuron disease progression using electronic health record data with a cloud computing big data approach. In: IEEE Conference on Computational Intelligence in Bioinformatics and Computational Biology, pp. 1–6 (2014)
14. Kumar, M., Pachori, R.B., Acharya, U.R.: An efficient automated technique for CAD diagnosis using flexible analytic wavelet transform and entropy features extracted from HRV signals. Expert Syst. Appl. **63**, 165–172 (2016)
15. Kumar, M., Pachori, R.B., Acharya, U.R.: Characterization of coronary artery disease using flexible analytic wavelet transform applied on ECG signals. Biomed. Signal Process. Control **31**, 301–308 (2017)
16. Liu, W., Pokharel, P.P., Príncipe, J.C.: Correntropy: properties and applications in non-Gaussian signal processing. IEEE Trans. Signal Process. **55**, 5286–5298 (2007)
17. McGill, K.C., Lateva, Z.C., Marateb, H.R.: EMGLAB: an interactive EMG decomposition program. J. Neurosci. Methods **149**, 121–133 (2005)
18. McKight, P.E., Najab, J.: Kruskal-Wallis Test. Corsini Encyclopedia of Psychology (2010)
19. Mishra, V.K., Bajaj, V., Kumar, A.: Classification of normal, ALS, and myopathy EMG signals using ELM classifier. In: 2nd International Conference on Advances in Electrical, Electronics, Information, Communication and Bio-Informatics (AEEICB), pp. 455–459 (2016)
20. Mishra, V.K., Bajaj, V., Kumar, A., Singh, G.K.: Analysis of ALS and normal EMG signals based on empirical mode decomposition. IET Sci. Meas. Tech. **10**, 963–971 (2016)
21. Nikolic, M.: Detailed analysis of clinical electromyography signals: EMG decomposition, findings and firing pattern analysis in controls and patients with myopathy and amytrophic lateral sclerosis, Faculty of Health Science, University of Copenhagen, Ph.D. thesis (2001)
22. Nikolic, M., Krarup, C.: EMGTools, an adaptive and versatile tool for detailed EMG analysis. IEEE Trans. Biomed. Eng. **58**, 2707–2718 (2011)
23. Pal, P., Mohanty, N., Kushwaha, A., Singh, B., Mazumdar, B., Gandhi, T.: Feature extraction for evaluation of muscular atrophy. In: IEEE International Conference on Computational Intelligence and Computing Research (ICCIC), pp. 1–4 (2010)
24. Sharma, R., Kumar, M., Pachori, R.B., Acharya, U.R.: Decision support system for focal EEG signals using tunable-Q wavelet transform. J. Comput. Sci. **20**, 52–60 (2017)
25. Sharma, R.R., Pachori, R.B.: Time-frequency representation using IEVDHM-HT with application to classification of epileptic EEG signals. IET Sci. Meas. Tech. **12**(1), 72–82 (2018)
26. Sharma, R.R., Pachori, R.B.: A new method for non-stationary signal analysis using eigenvalue decomposition of the Hankel matrix and Hilbert transform. In: Fourth International Conference on Signal Processing and Integrated Networks (SPIN 2017), Noida, India, pp. 484–488 (2017)

27. Xu, D., Erdogmuns, D.: Renyi's entropy, divergence and their nonparametric estimators. In: Information Theoretic Learning: Renyi's Entropy and Kernel Perspectives, pp. 47–102 (2010)
28. Xu, J.W., Paiva, A.R., Park, I., Principe, J.C.: A reproducing kernel Hilbert space framework for information-theoretic learning. IEEE Trans. Signal Process. **56**, 5891–5902 (2008)
29. Yousefi, J., Hamilton-Wright, A.: Characterizing EMG data using machine-learning tools. Comput. Biol. Med. **51**, 1–13 (2014)
30. Zwarts, M.J., Drost, G., Stegeman, D.F.: Recent progress in the diagnostic use of surface EMG for neurological diseases. J. Electromyogr. Kinesiol. **10**, 287–291 (2000)

Machine Learning Toward Infectious Disease Treatment

Tulika Bhardwaj and Pallavi Somvanshi

Abstract The emergence of infectious diseases poses a serious threat to human and animal health. This is evident from the sudden cases of hospital outbreaks due to the surveillance and evolution of zoonotic pathogens. The main reason is the development of resistance mechanism in infectious pathogens against broad-spectrum drugs. This leads to the mortality rate of 69% due to infectious diseases at global level. Although improvements have been made at next-generation epidemiological study level to combat such issues but shortfall still observed due to the gap between patients and governmental authorities assigned for treatment. Additionally, handling, analyzing and updating of large datasets is time consuming and labor intensive. To overcome such limitations, applied informatics was employed for the sorting of multipart disciplines of research and pathogenesis identification and treatment. To understand the underlying problem, mining of the diagnostic techniques was performed focused to execute the correct disease diagnosis in different symptoms from the patient. Data preprocessing enables improvement quality of data as redundant data requires continuous discrete mining for analysis. Graphical interfaces were utilized for the comparative analysis of the problem in n number of ways by random decisions and tree making processes. Categorization of the single problem into supervised, unsupervised and weakly supervised principles offers a complete set of appropriate outputs directed towards disease treatment. The whole process favors data preprocessing, data mining, and data analysis by employing various machine learning approaches, data interpretation by statistical platforms and data visualization. The complete chapter reviewed challenges, pathway, and opportunities provided by machine learning approaches toward infectious disease treatments.

Keywords Support vector machine · Fuzzy logic · Artificial neural network
Infectious disease · Machine learning

T. Bhardwaj · P. Somvanshi (✉)
Department of Biotechnology, TERI School of Advanced Studies, 10,
Institutional Area, Vasant Kunj, New Delhi 110070, India
e-mail: psomvanshi@gmail.com

© Springer Nature Singapore Pte Ltd. 2019
M. Tanveer and R. B. Pachori (eds.), *Machine Intelligence and Signal Analysis*,
Advances in Intelligent Systems and Computing 748,
https://doi.org/10.1007/978-981-13-0923-6_58

1 Introduction

The increasing burden of the prevalence of infectious pathogens raised the quest to identify new ventures for early diagnosis and treatment of diseases. The major is to identify the patients at early stages of the prevailing disease to provide preliminary medication. Similarity in diagnostic symptoms often leads to the gap in treatment procedure and subsequent deaths [1]. Post-genomic era provides advancement of molecular epidemiological techniques to identify novel drug and vaccine development. But lack of knowledge regarding medications results in increase in mortality rate approximately 70% among the population. Therefore, artificial intelligence strategies were applied to reduce such errors by keeping in mind all social, environmental and political parameters. Population setting in tribal and tropical regions possesses less knowledge about the government initiatives, diagnostic methods and treatment procedures for prevailing potent infections. Due to the availability of a large amount of data by third generation sequencing procedures, handling and analyzing such data became cost-effective and tedious task [2]. Therefore, machine learning platforms were combined to generate models depicting early treatment procedures and diagnosis.

Machine learning approaches rely on probabilistic framework assigning n number of assumptions to individual query and postulating output combining all assumptions. This learning can be classified as (a) supervised, (b) unsupervised, and (c) reinforced [3]. These platforms are classified on the basis of dependency of input variables on external variables to direct towards function. Further, reinforced learning focuses on reducing the error rate of future predictions by maximizing capabilities of machine in respect to environmental factors. This learning mimics "fitting" of human brain procedures by performing preprocessing of data and normalization. *Data preprocessing* involves the acquisition of information from raw data. This iterative process involves data cleaning, data integration, data transformation and data reduction to reduce its dimensionality. While *normalization* is the transformation of data equally into a range of defining variables (0 to 1 and -1 to 0) by computing several parameters, for example, Euclidean distance, Manhattan distance, etc. [4].

Several researchers explore multiple sclerosis disease course using machine learning platforms. Support Vector Machine (SVM) was utilized to maximize the margin that separates the two classes of data (worsening v/s non-worsening cases). Logistic regression analysis finds out best fit the log odds of binary variables predicting the rate of adverse symptoms [5]. The effectiveness of drug dose for treatment was analyzed by Li and his co-workers. They employ FCM clustering algorithm for efficient clustering of input into two-dimensional variables considering dosage and side effects respectively. Overcoming limitation of dealing with noisy and incomplete data in drug delivery processes, FCM clustering approach categorized drug concentration and pathogen population based on pattern recognition strategy. Multiple drug concentrations were subjected to clustered pathogen populations to evaluate their effects on n clusters [6]. Further, supervised learning algorithms were developed to classify the open-text records and assessing doctor's professional performances [7].

Additionally, pattern recognition strategies were employed for clustering the datasets into test and training clusters to study the effect of one on another using unsupervised learning and decision tree processes [8]. This review highlights challenges and opportunities towards treatment, classification of algorithms, and application of support vector machines, neural networks and fuzzy logic principles in infectious disease treatment.

2 Challenges and Opportunities Toward Infectious Disease Treatment

The introduction of next-generation sequencing technology provides the platform to future researchers to identify major cause of disease and treatment. The elucidation of pathogenesis and molecular epidemiology due to the advancement of research and development strategies assist in the development of drug and vaccines [9]. The optimum care of patients became the major focus of public health organizations due to the endemicity of virulent pathogens across the globe [10]. Therefore, nucleic acid detection technology was introduced for more efficient epidemiological and molecular studies. The contribution of genomics and proteomics towards drug discovery and vaccine development is well known, for example, development of polio vaccine, and meningitis-causing pneumococci, and recent Ebola vaccine are considered as landmarks [11]. Additionally, governments and several health authorities raise the concern of prevalence of virulent pathogens and assist in addressing people about infectious diseases at socio-political front, preliminary treatment and disease control [12]. This strategy was adopted by World Health Organization (WHO) to eliminate HIV/AIDS [13], polio [14], whooping cough [15], measles [16], and tuberculosis [17].

Despite significant advancement in sequencing technologies, persistence of infectious disease is a global challenge. Factors like ecological, social, environmental and microbial contributed towards upsurge increase in disease endemicity [18]. The continuous evolvement of genetic changes in pathogenic strains confers their adaptation in new environmental conditions [19, 20]. These genetic changes mainly include development of antibiotic and chemical resistant genes [21–23] with the evolution of new strain. This enhances pandemicity of infectious disease. Another major responsibility for the surveillance of disease is resistance development of pathogens [24] towards drug as in the case of *Escherichia coli, Pneumococcus*, and *Staphylococcus aures*. Therefore, continuous efforts were made by researchers in respect to novel drug and vaccine development. Besides these major contributing factors, the similarity in symptoms for more than one pathogenic strain often confuses doctors to provide accurate treatment [25]. Therefore, machine learning platforms were introduced to unwind underlying pathogen complexity for accurate diagnosis and treatment of infectious diseases.

Fig. 1 Classification of
machine learning algorithms

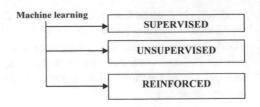

3   Supervised and Unsupervised Algorithms

3.1   Supervised Learning

This platform supports an association between training sets of input and output variables. The system minimizes the error in future predictions by assigning a deterministic function to each input with respect to its target values. A sequence of steps is involved in solving a problem using supervised learning platforms:

- Calculation of number of variables involved
- Selection of training set describing the problem
- Assigning the variables to the system-readable format to train the data
- Model validation by robustness and accuracy analysis.

The comparison between the model output and actual output enable adjustment of error in the connections between variables. This was performed by adjusting weights of underlying connections. This learning was further classified based on the behavior of target trained values like (a) *classification learning*, to find similarity between two elements in output space (b) *preference learning* to identify equality between two elements in output space (c) *function learning* for the optimization of a function for a given process.

3.2   Unsupervised Learning

For the data set used for training the system that does not contain target vectors, unsupervised learning was utilized. A cost function was assigned to each individual training set to be minimizing error rate during the learning process. This learning relies on estimation problems like clustering and dimensionality reductionism by making iterative decisions and future input predictions. For example, random clusters of input data were provided to the machine to learn a specific pattern to predict the output considering outliers as noise (Fig. 1).

4 Support Vector Machines

Support vector machines enable data analysis by performing classification and regression procedures. It acts a hyperplane to compute the largest minimum distance between training samples [26]. Support vector machines laid on the basis of supervised learning for general formulation of machines. Its ability depends on mainly two factors

1. minimizing error rate of training data
2. Computation of the integers (Vapnik–Chervonenkis dimensions) expressed as measuring power for classification functions [27].

SVM works as an abstract separator between two independent points at measuring distance in multidimensional environment. They perform classification based on the statistical evaluation overcoming the computational limitations provided by n number of tasks. They are also coined as "maximum margin classifiers" because they laid down on the principle of maximizing the margin between the classes and to minimizing the distance between the hyperplane points [28]. Hyperplane acts as a virtual separator which separates data based on largest distance analysis to classify the training data involved. The data points at one side of the plane are labeled as positive and on other side as negative. For n number of training datasets, classifiers enable the separation of the complete data on either side of the hyperplane [29]. This abstract separator is at maximal distance from the training points. Thus, the position of the hyperplane depends on the points closest to it. Therefore, they termed as support vectors and the perpendicular distance between support vectors and hyperplane is known as margin. *Optimal hyperplane* is defined as a separator with a maximal margin of separation between the two classes [30]. A nonlinear Kernel transformation function $\Phi(\cdot)$ was used for distance mapping. The binomial representation of hyperplane is

$$w \cdot \Phi(x) + b = 0$$

where $w \cdot \Phi(x)$ = product of data points and weights that determine their orientation
  b is the bias or offset of hyperplane from the origin.
  For the plane with positive class distribution,

$$w \cdot \Phi(xi) + b = 1$$

For the plane with positive class distribution,

$$w \cdot \Phi(xi) + b = 1$$

To overcome the limitation of constructive explicitly provided by Eigenvalues, various kernel functions were used based on the decision-making process to mea-

Fig. 2 Support vector
machine classification [26]

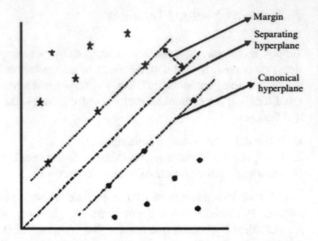

sure similarity measure between the distant objects. Some kernel functions used by
developers and researchers are as follows:

1. Polynomial kernel with degree d

$$K(X, Y) = (XT\ Y + 1)d$$

2. Radial basis function kernel with width σ

$$K(X,\ Y) = \exp(-\|X - Y\|2/(2\sigma 2))$$

   Closely related to radial basis function neural networks
   The feature space is infinite-dimensional.
3. Sigmoid with parameter θ and

$$K(X, Y) = \tan\ h\left(kT^{Y}+\theta\right)$$

where, 'd', 'σ' and 'θ' are parameters [31] (Fig. 2).

5   Artificial Neural Network

Artificial neural network mimics biological neural network built by connecting several neurons. It acts as human brain which can learn output behavior based on various
inputs that they get from their environment [32]. An individual neuron involves three
mathematical functions: multiplication, addition, and activation to solve the problems by assigning weights to every input [33]. Each individual input follows the
mathematical computation steps of multiplication to their corresponding weights,

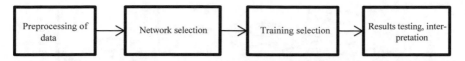

Fig. 3 Methodology for applying artificial neural networks approach

the addition of the individual participating inputs in connected pathway and transfer to an activation function (sigmoidal, linear, hyperbolic, tangent, or radial) [34]. Random arrangements of the artificial neurons in different patterns reflect the diversity in topological conformations of the artificial networks. The fine tuning of the neural network to solve a problem in called artificial neural network training. This training affects the accuracy and speed of the problem-solving efficiency of the model by employing both supervised and unsupervised learning platforms. In case of supervised learning, an external source was utilized to obtain desired output and then network weights were optimized by target matching. On the contrary, a self-organizing network identifies the silent features of input data set by assigning weights on random assumptions [35].

An artificial neuron reflects nodes arranged in three-layered networks: input layer, hidden layer and output layer. Connections between different nodes direct informational flow throughout the network in three ways (a) feed-forward neural network, (b) cascade-forward neural network, and (c) recurrent or feedback neural network [36]. The whole problem-solving process using artificial neural network platform involves four procedural steps: data preprocessing, network selection, training selection and testing and interpretation of results [37] (Fig. 3).

Li et al. elucidate the application of combined artificial neural networks toward the identification of infectious disease outbreaks, providing effective measures to reduce the mortality rate. The complete model integrates data from schools, hospitals, pharmacies and internet and combined BP artificial neural network modes, SIR model and the complex network model for forecasting preventive measures for disease attack [38]. Additionally, an artificial neural network approach was applied to classify the performance of medical students with their consent engaged in solving multiple disciplines enabling the students to discriminate between different infectious disease symptoms [39]. The complete model comprised of input layer, hidden layer, and output layer for information flow (Fig. 4).

Several researchers attempt to generate a model estimating the spread of endemic diseases. The assessment of several parameters affecting the surveillance of infection was described as control (vaccination) as a function of a number of infective individuals. This approach was used for identifying the patients suffering from active pulmonary tuberculosis. A group of 563 people was selected for assessment of prevailing tuberculosis cases. A general regression neural network (GRNN) was used to compute c-index, which is equivalent to the area under the receiver operating characteristic curve of predictive accuracy for the prevailing symptom [40]. Wang and co-workers developed a three-layered feed-forward back-propagation artificial neural network model (FFBPNN) to predict the weekly number of infectious

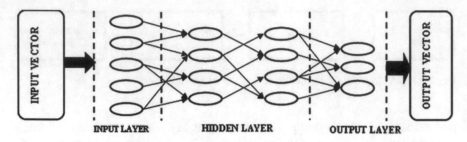

Fig. 4 Representation of basic feed-forward neural network [26]

diarrhea. They use meteorological factors as an input variable to detect infective cases. Besides infectious disease treatment, several neural network strategies were employed for cardiovascular [41], urological [42] and psychiatric disorders [43].

6 Fuzzy Logic

Fuzzy logic concepts deal with the concepts of partial truth whose value lies between 0 and 1. When value ranges from complete true to complete false, fuzzy logic expressions were used [44]. Fuzzification is the process involving all possible input values for individual decision to obtain a crisp output. The whole process involves (a) fuzzify all input variables (b) execution of all possible rules to obtain output (c) obtain output by de-fuzzification of fuzzy outputs. Such decisive outputs employ boolean operators (AND, OR, NOT) and logical operators (IF-THEN ELSE) for n number of decisions [45].

A detailed review on application of fuzzy knowledge-based system is combined to make diagnosis of tropical infectious diseases, asthma treatment etc. The detailed knowledge was mined from medical specialist in internal medicine which provides a crisp output dependent on certain related factors of surrounding environment. Researchers globally perform several studies under specific ethical considerations to connect the surveillance of disease might be due to the similarity in the symptoms of disease. Therefore, an expert system was developed combining fuzzy logic and certainty factors with the object tropical infectious diseases diagnosis and treatment. This knowledge-based expert system overcome vagueness of redundant data like physical symptoms and blood examination reports. The output contains percentage represents diagnosis of several diseases based on expert advice [46]. A knowledge acquisition process using a semantic network was developed for asthma treatment. A comparative analysis was performed between 53 individuals each (patients with asthma and non-asthmatic patients) with a cut-off value of 0.7 the level of accuracy of 100 and 94% response rate. The complete used is de-fuzzification centroid method. A knowledge representation system was developed to eliminate this chronic disorder at initial stages by categorizing data into two datasets (A and B). Each dataset consists of 6 and 8 modules respectively including symptoms, allergic rhinitis, environmental and genetic factors etc. A defuzziied system was utilized for assessment of asthma possibility in individual [47]. To avoid the successive rate of dengue cases, fuzzy logic approach was utilized to develop an expert system for the patients diagnosed with dengue to get further notification regarding hospital for diagnosis and treatment. Fuzzy logic based system acts as an interface triggering a set of questions (symptom-based questions) for the user. A cumulative percentage was analyzed for each question with the suggestion on the last page regarding patient's status [48].

7 Future Prospects

The underlying complexity of disease pathogenesis poses a serious threat to the public health system. The surveillance of pathogens, sudden hospital outbreaks, and lack of diagnostic techniques increase the burden of health regulation authorities globally. Machine learning techniques enable analysis and interpretation of large biological data towards précised treatment [49]. Several models were developed based on mathematical and statistical platforms to predict future diagnostic and treatment of infectious diseases [50, 51]. They provide accurate treatment than basic statistical technique based on large training set analysis. Combinations of such machine learning platforms enable researcher and patients to avail preliminary treatment in case of pathogen attack to combat the prevalence of diseases.

Ethics Approval and Consent to Participate Not Applicable

References

1. Schaepe, K.S.: Bad news and first impressions: patient and family caregiver accounts of learning the cancer diagnosis. Social Sci. Med. (1982) **73**(6), 912–921 (2011)
2. Muir, P., Li, S., Lou, S., Wang, D., Spakowicz, D.J., Salichos, L., et al.: The real cost of sequencing: scaling computation to keep pace with data generation. Genome Biol. **17**, 53 (2016)
3. Marblestone, A.H., Wayne, G., Kording, K.P.: Toward an Integration of Deep Learning and Neuroscience. Front. Comput. Neurosci. **10**, 94 (2016)
4. Siebert, J.C., Wagner, B.D., Juarez-Colunga, E.: Integrating and mining diverse data in human immunological studies. Bioanalysis **6**(2), 209–223 (2014)
5. Zhao, Y., Healy, B.C., Rotstein, D., Guttmann, C.R.G., Bakshi, R., Weiner, H.L., et al.: Exploration of machine learning techniques in predicting multiple sclerosis disease course. PLoS ONE **12**(4), e0174866 (2017)
6. Li, Y., Lenaghan, S.C., Zhang, M.: A data-driven predictive approach for drug delivery using machine learning techniques. PLoS ONE **7**(2), e31724 (2012)
7. Gibbons, C., Richards, S., Valderas, J.M., Campbell, J.: Supervised machine learning algorithms can classify open-text feedback of doctor performance with human-level accuracy. J. Med. Internet Res. **19**(3), e65 (2017)

8. Bhaskar, H., Hoyle, D.C., Singh, S.: Machine learning in bioinformatics: a brief survey and recommendations for practitioners. Comput. Biol. Med. **36**, 1104–1125 (2006)
9. Muldrew, K.L.: Molecular diagnostics of infectious diseases. Curr. Opin. Pediatr. **21**(1), 102–111 (2009)
10. Rweyemamu, M., Kambarage, D., Karimuribo, E., et al.: Development of a one health national capacity in Africa: the Southern African centre for infectious disease surveillance (SACIDS) one health virtual centre model. Curr. Topics Microbiol. Immunol. **366**, 73–91 (2013)
11. Caliendo, A.M., Gilbert, D.N., Ginocchio, C.C., et al.: Tests, better care: improved diagnostics for infectious diseases. Clin. Infect. Dis. **57**(3), 139–170 (2013)
12. Calmy, N.F., Hirschel, B., et al.: HIV viral load monitoring in resource-limited regions: optional or necessary? Clin. Infect. Diseases **44**(1), 128–134 (2007)
13. Pereira, C.F., Paridaen, J.T.: Anti-HIV drug development—an overview. Curr. Pharm. Des. **10**(32), 4005–4037 (2004)
14. Collett, M.S., Neyts, J., Modlin, J.F.: A case for developing antiviral drugs against polio **79**(3), 179–87 (2008)
15. Altunaiji, S., Kukuruzovic, R., Curtis, N., Massie, J.: Antibiotics for whooping cough (pertussis).Cochrane Database Syst. Rev. **1**, CD004404 (2005)
16. Plemper, R. K., Snyder, J. P.: Measles control—can measles virus inhibitors make a difference? Curr. Opin. Investig. Drugs (London, England: 2000) **10**(8), 811–820 (2009)
17. Swindells, S.: New drugs to treat tuberculosis. F1000 Med. Rep. **4**,12 (2012)
18. Klein, E.Y.: Antimalarial drug resistance: a review of the biology and strategies to delay emergence and spread. Int. J. Antimicrob. Agents **41**(4), 311–317 (2013)
19. Bryant, J., Chewapreecha, C., Bentley, S.D.: Developing insights into the mechanisms of evolution of bacterial pathogens from whole-genome sequences. Future Microbiol. **7**(11), 1283–1296 (2012)
20. Davies, J., Davies, D.: Origins and evolution of antibiotic resistance. Microbiol. Mol. Biol. Rev. MMBR **74**(3), 417–433 (2010)
21. Bhardwaj, T., Somvanshi, P.: Pan-genome analysis of clostridium botulinum reveals unique targets for drug development. Gene **623**, 48–62 (2017)
22. Venancio, T.M., Bellieny-Rabelo, D., Aravind, L.: Evolutionary and biochemical aspects of chemical stress resistance in *Saccharomyces cerevisiae*. Front. Genet. **3**, Article 47 (2012)
23. Khan, S., Somvanshi, P., Bhardwaj, T., Mandal, R.K., Dar, S.A., et al.: Aspartate-β-semialdeyhyde dehydrogenase as a potential therapeutic target of Mycobacterium tuberculosis H37Rv: evidence from in silico elementary mode analysis of biological network model. J. Cell. Biochem. **119**(3), 2832–2842 (2018). https://doi.org/10.1002/jcb.26458
24. Meyer, W.G., Pavlin, J.A., Hospenthal, D., et al.: Antimicrobial resistance surveillance in the AFHSC-GEIS network. BMC Public Health **11**(2) Article 8 (2011)
25. Fauci, A.S., Morens, D.M.: The perpetual challenge of infectious diseases. N. Engl. J. Med. **366**(5), 454–461 (2012)
26. Osama, K., Mishra, B.N., Somvanshi, P.: Machine Learning Techniques in Plant Biology. The Omics of Plant Science. Springer Publications, Plant Omics (2015)
27. Haykin, S.: Neural Networks: A Comprehensive Foundation, Fourth Indian Reprint. Pearson Education, Singapore (2003)
28. Ghumbre, S., Patil, C., Ghatol, A.: Heart disease diagnosis using support vector machine. In: Proceedings of the International Conference on Computer Science and Information Technology (ICCSIT '11), Pattaya, Thailand (2011)
29. Bhatia, S., Prakash, P., Pillai, G.N.: SVM based decision support system for heart disease classification with integer-coded genetic algorithm to select critical features. In: Proceedings of the World Congress on Engineering and Computer Science, San Francisco, USA, pp. 34–38 (2008)
30. Xiaoqing, G., Ni, T., Wang, H.: New fuzzy support vector machine for the class imbalance problem in medical datasets classification. Sci. World J. **2014** (2014)
31. Janaradanan, P., Heena, L., Sabika, F.: effectiveness of support vector machines in data mining. J. Commun. Softw. Syst. **11**(1) (2015)

32. Karim, M.N., Yoshida, T., Rivera, S.L., Saucedo, V.M., Eikens, B., Oh, G.-S.: Global and local neural network models in biotechnology: application to different cultivation processes. J. Ferment. Bioeng. **83**(1), 1–11 (1997)
33. Krenker, A., Bešter, J., Ko, A.: Introduction to the arti- ficial neural networks. In: Suzuki, K. (ed.) Artificial Neural Networks-Methodological Advances and Biomedical Applications, pp. 3–18. Carotia, Intech, Rijeka (2011)
34. Widrow, B., Hoff, M.: Adaptive switching circuits. 1960 IRE WESCON convention record, vol. 4, pp. 96–104. IRE, New York (1960)
35. Prasad, V., Gupta, S.D.: Applications and potentials of artificial neural networks in plant tissue culture. In: Gupta, S.D., Ibaraki, Y. (eds.) Plant Tissue Culture Engineering, pp. 47–67. Springer, Netherlands (2006)
36. Mandic, D.P., Chambers, J.: Recurrent Neural Networks for Prediction: Learning Algorithms, Architectures and Stability. Wiley, Chichester/New York (2001)
37. Yang, Z.R.: A novel radial basis function neural network for discriminant analysis. IEEE Trans. Neural Netw. **17**, 604–612 (2006)
38. Li, C.Y., Liang, G.Y., Yao, W.Z., et al.: Integrated analysis of long non-coding RNA competing interactions reveals the potential role in progression of human gastric Cancer. Int. J. Oncol. **248**, 1965–1976 (2016)
39. Stevens, R.H., Lopo, A.C.: Artificial neural network comparison of expert and novice problem-solving strategies. In: Proceedings of the Annual Symposium on Computer Application in Medical Care, pp. 64–68 (1994)
40. El-Solh, A.A., Hsiao, C.B., Goodnough, S., Serghani, R.N.J., Grant, B.J.B.: Predicting active pulmonary tuberculosis using an artificial neural network. Chest **116**, 968–973 (1999)
41. Narain, R., Saxena, S., Goyal, A.K.: Cardiovascular risk prediction: a comparative study of Framingham and quantum neural network based approach. Patient Prefer. Adher. **10**, 1259–1270 (2016)
42. Anagnostou, T., Remzi, M., Djavan, B.: Artificial neural networks for decision-making in urologic oncology. Rev. Urol. **5**(1), 15–21 (2003)
43. Cordes, J.S., Mathiak, K.A., Dyck, M., Alawi, E.M., Gaber, T.J., Zepf, F.D., et al.: Cognitive and neural strategies during control of the anterior cingulate cortex by fMRI neurofeedback in patients with schizophrenia. Front. Behav. Neurosci. **9**, 169 (2015)
44. Zadeh, L.A.: Fuzzy sets. Inf. Control **8**(3), 338–353 (1965)
45. Zaitsev, D.A., Sarbei, V.G., Sleptsov, A.I.: Synthesis of continuous-valued logic functions defined in tabular form. Cybern. Syst. Anal. **34**(2), 190–195 (1998)
46. Prihatini, P.M., Putra, I.K.G.D.: Fuzzy knowledge-based system with uncertainty for tropical infectious disease diagnosis. IJCSI Int. J. Comput. Sci. Issues **9**(4), 3 (2012)
47. Zarandi, F.M.H., Zolnoori, M., Moin, M., Heidarnejad, H.: A fuzzy rule-based expert system for diagnosing asthma. Trans. E. Ind. Eng. **17**, 129–142 (2010)
48. Razak, T.R.B., Ramli, M.H., Wahab, R.A.: Dengue notification system using fuzzy logic. In: 2013 International Conference on Computer, Control, Informatics and Its Applications (2013)
49. Gago, J., Landín, M., Gallego, P.P.: Artificial neural networks modeling the in vitro rhizogenesis and acclimatization of Vitis vinifera L. J. Plant Physiol. **167**, 1226–1231 (2010)
50. Goswami, N.D., Pfeiffer, C.D., Horton, J.R., Chiswell, K., Tasneem, A., Tsalik, E.L.: The state of infectious diseases clinical trials: a systematic review of clinicaltrials.gov. PLoS ONE **8**(10), e77086 (2013)
51. Wang, Y., Gu, J., Zhou, Z.: Artificial neural networks for infectious diarrhea prediction using meteorological factors in Shanghai 2015. Appl. Soft Comput. 280–290 (2015)

Online Differential Protection Methodology Based on DWT for Power Transmission System

Sunil Singh, Shwetank Agrawal and D. N. Vishwakarma

Abstract This paper presents an online methodology for transmission line relaying based on discrete wavelet transform (DWT) and differential energy estimation. Irrespective of traditional current differential relaying, the energy level of wavelet coefficients of differential and average currents are used as the discriminating factor in the proposed scheme. First, the wavelet coefficients of differential current (I_Δ) and average current (I_{avg}) are acquired by applying DWT. Thereafter, the energy of the wavelet coefficients (both I_Δ and I_{avg}) are computed, as operating and restraining factors. Finally, the decision regarding issuing of tripping signal is taken by comparing the operating and restraining factors. In order to effectuate the competency of the proposed methodology, it has been tested on simulated transmission system designed in MATLAB. The results of various test cases indicate that the present approach is capable of providing fast and efficient online protection to power transmission system. It has also been observed that developed methodology remains unaffected for external faults and power swings condition, which arises due to emergent load changes in the system.

Keywords Differential protection · Discrete wavelet transform · Power transmission line protection · Power swings

1 Introduction

Fast and pertinent fault detecting mechanism is essential for preserving the security and reliability of modern power network. Both security and reliability are directly related to protection mechanism used. Accurate and competent relaying system not only mitigates the outage period, but also helps in preserving system stability with least fault damages. The differential relaying technique is significantly applied in

S. Singh (✉) · S. Agrawal · D. N. Vishwakarma
Department of Electrical Engineering, Indian Institute of Technology (BHU),
Varanasi, Varanasi, India
e-mail: sksingh.rs.eee13@itbhu.ac.in

© Springer Nature Singapore Pte Ltd. 2019
M. Tanveer and R. B. Pachori (eds.), *Machine Intelligence and Signal Analysis*,
Advances in Intelligent Systems and Computing 748,
https://doi.org/10.1007/978-981-13-0923-6_59

power system and is principally based on the assessment of current at both ends of the protected section. The differential relaying management adjudges the spot of the fault, i.e., whether it is within protected section or not [1]. However, the conventional differential relaying approach easily gets affected by line charging and other abnormal phenomenon in the network. It leads to abnormal functioning of the relaying system and causes outages of supply. These limitations of conventional differential relaying are outperformed by application of signal processing and intelligent algorithm while designing the protection mechanism. Several schemes are already developed for distance relaying based on traveling wave theory, DWT, and neural network [2]. However, aforementioned approaches require extensive training data and time. Some authors also have applied differential relaying-based protection of transmission network [3–7]. In [3], a new approach based on differential energy is discussed for the protection of compensated transmission network. In [4], a differential mechanism-based protection of solar-based microgrid is explained. In [5], a noble power differential scheme has been discussed for a wide protection of transmission system. It fundamentally involves the assessment of real power at both ends of the protected section. If the differential power is more than the prespecified value, relay detects an internal fault. In [6], power differential relaying is explained for protecting transmission network. The key merits of the power differential protection approaches are it is independent of current charging phenomena; it reduces the data samples requirement and computation time. However, it is incompetent in identifying the particular faulty phase [7]. In [8], a differential scheme is proposed for line protection wherein wave pattern of current at two ends is used as the deciding factor for internal and external fault. However, it is only limited to 18–30 km lines. A differential relaying for transmission lines based on net energy fed into the protected section is explained in [9]. But the approach is quite complex and involves various assumptions. In [10], the authors have explained a new distance relaying approach for lines based on differential equation and equal transfer process. In [11] Hilbert transform and differential equations-based algorithm is presented for transmission line protection. However, it involves various presumptions and only tested for L-G and L-L faults. In [12], an advance current differential relaying for transmission line is explained. The proposed scheme is capable of overcoming the affects of distributive capacitive currents in long lines. In [13] also, a novel differential protection scheme is proposed which is competent of averting the impact of distributive capacitive currents. In [14], a fuzzy inference-based current differential scheme is presented for protection of high-voltage lines.

This paper proposes an online approach for transmission system relaying using differential phenomena and DWT. Currents samples of all phases are measured at both ends of the protected section. Then the value of differential (I_Δ) and average (I_{avg}) current in the protected section has been evaluated. There upon I_Δ and I_{avg} signals are decomposed using DWT for capturing wavelet coefficients. The energy of different levels of coefficients has been computed as operating and restraining factors. The tripping or no action decision is taken by comparing the operating and restraining factors. The feasibility of present approach has been tested on MATLAB-simulated test transmission system. The outcomes of the test cases reveal the competency of developed model in discriminating faulty conditions properly in the transmission network and issuing the tripping signal in a very little span of time duration. The proposed scheme has outperformed the abnormalities of traditional differential relaying schemes.

2 DWT-Based Differential Protection Mechanism

The principle associated with differential protection mechanism is nothing but a simple application of Kirchhoff's law and is normally applied as unit-type protection. In the traditional approach, the difference current is computed between two ends (entering and leaving) of protected element as operating factor, while average current is estimated as restraining factor. The relay gets actuated for any internal faults, i.e., when operating factor is more than restraining factor.

2.1 Discrete Wavelet Transform

Over the past few years, signal processing (SP) specifically fast Fourier Transform (FFT) and continuous wavelet transform (CWT)-based approaches are significantly applied in power system problems. SP techniques help in more realistic and reliable utilization of acquired signals in various power system applications. The CWT technique is a well-competent alternative of FFT as it outperforms prime limitations of FFT. During nonstationary (transient signals) signals analysis, CWT is preferred over FFT mainly because of its ability of providing better characteristics realization of signal. When CWT is computed at a discrete set of scale and translation, it is termed as DWT. The expression of DWT is given in Eq. (1). Daubechies (db4) is widely used by various researchers as mother wavelet while developing protection mechanism for the transmission network [15–17]. The db4 mother wavelet is pertinent for analyzing transients in transmission system [18, 19]. Hence, in the proposed scheme, the same db4 mother wavelet is opted for decomposing the differential and average current waveforms.

$$DWT_{(m,n)} = \frac{1}{\sqrt{a_0^m}} \sum f(t) g_{m,n} \left(\frac{t - n b_0 a_0^m}{a_0^m} \right) \tag{1}$$

where g(t) is mother wavelet; f(t) is the input signal, and a_0 and b_0 are fixed constant and $a_0 > 1$; $b_0 > 0$; m, n $\in$ Z where Z is set of integers. The differential mechanism combined with DWT is applied in the present approach of transmission line online relaying, as shown in Fig. 1. The current at both ends of the protected section are captured using current transformer. In the present approach, the differential and average currents are not directly used as actuating factors for relays like in traditional differential relaying mechanism. Instead of that, the energy level of wavelet

Fig. 1 DWT-based differential relaying management system

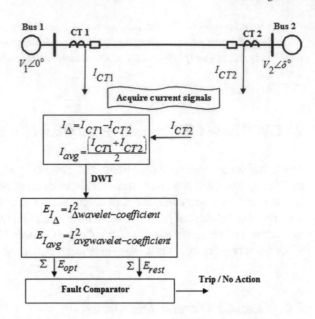

coefficients of differential and average currents is used as actuating factor. I_Δ and I_{avg} are computed using Eqs. (2, 3) and are decomposed using db4 mother wavelets for extracting wavelet coefficients.

$$I_\Delta = I_{CT1} - I_{CT2} \tag{2}$$

$$I_{avg} = \frac{(I_{CT1} + I_{CT2})}{2} \tag{3}$$

$$I_{\Delta wavelet-coefficients} = DWT(I_\Delta) \tag{4}$$

$$I_{avgwavelet-coefficients} = DWT(I_{avg}) \tag{5}$$

Thereupon, the energies of different level of coefficients of I_Δ and I_{avg} are estimated as operating and restraining factors.

$$E_{I_\Delta} = I_{\Delta wavelet-coefficient}^2 \tag{6}$$

$$E_{I_{avg}} = I_{avgwavelet-coefficient}^2 \tag{7}$$

The operating (E_{opt}) factor is addition of all the levels (l) of square of difference signal coefficients except the power frequency level as power frequency is not fault, whereas the restraining (E_{rest}) factor is addition all the levels of square of average signal coefficients which are computed using sliding data window.

$$E_{opt} = \sum_{l \neq fundamental} E_{I\Delta} \tag{8}$$

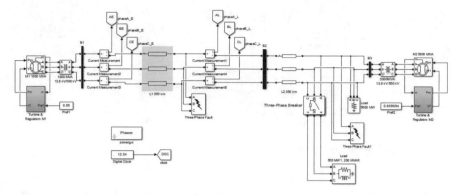

Fig. 2 Test-simulated system

$$E_{rest} = \sum_{l \neq fundamental} E_{lavg} \qquad (9)$$

The identification of internal fault and tripping of the circuit breaker is accomplished by comparing operating and restraining factor.

$$E_{opt} \succ \gamma E_{rest} \text{ Issue trip or alarm} \qquad (10)$$

$$E_{opt} \prec \gamma E_{rest} \text{ No action} \qquad (11)$$

3 Case Study

A test transmission network model of 100 MVA; two sections of 350 km have been simulated in MATLAB for ascertaining the feasibility of proposed online differential-based protection scheme. Figures 2 and 3 represents the simulated model and subsystems used for test cases. The positive and zero sequence resistance, inductance and capacitance of considered test model is 0.01755; 0.2758 Ω/km, 0.8737 e-3; 3.220 e-3 H/km, 13.33 e-9; 8.297 e-9 F/km, respectively. The proposed online methodology of differential protection is applied for the first section of line, i.e., 350 km. The db4 mother wavelet is applied for decomposing the differential and average current signal. The sampling rate is fixed as 3.2 kHz. The proposed scheme is evaluated for all presumable kinds of fault conditions, along with a different location. The power swing situation is also tested in the model. Different types of fault are inserted at different instance of time to visualize the ability of the proposed relaying scheme to identify the fault condition and generate trip signal only for internal faults. Flowchart of the proposed methodology is demonstrated in Fig. 4.

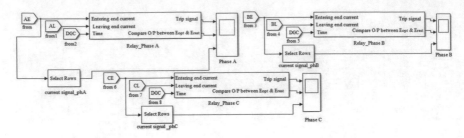

Fig. 3 Subset relaying simulation

The algorithm for online differential approach-based relaying in transmission system:

Step1 Capture data samples from both section of line, i.e., current at both ends.
Step2 Compute the differential and average currents values.
Step3 Find the 'N' samples of average and difference current, respectively, and save it in different variables.
Step4 Apply DWT to 'N' samples of average and difference current for realizing wavelet coefficients.
Step5 Evaluate the level energy by using above-mentioned equations
Step6 Thereafter, find operating factor (E_{opt}) and restraining factor (E_{rest})
Step7 Check if (E_{opt}) is less than (E_{rest}), then go to step3 otherwise go to next step.
Step8 Generate trip signal.

4 Result and Discussion

Different test conditions (like internal and external faults and emergent load changes) for all types of faults have been considered during assessment of feasibility of proposed online protection methodology. Various fault situations, at different instances of time are simulated in MATLAB, in order to evaluate the selectivity and sensitivity of proposed approach. Figure 5 shows the circumstances of L-G fault (phase-A) occurrence at two sections of test system (i.e., internal and external) and power swings condition (i.e., emergent load change). The three situations are simulated at different time instances. The lower window of Fig. 5 shows phase current; middle window shows the difference of E_{opt} and E_{rest}; whereas, the upper one has demonstrated the status of trip circuit. Figures 6, 7 and 8 are representing the magnifying version of Fig. 5 for obtaining the distinct view of different situations, i.e., internal, external fault, and power swings. It has been observed from Fig. 6 that, an internal fault at 0.5 s is identified as differential energy is positive, i.e., E_{opt} is greater than E_{rest}. The tripping signal is issued in 10.042 ms. Figure 7 has demonstrated the condition of power swing arises due to emergent load change at 1.0 s in the system. It has been observed that, although there is fluctuation in phase 'A' current waveform, but

Fig. 4 Flowchart of the
proposed relaying

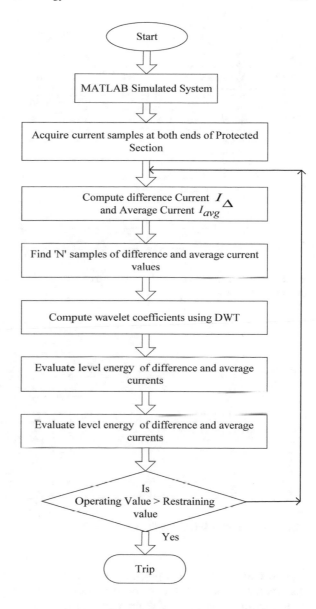

no tripping is assigned as the differential energy is zero, i.e., E_{opt} is equal to E_{rest}. It certifies that the proposed online protection approach is well competent of discriminating between faults and power swing conditions in the transmission line. In Fig. 8, the condition of external fault is depicted. An external fault is created at 1.5 s in the designed transmission system. It can be traced by current waveform pattern in phase 'A'. However, no tripping signal is generated during this instance also, as it has been observed that the differential energy is negative, i.e., E_{opt} is less than

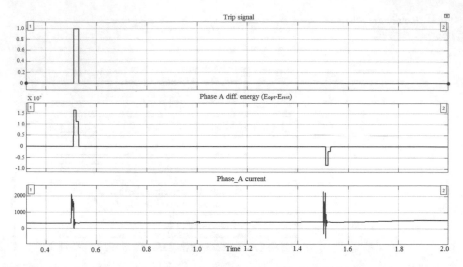

Fig. 5 Abnormatilies created at different instances of time, i.e., L-G fault (phase-a) internal, external, and power swing

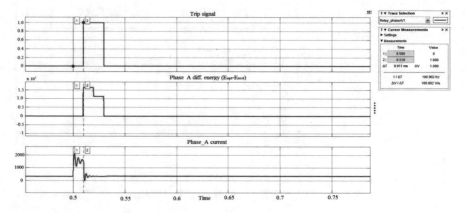

Fig. 6 L-G (phase-a) internal fault case

E_{rest}. It reasserted that the proposed protection methodology is also competent of discriminating faults within the protected section or outside of protection section.

Similarly, others types of faults such as L-L, LLG, 3-phase also have been simulated and the corresponding outcomes of the proposed scheme along with relaying functioning time is demonstrated in Table 1. T-Trip; R-Restrain.

Figure 9 separately represents the relay functioning time for different kinds of faults. From the results of different cases, it has been observed that the designed methodology of protection is competent in discriminating the faults (i.e., internal and external) and abnormal situations like power swings, which is one of the utmost important factors considered while designing any relaying scheme.

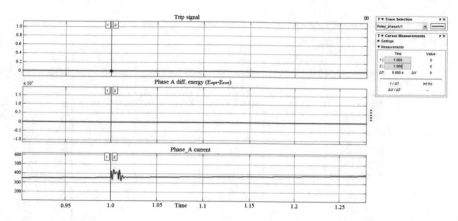

Fig. 7 Power swing case

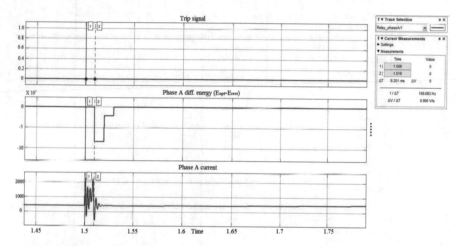

Fig. 8 L-G (phase-a) external fault case

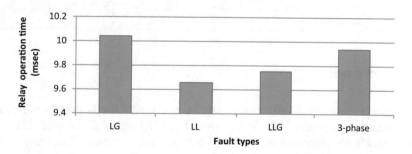

Fig. 9 Relay functioning time for different kinds of faults

Table 1 Output of the protection scheme for different abnormalities in line

S. No	Relay performance					
	Different condition	Instance at (s)	Relay-operation time (m s)	Phase-A relay	Phase-B relay	Phase-C relay
1	Internal fault SLG phasc-A	0.5	10.042	T	R	R
2	External fault SLG phase-A	1.5	9.407	R	R	R
3	Internal fault LL phases-AB	0.7	9.658	T	T	R
4	External fault LL phases-AB	1.4	8.979	R	R	R
5	Internal fault LLG phases-AB	0.4	9.753	T	T	R
6	External fault LLG phases-AB	1.5	9.551	R	R	R
7	Internal Fault 3-phase fault phases-ABC	0.6	9.939	T	T	T
8	External fault LLLG phases-ABC	0.62	9.706	R	R	R
9	Power swing case	0.8	9.315	R	R	R

5 Conclusion

This paper proposes an online methodology based on DWT for differential protection of transmission network. The energy level of coefficients of differential and average current in the protected section is primely applied as the discriminating factor. The tripping decision is taken on the basis of differential energy, i.e., E_{opt}-E_{rest}. The strength and feasibility of the proposed scheme have been appraised using MATLAB simulation. By observing the acquired results of different test cases, it has been concluded that the designed methodology of protection is competent in discriminating the faults (i.e., internal and external) and abnormal operating conditions like power swings. The prime attribute of the proposed scheme is it is competent of providing online protection to power transmission system. The application of DWT-based differential energy estimation has mitigated the direct dependency of differential relaying management only on current values. In the proposed approach, there is no need of extensive training data and time (which is the prime limitation of any intelligent-technique-based approaches reported in the literature) as it does not involve any training mechanism. Very less relay operating period is another worthy feature of the proposed scheme, which makes it apt for online protection application.

References

1. Ram, B., Vishwakarma, D.N.: Power System Protection and Switchgear. Tata Mc-Graw Hill Publication Company Limited, New Delhi (2005)
2. Singh, S., Vishwakarma, D.N.: Intelligent techniques for fault diagnosis in transmission lines-an overview. In: International Conference on Recent Developments in Control, Automation and Power Engineering (RDCAPE), pp. 129–134 (2015)
3. Samantaray, S.R., Tripathy, L.N., Dash, P.K.: Differential energy based relaying for thyristor controlled series compensated line. Int. J. Electr. Power Energy Syst. **43**, 621–629 (2012)
4. Abdulwahid, A.H., Wang, S.: Application of differential protection technique of solar photovoltaic based microgrid. Int. J. Control Autom. **9**(1), 371–386
5. Namdari, F., Jamali, S., Crossley, P.A.: Power differential based wide area protection. Electr. Power Syst. Res. **77**, 1541–1551 (2007)
6. Kawady, T.A., Taalab, A.M.I., Ahmed, E.S.: Dynamic performance of the power differential relay for transmission line protection. Electr. Power Energy Syst. **32**, 390–397 (2010)
7. Aziza, M.M.A., Zobaa, A.F., Ibrahima, D.K., Awad, M.M.: Transmission lines differential protection based on the energy conservation law. Electr. Power Syst. Res. **78**, 1865–1872 (2008)
8. Das, C.: Microprocessor based differential protection of a feeder depending on phase comparison of current. Int. J. Emerg. Technol. Adv. Eng. **3**(5) (2013)
9. Wen, M., Chen, D., Yin, X.: An energy differential relay for long transmission lines. Electr. Power Energy Syst. **55**, 497–502 (2014)
10. Wen, M., Chen, D., Yin, X.: A novel fast distance relay for long-transmission lines. Electr. Power Energy Syst. **63**, 681–686 (2014)
11. Liu, X., He, Z.: Transmission lines distance protection using differential equation algorithm and hilbert-huang transform. J. Power Energy Eng. **2**, 616–623 (2014)
12. Xu, Z.Y., et al.: A current differential relay for a 1000-kV UHV transmission line. IEEE Trans. Power Deliv. **22**(3) (2007)
13. Qiao, Y., Qing, C.: A novel current differential relay principle based on marti model. In: 10th IET International Conference on Developments in Power System Protection, pp 1–6 (2010). https://doi.org/10.1049/cp.2010.0311
14. Rebizant, W., Solak, K. Wiszniewski, A., Klimek, A.: Fuzzy inference supported current differential protection for HV transmission lines. In: 10th IET International Conference on Developments in Power System Protection, pp 1–5 (2010). https://doi.org/10.1049/cp.2010.0278
15. Dubey, R., Samantaray, S.R., Tripathy, A., Babu B.C., Ehtesham, M.: Wavelet based energy function for symmetrical fault detection during power swing. In: Student Conference on Engineering and Systems, pp 1–6. (2012). https://doi.org/10.1109/sces.2012.6199019
16. Jamil, M., Kalam, A., Ansari, A. Q., Rizwan, M.: Generalized neural network and wavelet transform based approach for fault location estimation of a transmission line. Appl. Soft Comput. 322–332 (2014)

17. Yadav, A., Swetapadma, A.: A single ended directional fault section identifier and fault locator for double circuit transmission lines using combined wavelet and ANN approach. Electr. Power Energy Syst. **69**, 27–33 (2015)
18. Eristi, H.: Fault diagnosis system for series compensated transmission line based on wavelet transform and adaptive neuro-fuzzy inference system. Measurement **46**, 393–401 (2013)
19. Ekici, S.: Support vector machines for classification and locating faults on transmission lines. Appl. Soft Comput. **12**, 1650–1658 (2012)

A Review on Magnetic Resonance Images Denoising Techniques

Abhishek Sharma and Vijayshri Chaurasia

Abstract Medical image denoising is a very important and challenging area in the field of image processing. Magnetic resonance imaging is a very popular and most effective imaging technique. During the acquisition, MR images get affected by random noise which could be modeled as Gaussian or Rician distribution. In the past few decades, a wide variety of denoising techniques have been proposed. This paper presents a survey of advancements proposed for the denoising of magnetic resonance images. The performance of most significant image denoising domains has been analyzed qualitatively as well as quantitatively on the basis of mean square error and peak signal-to-noise ratio.

Keywords Magnetic resonance images · Rician noise · Gaussian noise
Mean square error · Peak signal to noise ratio

1 Introduction

Digital image processing is the most important and widely used domain in the field of research. It has applications in many areas like computer vision, remote sensing, communication, defense and medical field. With the software development in image processing, new techniques and algorithms have been introduced in the past few decades, which used before treatment, during treatment and after the treatment images. Some processes like segmentation, registration, visualization and simulation are main key components of medical image processing. X-ray, computed tomography (CT), ultrasound and magnetic resonance imaging (MRI) are most useful terms of image processing. These different types of imaging techniques are generally known as modalities [1].

A. Sharma (✉) · V. Chaurasia
MANIT, Bhopal, Madhya Pradesh, India
e-mail: abhishektit09@gmail.com

V. Chaurasia
e-mail: vchaurasia@manit.ac.in

© Springer Nature Singapore Pte Ltd. 2019
M. Tanveer and R. B. Pachori (eds.), *Machine Intelligence and Signal Analysis*,
Advances in Intelligent Systems and Computing 748,
https://doi.org/10.1007/978-981-13-0923-6_60

X-ray diagnoses bone degeneration, fractures and dislocation [2]. Ultrasound use high-frequency sound waves for imaging human body structure and observe distinctive patterns of echoes. Ultrasound does not use ionizing radiation which damage body tissues. The computer tomography (CT) is a 3D view of X-ray radiography. It uses contrast agent to artificially increase the contrast in the tissues of human body to get better image quality [2]. Cancer, abnormal chest X-rays and bleeding in the brain because of injury is better shown in CT scan, but tendons and ligaments cannot be shown very well. These all are best seen by MRI like spinal cord of knees and shoulders. MRI images produce clearer differences between normal and abnormal tissue than CT images [3]. The MRI technique does not use high radiation like X-ray images; therefore it is the "safe method".

2 Noise Model for Medical Images

During acquisition or transmission, MRI images are mostly corrupted by noise. It may be due to imperfect instrument, susceptibilities between neighboring tissues, interference, rigid body motion and compression. The magnitude MRI images are best explained by a Gaussian distribution [4].

Magnitude images calculated using magnitudes of each pixel one by one from real and imaginary images. Because of nonlinearity of calculations, noise cannot be represented as Gaussian distribution. Now probability distribution for measured pixel intensity can be shown as

$$p_M(M) = \frac{M}{\sigma^2} e^{-\frac{A^2+M^2}{2\sigma^2}} I_0\left(\frac{A \cdot M}{\sigma^2}\right) \tag{1}$$

where A is represent original pixel intensity and M as measured pixel intensity.

I_0 = Modified zeroth-order Bessel function of the first kind and

σ = standard deviation of the Gaussian noise in the real and the imaginary images.

It is observed that the distribution of noise is considered as Rician for the range of $1 \leq \frac{A}{\sigma} \leq 3$. For $\frac{A}{\sigma} \geq 3$, this cannot be Rician but approximated towards Gaussian distribution. For $\frac{A}{\sigma} = 0$ or $A = 0$, Rician distribution is known as Rayleigh distribution, where no image signal is and only noise is present in MRI image.

$$\frac{A}{\sigma} = \begin{cases} \frac{A}{\sigma} = 0 \text{ Rayleigh distribution} \\ 1 \leq \frac{A}{\sigma} \leq 3 \text{ Rician distribution} \\ \frac{A}{\sigma} \geq 3 \text{ Gaussian distribution} \end{cases} \tag{2}$$

3 MRI Denoising Methods

The evaluations of modern denoising techniques in medical imaging have become very advanced, analytic and complex. During the noise removal of MRI images, fine details of image and other parameters should not get effected by noise removal process. Here, different post-acquisition noise removal algorithms are discussed. The noise removal techniques can be divided into three main categories, i.e., filtering domain, transform domain and statistical domain [5]. These domains can be further classified in subcategories such as filtering process can be linear or nonlinear. Transform domain can be classified as wavelet transform, curvelet transform or contourlet transform. Statistical domain can be based on maximum likelihood, LMMSE (Linear Minimum Mean Square Error) or error-based estimation.

3.1 Transform Domain

Wavelet-domain filtering is very important and effective for spatial variations in the signal behavior. It also preserves important edges and other fine details while removing noise [5]. Weaver et al. [6] and Nowak [7] developed a denoising method based on wavelet filter to reduce Rician noise. Wavelet packets are better concepts to represent signals more accurately. Wood and Johnson [8], Zaroubi and Goelman [9] and Alexander et al. [10] denoising algorithms based on Fourier transform preserve edges in a better way using Haar wavelet. Some other methods are Bao and Zang [11] based on adaptive multiscale products, Placidi et al. [12] based on the calculation of the correlation factor, Yu and Zhao [13] proposed an iterative scheme based on wavelet shrinkage, Tan and Shi [14] based on multi-wavelets and Wu et al. [15] based on Rayleigh background noise removal.

To deal with high-dimensional data and sharp edges, which cannot be described well by wavelet, curvelet transform-based method [16, 17] is a better option. First, it compute all thresholds and norms of curvelet, then apply curvelet transform by hard thresholding on curvelet coefficient and finally inverse curvelet transform to reconstruct denoised image. The contourlet transform [18] represents contours and fine details more specifically (Fig. 1).

3.2 Statistical Approach

Maximum likelihood (ML)-based approaches [19, 20] are suggested to remove Rician noise. L. He and I.R. Greenshields combine nonlocal and ML concept to estimate Rician noise and Rajan et al. [21, 22] uses local variance for ML-based estimator. Some statistical methods [23, 24] use Linear Minimum Mean Square Error (LMMSE) estimator for Rician noise, which uses local variance, the local mean, and

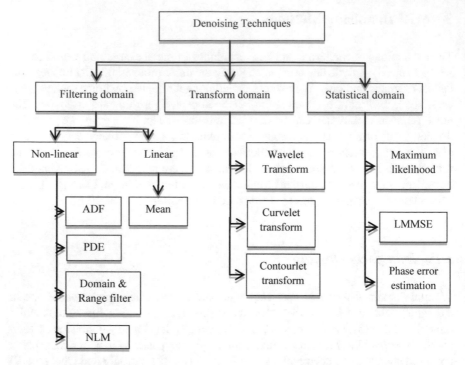

Fig. 1 Classification of denoising methods

the local mean square value. Phase error estimation [25] is another concept which uses a series of nonlinear filters iteratively.

3.3 Filtering Domain

Filters are of two types, i.e., linear filters and nonlinear filters. Linear filters are spatial filters and temporal filters proposed by McVeigh et al. [26]. These filters are very effective to remove the Gaussian noise from MRI images, therefore, also known as Gaussian filter. But these filters introduce blurring into edges because it averages pixels of nonsimilar patterns. To overcome this problem of blurring edges, nonlinear filters were introduced like Perona and Malik [27] developed a multiscale smoothing algorithm with edge preservation called as Anisotropic Diffusion Filter. It is based on second-order partial differential equation (PDE). Tomasi and Manduchi [28] proposed an improved filter, which does not use PDE and can be implemented using single iteration. It is the combination of two Gaussian filters (Domain and Range filter) known as bilateral filter. But all these methods are based on small filtering window and mainly concentrate on local pixels values for filtering. Another nonlinear filter is nonlocal means (NLM) filter, which is based on providing different weights to neighbors and taking average of those pixels.

Some enhancement of NLM filters were introduced like fast nonlocal means [29], optimized blockwise nonlocal means denoising filter [30, 31], adaptive NLM techniques [32] and maximum likelihood methods [33]. These filters are very fast due to automatic estimation of specially varying noise and improves computational performance.

4 Result Analysis

The analysis of different denoising algorithms is based on quality measure like peak signal-to-noise ratio (PSNR) [34], mean squared error (MSE) [34].

PSNR defines the performance of algorithm and it is represented mathematically as

$$PSNR = 10\log\left[\frac{255^2}{\frac{1}{MN}\sum_{i=0}^{M-1}\sum_{j=0}^{N-1}\left(I(i,j)-\hat{I}(i,j)\right)^2}\right] \tag{3}$$

And, mean square error (MSE) is defined as

$$MSE = \frac{1}{MN}\sum_{i=0}^{M-1}\sum_{j=0}^{N-1}\left(I(i,j)-\hat{I}(i,j)\right)^2 \tag{4}$$

where $M \times N$ is size of the image, $I(i,j)$ represents original image, and $\hat{I}(i,j)$ represents restored image.

Another quality measures is structural similarity index matrix (SSIM), which specifies the human visual system.

$$SSIM(x,y) = \frac{\left(2\mu_x\mu_y + c_1\right)\left(2\sigma_{xy} + c_2\right)}{\left(\mu_x^2 + \mu_y^2 + c_1\right)\left(\sigma_x^2 + \sigma_y^2 + c_2\right)} \tag{5}$$

The parameters μ_x and μ_y represent mean value of images x and y. σ_x and σ_y are standard deviation of images x and y. σ_{xy} represents covariance of x and y. Constant C_1 and C_2 are $C_1 = (K_1 L)^2$ and $C_2 = (K_2 L)^2$, where L is dynamic range and $K_1 = 0.01$, $K_2 = 0.03$. For 8-bit gray images, the value of L is 255.

Figure 2 showing the simulated T1 phantom images and image effected by Rician noise and then denoised using NLM and ANLM filter. In Fig. 3, T1 image is used form BrainWeb database [35]. It shows the noisy image corrupted by Rician noise and then restored through different denoising techniques (AD, TV, and Optimized Blockwise NLM). It is clear from the observation that the homogeneity of white

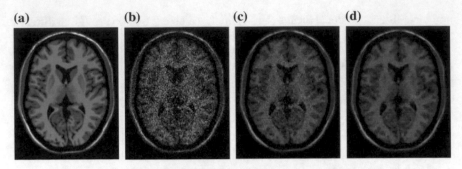

Fig. 2 **a** T1 MR images, **b** Noisy image, **c** Filtered by NLM, **d** Filtered by ANLM [31]

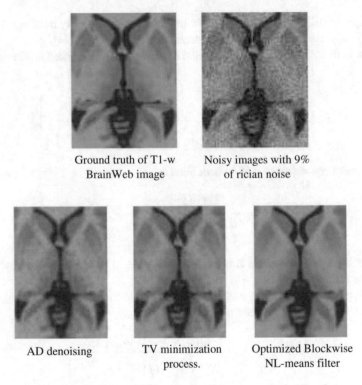

Ground truth of T1-w   Noisy images with 9%
BrainWeb image   of rician noise

AD denoising   TV minimization   Optimized Blockwise
                      process.   NL-means filter

Fig. 3 Comparison with AD, TV, and NL-means denoising on synthetic T1-w images

matter is very good when it denoised by the Optimized Blockwise NLM algorithm. Hence, it is clear that NLM denoised technique better preserves the high-frequency components of the image corresponding to anatomical structures while denoising process.

Table 1 shows the quantitative analysis based on MSE and Table 2 shows PSNR of different denoising techniques like Anisotropic Diffusion (AD), Total Variation (TV), NLM, Oracle discrete cosine transform (ODCT) and transform-domain collaborative

Table 1 Mean square error (MSE)

Sr. no.	Technique	Noise density (%)		
		3%	9%	15%
1	Noisy [30]	20.56	159.61	460.34
2	AD [30]	14.55	47.10	76.39
3	TV [30]	12.39	46.56	81.86
4	NLM [30]	8.18	33.34	78.17
5	ODCT [36]	3.28	11.88	22.03
6	BM4D [36]	2.53	9.52	17.02

Table 2 Peak signal-to-noise ratio (PSNR) in dB

Sr. no.	Technique	Noise density (%)		
		3%	9%	15%
1	Noisy [30]	35.00	26.10	21.50
2	AD [30]	36.50	31.40	29.30
3	TV [30]	37.20	31.45	29.00
4	NLM [30]	39.00	32.90	29.20
5	ODCT [36]	42.96	37.38	34.70
6	BM4D [36]	44.09	38.34	35.82

Table 3 Structural similarity index matrix (SSIM)

Sr. no.	Technique	Noise density (%)		
		3%	9%	15%
1	Noisy [30]	0.81	0.43	0.25
2	AD [30]	0.85	0.65	0.62
3	TV [30]	0.88	0.65	0.60
4	NLM [30]	0.95	0.79	0.62
5	ODCT [36]	0.97	0.90	0.83
6	BM4D [36]	0.98	0.92	0.88

filtering algorithm (BM4D). The results show that BM4D method produces much better results than the other state-of-the-art methods. Table 3 shows the comparison of SSIM which shows that ODCT and BM4D produced best results among all other methods.

5 Conclusion

An efficient filter should be capable of removing the maximum noise and restore the image while keeping the image structure and fine details unaltered. It should not only fulfill the quantitative criteria, but also preserve the visual quality of images. Wavelet-domain filters use threshold to detect noise but add some artifacts into the image because of which detection process gets affected. Curvelet transform does not work well in smooth region. Contourlet transform has high complexity. Nonlocal mean filter removes these drawbacks but introduce heavy computational load. With some improvements in NLM computational time can be reduced and improve the denoising efficiency. ABM4D method is very good for Gaussian noise and Rician noise because it preserves the image details much better and increases the PSNR.

References

1. Angenent, S., Pichon, E., Tannenbaum, A.: Mathematical methods in medical image processing. Bull. (New Ser.) Am. Math. Soc. (2005)
2. Roentgen, W.C.: Ueber eine neue Art von Strahlen. Ann. Phys. **64**, 1–37 (1898)
3. http://www.ctscaninfo.com/mrivsctscan.html
4. Gudbjartsson, H., Patz, S.: The Rician distribution of noisy MRI data. Magn. Reson. Med. **34**(6), 910–914 (1995)
5. Mohana*, J., Krishnavenib, V., Guoca, Y.: A survey on the magnetic resonance image denoising methods. Biomed. Signal Process. Control **9**, 56–69 (2014)
6. Weaver, J.B., Xu, Y., Healy, D.M., Cromwell, L.D.: Filtering noise from images withwavelet transforms. Magn. Reson. Imaging **21**, 288–295 (1991)
7. Nowak, R.D.: Wavelet-based Rician noise removal for magnetic resonance imaging. IEEE Trans. Image Process. **8**, 1408–1419 (1999)
8. Wood, J.C., Johnson, K.M.: Wavelet packet denoising of magnetic resonanceimages: importance of Rician noise at low SNR. Magn. Reson. Med. **41**, 631–635 (1999)
9. Zaroubi, S., Goelman, G.: Complex denoising of MR data via wavelet analysis:application for functional MRI. Magn. Reson. Imaging **18**, 59–68 (2000)
10. Alexander, M.E., Baumgartner, R., Summers, A.R., Windischberger, C., Klarhoefer, M., Moser, E., Somorjai, R.L.: A wavelet-based method for improvingsignal-to-noise ratio and contrast in MR images. Magn. Reson. Imaging **18**, 169–180 (2000)
11. Bao, P., Zhang, L.: Noise reduction for magnetic resonance images via adap-tive multiscale products thresholding. IEEE Trans. Med. Imaging **22**, 1089–1099 (2003)
12. Placidi, G., Alecci, M., Sotgiu, A.: Post-processing noise removal algorithm formagnetic resonance imaging based on edge detection and wavelet analysis. Phys. Med. Biol. **48**, 1987–1995 (2003)
13. Yu, H., Zhao, L.: An efficient denoising procedure for magnetic resonance imaging. In: Proceedings of IEEE 2nd International Conference on Bioinformatics and Biomedical Engineering, pp. 2628–2630 (2008)
14. Tan,L., Shi, L.: Multiwavelet-based estimation for improving magnetic reso-nance images. In: Proceedings of IEEE Conference, pp. 1–5 (2009)
15. Wu, Z.Q., Ware, J.A., Jiang, J.: Wavelet-based Rayleigh background removal inMRI. IEEE Electron. Lett. **39**, 603–605 (2003)
16. Starck, J.L., Candes, E.J., Donoho, D.L.: The curvelet transform for image denois-ing. IEEE Trans. Image Process. **11**, 670–684 (2002)

17. Ma, J., Plonka, G.: Combined curvelet shrinkage and nonlinear anisotropic dif-fusion. IEEE Trans. Image Process. **16**, 2198–2206 (2007)
18. Do, M.N., Vetterli, M.: The contourlet transform: an efficient directional mul-tiresolution image representation. IEEE Trans. Image Process. **14**, 2091–2106 (2005)
19. Jiang, L., Yang, W.: Adaptive magnetic resonance image denoising using mixture model and wavelet shrinkage. In: Sun, C., Talbot, H., Ourselin, S., Adriaansen, T. (eds.) Proceedings of VIIth Digital Image Computing: Techniques and Applications, The University of Queensland, Sydney, Australia. St. Lucia, Australia, pp. 831–838 (2003)
20. Sijbers, J., Poot, D., den Dekker, A.J., Pintjenst, W.: Automatic estimation of thenoise variance from the histogram of a magnetic resonance image. Phys. Med. Biol. **52**, 1335–1348 (2007)
21. Rajan, J., Poot, D., Juntu, J., Sijbers, J.: Noise measurement from magnitude MRI using local estimates of variance and skewness. Phys. Med. Biol. **55**, 441–449 (2010)
22. Rajan, J., Jeurissen, B., Verhoye, M., Audekerke, J.V., Sijbers, J.: Maximum like-lihood estimation-based denoising of magnetic resonance images usingrestriced local neighborhoods. Phys. Med. Biol. **56**, 5221–5234 (2011)
23. Aja-Fernández, S., Alberola-López, C., Westin, C.F.: Noise and signal estimationin magnitude MRI and Rician distributed images: a LMMSE approach. IEEETrans. Image Process. **17**, 1383–1398 (2008)
24. Golshan, H.M., Hasanzedeh, R.P.R., Yousefzadeh, S.C.: An MRI denoising method using data redundancy and local SNR estimation. Magn. Reson. Imaging **31**, 1206–1217 (2013)
25. Tisdall, D., Atkins, M.S.: MRI denoising via phase error estimation. Proc. SPIE **5747**, 646–654 (2005)
26. McVeigh, E.R., Henkelman, R.M., Bronskill, M.J.: Noise and filtration in magnetic resonance imaging. Med. Phys. **12**, 586–591 (1985)
27. Perona, P., Malik, J.: Scale-space and edge detection using anisotropic diffusion. IEEE Trans. Pattern Anal. Mach. Intell. **12**, 629–639 (1990)
28. Tomasi, C., Manduchi, R.: Bilateral filtering for gray and color images. In: Presented at the 6th International Conference Computer Vision, Bombay, India, pp. 839–846 (1998)
29. Coupe, P., Yger, P., Barillot, C.: Fast non local means denoising for 3D MR images. In: Larsen, R., Nielsen, M., Sporring, J. (eds.) MICCAI 2006. LNCS, vol. 4191, pp. 33–40 (2006)
30. Coupe, P., Yger, P., Prima, S., Hellier, P., Kervrann, C., Barillot, C.: An optimized blockwise nonlocal means denoising filter for 3-D magnetic resonance images. IEEE Trans. Med. Imaging **27**(4) (2008)
31. Manjon, J.V., Coupe, P., Buades, A., Louis Collins, D., Robles, M.: New methods for MRI denoising based on sparseness and self-similarity. Med. Image Anal. **16**(1), 8–27 (2012)
32. Coupe, P., Manjon, J.V., Robles, M., Collins, D.L.: Adaptive multiresolution non-local means filter for three-dimensional magnetic resonance image denoising. IET Image Process. **6**(5), 558–568 (2012)
33. Rajan, J., den Dekker, A.J., Sijbers, J.: A new non-local maximum likelihood estimation method for Rician noise reduction in magnetic resonance images using the Kolmogorov–Smirnov test. Signal Process. **103**, 16–23 (2014)
34. Kim, D.W., Kim, C., Kim, D.H., Lim, D.H.: Rician nonlocal means denoising for MR images using nonparametric principal component analysis. EURASIP J. Image Video Process. (2011)
35. http://brainweb.bic.mni.mcgill.ca/brainweb
36. Maggioni, M., Katkovnik, V., Egiazarian, K., Foi, A.: Nonlocal transform-domain filter for volumetric data denoising and reconstruction. IEEE Trans. Image Process. **22**(1) (2013)

Detection and Analysis of Human Brain Disorders

Deeksha Tiwari and Hem Chandra Jha

Abstract In this review, we noted that any single technique available till date is not sufficient enough to provide us with all the answers. Fortunately, the strengths and weakness of available techniques complement each other to a great extent. Thus, for obtaining desired information, data from different techniques can be poled together leading to the introduction of a variety of multimodal imaging technology. The direct comparison of the acquired information has contributed greatly to the understanding of different neurodegenerative disorders and better application of imaging techniques to it. This has also led to use of imaging technology as disease monitoring application. Disease detection and subsequent progression can be effectively monitored using such multimodal imaging techniques. There's a huge scope for development of new and improvements to be incorporated in existing techniques for enhancing the detection probability of neurodegeneration in early stages. Early detection might help in improvement of therapeutics and treatment available.

1 Introduction

Neurodegenerative disorders, a class of human brain disorders involve progressive degradation of the basic unit of nervous system, i.e., the neurons. Interestingly, if a neuron dies it is irreplaceable in an adult brain, only the damage can be repaired up to some extent. Thus, these diseases as they progress render the patient incapable of doing even his daily chores, and making him dependent on others for very basic activities like eating, bathing, etc. Different neurodegenerative diseases show different pathophysiology, depending on the location of damage in the brain. For instance, it is now well known that Alzheimer's disease affects hippocampus, resulting in memory impairment; Parkinson's disease affects motor areas of the brain, causing motor dysfunction, etc.

D. Tiwari · H. C. Jha (✉)
Department of Biosciences & Biomedical Engineering, Indian Institute of Technology Indore, Simrol Campus, Indore 453552, India
e-mail: hemcjha@iiti.ac.in

© Springer Nature Singapore Pte Ltd. 2019
M. Tanveer and R. B. Pachori (eds.), *Machine Intelligence and Signal Analysis*, Advances in Intelligent Systems and Computing 748, https://doi.org/10.1007/978-981-13-0923-6_61

The major obstacle in treating the neurodegenerative disorders is our inefficiency in diagnosing these diseases at early stage. Pathological symptoms that form the basis of current diagnosis appear only after the disease has already progressed to advanced stage. It becomes virtually impossible to treat such advance cases of neurodegeneration as the damage that has been done is irreversible. At present, only disease modification therapies are available in the market for some of these diseases that slow down the disease progression but could not cure it. The only possible way to treat neurodegeneration is to stop the disease from progressing to advance levels after its early diagnosis. This review tries to summarize and compare currently available techniques for analysis and detection of human brain disorders (in particular, Alzheimer's disease and Huntington's disease) based on their utility and limitations.

2 Current Diagnostic Techniques Available

Currently available techniques for diagnosis of neurodegeneration are only there for helping the clinician to rule out any other possible cause of symptoms. The diagnosis is primarily based on biomedical imaging techniques, CSF and blood biomarkers screening, and genetic risk profiling accompanied by complete psychological and cognitive examination by trained medical professionals.

The clinician first establishes the possibility of neurodegeneration by carefully examining the patient's medical history and thorough physical examination that might indicate some vital signs of neuronal damage. Certain lab tests like CSF and blood tests are then done to corroborate the findings. Thereafter, neuroimaging is done to confirm the positive results as found by prior tests. We tried to list some neuroimaging techniques that are employed in clinics at present.

Neuroimaging Techniques are broadly characterized as structural and functional neuroimaging. Structural neuroimaging techniques such as MRI and CT scans are used for the diagnosis of gross intracranial diseases and injuries whereas; functional neuroimaging is used to localize the brain activity related to any stimulus inside the brain anatomically and include fMRI and PET scans. Different neuroimaging techniques are employed for obtaining different information about the brain anatomy and functioning in normal and pathological conditions.

i. **Computer-Assisted Tomography (CAT/CT) Scan**:
   As the name suggests computer-assisted tomography scan is a radiographic technique that utilizes combination of computer-processed multiple X-ray images. Several two-dimensional radiographic images are taken around a single axis of rotation at different angles. These images are then computed with digital geometry processing techniques to generate cross-sectional three-dimensional image of the anatomy being scanned. CT scan produces the images based on the relative amount and quickness of X-rays being absorbed by the tissue or a particular organ of the body. It can reveal physical changes in the brain including atrophy, widened indentation in tissue and enlargement of the brain ventricles.

Sometimes to enhance detection of abnormalities in the brain tissue, gadolinium-based or iodine-based contrast materials could be given to the patient based on his clinical history.

ii. **Magnetic Resonance Imaging (MRI)**: MRI is the most widely used imaging technique that operates on the principle of NMR (Nuclear Magnetic Resonance). Certain atomic nuclei are able to absorb and emit radio energy when placed in external magnetic field, which is then received by antennas placed near the structure being examined. Majority of MRI machines used in clinical settings uses 1.5 T magnets. In MRI of biological organisms, hydrogen is used to generate these signals.

iii. **Positron Emission Tomography (PET) Scan**: PET scanning technique utilizes the principle of differential metabolism of compounds in body. Radioactive isomer of a biologically active molecule is administered to the patient several hours before test. This radiotracer is absorbed by the organs and tissues of the body and their subsequent decay signals, as captured by the detector indicate biochemical changes in real time (Table 1).

1. **Alzheimer's disease**:

1.1 Role of imaging in Alzheimer's Disease Prognosis and Diagnosis:

Current definitive diagnosis of AD available is only postmortem detection of Amyloid-β plaques and neurofibrillary tangles in the brain tissue. Establishing AD in clinical setting requires thorough examination of patient's medical history, neurophysiological and physical evaluation in addition to several lab tests. Neuroimaging has although historically been used to rule out any potential surgically treatable cause of cognitive decline. But, recently its application is expanded to diagnosis in providing positive support and monitoring disease progression as well as detection of structural and functional changes in brain anatomy. It is being used to exclude non-AD dementia pathologies (vascular and frontotemporal dementia) in clinical settings. Interestingly, only a fraction of patients with MCI (Mild Cognitive Impairment) develops clinical AD over 5–10 years, whereas others may revert back to normal cognition eventually. Henceforth, prediction of probability to develop clinical symptoms of AD early in the disease progression based on neuroimaging is of great value. With advent of new disease-modifying therapies the importance of early diagnosis with noninvasive techniques like neuroimaging is increasing, as these therapies work best when initiated early in prodromal stages (Figs. 1 and 2).

1.2 Structural MRI in AD

● Utility of Structural MRI in AD:

Structural MRI having biggest strength in its easy availability is primarily done to assess atrophy and changes in brain tissue characteristics of AD patients clinically. T_1-weighed volumetric sequences are used to visualize progressive cerebral atrophy.

Table 1 Comparative study of different imaging techniques

Techniques	Methodology	Spatial resolution	Temporal resolution	Outcome measured	Primary usage	Advantages	Disadvantages
CT	X-ray generator and detector rotates around generating multiple images in each plane			Morphology	Diagnostic purpose	Rapid, Noninvasive	Exposure to radiation
Structural MRI	Utilizes radio waves and magnetic field to generate images of the body based on variable alignment of hydrogen ions in the tissue	~2 mm	4–8 s	Whole or regional brain volumes are measured	Diagnostic and morphology	Noninvasive, precise, no radiation exposure	Links between structure and function are indirect, expensive and relatively difficult to use
fMRI	Magnetic field fluctuations produced locally by changes in proportions of oxyhaemoglobin and deoxyhaemoglobin in blood which approximate neural activity	<5 mm	200 ms–1 s	Evaluate location and strength of neuronal activation in response to task or treatment	Diagnostic; Neuronal activity	Noninvasive; no radiation exposure; functions can be directly linked with neural response	BOLD signals cannot resolve increase/decrease in activation at neurochemical level; temporal delay between stimulus and output; difficult interpretation
PET/SPECT	Radioactive decay (Positron emission in PET and gamma emission in SPECT) of ingested radioisotope used to detect in vivo molecular changes	2–3 mm	40 s	Metabolic increase/decrease in specific brain regions where ligand has bound	Diagnostic; areas of biochemical activation	Detection of subcortical neurons; real time monitoring of molecular changes in brain	Radioactivity exposure; quite expensive; difficult to use

Fig. 1 Alzheimer's
detection techniques

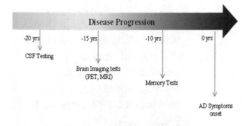

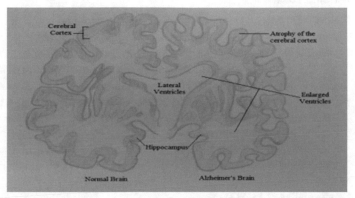

Fig. 2 Comparative sketch showing atrophy in brain of an Alzheimer's disease patient and a normal brain

Anatomically the entorhinal cortex of medial–temporal lobe is first to be affected by atrophy [1], closely followed by the hippocampus, parahippocampus, and amygdala [2, 3]. Other structures associated with limbic system such as posterior cingulate are also affected early on. Gradually temporal neocortex and then all neocortex associated areas also develop atrophy usually in a symmetrical manner. Histopathological studies that have established the stages of disease progression and spread of neurofibrillary tangles coincide with this sequence of progression of atrophy on MRI [4]. The atrophy is widespread in whole brain by the time clinical diagnosis is made [5, 6]. Total brain volume shrinks down by 6% with rate of 0.3% loss in 2 years up till diagnosis is made, with entorhinal volumes being reduced by 20–30% and hippocampal volumes by 15–25% even in mildly affected individuals [2, 7, 8].

Medial temporal atrophy assessment with the help of MRI has been proved to have positive predictive value with sensitivity and specificity of 50–70% for differentiating between individuals who might develop AD in future and who will not. And thus, medial temporal lobe atrophy is proposed to be included in diagnosis criteria as one of the biomarkers of prodromal AD at a pre-dementia stage [9]. Although severity of hippocampal atrophy is greatest in case of frontotemporal dementia (FTD) anteriorly followed by AD and then dementia with Lewy body (DLB) and vascular dementia (VaD), it is considered to be a characteristic feature of DLB and VaD [10, 11]. Therefore, for differential diagnosis of AD overall imaging pattern and other features of the dementia needs to be taken into account. For example, on MRI focal frontal/temporal lobar atrophy would suggest diagnosis of FTD, whereas marked signal changes in white matter may point to VaD [2, 10–12].

- Limitations:

Structural MRI cannot pick up the histopathological markers of AD (β-Amyloid plaques and neurofibrillary tangles), and hence lack molecular specificity. Neuronal damage gives rise to cerebral atrophy nonspecifically, certain loss patterns are characteristic feature of non-AD pathologies. Whereas, atrophy patterns not being entirely specific to AD often overlaps with other diseases as well. Also, volume changes as seen on MRI may be the result of certain factors other than progressive neuronal loss.

Moreover, performing MRI in severely affected individuals and those with claustrophobia may not be possible instead CT would be more feasible.

1.3 Functional MRI in AD

- Utility of functional MRI in AD:

fMRI exploits the changes in blood flow and volume including oxyhemoglobin/deoxyhemoglobin ratio in blood measured as blood oxygen level-dependent (BOLD) MR signals that indirectly indicates neuronal activity [11]. It can be used to test the functional integrity/connectivity (Fc-MRI) of neuronal networks in resting state or for comparative study of a cognitive task in a typical condition (encoding new information) versus control condition (recalling existing information in familiar conditions). Both the techniques have the potential to be used for detection of AD-related early brain dysfunction and therapeutic response monitoring for short time period.

By far, the fMRI findings on subjects clinically diagnosed with AD have been consistent to show decrease in hippocampal activity during formation of new memory [13–16]. And as a probable compensatory mechanism for hippocampal dysfunction in early AD pathology setting prefrontal cortex has been reported to be hyperactive according to some studies [14], particularly during successful memory trials. fMRI may be of greatest use for evaluating novel pharmacological strategies for treatment of AD whether during cognitive paradigms or resting state. It is now being integrated with ongoing Phase II and Phase III trials for obtaining valuable information about its potential utility in clinical settings.

- Limitations:

  fMRI being sensitive to head motion is difficult to be performed on patients with severe neurodegenerative dementias. In such cases, resting state MRI is more feasible. Whereas, in patients with more severe cognitive impairment who are not able to perform the cognitive tasks the greatest advantage of fMRI, i.e., task fMRI activation studies is lost.

1.4 FDG-PET in AD:

The brain is one of the most important organs in body and its proper functioning is dependent on adequate supply of blood through intricate network of vessels, which carries oxygen and nutrients with it. Glucose being the primary and almost exclusive source of energy for brain, its analogue FDG is a suitable indicator of brain metabolism when labeled with Fluorine-18 (half-life of 110 min) which can be easily detected on PET scans. Brain utilizes most of its energy budget on maintenance of intrinsic, resting (task independent) activity, which in cortex is largely maintained by glutaminergic synaptic signaling [17]. FDG uptake on autopsy exams strongly coincides with the levels of synaptic vesicle protein synaptophysin [14], indicating FDG as a valid biomarker of overall brain metabolism. And thus, FDG-PET is mainly utilized primarily to depict synaptic activity.

- Utility:

  FDG hypometabolism appears to precede appearance of cognitive symptoms and hence FDG has emerged as a robust biomarker for cognitive impairment and neurodegeneration. FDG can be used as a good predictive biomarker for rate of progressive cognitive decline in individuals who progress from MCI to develop AD in later stages of life. FDG hypometabolism highly coincides with the level of cognition among normal, preclinical, prodromal and established AD patients. However, cognitive levels and intellect alters the results toward showing lesser strength of correlation. An intelligent AD patient may show mild clinical symptoms, but severe hypometabolism [14, 15].

  Amyloid-β accumulation appears to change FDG metabolism. Thus, it is possibly indicating the intermediate stage between the initiation and subsequent development of neurodegeneration [14]. The ultimate development of AD can be predicted by FDG hypometabolism that actually occurs before impairment [15, 18] and brain volume loss has also been reported in cognitively normal individuals who go on to develop AD [15].

- Limitations:

  The major limitation of FDG-PET is it not being cost-effective and easily available diagnostic technique. Although at lower levels than known significant risk, FDG involves exposure to radioactivity, through intravenous access. FDG retention in brain is also nonspecific indicator of brain metabolism as it can be altered by a variety of reasons including ischemia or inflammation. Thus, FDG can be an indirect indicator of AD pathologies in the brain and results need to be substantiated with other pathological findings in clinical settings.

1.5   Amyloid PET in AD:

Amyloid-β PET is primarily an indicator of amyloid pathologies in brain rather than a definitive proof of clinical diagnosis of AD. It strongly correlates with postmortem findings of amyloid deposition in the brain of AD patients. Substrate for amyloid-β imaging in vivo for all the known tracers by far is Amyloid-β in beta sheet confirmation that forms the fibrillary tangles conformation [15]. Although, amyloid-β may exist in the form of monomer, oligomer or fibrillar Aβ affinity of tracers have been found to be more with beta sheets indicating a strong signal correlation between amyloid PET signal and oligomer concentration in brain based on equilibrium between all three forms. The most widely evaluated amyloid tracers is PiB (Pittsburgh compound B). Other tracers available are florbetaben and fluorbetapir (fluorine-18 labeled tracers).

- Utility:

Neurofibrillary tangles have been claimed to be accurately traced with amyloid tracers but there is no validatory proof available for the same. On the contrary, some studies suggest that amyloid tracers do not bind to neurofibrillary pathologies [19, 20].

The biggest strength of amyloid imaging is that is has allowed the Amyloid-β content determination to be moved from pathology laboratories to clinical settings and is now widely applicable as a substantial proof for amyloid pathologies. It can successfully and specifically detect cerebral β-amyloidosis giving negative signals with conformed cases of prion amyloid [15], pathologically confirmed pure α-synucleinopathies [18], as well as cases of pure tauopathies in semantic dementia [15].

Obvious advantage of amyloid PET is the regional information it provides even with the continuously varying biological changes. The quantitative data obtained by amyloid PET is most regionally specific among entire clinical spectrum.

- Limitations:

Cost and availability remains the major obstacle in ensuring widespread use of amyloid PET in clinics. When only presence or absence of amyloid deposits in the brain is a concern, CSF measurements of Aβ42 can provide the information. Amyloid PET is not a good surrogate marker of diseases progression in clinical stages as amyloid deposition is an early event in the AD pathogenesis [15]. Moreover, amyloid imaging provides much more of a binary diagnostic readout than techniques like MRI and FDG-PET indicating identical readout in case of absence of pathology and negative scan as well regardless of non-etiology of the dementia.

3 Conclusion

In this review we conclude that, there is a lot of scope of improvement in application of imaging techniques in the field of detection and analysis of human brain disorders, especially in multimodal imaging. Future of diagnosis and disease improvement therapies relies greatly on imaging techniques.

Acknowledgements This project was supported by Council of Scientific and Industrial Research grant no 37(1693)/17/EMR-II and Department of Science and Technology as Ramanujan Fellowship grant no. SB/S2/RJN-132/20/5. We are thankful to Ministry of Human Resource and Development for fellowship to Deeksha Tiwari. We are also thankful to Indian Institute of Technology Indore, for providing the resources. The funding organization has not played any role in study design, the decision to publish, or the preparation of the manuscript. We appreciate our lab colleagues for insightful discussions and advices.

References

1. Scahill, R.I., Schott, J.M., Stevens, J.M., et al.: Mapping the evolution of regional atrophy in Alzheimer's disease: unbiased analysis of fluid-registered serial MRI. Proc. Natl. Acad. Sci. **99**, 4703–4707 (2002). https://doi.org/10.1073/pnas.052587399
2. Chan, D., Fox, N.C., Scahill, R.I., et al.: Patterns of temporal lobe atrophy in semantic dementia and Alzheimer's disease. Ann. Neurol. **49**, 433–442 (2001)
3. Killiany, R.J., Hyman, B.T., Gomez-Isla, T., et al.: MRI measures of entorhinal cortex vs hippocampus in preclinical AD. Neurology **58**, 1188–1196 (2002)
4. Braak, H., Braak, E.: Neuropathological stageing of Alzheimer-related changes. Acta Neuropathol. **82**, 239–259 (1991)
5. Chan, D., Janssen, J.C., Whitwell, J.L., et al.: Change in rates of cerebral atrophy over time in early-onset Alzheimer's disease: longitudinal MRI study. Lancet **362**, 1121–1122 (2003). https://doi.org/10.1016/S0140-6736(03)14469-8
6. Ridha, B.H., Barnes, J., Bartlett, J.W., et al.: Tracking atrophy progression in familial Alzheimer's disease: a serial MRI study. Lancet Neurol. **5**, 828–834 (2006). https://doi.org/10.1016/S1474-4422(06)70550-6
7. Dickerson, B.C., Goncharova, I., Sullivan, M.P., et al.: MRI-derived entorhinal and hippocampal atrophy in incipient and very mild Alzheimer's disease. Neurobiol. Aging **22**, 747–54
8. Schuff, N., Woerner, N., Boreta, L., et al.: MRI of hippocampal volume loss in early Alzheimer's disease in relation to ApoE genotype and biomarkers. Brain **132**, 1067–1077 (2008). https://doi.org/10.1093/brain/awp007
9. Dubois, B., Feldman, H.H., Jacova, C., et al.: Research criteria for the diagnosis of Alzheimer's disease: revising the NINCDS-ADRDA criteria. Lancet Neurol. **6**, 734–746 (2007). https://doi.org/10.1016/S1474-4422(07)70178-3
10. Barber, R., Ballard, C., McKeith, I.G., et al.: MRI volumetric study of dementia with Lewy bodies: a comparison with AD and vascular dementia. Neurology **54**, 1304–1309 (2000). https://doi.org/10.1212/WNL.54.6.1304
11. Burton, E.J., Barber, R., Mukaetova-Ladinska, E.B., et al.: Medial temporal lobe atrophy on MRI differentiates Alzheimer's disease from dementia with Lewy bodies and vascular cognitive impairment: a prospective study with pathological verification of diagnosis. Brain **132**, 195–203 (2009). https://doi.org/10.1093/brain/awn298
12. Scheltens, P., Leys, D., Barkhof, F., et al.: Atrophy of medial temporal lobes on MRI in probable Alzheimer's disease and normal ageing: diagnostic value and neuropsychological correlates. J. Neurol. Neurosurg. Psychiatry **55**, 967–972 (1992)

13. Kwong, K.K., Belliveau, J.W., Chesler, D.A., et al.: Dynamic magnetic resonance imaging of human brain activity during primary sensory stimulation. Proc. Natl. Acad. Sci. USA **89**, 5675–5679 (1992)
14. Frisoni, G.B., Fox, N.C., Jack, C.R., et al.: The clinical use of structural MRI in Alzheimer disease. Nat. Rev. Neurol. **6**, 67–77 (2010). https://doi.org/10.1038/nrneurol.2009.215
15. Logothetis, N.K., Pauls, J., Augath, M., et al.: Neurophysiological investigation of the basis of the fMRI signal. Nature **412**, 150–157 (2001). https://doi.org/10.1038/35084005
16. Small, S.A., Perera, G.M., DeLaPaz, R., et al.: Differential regional dysfunction of the hippocampal formation among elderly with memory decline and Alzheimer's disease. Ann. Neurol. **45**, 466–472 (1999)
17. Rabinovici, G.D., Furst, A.J., O'Neil, J.P., et al.: 11C-PIB PET imaging in Alzheimer disease and frontotemporal lobar degeneration. Neurology **68**, 1205–1212 (2007). https://doi.org/10.1212/01.wnl.0000259035.98480.ed
18. Ogawa, S., Lee, T.M., Nayak, A.S., Glynn, P.: Oxygenation-sensitive contrast in magnetic resonance image of rodent brain at high magnetic fields. Magn. Reson. Med. **14**, 68–78 (1990)
19. Klunk, W.E., et al.: The binding of 2-(4'- methyl aminophenyl) benzothiazole to post-mortem brain homogenates is dominated by the amyloid component. J. Neurol. **23.6**, 2086–2092 (2003)
20. Ikonomovic. M.D., Klunk, W.E., et al.: Post-mortem correlates of in vivo PiB-PET amyloid imaging in a typical case of Alzheimers's disease. Brain **131** (6), 1630–1645 (2008)

Experimental Analysis on Effect of Nasal Tract on Nasalised Vowels

Debasish Jyotishi, Suman Deb, Amit Abhishek
and Samarendra Dandapat

Abstract In almost every language across the globe nasalised speech is present. Our work is motivated by the fact that nasalised speech detection can improve the speech recognition system. So, to analyse the nasalised speech better, we have designed a device to separate nasal murmur from oral speech, when nasalised speech is spoken. Speech data of different speakers are collected and analysed. Nasalised vowels are analysed first and it has been found that an additional formant is consistently being introduced between 1000 and 1500 Hz. Using various signal processing techniques we analysed different nasalised vowels and found that nasal murmur produced, is invariant irrespective of the nasalised vowels and so is the nasal tract. Nasalisation is being produced in speech by coupling of nasal tract with oral tract. So, when effect of coupling is analysed experimentally, it came out to be addition.

1 Introduction

Nasalised vowels or consonants contribute to the vocabulary of almost every language, although their impact of contribution varies widely across the languages. In English language nasalisation is generally introduced by nasal consonants, /m/ and

D. Jyotishi (✉) · S. Deb · A. Abhishek · S. Dandapat
Department of Electronics and Electrical Engineering, Indian Institute
of Technology Guwahati, Guwahati 781039, India
e-mail: debasish.jyotishi07@iitg.ernet.in

S. Deb
e-mail: suman.2013@iitg.ernet.in

A. Abhishek
e-mail: amit.abhishek@iitg.ernet.in

S. Dandapat
e-mail: samaren@iitg.ernet.in

© Springer Nature Singapore Pte Ltd. 2019   727
M. Tanveer and R. B. Pachori (eds.), *Machine Intelligence and Signal Analysis*,
Advances in Intelligent Systems and Computing 748,
https://doi.org/10.1007/978-981-13-0923-6_62

/n/. But in Indian language we can deliberately introduce nasality with the help of 'Matra', which are present in scripts, and also through nasal consonants. Thus for speech recognition system, it becomes increasingly difficult to distinguish words, only differed by introduction of nasality. This issue can easily be handled through detection of nasality. For that purpose, different characteristics of nasalised vowels and nasal consonants should be studied.

There are few works in the literature dealing with the analysis of nasalised vowels and nasal consonants. Glass and Zue [1] have done extensive statistical analysis on nasalised vowels in the pursuit of acoustic cues. But, these statistical parameters have limitations for detecting nasalised speech accurately. Vijaylaxmi and Reddy [2] have analysed hypernasal speech, which has limitations in case of nasalised speech.

To understand the properties of nasalised vowels better, we have designed a device to separate nasal murmur from oral vowel, when a nasalised vowel is spoken. Speech data collected using the device are used to study the variation of nasalisation when different nasalised vowels are spoken [3]. We have also studied the effect of nasal tract and vocal tract coupling on nasalised speech signal. Analysis on formants of nasalised vowel is also done in this paper.

The rest of the paper is organised as follows. Section 2 presents the database description. Section 3 comprises of analysis on oral and nasalised vowel. In Section 4, different signal processing techniques are used to analyse variability of nasal tract across different oral vowel. Finally, we conclude the work in Sect. 5.

2   Experimental Setup and Database Description

When nasalised speech is spoken the nasal tract gets coupled with the oral tract and the radiation occurs both at nostril and lips. This work aims to study the respective effect of nasal tract and vocal tract for different nasalised sounds. For this purpose, a device is designed to separate nasal murmur and oral sounds during the utterance of a nasalised vowel. This purpose is achieved by designing an acoustically closed chamber for the sound produced from mouth. The nasal murmur produced stay outside of the chamber, as the nose is placed outside during speaking. Figure 1 shows the recording of a speaker using the device. For simultaneous recording of both nasal murmur and oral sounds, we have used two microphones with two-channel recording facility. One microphone is placed near nose to record nasal murmur, while another one microphone is placed near mouth to record oral speech. Figure 2 shows the inner part and outer part of our designed device with two microphones.

Speech data are collected two times from a person, once using the device and other time without using the device. Speakers are asked to utter three vowels(/e/, /i/ and /u/), their nasalised counterparts and the word 'summer'. The nasal consonant part of the word 'summer' is extracted. The speech data are recorded using Audacity software, at 8 kHz sampling frequency.

All the participants have consented for use of the speech data for research purpose and also for publications.

Fig. 1 Picture of a speaker with the device

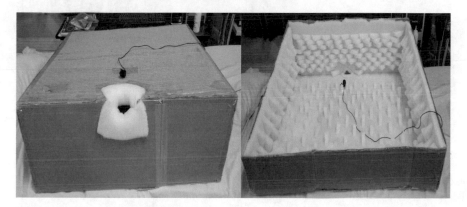

Fig. 2 Picture of inner part and outer part of the device

3 Acoustic Analysis

In the literature [1, 2, 4], it is found that nasalised vowels have in general three cues, i.e. (i) introduction of extra formants (ii) presence of antiformants and (iii) broadening of bandwidth of the first formant. In this section, we will focus on the analysis of extra formant introduction in case of nasalised vowels.

3.1 Acoustic Analysis of Oral Speech

Speech data considered for this study are the oral vowels recorded without using the device. Linear prediction (LP) spectrum of oral vowel is analysed and formants are

Table 1 Formant location of oral vowels

Formants	/e/	/i/	/u/
F1	384	300	370
F2	2290	2514	848
F3	2924	3208	1915

Fig. 3 Pdf of formants of oral vowel /i/

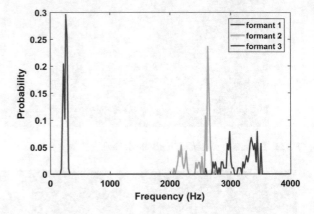

Fig. 4 Pdf of formants of oral vowel /u/

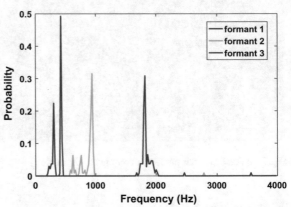

extracted. A 20 ms window and LP of order 12 are used for this purpose. The formant information extracted from oral vowels are tabulated in Table 1.

From Table 1 we can observe that the first formant of vowel /e/, /i/ and /u/ is coming out to be 384, 300, 370. The result is consistent with the existing literature [5]. For better visualisation of formant variation, we have drawn the pdf plot of formants. Pdf plots of formants of phoneme /i/, /u/ and /e/ are given in Fig. 3, Fig. 4 and Fig. 5 respectively.

Fig. 5 Pdf of formants of oral vowel /e/

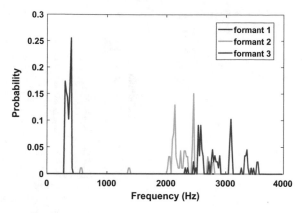

Table 2 Formant location of nasalised vowels

Formants	/e/	/i/	/u/
F1	450	332	336
F2	1128	1300	816
F3	2298	2550	1644
F4	2824	3159	2581

Fig. 6 Pdf of formants of nasalised vowel /i/

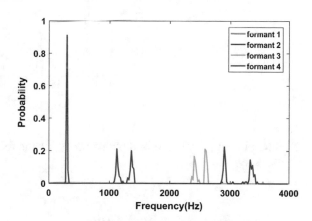

3.2 Acoustic Analysis on Nasal Speech

Speech data considered for the present study are the nasal vowels recorded without using the device. Linear prediction (LP) spectrum of nasalised vowels is analysed and formants are extracted. The formant informations from nasal vowels are tabulated in Table 2.

From the Table 2, it is observed that there is an introduction of extra formant between 1000 Hz to around 1500 Hz across all the nasalised vowels. This is an acoustic cue which is present in a nasalised vowel. For visual analysis of the fact,

Fig. 7 Pdf of formants of
nasalised vowel /u/

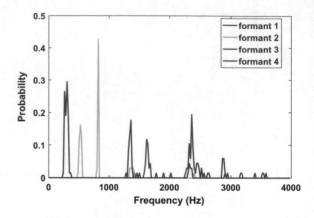

Fig. 8 Pdf of formants of
nasalised vowel /e/

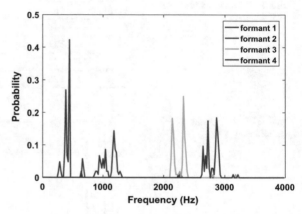

we have plotted pdf of nasalised phoneme /i/, /u/ and /e/, in Fig. 6, Fig. 7 and Fig. 8 respectively.

4 Signal Processing Techniques

Oral vowels are being produced by resonance along the oral cavity. So, for producing different oral vowels we change the oral tract. Thus different oral vowels have different formant frequency, different time domain and frequency domain response. While nasal murmur is introduced only by lowering the velum. Using the device the nasal murmur and the oral vowels are separated for every nasalised vowels. In this section, this nasal murmur and oral vowels separated by the device is being used for analysis purpose. Different signal processing techniques are used for this purpose. We have assumed that the vocal tract and nasal tract are speaker dependent. So the signal processing techniques are applied on person specific data.

4.1 Formant Detection

A number of studies have been reported for formant detection. Cepstral smoothed spectra and linear prediction spectra [5] are mostly used by researchers for formant detection. Lim [6] has proposed a spectral root homomorphic deconvolution for formant detection. Murthy and Yagnarayana [7] have proposed an algorithm to extract formant using group delay method. Yagnarayana [8] has shown that linear prediction phase spectra has magnitude information and thus formants can be extracted from the phase spectra. In this analysis, we have used linear prediction phase spectra for formant extraction.

Three formants are extracted from oral vowel, nasal murmur of all the nasalised vowels and nasal consonants of word 'summer'. Then, their pdf distribution is plotted. Figure 9 shows the pdf of three formants of oral vowels and Fig. 10 shows the pdf of nasal murmur and nasal consonants combined.

From the Figs. 9 and 10, it can be observed that while oral formants are distributed all across the frequency ranges, the nasal formants are well localised. This shows that

Fig. 9 Pdf of formants of oral vowels

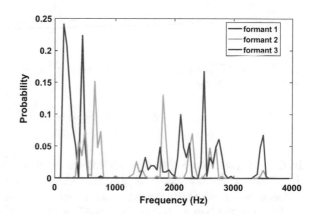

Fig. 10 Pdf of formants of nasal vowels

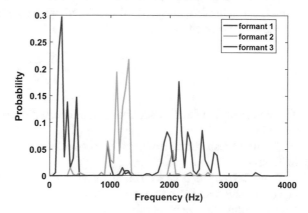

as a person speaks different nasalised vowels the vocal tract takes different shapes, while the nasal tract remains same and the input to nasal tract also remains same. Nasal consonants or nasal murmur are produced through nasal tract and by closing the oral tract. Figure 10 pdf plot includes formants from nasal murmur of nasalised vowels and nasal consonants of word 'summer'. From the figure, we can observe that the formant location of both nasal murmur and nasal consonants are nearly same. The nasal murmur data collected here, are from different nasalised vowels. Formant location of these speech data is same as the nasal consonant produced by closing the oral tract. So, this analysis shows that the effect of coupling is addition.

4.2 Log-Spectral Distance

Log-spectral distance of two speech measures the distortion between them. Log-spectral distance appears to be closely tied with the subjective assessment of sound differences. Log-spectral distance is defined as [5]

$$d(S, S')^P = (d_P)^p = \int_{-\pi}^{\pi} |V(\omega)|^p \frac{d\omega}{2\pi}$$

here 'S' and 'S'' corresponds to LP spectra of two different speech signals and 'p' is the order of distance and

$$V(\omega) = log S(\omega) - log S'(\omega)$$

In this analysis, rms log-spectral distortion is used. So value of p taken is 2. In this analysis, distance between different oral sounds and also, the corresponding distance between nasal murmurs were found out. Log-spectral distance between nasal consonant, extracted from the word 'summer' and the nasal murmur extracted from the nasalised sounds is also calculated. As we have discrete frequency, so the formulae is modified as below,

$$d(S, S')^P = (d_P)^p = \frac{1}{2\pi} \sum_{-\pi}^{\pi} |V(\omega)|^p$$

The mean result of the distance obtained between different nasal murmur is tabulated in Table 3.

The mean result of the distance obtained between different oral vowels is tabulated in Table 4.

From Tables 3 and 4, we can see that the log-spectral distance across the oral vowels are high and also varying widely. While the log-spectral distance across the nasal murmur of different nasalised vowels are almost same and also low. This shows

Table 3 Log-spectral distances among nasal murmurs

Vowel	Nasal consonants	/i/	/u/	/e/
/i/	153	–	113	114
/u/	159	113	–	169
/e/	176	114	169	–

Table 4 Log-spectral distances among oral vowels

Vowel	/i/	/u/	/e/
/i/	–	492	232
/u/	492	–	387
/e/	232	387	–

that while the oral tract changes widely for production of different vowels, the nasal tract as well as the input to the nasal tract remains almost same. The log-spectral distance between nasal murmur and nasal consonants is also low and nearly invariant. This shows that the nasal tract coupling has an effect of addition.

4.3 Cross Correlation

Correlation between two signals show the similarity between two signals. Cross correlation between two signals is defined as [9]

$$\rho_{xy}(l) = \frac{r_{xy}(l)}{\sqrt{r_{xx}(0).r_{yy}(0)}}$$

where

$$r_{xy}(l) = \sum_{-\infty}^{\infty} x(n)y(n-l).$$

As the correlation between two signals increase, the correlation coefficients approach towards one. Similarly, as the correlation between signals decrease correlation coefficients approach towards zero. And correlation coefficients of uncorrelated signals is zero.

Here we have used the speech data, collected using the device to calculate two types of cross-correlation result. One type of cross-correlation result is obtained using different oral sounds, while the other type of cross correlation result is obtained using respective nasal murmur.

The mean result of the cross correlation obtained between different oral vowels is tabulated in Table 4.

From Tables 5 and 6, we can see that the cross correlation across the oral vowels are low. While the cross correlation across the nasal murmur of different nasalised

Table 5 Correlation among nasal murmurs

Vowel	/i/	/u/	/e/
/i/	–	0.8535	0.7841
/u/	0.8535	–	0.8345
/e/	0.7841	0.8345	–

Table 6 Correlation among oral vowels

Vowel	/i/	/u/	/e/
/i/	–	0.7226	0.6997
/u/	0.7226	–	0.7365
/e/	0.6997	0.7365	–

vowels is high. This time domain analysis shows that the time domain signal for different oral vowels are different and hence have low correlation. But, the time domain signal of nasal murmur produced from different nasalised vowel is nearly the same and highly correlated.

5 Conclusion

In the literature, few studies have been performed to understand the characteristics of nasality. But no research has been carried out on the effect of nasal tract for different nasalised vowels. So we designed a device to separate nasal murmur and oral sounds. Data collected using the device helped us in analysing the effect of nasal tract for different nasalised vowels. The acoustic analysis of nasalised vowels showed that there is an extra formant from 1000 Hz to around 1500 Hz across all the nasalised vowels. This can be considered as an acoustic cue for nasalised vowels. Different signal processing techniques; formant detection, log-spectral distance and cross correlation are used in this work to analyse the effect of nasal tract on different nasalised vowels. From this analysis we observed that, while speaking different nasalised vowels the oral tract changes its shape but the nasal tract remains invariant. And also from the experiment, we came to know that the effect of coupling of nasal tract with the vocal tract has an effect of addition for speech signals.

References

1. Glass, J., Zue, V.: Detection of nasalized vowels in American english. In: Acoustics, Speech, and Signal Processing, IEEE International Conference on ICASSP'85, vol. 10, pp. 1569–1572. IEEE (1985)
2. Vijayalakshmi, P., Ramasubba Reddy, M., O'Shaughnessy, D: Acoustic analysis and detection of hypernasality using a group delay function. IEEE Trans. Biomed. Eng. **54**(4), 621–629 (2007)
3. Vijayalakshmi, P., Ramasubba Reddy, M.: Analysis of hypernasality by synthesis. In: INTER-SPEECH (2004)

4. Hawkins, S., Stevens, K.N.: Acoustic and perceptual correlates of the non-nasal-nasal distinction for vowels. J. Acoust. Soc. Am. **77**(4), 1560–1575 (1985)
5. Rabiner, L.R., Schafer, R.W.: Digital Processing of Speech Signals, vo. 100. Prentice-Hall, Englewood Cliffs (1978)
6. Lim, J.: Spectral root homomorphic deconvolution system. IEEE Trans. Acoust. Speech Signal Process. **27**(3), 223–233 (1979)
7. Murthy, H.A., Yegnanarayana, B.: Formant extraction from group delay function. Speech Commun. **10**(3), 209–221 (1991)
8. Yegnanarayana, B.: Formant extraction from linear-prediction phase spectra. J. Acoust. Soc. Am. **63**(5), 1638–1640 (1978)
9. Proakis, J.G., Manolakis, D.G.: Digital Signal Processing, 3rd edn. Prentice-Hall, Englewood Cliffs (1996)

A Fast Adaptive Classification Approach Using Kernel Ridge Regression and Clustering for Non-stationary Data Stream

Chandan Gautam, Raman Bansal, Ruchir Garg, Vedaanta Agarwalla
and Aruna Tiwari

Abstract Classification on non-stationary data requires faster evolving of the model while keeping the accuracy levels consistent. We present here a faster and reliable model to handle non-stationary data when a small number of labelled samples are available with the stream of unlabelled samples. An active learning model is proposed with the help of supervised model, i.e. Kernel Ridge Regression (KRR) with the combination of an unsupervised model, i.e. K-means clustering to handle the concept drift in the data efficiently. Proposed model consumes less time and at the same time yields similar or better accuracy compared to the existing clustering-based active learning methods.

Keywords Kernel · Non-stationary data · Concept drift · Multi-class classification

1 Introduction

Multi-class classification is a long researched topic and many different techniques are there to solve the problem for both stationary as well as non-stationary environment [1]. In a stationary environment, model is trained once for all the training samples and there is no need to adapt the model to any new sample. This assumes that data does not show any concept drift. But this is not always the case as the source of data is not always immutable. For example, human behaviour changes over the course of time, so the data associated with it also drifts. Shopping trends of a consumer changes with his salary, fashion, innovation in the market, etc. A college graduate might have got a good job, now we may classify him as a rich customer. This is an

C. Gautam (✉) · R. Bansal · R. Garg · V. Agarwalla · A. Tiwari
Indian Institute of Technology Indore, Indore, Madhya Pradesh, India
e-mail: chandangautam31@gmail.com

© Springer Nature Singapore Pte Ltd. 2019
M. Tanveer and R. B. Pachori (eds.), *Machine Intelligence and Signal Analysis*,
Advances in Intelligent Systems and Computing 748,
https://doi.org/10.1007/978-981-13-0923-6_63

example of change in data characteristics and this concept is called concept drift. Brain– computer interface (BCI) is also one of the most popular kind of examples where non-stationarity occurs in the data [2].

Non-stationary data possesses concept drift that requires automatic updates of model; else the accuracy suffers. The model needs to be evolved consistently [3]. New data that arrives might or might not have label information with it. When the model learns from a pool of unlabelled samples, then this strategy is called active learning [4]. In some cases, obtaining the labels of new data has some cost associated with it as a human annotator is required to label the unlabelled samples for retraining the model [5] and most of the models follow this strategy. Directly using the unlabelled samples without the help of human annotator is still a challenging task in this area [6]. In this paper, we want to minimize the cost of labelling of unlabelled samples. Our proposed model just takes small size of labelled samples initially for training and afterwards, user interference is not required for retraining or adaption of the model. For the purpose of initial training, we need data with labelled samples to initialize our model. Thus, it is completely supervised in the initial phase. Thereafter, a clustering algorithm helps the model to identify the drift from the unlabelled samples and also prepares the set of new labelled samples for retraining the classifier. Hence, model is continuously evolved to handle this drift in data using the predicted label not the actual label. Thus our approach is supervised in the start and becomes unsupervised after the initialization phase. Some other researchers also follow similar strategies and have just used initial samples for training the model [7]. However, their proposed model consumes a lot of time during training, which can be a bottleneck in the real scenario as data comes in a stream and the model needs to adapt the new data as fast as possible. We compare our proposed model from the methods proposed by [7] and obtain similar or better accuracy with significant improvements in time.

With the advent of big data applications, extremely fast level of active learning with self-adaption of non-stationary environment is required for classification, which is not addressed by current research. Training of neural networks and SVM requires huge amount of computation time and power. This paper aims to tackle that problem by introducing a faster adaptable model in online learning. In this paper, we propose a faster approach to multi-class classification for non-stationary datasets. Proposed algorithm is further tested on 16 datasets to show that our model is significantly faster than existing models while keeping the accuracy level similar or in some cases even higher.

Rest of the paper is as follows: Sect. 2 discusses KRR based multi-class classification. Proposed model and their experimental results are discussed in the Sects. 3 and 4.

2 Kernel Ridge Regression (KRR) for Multi-Class Classification

KRR uses the kernel method and enhances the ridge regression for non linear cases. Such a method has been well applied in problems relating to regression tasks and achieved credible accuracy in the past. However, it is not much explored for the multi-class classification task, especially for the non-stationary environment. KRR is discussed below for the multi-class classification task and named as **KRRMC**.

Class-codec is used to describe the output target $\mathbf{Y}$ as a 0–1 encoding. So, for M number of classes and N samples from the data stream, $\mathbf{Y}$ should be a NxL matrix obtained as follows:

$$\mathbf{Y}_{ij} = \begin{cases} 1 & ith \text{ sample is from } jth \text{ class} \\ 0 & \text{otherwise} \end{cases} \tag{1}$$

The kernel method that enables faster calculation of the projection of samples $\mathbf{x}_i$ in a higher dimension (also known as Hilbert space) is enabled by the calculation of the kernel matrix $\mathbf{K}$ as follows:

$$\mathbf{K}_{ij} = k(\mathbf{x}_i, \mathbf{x}_j) = \phi(\mathbf{x}_i)^T \phi(\mathbf{x}_j) \tag{2}$$

A classical linear regression problem is formulated as:

$$min_{\mathbf{w}} \sum_k (\mathbf{w}^T \mathbf{x}_k - \mathbf{t}_k)^2 + \lambda \|\mathbf{w}\|_2^2 \tag{3}$$

Here, the regularization parameter is denoted by λ and it helps in controlling the model complexity. Above minimization problem can be solved by employing Karush–Kuhn–Tucker (KKT) theorem [8] and obtained the solution as follows:

$$\mathbf{w} = (\mathbf{X}^T \mathbf{X} + \lambda \mathbf{I})^{-1} \mathbf{X}^T \mathbf{Y} \tag{4}$$

Here matrix $\mathbf{I}$ denotes an identity matrix, vector $\mathbf{Y}$ denotes a vector where each element is the target output y_k for sample x_k and and matrix $\mathbf{X}$ has one sample per row $\mathbf{x}_i$.

The solution of $\mathbf{w}$ can be expressed as a linear combination of the samples in the Hilbert space using the representer theorem as follows:

$$\mathbf{w} = \sum_i \alpha_i \phi(\mathbf{x}_i) \tag{5}$$

Using the above equation for **w** and substituting in Eq. 5, we get the following reformulated function for Kernel Ridge Regression:

$$min_\alpha \|\mathbf{Y} - \mathbf{K}\alpha\|^2 + \lambda\alpha^T\mathbf{K}\alpha \tag{6}$$

Calculating the closed for solution for the above equation yields:

$$\alpha = (\mathbf{K} + \lambda\mathbf{I})^{-1}\mathbf{Y} \tag{7}$$

3   Proposed Model

We propose a model, that handles the concept drift of the data stream by combining a supervised and an unsupervised learning approach. The base multi-class classification algorithm for our model is kernel ridge regression, which is described in the above Sect. 2. Further, this model is assisted by a clustering algorithm for getting the labelled data from the stream of unlabelled data for retraining the KRRMC. Hence, the proposed model can be described in three phases: 1. Initialization 2. Active Learning for the unlabelled stream 3. Retraining of the classifier KRRMC.

3.1   Initialization

Initial training step involves training the model with a batch of labelled data in order to define the problem for multi-class classification. This will enable the model to know the number of classes and their representation in the feature space. The number of classes are also assumed to be fixed through the duration of the runtime. Let this number of classes be c. Such an assumption is often used in many works for such a domain. We also assume that though the data stream is non-stationary, the characteristics of the environment changes gradually. Note that such a gradual change in data characteristic should be much more harder to detect than a abrupt change in data characteristic (Fig. 1).

3.2   Active Learning for the Unlabelled Stream

We use an **Active Learning** approach for our model. This infers that small size of the labelled samples is only available to train and rest of the upcoming samples in the stream are completely unlabelled. Therefore, we do not use the actual labels of the newly obtained unlabelled samples and thus only the labels obtained as an output of the proposed model is used to adapt the model to the varying characteristic of the data for the better classification of the upcoming samples in the stream. In the case for

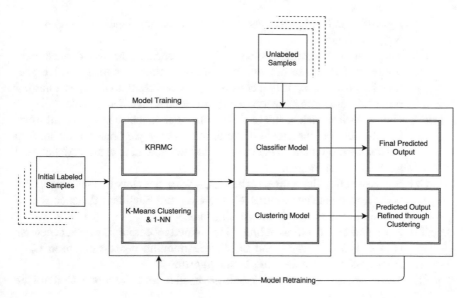

Fig. 1 Flow diagram of the proposed model

active learning, we have a teacher that helps us to obtain the labels of the unlabelled samples for further model training. Such a approach is often used in areas where gathering the actual labels is an expensive task. Here, we use a clustering algorithm as a teacher to obtain the new sample labels. A clustering method would be required for labelling the unlabelled samples for retraining. K-means clustering algorithm [9] is a simple and efficient algorithm, which is used for the experimentation. The latter uses a clustering algorithm to relate the previously known information with the new clusters formed to obtain a mapping between the cluster and the labels. Each sample in the cluster is assigned the corresponding class label regardless of the multi-class classification model's output. Following steps are required for labelling the unlabelled data stream for retraining:

step I   Initially, we divide our dataset into k number of clusters where $k \leq c$. Let C_0 denote this initial set of cluster centers:

$$C_0 = \{C_{01}, C_{02}, \ldots, C_{0k}\} \tag{8}$$

step II If the number of class is equal to number of clusters (Unimodal case) then calculate the cluster centres by taking the mean of data of each class from the given set of initial labelled data and assign a class label to each cluster centre with their corresponding class of the samples whose mean is calculated for obtaining the cluster centres.

step III If number of class is **not** equal to number of clusters (Multi-modal case) then calculate the cluster centres by directly employing K-means clustering algorithm and assign the label to the centre using 1-nearest neighbor (1-NN) algorithm.

step IV   The clustering algorithm is run again after reaching the user-defined maximum pool length to obtain a new set of k cluster centres, C_1. Here, at time t, clustering algorithm considers previous centres (C_{t-1}) as a starting point for performing clustering algorithm on unlabelled samples.

step V   Now associate each centre in C_1 to a class using centre in C_0 and information available in the previous stage. In this way, we obtain a relationship between the old data characteristics and the new data characteristics, and a label of each sample.

step VI   For associating the centres at time t, C_t, with class labels, 1-NN algorithm is employed between centres, C_t and C_{t-1} and Euclidean distance is used as a similarity measurement for this task.

step VII Now, new set of centres at time t is obtained by combining current centres (C_t) and median of C_t and their corresponding nearest centre in C_{t-1}. Again assign the label using 1-NN algorithm.

step VIII These newly obtained class labels are used to train the model to adapt the new conditions.

Thus our algorithm goes in a cycle of classification (supervised) and clustering (unsupervised). Previously labelled data is used to build the classification model whose labels have been obtained from the clustering algorithm.

3.3   Retraining of the Classifier KRRMC

The trained classification algorithm is run on the incoming data stream and the predicted results of these unlabelled samples are used as an index to a bucket where these samples are pooled. Thus each bucket will correspond to one and only one class. After a user-defined window (300 in our experiments) of newly arrived unlabelled samples have been fed to the classification algorithm, the process comes to a halt. We take a moment to discuss how to fix the maximum pool length. A shorter pool size will result in faster adaptation of the model to concept drift but will be associated with higher computational cost. A larger pool size may miss certain gradual drifts leading to a abrupt change in characteristic of data and thus poor efficiency of the model. The proposed model needs to retrain after receiving new unlabelled samples up to a number of maximum pool size. KRRMC will be retrained by the labels samples provided after refinement by the clustering algorithms.

In the next section, the proposed model is experimented with and without clustering to analyse the impact of clustering algorithm on the model. **KRRMC** is a classification algorithm without clustering, i.e. it uses no clustering algorithm as a basis to obtain labels and completely relies on the classification model's output labels as the correct ones for further training. The proposed model **KRRMC_Clust** uses a clustering algorithm to relate the previously known information with the new clusters formed to obtain a mapping between the cluster and the labels. Thus, it is a **supervised with unsupervised** model that enables the use of active learning for multi-class classification.

4 Experimental Evaluation

For testing purpose, 16 non-stationary datasets have been used. As these datasets have incremental and gradual concept drift, therefore, these are suitable for testing online algorithms that involves concept drift. These datasets are collected from various papers as mentioned in Table 1. All experiments has been conducted on MATLAB 2016b with 4 GB of RAM and Intel i5 processor. For testing purposes, average accuracy is taken as total accuracy. For initial training, 50 initial labelled samples were taken and pool size was taken as 300. Proposed model is evaluated in term of accuracy as well as time in the following section.

4.1 Performance Analysis

In this section, we compared our proposed model with the methods proposed in [7] i.e. SCARGC(1NN), SCARGC(SVM).

Table 2 provides the accuracy of different underlying algorithms, namely KRRMC, KRRMC_Clust and SCARGC(1NN, SVM). The number of data points in these

Table 1 Datasets description

Dataset name	No. of classes	No. of attributes	Drift	No. of samples
1CDT [7]	2	2	400	16,000
2CDT [7]	2	2	400	16,000
1CHT [7]	2	2	400	16,000
2CHT [7]	2	2	400	16,000
4CR [7]	4	2	400	144,400
4CRE-V1 [7]	4	2	1,000	125,000
4CRE-V2 [7]	4	2	1,000	183,000
5CVT [7]	5	2	1,000	40,000
1CSurr [7]	2	2	600	55,283
4CE1CF [7]	5	2	750	173,250
FG-2C-2D [10]	2	2	2,000	200,000
UG-2C-2D [11]	2	2	1,000	100,000
UG-2C-3D [11]	2	3	2,000	200,000
UG-2C-5D [11]	2	5	2,000	200,000
MG-2C-2D [11]	2	2	2,000	200,000
GEARS-2C-2D [11]	2	2	2,000	200,000

Table 2 Mean Accuracy of the proposed and existing classifiers

	SCARGC (1NN)	SCARGC (SVM)	KRRMC	KRRMC_Clust	Best k
1CDT [7]	99.75	99.63	99.72	99.72	2
2CDT [7]	90.90	91.01	53.27	91.24	2
1CHT [7]	99.24	99.18	98.99	99.20	2
2CHT [7]	85.98	85.83	52.78	86.06	2
4CR [7]	99.95	99.53	99.98	99.95	4
4CRE-V1 [7]	97.45	97.54	33.44	97.60	4
4CRE-V2 [7]	91.89	92.03	36.63	92.09	4
5CVT [7]	85.20	85.23	36.38	85.69	5
1CSurr [7]	94.29	94.80	67.59	94.50	4
4CE1CF [7]	94.09	92.40	95.32	94.14	5
FG-2C-2D [10]	95.18	95.34	95.22	95.33	4
UG-2C-2D [11]	95.55	95.64	95.55	95.63	2
UG-2C-3D [11]	94.77	94.98	93.54	95.03	2
UG-2C-5D [11]	91.01	91.72	78.73	91.68	2
MG-2C-2D [11]	92.77	92.86	69.42	92.88	4
GEARS-2C-2D [11]	95.89	95.16	95.23	95.28	2

datasets ranges from 16,000 to 2,00,000, while the total number of classes ranges from 2 to 5. KRRMC_Clust outperforms the other three models for more than half of the datasets. Such an improvement is a result of better classification ability and better label association through clustering technique which helps in cases where the classes overlaps. In such cases of classes overlapping, the model needs a smooth boundary of partition that is provided by the clustering algorithm as opposed to the overfitting done by the pure classification models. For calculating accuracy and plotting their behaviour, divided the datasets in the batch of 100 and calculate the average accuracy over hundred batches, which is taken as final accuracy of the model. Accuracy over period of time also gives us the idea of drift. A large fall in accuracy represents an instantaneous drift. Figure 2 represent the accuracy over multiple steps of online learning. Performance of the classifiers are discussed on these datasets by dividing them in the following category:

(i) If separation boundary stays static overtime or one class is static over time while data of other class is non-stationary in the data then classifier can easily draw a

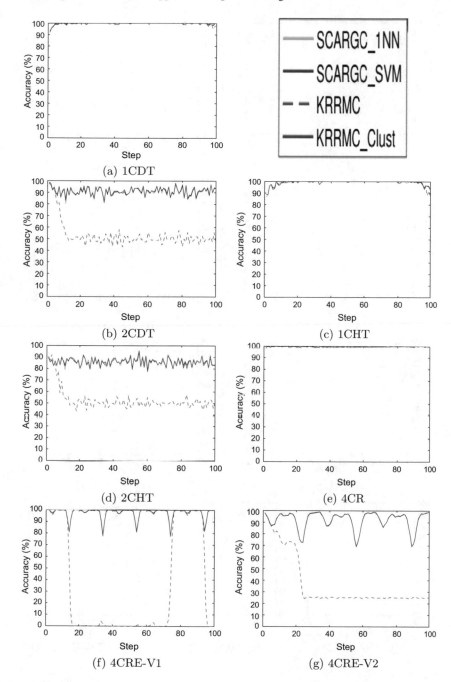

Fig. 2 Performance over 16 datasets in 100 steps

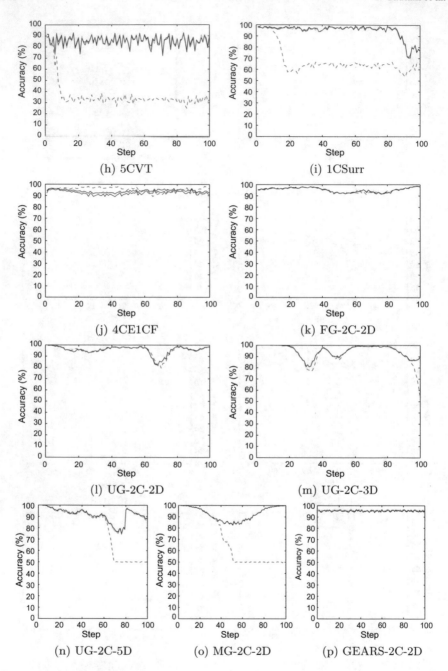

(h) 5CVT

(i) 1CSurr

(j) 4CE1CF

(k) FG-2C-2D

(l) UG-2C-2D

(m) UG-2C-3D

(n) UG-2C-5D

(o) MG-2C-2D

(p) GEARS-2C-2D

Fig. 2 (continued)

separation boundary. 1CDT, 1CHT, GEARS-2C-2D and 4CE1CF are such type of datasets. The accuracy of all the models, including those of pure classification i.e. KRRMC and those of combination of classification with clustering i.e. KRRMC_Clust, perform similar as the labels yield by KRRMC are not required to rectify by the clustering algorithm.

(ii) When data of both classes move simultaneously then KRRMC needs assistance of clustering algorithm. 2CDT, 2CHT, 4CRE-V2, 1CSURR etc. are such type of datasets. Therefore, KRRMC doesn't perform well on these datasets, however, KRRMC_Clust have performed well on these datasets and outperform KRRMC as well as other two existing classifiers as you can see from Fig. 2 and Table 2.

(iii) Some datasets contain abrupt drift like 4CRE-V1 and 5CVT. Proposed as well as existing model adapts these drifts easily however, KRRMC is not able to adapt the abrupt drift.

(iv) When data contains Unimodal Gaussian distribution. UG-2C-2D, UG-2C-3D, UG-2C-5D dataset is a Unimodal Gaussian dataset where Gaussian clusters of corresponding dimension rotate around a common axis and the distance between them keeps on varying. They also overlap at certain intervals in time. All models give comparable results with the highest being in SCARGC (SVM). But a point to note here is the significant decrease in time consumption by our models accompanied by similar accuracy. It is to be noted that KRRMC also perform similar to other classifiers in the case of unimodal Gaussian distribution.

(v) When data contains multi-modal Gaussian distribution. MG-2C-2D dataset is a multi-modal Gaussian dataset, which evolves over time, so we require a higher value of k, i.e. the number of clusters to achieve higher efficiency of classification. For experimentation, we set the value of k to 4. Due to its evolving multi-modal nature, we observe a significant increase in accuracy in models using clustering algorithms. Comparable accuracy is obtained in models of KRRMC_Clust, SCARGC (1NN, SVM) with the highest being in the proposed model KRRMC_Clust.

4.2   Efficiency Analysis

We have presented results on different number of clusters in each dataset, which is taken as either equal or more than the number of classes presents in the initial data. However, apart of multi-modal Gaussian distribution cases, if cluster number is equal to number of classes then that would be enough to adapt the drift in the dataset. In streaming datasets, training time is always a crucial factor as model needs to retrain multiple times for adapting the present drift. As you can see from Tables 2 and 3, our proposed model KRRMC_clust is more than 1000 times faster than existing approaches while giving consistent accuracy over all datasets. Table 3 presents the comparison of total time taken by SCARGC(1NN), SCARGC(SVM), KRRMC and KRRMC_Clust during training and retraining the model. For the fair

Table 3 Time (in Second) consumed by the classifiers

	SCARGC (1NN)	SCARGC (SVM)	KRRMC	KRRMC_Clust
1CDT [7]	8.71	1.32	0.00045	0.00046
2CDT [7]	6.49	1.45	0.00048	0.00046
1CHT [7]	6.48	1.16	0.00048	0.00044
2CHT [7]	6.69	1.44	0.00046	0.00046
4CR [7]	60.59	22.99	0.00059	0.00067
4CRE-V1 [7]	51.25	9.50	0.00058	0.00059
4CRE-V2 [7]	80.60	18.41	0.00063	0.00074
5CVT [7]	9.95	2.57	0.0004	0.00047
1CSurr [7]	23.19	6.21	0.00052	0.0005
4CE1CF [7]	71.52	38.18	0.00068	0.0006
FG-2C-2D [10]	78.77	17.15	0.00067	0.00073
UG-2C-2D [11]	41.74	7.86	0.00057	0.00056
UG-2C-3D [11]	85.09	16.02	0.00067	0.00072
UG-2C-5D [11]	76.87	23.24	0.00063	0.0006
MG-2C-2D [11]	85.71	17.86	0.00072	0.00063
GEARS-2C-2D [11]	80.05	13.82	0.00073	0.00073

comparison of training time, we executed existing and proposed approach in the same environment.

5 Conclusion

This paper presents an adaptable and faster classification model based on KRR and K-means clustering for non-stationary environment. KRR provides model a speed up and global learning capability, while K-means clustering helps in active learning and prepare the samples to train on KRR for classification. We performed a wide experimental evaluation over 16 non-stationary datasets. In all the datasets, proposed model either performed better or similar compared to the existing methods, however, it outperformed all other existing methods by high margin (more than 1000 times) in time taken. In future work, selection of a number of clusters need to be automated.

References

1. Ditzler, G., Roveri, M., Alippi, C., Polikar, R.: Learning in nonstationary environments: a survey. IEEE Comput. Intell. Mag. **10**(4), 12–25 (2015)
2. Das, A.K., Sundaram, S., Sundararajan, N.: A self-regulated interval type-2 neuro-fuzzy inference system for handling nonstationarities in eeg signals for bci. IEEE Trans. Fuzzy Syst. **24**(6), 1565–1577 (2016)
3. Polikar, R., Upda, L., Upda, S.S., Honavar, V.: Learn++: an incremental learning algorithm for supervised neural networks. IEEE Trans. Syst. Man Cybern. Part C (Appl. Rev.) **31**(4), 497–508 (2001)
4. Aggarwal, C.C., Kong, X., Gu, Q., Han, J., Yu, P.S.: Active learning: a survey (2014)
5. Settles, B.: Active learning literature survey. University of Wisconsin, Madison **52**(55–66), 11 (2010)
6. Krempl, G., Žliobaite, I., Brzeziński, D., Hüllermeier, E., Last, M., Lemaire, V., Noack, T., Shaker, A., Sievi, S., Spiliopoulou, M., et al.: Open challenges for data stream mining research. ACM SIGKDD Explor. Newslett. **16**(1), 1–10 (2014)
7. Souza, V.M.A., Silva, D.F., Gama, J., Batista, G.E.A.P.A: Data stream classification guided by clustering on nonstationary environments and extreme verification latency. In: Proceedings of SIAM International Conference on Data Mining (SDM), pp. 873–881 (2015)
8. Gill, P.E., Murray, W., Wright, M.H.: Practical Optimization (1981)
9. Pelleg, D., Moore, A.W., et al.: X-means: extending k-means with efficient estimation of the number of clusters. ICML **1**, 727–734 (2000)
10. Ditzler, G., Polikar, R.: Incremental learning of concept drift from streaming imbalanced data. IEEE Trans. Knowl. Data Eng. **25**(10), 2283–2301 (2013)
11. Dyer, K.B., Capo, R., Polikar, R.: Compose: a semisupervised learning framework for initially labeled nonstationary streaming data. IEEE Trans. Neural Netw. Learn. Syst. **25**(1), 12–26 (2014)

Automatic Segmentation of Intracerebral Hemorrhage from Brain CT Images

Anjali Gautam and Balasubramanian Raman

Abstract Intracerebral hemorrhage (ICH) diagnosis is a neurological deficit that can occur in the patients suffering from high blood pressure and head trauma. Manual segmentation of ICH is tedious and operator dependent, therefore the purpose of this study is to present a robust fully automated system for hemorrhage detection from Computed Tomography (CT) scan images. The proposed method is based on White Matter Fuzzy c-Means (WMFCM) clustering and wavelet-based thresholding. The suggested method starts with the removal of components which do not look like brain tissues including skull by using a new WMFCM technique. After brain extraction, a new segmentation technique based on wavelet thresholding is used for detection and localization of hemorrhagic stroke. The proposed segmentation method is fast and accurate where standard evaluation metrics like dice similarity coefficients, Jaccard distance, Hausdorff distance, precision, recall, and F1 score are used to measure the accuracy of the proposed algorithm. Our method is demonstrated on a dataset of 20 brain computed tomography (CT) images suffered ICH and results obtained are compared with the ground truth of images. We found that our method can detect ICH with an average dice similarity of 0.82 and perform better as compared to standard fuzzy c-means (FCM) and spatial FCM (SFCM) clustering methods.

Keywords Computed tomography (CT) · Intracerebral hemorrhage (ICH) Segmentation · Fuzzy c-means

A. Gautam (✉) · B. Raman
Department of Computer Science and Engineering,
Indian Institute of Technology Roorkee, Roorkee, India
e-mail: anga3.dcs2015@iitr.ac.in

B. Raman
e-mail: balarfma@iitr.ac.in

© Springer Nature Singapore Pte Ltd. 2019   753
M. Tanveer and R. B. Pachori (eds.), *Machine Intelligence and Signal Analysis*,
Advances in Intelligent Systems and Computing 748,
https://doi.org/10.1007/978-981-13-0923-6_64

1 Introduction

Hemorrhagic stroke is a life threatening medical emergency, which may result from weakened vessel that ruptures and bleeds into the surrounding brain. In the United States, stroke is the fifth leading cause of death where 13% of all the stroke cases are hemorrhagic and in India it accounts for 22% of all stroke cases [1, 2]. ICH may be caused by different medical conditions such as sudden weakness, severe headache, trouble in swallowing, vision problem, and paralysis of body parts specially one side of body, etc. Death cases due to hemorrhagic stroke can be avoided by providing proper medical treatment which include stroke identification [3]. For this, physicians suggest Computed Tomography (CT) or Magnetic Resonance Imaging (MRI) scans of the patients. CT is widely used in hospitals for the initial identification of the stroke and if some abnormality is found, then, in that case, MRI scans are required to get the high-quality images of the brain regions. There are different medical imaging techniques such as nuclear medicine, positron emission tomography (PET)-CT, ultrasound or sonography, X-rays, endoscopy, elastography, tactile imaging, thermography, and biomarkers which are used by the radiologist for proper identification of brain and are not limited to only CT and MRI scans. After medical imaging, manual examination of each and every slice of the CT or MRI scans are analyzed by medical experts. Manual examination of imaging data is very time consuming, drowsy process and completely operator dependent and due to this procedure patients have to suffer that may even cause death [4]. Hence, reliable algorithms are required for the automatic segmentation and identification of different lesion characteristics, that can help doctors to start their treatment as early as possible. Segmentation of brain images has been performed by using different techniques like image segmentation, classification, graph based, and pattern recognition. CT scan images have been used in this study, where skull and hemorrhagic stroke appears white. Therefore, to delineate only the stroke region, proper segmentation algorithm should come into existence.

The organization of paper is as follows: first, we discuss the previous work related to hemorrhage segmentation in Sect. 2 followed by proposed methods and procedures used for stroke lesion identification by using fuzzy clustering and wavelet thresholding are given in Sect. 3, after that results obtained and conclusions are presented.

2 Related Work

In image processing, segmentation is considered as a very difficult problem and recent research finds that automatic system can give more accurate segmentation results that can be accepted in medical practice. Accurate identification of abnormality is very important in medical terminology for better analysis of normal or abnormal body parts. In medical diagnosis, manual examination of medical images is widely used and this process is very time consuming. Hence, an automated system should come into existence that gives results near to manual segmentation.

Various methods have been developed for semi-automatic and automatic segmentation of the region of interest. Among them fuzzy clustering-based segmentation is extensively used in many researches where image is divided into different meaningful regions. In 1984, fuzzy c-means clustering was proposed by Bezdek et al. [5] after that many variations of FCM was developed based on the requirement of segmentation where spatial and local information of the image pixel is embedded with FCM. Spatial fuzzy c-means (SFCM) clustering was proposed by Chuang et al. [6] where they incorporate spatial information into the membership function of FCM.

Many segmentation methods have been developed for the detection of ICH using semiautomated or automated methods. Phillips et al. [7] have used FCM-based segmentation technique to differentiate brain tissue from MR images like tumor and hemorrhage, gray matter, normal white matter, edematous white matter and air from MR scan images of hemorrhagic glioblastoma multiforme.

Liang et al. [8] have used different imaging techniques to detect ICH with susceptibility like gradient-recalled echo (GRE) imaging, GRE-type single-shot echo-planar imaging (GRE-EPI), turbo spin-echo (TSE) imaging, spin-echo-type single-shot echo-planar imaging (SE-EPI), half-Fourier single-shot turbo spin-echo (HASTE) imaging, and segmented HASTE (s-HASTE) imaging. CT imaging was also used by many researcher, Liu et al. [9] in their work have presented automated method to detect only the hemorrhage slices from multi-slice CT scan images. They used Haralick texture descriptors as the feature extraction model and support vector machine was used for hemorrhage detection.

Loncaric et al. [10] used K-means histogram-based clustering to determine three characteristic brain image values for background (noise), brain tissue and skull. After that they performed mapping of image-gray values through the mapping determined by fourth-order polynomial where value for background, brain tissue and skull remains constant, i.e., 30, 90, and 230, respectively, to characterize brain region more accurately. After that skull removal was done by morphological operations followed by region growing to detect the primary region of ICH. Cosic and Loucaric [11] have proposed a method to delineate the ICH lesion by implementing fuzzy clustering and expert system-based labeling. However, two methods [10, 11] perform good but due to some constant values and rules on the particular image dataset, their method cannot be universally accepted. Chan [12] has detected acute intracranial hemorrhage based on thresholding, morphological operations and knowledge-based classification system.

Bardera et al. [13] have presented a semiautomatic method to segment hematoma and edema by combining a region growing and a level set segmentation technique. Bhadauria et al. [14, 15] have combined active contour and FCM clustering method to segment stroke lesion from CT images. However, Shahangian and Pourghassem [16] developed a modified version of distance regularized level set evolution method [17]. Muschelli et al. [18] worked on 3D CT scan images for the segmentation of hemorrhage lesion using logistic regression, generalized additive model (GAM), and random forest classifier.

Several other methods have been proposed for the detection of ICH, but those were very time consuming methods. Hence, to tackle this time issue a good computer software should come into existence that can easily, accurately detect ICH and make a great impact in clinical practice.

3 Methods and Procedures

In this section, we have discussed the methodologies used for the delineation of hemorrhagic stroke from CT scan images. CT image dataset have been taken from Himalayan Institute of Medical Sciences Dehradun, India. All the CT images were acquired from Siemens SOMATOM Sensation 64 CT scanner, where each slice is having a resolution of 512×512. Figure 1 depicts the block diagram of the proposed framework and detailed description are given in subsections given below and also in Algorithm 1. This section is divided into two parts, i.e., skull removal and wavelet-based thresholding. The first subsection describes the brain extraction method and in other subsection wavelet-based thresholding technique is described which was used as the new segmentation method to remove most of the brain structure which was followed by Gaussian filtering and morphological operations.

3.1 Skull Removal

The whole structure of the head, i.e., the internal (brain) and the external (skull and skin of head) parts along with the head fixation device usually made of iron is displayed in the CT scan image. In case of ICH, only internal structure of the brain is required for diagnosis and hence all the other components present in the images are not required. In this subsection, we will discuss the skull removal using White Matter Fuzzy c-Means (WMFCM) clustering technique, FCM was developed by Dunn in 1973 [19] and improved by Bezdek in 1984 [5]. Fuzzy clustering is an unsupervised method of segmentation and widely used by many algorithms as discussed in related work. Image segmentation is a procedure where an image is divided into meaningful regions. In FCM, entire image is partitioned into C crisp maximally connected regions R^n where each pixel participates in many memberships

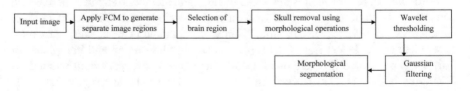

Fig. 1 Block diagram of the proposed system framework

such that each R^n is homogeneous with respect to some criteria. Image partitioning into different regions can be obtained by iterative optimization of the FCM objective function, given as:

$$J_m = \sum_{i=1}^{N} \sum_{j=1}^{C} u_{ij}^m \parallel x_i - c_j \parallel^2, 1 \leq m < \infty \tag{1}$$

where m is the weighting exponent greater than or equal to 1, x_i is the ith data of region R^n, c_j is the center of cluster j, and $\parallel * \parallel$ is norm on R^n to measure the similarity between data and cluster center C ($2 \leq C \leq N$) where N represents the number of data points in R^n. u_{ij} is the degree of membership of x_i in cluster j, the overall sum of u_{ij} is equal to 1 and can be expanded as given in (2):

$$u_{ij} = \frac{1}{\sum_{k=1}^{C} \left(\frac{\parallel x_i - c_j \parallel}{\parallel x_i - c_k \parallel} \right)^{\frac{2}{m-1}}} \tag{2}$$

where c_j is defined as

$$c_j = \frac{\sum_{i=1}^{N} u_{ij}^m \cdot x_i}{\sum_{i=1}^{N} u_{ij}^m} \tag{3}$$

and $\parallel x_i - c_j \parallel^2$ is the squared Euclidean distance between data point x_i and cluster center c_j. The iteration of the objective function will stop when $max_{ij} \left| u_{ij}^{(k+1)} - u_{ij}^{(k)} \right| < \varepsilon$, where ε is a termination criterion between 0 and 1, whereas k represents iteration steps, here m is 2 and j is 1, 2 and 3. After fuzzy clustering, the cluster that have hemorrhagic lesion will be considered for morphological operations to remove unwanted regions. Skull removed sample images are shown in Fig. 2 and the details of WMFCM are given in Algorithm 1.

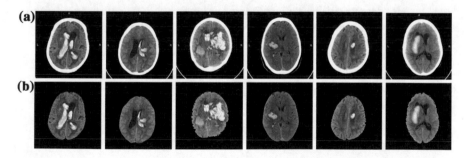

Fig. 2 Skull-removed images using fuzzy clustering

Algorithm 1: Fuzzy clustering and wavelet based ICH segmentation

---
Input : Load CT image I from the dataset
Output: ICH Localization and Extraction from CT scan image

1 **Function** $I=WMFCM\ (I,C)$
2   Select the initial clustering centers c_j and set the termination criterion ε to a low value
3   Randomly initialize fuzzy partition matrix u_{ij} for each c_j
4   Update u_{ij} using (2)
5   New clustering centers are generated using (3)
6   Stop clustering when $max_{ij}\left|u_{ij}^{(k+1)} - u_{ij}^{(k)}\right| < \varepsilon$ otherwise goto step 4
7   Consider the cluster which likely to have a brain region
8   Apply morphological operations to remove remaining parts of skull and head fixation device.
9 **Function** $I_{CH} = WT(I)$
10   Discrete wavelet transform is used such that: $Approx_1 = dwt(I)$
11      $\eta_T(Approx_1)$ is generated using (4) where $T = 0.8$
12      $Approx_2 = dwt(\eta_T(Approx_1))$
13      $\psi_T(Approx_2)$ is obtained by (5) with $T = 1$
14      $Approx_3 = dwt(HardApprox_2)$
15      $\eta_T(Approx_3)$ is obtained using (4), here $T = 0.8$
16   Three level wavelet reconstruction with wiener filtering, results in W_{Rec}
17   Apply Gaussian filtering on W_{Rec}
18   Morphological closing operation is applied for ICH segmentation.

3.2 ICH Segmentation

Segmentation of ICH is done using wavelet thresholding in order to reduce the time complexity of stroke detection. Wavelet transform is widely used in medical image analysis, it decomposes the signal into its "wavelets" which is obtained from a single prototype wavelet called mother wavelet, translated by the amount t with a frequency of s, where high-pass and lowpass filter are used to break the signal into different frequency bands [20]. In the proposed method, we have used three-level wavelet decomposition with soft and hard thresholding. Soft thresholding (or shrinkage function) shrinks the argument towards zero by the threshold T [21].

$$\eta_T(x) = sgn(x)(|x| - T)_+ \qquad (4)$$

where $(a)_+ = 0$ if $a < 0$ otherwise $(a)_+ = a$.

Hard-threshold function is given as

$$\psi_T(x) = \begin{cases} x \ if \ |x| > T, \\ 0 \ \ otherwise \end{cases} \qquad (5)$$

where input is stored if it is greater than the threshold T; otherwise, it is set to zero [22].

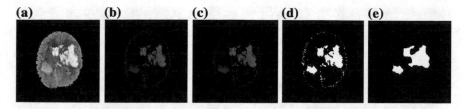

Fig. 3 ICH Segmentation procedure: **a** Skull-removed CT image, **b** Wavelet thresholding, **c** Gaussian Filtering, **d** Image binarization, and **e** Final segmentation result using morphological operations

In the proposed method, wavelet thresholding used as a procedure to remove most of the brain parts except white matter in the image as shown in Fig. 3. We have used three-level decomposition, in first and third decomposition, soft thresholding with 0.8 threshold is applied to remove noise from the image whereas in second decomposition hard thresholding with value of 1 is used to keep image element greater than threshold value. Then, wavelet reconstruction is applied with Wiener filtering in all the reconstructed level that can results in an image having elements close to the white matter. Results obtained after wavelet transformation are followed by Gaussian filtering to remove all the remaining noise.

Segmentation of hemorrhage from the resulted image obtained from the above method can be achieved by morphological operations. In this paper, we have used morphological closing operation to get the results close to the ground truth of input image. Closing is a dilation followed by erosion where same structuring element (disk) is used for erosion and dilation. The final segmentation results are shown in Fig. 3e.

4 Results

The results obtained using our proposed method are shown in Fig. 4, where input CT scan image indicates in (a); ground truth (GT) of the ICH obtained from the medical expert are overlaid on the original image and shown in (b); (c) indicates the results obtained from the proposed method overlaid on the original image (a). Ground truth of the ICH and proposed results are represented as green and red color respectively. Table 1 shows the accuracy of our method in comparison with the ground truth of lesion by using standard evaluation metrics like dice similarity coefficient (DSC), Jaccard distance (JD), Hausdorff distance (HD), precision, recall, and F1 score. DSC and Jaccard similarity give the volume overlap of segmentation results with different formula, if their values are close to 1, it represents perfect segmentation; HD measures the maximum surface distance and its value should be nearer to 0 for accurate segmentation, precision is the positive predictive value and recall is the sensitivity. Overall accuracy of the result can be obtained by combining precision and recall which is known as F1 score.

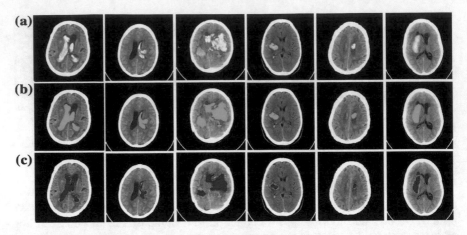

Fig. 4 ICH CT scan images with their ground truth and results obtained by proposed method (Note: green color is for the ground truth and red color for our obtained results)

Supposed two sets of pixels from image I_1 (ground truth) and I_2 (segmentation), then DSC can be calculated as

$$DSC = \frac{2\left|I_1 \bigcap I_2\right|}{|I_1| + |I_2|} \qquad (6)$$

Jaccard distance is given by

$$JD = \frac{\left|I_1 \bigcap I_2\right|}{|I_1| + |I_2| - \left|I_1 \bigcap I_2\right|} \qquad (7)$$

Similarly, maximum of all surface distances can be given by HD (mm) as

$$HD(I_1, I_2) = max\left\{\frac{1}{N_a}\sum_{a \in I_1} min_{b \in I_2}d(a, b), \frac{1}{N_b}\sum_{b \in I_2} min_{a \in I_1}d(b, a)\right\} \qquad (8)$$

The precision (also called positive predictive value) and recall (also known as sensitivity) are defined as

$$precision = \frac{TP}{TP + FP} \qquad (9)$$

$$recall = \frac{TP}{TP + FN} \qquad (10)$$

F1 score is also calculated to check the accuracy of segmentation results obtained after implementing the proposed method. It is similar to dice similarity coefficient and given as

$$F_1 score = 2.\cfrac{1}{\cfrac{1}{recall} + \cfrac{1}{precision}} \tag{11}$$

$$F_1 score = 2.\frac{precision \times recall}{precision + recall} \tag{12}$$

Table 1 shows the overall accuracy of all CT images as the mean of evaluation metrics where DSC is 0.82, for JD we achieved 0.71 accuracy, in terms of HD it is 1.85, precision, recall, and F1 score obtained are 0.96, 0.74, and 0.82 respectively. Average accuracy in terms of standard deviation as DSC is 0.10, 0.13 is achieved using JD, HD obtained here is 3.19, precision, recall, and F1 score obtained are 0.06, 0.15, and 0.10 respectively.

We have also compared our results with FCM and SFCM shown in Fig. 5 and we found that our method is able to detect the brain hemorrhagic lesions more accurately

Table 1 Results of input CT image data with an average mean and standard deviation of the evaluation metrics

CT Image	DSC	JD	HD	Precision	Recall	F1 score
1	0.93	0.86	0.61	0.98	0.88	0.93
2	0.79	0.64	0.94	0.92	0.68	0.79
3	0.89	0.79	0.81	0.96	0.82	0.89
4	0.84	0.73	0.47	0.96	0.75	0.84
5	0.96	0.92	0.07	0.94	0.98	0.96
6	0.89	0.79	0.55	0.99	0.80	0.89
7	0.90	0.82	0.37	0.99	0.82	0.90
8	0.84	0.72	0.55	0.99	0.73	0.84
9	0.90	0.81	1.6	0.98	0.83	0.90
10	0.87	0.78	0.40	0.99	0.78	0.87
11	0.76	0.62	3.60	0.93	0.64	0.76
12	0.84	0.73	0.98	0.98	0.74	0.84
13	0.68	0.51	2.17	1	0.51	0.68
14	0.87	0.77	1.05	0.97	0.79	0.87
15	0.85	0.74	1.78	0.98	0.75	0.85
16	0.52	0.35	14.83	1	0.35	0.52
17	0.87	0.78	0.51	0.99	0.78	0.87
18	0.75	0.60	3.23	0.93	0.63	0.75
19	0.85	0.73	1.07	0.74	0.99	0.85
20	0.72	0.56	1.39	1	0.56	0.72
Mean	**0.82**	**0.71**	**1.85**	**0.96**	**0.74**	**0.82**
Std.	**0.10**	**0.13**	**3.19**	**0.06**	**0.15**	**0.10**

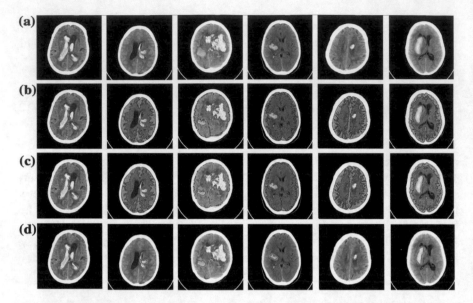

Fig. 5 Segmentation result comparision of **a** Input image, using **b** standard FCM, **c** SFCM and **d** Proposed Method (Note: Green color represents ground truth of stroke lesion and red indicates segmentation result)

from all the CT cases as shown in Fig. 5d whereas FCM and SFCM are not able to recognize the lesions as indicated in Fig. 5b, c.

5 Conclusion

Hemorrhage detection from CT scan image provides useful information to physicians which results in a improved computational aid in the diagnosis of patients. We have developed an automated system for detection and localization of ICH in CT scan images. The method is based on White Matter Fuzzy c-Means (WMFCM) clustering approach to extract brain from the input image which is then followed by the wavelet based thresholding (WT) technique to determine white matter objects, i.e., hemorrhage from the image by reducing the contrast of other brain region. We found that the proposed method is successful in detection and segmentation of hemorrhage by taking the advantage of wavelet transform. Six evaluation metrics were used to measure the accuracy of proposed method and in terms of average dice similarity it is 0.82 when compared with ground truth of the images. This shows that our segmentation accuracy is very high and better than FCM and SFCM clustering methods. Future work can also be carried out to get highly accurate results for hemorrhage lesion segmentation.

Acknowledgements We thank Dr. Shailendra Raghuwanshi, Head of Radiology Department, Himalayan Institute of Medical Sciences, Dehradun, Uttarakhand, India for providing hemorrhagic stroke CT image dataset.

References

1. Impact of Stroke (Stroke statistics) (2017). http://www.strokeassociation.org/STROKEORG/AboutStroke/Impact-of-Stroke-Stroke-statistics_UCM_310728_Article.jsp
2. Banerjee, T.K., Das, S.K.: Fifty years of stroke researches in India. Ann. Indian Acad. Neurol. **19**(1), 1–8 (2016)
3. Heit, J.J., Iv, M., Wintermark, M.: Imaging of intracranial hemorrhage. J. Stroke **19**(1), 11 (2017)
4. Anbeek, P., Išgum, I., van Kooij, B.J., Mol, C.P., Kersbergen, K.J., Groenendaal, F., Viergever, M.A., de Vries, L.S., Benders, M.J.: Automatic segmentation of eight tissue classes in neonatal brain MRI. PLoS One **8**(12), e81895 (2013)
5. Bezdek, J.C., Ehrlich, R., Full, W.: FCM: the fuzzy c-means clustering algorithm. Comput. Geosci. **10**(2–3), 191–203 (1984)
6. Chuang, K.S., Tzeng, H.L., Chen, S., Wu, J., Chen, T.J.: Fuzzy c-means clustering with spatial information for image segmentation. Comput. Med. Imaging Graph. **30**(1), 9–15 (2006)
7. Phillips, W.E., Velthuizen, R.P., Phuphanich, S., Hall, L.O., Clarke, L.P., Silbiger, M.L.: Application of fuzzy c-means segmentation technique for tissue differentiation in MR images of a hemorrhagic glioblastoma multiforme. Magn. Reson. Imaging **13**(2), 277–290 (1995)
8. Liang, L., Korogi, Y., Sugahara, T., Shigematsu, Y., Okuda, T., Ikushima, I., Takahashi, M.: Detection of intracranial hemorrhage with susceptibility-weighted MR sequences. AJNR Am. J. Neuroradiol. **20**(8), 1527–1534 (1999)
9. Liu, R., Tan, C.L., Leong, T.Y., Lee, C.K., Pang, B.C., Lim, C.T., Tian, Q., Tang, S., Zhang, Z.: Hemorrhage slices detection in brain ct images. In: Proceedings of the Nineteen International Conference on Pattern Recognition, pp. 1–4. IEEE (2008)
10. Loncaric, S., Dhawan, A.P., Broderick, J., Brott, T.: 3-D image analysis of intra-cerebral brain hemorrhage from digitized CT films. Comput. Methods Programs Biomed. **46**(3), 207–216 (1995)
11. Cosic, D., Loucaric, S.: Computer system for quantitative: analysis of ICH from CT head images. In: Proceedings of the 19th Annual International Conference of Engineering in Medicine and Biology Society, pp. 553–556. IEEE (1997)
12. Chan, T.: Computer aided detection of small acute intracranial hemorrhage on computer tomography of brain. Comput. Med. Imaging Gr. **31**(4), 285–298 (2007)
13. Bardera, A., Boada, I., Feixas, M., Remollo, S., Blasco, G., Silva, Y., Pedraza, S.: Semi-automated method for brain hematoma and edema quantification using computed tomography. Comput. Med. Imaging Gr. **33**(4), 304–311 (2009)
14. Bhadauria, H.S., Singh, A., Dewal, M.L.: An integrated method for hemorrhage segmentation from brain CT imaging. Comput. Electr. Eng. **39**(5), 1527–1536 (2013)
15. Bhadauria, H.S., Dewal, M.L.: Intracranial hemorrhage detection using spatial fuzzy c-mean and region-based active contour on brain CT imaging. SIViP **8**(2), 357–364 (2014)
16. Shahangian, B., Pourghassem, H.: Automatic brain hemorrhage segmentation and classification algorithm based on weighted grayscale histogram feature in a hierarchical classification structure. Biocybern. Biomed. Eng. **36**(1), 217–232 (2016)
17. Li, C., Xu, C., Gui, C., Fox, M.D.: Distance regularized level set evolution and its application to image segmentation. IEEE Trans. Image Process. **19**(12), 3243–3254 (2010)
18. Muschelli, J., Sweeney, E.M., Ullman, N.L., Vespa, P., Hanley, D.F., Crainiceanu, C.M.: PItcH-PERFeCT: primary intracranial hemorrhage probability estimation using random forests on CT. Neuroimage Clin. **14**, 379–390 (2017)

19. Dunn, J.C.: A fuzzy relative of the ISODATA process and its use in detecting compact well-separated clusters. J. Cybern. **3**(3), 32–57 (1973)
20. Antonini, M., Barlaud, M., Mathieu, P., Daubechies, I.: Image coding using wavelet transform. IEEE Trans. Image Process. **1**(2), 205–220 (1992)
21. Donoho, D.L.: De-noising by soft-thresholding. IEEE Trans. Inf. Theory **41**(3), 613–627 (1995)
22. Chang, S.G., Yu, B., Vetterli, M.: Adaptive wavelet thresholding for image denoising and compression. IEEE Trans. Image Process. **9**(9), 1532–1546 (2000)

Author Index

Printed in the United States
By Bookmasters